T0191638

Lecture Notes
in Business Information Processing   **446**

More information about this series at https://link.springer.com/bookseries/7911

Renata Guizzardi · Jolita Ralyté ·
Xavier Franch (Eds.)

Research Challenges in Information Science

16th International Conference, RCIS 2022
Barcelona, Spain, May 17–20, 2022
Proceedings

 Springer

Editors
Renata Guizzardi (iD)
University of Twente
Enschede, The Netherlands

Jolita Ralyté (iD)
University of Geneva, CUI
Carouge, Switzerland

Xavier Franch (iD)
Polytechnic University of Catalonia
Barcelona, Spain

ISSN 1865-1348 ISSN 1865-1356 (electronic)
Lecture Notes in Business Information Processing
ISBN 978-3-031-05759-5 ISBN 978-3-031-05760-1 (eBook)
https://doi.org/10.1007/978-3-031-05760-1

This Springer imprint is published by the registered company Springer Nature Switzerland AG
The registered company address is: Gewerbestrasse 11, 6330 Cham, Switzerland

Preface

We are pleased to welcome you to the proceedings of the 16th edition of the International Conference on Research Challenges in Information Science (RCIS 2022), which took place in Barcelona from May 17–20, 2022.

RCIS aims to bring together scientists, researchers, engineers and practitioners from a wide range of information science fields and to provide opportunities for knowledge sharing and dissemination. This year, RCIS 2022 has chosen the theme "Ethics and Trustworthiness in Information Science" be at the center of its reflections and discussions. The way information is captured, interpreted, transformed and used raises important ethical issues, such as who can access the information and to what extent, how bias can be avoided when interpreting and using information, and how fair are decisions made based on particular information artifacts. Moreover, trust is crucial in such contexts, as people become increasingly dependent on how information is handled. The more we need information to make critical decision, the greater the need to develop ethical and trustworthy information systems and services.

The conference program included three engaging keynote speeches. "The FAIR Guiding Principles in Times of Crisis", by Prof. Dr. Barend Mons, from the Human Genetics Department of Leiden University Medical Center, in the Netherlands; "We Are Open. The Door Is Just Very Heavy: New Challenges in Information Science or How Can IS Help with Fairness, Diversity, Non-discrimination?", by Prof. Dr. Florence Sèdes, from IRIT—Université Toulouse III—Paul Sabatier, France; and "Engineering the New Fabric of the Information Infrastructure—IoT, Edge, and Cloud", by Prof. Dr. Schahram Dustdar, from the Distributed Systems Group, TU Wien, Austria.

A total of 100 papers were submitted to the main track of the conference, from which 35 have been accepted as full papers and included in these proceedings. For paper review and selection, we relied on a Program Committee composed of 58 members, in addition to a Program Board of 15 members. Each of the submitted papers went through a thorough review process with at least three reviews from the Program Committee. The papers with no clear decision were discussed online. The discussions were moderated by Program Board members, who concluded by writing meta-reviews. The final decisions were taken during the Program Board meeting. We are deeply thankful to the Program Committee and Program Board for their fair, competent and active contribution in selecting papers for publication in the RCIS 2022 proceedings.

Besides the main track, RCIS 2022 hosted three other tracks accepting short papers: Forum (presenting research challenges, novel ideas, and tool demonstrations), Research Projects (presenting initial or intermediate project results), and Doctoral Consortium (describing PhD research projects).

The Forum track, chaired by Estefanía Serral Asensio and Marcela Ruiz, received 11 submissions from which 5 were accepted. Furthermore, 13 additional papers were accepted from the papers invited from the main conference track, leading to a total of 18 forum papers. The Doctoral Consortium track, chaired by Andreas Lothe Opdahl and Marko Bajec, attracted 10 submissions out of which 6 were accepted. Finally, the Research Projects track, chaired by Alessandra Bagnato and Lidia López, received 16 submissions, which are published in an online volume of CEUR-WS proceedings. We express our warm thanks to the track chairs for their involvement in soliciting and selecting papers for each of these tracks.

In addition, RCIS 2022 Program also included the presentation of three interesting tutorials: "Information Security & Risk Management: Trustworthiness and Human Interaction", by Nicholas Fair, Stephen Phillips, Gencer Erdogan and Ragnhild Halvorsrud; "The Challenge of Collecting and Analyzing Information from Citizens and Social Media in Emergencies: The Crowd4sdg Experience and Tools", by Barbara Pernici, Carlo Alberto Bono, Mehmet Oguz Mulayim and Jose Luis Fernandez-Marquez; and "Information Science Research with Machine Learning: Best Practices and Pitfalls" by Andreas Vogelsang. We thank the Tutorial chairs, Cinzia Cappiello and Paola Spoletini, for their great work in attracting and selecting the tutorials.

This year, for the first time, RCIS invited workshop submissions. Three exciting workshops have been selected, namely: "The 3rd International Workshop on Quality and Measurement of Software Model-Driven Development" (QUAMES 2022), organized by Beatriz Marín, Giovanni Giachetti and Estefanía Serral Asensio; "The 1st International Workshop on Cyber-Physical Social Systems for Sustainability: Challenges and Opportunities", organized by Isabel Sofia Sousa Brito, Nelly Condori-Fernandez and Leticia Duboc; and "Ethical, Social and Environmental Accounting of Conferences: the Case of RCIS", organized by Sergio España, Vijanti Ramautar and Quang Tan Le. We are very grateful to the Workshop chairs, João Araújo and Jose Luis de la Vara, for their work in attracting and selecting the workshops.

Overall, organizing RCIS 2022 was a great pleasure, since we had an exceptional organizing committee chaired by Carme Quer. We thank them for their engagement and support. We are also grateful to our publicity chairs Joaquim Motger, Tiago Sales, and Ben Roelens, and the webmaster Carles Farré who ensured the visibility of the conference. Special thanks go to the Proceedings chair, Rébeca Deneckère, who carefully managed the proceedings preparation process, and Christine Reiss and Ralf Gerstner from Springer who assisted us in its production.

Finally, we thank all the authors for sharing their work and findings in the field of Information Science and all the attendees for their lively participation and constructive feedback to the authors. We hope you will enjoy reading this volume.

April 2022

Renata Guizzardi
Jolita Ralyté
Xavier Franch

Organization

General Chair

Xavier Franch — Universitat Politècnica de Catalunya, Spain

Program Committee Chairs

Renata Guizzardi — Twente University, The Netherlands
Jolita Ralyté — University of Geneva, Switzerland

Organizing Chair

Carme Quer — Universitat Politècnica de Catalunya, Spain

Forum Chairs

Marcela Ruiz — Zurich University of Applied Sciences, Switzerland
Estefanía Serral Asensio — Katholieke Universiteit Leuven, Belgium

Doctoral Consortium Chairs

Marko Bajec — University of Ljubljana, Slovenia
Andreas L. Opdahl — University of Bergen, Norway

Tutorial Chairs

Paola Spoletini — Kennesaw State University, USA
Cinzia Cappiello — Politecnico di Milano, Italy

Workshop Chairs

João Araujo — Universidade Nova de Lisboa, Portugal
Jose Luis de la Vara — University of Castilla-La Mancha, Spain

Research Project Chairs

Alessandra Bagnato — Softeam, France
Lidia López — Barcelona Supercomputing Center, Spain

Publicity Chairs

Joaquim Motger	Universitat Politècnica de Catalunya, Spain
Tiago Sales	Free University of Bozen-Bolzano, Italy
Ben Roelens	Open University, The Netherlands

Journal First Chair

Silverio Martínez-Fernández	Universitat Politècnica de Catalunya, Spain

Procceedings Chair

Rebecca Deneckere	Université Paris 1 Panthéon - Sorbonne, France

Steering Committee

Saïd Assar	Institut Mines-Telecom Business School, France
Marko Bajec	University of Ljubljana, Slovenia
Pericles Loucopoulos	Institute of Digital Innovation and Research, Ireland
Haralambos Mouratidis	University of Essex, UK
Selmin Nurcan	Université Paris 1 Panthéon - Sorbonne, France
Oscar Pastor	Universidad Politècnica de Valencia, Spain
Jolita Ralyté	University of Geneva, Switzerland
Colette Rolland	Université Paris 1 Panthéon - Sorbonne, France
Jelena Zdravkovic	Stockholm University, Sweden

Program Board

Saïd Assar	Institut Mines-Telecom Business School, France
Marko Bajec	University of Ljubljana, Slovenia
Jennifer Horkoff	University of Gothenburg, Sweden
Pericles Loucopoulos	Institute of Digital Innovation and Research, Ireland
Nikolay Mehandjiev	The University of Manchester, UK
Haralambos Mouratidis	University of Essex, UK
Selmin Nurcan	Université Paris 1 Panthéon - Sorbonne, France
Andreas L. Opdahl	University of Bergen, Norway
Oscar Pastor	Universidad Politècnica de Valencia, Spain
Anna Perini	Fondazione Bruno Kessler Trento, Italy
Colette Rolland	Université Paris 1 Panthéon - Sorbonne, France
Samira Si-Said Cherfi	CNAM, France
Monique Snoeck	Katholieke Universiteit Leuven, Belgium
Pnina Soffer	University of Haifa, Israel
Jelena Zdravkovic	Stockholm University, Sweden

Main Track Program Committee

Nour Ali	Brunel University, UK
Raian Ali	Hamad Bin Khalifa University, Qatar
Carina Alves	Universidade Federal de Pernambuco, Brazil
Daniel Amyot	University of Ottawa, Canada
João Araujo	Universidade Nova de Lisboa, Portugal
Fatma Başak Aydemir	Boğaziçi University, Turkey
Dominik Bork	TU Wien, Austria
Mario Cortes-Cornax	Université Grenoble Alpes, France
Fabiano Dalpiaz	Utrecht University, The Netherlands
Maya Daneva	University of Twente, The Netherlands
Rebecca Deneckere	Université Paris 1 Panthéon - Sorbonne, France
Chiara Di Francescomarino	Fondazione Bruno Kessler, Italy
Sophie Dupuy-Chessa	Université Grenoble Alpes, France
Hans-Georg Fill	University of Fribourg, Switzerland
Andrew Fish	University of Brighton, UK
Ines Gam	Université Tunis Manar, Tunisia
Mohamad Gharib	University of Tartu, Estonia
Cesar Gonzalez-Perez	Incipit CSIC, Spain
Giancarlo Guizzardi	University of Twente, The Netherlands
Fayçal Hamdi	CNAM, France
Felix Härer	University of Fribourg, Switzerland
Mirjana Ivanovic	University of Novi Sad, Serbia
Haruhiko Kaiya	Kanagawa University, Japan
Christos Kalloniatis	University of the Aegean, Greece
Maria Karyda	University of the Aegean, Greece
Evangelia Kavakli	University of the Aegean, Greece
Marite Kirikova	Riga Technical University, Latvia
Elena Kornyshova	CNAM, France
Tong Li	Beijing University of Technology, China
Lidia López	Barcelona Supercomputing Center, Spain
Andrea Marrella	Sapienza University of Rome, Italy
Patricia Martin-Rodilla	Incipit CSIC, Spain
Massimo Mecella	Sapienza University of Rome, Italy
Giovanni Meroni	Politecnico di Milano, Italy
Marco Montali	Free University of Bozen-Bolzano, Italy
Denisse Muñante	ENSIIE & SAMOVAR, France
John Mylopoulos	University of Ottawa, Canada
Kathia Oliveira	Université Polytechnique Hauts-de-France, France
George Papadopoulos	University of Cyprus, Cyprus
Geert Poels	Ghent University, Belgium
Henderik A. Proper	LIST, Luxembourg
Gil Regev	EPFL, Switzerland
Patricia Rogetzer	University of Twente, The Netherlands
Marcela Ruiz	Zurich University of Applied Sciences, Switzerland

Maribel Yasmina Santos	University of Minho, Portugal
Rainer Schmidt	Munich University of Applied Sciences, Germany
Florence Sèdes	Université Toulouse III Paul Sabatier, France
Estefanía Serral Asensio	Katholieke Universiteit Leuven, Belgium
Anthony Simonofski	Katholieke Universiteit Leuven, Belgium
Dimitris Spiliotopoulos	University of the Peloponnese, Greece
Paola Spoletini	Kennesaw State University, USA
Erick Stattner	University of the French West Indies, France
Angelo Susi	Fondazione Bruno Kessler, Italy
Eric-Oluf Svee	Stockholm University, Sweden
Ernest Teniente	Universitat Politècnica de Catalunya, Spain
Nicolas Travers	De Vinci Research Center, France
Juan Carlos Trujillo	University of Alicante, Spain
Jean Vanderdonckt	Université catholique de Louvain, Belgium
Yves Wautelet	Katholieke Universiteit Leuven, Belgium
Hans Weigand	Tilburg University, The Netherlands

Forum Program Committee

Claudia P. Ayala	Universitat Politècnica de Catalunya, Spain
Fatma Başak Aydemir	Boğaziçi University, Turkey
Judith Barrios Albornoz	University of Los Andes, Colombia
Cinzia Cappiello	Politecnico di Milano, Italy
Jose Luis de la Vara	University of Castilla-La Mancha, Spain
Abdelaziz Khadraoui	University of Geneva, Switzerland
Manuele Kirsch-Pinheiro	Université Paris 1 Panthéon - Sorbonne, France
Elena Kornyshova	CNAM, France
Emanuele Laurenzi	FHNW, Switzerland
Dejan Lavbič	University of Ljubljana, Slovenia
Giovanni Meroni	Politecnico di Milano, Italy
Patricia Martin-Rodilla	Incipit CSIC, Spain
Beatriz Marín	Universitat Politècnica de València, Spain
Selmin Nurcan	Université Paris 1 Panthéon - Sorbonne, France
Vik Pant	University of Toronto, Canada
George Papadopoulos	University of Cyprus, Cyprus
Francisca Pérez	Universidad San Jorge, Spain
Iris Reinhartz-Berger	University of Haifa, Israel
Gianluigi Viscusi	Imperial College Business School, UK
Manuel Wimmer	Johannes Kepler University Linz, Austria

Doctoral Consortium Program Committee

Fabiano Dalpiaz	Utrecht University, The Netherlands
Maya Daneva	University of Twente, The Netherlands
Hans-Georg Fill	University of Fribourg, Switzerland
Giancarlo Guizzardi	University of Twente, The Netherlands

David Jelenc	University of Ljubljana, Slovenia
Nikolay Mehandjiev	The University of Manchester, UK
Oscar Pastor	Universidad Politècnica de Valencia, Spain
Gil Regev	EPFL, Switzerland
Pnina Soffer	University of Haifa, Israel
Angelo Susi	Fondazione Bruno Kessler, Italy
Eric-Oluf Svee	Stockholm University, Sweden
Juan Carlos Trujillo	University of Alicante, Spain
Jean Vanderdonckt	Université catholique de Louvain, Belgium
Tanja Vos	Universidad Politècnica de València, Spain
Hans Weigand	Tilburg University, The Netherlands

Research Project Program Committee

Nelly Condori-Fernández	Universidade da Coruña, Spain
Marilia Curado	University of Coimbra, Portugal
Tolga Ensari	Arkansas Tech University, USA
Davide Fucci	University of Hamburg, Germany
Antonio Garcia-Dominguez	Aston University, UK
Filippo Lanubile	University of Bari, Italy
Marc Oriol Hilari	Universitat Politècnica de Catalunya, Spain
Dimitrios Soudris	National Technical University of Athens, Greece
Alin Stefanescu	University of Bucharest, Romania
Dalila Tamzalit	University of Nantes, France
Tanja Vos	Universidad Politècnica de València, Spain

Additional Reviewers

Simone Agostinelli	Daniel Lehner
Markus Fasnacht	Sabine Sint
Jose David Mosquera Tobón	Antoine Clarinval
Syed Juned Ali	Ana León
Jérôme Fink	Evangelia Vanezi
Fabian Muff	Simon Curty
Faten Atigui	Katerina Mavroeidi
Mattia Fumagalli	Katerina Vgena
Michail Pantelelis	Francesca De Luzi
Fabio Azzalini	Christos Mettouris
Ana Lavalle	Sourav Debnath
Argyri Pattakou	Claudenir Moraes
Dario Benvenuti	

Keynotes

Engineering the New Fabric of the Information Infrastructure - IoT, Edge, and Cloud

Schahram Dustdar 🄳

Distributed Systems Group, TU Wien, Vienna, Austria
dustdar@dsg.tuwien.ac.at

Keynote Abstract

As humans, things, software and AI continue to become the entangled fabric of distributed systems, systems engineers and researchers are facing novel challenges. In this talk, we analyze the role of IoT, Edge, and Cloud, as well as AI in the co-evolution of distributed systems for the new decade. We identify challenges and discuss a roadmap that these new distributed systems have to address. We take a closer look at how a cyber-physical fabric will be complemented by AI operationalization to enable seamless end-to-end distributed systems.

"We Are Open. The Door is Just Very Heavy": New Challenges in Information Science or How can IS Help with Fairness, Diversity, Non-Discrimination?

Florence Sèdes (iD)

IRIT, Paul Sabatier University, Toulouse, France
florence.sedes@irit.fr

Keynote Abstract

Laws and administrative rules have been addressing disability and accessibility, through quotas and financial penalties, or students? social criteria assessment on French national ranking platforms. Such official measures enable minorities and other discriminated groups to be represented.

As half of the humanity does not constitute a minority, no quota policy is sup-posed to be applied, leaving gender imbalance as a potential issue. Academics have launched various inquiries and studies on this issue (Mothers in Science, UNESCO I'd Blush if I Could, etc.), that is crucial because of the female under-representation in scientific fields in STEM (Science, Technology, Engineering, and Mathematics).

"Gender equality paradox" names the observation that the under-representation of women in scientific fields (in particular, those related to STEM) is stronger in the most developed countries. As one says: "We are open. The door is just very heavy"[1]. The door is not closed but women must get a foot in the door to be invited, integrated, accepted, and hired. Women fight to break the glass ceiling in many professions. At the same time, as they can encounter obstacles, the sticky floor restricts them to relative non-strategic positions and prevents their scientific career from really taking off.

The tooth paste tube phenomenon is also invoked as a metaphor that illustrates how few female colleagues must assume the committees and representations for the community to comply with F/M quotas when required for hiring or mediating, for instance. The consequence can be an over-solicitation and over-impact on the communities. The facts are horrendous, and actions have hence to be taken to reach a global balance with diversity, inclusiveness and fair cohabitation without discrimination. All the launched initiatives testify this awareness. As a warning underlies the need to support these actions, one must take care of a negative effect on under-represented members by (unfairly) reinforcing their impostor syndrome.

---

[1] https://me.me/i/we-are-open-the-door-is-just-very-heavy-when-20430464.

Inside the science itself, and not only in terms of population or scientometrics, but a new phenomenon also arises with the evolution of topics, technics, and tools. One striking example is the risk with gender and minority "invisibilization" in AI, as the social representation biases are emphasized (few entries in Wikipedia, Matilda effect, lack of historical figures and illustration, rare "role models" in science,...) with unbalanced learning data. The dual issue is how rule-based systems and by whom, if any bias here also, are encoded to be aware of all the diversity any decision implies.

Contents

Business Process Mining

Digital Transformation and Smart Life

Conceptual Modelling and Ontologies

Requirements Engineering

Model-Driven Engineering

Doctoral Consortium

Tutorials

Data Science and Data Management

Personal Data Markets: A Narrative Review on Influence Factors of the Price of Personal Data

Julia Busch-Casler(✉) [ID] and Marija Radic [ID]

Fraunhofer IMW, Neumarkt 9-19, 04109 Leipzig, Germany
{julia.busch-casler,marija.radic}@imw.fraunhofer.de

Abstract. Personal data has been described as the "the new oil of the Internet." The global data monetization market is projected to increase to USD 6.1bn by 2025, and the success of giants like Facebook or Google speaks for itself. Almost all companies create, store, share and/or use personal data i.e. information from or about individuals. While the current assumption is that data subjects voluntarily share their data in exchange for a "free" service, the awareness of the value of personal data and data sovereignty is growing amongst consumers, businesses, and regulators alike. However, there is currently no consensus on which factors influence the value of personal data and how personal data should be priced regarding self-determination and data sovereignty. With this narrative review, we answer the following research question: Which factors influence the pricing of personal data? We show that research on the subject is diverse and that there is no consensus on the optimal pricing mechanism. We identify individual privacy and risk preferences, informational self-determination, sensitivity of data and data volume and inferability as most prevalent influence factors. We underline the need to establish ways for data owners to exercise data sovereignty and informed consent about data usage.

Keywords: Pricing personal data · Value of personal data · Data products · Data markets · Data monetization

1 Introduction

Personal data has been described as the "the new oil of the Internet and the new currency for the digital world" [1] and is created, stored, shared, and used by almost all companies. Market studies expect the overall global data monetization market to increase from USD 2.3bn in 2020 to USD 6.1bn in 2025 [2]. A sharp increase of data creation and low cost of storage are main drivers. Companies such as Facebook and Google have created massive data environments and developed their business models around the voluntary sharing of personal data by individuals in exchange for using their services [3, 4].

The OECD defines personal data as "any information relating to an identified or identifiable individual (data subject)" [5]. It shows traits of a public good rather than a commercial good as it is difficult to exclude other parties from using it in an effective

© The Author(s), under exclusive license to Springer Nature Switzerland AG 2022
R. Guizzardi et al. (Eds.): RCIS 2022, LNBIP 446, pp. 3–19, 2022.
https://doi.org/10.1007/978-3-031-05760-1_1

way and usage by one party does not prevent other parties from using it as well [3, 5, 6]. Data owners on the other hand have a need for privacy [7]. With the collection of personal data happening continuously and through a multitude of interconnected devices and modalities, there is an abundance of data and it is challenging to introduce effective control mechanisms on who transfers and uses the data [3]. Additionally, at least in Europe, privacy and the right for information to remain private is of high value and are protected by complex laws and regulations such as the General Data Protection Regulation [8] or new regulations aiming to increase data sovereignty [9, 10]. The current business assumption is that users are willing to share their data in exchange for the (free) service they receive. However, studies such as Schwartz [11], have shown that users are often not fully aware that they are paying for the service with their data and that their data may be collected and sold to data brokers who in turn sell data bundles to different data users along a data value chain. Other studies such as Sindermann et al. [12] investigate whether this influences customers willingness to pay for Social Media and show that only a minority supports a monetary payment model. As consumers learn more about how their data is used, this duality will eventually have consequences for companies' business models [3, 12]. Thus, it will be necessary for companies to attach a monetary value to the personal data shared with them and allow data producers/owners to actively consent to the use of their personal data in return for monetary or non-monetary compensation and develop a data market [6]. This can be facilitated through a data broker who transfers data from data owner to data user and monetary or non-monetary compensation from data user to data owner. The price of data depends highly on a diverse range of factors, some inherent in the data and some based on data context and the data subject. Combined with the need to account for the value attributed to privacy by data subjects, this adds an extra layer of complexity to pricing models as the value of the privacy of the data owner may exceed the value companies are willing to pay. The question of how to price personal data remains open despite multiple calls for research concerning this topic [6, 13]. Literature reviews in this area focus on pricing models [14, 15] or on the value of privacy [7] and have been published some 5 years ago. A more current review was undertaken by Wdowin and Deepeeven [16] as part of a research project and describes some factors related to the value of data. However, there is no detailed description of the search process and which criteria were chosen to define papers included in the review. There is currently no consensus in the literature on which factors influence the value of personal data. Given the omnipresence of personal data across divergent research fields and the regulatory efforts focusing on data sovereignty, a narrative synthesis of current findings is needed to provide a sound basis for research, model development and decision makers to answer the research question: Which factors influence the pricing of personal data?

We contribute to the discussion on pricing personal data in the following respects: (1) we identify relevant influence factors for pricing personal data and (2) provide a structure and categories for further qualitative and quantitative analysis of the subject.

2 Research Approach and Sample

2.1 Research Approach

We perform a narrative review [17] to conceptually integrate different fields of research on influence factors of the price of personal data. We use the following keywords: pricing of data, data markets, value of data, data valuation, data monetization, economics of personal data, pricing personal data and worth of privacy. We search EBSCO Business Source Complete as well as eLib, a specific directory which encompasses Web of Science, Scopus, Tema, Springer Link, Science Direct and other open access directories. We search for the keywords in TITLE/ABSTRACT of publications between 2001 and 2021. We find 1,535 papers in total. We screen the title and abstract based on (a) the topic of pricing privacy, pricing personal data or personal data markets and (b) peer review. We exclude papers that are out of scope, e.g. focusing on company data, pricing goods or services using personal data or using personal data for price discrimination. We exclude journal editorials or short summaries of conference papers. Overall, we exclude 1383 papers. We initially include 152 papers in our sample. We perform forward/backward referencing and include 8 additional in-scope papers respectively grey literature. Following the PRISMA [18] recommendations, we then perform a full text screening and exclude an additional 107 papers (1 duplicate, 106 papers out of scope based on the above mentioned criteria) from our final sample. Our overall sample consists of 54 papers. We extract relevant influence factors using an inductive content analysis [19] to identify the major themes in the literature. Two researchers perform the initial open coding independently and then perform multiple rounds of clustering and categorization to determine the main influence factors on the price of personal data. We derive a description for each factor and analyze the frequency of occurrence within our sample. We count each factor once per paper to determine the frequency.

2.2 Sample Description

The papers in the sample are published between 2005 and 2021 (see Fig. 1). The topic has become more frequently analyzed in the literature since 2014, highlighting its importance for current political and public discussion. There is a large spread of publications amongst different scientific outlets. About 60% of the analyzed papers were published in different scientific journals, while 30% were published in conference proceedings and the remaining 9% were published as grey literature reports. The topic of pricing personal data lies at an interface between business, economics, and information technology research. This is reflected in the publication outlets. The journals differed widely within the sample, with Electronic Markets being the most used outlet (3 publications), followed by Computer Law & Security Review (2 publications) and the IEEE Internet of Things Journal (2 publications). The remaining journals published one paper on the subject.

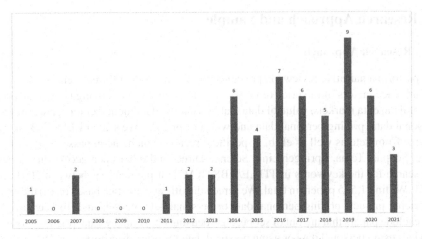

Fig. 1. Papers according to their year of publication

3   Results

3.1   General Results

Overall, the analyzed papers portrait a diverse stream of research. We classified them into eight categories: case study; commentary; data market model; data market model, technical[1]; data pricing model; experiment; literature review and report. The results are provided in Fig. 2. The appendix provides an overview of the core topics covered.

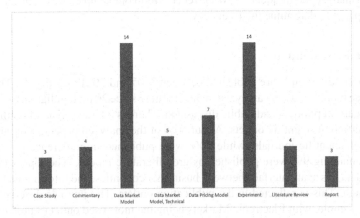

Fig. 2. Papers according to their content

The two most prevalent categories are theoretical data market models (14 papers + 5 papers on algorithm-based data markets) and experiments (14 papers). Data market

---

[1] The category "Data Market Model, Technical" refers to papers focusing on algorithms and technical solutions for data markets.

models focus on the development of different theoretical scenarios for establishing a personal data market. Most of the papers are theoretical in nature and deal with specific game-theoretical [20, 21] or auction-based [22] approaches for market creation. Some underline their findings with simulations based on real life data sets [23, 24]. Algorithm-based models show different ways to technologically facilitate data market settings and highlight the difficulties incorporating real-life influence factors such as anonymization and noise as well as profit-maximization calculations into an algorithm [25–27]. The experiments focus mainly on (a) eliciting willingness to pay for e.g. keeping personal data such as social media data [3, 28, 29] or preferences private [30] and (b) willingness to accept money for e.g. social media data [31] or location data [32, 33] from the participants[2]. A notable finding is a gap between WPT and WTA-values with the average WTA-value being significantly higher than the WTP to protects ones privacy [35–37]. We further find several papers on what we call data pricing models, which focus on different pricing methods and ways to elicit prices for personal data. Literature reviews in this area focus on pricing models [14, 15] or on the value of privacy [7] and have been published some 5 years ago. A more current review was undertaken by Wdowin and Deepeeven [16] as part of a research project and describes some factors related to the value of data. However, there is no detailed description of the search process and which criteria were chosen to define papers included in the review. The few case studies found in our research focus on the complexity of choice and value estimation for personal data.

3.2 Influence Factors on Pricing Personal Data

We find a multitude of influence factors on the price of personal data. We categorize them into four overarching categories: (1) data properties, (2) data context, (3) perceptions of data owner and (4) perceptions of data user. A detailed overview of the subsumed influence factors, their description and frequency of occurrence across the sample can be found in the respective Tables 1, 2, 3 and 4. "Data properties" refer to the inherent properties of a personal data dataset. This category represents general properties of data that are not limited to personal data, but apply to all types of data. We included this category as it lists important factors for pricing data. Within this category, sensitivity of data was found most frequently in the literature, followed by the data content, volume and coverage of data, quality data and level of data aggregation. The remaining factors were mentioned less frequently.

"Data context" refers to the environment and background of the personal data dataset. Within this category, data volume and inferability were found most often in the literature, often referring to arbitrage situations. Further, cost and length of data storage and cost of data gathering were discussed in the literature, while the remaining factors occurred less often.

The category "perceptions of data owner" focuses on the preferences and views of the data owner and the related willingness to consider selling personal data in general. Individual privacy and risk preferences and informational self-determination were found

---

[2] Willingness to pay (WTP) refers to the maximum amount of money an individual would be willing to pay to secure a specific change, while willingness to accept (WTA) refers to the minimum amount a person would be willing to accept to forego said change [34].

Table 1. Influence factors within the category "Data properties" and frequency of occurrence

Influence factor	Description	Frequency
Sensitivity of data	Refers to the level of identifiableness/anonymity and confidentiality and the associated loss of privacy in case of a data breach	18
Data content	Refers to the subject matter of the data	17
Level of data aggregation	Refers to the level of noise being added to a data set	12
Temporal coverage of data	Refers to the age and topicality of data	10
Volume and coverage of data	Refers to the amount and content of a specific dataset in terms of spatial coverage, granularity and generality of data	9
Uniqueness of data	Refers to the level of rarity/exclusivity of the data set	5
Data format	Refers to the data format and the linked interoperability, structure and resulting linkability of the data set with existing data	5
Intangibility of data	Refers to the inherent property that data is not a tangible asset	5
Completeness of data	Refers to the proportion of reality represented by the data (e.g. missing data points, biases in data collection)	4
Origin of data	Refers to the way the data was initially created	2
Filtering	Refers to the potential and level of filtering that is possible	1

Table 2. Influence factors within the category "Data context" and frequency of occurrence

Influence factor	Description	Frequency
Data volume and inferability	Refers to the amount of data available in general	17
Cost and length of data storage	Refers to the resources (e.g. hardware, financial, security and risk, time) used for data storage	12
Cost of data gathering	Refers to the resources (e.g. human, financial, time) used for data gathering	10

(continued)

Table 2. (*continued*)

Influence factor	Description	Frequency
Data gathering	Refers to the method (volunteered, surrendered, observed, inferred) and level of precision of data gathering	9
Level in the data value chain	Refers to the (commercial) potential and need for data refinement or aggregation for a specific purpose	9
Initial data owner	Refers to the profile and classification of the initial owner of the data	8
Frequency and type of usage	Refers to the delivery cadence of data and its usage	8
Data gatherer	Refers to the institution, person or device responsible for gathering data	5
Level of ownership	Refers to the accessibility and usage restrictions associated with the data	5
Culture	Refers to the societal context of data and the related cultural norms on e.g. sharing	5
Socio-economic impact	Refers to the social and economic consequences of data usage	4
Time of data access	Refers to the time and length of data access	1
Credibility	Refers to the level of verification of "correctness" of data	1

most frequently in the literature and are the most frequent influence factors mentioned overall.

Table 3. Influence factors within the category "Perceptions of data owner" and frequency of occurrence

Influence factor	Description	Frequency
Individual privacy and risk preferences	Refers to the individual's rationales and preferences on privacy (loss) and associated risk taking	26
Informational self-determination	Refers to the awareness of the value and ownership of data and the ability to act accordingly	26
Trust in data market	Refers to the perceived transparency, fairness, quality and morality of a specific data market by the data owner	12

(*continued*)

Table 3. (*continued*)

Influence factor	Description	Frequency
Trust in data user	Refers to the perceived level of transparency, fairness and morality of the data user and the post-sale use of the data	9
Service level offered to data owner	Refers to the perceived service level of the data market place offered to the owner of the data	8

The category "perceptions of data user" focuses on the preferences and views of the data user. Within this category, the trust in the data market and the individual utility of the purchased data for the data user were most frequently mentioned.

Table 4. Influence factors within the category "Perceptions of data user" and frequency of occurrence

Influence factor	Description	Frequency
Trust in data market	Refers to the perceived transparency, fairness, quality and morality of a specific data market by the data user	12
Individual utility of purchased data	Refers to the perceived expected utility of the purchased data for the data user	10
Service level offered to data user	Refers to the perceived service level of the data market place offered to the user	8

Overall, individual preferences on risk and privacy, informational self-determination, sensitivity of data and data volume and inferability were identified as the most mentioned influence factors in the literature.

4   Synthesis of Findings

Our main finding is that while there already is a body of literature concerned with pricing personal data, the research base is heterogenous. Research is spread out amongst fields of science, journals, and research streams. It seems that there is not yet a consensus on key questions and a lack of overarching frameworks on the subject. This diverse research base, however, highlights the importance of the topic.

We have identified two main research aspects: (1) theoretical and what we call technical data market models and (2) experiments to elicit willingness to accept money/willingness to pay for privacy in exchange for data. Market models are focused on theoretical aspects of pricing personal data and are utilizing game theoretical and profit maximization for developing narrow scope data market models. Technical data market models provide algorithms for different pricing mechanisms such as query-based pricing. While most papers try to validate their models with real life data sets, no paper provides insights from a real-life application of their model, which leads us to assume that data access is limited and that companies already operating data markets are unwilling to disclose their models as they are presumably core to their respective business models. Experiments focus on very narrow experimental settings, often with students as their subjects and are mainly focused on social media data (likes, shares, and general personal information) and a few using location data. To our knowledge, there are no experiments focusing on pricing more sensitive personal data such as electronic health records and very few studies with demographically diverse participants. Particularly when considering more sensitive personal data it would be interesting to gather information from a broad demographic including diverse age groups, educational backgrounds and levels of digital aptitude, as experiments show (1) irrationality concerning ones data (e.g. [35]) and (2) for parts of the participants a plain refusal to partake in data pricing (e.g. [38]). Additionally, there are some studies on data pricing models and very few case studies and literature reviews on the subject. While each of the streams has created a significant insight into the topic, it would be most useful to combine them to merge the aspects of (irrational) decision making of data subjects with more economical and algorithm-based thinking into a more practical data market model/algorithm, as has been attempted by e.g. Biswas et al. [39].

Looking at the factors that influence the price of personal data, we develop four categories of influence factors: (1) data properties, (2) data context, (3) perceptions of data owner and (4) perceptions of data user. All categories include several factors which we also rank by occurrence in the papers of the literature review. The most prevalent factors are individual privacy and risk preferences and informational self-determination. This is not surprising, as those factors are what differentiates personal data from e.g. company data and should thus have a strong impact on the price of personal data. Factors relating to trust, while still important, occur much less frequently, despite seemingly being a factor in increasing market participation by data owners. Market models thus may exclude an important factor in their setups if they exclude trust-creating mechanisms, Sensitivity, data content, and data volume and inferability are further factors that frequently appear in the literature. Those factors are rather generic and certainly applicable not only to the price of personal, but also data in general. Pricing also seems to be dependent on cost and length of data storage and cost of data gathering, which is not surprising since the utility derived from the data needs to outweigh the cost and since most data consumers operate within a restricted budget. Other factors such as culture, ownership level or origin of data appear infrequently and seem to be of less importance. Relating to culture, this is interesting since there are significant differences in how cultures approach the topic of privacy [40]. The most frequently used pricing method is market pricing. Other pricing concepts, such as option, per query or auction models appear much less frequently. This may be due to their perceived complexity in setup and operation.

5 Conclusion and Limitations

We conduct a narrative literature review and show a diverse array of research streams and questions that are related to the topic of pricing personal data and emphasize its importance. Due to the broad frame an aggregation of the results can only be done in a limited way. We believe that the resulting influence factors of our work are a valuable contribution to the scientific discussion and model development in future research. Our qualitative analysis of influence factors based on the underlying literature shows that the influence factors of the price of personal data can be classified into four categories. We provide a first description and ranking of the factors based on our literature review. These factors are only a starting point to researching the influence factors for pricing personal data and need further verification through empirical analyses such as a quantitative study or structured equation modelling. We aim to validate these factors empirically in future projects and to develop an operational model to quantify a price for personal data.

Acknowledgements. This research was partially funded by the German Federal Ministry of Education and Research (BMBF) within the scope of the research project DaWID (Platform for value determination and self-determined data release; funding reference number: 16SV8383).

Appendix - Overview of Included Papers

#	Paper	Classification	Key issue explored
13	Feijóo et al. [4]	Case Study	Case study on estimation of personal data
18	Hacker & Petkova [41]	Case Study	Case study on active choice of using data as currency
20	Holt et al. [42]	Case Study	Case study on value of data in stolen data markets
21	Jentzsch [43]	Commentary	Commentary on the difficulties of valuing personal data
37	Perera et al. [44]	Commentary	Commentary on the challenges of privacy protection in IoT
39	Raskar et al. [45]	Commentary	Commentary on challenges of data pricing and data markets
44	Sidgman & Crompton [6]	Commentary	Theoretical commentary on current challenges and research opportunities
4	Bataineh et al. [23]	Data Market Model	Two-sided data market model with experimental comparison based on real life data set
8	Choi et al. [46]	Data Market Model	Theoretical data market model with consumer consent for data collection

(*continued*)

(*continued*)

#	Paper	Classification	Key issue explored
12	Dimakopoulos & Sudaric [47]	Data Market Model	Theoretical data market with platform competition
16	Gkatzelis et al. [48]	Data Market Model	Theoretical data market model for unbiased data samples
22	Jiao et al. [22]	Data Market Model	Data market model with Bayesian profit maximization auction
23	Lei Xu et al. [24]	Data Market Model	Theoretical data market model with privacy and learning policies in a multi-armed bandit model
24	Li & Raghunathan [49]	Data Market Model	Data market model when purpose of data use is unclear
34	Niyato et al. [20]	Data Market Model	Theoretical data market model for optimal big data pricing with simulation
36	Oh et al. [50]	Data Market Model	Theoretical data market model between broker and service provider under profit maximization and respect for privacy protection and valuation
38	Radhakrishnan & Das [51]	Data Market Model	Theoretical data market model for smart grid data
45	Spiekermann et al. [52]	Data Market Model	Theoretical data market model focusing on challenges of personal data markets
46	Spiekermann & Novotny [53]	Data Market Model	Theoretical data market model focusing on operating principles
49	Tian et al. [54]	Data Market Model	Theoretical data market model based on optimal contract-based mechanisms
50	Wang et al. [21]	Data Market Model	Theoretical data market model with data owners exhibiting informed consent in a Nash equilibrium with a non-trusted data collector
3	Balazinska et al. [55]	Data Market Model, Technical	Technical data market model with query-based pricing

(*continued*)

(continued)

#	Paper	Classification	Key issue explored
11	De Capitani Di Vimercati et al. [27]	Data Market Model, Technical	Technical data market model focusing on including privacy issues in a cloud setting
25	Li et al. [56]	Data Market Model, Technical	Technical data market model with query-based pricing
30	Nget et al.[25]	Data Market Model, Technical	Technical market model and simulation of query-based pricing mechanism
53	Yang & Xing [26]	Data Market Model, Technical	Algorithm for personal data pricing with multi-level privacy division
7	Biswas et al. [39]	Data Pricing Model	Theoretical model to induce data provider to accurately report privacy price within differential privacy
29	Mehta et al. [57]	Data Pricing Model	Theoretical data pricing model with price-quantity schedule and approximation scheme for data seller
32	Niu et al.[58]	Data Pricing Model	Technical pricing model for trading aggregate statistics over private correlated data
33	Niu et al. [59]	Data Pricing Model	Algorithm for personal data pricing with reverse price constraint
42	Shen et al. [60]	Data Pricing Model	Data pricing model for Big Personal Data based on tuple granularity
43	Shen et al. [61]	Data Pricing Model	Data pricing model based on data provenance
54	Zhang et al. [62]	Data Pricing Model	Data pricing with privacy concern introducing privacy cost concept
1	Acquisti et al. [35]	Experiment	Experiment on WTP/WTA money for private data and privacy
5	Bauer et al. [29]	Experiment	Survey-based experiment on value of Facebook user information from user perspective

(continued)

(continued)

#	Paper	Classification	Key issue explored
6	Benndorf & Normann [31]	Experiment	Experiment to extraxt WTA money with take-it-or-leave-it offers
10	Danezis et al. [33]	Experiment	Experiment on WTA money for location tracking
15	Frik & Gaudeul [30]	Experiment	Method and experimental validation for elicitating the implicit value of privacy under risk
17	Grossklags & Acquisti [37]	Experiment	Experiment on WTP/WTA money for private data and privacy
19	Hann et al. [63]	Experiment	Conjoint analysis to estimate individual's utility of mitigate privacy concerns
26	Lim et al. [64]	Experiment	Discrete choice experiment to estimate value of types of personal information leakage
27	Mahmoodi et al. [28]	Experiment	Experiment quantifying WTP for different levels of privacy on social media platforms & analysis of psychological factors (ongoing)
31	Nielsen [38]	Experiment	Experiment showing lay peoples reaction to data markets is diverse and shows unwillingness to participate in data market
41	Schomakers et al. [65]	Experiment	Mixed-method study on data sharing and privacy preferences in data markets
47	Spiekermann & Korunovska [3]	Experiment	Experiment on WTP/WTA money for private data and privacy
48	Staiano et al. [32]	Experiment	Living lab experiment focusing on pricing and correlated behaviour patterns
52	Winegar & Sunstein [36]	Experiment	Survey-based experiment on the disparity of WTP and WTA money to give up privacy
2	Acquisti et al. [7]	Literature Review	Literature review on privacy

(continued)

(*continued*)

#	Paper	Classification	Key issue explored
14	Fricker & Maksimov [15]	Literature Review	Literature review on pricing of data products
28	Malgieri & Custers [14]	Literature Review	Literature review on pricing of personal data
51	Wdowin & Diepeveen[16]	Literature Review	Literature review on value of personal data
9	Coyle et al. [66]	Report	Policy Recommendation for capturing data value
35	OECD [5]	Report	Overarching report on pricing personal data
40	Rose et al. [67]	Report	Report on value of digital identity based on EU survey

References

1. Kuneva, M.: Keynote speech. Roundtable on online data collection, targeting and profiling, 31 March 2009
2. MarketsandMarkets: Data Monetization Market (2020). https://www.marketsandmarkets.com/Market-Reports/data-monetization-market-127405959.html. Accessed 23 Sep 2020
3. Spiekermann, S., Korunovska, J.: Towards a value theory for personal data. J. Inf. Technol. (2017). https://doi.org/10.1057/jit.2016.4
4. Feijóo, C., Gómez-Barroso, J.L., Voigt, P.: Exploring the economic value of personal information from firms' financial statements. Int. J. Inf. Manag. (2014). https://doi.org/10.1016/j.ijinfomgt.2013.12.005
5. OECD: Exploring the Economics of Personal Data: A Survey of Methodologies for Measuring Monetary Value (2013). https://doi.org/10.1787/5k486qtxldmq-en
6. Sidgman, J., Crompton, M.: Valuing personal data to foster privacy: a thought experiment and opportunities for research. J. Inf. Syst. (2016). https://doi.org/10.2308/isys-51429
7. Acquisti, A., Taylor, C., Wagman, L.: The economics of privacy. J. Econ. Lite. (2016). https://doi.org/10.1257/jel.54.2.442
8. European Commission: EU data protection rules (2020). https://ec.europa.eu/info/law/law-topic/data-protection/eu-data-protection-rules_en. Accessed 27 July 2020
9. Madiega, T.: Digital sovereignty for Europe. European Union (2020)
10. European Commission: A European Strategy for data (2021). https://digital-strategy.ec.europa.eu/en/policies/strategy-data. Accessed 13 Jan 2022
11. Schwartz, P.M.: Property, privacy, and personal data. Harv. Law Rev. (2004). https://doi.org/10.2307/4093335
12. Sindermann, C., Kuss, D.J., Throuvala, M.A., Griffiths, M.D., Montag, C.: Should we pay for our social media/messenger applications? Preliminary data on the acceptance of an alternative to the current prevailing data business model. Front. Psychol. (2020). https://doi.org/10.3389/fpsyg.2020.01415
13. Currie, C.S.M., Dokka, T., Harvey, J., Strauss, A.K.: Future research directions in demand management. J. Revenue Pricing Manag. (2018). https://doi.org/10.1057/s41272-018-0139-z

14. Malgieri, G., Custers, B.: Pricing privacy–the right to know the value of your personal data. Comput. Law Secur. Rev. (2018). https://doi.org/10.1016/j.clsr.2017.08.006
15. Fricker, S.A., Maksimov, Y.V.: Pricing of data products in data marketplaces. In: Ojala, A., Holmström Olsson, H., Werder, K. (eds.) ICSOB 2017. LNBIP, vol. 304, pp. 49–66. Springer, Cham (2017). https://doi.org/10.1007/978-3-319-69191-6_4
16. Wdowin, J., Diepeveen, S.: The value of data: literature review, Cambrige, UK (2020)
17. Cook, D.J., Mulrow, C.D., Haynes, R.B.: Systematic reviews: synthesis of best evidence for clinical decisions. Ann. Intern. Med. (1997). https://doi.org/10.7326/0003-4819-126-5-199 703010-00006
18. Moher, D., Liberati, A., Tetzlaff, J., Altman, D.G.: Preferred reporting items for systematic reviews and meta-analyses: the PRISMA statement. BMJ (Clin. Res. Ed.) (2009). https://doi.org/10.1136/bmj.b2535
19. Elo, S., Kyngäs, H.: The qualitative content analysis process. J. Adv. Nurs. **62**, 107–115 (2008)
20. Niyato, D., Alsheikh, M.A., Wang, P., Kim, D.I., Han, Z.: Market model and optimal pricing scheme of big data and Internet of Things (IoT). In: 2016 IEEE International Conference on Communications (ICC). 2016 IEEE International Conference on Communications (ICC), pp. 1–6 (2016). https://doi.org/10.1109/ICC.2016.7510922
21. Wang, W., Ying, L., Zhang, J.: The value of privacy: strategic data subjects, incentive mechanisms and fundamental limits. In: Proceedings of the SIGMETRICS 2016: Proceedings of the 2016 ACM SIGMETRICS International Conference on Measurement and Modeling of Computer Science (2016)
22. Jiao, Y., Wang, P., Niyato, D., Abu Alsheikh, M., Feng, S.: Profit maximization auction and data management in big data markets. In: 2017 IEEE Wireless Communications and Networking Conference (WCNC). 2017 IEEE Wireless Communications and Networking Conference (WCNC), pp. 1–6 (2017). https://doi.org/10.1109/WCNC.2017.7925760
23. Bataineh, A.S., Mizouni, R., Barachi, M.E., Bentahar, J.: Monetizing personal data: a two-sided market approach. Procedia Comput. Sci. (2016). https://doi.org/10.1016/j.procs.2016.04.211
24. Xu, L., Jiang, C., Qian, Y., Zhao, Y., Li, J., Ren, Y.: Dynamic privacy pricing: a multi-armed bandit approach with time-variant rewards. IEEE Trans. Inf. Forensics Secur. (2017). https://doi.org/10.1109/TIFS.2016.2611487
25. Nget, R., Cao, Y., Yoshikawa, M.: How to balance privacy and money through pricing mechanism in personal data market (2017)
26. Yang, J., Xing, C.: Personal data market optimization pricing model based on privacy level. Information **10**, 123 (2019)
27. Di Vimercati, S.D.C., Foresti, S., Livraga, G., Samarati, P.: Toward owners' control in digital data markets. IEEE Syst. J. 1–8 (2020). https://doi.org/10.1109/JSYST.2020.2970456
28. Mahmoodi, J., et al.: Internet users' valuation of enhanced data protection on social media: which aspects of privacy are worth the most? Front. Psychol. (2018). https://doi.org/10.3389/fpsyg.2018.01516
29. Bauer, M., et al.: Using crash outcome data evaluation system (CODES) to examine injury in front vs. rear-seated infants and children involved in a motor vehicle crash in New York State. Inj. Epidemiol. **8**, 1–10 (2021)
30. Frik, A., Gaudeul, A.: A measure of the implicit value of privacy under risk. J. Consum. Mark. (2020). https://doi.org/10.1108/JCM-06-2019-3286
31. Benndorf, V., Normann, H.-T.: The willingness to sell personal data. DICE discussion paper, vol. 143. DICE, Düsseldorf (2014)
32. Staiano, J., Oliver, N., Lepri, B., de Oliveira, R., Caraviello, M., Sebe, N.: Money walks: a human-centric study on the economics of personal mobile data. In: Proceedings of the 2014 ACM International Joint Conference on Pervasive and Ubiquitous Computing, 13 September 2014, pp. 583–594 (2014). https://doi.org/10.1145/2632048.2632074

33. Danezis, G., Lewis, S., Anderson, R.J.: How much is location privacy worth? In: WEIS
34. Hanemann, W.M.: Willingness to pay and willingness to accept: how much can they differ? Am. Econ. Rev. **81**, 635–647 (1991)
35. Acquisti, A., John, L.K., Loewenstein, G.: What is privacy worth? J. Leg. Stud. (2013). https://doi.org/10.1086/671754
36. Winegar, A.G., Sunstein, C.R.: How much is data privacy worth? A preliminary investigation. SSRN J. (2019). https://doi.org/10.2139/ssrn.3413277
37. Grossklags, J., Acquisti, A.: When 25 cents is too much: an experiment on willingness-to-sell and willingness-to- protect personal information. In: Proceedings of the Workshop on the Economics of Information Security Proceedings 2007 (2007)
38. Nielsen, A.: Measuring lay reactions to personal data markets. In: Proceedings of the 2021 AAAI/ACM Conference on AI, Ethics, and Society (AIES 2021) (2021). https://doi.org/10.1145/3461702.3462582
39. Biswas, S., Jung, K., Palamidessi, C.: An incentive mechanism for trading personal data in data markets. In: Cerone, A., Ölveczky, P.C. (eds.) ICTAC 2021. LNCS, vol. 12819, pp. 197–213. Springer, Cham (2021). https://doi.org/10.1007/978-3-030-85315-0_12
40. Cockcroft, S., Rekker, S.: The relationship between culture and information privacy policy. Electron. Mark. **26**(1), 55–72 (2015). https://doi.org/10.1007/s12525-015-0195-9
41. Hacker, P., Petkova, B.: Reining in the big promise of big data: transparency, inequality, and new regulatory frontiers. Northwestern J. Technol. Intellect. Prop. **15**(1), 6–42 (2017)
42. Holt, T.J., Smirnova, O., Chua, Y.T.: Exploring and estimating the revenues and profits of participants in stolen data markets. Deviant Behav. (2016). https://doi.org/10.1080/01639625.2015.1026766
43. Jentzsch, N.: Monetarisierung der privatsphäre: welchen preis haben persönliche daten? DIW Wochenbericht (2014)
44. Perera, C., Ranjan, R., Wang, L.: End-to-end privacy for open big data markets. IEEE Cloud Comput. **2**(4), 44–53 (2015). https://doi.org/10.1109/MCC.2015.78
45. Raskar, R., Vepakomma, P., Swedish, T., Sharan, A.: Data markets to support AI for all: pricing, valuation and governance (2019)
46. Choi, J.P., Jeon, D.-S., Kim, B.-C.: Privacy and personal data collection with information externalities. J. Public Econ. (2019). https://doi.org/10.1016/j.jpubeco.2019.02.001
47. Dimakopoulos, P.D., Sudaric, S.: Privacy and platform competition. Int. J. Ind. Organ. (2018). https://doi.org/10.1016/j.ijindorg.2018.01.003
48. Gkatzelis, V., Aperjis, C., Huberman, B.A.: Pricing private data. Electron. Mark. **25**(2), 109–123 (2015). https://doi.org/10.1007/s12525-015-0188-8
49. Li, X.-B., Raghunathan, S.: Pricing and disseminating customer data with privacy awareness. Decis. Support Syst. (2014). https://doi.org/10.1016/j.dss.2013.10.006
50. Oh, H., Park, S., Lee, G.M., Choi, J.K., Noh, S.: Competitive data trading model with privacy valuation for multiple stakeholders in IoT data markets. IEEE Internet Things J. **7**(4), 3623–3639 (2020). https://doi.org/10.1109/JIOT.2020.2973662
51. Radhakrishnan, A., Das, S.: Data markets for smart grids: an introduction. In: 2018 IEEE Innovative Smart Grid Technologies - Asia (ISGT Asia). 2018 IEEE Innovative Smart Grid Technologies–Asia (ISGT Asia), pp. 1010–1015 (2018). https://doi.org/10.1109/ISGT-Asia.2018.8467818
52. Spiekermann, S., Acquisti, A., Böhme, R., Hui, K.-L.: The challenges of personal data markets and privacy. Electron. Mark. **25**(2), 161–167 (2015). https://doi.org/10.1007/s12525-015-0191-0
53. Spiekermann, S., Novotny, A.: A vision for global privacy bridges: technical and legal measures for international data markets Comput. Law Secur. Rev. (2015). https://doi.org/10.1016/j.clsr.2015.01.009

54. Tian, L., Li, J., Li, W., Ramesh, B., Cai, Z.: Optimal contract-based mechanisms for online data trading markets. IEEE Internet Things J. **6**(5), 7800–7810 (2019). https://doi.org/10.1109/JIOT.2019.2902528

55. Balazinska, M., Howe, B., Suciu, D.: Data markets in the cloud: an opportunity for the database community. Proc. VLDB Endow. **5**(12), 1962–1965 (2011)

56. Li, C., Li, D.Y., Miklau, G., Suciu, D.: A theory of pricing private data. ACM Trans. Database Syst. (2014). https://doi.org/10.1145/2691190.2691191

57. Mehta, S., Dawande, M., Janakiraman, G., Mookerjee, V.: How to sell a dataset? Pricing policies for data monetization. SSRN J. (2019). https://doi.org/10.2139/ssrn.3333296

58. Niu, C., Zheng, Z., Wu, F., Tang, S., Gao, X., Chen, G.: Online pricing with reserve price constraint for personal data markets (2019)

59. Niu, C., Zheng, Z., Wu, F., Tang, S., Gao, X., Chen, G.: ERATO: trading noisy aggregate statistics over private correlated data. IEEE Trans. Knowl. Data Eng. (2019). https://doi.org/10.1109/TKDE.2019.2934100

60. Shen, Y., Guo, B., Duan, X., Dong, X., Zhang, H.: A pricing model for big personal data. Tsinghua Sci. Technol. **21**(5), 482–490 (2016). https://doi.org/10.1109/TST.2016.7590317

61. Shen, Y., et al.: Pricing personal data based on data provenance. Appl. Sci. Basel **9**(12) (2019). https://doi.org/10.3390/app9163388

62. Zhang, Z., Song, W., Shen, Y.: A reasonable data pricing mechanism for personal data transactions with privacy concern. In: U, L.H., Spaniol, M., Sakurai, Y., Chen, J. (eds.) APWeb-WAIM 2021. LNCS, vol. 12859, pp. 64–71. Springer, Cham (2021). https://doi.org/10.1007/978-3-030-85899-5_5

63. Hann, I.-H., Hui, K.-L., Lee, S.-Y.T., Png, I.P.L.: Overcoming online information privacy concerns: an information-processing theory approach. J. Manag. Inf. Syst. (2007). https://doi.org/10.2753/MIS0742-1222240202

64. Lim, S., Woo, J., Lee, J., Huh, S.-Y.: Consumer valuation of personal information in the age of big data. J. Assoc. Inf. Sci. Technol. (2018). https://doi.org/10.1002/asi.23915

65. Schomakers, E.-M., Lidynia, C., Ziefle, M.: All of me? Users' preferences for privacy-preserving data markets and the importance of anonymity. Electron. Mark. **30**(3), 649–665 (2020). https://doi.org/10.1007/s12525-020-00404-9

66. Coyle, D., Diepeveen, S., Wdowin, J., Kay, L., Tennison, J.: The value of data: policy implications, Cambridge, UK (Februaray 2020). https://www.nuffieldfoundation.org/project/valuing-data-foundations-for-data-policy

67. Rose, J., Rehse, O., Robe, B.: The value of our digital identity (2012). https://www.bcg.com/publications/2012/digital-economy-consumer-insight-value-of-our-digital-identity.aspx. Accessed 29 Apr 2020

What's in a (Data) Type? Meaningful Type Safety for Data Science

Riley Moher[✉], Michael Gruninger, and Scott Sanner

Department of Mechanical and Industrial Engineering, University of Toronto,
Ontario M5S 3G8, Canada
Riley.moher@mail.utoronto.ca

Abstract. Data science incorporates a variety of processes, concepts, techniques and domains, to transform data that is representative of real-world phenomena into meaningful insights and to inform decision-making. Data science relies on simple datatypes like strings and integers to represent complex real-world phenomena like time and geospatial regions. This reduction of semantically rich types to simplistic ones creates issues by ignoring common and significant relationships in data science including time, mereology, and provenance. Current solutions to this problem including documentation standards, provenance tracking, and knowledge model integration are opaque, lack standardization, and require manual intervention to validate. We introduce the meaningful type safety framework (MeTS) to ensure meaningful and correct data science through semantically-rich datatypes based on dependent types. Our solution encodes the assumptions and rules of common real-world concepts, such as time, geospatial regions, and populations, and automatically detects violations of these rules and assumptions. Additionally, our type system is provenance-integrated, meaning the type environment is updated with every data operation. To illustrate the effectiveness of our system, we present a case study based on real-world datasets from Statistics Canada (StatCAN). We also include a proof-of-concept implementation of our system in the Idris programming language.

Keywords: Data science · Dependent types · Type safety · Data provenance · Meaningful types

1 Introduction

Data science is a delicate task involving the analysis and manipulation of heterogenous datasets that represent real-world phenomena embedded with assumptions, rules, and interpretations. This data, however, is ultimately represented using simple datatypes like strings or integers, which fail to typify the real-world concepts the data represents. Whether comparing net and gross profit, averaging populations of overlapping regions, or summing measurements from different time periods, real-world phenomena are much more than the simple datatypes that represent them. With data science becoming an increasingly integral part of many industries, important decisions are being made based on results produced by data scientists. These decisions can have significant consequences, the wrong decision made by a hospital administrator puts human lives at stake, the decisions of policy-makers can have far-reaching impacts to our society; ensuring data science is correct and verifiable will have significant benefits.

To solve datatype issues, existing work supplements simple datatypes with external information in an informal, ad-hoc, and highly manual manner. These approaches reduce the nuanced real-world rules and interpretations of data to scattered informal knowledge and documentation, provenance records, and prohibitively complex knowledge models. This leads to an environment where results become extremely difficult to validate and verify, especially throughout the long and complex process that is the modern data science pipeline.

In this paper, we present the meaningful type safety (MeTS) framework, where type checking equates to validating the rules and assumptions of the real-world phenonema the data represents. Rather than use strings or floats, MeTS relies on semantically-rich types like population, time intervals, and geospatial regions. We do this through powerful, expressive, and decidable dependently-typed programming, based on axioms of formal knowledge models. Our approach, contrary to the current state, is formal, automatic, and easily verifiable.

This paper consists of a detailed presentation of the issue of datatypes and how our framework can be applied to solve it, including a case study of StatCAN census data to make our framework concrete. To provide salient motivation, we identify specific scenarios where errors of real world rules and interpretations frequently occur in data science, and identify how current approaches fail to address these scenarios. The most significant portion of this paper is the specification of MeTS, presented in a formal syntax based on Martin-Löf type theory [18], and also instantiated for the StatCAN census data with a proof-of-concept implementation in the Idris programming language. Finally, we discuss how our work fits into a broader research program including a correspondence theorem and a focus on tractability.

2   Datatypes Fail to Typify Data

This section will detail three prominent and significant classes of datatype problems including those associated with time, mereology, and data provenance. Datatype errors are a result of the fact that one simple datatype may be used to represent multiple very different real-world concepts. Both a person's age and the population of a city could be represented by an integer, despite ages and city populations having very different rules and interpretations. This leads to many different kinds of problems, from something seemingly simple like adding feet to metres, to more complex errors like geospatial mereology changing as a function of time.

2.1   Time

Time is an especially important concept, it is an ever-present factor in data which contains many implicit rules and assumptions. For example, training a machine learning model on observations that occurred later than test-set observations would be predicting the past from the future, producing a non-sensical, anti-causal model. Furthermore, the real-world interpretation of data may change as a function of time, while labels do not. Consider comparing the price of a home in 1967 to that of a home in 2021, the relationship of purchasing power and time must be reconciled.

Time issues can be further exacerbated through the layering of time with other metadata like units or labels. Consider financial analysis on the Frankfurt stock exchange; even ignoring inflation, a quote from 1960 would be given in Marks, while one from 2004 would be given in Euros. The real-world rules and assumptions of time in a given dataset, even if well-understood, lack a formal means of integration or validation, a human still must interpret column labels, consult documentation, or manually review code. So long as manual intervention is necessary to verify and validate results, errors are bound to occur, and negative consequences are bound to follow.

2.2   Mereology

One of the most fundamental relationships that exists in data is mereology, the relationship of parts to a whole. The seminal work of Winston et al. [22] and the myriad of work that built upon it, especially recent developments in the knowledge modelling community [4,12], illustrates that mereology is a complex concept. In data science, mereology manifests itself in many ways, from categories and sub-categories to overlapping time intervals, data scientists must be aware of these relationships to preserve the integrity of interpretation. For example, consider a data scientist working with COVID-19 vaccination data: they must understand that "individuals with two vaccine doses" is a subset of "individuals who have received a single dose", and that both are part of the larger whole of "eligible individuals". Understanding these relationships is crucial to avoiding errors like double-counting and preserving the integrity of results. Furthermore, time may complicates these issues further, "eligible population" may change to encompass new age groups, booster shots and a new 3+ dose category could alter the definition of "fully vaccinated", etc.

Underlying mereological relationships are not formally modelled or integrated in data science pipelines, so catching these kinds of errors requires careful and laborious manual review. In the data science pipeline, simple datatypes reduce concepts with mereological relationships to formats that cannot capture this reality.

2.3   Provenance

Just as time alters the rules and interpretation of data, so do the operations data scientists perform in their analyses. Even assuming a complete formal understanding of a dataset, the interpretation of that dataset will throughout the data pipeline. For example, the average of city populations is no longer itself a "city population"; a new entity has been derived, with its own real-world rules and interpretation. Additionally, just like time, provenance adds a layer of complexity in conjunction with the previously mentioned issues.

Consider a relevant example of provenance issues involving physical units, whereby types consist of physical units; going beyond floats and integers to kilograms or feet. Even with this more complex representation, ensuring the integrity of the data's real-world interpretation is more than ensuring: `Unit X == Unit Y`. Take for example data consisting of individual's height measurements: if Bob and Alice's heights are both measured in centimetres, $height_{bob} + height_{alice}$ is completely type-safe from a unit perspective, but the result of this computation has no meaningful interpretation; it is no longer a height. Alternatively, $population_{regionA} + population_{regionB}$ does have a meaningful interpretation: it makes sense to speak of "total populations", but not "total heights". Even with more robust datatypes, it is not enough to enforce only equal types for any given operation, operations each have their own preconditions, and may produce entirely new datatypes as a result.

3   Current Approaches

Data scientists utilize many tools and methodologies to address the problems of datatypes, but none of them sufficiently address datatype issues like time, mereology, or provenance. These limitations result from various factors inherent in existing methods, like the necessity of manual effort, informal representations, poor standardization, and ad-hoc applications. This section will critically examine these approaches in three categories: documentation, provenance tracking, and knowledge representation techniques.

3.1   Documentation Standards

Documentation is a tool intended to ensure quality, reduce risk, save time, and encourage knowledge sharing. However, effective utilization of documentation for data science is a challenging task. The most significant issues of using documentation to address datatype issues is their informal nature, their inability to properly address provenance, and their lack of standardization.

One of the most significant recent developments for data science documentation is Datasheets for Datasets [7], an effort to create a gold-standard for machine learning dataset documentation. These datasheets consist of answers to curated questions about a wide range of dataset information, like their motivation, collection process, pre-processing steps, and maintenance. While encouraging ML practitioners to be well-informed about their data is a positive intent, this work cannot avoid the inherent limitations of documentation. Even with a complete datasheet, there is no guarantee that the answers supplied to the set of questions will be of sufficient detail, contain sufficient contextual information, or be unambiguous. The ambiguity of natural langauge is problematic for data science, without any formal specification beyond natural language, the issues of semantic heterogeneity are unavoidable. Moreover, while the authors do provide example datasheets for well-known datasets, they do not evaluate these examples or provide any means of evaluation; there are no quality standards for datasheets.

An additional issue with documentation-based approaches is a failure to sufficiently address provenance. With the interpretation of data changing as the result of data transformations, the information contained in the documentation must change accordingly. This will require an unrealistic degree of manual effort to update the documentation and maintain effective versioning. The last major issue of documentation approaches is a lack of standardization. Even for Datasheets for Datasets, the authors acknowledge that enterprises use unique implementations of their work, like Google's Model Cards [19] or IBM's factsheets [15]. This lack of standardization places unnecessary burdens on data scientists which hinders knowledge-sharing and interoperability. With no formal semantics, quality criteria, provenance-awareness, or standardization, documentation-based approaches cannot adequately address the datatype issues.

3.2 Provenance Tracking

Provenance tracking provides information about the origins, history and derivation of data. For an excellent survey on data provenance work, we refer the reader to [14]. While provenance tracking is mature, formal, and well-understood within the data modelling and engineering community, provenance tracking alone does not sufficiently address the nuances of datatype issues. One example of a formal representation of provenance is in Buneman et al.'s Graph Model of Data and Workflow provenance [1], which models workflow provenance through directed acyclic graphs called provenance graphs. As opposed to documentation-based approaches, these approaches have formal semantics and are machine-readable. However, despite this formality, this provenance information still must be interpreted with respect to the specific real-world phenomena modelled in the data. Provenance tracking along does not integrate real-world knowledge with provenance information, and fails to automatically detect datatype errors. While provenance-based approaches are preferable to manually reviewing code, it still lacks the additional semantics necessary to enforce meaningful data science.

3.3 Knowledge Representation

Knowledge Representation is a rich discipline containing many models which represent complex rules and interpretations from the real-world in a very formal way. While there are many widely accepted and useful knowledge models, such as ontologies for units of measure [11,21], time [13], and processes [10], current work does not sufficiently integrate and apply them for general-purpose data science. Even given an agreed-upon ontological representation of real-world concepts, there is no consensus on how to integrate this model into the data science pipeline, and no guarantee on the tractability or scalability of these methods applied to data science.

Context interchange technology (COIN) [6,8,17] identifies a similar problem identified in this paper, that of semantic heterogeneity between datasets, and addresses it using a general knowledge model to translate between more specific models, or "contexts". The issue with this approach is that COIN accounts only for a single data science operation: projection. The user queries some data and it is converted into their specific context for them; it does not support the interpretation of addition, multiplication, averages, etc. Furthermore, COINs contexts are specified weakly, with just two relations: is_a and attribute. COIN's shortcomings are inherent to description logic-based (DL) approaches, as even with additional properties, DL lacks the expressiveness to integrate real-world rules into various operations.

Foundational Ontologies for Units of Measure (FOUnt) [11] is a collection of ontologies that model rules for combining units of measure with respect to the physical objects and processes they describe. While these ontologies are more expressive than those of COIN, they also provide limited coverage of data science operations; only specific instances of addition, subtraction, and division. In addition, FOUnt lacks consideration of provenance; the axioms of units do not change as specific operations are performed unless an entirely new unit is derived (like density from mass and volume). The other issue with higher-order logic-based approaches like FOUnt is complexity; reasoning in a data science application would require an enormous volume of instantiations and undecidable reasoning with a theorem prover.

Real-World Types (RWT) [23,24] is a framework with a motivation very similar to that of COIN and this paper; it identifies a mismatch between machine representation and the real-world entities they represent. Unlike COIN and FOUnt, however, RWT is much more informal in specifying types, they are essentially annotations of OOP objects, consisting of a name, a natural language definition, possible values, and references. Furthermore, the RWT framework relies on OOP naming conventions to identify entities and generate candidates for real-world types, and requires manual review to identify errors. This approach does not rely on any formal model or logic to detect errors or enforce rules, being more akin to a low-level and object-oriented version of an approach like datasheets for datasets [7]. RWT's treatment of real-world rules as add-ons to objects speaks to the larger issue of a reliance on objects as an alternative to simple datatypes. The object-oriented-paradigm can perform many run-time checks to verify properties of data and attempt to enforce real-world rules. The specification of real-world rules themselves, however, must be formally specified in some other format, as annotations, an ontology, etc. Furthermore, reliance on object implementations to ensure the integrity of real-world rules entangles the hides the ontological commitments of programmers behind implementation decisions.

Other knowledge modelling approaches involve utilizing upper ontologies, those that model real-world concepts generally, in conjunction with conceptual modelling languages like UML. For instance, in [2], Albuquerque et al. employ semantic reference spaces to ontologically ground UML datatypes. The reference structures used to specify the meaning of datatypes consist of a taxonomy for measurement dimensions like ordinal dimensions or interval dimensions. Consider the datatype for the temperature in Toronto, $10°$ C is an integer interval dimensions which is associated to the measurable quality `Outside Temperature`, of the `Toronto` city. Knowing that this quantity is an interval dimension, I know I can compare it to other temperatures through relations like `hotterThan`, and knowing it is a characterization of a place, I could limit these comparisons to other cities, to answer questions like "Is Chicago hotter than Toronto?". However, when one considers the data science application, this approach provides no semantic information to address which operations could be performed on this quantity; can I add together the temperature of Toronto and Chicago? The application of these taxonomy-centric solutions often focus on providing explanations for how quantities may be associated to concepts, but do not provide solutions for the complex issues encountered when combining and manipulating datatypes in the data science pipeline.

An additional form of knowledge modelling is conceptual modelling in the database community, in approaches like the entity relationship (ER) model [5], temporal ER models [9], and semantic data models (SDM) [20]. While these approaches provide a more formal means of understanding and documenting database schema, they, like knowledge modelling approaches in general, lack a focus on data science specifically. This kind of conceptual modelling establishes understandings of concepts and their relationships for well-organized and collected data. Data science however, often deals with found data, and must integrate semantically heterogeneous data from different sources to enable unambigious sharing. For example, a demographics database will likely have a singular strict definition for "family income", whereas data science tasks may require reconciliation of "family income" as defined by StatCAN, and varying definitions defined by other political entities or organizations. Provenance is another important concern, as we have established, the interpretations of concepts are not static as a consequence of data science operations. While knowledge models do make steps in the right direction in terms of formality, their lack of a data science-focused approach means they cannot adequately solve datatype issues.

4   Case Study: StatCAN Census Data

Now having a comprehensive understanding of datatype issues including how current work fails to sufficiently address them, we provide further, concrete motivation drawn from Statistics Canada (StatCAN) census data. These datasets model a variety of demographic information for various geographic areas throughout Canada, and are made publicly available[1]. We base our analysis on population data divided by geographic divisions of various levels[2] from 2011

---

[1] StatCAN's full range of Census data can be accessed at https://www12.statcan.gc.ca/census-recensement/2021/dp-pd/index-eng.cfm.

[2] The specific datasets we referenced can be accessed by their StatCAN catalogue numbers: 98-401-X2016041, 98-401-X2016043, 98-401-X2016066.

and 2016. These datasets allow us to provide specific and concrete examples of the aforementioned datatype problem classes and can also be used to evaluate our framework. Furthermore, concepts like geospatial regions, time, and mereology are not only integral to these datasets, but are commonly represented across many diverse datasets.

4.1 Example Operations

In order to provide specific and concrete instances of how the problems of datatypes occur within data science, we exemplify datatype errors through operations a data scientist may perform on census data. These operations may be addition, subtraction, or averages, among other common operations, and are performed on population quantities defined for a specific geospatial region and timepoint. While this case study is on population data, the embedded concepts like time, geospatial regions, and mereology are widely applicable to general datasets. Each of these example computations violates real-world assumptions or rules associated with census data, and would produce a result that, while a quantity could be computed, would have no meaningful interpretation. These examples form a foundation for the problem of providing precise explanations for how and why real-world rules are violated through datatype errors.

Disjointedness. The following two example computations violate real-world rules of temporal and geospatial disjointedness, respectively.

– Computing the median of non-disjoint populations (campbelton and each of its parts in bordering provinces)

```
median(Toronto, ... Campbelton, Campbelton (NB Part), Campbelton (Quebec Part))
```

– Computing the sum of populations over disjoint time periods

```
sum(Toronto2016, Hamilton2016, ... Guelph2011)
```

Each of these errors stems from disjointedness, with the first example, adding Campbelton's population together with that of its parts, is a case of double-counting overlapping regions. The second example contains addition over populations which vary over both time *and* geospatial regions, producing a nonsensical result.

Provenance. The following two examples illustrate how data science operations produce new kinds of data, forming new rules.

– Computing the average of two average populations

```
avg(avg(Toronto, Hamilton), avg(Guelph, Kitchener))
```

– Computing the average of a population change over time, and a population difference in regions

```
avg((Toronto2016 - Toronto2011), (Guelph 2011 - Hamilton2011))
```

For the first example, while an average of averages could have an interpretation, it is not the correct means of obtaining an average over the four regions (which is true not just of demographics data but for any data). With the second example, an average of differences seems like an average over the same kind of quantity. However, the key distinction is what these differences are aggregated 'over': change over time is not comparable to a difference over geospatial regions.

5   Meaningful Type Safety Framework (MeTS)

In this section, we provide a high-level overview of the framework designed to model real-world concepts and rules like those contained in StatCAN Census data and detect errors like those given in the example operations. The Meaning-ful Type Safety framework (MeTS) rejects simple datatypes and elevates type safety to a meaningful result, one which is derived from specific representations of real-world concepts and their associated rules and interpretations. The complete and production-ready implementation of MeTs would include several compo-nents, including a method of integration with existing data science tools, and a complete correspondence theorem, both of which fit into the broader research program of MeTS (discussed in Sect. 8). The focus of this chapter, however, is the technical heart of MeTS, namely the type-checking mechanism. In keeping with a focused scope, we present the MeTS architecture as two primary components: the interface component, and the typing component, depicted in Fig. 1.

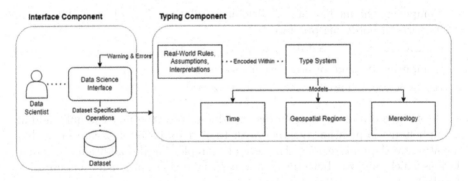

Fig. 1. Architecture for the meaningful type safety framework (MeTS)

5.1   Interface Component

The interface component is how the data scientist interacts with MeTS, by performing operations on some dataset(s) and receiving corresponding feed-back from the type system. In the context of this paper, the interface com-ponent exists within our proof-of-concept implementation (further elaborated in Sect. 6.4) using the Idris programming language [3]. However, it should be noted that *we do not expect data scientists to learn dependently-typed functional programming*, a production-ready interface would be implemented through common data science tools like pandas, R, or Tableau for familiarity and ease of use, with type-checking still performed in a dependently-typed language. Operations that a data scientist performs would be done the same as they currently are, while real-world concepts would be specified at the beginning stage of the data pipeline. This initial specification would associate a dataset with concepts formally modeled in the type system, such as time, geospatial regions, or mereological relationships, thereby pairing the dataset with its real-world interpretation and rules. This pairing will then enable the automatic detection of real-world rule violations for any downstream operations involving the dataset. This pairing is trivial in our proof-of-concept implementation since the specification of the dataset(s) is already in Idris.

5.2 Typing Component

The typing component is the primary reasoning mechanism of the framework, based on a dependently-typed representation of the real-world rules and assumptions of the dataset(s) being modelled. The type system has the goal of implementing the kinds of real-world checks that would normally be done manually, through type-checking. This elevates the notion of type-safety from a trivial check to guaranteeing a level of interpretability. It should be noted that the real-world rules and assumptions encoded within this typing component are not arbitrary and are rooted in formal logic models, namely first order logic ontologies. The specific relationship between these ontologies and the typing system is outside the scope of this paper and encompassed in the correspondence theorem, as described in Sect. 8.1. The specifics of the dependently-typed program component will be presented in complete detail in Sect. 6.

6 Type System

The type system of MeTS uses dependent type theory and expression tree representations to construct preconditions for a wide range of data science operations. For the purpose of this paper, MeTS is presented in a simplified syntax based on Martin-Lof's intuitionistic type theory [18], and basic functional programming. MeTS is presented here with specific type-checking functionality given for time, geospatial entities, and data provenance. Additionally, the type system is evaluated with respect to the previously presented example operations (with a corresponding proof-of concept implementation in the Idris programming language [3]).

6.1 Syntax

Since dependent type theory is not typically associated with data science, our presentation of the MeTS type system assumes little to no prior experience with dependent types or functional programming (for a more thorough introduction to these topics, we refer the reader to [16]).

To illustrate our function syntax, consider the basic factorial function:

```
factorial : Integer → Integer
factorial(0) = 1
factorial(n) = n*(factorial(n - 1))
```

Three syntactic constructs should be observed here: typing declarations, function notation, and pattern matching. We use the : operator to denote typing declarations, it can be read as "has the type". The → operator denotes a function type, a function taking one argument will have one →, infixed between the input type and the output type, a function taking n arguments will have $n \rightarrow$ symbols. Lastly, the actual function definition utilizes pattern matching, where specific patterns for the arguments of the function are given as a 'blueprint' for the function to follow. In our example, the first pattern is given as a base-case, and all other possible inputs to our function can be computed recursively. It should also be noted that MeTS utilizes only total functions, that is, functions will have patterns for all possible value of the input type(s).

Dependent Type Theory. The central idea of dependent types is that types may depend on values or other types. Dependent type theory has two new type constructs, the dependent function type and the dependent pair type. The dependent function type constructs a type from some parameter, the canonical example of which is vectors of length **n**, with **n** being a type parameter of the natural number type, denoted $n : \mathbb{N}$. In the MeTS framework, we utilize the dependent pair type, since it provides a method of enforcing operation preconditions analogous to real-world rules.

The basic example of a dependent pair type describes pairs of values where the type of the second element depends on the value of the first. In general, we can write a dependent pair type as:

$$\sum_{a:\mathbb{A}} B(a)$$

Firstly note that the typical meaning of the $\sum$ operator is interpreted differently when read in a typing statement, it denotes a dependent pair type here, also sometimes called a sigma type for this reason. If $(a, b) : \sum_{a:\mathbb{A}} B(a)$, then we can say $a : A$, and $b : B(a)$. That is, b's type depends on the value of a. With this construct, we can also enforce relationships within a tuple, consider the following type:

$$\sum_{m:\mathbb{N}} \sum_{n:\mathbb{N}} ((m < n) = True)$$

This type describes tuples (m, n) of natural numbers where the first element is less than the second. Values inhabiting this type would consist of a triple containing: the two elements m and n, and a proof of $m < n = True$. Given that the definition of the $<$ operator is total, this expression simply reduces to $True = True$ and a proof of the above can be done via reflection, type-checking is decidable for total functions. Types of this form are essential to our typing framework: functions with arguments of this form are analogous to preconditions for those functions.

Alternate Notations. We have chosen a notation similar that of Martin-Lof type theory since it is the seminal notation, and because the it is compact and simple to follow. Other notations are sometimes used in more application-oriented work, as in many functional programming papers, and may have minor variations. Table 1 shows the dependent pair types written in some of these alternate notations, as well as how these types could be written in Idris syntax.

Table 1. Differences in dependent type theory notations

Notation	Examples
Martin-Lof Type Theory	$\sum_{a:A} B(a)$
	$\sum_{m:\mathbb{N}} \sum_{n:\mathbb{N}} ((m < n) = True)$
Dependently-Typed Lambda Calculus	$\sum a : A.B(a)$
	$\sum m : \mathbb{N}. \sum n : \mathbb{N}.((m < n) = True)$
Dependently-Typed Lambda Calculus (Alt)	$\exists a : A.B(a)$
	$\exists m : \mathbb{N}.\exists n : \mathbb{N}.((m < n) = True)$
Idris	`(a : A) -> (b : a -> Type) -> Type`
	`mn:(Nat, Nat)**((fst mn < snd mn)=True)`

6.2   Expression Trees

To properly account for the effects of provenance, MeTS incorporates this information in type-checking through an expression tree construct. Essentially, types are a combination of base types and the operations that have been performed, in a recursively-defined tree structure:

$$\mathbb{B} :: = Population \mid Time \mid GeoEntity \mid ...$$
$$\mathbb{O} :: = Sum \mid Sub \mid Mult \mid Div \mid Avg \mid ...$$
$$\mathbb{E} :: = Atom(\mathbb{B}) \mid Over(\mathbb{O}, (List\ \mathbb{E}))$$

That is, expressions are either an atom (base type) or some operation 'over' a set of expressions. This structure allows us to define type-checking functions, preconditions, that traverse these trees, an essential component of the primary mechanism behind MeTS.

6.3   Preconditions

Preconditions are defined by typing operation functions with dependent pair types consisting of the actual operands and a proof that they satisfy some set of preconditions. The general form is given by:

$$\text{operation} : \sum_{\text{operands} : \text{(List } \mathbb{E})} (\text{opPreconditions(operands)=True}) \rightarrow \text{result}:\mathbb{E}$$

The functions that evaluate preconditions, `opPreconditions` are structured in a way that facilitates re-use and sharing of preconditions. For example, the precondition definition function for population sums is given by:

```
popSumPreconditions : List Population → Bool
popSumPreconditions(ps) = disjointRegions(ps) & allSameTime(ps) &
allMeasured(ps)
```

The preconditions for a sum over populations contains more general precondition functions defined over geospatial regions, time, and provenance, that can be referenced in other operation preconditions. These specific preconditions model the real-world assumptions of a sum of populations, that it is a count of individuals over disjoint areas, at some consistent point in time. `allMeasured` is an example of a provenance-based precondition, its truth value is based on the the structure of the expression tree. Intuitively, the statement "the total population of Toronto and Hamilton 125.3 people", is nonsensical, a total population should never result in a non-whole number. Analogously, a sum of populations should never be performed over estimated or aggregated populations. The `allMeasured` function is the formal representation these real-world rules, and acts as a method of catching a violating operation before it happens, ensuring type-safe operations are meaningful.

6.4   Implementation

For the purposes of demonstrating MeTS, we implemented the type system in the dependently-typed programming language Idris [3][3]. Idris was chosen for its simple to read syntax, and its decidable type-checking. Idris type-checking is decidable since Idris can enforce totality of functions, and ensure that any functions involved in a type-checking operation (like precondition functions) are total, and therefore, decidable. It is important to note that our framework is independent of the implementation language chosen, part of the reason why we do not present our framework in Idris syntax, but a more general dependently-typed notation.

7   Application and Evaluation on StatCAN Data

In order to deliver on the original motivation for meaningful types, we demonstrate MeTS in action on the previously mentioned StatCAN data. This section references the specific type-checking mechanisms that detect violations of real-world rules including disjointedness and provenance, and how this form of type-checking can be abstracted to general problems.

---

[3] Our proof-of-concept implementation with all the example census computations and more is available at https://github.com/riley-momo/Meaningful-Type-Safety-For-Data-Science.

7.1 Disjointedness

Disjointedness is the lack of any overlap or parthood, modelling disjointedness requires a model of mereology, which we have implemented in MeTS for time and geospatial regions, as well as defining a general mereological interface. Take the example computation, performing addition over Campbelton's population, and that of Campbelton's provincial parts. This is not a type safe operation since addition over populations assumes geospatial disjointedness, to avoid doublecounting. Correctly identifying this error requires 3 elements: modelling the knowledge that the provincial parts are parts of the whole, a definition of disjointness for geospatial entities, and a precondition for the addition operator that the geospatial regions of the operands must be disjoint from one another. As for the first element, we leverage pattern-matching in a general mereological partOf function to accomplish this:

```
partOf : GeoEntity → GeoEntity → Bool
partOf(Campbelton (Quebec Part),Campbelton) = True
partOf(Campbelton (NB Part), Campbelton) = True
...
```

For this geospatial disjointedness, we use a qualitative mereology, as would be done in a formal logic; pattern matching here is analogous to relations in the A-box. For temporal mereology however, we do not need to rely on A-box style relations of parts; it is an arithmetic computation.

```
partOf : Time → Time → Bool
partOf(Interval(x1, x2), Interval(y1, y2)) = x1 <= y2 & y1 <= x2
...
partOf(t1, t2) = t1 == t2
```

This is a very obvious example of MeTS efficiency over reasoning, as this kind of quantitative reasoning in a formal logic would either require an ontology of time and processes, or rely on SMT solvers, as opposed to our comparatively simple use of functional programming techniques. For the definition of disjointedness in general, we define it in terms of the partOf relation:

```
disjoint(x,y) = not (partOf(x, y) OR partOf(y, x))
```

Finally, we require a precondition function which ensures all the geospatial regions of a set of populations are disjoint from one another, as a pre-condition for arithmetic sum of populations.

```
disjointRegions : (List Population) → Bool
disjointRegions(ps) = disjointList (getRegions ps)

popSumPreconditions : (List Population) → Bool
popSumPreconditions(ps) = disjointRegions(ps) & ...
```

$$\texttt{populationSum} : \left(\sum\nolimits_{ps:(\text{List Population})} (\texttt{popSumPreconditions(ps)} = \texttt{True})\right) \rightarrow \texttt{Population}$$

Note that for the purpose of this paper, we omit some basic function definitions and pattern matching (they are fully implemented in our proof-of-concept implementation), since function definitions like disjointList are basic exercises in functional programming, and not relevant to our key contributions. With these definitions, the sum of populations guarantees all the geospatial regions of the populations are disjoint, and thus ensures no errors of double-counting due to geospatial overlap. Furthermore, an addition operation requiring disjointedness over one of its 'stratifiers' is not specific to populations. Summing the mass of physical objects, for example, similarly assumes they do not occupy the same space; the methodology of type-checking is the same in both cases.

7.2 Provenance

Integrating provenance with type-checking is where the power of dependent typing is most evident. Revisiting the earlier example operation, an average of averages should not be type-safe. To correctly type this, there are two important components: the way operations derive new expression trees, and how precondition functions distinguish between different expression trees. Returning the correct expression trees is straightforward, we need only ensure the right-hand side of our pattern matching expressions include the correct operator:

```
populationSum : (∑ps:(List Population) (popSumPreconditions(ps) = True)) → Population
populationSum(ps) = Over(Sum, ps)
```

```
populationAvg : (∑ps:(List Population) (popAvgPreconditions(ps) = True)) → Population
populationAvg(ps) = Over(Avg, ps)
```

The next component is incorporating patterns of expression trees into precondition functions. For example, a precondition for population trees which have had an average or median operation performed on them:

```
aggregatePop : Population → Bool
aggregatePop(Over(Avg, ps)) = True
aggregatePop(Over(Median, ps)) = True
aggregatePop(_) = False
```

Precondition functions of this form allow us to make distinctions between kinds of data based on transformations that have been applied to them. This, in conjunction with list comprehension functions, like all, none, atLeastOne, etc., can be incorporated into operation preconditions to give us fine-grained control over which sequences of operations produce meaningful results for the given data types.

Where the expressive power of preconditions becomes even more evident is in the combination of mereological and provenance preconditions. Consider the earlier example computation, avg((Toronto2016 - Toronto2011), (Guelph 2011 - Hamilton2011)). An explanation for why this computation is not meaningful requires both a mereological explanation and a provenance-based one. While both operands are *population differences* (provenance tree head is the same operation), they are different kinds of differences (mereological distinction). The complete preconditions for population average provide insight into how this is type checked:

```
avgPreconditions : List Population → Bool
avgPreconditions(ps) = ((allSameRegions(ps)∧differingTime(ps))∨
                (allSameTime(ps)∧(disjointRegions ps)))∧
```

```
(all(measuredPop, ps)∨all(scaledPop, ps)∨
 all(differencePop, ps)∨all(ratioPop, ps))
```

In order to compute a population average, the population values we are averaging over must belong to the type described by: all possible values of ps such that popAvgPreconditions(ps) = True. In our example, the provenance portion of the preconditions (after the ∧) are satisfied; both operands are population differences. However, our mereological component is not satisfied; the operands vary over both time and regions. Since we cannot generate a proof of our example operands satisfying this precondition, it is not type-safe, and thusly not a meaningful computation.

8   Research Program

While the type system of the MeTS framework is the primary mechanism for meaningful data science, its broader impact is realized when placed within the greater research program of MeTS. This research program expands upon the MeTS framework by modelling its correspondence to formal knowledge models and by extending the implementation with tractability in mind.

8.1   Correspondence Theorem

In order to model concepts like time and mereology, fundamental assumptions about their semantics must be made. MeTS is no different, because behind type-checking functions and preconditions are ontological commitments. It is therefore important to make these commitments clear and transparent; no specific interpretation should be forced upon data scientists when implementing the framework. Furthermore, the ontological commitments should not be so entangled with the type system that it becomes impossible to adjust these commitments without significant re-engineering. This modular and ontologically-agnostic approach not only increases transparency, and supports more domains and applications, it also has broader implications for the knowledge modelling community.

The correspondence theorem of the MeTS framework intends to prove the soundness and completeness of a given MeTS program with respect to some formal ontology(ies). We have already developed FOL ontologies for census data which correspond to the implicit modelling commitments in the exemplary StatsCAN type system, including a method of proving the soundness of a MeTS type environment with repsect to some FOL theory. MeTS is not intended to be used as a 'black-box' set of dependently-typed programs, but as a method of leveraging knowledge model(s) through a dependent type system for data science. This modular and ontologically-agnostic approach not only increases transparency, and supports more domains and applications, it also enables the exploration of alternative methods of reasoning within the knowledge modelling community.

8.2 Tractability

With data becoming increasingly available and used in many industries, data science tools should facilitate a fast and efficient data science pipeline. Considering the expressive power of MeTS and its reliance on programming paradigms usually not associated with data science, tractability is a critical factor. One significant benefit of MeTS is the expressive power it achieves while remaining decidable. However, decidability itself does not guarantee adoption, a production-ready implementation of MeTS should also prioritize integration within existing data science infrastructure. In our proof of concept implementation, the interface component and typing component of MeTS are both specified in idris, a general-purpose dependently typed programming langauge. However, we should not require data scientists to learn dependent types and functional programming in order to employ MeTS into their existing workflows. In order for the vision of MeTS to be completely realized and to promote adoption in the data science community, MeTS should include efficient out-of-the-box integration with common data science tools like pandas, Tableau, etc.

9 Conclusion

The current state of data science relies heavily on datatypes which lack the expressiveness to accurately model the real world phenomena they represent. Simple datatypes fail to typify data because their simplistic representation cannot capture the concepts which underlie data including time, mereology, and provenance. Existing work addresses this issue by supplementing datatypes with external documentation, provenance tracking, and knowledge models in informal and ad-hoc ways that require manual effort and are prohibitively complex. Conversely, MeTS embeds the rules, assumptions, and interpretations of real-world concepts in types, elevating type safety from a trivial check to a meaningful result. The mechanism behind the MeTS type system is dependent types and expression trees, which integrate temporal, mereological, and provenance information into preconditions for fundamental data science operations. Operations involving StatCAN census data demonstrates concrete examples of how MeTS enforces real-world rules of provenance, geospatial and temporal disjointedness through type-checking. The MeTS framework is also part of a larger research program including a correspondance theorem with broader impacts for knowledge modelling and type theory, as well as a focus on tractability including integration with common data science tools. Ultimately, the MeTS framework is a significant step towards transparent, verifiable, and meaningful data science.

References

1. Acar, U.A., Buneman, P., Cheney, J., Van den Bussche, J., Kwasnikowska, N., Vansummeren, S.: A graph model of data and workflow provenance. In: TaPP (2010)

2. Albuquerque, A., Guizzardi, G.: An ontological foundation for conceptual modeling datatypes based on semantic reference spaces. In: IEEE 7th International Conference on Research Challenges in Information Science (RCIS), pp. 1–12. IEEE (2013)

3. Brady, E.: Idris, a general-purpose dependently typed programming language: design and implementation. J. Funct. Program. **23**(5), 552–593 (2013)

4. Canavotto, I., Giordani, A.: An extensional mereology for structured entities. Erkenntnis 1–31 (2020)

5. Chen, P.P.S.: The entity-relationship model-toward a unified view of data. ACM Trans. Database Syst. (TODS) **1**(1), 9–36 (1976)

6. Firat, A.: Information integration using contextual knowledge and ontology merging. Ph.D. thesis, Massachusetts Institute of Technology (2003)

7. Gebru, T., Morgenstern, J., Vecchione, B., Vaughan, J.W., Wallach, H., Daumé III, H., Crawford, K.: Datasheets for datasets. arXiv preprint arXiv:1803.09010 (2018)

8. Goh, C.H.: Representing and reasoning about semantic conflicts in heterogeneous information systems. Ph.D. thesis, Massachusetts Institute of Technology (1997)

9. Gregersen, H., Jensen, C.S.: Temporal entity-relationship models-a survey. IEEE Trans. Knowl. Data Eng. **11**(3), 464–497 (1999)

10. Grüninger, M.: Ontology of the process specification language. In: Staab, S., Studer, R. (eds.) Handbook on Ontologies. International Handbooks on Information Systems. Springer, Berlin, Heidelberg (2004). https://doi.org/10.1007/978-3-540-24750-0_29

11. Grüninger, M., Aameri, B., Chui, C., Hahmann, T., Ru, Y.: Foundational ontologies for units of measure. In: FOIS, pp. 211–224 (2018)

12. Grüninger, M., Chui, C., Ru, Y., Thai, J.: A mereology for connected structures. In: Formal Ontology in Information Systems, pp. 171–185. IOS Press (2020)

13. Grüninger, M., Li, Z.: The time ontology of allen's interval algebra. In: Proceedings of the 24th International Symposium on Temporal Representation and Reasoning (TIME 2017). Schloss Dagstuhl-Leibniz-Zentrum fuer Informatik (2017)

14. Herschel, M., Diestelkämper, R., Lahmar, H.B.: A survey on provenance: what for? what form? what from? VLDB J. **26**(6), 881–906 (2017)

15. Hind, M., et al.: Increasing trust in ai services through supplier's declarations of conformity **18**, 2813–2869 (2018). arXiv preprint arXiv:1808.07261

16. Löh, A., McBride, C., Swierstra, W.: A tutorial implementation of a dependently typed lambda calculus. Fund. Inform. **102**(2), 177–207 (2010)

17. Madnick, S., Zhu, H.: Improving data quality through effective use of data semantics. Data Knowl. Eng. **59**(2), 460–475 (2006)

18. Martin-Löf, P., Sambin, G.: Intuitionistic Type Theory, vol. 9. Bibliopolis, Naples (1984)

19. Mitchell, M., et al.: Model cards for model reporting. In: Proceedings of the Conference on Fairness, Accountability, and Transparency, pp. 220–229 (2019)

20. Peckham, J., Maryanski, F.: Semantic data models. ACM Comput. Surv. (CSUR) **20**(3), 153–189 (1988)

21. Rijgersberg, H., Van Assem, M., Top, J.: Ontology of units of measure and related concepts. Semant. Web **4**(1), 3–13 (2013)

22. Winston, M.E., Chaffin, R., Herrmann, D.: A taxonomy of part-whole relations. Cogn. Sci. **11**(4), 417–444 (1987)

23. Xiang, J., Knight, J., Sullivan, K.: Real-world types and their application. In: Koornneef, F., van Gulijk, C. (eds.) SAFECOMP 2015. LNCS, vol. 9337, pp. 471–484. Springer, Cham (2015). https://doi.org/10.1007/978-3-319-24255-2_34
24. Xiang, J., Knight, J., Sullivan, K.: Is my software consistent with the real world? In: 2017 IEEE 18th International Symposium on High Assurance Systems Engineering (HASE), pp. 1–4. IEEE (2017)

Research Data Management in the Image Lifecycle: A Study of Current Behaviors

Joana Rodrigues[1,2](✉) [iD] and Carla Teixeira Lopes[1,2] [iD]

[1] Faculty of Engineering of the University of Porto,
Rua Dr. Roberto Frias, 4200-465 Porto, Portugal
`joanasousarodrigues.14@gmail.com, ctl@fe.up.pt`
[2] INESC TEC, Rua Dr. Roberto Frias, 4200-465 Porto, Portugal

Abstract. Research data management (RDM) practices are critical for ensuring research success. Data can assume diverse formats and data in image format have been understudied in RDM. To understand image management habits in research, we have conducted semi-structured interviews with researchers from four research domains. Most researchers do not formally manage their images, nor do they develop RDM plans. They assume that image management is not a topic discussed at project meetings. In turn, they tend to perform some individual practices, depending on the context and their own opinion, such as creating captions to describe the images and organizing and storing the images in specific locations. However, they see these habits as necessary and admit that they will start to do so in a formal and collaborative way with the working group. These results provide valuable information on practical aspects of the use and production of images in research.

Keywords: Research data management · Images · Research · Image life cycle

1 Introduction

Images are dominant in communication, and their abundance, diversity of origins, and variety of holders create a multiplicity of practices regarding description, interpretation, and systematic use [15]. Many authors talk about the role of human behaviors in the life cycle of images [7]. The awareness and education of others for the practices of image record description is essential, commenting on the importance of eliminating description mistakes associated with automatic systems [4].

Research Data Management (RDM) is gaining the attention of researchers, who often ask for support in the process [2]. Tasks as an adequate data organization, rigorous description, storage, and sharing allow data to keep their meaning and facilitate their interpretation [7]. Data management is also fundamental in the communication between researchers and the scientific community, where images can play a central role in the "teaching-learning" process [10].

© The Author(s), under exclusive license to Springer Nature Switzerland AG 2022
R. Guizzardi et al. (Eds.): RCIS 2022, LNBIP 446, pp. 39–54, 2022.
https://doi.org/10.1007/978-3-031-05760-1_3

When we talk about research data, a dictionary in the area of RDM tells us that they are "Facts, measurements, recordings, records, or observations about the world collected by scientists and others, with a minimum of contextual interpretation. Data may be in any format or medium taking the form of writings, notes, numbers, symbols, text, images, films, video, sound recordings (...)" [6]. In its turn, image is the visual representation of something. It can be captured through various instruments or obtained through manual techniques (such as drawings, paintings, engravings and illustrations). Images can be in an analog or digital format. Examples of images are photographs, microscopic images, medical images (e.g., x-ray image), paintings, illustrations, videos (moving images), computer-made images, engravings, graphics, drawings, maps.

The lack of knowledge about image management in research motivates this study. Images are fundamental pieces in the research process, facilitating interpretation and analysis. For this reason, it is essential to know how researchers deal with images and to make them aware of the value of images, motivating its adequate management.

The capture and production of images are increasingly facilitated by the emergence of technologies that allow you to obtain an image in seconds. With a simple smartphone, we can capture or produce an image quickly and easily. In addition, technological devices also allow, in a fraction of a second, an image to be broadcast to any part of the world. This scenario of image empowerment leads to more images being used and produced in the context of research. However, the formal management practices of these images are not aligned with their exponential growth. How do researchers deal with their images? Images are organized, analyzed, described, shared according to previously established methods? This work arises to clarify what happens with data in image format. We conducted interviews with researchers from different research domains addressing how they organize and manage the images they use or produce in their projects. Results contribute to the definition of a current scenario, leading to proposing the conditions and the necessary steps to be followed.

2   Literature Review

An unprecedented growth characterizes the volume of data produced in research, as powerful computational capabilities are available to even small research groups: the so-called long-tail of science [9]. Usually, small groups or individual researchers have minimal resources to ensure the long-term availability of their data [5]. As such, they need good research data management practices supported by practical tools so that the datasets they produce can be made available to others [18]. Make data available is especially important as more research funding agencies adhere to the European Commission's Guidelines on FAIR Data Management in Horizon 2020, which advocate for a set of principles to make data Findable, Accessible, Interoperable, and Reusable [14].

Palmer et al. [13] analyze the importance of creating a structure of principles and processes that help articulate and support data description. One point of consensus was the crucial role of images, a typology that constitutes a social object, with materiality and associated ideas [1,12]. For researchers, images have a double purpose: they function as metadata, providing context, and as a vital medium to record the object of study [1]. This approach leaves room for positioning an image in RDM as a fundamental asset in describing and interpreting specific domains.

To the best of our knowledge, only one work studied the habits of researchers managing images in the research process. Fernandes et al. [8] created a questionnaire for this purpose and analyzed the collected responses. The authors concluded that researchers do not have the practice of managing and organizing the images they use or produce in their projects. Unfamiliarity, lack of standardized practices, and time constraints are the main reasons for these gaps in image management.

3   Methodology

To further investigate the responses and conclusions of Fernandes et al. [8], we have conducted interviews with researchers from different domains. Although the questionnaires allow reaching a more significant number of people, the interviews allow more profound analysis and validation of the results, provided by the direct contact that the interviews allow. With the results obtained in both works, we can develop a holistic view of the conclusions, facilitating a comprehensive and general understanding of the phenomena.

The methodology used in this work is a qualitative method of investigation through the case study, namely exploratory research.

Exploratory research aims to improve familiarization with the case under study, so this method will investigate the topic so that it is possible to reflect on it and improve its understanding [16]. The exploratory research aims to generate ideas and hypotheses that complement this theme and help the work be developed, namely in identifying a set of habits and behaviors related to data management in image format. The case study is an integral part of the qualitative method of investigation, which aims to specifically analyze the topic in question and, in this way, create in-depth clarification about it.

In this case, qualitative research relies on primary sources in the field of empirical materials. For this, semi-structured interviews are carried out. We recruited, by convenience, 15 researchers in 4 different domains: 4 researchers in Life and Health Sciences (LHS), 5 in Exact Sciences and Engineering (ESE), 3 in Natural and Environmental Sciences (NES), and 3 in Social Sciences and Humanities (SSH). These are four main research domains used by the national funding agency for science, research, and technology in the country of this study.

For recruitment, the choice fell on some researchers who had already participated in other projects and partnerships, as it is more agile, faster, and more likely to be successful. Table 1 shows the profile of the fifteen researchers interviewed. To maintain the confidentiality of the participants, we do not disclose all the information about the participants.

Table 1. Researchers profile

Research	Specific work area	Academic qualifications	Position in the institution	Age group
ESE1	Data science	PhD	Senior Researcher	30–40
ESE2	Enterprise systems engineering	MSc	Researcher	30–40
ESE3	Human-centered computing	PhD student	Research Assistant	30–40
ESE4	Enterprise systems engineering	MSc	Researcher	20–30
ESE5	Electrical and computer engineering	PhD	Senior Researcher	40–50
LHS1	Genome editing	PhD student	Research Assistant	30–40
LHS2	Animal health	PhD	Researcher	30–40
LHS3	Biological an molecular science	MSc	Researcher	20–30
LHS4	Reproductive genetics	PhD	Associate Researcher	20–30
NES1	Biology	PhD student	Researcher	20–30
NES2	Biodiversity	PhD	Post-Doc Researcher	30–40
NES3	Biology	PhD	Researcher	40–50
SSH1	Psychology	PhD	Senior Researcher	30–40
SSH2	Psychology	PhD	Researcher	40–50
SSH3	Anglo-American studies	PhD student	Researcher	20–30

The interviews aim to give credibility and depth to this study, as they make it possible to indicate results based on real practices. Interviews can be divided into three types: informal conversation, interview guide, and semi-structured [11]. The choice was made for the latter since it does not require a rigid, fixed, and standardized protocol, that is, the questions do not have to be asked in an orderly manner and with a standardized formulation. So, this interview model allows the elaboration of a set of pre-defined questions, however, the guide is adaptable according to the direction of the dialogue between the interviewer and the interviewee [11].

Interviews were semi-structured by the phases of the images life cycle: Planning, Creation/Compilation, Quality assurance, Processing/Analysis, Description, Storage, and Sharing [3,17,19], as you can see in Fig. 2. This choice was because this life cycle includes all the phases that we consider essential in managing data in image format. We intended to balance the importance given to all stages, showing that the success of image data management lies in the joint use of all these phases of the life cycle.

The interview script includes fourteen questions that are available in the INESC TEC data repository[1]. Each set of questions was developed in order to understand the behaviors in each of the phases of the cycle, so that it was possible to concretely determine the actions, without mixing lifecycle stages. Due to COVID-19, the interviews took place remotely. The sessions were recorded in video and audio for later analysis. Data were anonymized, and all recordings will be deleted when the study is complete. The interviews took place from May to October 2020. For the development of the interviews, all the researchers read and filled out a document called "Informed, Free and Clarifies Consent to Participate in a Research Projet according to the Declaration of Helsinki and the Convention of Oviedo", where they attested that they agreed to participate

---

[1] https://doi.org/10.25747/hcrd-ht83.

in the study and that they were guaranteed confidentiality, the exclusive use of the data collected and anonymity, promising never to public the identification of participants.

To process and analyze the data from the interviews, we began by transcribing the interviews (after all interviews have already been done). After that a table was prepared for each question, where the identifier of each interviewee was placed in the first column (e.g., ESE1), in the second column, it was described o the research domain and in the third column a response summary. The summary is composed of topics that include the main aspects of what was said, in order to make clear the specific behavior in that situation. Below, in Fig. 1, we present a summary of our methodological approach.

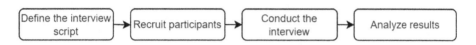

Fig. 1. Our methodological approach

4   Results

In this section, we summarize the findings obtained in the context of the interviews, organized by the different phases of the life cycle of the images. Figure 2 visually represents the topics examined in the results.

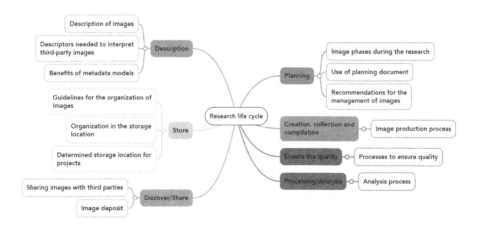

Fig. 2. Mind-map of the topics examined in the results

4.1 Planning

We asked if interviewees plan the phases through which the image goes, produce a document establishing guidelines for the management of the images, and if and what recommendations they would like to have for image management.

Image Phases During the Research

It was clear that most researchers do not formally plan the management of the images. However, in the research routines, they carry out practices that indicate informal planning. We identified sequences of actions in the interviews. Table 2 shows these sequences, columns identify the researchers, along with their research domains, and rows represent phases of image management. Cells' values indicate the order of the stage in the overall sequence.

Table 2. Image management planning cycles identified by researchers

	ESE1	ESE2	ESE3	LSH1	LSH2	NES1	SSH1
Collect	1				1		
Select	2						
Store	3				2	2	1
Create		1		1		1	
Treat		2		2			
Use		3			3		
Share		4				4	3
Analyze			1			3	
Investigate			2				
Design			3				
Assess quality				3			
Refine				4			
Index				5			
Publish				6			
Back up							2

As previously mentioned, in most cases, formal planning of image management does not exist. The images go through the cycle that the researchers, at that moment, considered appropriate. Sometimes, the image production is planned and described in advance. Other times, it appears in meetings, and then they are used for various purposes. Some researchers refer that when there is planning, the image is produced and worked collaboratively. Data observation is one of the stages that typically occurs when there is planning. In these cases, the image has a prominent role since it allows the interpretation and further processing of the data. For some researchers, if an image is reused, the image quality check is planned, and only then is it used for the research. One of the researchers said that the evaluation of this quality was also carried out through metadata, as it allows to verify the information and better interpret the image. Only one of the interviewees said that planning always takes place.

Of the seven cases presented in the Table 2, it is interesting to highlight the researcher designated by SSH1. This is the one that most significantly differs from the sequences of the other researchers. This researcher, from the domain of psychology, mentioned that his practices are neither formal nor pre-established, but carried out according to what he considers to be "the most qualified method for treating images". In the absence of a structured procedure, this researcher stores your images, then prepares a back-up, as it believes that "data safeguarding is essential and its eventual loss could constitute a rupture in the research process" and, finally, shares the data with third parties, particularly in data repositories.

Use of Planning Document

Only two researchers said they always have a planning document, as it ensures that the team knows what has been defined and approved. Yet, the predominance is the absence of a document including standard guidelines for image management. Many say they don't feel the need for it. One of the researchers said that he knows only a few open-access guidelines and tools mainly aimed at metadata. Another researcher stated that the non-use of this document happens because there is no such culture in science, as it is not a habit in the context of research projects. Most researchers assume that they never thought about the possibility of developing this type of guide, but they consider it an advantage. One of the researchers notes everything planned and accomplished, but not in the form of a guiding document. There were no significant differences by research domain.

Recommendations for the Management of Images

Some researchers answered recommendations about image management are essential but do not know precisely at what stage they may be relevant. One stated that recommendations are important mainly at an early stage of the project. Another said recommendations are vital in the planning phase to affect the quality of the production and use of the images. Another researcher noted that recommendations are not essential and may stop being applicable as projects change. However, many researchers said these recommendations would be helpful. Most of them highlighted that a document with universal recommendations to guide researchers to prepare and publish their data clearly would be advantageous. They also spoke of the importance of these recommendations providing practical examples. Many researchers said that these recommendations could safeguard characteristics of the use of images and their benefits. Also, these recommendations can make researchers able to more quickly and easily assess the images they use or produce.

4.2 Creation/Compilation

At this stage, it was asked how the image production process develops during the investigation.

Image Production Process

One of the ESE researchers said that he produces many photographs, maps, and graphs. The process involves documenting the code, rerunning the code, and generating the image again. Other ESE researchers said that this process could vary, but, usually, existing images are chosen and treated. The image reference (name that clearly identifies the image) for the future is always kept. The researchers also mentioned that images can come in two forms: 1) they are produced spontaneously when immediate need arises 2) they are planned in advance. In several cases, the images are produced collaboratively in this research domain.

In the domain of LHS, all the researchers said that the production process depends on the type of image. Production processes change with the types of image. Graphics are usually associated with data collection, data organization, image creation, image refinement, and treatment. Graphical abstract and infographics - root creation. Western plots - created photography, simple editing, and comparison of the bands. The creation of histograms, boxplots, funnel plots, and different types of graphs were also mentioned. Researchers also create explanatory diagrams and flowcharts, usually created with software.

The NES researchers mainly produce images in geographic information systems, where they create the map and add informational elements. However, this process depends on the type of investigation. This issue is often seen as a part of the research, where data collection and analysis processes are also considered. Satellite images are usually downloaded and pre-processed to identify defects, apply corrections, and georeference the image.

The SSH researchers say the images can be produced or reused, depending on the type of research and its purpose. In sociocultural projects, as a rule, images are produced. When dealing with historical or conceptual projects, the images are reused from repositories, articles already published, or past projects. Regarding production, all the researchers in the area mentioned the importance of anonymization. For this reason, the production process consists of collecting, blurring the characteristics of identification, storage, and use.

4.3 Quality Assurance

In this section of the interview, we asked about the existence of processes to ensure the quality and the activities inherent to this.

Processes to Ensure Quality

Most researchers say there is little or no activity to guarantee or verify the quality of the images. One of the researchers explicitly stated that there is no such verification, but said that if the image "looks good" and reflects what was intended, it is approved. One of the researchers noted that this verification occurs mainly in western plots. For some interviewees, verification only occurs when they fail to perceive the original image. Some researchers claim that, although images are critical to projects, they are not published and shared most of the time, so a raw image is sufficient to analyze what is intended. One of the researchers said he checked the quality when it comes to videos and photographs, as the details of these images can be critical. Some researchers mentioned adjustments to the code, cropping of images, treatment of image noise, care with contrasts, verification of the dimensions and source of images, aesthetic validation of the image, and verification of spatial resolution, temporal resolution and georeferencing.

4.4   Processing/Analysis

For this topic, we asked about the image analysis process, namely how information is extracted from the images.

Analysis Process

Most interviewees do not have a pre-established method of analyzing/processing the images. Most researchers said they did not analyze the images. Some researchers claim to do this frequently with methods that depend on the images and the project. One researcher stated that this analysis/processing depends on who performs it, as he will check what will be necessary according to the type of image. One researcher that works exclusively with maps said that this type of image is already the final product. Researchers identified the semi-automatic classification of images, the geostatistical processes, the extraction of direct or indirect information from the image, content analysis (especially for videos and photographs), information extraction (for graphics), visualization of the data, and the annotation of the image results as analysis/processing methods. Researchers also mentioned visual analysis, measuring distances, checking details, and cutting out irrelevant parts.

4.5   Description

We asked interviewees how they describe images, which information they need to understand and use third-party images, and the benefits and interest of a metadata model for image description.

Description of Images

Few researchers describe their images. The description is not a frequent practice, and its realization depends on the context, the image, and the researcher
who produces it. For one of the researchers, "less is more", that is, too much
description can affect the purpose of the description and confuse whoever tries
to interpret it. We identified the following examples of description: 1) description through technical details (for those who already know the context of the
image) and general details (for those who do not know), 2) description through
captions (most mentioned), 3) elaboration of metadata (most used to describe
components of maps, production periods, dates and the purpose of the image), 4)
description through annotations in a laboratory notebook. For researchers who
describe images, the advantages include sharing metadata/captions for better
interpretation and dissemination of the study.

Descriptors Needed to Interpret Third-party Images

The most mentioned descriptors are the source, rights of use, author, place
of origin, title, context, characteristics of obtaining and processing the image,
caption, scale, related resources, methodology, description, date, production
method, treatment and analysis processes, spatial resolution, temporal resolution, abstract, keywords, and the area of study. Elements of authorship, date,
context (abstract, description, keywords), methodology, production techniques,
and study area were the most mentioned and are transversal to all research
domains. Some elements vary according to the particularities of the research
domain.

Benefits of Metadata Models

Researchers see metadata models as very important. Only one researcher said
he was not sure, as he did not know whether the outcomes would compensate
for the time spent in the description. The other researchers responded positively
to the possibility of using metadata models, stating that they will facilitate the
storage of relevant information about an image. In addition, the metadata model
would serve as a recommendation to describe, a description standard to follow,
and would facilitate description for researchers. One researcher said that the
model should be adapted to the type of investigation. Another researcher thinks
it is difficult for a model to reach a balance of descriptors. On the one hand,
there are general recommendations that have general applicability and are very
relevant. On the other hand, a generic model can leave out important details,
such as the context and the scientific area.

  Almost all researchers mention that it is helpful to have guidelines for describing images. They also claim that it can be very advantageous to have a cross-
cutting model for multidisciplinary research groups to harmonize and create
common communication points. They highlight the importance of a core of
generic and specific metadata depending on the area. Researchers stress that
the metadata model can function as a working tool and help members of the

research group follow the same working methods, avoiding deviations. The metadata model can also stimulate the publication and sharing of data in repositories and directories.

4.6 Storage

We asked researchers whether they define guidelines for the organization of the images, what is the organization of the images in the storage location and if there is some central location where the images are stored.

Guidelines for the Organization of Images

No researcher defines guidelines for organizing their images. One said that he does not do it formally but pays special attention to the formats, the way they are organized, and the place where they are stored. Another admitted that this theme is a general failure, mainly because each researcher stores his own images, which is not safe and does not provide access to other research group members. Many researchers admit to having their method of organization and storage, but it is not formal or discussed in the working group. Everyone says they give importance to the way they organize the data, but it is not a procedure stipulated by the project's guidelines.

Organization in the Storage Location

All researchers claim to organize the images into project folders and subfolders on the computer. Some affirm organize projects folders by content (separate the content into categories that they create). Some researchers divide the raw data from the processed data. Others have folders for analysis and folders for scientific articles, each having different versions of images.

Determined Storage Location for Projects

No ESE researcher has a storage location, except one researcher who said that the images are stored on the computer and accessible via the internal network. In the LSH domain, only one researcher claims to have a storage location for the project's images. This researcher said that the existence of this location depends on the project. If it is a small project, all the images are on the computers; if it is a large project, a folder is created in the cloud for everyone to access. Two of the NES researchers say that this central location exists. One said that it is on the computer or the external disk and that, sometimes, it is in both places at the same time. Another stated that they have a Network-Attached Storage (NAS), a device dedicated to the storage of network data, allowing homogeneous access to data. For the SSH domain, only one of the researchers stated that this location exists. He said that they have an external disk where they store all the data. This disk is kept in the group's laboratory and accessible to project members. Researchers also have a copy of this data on their computers.

4.7 Sharing

In this last phase, we asked with who the interviewees usually share their images, what motivates sharing, and if they ever made an image deposit.

Sharing Images with Third Parties

Some researchers share their images with group colleagues in presentations to colleagues, at conferences, in articles, and on posters. Researchers also share the images with people associated with the project for specific tasks. Some share images with colleagues in the group to receive feedback on the image, to know if it is readable if it makes sense and if it is reliable. One researcher never shares the original images. Another said that it only shares images in articles. Most researchers share images after being treated because others may not easily interpret a raw image. One of the researchers says he is careful to accompany the image of a caption with the context whenever he shares it. The reuse of data is one of the main reasons for sharing and the possibility of new collaborations, sharing new knowledge. Some researchers point to confidentiality and sensitive data as barriers to sharing. The fact that images are not central to the project also limits sharing.

Image Deposit

Only some researchers have already deposited their images in a joint deposit that also included other types of data. One of the researchers said it was unusual and that he normally didn't hear about it. Another researcher deposits data, but never deposited only images. This researcher said he did not see any relevance in deposit the individual images. One of the researchers said he knew that depositing is one of the data management requirements but has not yet done so with the images. One of the researchers said that sometimes deposit their data but considers that it is more beneficial to share the information that gave rise to the images and the results than the image itself. A part of the interviewees claim to see no use in depositing the images in isolation; it would only make sense accompanied by textual information. These researchers say that they only make deposits if it is a requirement of the conferences or scientific journals. Deposit normally occurs in repositories.

5   Discussion

We interviewed a group of researchers about how they proceed and see the management of imagery data. We realize that all researchers produce or use images, but these are not always essential parts of the study. Therefore, many researchers are not concerned with its management. The results coincide with those collected by Fernandes et al. [8], where the generality said they did not have a guiding document or formal practices for this purpose. Although images are widely produced and used in research, there are no guidelines to direct researchers to standardized practices regarding images.

In the creation process, it is visible that the methods differ by research domain. In this study, the use of digital instruments is highlighted, prevalent in ESE, LHS, and NES, which was also confirmed in the previous survey [8]. At certain times, researchers pay attention to quality assurance, mainly in the domains of LSH and NES. However, there were no standard procedures to do so, varying with the type of study. In the questionnaire article [8], not all researchers perform this task, which also happens here. Regarding the processing/analysis phase, most researchers admit that it occurs according to the situations and context of the work, without any formal method or common practice in the work group. In the questionnaire article [8], it was identified the same.

Most researchers do not use metadata models. They consider it relevant, but it is not a common practice. The description is usually done through captions. They believe that when the image is reused, metadata is essential for interpretation, and the use of a metadata model for projects would bring benefits. The previous article [8] also shows that researchers do not use metadata models. However, in this work, we gathered more comprehensive answers regarding the descriptors that researchers consider relevant. All the researchers were able to find a justification for choosing a specific descriptor. For example, the keywords were mentioned several times, as they are essential for a preliminary analysis of the relevance of the data. Regarding storage, researchers do not have formal habits or standardized methods of organization. The computer prevails as a storage location, as it is easily accessible. The same was evident in the study of Fernandes et al. [8]. Finally, the researchers showed that no sharing habits. Sharing demonstrations take place mainly in scientific articles and conferences. Internally within the project, sharing occurs frequently. They have also not been shown to have depositing practices in repositories or the like. The same results were obtained in the questionnaire article [8].

This work allowed us to confirm several results obtained previously with a questionnaire. The use of the interviews allowed close contact with the researchers. If the questionnaire makes it possible to collect factual information, the interviews were essential to determine each of the choices, obtain justifications for answers, understand the various topics and understand the researchers' predisposition to initiate image management practices.

We noticed that most researchers are aware of the RDM guidelines, but this is not usually discussed in group/project meetings. Researchers also do not yet assume that imagery data is amenable to management, for many researchers as images are complements to the work and, therefore, their management is not essential. For most researchers, the time spent is still not worth the effort in data management unless that is an obligation. It became clear that the existence of practical and easy-to-implement standards and models (metadata model for image description, for example) is welcomed by researchers, as it is a way of seeing the work done with little effort. It was clear that the type of image is one of the most significant differentiating factors for the kind of management done. The research domain brings particularities, but it does not always determine all differences. Sometimes, the type of images and the conditions leading to their acquisition are more differentiating than other characteristics.

During the research phase in the literature on the research life cycles, we realized that none of the options contemplated data security and privacy procedures as one of the main phases. However, issues related to data confidentiality, namely images produced or used in the research context, must be reflected. In fact, an image can indicate, for example, a place or a person, thus breaking the principles of confidentiality that are often desired.

The General Data Protection Regulation (GDPR), approved on April 27, 2016, launched new challenges to the various audiences that the sphere of privacy affects, from citizens, companies, organizations, educational institutions, or research centers. The protection of individuals concerning the processing of personal data is a fundamental right that needs to be defended, regardless of the context. With the validity of the Regulation, a solid and more coherent protection framework at the European level was demanded. Personal data are understood to be those that allow the identification of the data subject, such as name, identification number, location data, identifiers electronically, and data related to physiology, genetics, mental health, economic, cultural, or social situation. However, if an image shows or indicates any of these cases, it may compromise the protection of the author's data or the actors highlighted in the image.

In the context of the research, some guidelines have already emerged to prevent this problem. Data Management Plans (DMP) and Privacy Impact Assessments (PIA) are very useful tools in data protection. The DMP allows structuring and organizing data from research projects, playing a central role in the development of good research practices, contemplating privacy actions to be applied to data. The PIA was introduced through the General Data Protection Regulation (article 35). The PIA aims to prepare a document with guidelines that intend to direct and control privacy risks. This document must foresee the risks and establish concrete scenarios associated with the research data, in order to anticipate problems and define strategies for the success of the management.

Therefore, it would be important to start a reflection on the clear introduction of a privacy phase in the research life cycles, so that practices related to this topic are mandatory for researchers who want to fully comply with security and data confidentiality.

6   Conclusion and Future Work

There is still a way to go in the sphere of research data management. Researchers tend to agree that the RDM is vital in science, but a lack of knowledge and time is often a hindrance to progress. Within the various types of data, the text is the one we see most often associated with data management practices. However, the rest of the typologies should not be overlooked. Images, for example, can be valuable instruments in scientific production. Their capacity for representation can help researchers. Besides, the use or production of images is not limited to a domain of research. Allied to this, the enormous technological evolution and the new capture devices that allow to obtain and spread an image quickly.

It is visible that when confronted with this reality, researchers realize what habits they have regarding images. However, they do not usually set guidelines and plan methods for managing their images. Most of the behaviors highlighted in this work mainly occur informally and without conformity of the working group. The path is to continue to reflect on these issues, confront researchers with these issues and motivate data management practices to become recurrent. Once implemented in the working group, they are easily adapted to the various projects, bringing significant benefits.

The work that we have been developing in the field of RDM in INESC TEC, leads us to believe that there is growing knowledge and interest in developing image data management practices, on the part of the research community. First, more and more researchers look to us to help them in the process of organization (through the development of data management plans), of description (through the use of metadata models), and of deposit/sharing (through the choice of repository and assistance in the filing process). Second, more researchers are concerned about reuse issues of their images, especially authorship issues, showing interest in knowledge about license and rights issues. The possibility of attributing DOI to the data has been an incentive to share. Finally, researchers almost always admit that a data management policy would be very helpful. This policy does not arise for lack of time or in-depth knowledge. In fact, some practices are beginning to be applied in projects, in order to initiate, even if superficially, behaviors that lead to good image management. Therefore, we believe that this study is necessary for the definition of best practices of research data management, namely image management, as it allows establishing a scenario and, from there, define a set of guidelines and offer tools for the image management. In addition, this study makes it possible to strengthen the existing state of the art, through analyzes of real practices.

In this sense, future works include the development of a metadata model for the description of data in image format. Controlled vocabularies will be associated with some of the elements of the model, as we believe that their use will improve the description and facilitate the researchers' work. This model will emerge from a development and evaluation process that seeks to meet the expectations of the scientific community, regardless of its domain, always in line with good RDM practices. This model is already in an advanced stage of development and will be presented shortly. Additionally, we will propose a guidance document where researchers can find tips for managing their images.

Finally, talk about the challenges that images bring to RDM. Information systems are better equipped to handle text. Nowadays, technology allows extracting automated the content of a text with some ease, but from an image, it is more difficult. Therefore, it is essential to have good description mechanisms that contribute to the search/findability of data in image format.

Acknowledgement. Joana Rodrigues is supported by research grant from FCT - Fundação para a Ciência e Tecnologia: PD/BD/150288/2019. Thanks to Miguel Fernandes who conducted the initial interviews and was involved in the construction of the interview script.

References

1. Abell, C.: The epistemic value of photographs. In: Abell, C., Bantinaki, K. (eds.) Philosophical Perspectives on Depiction, pp. 82–103. Oxford University Press (2010). https://doi.org/10.4324/9780203790762
2. Aguiar Castro, J., Amorim, R., Gattelli, R., Karimova, Y., Rocha da Silva, J., Ribeiro, C.: Involving data creators in an ontology-based design process for metadata models. In: Developing Metadata Application Profiles, pp. 181–213 (2017). https://doi.org/10.4018/978-1-5225-2221-8.ch008
3. Ball, A.: Review of Data Management Lifecycle Models. Technical report, Bath, UK (2012). https://purehost.bath.ac.uk/ws/files/206543/redm1rep120110ab10.pdf
4. Banks, M.: Using Visual Data in Qualitative Research, 1st edn. SAGE Publications, London (2007)
5. Briney, K.: Data Management for Researchers: Organize, Maintain and Share your Data for Research Success. Pelagic Publishing, Exeter, UK (2015)
6. CASRAI: Data (March 2022). https://casrai.org/term/data/
7. Cox, A.M., Tam, W.W.T.: A critical analysis of lifecycle models of the research process and research data management. Aslib J. Inf. Manag. **70**(2), 142–157 (2018). https://doi.org/10.1108/AJIM-11-2017-0251
8. Fernandes, M., Rodrigues, J., Lopes, C.T.: Management of research data in image format: an exploratory study on current practices. In: Proceedings of the International Conference on Theory and Practice of Digital Libraries–Digital Libraries for Open Knowledge, pp. 212–226 (2020). https://doi.org/10.1007/978-3-030-54956-5_16
9. Heidorn, P.B.: Shedding light on the dark data in the long tail of science. Libr. Trends **57**(2), 280–299 (2008)
10. Huds, D.: The impact of photography on society. Our Times (2019). https://ourpastimes.com/the-impact-of-photography-on-society-12377030.html
11. Kallio, H., Pietilä, A.M., Johnson, M., Kangasniemi, M.: Systematic methodological review: developing a framework for a qualitative semi-structured interview guide. J. Adv. Nurs. **72**, 2954–2965 (2016). https://doi.org/10.1111/jan.13031
12. Lacerda, A.: A fotografia nos arquivos: produção e sentido de documentos visuais. História, Ciências, Saúde (1) (2012)
13. Palmer, C., et al.: Site-based data curation based on hot spring geobiology. PLoS ONE **12**(7) (2017)
14. E.C.D.G., Innovation for Research: Guidelines on fair data management in horizon 2020 (2016)
15. Sandweiss, M.A.: Image and artifact: the photograph as evidence in the digital age. J. Am. History **94**(1), 193–202 (2007). https://doi.org/10.2307/25094789
16. Saunders, M., Lewis, P., Thornhill, A.: Research Methods for Business Students, 6th ed. Pearson Education Limited, New York (2012)
17. Structural Reform Group: DDI Version 3.0 Conceptual Model (2004). https://ddialliance.org/sites/default/files/Concept-Model-WD.pdf
18. Tenopir, C., et al.: Data sharing by scientists: practices and perceptions. PLoS ONE **6**(6), 1–21 (2011). https://doi.org/10.1371/journal.pone.0021101
19. UK Data Archive: Research data lifecycle (2019). https://www.ukdataservice.ac.uk/manage-data/lifecycle.aspx

Information Search and Analysis

Towards an Arabic Text Summaries Evaluation Based on AraBERT Model

Samira Ellouze[✉] and Maher Jaoua

ANLP Research Group, MIRACL Laboratory, University of Sfax, Sfax, Tunisia
ellouze.samira@gmail.com, maher.jaoua@gmail.com

Abstract. The evaluation of text summaries remains a challenging task despite the large number of studies in this field for more than two decades. This paper describes an automatic method for assessing Arabic text summaries. In fact, the proposed method will predict the "Overall Responsiveness" manual score, which is a combination of the content and the linguistic quality of a summary. To predict this manual score, we aggregate, with a regression function, three types of features: lexical similarity features, semantic similarity features and linguistic features. Semantic features include multiple semantic similarity scores based on Bert model. While linguistic features are based on the calculation of entropy scores. To calculate the similarity between a candidate summary and a reference summary, we begin by doing an exact match between n-grams. For the unmatched n-grams, we present them as Bert vectors, and then we compute the similarity between Bert vectors. The proposed method yielded competitive results compared to metrics based on lexical similarity such as ROUGE.

Keywords: Summary evaluation · Arabic summary · AraBERT model · Contextual word embedding

1 Introduction

The evaluation of text summaries plays an important role in the process of developing summarization systems. Several studies have focused on the evaluation of the content of the summary (Lin 2004; Giannakopoulos and Karkaletsis 2011; Cabrera-Diego and Torres-Moreno 2018), others are interested by the linguistic quality evaluation of the summary (Pitler and Nenkova 2008; Pitler et al. 2010; Dias et al. 2014; Ellouze et al. 2016; Xenouleas et al. 2019). Some other studies have investigated in the assessment of both the content and the linguistic quality of the text summary (Lin et al. 2012; Ellouze et al. Ellouze et al. 2017a; Wang et al. 2020). All previous cited works are dedicated to evaluation summaries in English language. While the existing automatic evaluation methods for Arabic summary are rare (Elghannam and El-Shishtawy 2015; Ellouze et al. 2017b). Predicting the quality of an Arabic summary is a challenging task that has taken, until now, only the lexical level of similarity evaluation.

In this paper, we focus on the evaluation of the Arabic text summary, especially on predicting the overall responsiveness manual score. This score is a combination of

R. Guizzardi et al. (Eds.): RCIS 2022, LNBIP 446, pp. 57–69, 2022.
https://doi.org/10.1007/978-3-031-05760-1_4

content and linguistic quality evaluation. To achieve this goal, we proposed a method that aggregates several linguistic qualities, semantic similarity, and lexical similarity features. This method solves two common defects in lexical similarity-based methods. First those methods often fail to match paraphrases. Second, they penalize the ordering changes between sentence words in reference summary and others in system summary.

We study the correlation between lexical similarity scores (i.e., ROUGE, AutoSum-mENG, MeMoG and Simetrix scores) and the overall responsiveness score. This study shows that most lexical similarity scores have lower correlation with the manual score. Then, we propose a new method for evaluating text summary that relies on the study of similarity between a system summary and a reference summary (manual summary) by presenting n-grams with the AraBERT model as vectors. Next, we compare the AraBERT vectors of system summary with the AraBERT vectors of reference summary. In addition, we have integrated multiple character and word entropy scores to capture the linguistic quality of a summary.

The main contribution of this paper is the introduction of a novel method of Arabic text summary evaluation based on three types of features: lexical similarity features, semantic similarity features and linguistic quality features. In fact, the motivation of combining lexical similarity with semantic similarity features is based on the idea that when a human judge (evaluator) observes a system summary, normally he will begin by observing the exact match between words in this summary and words in reference summary or source documents. Then he goes on to check for the presence of semantic similarity.

The second contribution is that we have calculated scores based on embedding n-grams as follows: First, we have extracted multiple lengths of n-grams from system and reference summaries then we have presented them with AraBERT as embedding n-grams. Moreover, we have calculated the cosine similarity between embedding n-grams from system summary and reference summary. Finally, when we calculate the similarity score, we include only cosine similarity that is greater than or equal to a threshold. We tested several thresholds of cosine similarity which varies from 0.5 to 0.8, for each length of an embedding n-gram. In fact, we have chosen that the lower threshold will be 0.5, because we have observed that the embedding n-grams having cosine similarity lower that 0.5 don't seem to be similar enough.

2 Related Works

In the past two decades, there has been a lot of effort done by researchers to propose new methods for automatic text evaluation. The first methods are based on the lexical level of similarity between system summary and reference summary. The most known metric is ROUGE (Lin 2004) metric. It is relying on comparing word n-grams of system summary with n-grams of reference summary. The final score is obtained by computing a recall, precision, or f-measure score based on n-grams overlap. Another metric based on character n-grams, has been proposed by Giannakopoulos and Karkaletsis (2011). In fact, a summary is represented by a graph where the nodes are the n-grams of characters, and the edges are the relation between them. The estimation of the similarity degree is performed by comparing the graph of the system summary with the graph of reference summary.

Other methods are based on syntactic similarity such as BE (Hovy et al. 2006) metric. This name comes from the fact that the calculation of this metric is based on the decomposition of each sentence into minimal units called "Basic Elements" (BE). Then a score of similarity is computed by comparing the BE of the system summary with the BE of reference summary.

A few years later, some methods based on semantic similarity were appeared. The first metric that deals with semantic similarity is BEwTE (Tratz and Hovy 2008). This metric is an extension of the BE which improves the correspondence between the BEs of the system summary and the BEs of the reference summary, thanks to the use of semantic relations (synonymy, abbreviation, meronym, holonym, etc.) which can be the subject of correspondence between two BEs.

More recently, Zhang et al. (2020) have proposed a method for evaluating the content of a summary by computing a similarity score for each token in the system summary with each token in the reference summary. But, instead of the exact match used by ROUGE metric, the authors compute a similarity using contextual embeddings such as Bert (Devlin et al. 2019), XLNet (Yang et al. 2019), etc.

All previous works have focused on the evaluation of English text summaries. Now we take a peek on the two existing works that have proposed method to evaluate text summaries on Arabic language. We begin by the metric proposed by Elghannam and El-Shishtawy (2015), who evaluate summaries based on key phrases overlap. In fact, after the extraction of key phrases from system summary and reference summary, the metric counts the matched key-phrases between the candidate (system) summary and the reference summary. Later Ellouze et al. (2017b) have merged lexical similarity scores with linguistic features using a regression model. Lexical similarity scores include ROUGE, AutoSummENG, NPowER-ed, while linguistic features include syntactic-based features. Until now, there is no Arabic language summary evaluation metric that supports semantic similarity.

3   Challenges of Arabic Language Processing

Arabic is the most widely spoken Semitic language that is the native language for more than 330 million people. In fact, it has a complex linguistic structure (Attia 2008), which made Arabic NLP applications faced with several complex problems related to the nature and structure of this language (Farghaly and Shaalan 2009). The main phenomena that cause those problems are the words agglutination, the absence of short vowels, the pro-drop phenomenon (Farghaly 1982), and the very broad vocabulary (comprising approximately 12.3 million words). The agglutination phenomenon presents the major problem because a single word can represent an entire sentence such as the Arabic word "أتتذكرونهم؟" (ātatadakarūnahum) which means "Do you remember them?". In addition, the absence of short vowels causes the problem of homograph words that have the same spelling but different meaning, such as the word " رجل" without short vowels can have various meaning among the meanings of this word we cite, " رَجُل" (raǧul) which means "man", " رِجْل" (riǧl) which means "leg", " رُجْل" (ruǧl) which means "men", etc. The other linguistic phenomenon that complicates Arabic NLP applications is the pro-drop phenomenon which means subject pronouns are omitted while semantic information remains. More precisely, for a human an omitted subject pronoun can be deduced from the previous sentences. But for a system it is so hard to determine it.

The fact that a wide number of people use this language makes the construction of specific NLP applications for Arabic become a necessity. But the problems related to the processing of this language make this task very difficult and challenged.

4   Proposed Method

We propose an automatic evaluation method for Arabic text summary. Our method relies on the aggregation of different types of features using a regression model to predict overall responsiveness. In fact, we combine lexical similarity features, semantic similarity features and linguistic quality features. In the first step of our method, we begin by computing all features that we will include in the step of predictive model construction. After the construction of the predictive model, we will use it in the calculation of a new system summary score. In the following subsection, we will present each type of features and how we have obtained them.

4.1   Features

Lexical features: in this type of features, we have studied the correlation between overall responsiveness score and many existing scores that are based on lexical similarity such as ROUGE, AutoSummENG, MeMoG, NPowER-ed (Giannakopoulos and Karkaletsis 2013) and Simetrix (Louis and Nenkova 2013) (KLSummaryInput, KLInputSummary, unsmoothedJSD, unsmoothedJSD, smoothedJSD, unigramProb and multinomialProb.). This study will give an idea about the performance of each score and the best score that correlate the best with the human score.

Table 1. Pearson correlation between overall responsiveness and lexical scores on multiling2011 and multiling2013 for Arabic language

Score	Correlation	Score	Correlation
ROUGE-1	−0.1071	Memog	−0.0714
ROUGE-2	−0.0993	KLSummaryInput	0.0695
ROUGE-3	−0.1021	KLInputSummary	0.1039
ROUGE-4	−0.0937	unsmoothedJSD	0.1363
ROUGE-5	−0.0930	smoothedJSD	0.1152
AutoSummENG	−0.0730	unigramProb	0.2024
NPowER	−0.0726	multinomialProb	0.0969

From Table 1 we notice that ROUGE scores do not correlate with manual score in Arabic language, while until now the most summarization system used ROUGE scores to evaluate its performance. In addition, we notice that the best lexical score is the unigram

Probability Simetrix score with 0.2 correlation with Overall responsiveness. In addition, we find that most Simetrix scores have a better correlation with Overall responsiveness, compared to other lexical scores. For this reason, we include all the six Simetrix scores as features to find the best predictive model of the manual score.

Semantic Similarity Features: AraBERT based features.

Analogously to common lexical similarity-based methods such as ROUGE, we compute a similarity score for each n-gram type (i.e., unigrams, bigrams, trigrams, four grams and fivegrams) in the candidate summary with each n-gram type in the reference sentence. Nevertheless, instead of exact matches of n-grams, we compute n-gram similarity using the AraBERT model. Bert is a neural network language models for multiple language, while Arabert is a pre-trained BERT, which was trained on a large Arabic corpus (Antoun et al. 2020). The corpus used in pre-training includes 70 million sentences of news articles from several Arab newspapers and with different topics. The size of this corpus is around 24 GB. AraBERT has produced better results on various Arabic NLP tasks such as sentiment analysis and Named Entity Recognition. From AraBERT, we can extract contextualized word embeddings which means that we can eliminate most disambiguates related to the absence of short vowels and homograph phenomenon when we calculate the similarity between two embedding n-grams.

Now we explain how we used AraBERT to detect semantic similarity. First, we have used n-grams of different length from unigrams to five grams. For each reference and system summary, we extract the list of n-grams of a specific length. Then, we transform each n-gram to an embedding n-gram using AraBERT. After, we match embedding n-grams from system summary with embedding n-grams from reference summary using different thresholds of cosine similarity. The threshold varies from 0.5 to 0.8. For example, for a threshold of 0.5, we match embedding n-grams that have cosines similarity (CS) value equal or great to 0.5. With more details, first, for a given embedding n-gram from the system summary, we use a greedy matching, we retain the best cosine similarity value between this embedding n-gram and each embedding n-gram from a reference summary. For an embedding n-gram from candidate summary ENs_i and a set of n-grams from reference summary $R = \{ENr_1, ENr_2, \ldots, ENr_m\}$ with m is the number of embedding n-grams in the reference summary, we will retain in this step the match that has the maximum cosine similarity using this formula:

$$CS_{score} = Max_{j \in [1,m]}(ENs_i, ENr_j)$$

Second, this match will be permanently retained only if its cosine similarity is equal or greater to a threshold. In our method, we define the function that maintains a CS_{score} between two embedding n-gram as:

$$f(EN_i, EN_j) \begin{cases} 0 \text{ if } CS_{score} < c \\ CS_{score} \text{ if } CS_{score} \geq c \end{cases}$$

where EN_i is an embedding n-gram from the system summary and EN_j is an embedding n-gram from the reference summary. c is a threshold of cosine similarity.

For each threshold, two final similarity scores between a system summary and reference are calculated, one is oriented recall and the second is oriented precision. We define the two final similarity scores as:

$$Similarity\_Score_{recall} = \frac{\sum_{i=1}^{n} \sum_{j=1}^{m} f\left(EN_i, EN_j\right)}{m}$$

$$Similarity\_Score_{precision} = \frac{\sum_{i=1}^{n} \sum_{j=1}^{m} f\left(EN_i, EN_j\right)}{n}$$

Where n and m are the total number of embedding n-grams in respectively system summary and reference summary.

Due to the presence of three reference summaries, we have for each system summary three $Similarity\_Score_{recall}$ and three $Similarity\_Score_{precision}$. To get one score to each summary, we adopted two methods, the average of the three scores and the maximum of the three scores.

$$Max\_Similarity\_score_x = max_{i \in [1..3]}(Similarity\_score_{x_i})$$

$$AVG\_Similarity\_Score_x = \frac{\sum_{i=1}^{3} Similarity\_Score_{x_i}}{3}$$

where x refers to recall or precision score and i refers to one of the reference summaries.

We recall that the threshold varies from 0.5 to 0.8 and the length of the n-grams varies from one to five grams. So As a final score for a system summary on a given threshold and a given n-gram length; we have: $AVG\_Similarity\_Score_{precision}$, $AVG\_Similarity\_Score_{recall}$, $Max\_Similarity\_score_{precision}$ and $Max\_Similarity\_score_{recall}$. As consequence, we calculated 20 scores for each threshold. And in total we have obtained 80 scores based on embedding n-grams similarity. We include all those scores in the step of the construction of the predictive model.

Linguistic Quality Features: for linguistic quality, we have calculated n-grams entropy features of words and of characters: the word unigram entropy, the normalized word unigram entropy, bigram entropy, the normalized word bigram entropy, the character unigram entropy, the normalized character unigram entropy, the character bigram entropy, the normalized character bigram entropy of a given system (candidate) summary. The motivation of calculating entropy is that it has an effect in text readability and comprehension. Moreover, according to (Bentz et al. 2017) word n-grams entropy incorporates dependencies between words. The entropy of a summary S is defined as:

$$H(S) = -\sum_{w \in \Omega} P(w)\log_2 P(w)$$

Where S is the system summary, Ω is a finite set that contains n n-grams presented in a system summary. w is an n-gram of a word or character from Ω, P is a probability distribution of w on Ω. The n-grams entropy can be normalized by dividing it by the number of n-grams in Ω. The normalized entropy is defined as:

$$H_n(S) = -\sum_{w \in \Omega} \frac{p(w)\log_2 P(w)}{\log_2 n}$$

Normalizing the entropy by $\log_2 n$ gives $H_n(S) \in [0,1]$.

We obtain 8 entropy scores that will be used in the model construction phase.

4.2 Model Construction

In this step, we will aggregate the previously computed features, by trying several regression classifiers such as LinearRegression, PoissonRegressor, GaussianProcessRegressor, Ridge, SVR, ARDRegression and HuberRegressor that exist in the scikit-learn (Pedregosa et al. 2011) python module. But, before testing any algorithm, we will select the best features that can contribute the maximum in the optimum built model, by using Recursive Feature Elimination (RFE) (Guyon et al. 2002) algorithm. In fact, this algorithm has proven its efficiency in selecting adequate features to predict the targeted variable. It incorporates an external estimator that assigns weights to features, in our case we use RandomForest as an external estimator. The purpose of recursive feature elimination (RFE) is to pick features by evaluating smaller and smaller sets of features in a recursive manner. This process is executed recursively on the pruned set until the appropriate number of features to select is attained. We have given all the features presented in the last subsection as an entry to RFE. To test cited classifiers, we have used two methods. The first by dividing our dataset into a training data set and a test data set. The second using cross-validation method. Then, we adopted the one that best correlates with the "Overall responsiveness".

5 Dataset

In our experiment, we use as dataset the Arabic portion of the Multiling 2011 and 2013 summarization task datasets: TAC 2011 MultiLing Pilot 2011 data set (Giannakopoulos et al. 2011) and the MultiLing 2013 data set (Giannakopoulos 2013). MultiLing Pilot 2011 and MultiLing 2013 data set, contain 10 and 15 collections of newspaper articles, respectively that belongs to different topics. For each topic, there are 10 newspaper articles. In addition, the two data sets involve system summaries and three model summaries for each collection.

Also, three Overall responsiveness manual scores are provided for each system summary. Each one is obtained by comparing the candidate summary with a reference summary. To obtain one manual score for each system summary, we have computed the average of the three scores. We have used this score to build the predictive model and to calculate the correlation between this score and the scores obtained by our predictive model.

6 Results and Discussion

We have tested our method first on the summary level evaluation, in which we calculate the score of each summary. Then, on the system level evaluation, in which we averaged the scores of summaries that belong to the same summarization system.

As mentioned earlier, we evaluate our models, firstly using test sets and then using cross-validation. In the first evaluation, we have randomly divided our data set into training and testing sets: the training set represents the two thirds of the data set and the rest represent the test set. To evaluate the model produced by each classifier, we studied the Pearson, Spearman and Kendall correlation between the manual scores "Overall Responsiveness" and the scores produced by the model. Furthermore, we reported the "Root Mean Squared Error" (RMSE) measure for each model. This measure is based on the difference between the manual scores and the predicted ones. In the next section, we present the results obtained by the tested model in summary level evaluation.

6.1   Summary Evaluation Level

We begin by presenting the features selected by the "recursive feature elimination" algorithm that incorporates RandomForest estimator. This algorithm has selected 24 features from the three types of features presented below. Table 1 presents all those features.

Table 2. List of features selected by "recursive feature elimination" to predict the "Overall Responsiveness" score for Arabic text Summaries

Features type	List of features
Lexical similarity	Simetrix_KLInputSummary, Simetrix_KLSummaryInput, Simetrix_unsmoothedJSD, Simetrix_unigramProb, Simetrix_multinomialProb
Semantic similarity	$\text{avg\_unigram\_Similarity\_Score\_with\_0.5\_threshold}_{recall}$ $\text{avg\_unigram\_Similarity\_Score\_with\_0.5\_threshold}_{precision}$ $\text{avg\_unigram\_Similarity\_Score\_with\_0.6\_threshold}_{recall}$ $\text{max\_unigram\_Similarity\_Score\_with\_0.6\_threshold}_{recall}$ $\text{avg\_unigram\_Similarity\_Score\_with\_0.7\_threshold}_{precision}$ $\text{max\_unigram\_Similarity\_Score\_with\_0.7\_threshold}_{recall}$ $\text{avg\_unigram\_Similarity\_Score\_with\_0.8\_threshold}_{recall}$ $\text{avg\_unigram\_Similarity\_Score\_with\_0.8\_threshold}_{precision}$ $\text{max\_unigram\_Similarity\_Score\_with\_0.8\_threshold}_{recall}$ $\text{max\_unigram\_Similarity\_Score\_with\_0.8\_threshold}_{precision}$ $\text{avg\_bigram\_Similarity\_Score\_with\_0.7\_threshold}_{precision}$ $\text{max\_trigram\_Similarity\_Score\_with\_0.5\_threshold}_{recall}$ $\text{max\_trigram\_Similarity\_Score\_with\_0.5\_threshold}_{precision}$ $\text{max\_fivegram\_Similarity\_Score\_with\_0.5\_threshold}_{recall}$ $\text{max\_fivegram\_Similarity\_Score\_with\_0.6\_threshold}_{precision}$
Linguistic quality	Unigram_character_entropy, normalized_Unigram_character_entropy, Unigram_word_entropy, normalized_unigram_word_entropy

From Table 2 we notice the presence of all types of features in the selected features. This means that all those types have an impact on the prediction of the manual score. In the semantic similarity features type we remark the presence of different length of n-grams, but the semantic similarity features based on the embedding of unigrams are more present the others. In addition, five of the six Simetrix scores are among the selected features. This indicates the power of the Simetrix metric.

Now, we present the results produced by divers classifiers such as LinearRegression, SVR, GaussianProcessRegressor, etc. Each classifier has as input the 24 selected features. Table 3 shows the correlation of the score obtained by each classifier with the human score using cross-validation.

Table 3. Correlation (p-value between brockets) and RMSE for classifiers using cross-validation

Classifiers	Pearson	Spearman	Kendall	RMSE
LinearRegression	0.2416(0.0001)	0.2674(2.68e−05)	0.1840(4.78e−05)	0.6999
SVR using linear kernel	0.2693(2.34e−05)	0.3034(1.66e−06)	0.2086(4.02e−06)	0.6745
SVR using non-linear Kernel	0.2399(0.0001)	0.2508(8.52e−05)	0.1737(0.0001)	0.6701
Ridge using linear kernel	0.2416(0.0001)	0.2674(2.68e−05	0.1841(4.78e−05)	0.6999
Ridge using non-linear kernel	0.2649(3.21e−05)	0.2833(8.24e−06)	0.1976(1.25e−05)	0.6950
GaussianProcessRegressor	0.2415(0.0001)	0.2658(3.02e−05)	0.1835, 5.14e−05	0.6994
ARDRegression	0.2597(4.64e−05)	0.2740(1.66e−05)	0.1882(3.20e−05)	0.6704
HuberRegressor	0.2107(0.0010)	0.2476(0.0001)	0.1680(0.0002)	0.6868

Table 3 shows that all classifiers give low correlation which does not exceed 0.26. Whereas we notice that the model built with SVR classifier using linear kernel has the best Pearson, Spearman and Kendall correlation. And the model built with SVR classifier using nonlinear kernel has the least RMSE.

In what follows, we present the obtained results for each classifier using a test set instead of cross-validation. Table 4 shows the correlation of the score obtained by each classifier with human score using the test set validation method.

Table 4. Correlation (p-value between brockets) and RMSE for the obtained scores using each classifiers using test set

Classifiers	Pearson	Spearman	Kendall	RMSE
LinearRegression	0.4392(4.57e−05)	0.4518(2.59e−05)	0.3324(2.61e−05)	0.6733
SVR using linear kernel	0.4922(3.52e−06)	0.4639(1.46e−05)	0.3397(1.73e−05	0.6612
SVR using non-linear kernel	0.3246(0.0033)	0.3713(0.0006)	0.2643(0.0008)	0.7100
Ridge using linear kernel	0.4393(4.57e−05)	0.4518(2.59e−05)	0.332(2.61e−05)	0.6733
Ridge using non-linear kernel	0.4368(5.09e−05)	0.4240(8.85e−05)	0.3033(0.0001)	0.6702
GaussianProcessRegressor	0.4382(4.79e−05)	0.4553(2.19e−05)	0.3333(2.52e−05)	0.6710
ARDRegression	0.4263(8.05e−05)	0.3801(0.0005)	0.2703(0.0006)	0.6873
HuberRegressor	0.3348(0.0023)	0.2811(0.0115)	0.2029(0.0103)	0.6991

From the Table 4 we constate that also with the test set validation method, the result obtained by SVR classifier with linear kernel has the best Pearson, Spearman and Kendall correlation with Overall Responsiveness and the least RMSE. Besides, we clearly remark the weak correlations produced by all classifiers. In fact, perhaps this is because we do not have enough summaries to obtain a more robust model that can predict scores having a higher correlation with the manual score and a lower RMSE.

In addition, we noticed that the correlation using test set is better than the correlation using cross-validation. So, we retained the model built using SVR with the linear kernel model.

Now we compare the results obtained by the SVR model with linear kernel using a test set with multiple baselines such as ROUGE, AutoSummENG, Simetrix, etc. Table 5 presents the different correlations of each baseline or our experiments with overall responsiveness.

Table 5. Correlation with overall responsiveness scores, measured with Pearson, Spearman, and Kendall coefficients in summary level evaluation

Baseline/our experiments	Pearson	Spearman	Kendall
ROUGE-1	−0.1071	−0.0947	−0.0675
ROUGE-2	−0.0993	−0.1086	−0.0786
ROUGE-3	−0.1021	−0.1163	−0.0833
ROUGE-4	−0.0937	−0.1117	−0.0807
ROUGE-5	−0.0930	−0.1136	−0.0831
AutoSummENG	−0.0211	−0.0216	−0.0157
MeMoG	−0.0123	0.0141	0.0103
NPowER	−0.0175	−0.0159	−0.0122
Simetre_KLInputSummary	0.1039	0.0859	0.0581
Simetrix_KLSummaryInput	0.0695	0.0824	0.0558
Simetrix_unsmoothedJSD	0.1363	0.1362	0.0926
Simetrix_smoothedJSD	0.1152	0.1126	0.0769
Simetrix_unigramProb	0.2024	0.2318	0.1633
Simetrix_multinomialProb	0.0969	0.1325	0.0938
Retained model	0.4922	0.4639	0.3397
Model without lexical features	0.4115	0.3632	0.2590
Model without semantic features	0.3790	0.4253	0.2934
Model without linguistic features	0.4779	0.4475	0.3264

From Table 5, we observe that our retained model outperforms all baseline metrics in the three types of correlation with a difference of more than 0.2 in the correlation. In addition, Table 5 shows that the five Simetrix scores KLInputSummary, unsmoothedJSD,

unsmoothedJSD, smoothedJSD, unigramProb, have better correlation with manual score than other baseline metrics that have demonstrated its efficacy on the evaluation of English text summary.

Besides, we observe that the elimination of each type of features from the selected features, has decreased the correlation. This means that each type has an impact on the retained model. But we notice that the elimination of semantic similarity features type has the most important effect on the correlation. In fact, the Pearson correlation has decreased from 0.4922 to 0.3790.

6.2   System Evaluation Level

The evaluation of the efficiency of an automatic system summarization is an important task that comes after the evaluation of each summary produced by a system. In this section, we calculate the average scores produced by each system to obtain its overall score.

Table 6 shows the correlation of the scores produced by each baseline or experiment with the human scores "Overall responsiveness".

Table 6. Correlation with overall responsiveness scores, measured with Pearson, Spearman, and Kendall coefficients in system level evaluation

Baseline/our experiment	Pearson	Spearman	Kendall
ROUGE-1	−0.3241	−0.2860	−0.2111
ROUGE-2	−0.2628	−0.3186	−0.2229
ROUGE-3	−0.2530	−0.2905	−0.2111
ROUGE-4	−0.2458	−0.2887	−0.1994
AutoSummENG	−0.2557	−0.2913	−0.1759
MeMoG	−0.1511	−0.1746	0.0938
NPowER	−0.2072	−0.2317	−0.1408
Simetre_KLInputSummary	0.1897	0.2018	0.1056
Simetrix_KLSummaryInput	−0.0122	0.2106	0.1056
Simetrix_unsmoothedJSD	0.2179	0.2282	0.1173
Simetrix_smoothedJSD	0.1815	0.2185	0.1056
Simetrix_unigramProb	0.0724	−0.0263	−0.0117
Simetrix_multinomialProb	−0.2233	−0.1983	−0.1524
Retained model	0.7110	0.6064	0.4692

From Table 6, we notice that our retained model has the best correlation with overall responsiveness on system level evaluation. In addition, we observe that there is a huge correlation gap between our retained model and baselines which exceeds 0.4 for Pearson correlation.

7 Conclusion

In this article, we have presented a novel method for automatically evaluating text summaries. In this method, we used the AraBERT model to evaluate the semantic similarity between system summary and reference summary. In addition, we have combined semantic similarity with lexical similarity and linguistic quality features. The obtained results in the system and the summary levels of evaluation outperform the results obtained by baseline metrics.

Looking ahead, it is important to use a larger dataset of summarization systems and system (candidate) summaries to obtain a model that correlates more with the manual score. In addition, we aspire to include more evaluation systems that take into account paraphrasing. Also, we aim to study the semantic similarity match between different n-grams lengths.

References

Lin, C.Y.: Rouge: a package for automatic evaluation of summaries. In: Proceedings of the Workshop on Text Summarization Branches Out, Post-Conference Workshop of ACL, Barcelona, Spain, pp.74–81 (2004)

Giannakopoulos, G., Karkaletsis, V.: AutoSummENG and MeMoG in evaluating guided summaries. In: Proceedings of the Text Analysis Conference (TAC) (2011)

Cabrera-Diego, L.A., Torres-Moreno, J.: Summtriver: a new trivergent model to evaluate summaries automatically without human references. Data Knowl. Eng. **113**, 184–197 (2018)

Pitler, E., Nenkova, A.: Revisiting readability: a unified framework for predicting text quality. In: Proceedings of the Empirical Methods in Natural Language Processing (EMNLP), pp. 186–195 (2008)

Pitler, E., Louis, A., Nenkova, A.: Automatic evaluation of linguistic quality in multi-document summarization. In: Proceedings of the 48th Annual Meeting of the Association for Computational Linguistics, pp. 544–554 (2010)

de S. Dias, M., Feltrim, V.D., Pardo, T.A.S.: Using rhetorical structure theory and entity grids to automatically evaluate local coherence in texts. In: Baptista, J., Mamede, N., Candeias, S., Paraboni, I., Pardo, T.A.S., Volpe Nunes, M.D.G. (eds.) PROPOR 2014. LNCS (LNAI), vol. 8775, pp. 232–243. Springer, Cham (2014). https://doi.org/10.1007/978-3-319-09761-9_26

Ellouze, S., Jaoua, M., Belguith, L.H.: Automatic evaluation of a summary's linguistic quality. In: Métais, E., Meziane, F., Saraee, M., Sugumaran, V., Vadera, S. (eds.) NLDB 2016. LNCS, vol. 9612, pp. 392–400. Springer, Cham (2016). https://doi.org/10.1007/978-3-319-41754-7_39

Xenouleas, S., Malakasiotis, P., Apidianaki M., Androutsopoulos I.: Sum-QE: a BERT-based summary quality estimation model. In: Proceedings of the 2019 Conference on Empirical Methods in Natural Language Processing and the International Joint Conference on Natural Language Processing (EMNLP-IJCNLP), pp. 6004–6010 (2019)

Lin, Z., Liu, C., Ng, H.T., Kan, M.Y.: Combining coherence models and machine translation evaluation metrics for summarization evaluation. In: Proceedings of the Annual Meeting of the Association for Computational Linguistics (Volume 1: Long Papers), pp. 1006–1014 (2012)

Ellouze, S., Jaoua, M., Hadrich Belguith, L.: Mix multiple features to evaluate the content and the linguistic quality of text summaries. J. Comput. Inf. Technol. **25**(2), 149–166 (2017)

Wang, X., Liu, B., Shen, L., Li, Y., Gu, R., Qu, G.: A summary evaluation method combining linguistic quality and semantic similarity. In: Proceedings of 2020 International Conference on Computational Science and Computational Intelligence (CSCI), pp. 637–642 (2020). https://doi.org/10.1109/CSCI51800.2020.00113

Elghannam, F., El-Shishtawy, T.: Keyphrase based evaluation of automatic text summarization. Int. J. Comput. Appl. **117**(7), 5–8 (2015)

Ellouze, S., Jaoua, M., Hadrich Belguith, L.: Arabic text summary evaluation method. In: Proceedings of the International Business Information Management Association Conference-Education Excellence and Innovation Management through Vision2020: From Regional Development Sustainability to Global Economic Growth, pp. 3532–3541 (2017)

Attia, M.: Handling Arabic morphological and syntactic ambiguities within the LFG framework with a view to machine translation. Ph.D. dissertation, University of Manchester (2008)

Farghaly, A., Shaalan, K.: Arabic natural language processing: challenges and solutions. ACM Trans. Asian Lang. Inf. Process. **8**(4) (2009). https://doi.org/10.1145/1644879.1644881. Article 14, 22 pages

Farghaly, A.: Subject pronoun deletion rule. In: Proceedings of the English Language Symposium on Discourse Analysis (LSDA 1982), pp. 110–117 (1982)

Hovy, E., Lin, C., Zhou, L., Fukumoto, J.: Automated summarization evaluation with basic elements. In: Proceedings of the Conference on Language Resources and Evaluation, pp. 899–902 (2006)

Tratz, S., Hovy, E.: BEwTE: basic elements with transformations for evaluation. In: Proceedings of Text Analysis Conference (TAC) Workshop (2008)

Zhang, T., Kishore, V., Wu, F., Weinberger, K.Q., Artzi, Y.: BERTScore: evaluating text generation with BERT. In: Proceedings of the International Conference on Learning Representations (ICLR) (2020)

Devlin, J., Chang, M.W., Lee, K., Toutanova, K.: BERT: pre-training of deep bidirectional transformers for language understanding. In: The Proceedings of the 2019 Conference of the North American Chapter of the Association for Computational Linguistics: Human Language Technologies, Volume 1 (Long and Short Papers), pp. 4171–4186 (2019)

Yang, Z., Dai, Z., Yang, Y., Carbonell, J., Salakhutdinov, R., Quoc, V.Le.: XLNet: generalized autoregressive pretraining for language understanding. In: Proceedings of the International Conference on Neural Information Processing Systems, pp. 5753–5763 (2019)

Giannakopoulos, G., Karkaletsis V.: Summary evaluation: together we stand NPowER-ed. In: Proceedings of International Conference on Computational Linguistics and Intelligent Text Processing, vol. 2, pp. 436–450 (2013)

Bentz, C., Alikaniotis, D., Cysouw, M., Ferrer-i-Cancho, R.: The entropy of words—learnability and expressivity across more than 1000 languages. Entropy **19**(6), 275 (2017)

Pedregosa, F., et al.: Scikit-learn: machine learning in python. J. Mach. Learn. Res. **12**, 2825–2830 (2011)

Giannakopoulos, G., El-Haj, M., Favre, B., Litvak, M., Steinberger, J., Varma, V.: TAC 2011 multiling pilot overview. In: Proceedings of the Fourth Text Analysis Conference (2011)

Giannakopoulos, G.: Multi-document multi-lingual summarization and evaluation tracks in ACL'acl 2013 multiling workshop'. In: Proceedings of the MultiLing 2013 Workshop on Multilingual Multi-document Summarization, pp. 20–28 (2013)

Louis, A., Nenkova, A.: Automatically assessing machine summary content without a gold standard. Comput. Linguist. **39**(2), 267–300 (2013). https://doi.org/10.1162/COLI_a_00123

Antoun, W., Baly, F., Hajj, H.: AraBERT: Transformer-based model for Arabic language understanding. In: Proceedings of the 4th Workshop on Open-Source Arabic Corpora and Processing Tools, with a Shared Task on Offensive Language Detection, pp. 9–15 (2020)

Guyon, I., Weston, J., Barnhill, S., Vapnik, V.: Gene selection for cancer classification using support vector machines. Mach. Learn. **46**, 389–422 (2002). https://doi.org/10.1023/A:101248 7302797

A Recommender System for EOSC.
Challenges and Possible Solutions

Marcin Wolski[1]([⊠])(iD), Krzysztof Martyn[1](iD), and Bartosz Walter[1,2](iD)

[1] Poznań Supercomputing and Networking Center, Poznań, Poland
{marcin.wolski,krzysztof.martyn,bartek.walter}@man.poznan.pl
[2] Poznań University of Technology, Poznań, Poland

Abstract. European Open Science Cloud (EOSC) is a pan-European
environment providing researchers with a plethora or publicly-available,
open resources and services to help them conduct their research. Availability of publications, datasets, computational power, networks or storage allows researchers to concentrate on their research rather than the
technical infrastructures. However, the plenitude and diversity of items
offered in EOSC increases and becomes overwhelming for researchers
who expect guidance and support. Recommender systems allow them to
assign rankings to subject object, based on their value for specific end
users, inferred from diverse data about them, their behaviour or various
relationships between users and objects. In this paper we present architectural and functional challenges related to the EOSC Recommender
System that could substantially improve the experience of researchers
using EOSC offerings.

Keywords: EOSC · Recommender system · Recommendations · User
experience

1 Introduction

The constantly growing amount of available data with diverse value and credibility is becoming a real challenge for data consumers and researchers. They face
increasing difficulties in understanding the data and making conscious, informed
decisions based on it. This observation laid a foundation for the advent of Recommender Systems (RSs), focused on supporting the end-users in identifying and
extracting relevant information, considering the perspective of specific users.

Recommender systems are used for filtering large volumes of data objects by
rating each object based on the preferences extracted for each individual user.
The rating represents the value of that object for the user. In order to attain
that goal, an RS infers and learns user preferences from available information
and, based on them, predicts the rating of the given object.

Supported by the *EOSC Future* project, which is co-funded by the European Union
Horizon 2020 Programme call INFRAEOSC-03-2020 – Grant Agreement Number
101017536.

Several use cases for RS have been proposed in the literature and applied in practice [6]. Most frequently they include various extensions for the search and discovery processes, aimed at improving user experience by suggesting the items that are likely to be of a user's interest. RSs appeared to be very effective in that, in particular in large retail e-commerce malls or video streaming services, substantially increasing sales and revenue. According to McKinsey's report, 35% of the Amazon's sales can be attributed to well-addressed recommendations[1].

The success of RSs in commercial applications sparked interest in other domains as well, including health, literature or even cuisine. A similar phenomenon is currently being observed in many areas of scientific research: plenitude of papers, data sets, environments and tools make it virtually impossible for researchers to find them and then to create and effectively operate environments for pursuing science. The European Open Science Cloud (EOSC) is an example of a distributed collection of research-related services and resources that are offered via a uniform portal. It aims at creating a pan-European science hub that will support researchers in exchanging ideas and knowledge as well as fostering collaboration. However, the large number of currently offered objects of different kinds and delivered by various vendors prevents the researchers from using it in an effective way, which leaves the potential of the EOSC ecosystem largely untapped. In addition to that, contemporary research does not rely only on a corpus of publications, but also on other artifacts, resources, services and entities, like data sets, tools, infrastructures and projects. Identification of relevant and useful items, by addressing various preferences and constraints is a non-trivial task. However, it could significantly facilitate and improve the process of organizing and managing the research.

In this paper we envision functional and architectural assumptions, expectations and impediments for the EOSC Recommender System (EOSC-RS), outlining the challenges and issues that need to be overcome. In particular, we show how a research-oriented recommender system differs from its commercial counterparts, and propose adequate solutions.

In this paper we envision functional and architectural assumptions, expectations and impediments for the EOSC RS. At first, we present the EOSC environment taking into account its distinctive attributes such as diverse scientific ecosystem, multiple of user roles accessing EOSC and variety of existing data sources. Next, we outline challenges and issues for the EOSC RS with a particular attention on Big Data handling. Based on that analyse, we delve into details and describe architectural solutions designed to overcome the identified challenges. Finally lessons learnt are presented.

---

[1] https://www.mckinsey.com/industries/retail/our-insights/how-retailers-can-keep-up-with-consumers.

2   Overview of Recommender Systems

A RS is a fully functional software system that makes recommendations by matching the relevant content to the user preferences [8]. To identify items useful for a user, the RS should predict (or compare) the utility of some of them and then decide which items to recommend [23].

RSs work with two types of data: (a) *the user-item interactions*, e.g., ratings made by the user concerning an item or data about the user behaviour in the system, and (b) *attribute information* about the users and items, e.g., textual profiles or relevant keywords [1]. Methods that use the former approach are referred to as *collaborative filtering* methods, whereas methods that apply the latter one are called *content-based* recommender methods. Some RSs combine the above-mentioned aspects to create hybrid systems to utilize the advantages of various types of RSs to perform more robustly in a wide variety of settings.

2.1   Goals of a Recommender System

From the user's perspective, recommendations help improving the overall user satisfaction with the website. For example, a user who repeatedly receives relevant recommendations from a web portal would have better experience with it and is more likely to visit the site again. At the provider end, the recommendation process can provide insights into the user's needs to help customizing it to the user experience.

When properly applied, recommendations can also increase the user's loyalty, which can eventually have a direct impact on revenue of e-commerce and related business. Providing the user with an explanation of why a particular item is recommended is even more useful. For example, in case of YouTube, recommendations are provided in pre-defined sections, such as best matches based on previously watched movies.

There is a wide diversity in the types of products recommended by such systems. Some RSs, such as the one in Facebook, do not directly recommend products. They rather recommend social connections, which have an indirect benefit to the site by increasing its usability and advertising profits.

The most obvious operational goal of RS, named *relevance*, concerns recommending the items that are important to the user at hand. Users are more likely to consume items they find interesting. Although relevance is usually the primary operational goal of RS, others are considered as well. *Diversity* of items prevents the user from getting bored by repeated recommendations of similar items. RS frequently does so by creating a ranking based on sorting the quality scores which combine characteristics that are relevant in a given context. However, such an approach can lead to a sub-optimal solution, as the pointwise estimator ignores correlations between the items. For example, given that a medical data set related to a specific disease has already been shown on the page, it may now be less useful to show another medical data set related to the same disease. This is exacerbated by the fact that similar data sets tend to have similar quality scores.

RSs are truly helpful when the recommended item is something that the user has not seen in the past, but has some value to them. For example, popular movies of a preferred genre would rarely be new to the user. This feature, called *novelty*, helps to avoid repeating the recommendations of popular items. A related notion is that of *serendipity*, wherein the items recommended are somewhat unexpected by the user, and therefore there is a modest element of lucky discovery, as opposed to obvious recommendations. Serendipity is different from novelty in that the recommendations are truly surprising to the user rather than simply something they did not know about before.

2.2 Academic Recommender Systems

In the research environment, the RS may recommend research papers, articles, books, conferences, etc. They may also help to find related authors, who have similar research interests, or to recommend e-infrastructure resources (computational power, storage, network) to support scientific experiments [12]. The academic RSs emerged to overcome and replace the keyword-based search techniques which do not consider users' different interests and purposes [4]. In addition, it is often the case that researchers do not know how to express their requirements, resulting in using weak keywords that yield weak results [19]. Unlike them, RSs can consider researchers' interests, co-author relationship and citations scores to design the recommendation algorithms and provide the recommendation lists. The number of the results can be short and controllable to ensure that the RSs are personalised and effective.

Among all types of academic recommenders, the most popular are those focused on research publications [9,26], with a notable example of Google Scholar[2]. The recommendation of scientific articles aims at recommending relevant articles in correlation with the interests of a researcher or a group of researchers [11]. On top of that, in recent years many researchers have become interested in recommending scientific venues (such as conferences) [10].

Academic social networks provide millions of researchers with functionalities that allow them to promote their publications, find relevant articles and discover trends in their areas of interest [11]. However, the rapidity with which new articles are published and shared, especially on these academic social networks, generates a situation of cognitive overload and is therefore a major challenge for the researcher in search of relevant and recently published information. It is in this context that scientific article recommendation systems are used to filter the huge number of articles shared on these platforms.

2.3 Novel Approaches for Building Recommender Systems

In several RS architectures, we can distinguish two main components: one responsible for data processing, and another one whose task is creating final recommendations [2,5,13,20,25]. This distinction is useful for relieving the

---

[2] https://scholar.google.com.

websites that provide recommendations to the users from time-consuming computationally-exhaustive operations. Thanks to this, they can operate in real-time, delivering reliable recommendations shortly after receiving the request. One of the tasks performed during data processing is generating feature vectors – embedding for both resources and users [13, 16, 21, 24]. They are used as a numerical vector of an object that describes it in a standardized and compressed form. They make it possible to effectively and directly determine the similarity between objects. During creating recommendations, both content-based and collaborative filtering methods use object embedding as the basic set of resource and user characteristics.

Methods of data acquisition in recommendation systems are two-fold, depending on how the data is delivered to the system and its characteristics: static or dynamic [22]. Static data, which changes infrequently, is supplied to the system periodically, most often in the form of batches. This data may include resource descriptions, historical data, or some metadata. Therefore, the volume of such data is large, and it requires long processing. As a result, it usually takes place in the background, to avoid affecting the primary objective, i.e., serving recommendations.

The second type of incoming data is real-time, dynamic data, which may contain current logs from the system, query phrases, or filters applied by the user, together with the request for recommendations. This information changes rapidly and can have a significant impact on recommendations. As a result, their processing is subject to a strong time constraint so that the response is sent back in an acceptable time.

There is also an intermediate data type. The data that comes in real-time, but is not immediately required for generating recommendations, can be handled in the *nearline layer* [2].

3   A Recommender System for EOSC

3.1   About EOSC

EOSC is an ongoing effort to connect existing European e-infrastructures, integrate cloud solutions and provide a coherent point of access for various public and commercial e-infrastructure services [12]. The EOSC Portal is considered as a universal open hub for European researchers that enables them to access, use and reuse research products and data across various scientific disciplines. EOSC Portal recognizes different user roles, with each role having access to different sets of functionalities offered by the different components.

EOSC has been paving the way towards a federated system (a.k.a. *a system of systems* [17]) among service providers. Each single system targets a variety of stakeholders (research communities, funding bodies, individual researchers, providers, data stewards etc.), collects usage statistics and accounting information and offers personalized features within the scope of its provided services.

It is envisaged that thematic clouds and EOSC clusters (such as ESFRI[3] or Blue-Cloud[4]) would allow for the different science communities to engage with their own developed solutions, workflows, computing resources and accessing their own data through domain-based catalogues as community-based "virtual research environments".

3.2   EOSC Users

The Framework of Actors in EOSC identifies roles rather than job titles, as it is likely that in some cases the same person could play different roles (e.g., trainer and researcher, data scientist and research software engineer, etc.) [14]. The identified roles address the needs of stakeholders from the entire EOSC landscape, including research infrastructures, technology providers, service providers, data managers, researchers, policy makers (including funders), and citizens in general.

The EOSC Portal recognizes different user roles, with each role having access to different sets of functionalities offered by the different components. Unregistered users are able to exploit the core facilities of the platform, which consist in service search, browsing and comparison, as well as a subset of aggregate statistics and visualizations of services. Registered users obtain access to advanced functionality, including personalized views of the catalogue's content, as well as the ability to provide user feedback and access notifications and recommendations to/from the platform in several forms. Resource Provider users are responsible for registering and managing services in the platform, representing essentially the resource providing organization; consequently, they are granted access to additional functionality related to their role. Funders/Policy makers are users coming from policy and funding agencies, having access to dashboards and views of the catalogue related to the funding source. Finally, members of the onboarding team and administrators of the portal registry are responsible for the management of the onboarding process and the content in the catalogue.

3.3   Motivation for Improved Recommendations in EOSC

The idea of a Recommender System for EOSC (EOSC-RS) emerged in response to challenges identified in EOSC resulting from a large number of diverse items provided to the users. Currently, the EOSC users can access EOSC to find and then to use the relevant resources. In particular, they may want to choose and combine services from different service providers. However, in that case the researchers expected to know in detail the constraints related to the chosen services. Furthermore, they need to be technically skilled to order and assemble the resources together. Typically, this is not the case, which results in an information gap and sub-optimal selection of resources.

---

[3] https://eosc-portal.eu/news/esfri-clusters-and-thematic-clouds-share-their-positions-eosc.

[4] https://www.blue-cloud.org/.

EOSC-RS is expected to provide the users with enhanced capabilities for discovery of EOSC resources in a user-friendly, customized manner, addressing the data about them. EOSC-RS will support researchers visiting the EOSC portal by delivering them a personalized perspective for the available EOSC resources.

Currently, the EOSC portal offers public and commercial e-infrastructure services, including distributed and cloud computing resources, to enable researchers and other users to process and analyse data[5]. In addition, EOSC Research Products comprising the result of a research process (e.g., publications, datasets, research-supporting software, configurations and other products) are going to be unveiled and made available to the users. They are characterised/described by metadata to be used for citation, attribution, reuse, reproducibility, semantic linking, and findability. The concept of providing relevant recommendations in EOSC is outlined in Fig. 1

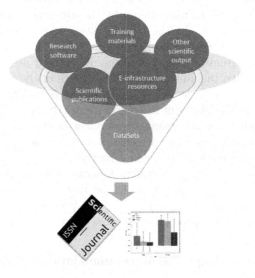

Fig. 1. Delivering relevant content in EOSC

3.4   User Demands for Recommendations in EOSC

EOSC from an end-user's perspective can be seen as a virtual, distributed repository of research data and related services, offering distributed and cloud computing resources to researchers. Such a diverse scientific ecosystem of EOSC can significantly complicate understanding relevant application scenarios for RS and learning users needs.

In general, three main scenarios for providing a personalized view of the EOSC resources and enhanced capabilities to explore EOSC have been identified.

---

[5] https://eosc-portal.eu/sites/default/files/EOSC-Enhance-WP2-JNP-D2.4-2020-11-26_v1.pdf.

These are (i) personalized recommendations, (ii) smart search and (iii) interactive workflows. They represent three main directions of applying the AI/ML to support the users in interacting with EOSC Portal.

Personalized recommendations support EOSC end-users visiting the portal in selecting services and resources based on their actual needs. The recommendation should encompass the user preferences (expressed explicitly or implicitly inferred from the data) extracted from the previous activity, relationships with other users/projects/organizations, or typical constraints, e.g., referring to the data residency etc.

Smart search supports personalized search across data, services and software available in EOSC. Typically, a search query provided by the user is incomplete or imprecise, and the responses are not properly ordered. The smart search will address the issue in two ways: (1) by suggesting related resources that could help the user to make more conscious and informed decision about the research setting s/he needs, (2) ordering the search results, so that the items more relevant to the specific user are listed on the top.

Interactive workflows support the discovery, composition, and execution of workflows obtained as a combination of resources compliant with the same framework. The multitude and heterogeneity of available solutions do not help the user making the right choice concerning elements of the research environment. Ideally, users should be guided and advised during the request processing with the available options, and typical constraints and possible solutions.

Those functionalities directly address the needs of researchers. The needs of other EOSC stakeholders, e.g., Providers, will be addressed indirectly, by using the data collected and extracted during provisioning the services targeted to Researchers.

Personalized recommendations have been found to have the highest value for the users, so they will be implemented in the first stage. Remaining approaches will be added successively.

4   Challenges for the EOSC Recommender System

The EOSC RS will recommend resources from various sources to provide researchers with a full set of tools. These can be e-infrastructure resources registered directly in the EOSC, e-learning resources from the EOSC Research Hub, or resources from external suppliers, e.g., publications, projects, datasets, software and other research products from OpenAire Research Graph (OARG) [18].

Moreover, the system should support various sources of user data, including information from Authentication&Authorization Infrastructures (AAIs) that provide basic information provided by the authorizing organization, e.g., affiliation or email address. Another source of data about the users are their interests, preferences, and current research topics. Along with this, indirect information about the user's preferences would also be collected by analyzing their behaviour while navigating the website, e.g., filters that were applied to the search results or the resources that have been clicked. Additional information could also be linking the users with authors of scientific works.

The challenges EOSC-RS faces can be presented as challenges in handling big data that are described via 5Vs: *volume, variety, velocity, value, veracity* and additionally: *variability* and *linkage* [15]:

- *Volume* – It includes over 150 million resources for recommendation coming from OARG, and the system history data that is processed offline. The total volume of data is roughly estimated to several TBs.
- *Variety* – EOSC-RS will aggregate different data sources with various:
  - data types:
    * *events* coming from users' navigation in the EOSC portal or the resource management events;
    * *database dumps* containing information from external data sources;
    * *data coming from APIs*, e.g., information with requests for recommendations.
  - file formats: jsonf, csv, gz and others;
  - structures: each type of data has a different structure and data meaning: publications, projects, datasets, software, other research products, authors, e-infrastructure resource, user events, resource management events;
  - unbalanced distribution of the number of particular types of resources.
- *Velocity* – One of the main goals of the recommendation systems is to increase the number of users and increase their activity in the EOSC Portal. This will be associated with a significant traffic growth and the EOSC-RS should cope with the several-fold increase in that.
- *Value* – Data must be processed, analyzed, and profiled to discover trends and features in data. Then the detected user- and resource features will be used in real-time for generating the recommendations.
- *Veracity* – OARG data comes from crawling and contains many defects, inconsistencies and inaccuracies. User preference information is only obtained indirectly. The actual user inquiries may also contain linguistic errors. The quality of both data and processing will have a direct impact on the user's satisfaction from using the EOSC-RS.
- *Variability* – Information about the user is collected continuously, e.g., even navigating the website enriches the user's profile with new data items. User preferences also may evolve over time. In addition, the user needs to be able customizing the recommendations by enabling and disabling certain types of recommendations, which will have a significant impact on the way user data is processed. On the other hand, new resources are also constantly added and OARG resource IDs are subject to change due to extracting new resources.
- *Linkage* – In order to create a consistent user profile, it is necessary to combine information coming from multiple sources: events, AAI, requests. There is no unambiguous mapping of EOSC users to the authors of publications, data sets from OARG.

Additional challenges associated with the EOSC-RS initiative include:

- Integration with existing components and projects.
- Making the system open for including attach new sources of data about resources in the future.

5 EOSC Recommender System – Architecture and Data Processing

We will try to solve the challenges discussed in the previous chapter by applying appropriate architectural solutions to the EOSC RS. The solution to the Volume- and Velocity issues is the application of an appropriate architecture that separates data from the processing into several, easily scalable modules. Variety-, Veracity- and Linkage-related challenges require the use of advanced preprocessing to clean and standardize the data included into the system. An explicit split of the data processing in the batch and event mode will allow for dealing with Variability of data. Modules analyzing the data with the use of artificial intelligence (AI) and machine learning (ML) at each stage of processing will allow for maximizing the Value of data.

The first and most important assumption is to build a system based on stateless microservices responsible for specific stages of data processing and preparation. With the use of message queues and load balancing, it will be possible to horizontally scale these functionalities that are critical for quick response of the system. A general diagram of the EOSC-RS architecture is shown in Fig. 2.

RS modules have been partitioned into three groups with respect to their operating mode: online, nearline, and offline.

5.1 Online Modules

Online modules are elements of the system that need to respond almost immediately, as they are responsible for handling user requests in the application. These are modules responsible for the preparation of final recommendations, smart search, chatbot, and they take part in project composition or interactive workflows. These modules use aggregated and pre-processed data from internal databases and also the information coming directly from the request.

The communication with the EOSC Portal has been divided in terms of the data processing modes. Direct communication with other EOSC systems, in particular with the EOSC Portal that includes the recommendation requests, has been designed as a REST API. This type of request is handled by the Online Engine, whose task is to quickly create a recommendation with an explanation and send it back to the EOSC Portal. Recommendations can be displayed in various places of the EOSC Portal, for example in the EOSC Marketplace, in the User Panel, or they can be sent to the user by other means of communication, e.g., by email. EOSC Marketplace allows for searching resources using various filters and queries that narrow down the set of resources displayed to the user. On the other hand, the User Panel displays many lists with personalized recommendations in several panels. In addition, it also allows the user for viewing his activity in the EOSC Portal and managing the settings, including the settings of personalized recommendations.

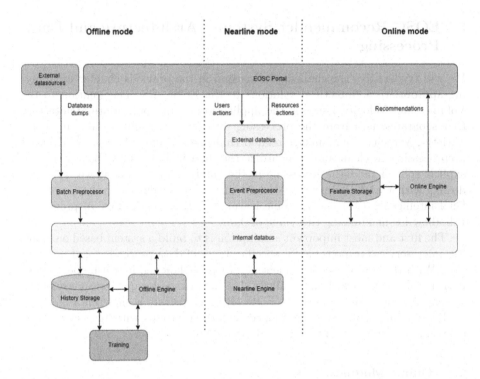

Fig. 2. EOSC recommender system general architecture

To create a recommendation, the Online Engine generates candidates for a recommendation, evaluates them using various methods, and then selects and aggregates all selected resources into one or more recommendation lists. Additionally, recommendations should be accompanied by explanations of where they come from and how they were created. Depending on the type of panel in which the recommendations will be displayed, the recommendations may be created with the use of content-based filtering or collaborative filtering.

While designing the ML/AI modules responsible for recommending resources in the scope of personalized recommendations or project composition we comply with the common architecture of recommender systems [3].

Modules are thus decomposed into the following three stages:

1. **Candidate generator** – This is the first stage of the recommender system's operation. EOSC Portal also sends a `RecomendationContext` with the recommendation request that includes the user ID, current query and filters, and information where to display the recommendation. Based on these data, a subset of the most promising resources is selected. If, after filtering, the set of candidates is too small, it should be extended with other resources based on the similarities between resources.
2. **Scoring** The main objective of this module is to score and select the most relevant candidates that can be recommended to the user. Various ML models

are used to provide scoring of the proposed resources by giving them a rating of relevance according to different objective functions. For each resource, a score or a utility value corresponding to the user's preferences is calculated. Possible scoring algorithms include ranking neural networks or ensembles of tree models. Models used in this step might utilize precomputed features describing users and resources obtained from the external data sources, such as the OARG. At inference, they make use of the feature representations of users and resources fetched from Feature Storage, as well as recommendations that have been prepared in advance. They take into account the user- and search context, as well as different constraints introduced by the user's search parameters. Because of this, multiple lists of highest-scoring recommended resources are prepared. Those lists can be used to provide recommendations in different areas of EOSC Portal. Additionally, each of the proposed items is provided with its corresponding explanation. Items recommended by different algorithms are intended to be complementary, and are further aggregated in the re-ranking step of the RS.

3. **Re-ranking/aggregator** In the third stage, the ranking system takes into account extra information to ensure supplementary recommendation qualities, such as diversity, novelty, fairness, and/or other aspects. It also aims to tackle the problem of a possibly large number of similar items provided by the scoring modules. For example, the system removes resources that have been explicitly disliked by the user earlier and also takes into account any resource recently introduced to EOSC Portal.

To speed up all calculations, the Online Engine uses a set of various databases, indexes, and tools from the Feature Storage. Its task is to store the information in a pre-prepared form so that the Online Engine does not have to perform time-consuming operations. Other tasks of Feature Storage include finding the most similar resources and users and storing cached recommendations.

The explanations contain a description of how individual recommendations have been generated. Their creation begins with the receipt of a request to create a recommendation. The first part is created during the generation of candidates and describes *how they were generated*, e.g., as resources similar to those previously viewed by this user. The next part of the explanation arises during the scoring process and explains *how the resources have been evaluated*. The explanation can be enriched with extra information at the re-ranking stage.

5.2   Nearline Modules

The second group consists of nearline modules that process incoming online data, like user or resource management events. Most of the live information reaching RS from external systems is not directly used in recommendation creating process, but serves to build knowledge about resources, users, and their relationships. So, for the nearline modules the response does not need to be real-time, and they can perform more complex calculations for live data, e.g., create profiles or recommendations that would be used in the future by the online modules.

Asynchronous communication, as well as the user and resource events coming from the entire EOSC system, are sent to the RS via the external data bus and go to the Event Preprocessor. The data here is filtered, analyzed, aggregated, and processed in such a way that it can be easily processed by ML algorithms. The preprocessor also deals with the correlation of data from various data sources and their unification so that they are consistent within the entire EOSC-RS.

From such prepared data, it is possible to create a set of features describing resources and users with the use of machine learning algorithms in the Nearline Engine module. Here, the representation of resources and users in the latent space is created, while the stream data is profiled, clustered and classified. Everything produced by this module is saved in the Feature Storage to be used later by the Online Engine.

5.3   Offline Modules

Processing of data in the batch mode or the analysis of the entire data set or historical data, training algorithms, creating indexes are examples of operations that require a long processing time due to the volume of the data or the complexity of calculations. Modules that would perform these operations in the offline mode are run cyclically, and their execution is not limited in time. Thanks to the separation of such tasks into separate microservices that use different resources and databases, it is possible to perform these tasks independently from the mainstream of data processing.

In this processing mode, the data is delivered in the form of database dumps from other EOSC components, containing, for example, information about EOSC users or e-infrastructure resources, and from external data sources such as OARG. Such raw data must be processed in a Batch Preprocessor where the data is cleaned, standardized, aggregated and correlated with existing data. All operations performed on this data, including analysis, profiling, discovering new knowledge, creating recommendations in advance, takes place in the Offline Engine. This module ensures the quality of recommendations and allows RS administrator to monitor the data processed in the system. If the distribution of data changes significantly, as a result of which the accuracy of recommendations decreases, it is possible to order re-training of the ML/AI algorithms and replacing them in the appropriate modules.

In this mode, there is also a History Storage that stores all historical data that is needed for training and the advanced processing methods that are not directly required during generating a recommendation.

The training module is responsible for training and updating ML/AI models that perform data analysis, profiling, and generate recommendations. The training procedure leverages data about newly recorded users' activities and changes in resources to improve the existing rules for ML/AI tasks in the EOSC Portal. It allows for providing more accurate analysis and recommendations for current and future users, based on a growing base of knowledge.

The training process can be triggered by system operator, or automatically by the Offline Engine. In this scenario, training is performed using historical data. It makes the training procedure time-consuming due to the complex optimization process on a large amount of data, but it also allows for detecting complex dependencies. Another possible scenario is to train the modules directly on the incoming data stream from the preprocessor event. It allows for training more quickly and providing the relevant recommendations for new users and resources without waiting for the cyclic update performed in the offline layer.

5.4   A Data Flow Example

1. After entering the relevant EOSC Portal website, the user triggers a recommendation request for individual panels on the website. In this request, the Recomendation Context object is sent, which includes all the most important information about the user, the search parameters, the page, and the panel where the recommendations will be displayed. This request must be handled in very short time (online), because the whole process takes part in direct interaction with the user. Therefore, all recommendations produced at this stage must come from fast algorithms using previously processed data or must be prepared in advance.
2. The requests from the EOSC Portal are directed to the Online Engine, more specifically to candidate generator, which will filter the list of all available resources and choose a subset of those resources that match the query parameters. From there, the selected candidates can be provided with new resources if the number of candidates is too low. On the other hand, if there are too many of them, candidates are filtered again using information about the user.
3. Selected candidates are then scored by a number of different scoring models that will evaluate each candidate for quality and relevance for the user. These assessments are prepared with the use of a previously processed user profile and additional information contained in the Feature Storage. At this stage, we can also use previously generated recommendations by the offline- and nearline models. All models, apart from generating recommendations, also provide an explanation of the proposed resource order.
4. Recommendations from many scoring modules are aggregated by the Reranking/Aggregator module and compiled into the final list of recommendations for every panel.
5. Finally, the Recommendations with explanations are sent back to the EOSC Portal, where they are presented to the user.
6. Recommendations are saved for reuse and RS quality assessment.

6   Discussion

The main challenges related to RS in EOSC are the complexity and the diversity of data sources that are available for generating recommendations. User needs and expectations are usually not expressed explicitly, and do not rely on one, uniform and well-structured source of data. Instead, the record of researcher's activity includes a variety of data: authored publications, but also the participation in research projects, implicit or explicit collaborations with other researchers, personal contacts, affiliated institutions etc. Yet another data can be extracted from the referenced data, e.g., papers that are cited by the researcher. In addition, also behavioural data about previous user's activity in EOSC portal is also potentially relevant: the viewed items, previously used or requested services deliver information about user interests that could be exploited in generating more adjusted recommendations. Moreover, data coming from various sources may refer to the same objects, opening also a question of the data consistency and credibility. As a result, the RS needs to discriminate among the available data sources and to choose the most reliable one.

The final success of RS largely depends on the users subjective perception of the quality and the adequacy of recommendations. A recommendation approach that works well in one scenario might not meet the expected objectives in another [7]. It means that several aspects play a role: (i) identification of relevant data sources, (ii) correct identification of their credibility, (iii) transparent process of generating the recommendations that is readable for the user, (iv) a selection of alternative recommendation generating schemas, and (v) the method of delivering recommendation to the users. In our work we addressed all of the points mentioned above. We identified the existing data sources in EOSC, determine their structure as well as the content of the data sets, with paying particular attention at the data quality and applicability for serving recommendations. An open, modular architecture of the RS should allow for extending it with new functions and adding alternative recommendation engines to help in managing the potentially increasing complexity of the EOSC environment. The future interoperability schema for EOSC-RS assumes the existence of various recommendation engines, and their interconnections via a common facade. The facade should determine the communication pattern with the available engines, and the way how the obtained recommendations are presented to the front-end layer (for example, through the aggregation of single results). Furthermore, the explicit and early recognition of the three types of recommendations with respect to data availability and computational complexity also will contribute to expected goals that is increased the user satisfaction and user experience, subject to timing constraints.

Similar limitations and challenges like for EOSC-RS are also present in other domains, and the proposed solution could be applicable in them as well. For example, medical decision-support systems also collect multi-dimensional and highly diverse information about patients. They could make a diagnosis based on a set of symptoms collected from various sources, and recommend a protocol of treatment that could be applied in this context. They could also suggest a set of periodic screening tests, based on the patient's history of therapy, the drugs they have been administered, and the available population statistical data. However, applying an RS in the healthcare domain is subject to legal concerns about the consequences of provided recommendations.

We learned a number of lessons from the analysis of the functionality and architecture of EOSC-RS. The first one concerns the openness of the architectural model to extensions and new data sources, concerning primarily the users, but also resources and services offered by the system. Another important issue concerns the data model. It is the user and their identity that connects various data sets. Unfortunately, they can be inaccurate and lead to inconsistencies, which will not only affect themselves, but also compromise the quality of recommendations. Then, there is a need for a comprehensive method of matching user data stored in EOSC with the references to that user in various external databases. The most promising option to correlate EOSC datasets with external databases is ORCID, which is the commonly used and accepted way for identifying a researcher and their scientific record. However, ORCID is only partially present in the currently available databases, and it might not be present in EOSC unless the user enters this information explicitly. Thus, the usefullness of ORCID is currently limited. As an alternative, user's personal data, e.g., name or email address, could be used to match the datasets, but this information is frequently unavailable in public database due to privacy issues.

Finally, the interviews with researchers who would be the prospective users of EOSC-RS indicated that there is no single set of services that satisfy all of them. On the contrary, they need different services and expect different support from the system, depending on the research area they represent and the seniority of their professional position.

All the findings indicate the interoperability as one of the core concerns and objectives for EOSC-RS. The need for smooth interaction, both internally, between various modules of the system, and externally, with third-party providers and other systems, appears vital for the development of a successful recommendation system.

7   Summary

In this paper we presented a functional analysis and architectural design of the Recommendation System for EOSC. EOSC is the main hub for European researchers with diverse services and resources. RS for EOSC will support the researchers in their scientific efforts such as searching, finding and composing digital objects or tools. Various RSs have been successfully implemented in many commercial applications. However, the research-related systems are usually simple and refer only to publications (for example Google Scholar). As a result, the process of generating the recommendations is much more complex and demanding. The concept of RS in EOSC differs in several aspects from typical commercial application. Diversity of resources and variety of users and their roles accessing the EOSC portal are the examples of distinctive attributes of EOSC.

The envisioned functionality and architecture of the EOSC RS address various challenges related to handling big data, in particular the volume and variety of data. Still, big data introduces a number of risks related to the development and operational deployment of RS that need to be properly managed and mitigated when it is feasible.

On the top of described architectural solutions, the paper shows the existing approaches for building RSs that can be applied in EOSC; it also determines the areas where additional research and development effort is required.

The development of the EOSC-RS is currently ongoing, based on a set of identified use cases for personalized recommendations. We plan to release the RS software to production in 2022. Our future work encompasses the further refinement of the EOSC-RS architecture and evaluation of the provided recommendations in a real-life environment.

References

1. Aggarwal, C.C.: Recommender Systems - The Textbook. Springer, Cham (2016). https://doi.org/10.1007/978-3-319-29659-3
2. Amatriain, X.: Big & personal: data and models behind Netflix recommendations. In: Proceedings of the 2nd International Workshop on Big Data, Streams and Heterogeneous Source Mining: Algorithms, Systems, Programming Models and Applications, pp. 1–6 (2013)
3. Niriksha, T.K., Surendiran, B., Muppana, M., Rajagopalan, N.: Analysis of sub-clustering in group recommender system **23**, 12 (2020)
4. Bai, X., Wang, M., Lee, I., Yang, Z., Kong, X., Xia, F.: Scientific paper recommendation: a survey. IEEE Access **7**, 9324–9339 (2019). https://doi.org/10.1109/ACCESS.2018.2890388
5. Baldominos, A., Saez, Y., Albacete, E., Marrero, I.: An efficient and scalable recommender system for the smart web. In: 2015 11th International Conference on Innovations in Information Technology (IIT), pp. 296–301. IEEE (2015)
6. Beel, J., Collins, A., Kopp, O., Dietz, L., Knoth, P.: Online Evaluations for Everyone: Mr. DLib's Living Lab for Scholarly Recommendations, pp. 213–219, April 2019. https://doi.org/10.1007/978-3-030-15719-7_27
7. Beel, J., Dinesh, S.: Real-world recommender systems for academia: the pain and gain in building, operating, and researching them [long version], April 2017
8. Beel, J., Gipp, B., Langer, S., Breitinger, C.: Research-paper recommender systems: a literature survey. Int. J. Digit. Libr. **17**(4), 305–338 (2015). https://doi.org/10.1007/s00799-015-0156-0
9. Beel, J., Gipp, B.: Google scholar's ranking algorithm: an introductory overview. In: Proceedings of the 12th International Conference on Scientometrics and Informetrics (ISSI 2009), vol. 1, pp. 230–241, Rio de Janeiro, Brazil (2009)
10. Beierle, F., Tan, J., Grunert, K.: Analyzing social relations for recommending academic conferences. In: HotPOST 2016 - Proceedings of the 8th MobiHoc International Workshop on Hot Topics in Planet-Scale mObile Computing and Online Social Networking, pp. 37–42 (2016). https://doi.org/10.1145/2944789.2944871
11. Boussaadi, S., Aliane, H., Abdeldjalil, O., Houari, D., Djoumagh, M.: Recommender systems based on detection community in academic social network. In: 2020 International Multi-Conference on: "Organization of Knowledge and Advanced Technologies" (OCTA), pp. 1–7 (2020). https://doi.org/10.1109/OCTA49274.2020.9151729
12. Budroni, P., Burgelman, J., Schouppe, M.: Architectures of knowledge: the European open science cloud. ABI Technik **39**, 130–141 (2019). https://doi.org/10.1515/abitech-2019-2006

13. Chamberlain, B.P., Hardwick, S.R., Wardrope, D.R., Dzogang, F., Daolio, F., Vargas, S.: Scalable hyperbolic recommender systems. arXiv preprint arXiv:1902.08648 (2019)
14. European Commission and Directorate-General for Research and Innovation: Digital skills for FAIR and Open Science: report from the EOSC Executive Board Skills and Training Working Group. Publications Office (2021). https://doi.org/10.2777/59065
15. Demchenko, Y., De Laat, C., Membrey, P.: Defining architecture components of the big data ecosystem. In: 2014 International Conference on Collaboration Technologies and Systems (CTS), pp. 104–112. IEEE (2014)
16. Kanakia, A., Shen, Z., Eide, D., Wang, K.: A scalable hybrid research paper recommender system for microsoft academic. In: The World Wide Web Conference, pp. 2893–2899 (2019)
17. Maier, M.W.: Architecting principles for systems-of-systems. Syst. Eng. J. Int. Council Syst. Eng. 1(4), 267–284 (1998)
18. Manghi, P., et al.: OpenAIRE research graph dump, December 2021. https://doi.org/10.5281/zenodo.5801283. A new version of this dataset is published every 6 months. The content available on the OpenAIRE EXPLORE and CONNECT portals might be more up-to- date with respect to the data you find here
19. Mitchum, R.: Unwinding the 'long tail' of science (2012). https://voices.uchicago.edu/compinst/blog/unwinding-long-tail-science/. Accessed 20 Jan 2021
20. Monsalve-Pulido, J., Aguilar, J., Montoya, E., Salazar, C.: Autonomous recommender system architecture for virtual learning environments. Appl. Comput. Inform. (2020)
21. Nisha, C., Mohan, A.: A social recommender system using deep architecture and network embedding. Appl. Intell. 49(5), 1937–1953 (2018). https://doi.org/10.1007/s10489-018-1359-z
22. Rabiu, I., Salim, N., Da'u, A., Osman, A.: Recommender system based on temporal models: a systematic review. Appl. Sci. 10(7), 2204 (2020)
23. Ricci, F., Rokach, L., Shapira, B.: Recommender systems: introduction and challenges. In: Ricci, F., Rokach, L., Shapira, B. (eds.) Recommender Systems Handbook, pp. 1–34. Springer, Boston (2015). https://doi.org/10.1007/978-1-4899-7637-6_1
24. Shi, B., et al.: DARES: an asynchronous distributed recommender system using deep reinforcement learning. IEEE Access 9, 83340–83354 (2021)
25. Tan, B., Liu, B., Zheng, V., Yang, Q.: A federated recommender system for online services. In: Fourteenth ACM Conference on Recommender Systems, pp. 579–581 (2020)
26. Xia, F., Liu, H., Lee, I., Cao, L.: Scientific article recommendation: exploiting common author relations and historical preferences. arXiv 2(2), 101–112 (2020). https://doi.org/10.1109/tbdata.2016.2555318

ERIS: An Approach Based on Community Boundaries to Assess Polarization in Online Social Networks

Alexis Guyot[✉], Annabelle Gillet, Éric Leclercq, and Nadine Cullot

Laboratoire d'Informatique de Bourgogne - EA 7534, University of
Bourgogne-Franche-Comté, Dijon, France
{alexis.guyot,annabelle.gillet,eric.leclercq,
nadine.cullot}@u-bourgogne.fr

Abstract. Detection and characterization of polarization are of major interest in Social Network Analysis, especially to identify conflictual topics that animate the interactions between users. As gatekeepers of their community, users in the boundaries significantly contribute to its polarization. We propose ERIS, a formal graph approach relying on community boundaries and users' interactions to compute two metrics: the community antagonism and the porosity of boundaries. These values assess the degree of opposition between communities and their aversion to external exposure, allowing an understanding of the overall polarization through the behaviors of the different communities. We also present an implementation based on matrix computations, freely available online. Our experiments show a significant improvement in terms of efficiency in comparison to existing solutions. Finally, we apply our proposal on real data harvested from Twitter with a case study about the vaccines and the COVID-19.

Keywords: Social networks · Polarization · Community boundaries · Community structure · Graph mining

1 Introduction

Online Social Networks (OSN) are large scale environments of exchanges and debates. Social Network Analysis (SNA) has diverse and numerous applications in domains such as sociology, politics, marketing, health, etc. The intrinsic characteristics of the large volume of data generated in OSN [3], such as the power law distribution, entail analysts to use algorithmic approaches to extract value.

SNA can benefit from the graph theory since graph structures are natural representations for OSN, where users can be represented by vertices and their interactions by edges. Communities of individuals form dense areas of nodes and can therefore be detected by algorithms like Louvain, Walktrap or Infomap for non-overlapping communities and SLPA, OSLOM or Game for overlapping ones [11].

© The Author(s), under exclusive license to Springer Nature Switzerland AG 2022
R. Guizzardi et al. (Eds.): RCIS 2022, LNBIP 446, pp. 88–104, 2022.
https://doi.org/10.1007/978-3-031-05760-1_6

Discussions about hot topics can lead to the creation of mutually antagonistic communities, with few individuals remaining neutral or holding an intermediate position. In social sciences, this phenomenon is called polarization [21]. Polarized communities negatively impact OSN by fostering social division, ideological isolation and misinformation spreading [4]. Detecting such communities is of major interest to proactively assist moderation and therefore avoid further escalation between users. Journalists could also benefit from this feature to identify areas in the network where fact checking could be needed. Moreover, detecting polarization allows a more precise understanding of individuals through their relationships. Domains such as politics or business intelligence could benefit from it to adapt their decision making and communication strategies.

In the literature, echo chambers are usually considered as the consequence of polarization [7]. However, only showing that a community is an echo chamber does not allow to conclude about its polarization. An echo chamber is a configuration in which one is exposed only to opinions that agree with their own [12]. So this phenomenon describes a global behavior within a community whereas polarization is also about relationships between communities [4]. Thus, the polarization is also carried by community members exposed to other communities and exposing the main topic of their community to the outside through their interactions. These members form the community boundaries. More formally, we can define a community boundary as the set of nodes having edges directed toward both the inside and the outside of the community [8]. In the literature, boundaries are fairly unexplored parts of communities. Nevertheless, the behaviors of boundary users have a significant impact on the strength of the polarization but also on the fragility of the echo chamber [10] as they contribute to the porosity of the boundary.

The major contributions of our work are: 1) a formal graph approach relying on community boundaries to unveil the polarization of networks created from the interactions between individuals; 2) two metrics to characterize the level of polarization and the porosity of boundaries; 3) an efficient algorithm based on matrix computations suitable for large volumes of data, and; 4) a case study on real data extracted from Twitter to experimentally validate our proposal.

The remainder of this paper is structured as follows. First, our method is positioned in relation to the state of the art in Sect. 2. In Sect. 3 we formally define the ERIS approach and propose an efficient algorithm to compute polarization metrics. The case study led on real data and validated by domain experts is described in Sect. 4. Finally, we draw conclusions of our work and open up perspectives for the future in Sect. 6.

2   Related Work

The problem of polarization in OSN was addressed back in 2011 [9]. The authors consider that echo chambers and polarized communities are the same. But with this assumption, interactions and relationships between communities, carried by community boundaries, are ignored. Moreover, the approach is an applied methodology which cannot be included in an automatic analytical workflow ready-to-use for domain experts.

Many works on social polarization use exploratory analyzes combining metrics from the graph theory with interpretations provided by Natural Language Processing (NLP) tools like sentiment analysis based on Naïves Bayes [2,20] or sentence embedding models based on Retweet-BERT [22]. These approaches best capture the semantics of discussions but require heavy involvement from the analyst, especially during the preprocessing step. Indeed, to set the stage for the NLP algorithms, many difficulties must be manually addressed like spelling approximations, abbreviations, slang, or ambiguities caused by humor, sarcasm or irony as discussed in [16,23,24]. Thus, they leave a large area for subjectivity and lack automatism.

Other approaches focus on the network structure with weakly-supervised strategies where just a limited amount of extra knowledge is used to initialize algorithms. In [25] an opinion score is manually assigned to seed users (*elites*) and then propagated to the other nodes of the network (*listeners*) in order to create two opposing groups and to assess their polarization degree. In [1], a similarity measure must be wisely chosen to create clusters of tweets (*assertions*) with the aim of unveiling polarized groups inside a network. To do so, they use a matrix factorization and an ensemble based gradient descent algorithm applied on the adjacency matrices of a bipartite source-assertion graph and a social influence graph. In any case the relevance of the results depends a lot on the extra knowledge brought, which must be revised for each dataset studied. Therefore, their automatism is limited. Moreover, they do not consider the relationships between polarized communities, meaning that two communities behaving as echo chambers but never interacting because they do not know each other could be considered as polarized.

Boundaries have a major impact on the polarization of their community by defining both how the community is exposed to the outside and how the outside is exposed to the community. A first non-supervised approach based on community boundaries was described by Guerra et al. in [17] with the aim of computing complementary metrics to be used besides cohesion and homophily metrics such as modularity. Antagonism between communities is assessed by measuring the involvement of users interacting with both the inside and the outside of their community. This approach does not need any *a priori* knowledge on the graph or on the individuals represented and can therefore be included in automatic analytical workflows designed for domain experts. However, the main limitation of this method is its specification on undirected and unweighted graphs whereas social interaction graphs usually are directed and weighted. Furthermore, only non-overlapping communities are handled whereas users of social networks more naturally belong to multiple communities [26].

As a conclusion, fully automatic methods are the best option to detect polarization. This property is indeed very important in SNA to allow domain experts like sociologists or decision-makers to use a method and to permit comparisons between datasets. Furthermore, the polarization of a community can be misinterpreted when interactions between communities are not considered, which can be avoided by examining the behavior of community boundaries.

3   The ERIS Method

In this section, we formally introduce the ERIS method and its metrics, *i.e.*, the *community antagonism* and the *porosity of boundaries*. Our approach relies on edge weighting and direction and handles overlapping communities. We also propose an efficient algorithm based on matrix computations to assess the metrics.

3.1   Formal Definitions

In the following definitions, a graph $G = (V, E)$ is composed of a set of vertices V and of a set of directed edges $E \subseteq V \times V$. An edge $e_{a,b} \in E$ connects a source $a \in V$ and a destination $b \in V$ with a weight $w(e_{a,b}) \in \mathbb{R}$. Communities are locally dense connected subsets of V.

Two communities (C_i, C_j) are polarized if they are mutually antagonistic. According to [17] and [4], a strong involvement from a boundary individual within the community, especially expressed by numerous interactions with the internal members, reveals a substantial emotional attachment to the community and its main topics. This attachment could easily lead to the expression of antagonism in response to a criticism, an attack or the broadcast of a negative opinion or information about these topics by another community.

The ERIS method consists in identifying, for each pair of communities (C_i, C_j): 1) the internal area $I_{i,j}$ of C_i, that is the set of vertices in C_i without any edge directed toward C_j, and; 2) the boundary area $B_{i,j}$ of C_i, that is the set of vertices in C_i with at least one edge directed toward $I_{i,j}$ and another one toward C_j. The method assesses the average antagonism expressed by the community C_i to the community C_j by measuring the involvement of the vertices in B_{ij}.

From the previous intuitive descriptions, we have established the following formal definitions:

$$I_{i,j} = \{v : v \in C_i, \nexists e_{v,n} \mid n \in C_j, i \neq j\} \tag{1}$$

$$B_{i,j} = \{v : v \in C_i, \exists e_{v,n_1} \mid n_1 \in C_j, \exists e_{v,n_2} \mid n_2 \in I_{i,j}, i \neq j\} \tag{2}$$

For each boundary, we consider the set of outgoing edges directed toward the other community (*external edges* or $EE_{i,j}$) as well as the set of edges directed toward the internal area $I_{i,j}$ of C_i (*internal edges* or $IE_{i,j}$):

$$EE_{i,j} = \{e_{s,d} : s \in B_{i,j} \land d \in C_j\} \tag{3}$$

$$IE_{i,j} = \{e_{s,d} : s \in B_{i,j} \land d \in I_{i,j}\} \tag{4}$$

We also consider $EE_{i,j}^v$ the external and $IE_{i,j}^v$ the internal edges of a vertex v as the subsets of edges, respectively included in $EE_{i,j}$ and $EI_{i,j}$, where v is the source of the edge:

$$EE_{i,j}^v = \{e_{v,d} : e_{v,d} \in EE_{i,j}\} \tag{5}$$

$$IE_{i,j}^v = \{e_{v,d} : e_{v,d} \in IE_{i,j}\} \tag{6}$$

The antagonism $A_{i,j}^v$ expressed by a vertex v is assessed as the weighted ratio of its internal edges' weights with the sum of its internal and external edges' weights. This value is compared to a null hypothesis, i.e., each node spreads its edges equally between internal nodes and nodes from the other community [17]:

$$A_{i,j}^v = \frac{\sum_{e \in IE_{i,j}^v} w(e)}{\sum_{e \in IE_{i,j}^v} w(e) + \sum_{e \in EE_{i,j}^v} w(e)} - 0.5 \tag{7}$$

Finally, the antagonism $A_{i,j}$ expressed by a boundary $B_{i,j}$ is the average antagonism expressed by its members:

$$A_{i,j} = \frac{1}{|B_{i,j}|} \sum_{v \in B_{i,j}} A_{i,j}^v \tag{8}$$

By assessing the antagonism values for each possible pair of communities in a graph, we obtain an asymmetrical matrix called the *antagonism matrix*, containing values ranging from -0.5 to 0.5. A community boundary with a value close to 0.5 should be considered as likely to be antagonistic toward the other community of the pair. Values on the lines of the antagonism matrix express how much the community heading the line is likely to express antagonism toward the communities heading the columns. Conversely, values on the columns indicate how much the community heading the column is likely to receive antagonism from the communities heading the lines.

Boundary vertices with negative antagonism values weaken the polarization of their community. Indeed, by interacting more with the outside than with the inside, they reduce the isolation of the community that leads to the creation of an echo chamber. As fellow members, they also seem more credible in the eyes of the others when they share more nuanced opinions about the main topics of the community [10]. Based on these ascertainments, we propose a novel metrics $P_{i,j}$ called the *porosity* of the boundary $B_{i,j}$, measuring the fragility of the boundary of C_i with C_j:

$$P_{i,j} = \frac{|NB_{i,j}|}{|B_{i,j}|} \times 100 \tag{9}$$

with $NB_{i,j} = \{v : v \in B_{i,j}, A_{i,j}^v < 0\}$ the subset of $B_{i,j}$ including all the vertices having negative antagonism values. Porosity values also can be represented inside an asymmetrical matrix called the *porosity matrix*.

We now illustrate the different sets and values presented in this subsection by applying the definitions on the toy example of Fig. 1. We focus our explanation

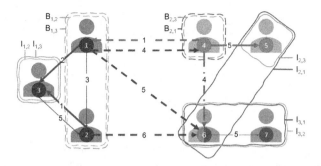

Fig. 1. Toy example with 3 communities C_1 (blue), C_2 (red) and C_3 (purple). Communities C_2 and C_3 overlap on vertex 6. For edges, solid lines are internal edges, dotted lines are external edges, thin lines are edges neither internal nor external. Note that $e_{4,6}$ is both internal and external. For areas, internal areas are surrounded by solid lines, boundary areas by dotted lines. (Color figure online)

on the community C_1. Both internal areas of C_1 with C_2 and C_3 include the same vertices, that is $I_{1,2} = I_{1,3} = \{3\}$. The same observation can be made with its boundary areas, $B_{1,2} = B_{1,3} = \{1,2\}$. External and internal edges of the pair (C_1, C_2) are $EE_{1,2} = \{e_{1,4}, e_{1,6}, e_{2,6}\}$ and $IE_{1,2} = \{e_{1,3}, e_{2,3}\}$.

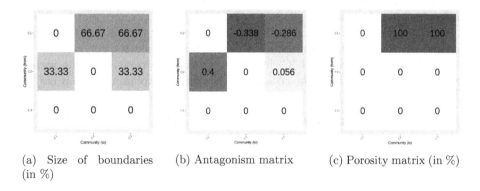

(a) Size of boundaries (in %)

(b) Antagonism matrix

(c) Porosity matrix (in %)

Fig. 2. Values calculated from the toy example

Figure 2 shows the antagonism and the porosity matrices obtained on the graph of the toy example. Figure 2a expresses the sizes of the different boundary areas as a percentage of community members belonging to the boundary. In Fig. 2b, the antagonism value $A_{1,2}$ expressed by the boundary of the community C_1 toward the community C_2 is equal to -0.338 and therefore does not reveal an antagonistic behavior. However, the boundary of the community C_2 is pretty likely to be antagonistic toward the community C_1 since its antagonism value $A_{2,1}$ is equal to 0.4. The matrix of Fig. 2a reveals that all the values equal to 0 in the antagonism matrix are default values resulting from empty boundaries. In

Fig. 2c, the porosity value $P_{1,2}$ is equal to 100, meaning that $B_{1,2}$ is very porous as its members interact more with C_2 than with $I_{1,2}$.

3.2 Algorithm

The adjacency matrix of a graph is at the core of the algorithm designed for ERIS, which uses a series of matrix computations to get the antagonism (M_{ANT}) and the porosity (M_{POR}) matrices. Naming conventions used in following paragraphs are listed in Table 1. We define ✦ as an operator computing element-wise multiplication between a vector of size N and each column of a matrix of size $N \times M$, resulting in a new matrix of size $N \times M$.

Table 1. Naming conventions

Symbol	Definition	Symbol	Definition
G	The graph to analyze	C	The set of communities in G
V	The set of vertices in G	$|V|$	The number of vertices in G
E	The set of edges in G	$|C|$	The number of communities in G

The inputs of the algorithm are the adjacency matrix M_A of G and a community membership matrix called M_C (Table 2). M_A is a square matrix of size $|V| \times |V|$ containing in each cell the weight of the edge whose source is the vertex heading the row and the destination is the vertex heading the column. M_C is a binary matrix of size $|V| \times |C|$ in which the value 1 means that the vertex heading the row belongs to the community heading the column, and 0 if not.

Table 2. Matrices used in Algorithm 1

Symbol	Size	Type	Name
M_A	$|V| \times |V|$	Int	Adjacency matrix
M_C	$|V| \times |C|$	Bin	Community membership matrix
M_{EE}	$|V| \times |C|$	Int	External edges weight matrix
M_I	$|V| \times |C|$	Bin	internals matrix
M_{IC}	$|V| \times |C|$	Bin	Current internals matrix
M_{IE}	$|V| \times |C|$	Int	Internal edges weight matrix
M_{BIE}	$|V| \times |C|$	Bin	Binary internal edges matrix
M_{VANT}	$|V| \times |C|$	Real	Vertices antagonism matrix
M_{ANT}	$|C| \times |C|$	Real	Antagonism matrix
M_{POR}	$|C| \times |C|$	Real	Porosity matrix

Algorithm 1. Matrix computations to assess the metrics of ERIS

Require: M_A, M_C
Ensure: M_{ANT}, M_{POR}
1: $M_{EE} \leftarrow M_A \times M_C$
2: $M_I \leftarrow (M_{EE} == 0)$
3: **for** $c = 1, \ldots, |C|$ **do**
4:    $M_{IC} \leftarrow M_C[,c] \blacklozenge (M_I \cdot \neg M_C)$
5:    $M_{IE} \leftarrow (M_C[,c] \blacklozenge (M_A \times M_{IC})) \cdot \neg M_I$
6:    $M_{BIE} \leftarrow (M_{IE}! = 0)$
7:    $M_{VANT} \leftarrow ((M_{IE}/(M_{IE} + M_{EE})) - 0.5) \cdot M_{BIE}$
8:    $M_{ANT}[c,] \leftarrow (M_C^T \times M_{VANT})/(M_C^T \times M_{BIE})$
9:    $M_{POR}[c,] \leftarrow 100 * (M_C^T \times (M_{VANT} < 0))/(\sum_{i=1}^{|V|} M_{BIE}[i,])$
10: **end for**

The initialization part of the algorithm consists in computing M_{EE}, an aggregated version of M_A grouped by community (line 1). The matrix contains the sum of the edges' weights whose source is the vertex heading the row and the destination is a vertex belonging to the community heading the column. This matrix is then used to extract M_I, a binary mask of M_{EE} in which the vertices belonging to at least one internal area of their communities are identified (line 2).

From these two common matrices, the main part of the algorithm computes for each community the antagonism and porosity values of its boundaries through four main steps:

- the detection of the internal areas of the current community c for each pair involving c (line 4);
- the aggregation of M_A to sum the weights of the edges directed toward the internal areas of c (line 5);
- the computation of the antagonism values for the vertices belonging to the boundaries of c (line 6);
- the computation of the antagonism and porosity values for the boundaries of c (lines 8–9).

An open source implementation of this algorithm in R is available on GitHub[1] to allow the use of ERIS on graphs built from real datasets.

4   Experimentations

We want to experimentally show the suitability of our method on real data from OSN. First, we verify its applicability on large graphs through an analysis of the algorithmic complexity in time achieved by our matrix computation based algorithm. Then, we explore the validity of our polarization metrics through a case study led on real data harvested from Twitter with the help of domain experts validating our results and interpretations.

---

[1] https://github.com/AlexisGuyot/ERIS.

4.1   Execution on Large Graphs

We compare the execution times of 3 algorithms aiming to measure the polarization of communities in graphs built from interactions between individuals:

- the matrix computation based algorithm of ERIS presented in the previous section (implemented in R);
- an iterative algorithm of ERIS proposed in a previous work [19] (implemented in R);
- the only algorithm of Guerra et al.'s method available online[2], not developed by the authors (implemented in Python).

We chose to compare ERIS with Guerra et al.'s method because both share a lot of common characteristics (no supervision needed, based on graph mining, etc.).

We have generated artificial graphs with decreasing sizes, ranging from 1 million to 500 vertices, based on a real graph extracted from a dataset harvested from Twitter. In this subsection, we do not take into consideration the semantics of the computed metrics, only the impact of the graph structure on the execution time.

For each algorithm, we have measured the elapsed time between the call of the function computing the antagonism matrices (the common metrics) and the return of a result[3]. For the Algorithm 1, this interval corresponds to lines 1 to 10. The three algorithms were run on a Dell PowerEdge R440 server with the following characteristics: Intel(R) Xeon(R) Bronze 3204 CPU @ 1.92 GHz, 6 cores, 128Go RAM.

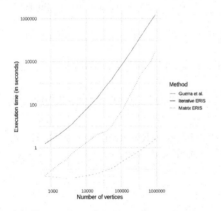
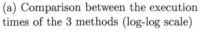
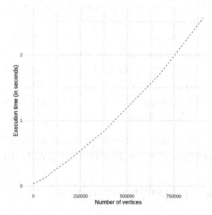

(a) Comparison between the execution times of the 3 methods (log-log scale)

(b) Focus on the execution times of the matrix computation based algorithm

Fig. 3. Execution times of the algorithms

Execution times of the three algorithms are compared on Fig. 3a. Figure 3b focuses on the execution times of the Algorithm 1 described in the last section. We can see that the matrix computation based algorithm of ERIS outperforms all the other implementations. For our biggest graph, the one with 1 million vertices directly extracted from the real corpus that we have harvested from Twitter, the matrix computation based version of ERIS took 2.5 s to compute the metrics. It is 12,828 times faster than the iterative version (32,070 s or almost 9 h) and 592,399 times faster than the algorithm of Guerra et al.'s method (1,528,389 s or more than 17 days).

Theoretically, the computational complexity for assessing both polarization metrics with the algorithm based on matrix computations is $\mathcal{O}(|V|^2|C|^2)$, as long as $|C| < \sqrt{\frac{|V|}{3}}$. Beyond this value, the order reaches $\mathcal{O}(|V||C|^3)$. However, in most of practical analyzes, the number of significant communities in a graph remains relatively small due to the resolution limit. Furthermore, domain experts also only require a small number of communities to left results open to interpretation. In these cases, $|C| \ll |V|$ and thus the computational complexity can be considered as $\mathcal{O}(|V|^2)$.

According to the previous theoretical and practical analyzes of the algorithmic complexity of our proposal, we can conclude that the matrix computation based algorithm of ERIS achieves our goals of applicability on large graphs and outperforms of several orders of magnitude the other algorithms available online to automatically assess polarization on graphs extracted from OSN.

4.2 Case Study on Real Data

We experimentally illustrate the interest of our approach on a real dataset about COVID-19 vaccines, which includes more than 18 millions tweets harvested in the context of the interdisciplinary project Cocktail[4] by the architecture Hydre [14] from December 1, 2020 to March 31, 2021 (120 days).

From this dataset, a directed graph of quotes[5] G_Q is extracted, in which the vertex representing an individual u has an outgoing edge directed toward the vertex representing the user n if u has already quoted at least twice the tweets of n. The weight w of an edge indicates the exact number of times u quoted n. Following [13], we do not consider isolated quotes as they can be random noise. The characteristics of G_Q are presented in Table 3.

---

[4] https://projet-cocktail.fr/.
[5] Quotes are retweets with additional comments.

Table 3. Characteristics of G_Q

Vertices count	24,591
Edges count	55,703
Average strength	4.46
Diameter	338
Power law exponent γ	2.27
Resolution limit	333
Significant community count	8
Modularity	0.59

We chose the quote as type of interaction to be consistent with the previous works led on polarization on Twitter. Indeed, the literature mainly agrees that retweets often imply endorsement [6] and thus not antagonism, and that the mention network is usually not polarized [9]. However, quotes are often used to twist a message out of its original context for humor and criticism purposes, leading to antagonistic responses [18]. Thus, quotes are the best type of interactions for community boundary approaches like ERIS to assess polarization.

G_Q is a scale-free network as the degree distribution of its vertices follows a power law of exponent $2 < 2.27 < 3$ [3]. As a result, modularity can be computed and therefore community detection algorithms based on the optimization of modularity, like Louvain [5], can be run. On G_Q, the previous algorithm revealed 8 significant communities, *i.e.*, having a largest size than the resolution limit of the graph [15] (Table 3). The overall modularity of G_Q is 0.59.

To better understand the communities and their relationships, the domain experts of the interdisciplinary project have manually assigned to each community a label related to its main topics by analyzing, for each, its 30 most used hashtags (top-hashtags). This labeling step revealed that the two biggest communities of G_Q gather respectively pro and anti-vaccine individuals. Table 4 lists the elements among the 30 top-hashtags of these last two communities used to infer the labels.

Table 4. Top-hashtags highlighting the main topics of the pro and anti-vaccine communities

Community	Top-hashtags (translated from French)
Pro-vaccines	Mutation, Lockdown3, Curfew, Schools, DigitalGreenCertificate, HealthDictatorship, Israel, IGetVaccinated, Pasteur
Anti-vaccines	Ivermectine, HealthDictatorship, IWillNotConfineMyself, Raoult, Hydroxychloroquine, AndTheTreatment, Plandemic, VeranResignation, TheStonesWillCryOut, GreatReset, Ethics, BeBraveWHO, IWillNotGetVaccinated

Since these two main topics are opposite, we expect polarization between the communities. Thus, the anti and pro-vaccine communities should be cohesive and closed communities and their mutual relationship should be antagonistic. To experimentally confirm this expectation, we apply our implementation of the Algorithm 1 on G_Q. The computed results are shown in Figs. 5 and 6. Figure 4 gives supplementary information about the sizes of the boundaries.

Values on the lines of the antagonism matrix (Fig. 5) express how much the community heading the line is likely to express antagonism toward the communities heading the columns. Conversely, values on the columns indicate how much the community heading the column is likely to receive antagonism from the communities heading the lines.

Columns related to the pro and anti-vaccine communities show that both do not receive much antagonism from the other communities. The community the most likely to be antagonistic with the pro-vaccine community is the anti-vaccine community (0.278) and *vice versa* (0.152). Lines related to these two communities show however that both are pretty likely to have antagonistic behaviors with all the communities. Two hypotheses might explain the lower values between the two communities in comparison with the others: 1) the anti and pro-vaccine boundaries are not very antagonistic with each other; 2) the lively debates between these communities lead some boundary members to communicate more with the outside than with the inside.

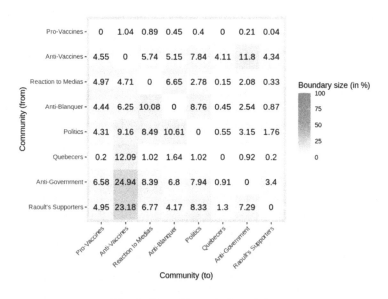

Fig. 4. Size of boundaries of G_Q

The matrix representing the porosity of boundaries (Fig. 6) allows to decide between the two previous hypotheses and shows that the second one is the more likely. Indeed, we see that 10% of the boundary members of the pro-vaccine community interact more with the anti-vaccine community than with the core members of their own community. For the anti-vaccine community, the equivalent value is around 5%. Thus, these contributions to the debates cause the decrease of the antagonism values for both boundaries. A possible interpretation for this observation is a need for these boundary users to convince their opponents to change their mind.

A deeper understanding of the behavior and roles of these two communities can also be achieved through the lines and columns of the porosity matrix. First, from a broader perspective, we can see that the values on the lines related to both communities are pretty low in comparison with the other ones, meaning that their boundaries do not interact much with the outside. So, anti and pro-vaccine communities are fairly closed communities. Furthermore, on the line related to the anti-vaccine community only, we can see that, even if the values are low, all the boundaries are nearly as porous. Therefore, the anti-vaccine community is almost equally exposed everywhere, which could reveal an additional need to control the debate and the image of the community.

Columns related to these two communities show that they both have a significant impact on the porosity of the other communities, pointing out the general interest of the individuals forming our corpus for the vaccination topic. Higher values in the column related to the anti-vaccine community reveal a trend for this

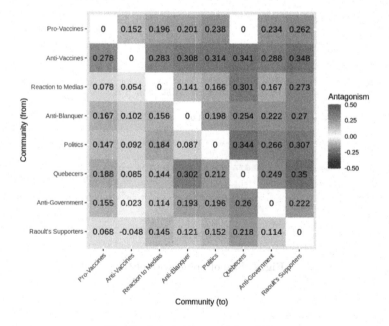

Fig. 5. Antagonism matrix of G_Q

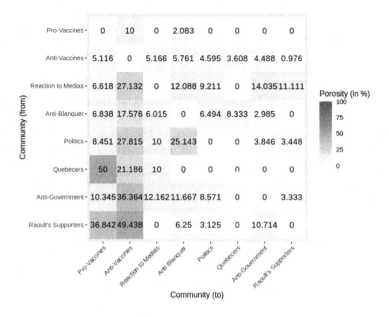

Fig. 6. Porosity matrix of G_Q

community to trigger a lot of reactions from the boundaries of the other communities. Because we are working with quotes, these reactions could be sarcastic, ironic or humorous, and therefore rather negative.

In brief, the metrics of the ERIS method describe a relationship likely to be antagonistic between two fairly closed communities. From this observation, we can conclude that, as expected, the pro and anti-vaccine communities are polarized in our corpus. The ERIS method successfully highlights the traces of polarization within a graph built from interactions between individuals in OSN.

5   Discussions

In this section, we discuss about threats to validity for our method and ideas to improve the interpretability of our metrics.

Before using ERIS, a well informed user should be careful about some points that could threaten its interpretations. First, one should make sure of the objectivity of the data harvesting process, to avoid bias in the data. The longer the time period covered by the dataset is, the more the information is diluted because of higher probabilities to find random or one-time interactions between individuals. There also may be a shift of attention to subjects. ERIS needs well defined communities, so the user should pay a close attention to the chosen algorithm. For example, the modularity of the graph should be high enough to ensure the realness of detected communities with methods like Louvain. Some communities also could be too small to be significant, so the resolution limit of the graph should be respected. Finally, some relationships do not carry antagonism. For example, sharing features like retweets usually imply endorsement. The user should therefore be well aware of the usual meaning of the chosen interaction before drawing any conclusion with the computed metrics.

If the results are not threatened by the previous points, further analytics can be led to achieve a better understanding of the detected polarization. First of all, keywords or hashtags can be used to characterize the source of the disagreement, for example by looking at the main topics of the boundaries. The impact of the different boundary users on the overall polarization of their community can be further investigated by computing their centrality inside the community and inside the whole graph. Finally, the detected polarization could be contextualized in time to gain insights on its first appearance and its evolution.

6   Conclusion

Social Network Analysis allows the extraction of value from interactions between individuals and communities of individuals. Discussions and debates about controversial topics can lead to the polarization of the communities of individuals, *i.e.*, their isolation inside closed and mutually antagonistic groups.

In this article, we propose ERIS, an automatic approach to assess polarization between pairs of communities inside graphs built from social interactions. The method analyzes the behavior of community boundaries, individuals acting as intermediaries between the inside and the outside of their community, to compute two metrics called the *community antagonism* and the *porosity of boundaries*.

Our formal definition of ERIS takes into consideration three major characteristics of graphs built from social interactions: the weighting, the edge direction and the possible presence of overlapping communities. We also propose an efficient algorithm based on matrix computations as well as an open source implementation in R freely available online.

By allowing a more precise description of the roles inside communities through the concepts of internal and boundary areas, the method could also be used to achieve several other objectives in future works. For example, a possible evolution of ERIS could improve discourse analyzes by exploiting these areas to comment the diffusion of topics from and toward the outside of a community.

Finally, boundary members of communities were identified as key elements to get rid of ideological echo chambers, created from the rejection of contradictory opinions [10]. As the ERIS method allows to detect both polarized pairs of communities and the individuals leading to the porosity of boundaries, we would like to investigate how ERIS could be used to build a tool to favor the depolarization of some targeted OSN communities.

Acknowledgement. This work is supported by ISITE-BFC (ANR-15-IDEX-0003) coordinated by G. Brachotte, CIMEOS Laboratory (EA 4177), University of Burgundy.

References

1. Al Amin, M.T., Aggarwal, C., Yao, S., Abdelzaher, T., Kaplan, L.: Unveiling polarization in social networks: a matrix factorization approach. In: IEEE INFOCOM 2017-IEEE Conference on Computer Communications, pp. 1–9. IEEE (2017)
2. Alamsyah, A., Adityawarman, F.: Hybrid sentiment and network analysis of social opinion polarization. In: 2017 5th International Conference on Information and Communication Technology (ICoIC7), pp. 1–6. IEEE (2017)
3. Barabási, A.L., Pósfai, M.: Network Science. Cambridge University Press, Cambridge (2016)
4. Baumann, F., Lorenz-Spreen, P., Sokolov, I.M., Starnini, M.: Modeling echo chambers and polarization dynamics in social networks. Phys. Rev. Lett. **124**(4), 048301 (2020)
5. Blondel, V.D., Guillaume, J.L., Lambiotte, R., Lefebvre, E.: Fast unfolding of communities in large networks. J. Stat. Mech. Theory Exp. **2008**(10), P10008 (2008)
6. Boyd, D., Golder, S., Lotan, G.: Tweet, tweet, retweet: conversational aspects of retweeting on twitter. In: 2010 43rd Hawaii International Conference on System Sciences, pp. 1–10. IEEE (2010)
7. Cinelli, M., Morales, G.D.F., Galeazzi, A., Quattrociocchi, W., Starnini, M.: The echo chamber effect on social media. Proc. Natl. Acad. Sci. **118**(9) (2021)
8. Clauset, A.: Finding local community structure in networks. Phys. Rev. E **72**(2), 026132 (2005)
9. Conover, M., Ratkiewicz, J., Francisco, M., Gonçalves, B., Menczer, F., Flammini, A.: Political polarization on twitter. In: Proceedings of the International AAAI Conference on Web and Social Media, vol. 5 (2011)
10. Donkers, T., Ziegler, J.: The Dual Echo Chamber: Modeling Social Media Polarization for Interventional Recommending. https://doi.org/10.1145/3460231.3474261. https://dl.acm.org/doi/10.1145/3460231.3474261
11. Fortunato, S.: Community detection in graphs. Phys. Rep. **486**(3–5), 75–174 (2010)
12. Garimella, K., De Francisci Morales, G., Gionis, A., Mathioudakis, M.: Political discourse on social media: echo chambers, gatekeepers, and the price of bipartisanship. In: Proceedings of the 2018 World Wide Web Conference, pp. 913–922 (2018)
13. Garimella, K., Morales, G.D.F., Gionis, A., Mathioudakis, M.: Quantifying controversy on social media. ACM Trans. Soc. Comput. **1**(1), 1–27 (2018)
14. Gillet, A., Leclercq, É., Cullot, N.: Évolution et formalisation de la Lambda Architecture pour des analyses à hautes performances – Application aux données de Twitter. Revue ouverte d'ingénierie des systèmes d'information (ISI) (Numéro 1), pp. 1–26 (2021)
15. Goldstein, M.L., Morris, S.A., Yen, G.G.: Problems with fitting to the power-law distribution. Eur. Phys. J. B-Condens. Matter Complex Syst. **41**(2), 255–258 (2004)
16. González-Ibánez, R., Muresan, S., Wacholder, N.: Identifying sarcasm in twitter: a closer look. In: Proceedings of the 49th Annual Meeting of the Association for Computational Linguistics: Human Language Technologies, pp. 581–586 (2011)
17. Guerra, P., Meira Jr, W., Cardie, C., Kleinberg, R.: A measure of polarization on social media networks based on community boundaries. In: Proceedings of the International AAAI Conference on Web and Social Media, vol. 7 (2013)

18. Guerra, P., Nalon, R., Assunçao, R., Meira, W., Jr.: Antagonism also flows through retweets: the impact of out-of-context quotes in opinion polarization analysis. In: Proceedings of the International AAAI Conference on Web and Social Media, vol. 11 (2017)
19. Guyot, A., Gillet, A., Leclercq, É.: Frontières des communautés polarisées: application à l'étude des théories complotistes autour des vaccins
20. Habibi, M.N., Sunjana: Analysis of Indonesia politics polarization before 2019 president election using sentiment analysis and social network analysis. Int. J. Modern Educ. Comput. Sci. 11(11) (2019)
21. Isenberg, D.J.: Group polarization: a critical review and meta-analysis. J. Pers. Soc. Psychol. 50(6), 1141 (1986)
22. Jiang, J., Ren, X., Ferrara, E.: Social media polarization and echo chambers: a case study of covid-19. arXiv preprint arXiv:2103.10979 (2021)
23. Joshi, A., Bhattacharyya, P., Carman, M.J.: Automatic sarcasm detection: a survey. ACM Comput. Surv. (CSUR) 50(5), 1–22 (2017)
24. McGlone, M.S.: Contextomy: the art of quoting out of context. Media Cult. Soc. 27(4), 511–522 (2005)
25. Morales, A.J., Borondo, J., Losada, J.C., Benito, R.M.: Measuring political polarization: Twitter shows the two sides of Venezuela. Chaos Interdisc. J. Nonlinear Sci. 25(3), 033114 (2015)
26. Xie, J., Kelley, S., Szymanski, B.K.: Overlapping community detection in networks: the state-of-the-art and comparative study. ACM Comput. Surv. (CSUR) 45(4), 1–35 (2013)

Business Process Management

Progress Determination of a BPM Tool with Ad-Hoc Changes: An Empirical Study

Lisa Arnold[(✉)], Marius Breitmayer, and Manfred Reichert

Institute of Databases and Information Systems, Ulm University, Ulm, Germany
{lisa.arnold,marius.breitmayer,manfred.reichert}@uni-ulm.do

Abstract. One aspect of monitoring business processes in real-time is to determine their current progress. For any real-time progress determination it is of utmost importance to accurately predict the remaining share still to be executed in relation to the total process. At run-time, however, this constitutes a particular challenge, as unexpected ad-hoc changes of the ongoing business processes may occur at any time. To properly consider such changes in the context of progress determination, different progress variants may be suitable. In this paper, an empirical study with 194 participants is presented that investigates user acceptance of different progress variants in various scenarios. The study aims to identify which progress variant, each visualised by a progress bar, is accepted best by users in case of dynamic process changes, which usually effect the current progress of the respective progress instance. The results of this study allow for an implementation of the most suitable variant in business process monitoring systems. In addition, the study provides deeper insights into the general acceptance of different progress measurements. As a key observation for most scenarios, the majority of the participants give similar answers, e.g., progress jumps within a progress bar are rejected by most participants. Consequently, it can be assumed that a general understanding of progress exists. This underlines the importance of comprehending the users' intuitive understanding of progress to implement the latter in the most suitable fashion.

Keywords: Business process monitoring · Empirical study · Progress determination · Progress visualisation · Real-time monitoring

1 Introduction

The quest to monitor business processes is very much in demand. Monitoring processes allows us to identify and address problems and errors at an early stage of process execution. A crucial task of business process monitoring is to determine the progress of running process. This is particularly challenging when facing ad-hoc changes during run-time, which might affect progress.

The unexpected addition or deletion of process instances as well as direct changes of the underlying process model itself are examples of such ad-hoc

© The Author(s), under exclusive license to Springer Nature Switzerland AG 2022
R. Guizzardi et al. (Eds.): RCIS 2022, LNBIP 446, pp. 107–123, 2022.
https://doi.org/10.1007/978-3-031-05760-1_7

changes, which have a direct impact on progress determination. As a result of this process flexibility, further investigation of how to determine and visualise the progress of a running business process is needed, and to understand which variants exist to cope with unexpected progress.

In this paper, we investigate how to cope with dynamic progress changes of flexible processes during run-time. In particular, we want to know whether there exists a generally accepted notation of progress and which type of visualising progress are accepted best by users. For this purpose, an empirical study to investigate the acceptance of different progress variants for business processes is conducted. Its aim is to determine which type of progress is accepted by the users of a business process monitoring system when facing process changes during run-time. The results of the study shall allow us to implement the most suitable variants of progress changes.

The remainder of this paper is structured as follows. Section 2 gives backgrounds on the process paradigm presumed by this paper, of object-aware business processes. In Sect. 3, the applied research methodology, and the design of the empirical study are described. Section 4 analyses, evaluates, and discusses the answers of the 194 participants of the study. Related work is discusses in Sect. 5, whereas Sect. 6 concludes the paper.

2 Backgrounds

The framework PHILharmonicFlows, which we have developed in recent years, enables implementing object-aware business processes, which allow for a high user flexibility as well allows ad-hoc process changes during run-time [1]. In object-aware processes the collection of process data is accomplished with data-driven form sheets. The latter are auto-generated from the process specification. Forms usually have to be designed manually for each activity while form generation is a built-in feature of PHILharmonicFlows due to its data-centric approach. Thereby, forms logic and enactment utilises the operational semantics of the dynamic aspects of the form, such as the next value that is required. The latter also allows for a high degree of flexibility, for example, a user is not forced to fill out the fields of a form in-order. User may choose their preferred order, e.g., by filling out the form from bottom to the top. Additionally, jumping back within a form allows users to check or correct previously filled fields [2].

Basic to our approach are business *objects*. Each object is defined by its *attributes* as well as its *lifecycle*, which describes object behaviour (i.e., the processing of object-related forms) during run-time [3]. Moreover, the relations between the various objects of a business process are manifested by a *relational process structure*. Possible cardinality constraints on these object relations are 1:n or n:m [4]. An object lifecycle is defined in terms of *states*. Each lifecycle has one start state and at least one end state, and an arbitrary number of intermediate states. Each state is represented by a *form sheet* and maybe refined by *steps*, which correspond to object attributes and hence represent the data input fields of the respective form. Finally, a *coordination process* controls the interactions between these multiple lifecycle processes making use the relation process structure [5].

In [6], we have defined four research questions to be investigated in the context of determining the progress of an object-aware business process. Moreover, in [7], we presented an approach that addresses Research Question 1 and 2. To answer Research Question 3, the challenges of determining the progress of large, dynamically evolving process structures, which consist of interacting loosely coupled, smaller processes that may be also subject to ad-hoc process changes must be investigated first. Regarding these challenges, a solution for progress determination with empirical evidence is provided.

Research Question 1 How can the progress of a single lifecycle process with its state-based view form be determined?

Research Question 2 How can the progress of the processing of a single state within a lifecycle process be measured?

Research Question 3 How can the progress of multiple, interacting (i.e., interrelated) lifecycles be determined?

Research Question 4 How does a coordination process affect the progress of an object-aware business process?

3   Research Methodology and Study Design

This section presents our research methods and the design of the empirical study. We combine the methodologies from [8–10] to investigate a real-world scenario for a process monitoring tool. Section 3.1 introduces the research questions addressed by this paper. Section 3.2 defines the data collection method. Section 3.3 presents the study design and Sect. 3.4 describes the used data analysis method.

3.1   Research Questions

For each business entity, a separate instance (with its own lifecycle instance) of an object type is created at run-time. In addition, changes of a lifecycle instance may be defined at run-time by adding new attributes, which are considered in the dynamically generated form sheets, i.e., new form sheets or input fields may be created at run-time. Moreover, form sheets and input fields may be deleted at run-time. In the context of such dynamic lifecycle and form changes as well needs to investigated how the progress of the overall process be adapted and what adaptations are accepted by different users or user groups. This leads us to the following research question:

Main Research Question: Does a generally suitable accepted understanding of progress exist?

This includes the following sub-research questions:

Sub-Research Question 1 Which variants of determining progress are rejected by most users?

Sub-Research Question 2 Are progress jumps or progress speed adjustments better accepted by users in the context of run-time changes?

Sub-Research Question 3 Which progress variant is most suitable?

3.2 Data Collection Method

The study is performed with an anonymous online questionnaire[1] leveraging the web-based tool `Google Forms` for data collection. Further, the study is available in German and English. The language options do not differ with respect to content or structure. The participants can choose their preferred language in the first step of the questionnaire. The questionnaire was available for the participants over a period of more than eight months.

3.3 Study Design

Demographic Questions: In the first section of the online questionnaire, demographic data of the participants are collected. The latter shall provide us with information about the participants. In addition, these data allow us to compare the results of the remaining study for different groups. In detail, the demographic questions refer to the participants' gender, age group, highest school or university degree, professional field, experience with reading or creation process models, experiences with process monitoring, and satisfaction with existing process monitoring tools (on a 5 point Likert-scale, from 1 *not at all satisfied* to 3 (*neutral*) to 5 *very satisfied*), if the participant has worked with any monitoring tools before.

Intuitive View of Progress: Following the demographic questions, each participant is presented an introductory video of a linear, continuous progress bar course of a lifecycle process i.e., sequentially passing the states of the lifecycle process without any deviations. However, such a progress course is generally not feasible due to unpredictable changes during lifecycle process execution. Therefore, alternatives need to be developed that reflect process changes in the progress bar of this process best. For this purpose, to each participant additional eight videos of progress bar courses are presented, which should be rated in terms of whether progress determination is properly presented to users and intuitively perceived by them. For these ratings, a 5 point Likert-scale ranging from 1 (*not at all true*) to 3 (*neutral*) to 5 (*completely true*) is used. The aim of this section of the questionnaire is to investigate whether there are progress variants which are rejected by users and therefore should be not used in the context of process monitoring.

Scenario View of Progress: In this section of the study, five intralogistic scenarios are described. Along these scenarios different challenges of determining the

---

[1] The questionnaire and the responses of the 194 participants are available via the following Researchgate link: https://www.researchgate.net/publication/358140443_Study_Results.

progress of a running business process are covered, e.g., when creating or deleting business object instances. The results shall help us to find the most suitable progress measurement for a business process monitoring system. Additionally, we want to evaluate our hypothesis that a generally accepted understanding of progress exists.

The following introduction is given before presenting the different intralogistic scenarios to the users: *Robots can support the intralogistic processes in an industrial environment, receiving goods. Robots can support human interactions when unloading, recognising and aligning packages. During the enactment of the process, its progress is displayed in an optimised way. Unexpected (i.e., dynamic) changes may occur during process execution e.g., the technical failure of a robot or the commissioning of a replacement robot. When a robot fails, the remaining robots must compensate this failure and perform additional work. In turn, this might lead to a delay of the total process. The situation is similar to the ad-hoc emergence of additional work.*

For each scenario, the participants are confronted with two videos describing different options for courses of the progress. In all five scenarios, Option A describes the progress with no jumps by speeding up or slowing down the progress depending on the respective ad-hoc changes. In turn, Option B enables progress with the same speed. If progress adjustments become necessary corresponding backward or forward jumps are made in the progress bar. In the following, the five scenarios are described.

Creation of Business Object Instances. During run-time an unexpected workload may occur. For example, the unexpected absence of staff (illness) or robots requires the redistribution of the workload among fewer staff or robots respectively. In this case, progress adjustments can be made by either jumping back in the progress bar or slowing down the remaining progress. Scenario 1 shall help us to find the most suitable solution when creating instances of a business object.

> ### Scenario 1: Reordering packages
>
> Before the start of each working day, the total amount of work is determined. Due to unexpected hoarding of toilet paper, as much toilet paper as possible is reordered and delivered during the working day. The ordered toilet paper is delivered by an unknown number of trucks throughout the afternoon. The current progress of the goods' delivery is determined each time a truck arrives by considering the additional work.

In Fig. 1, the progress determination according to Option A and B are shown. Both progress determinations are identical until 60%. Then the progress according to Option A grows slower until 80%, due to the unexpected delivery of toilet paper from the trucks. For Option B, with this delivery the progress determination drops immediately to 50% and is growing up to 80% with the same speed as before. When the progress reaches 80%, another unexpected truck load arrives in the intralogistic centre. Again, for Option A the progress grows slower until

100%, whereas for Option B the progress drops immediately to 60% and is growing up to 100% with the same speed as before.

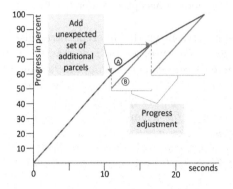

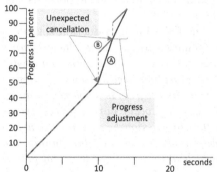

Fig. 1. Progress adjustments in Option A and B after the unexpected delivery of additional parcels (cf. Scenario 1).

Fig. 2. Progress adjustments in Option A and B after the cancellation of parcels (cf. Scenario 2).

Deletion of Business Object Instances. Besides the unexpected increase of the total workload, an unexpected decrease of the latter may occur as well, e.g., after cancelling orders or work assignments. Scenario 2 shall help us to find the most suitable solution for adjusting progress measures in such cases.

> **Scenario 2: Order cancellation**
>
> Due to a more favourable price development of a competitor's product, unexpected cancellations are made throughout the working day. Consequently, the overall workload is reduced.

In Fig. 2, for Option A and B the respective progress with respect to Scenario 2 is depicted. For both options the progress determination evolves the same way until reaching 50%. Then, for A it grows significantly faster up to 80%, due to a first unexpected order cancellation. For Option B, with the order cancellation the progress jumps immediately from 50% to 70%, and is then growing up to 80% with the same speed as before. When reaching 80%, another unexpected cancellation occurs. For Option A the progress determination is then growing even faster up to 100%, whereas for Option B with the cancellation the progress immediately jumps from 80% to 90%, and is growing up to 100% with the same speed as before.

Replacement of Business Object Instances. During process execution, multiple unexpected events might occur, which affect the overall progress determination.

> **Scenario 3: Robot failure and replacement by another robot**
>
> Assume that a robot fails when reaching 60% of the calculated progress due to technical reasons. Adding a replacement robot takes some time as the failed robot has to be removed and the new one needs some time to be ready for use. Assume further that is not possible to predict whether the replacement robot will be ready for use before completing the entire process.

In Fig. 3, for both options the respective progress determination is shown. The latter are the same until reaching 60%. Then, for Option A the progress is growing slower due to of the unexpected robot failure, whereas for Option B the progress drops immediately to 45%. Afterwards, for Option B the progress is growing from 45% to 70%, whereas for Option A the progress is growing from 60% to 70% in the same time. When activating the backup robot, the progress determination readopted adapts again. For Option A, the progress is growing significantly faster, whereas for Option B it jumps from 70% to 85% with the same speed as before.

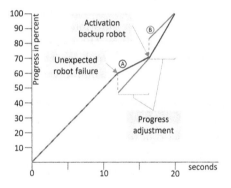
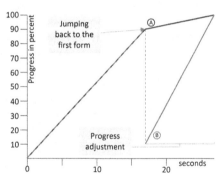

Fig. 3. Progress adjustments in Option A and B after the unexpected robot failure and provision of a backup robot (cf. Scenario 3).

Fig. 4. Progress adjustments in Option A and B when jumping back to a previous form (cf. Scenario 4).

Progress Jumps Within a Form. As shown in Sect. 2, in the PHILharmonicFlows approach the end-users interact with forms to collect data. After filling out the current form, it is possible to activate the next one. However, it is also possible to jump back to a previous form to review and update the input added previously with the respective form. The input data of all forms is preserved.

Scenario 4: Jumping back to a previous form

This scenario switches to the customer's perspective. During an order process, multiple forms are displayed to the customer one following the other. These forms and their order are as follows: *Shopping cart → Enter address → Payment method → Confirm → Done* (end of order process). Assume that before confirming the customer jumps back to the *Shopping cart* and re-checks it. In this context note that the input data of the previously filled forms *Enter address* and *Payment method* persists.

In Fig. 4, for both options the progress determination is shown. Both progresses evolve the same way until 90%. For Option A the progress is growing slower due to the backward jump in the form, whereas for Option B the progress immediately drops to 10%. After the backward jump, for Option B the progress is growing from 10% to 100% in the same time as it is growing from 90% to 100% in Option A.

Time- or Work-Based. In general, there are two metrics that can be used to determine the progress of a business process. First, the progress calculation can be based on the total execution time. Assume, for example, that a business process needs from its start five days upon completion. Accordingly, on the second day at 6am the progress reaches 25% (presuming a duration of 5 days with 24 h). Alternatively, progress calculation can be based on the completion of work packages, which results in progress of 20% on the second day at 6am as only the work of the first day has been completed so far (assuming working hours from 8am to 5 pm).

Scenario 5: Time- or work-based

If there are planned breaks during process execution in which no active work is done, this affects the progress determination of the process. Assuming that the following sequence of work packages and breaks, together with the time required for them, is given.
Work package A: 3 hours + 15 minutes break
Work package B: 1 hour + 15 minutes break
Work package C: 4 hours + 30 minutes break
Work package D: 2 hours

Figure 5 depicts the determination for progress time- and worked-based scenario. For Option A, there is no progress growth in the three breaks, whereas for Option B the progress is growing continuously from 0% to 100%.

Outlook Section of the Questionnaire: The last section of the questionnaire concludes the study by asking general questions about progress determination and presentation. The aim is to get insights into the accepted duration for progress determination and further fundamental monitoring components.

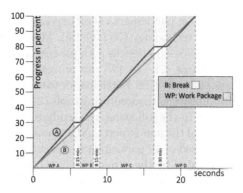

Fig. 5. Progress adjustments in Option *A* based on work and *B* based on time (cf. Scenario 5).

Ideally, progress determination is accomplished in real-time. However, the calculations take time, and the more comprehensive the process is, the greater the delay due to the continuous progress calculation will be. For this reason, we investigated what delay in progress determination is acceptable for the users of a process monitoring tool. For this investigation, we defined four process duration: *one hour, one day, one week,* and *one month.* For each process duration, the participants gave the maximal acceptable delay of progress determination.

Besides the progress determination, many more monitoring components exist, e.g., *Risk Management, Resource Management, Process Performance Measurement, Alarm Trigger and Error Warnings,* and *Numbers of running processes.* Therefore, the participants were asked which components shall be part of a process monitoring tool and which of them is the most important.

3.4   Data Analysis Method

All collected data are analysed and evaluated in a structured way to answer the research question defined in Sect. 3.1. For this purpose, the methodology from [8] is used. In a first step, the answers of all participants are analysed. In a second step, an expert group among the participants is defined based on the answers given in the demographic section of the questionnaire, e.g., experience in modelling processes. The answers of these experts are then compared with the ones of the first step. **Pearson** and **Spearman Correlation** are used to evaluate the correlation between the data of the questionnaire.

4   Evaluation

Demographic Questions: Overall, more male (110 | 56.7%) than female (84 | 43.3%) participants took part in the study. The majority of the participants are between 18 and 25 years old (152 | 78.4%); 36 participants are between 26 and 35 years old (36 | 18.6%). The number of participants with an age between 36 and

65 (6 | 3.0%) is negligible. Most participants work in economy and administration (89 | 45.9%) or information technology (76 | 39.2%). More than two third (133 | 68.6%) of the participants have experiences with process models or process modelling. Much fewer participants (28 | 14.4%) use process monitoring tools in the context of their professional activities. Participants with both modelling experience and professional familiarity with monitoring tools (24 | 12.8%) are considered as the experts in our study. In the following, the study results of the experts (E) are compared with the ones of the non-experts (NE) to enable a profound analysis. The non-experts are composed of all participants expect the experts. Finally, the satisfaction score of 3.3/5 rated by the experts is slightly above a neutral rating, which indicates that existing monitoring tools are not perceived as satisfaction by users.

Intuitive View of Progress: This section of the study deals with the intuitive view users have on the progress determination of a process. For this purpose, they are confronted with eight different progress variants, which they shall evaluate with respect to suitability and usability. Table 1 describes the considered progress determinations of Videos 1–8 as well as the average rating received from experts and non-experts, respectively. For Videos 1, 3, and 6 the scores of the non-experts and experts are almost the same (≤ 0.10). For Videos 2, 5, 7, and 8, the scores are slightly different (≤ 0.38). The biggest difference (0.50) occurs for Video 4.

In a nutshell, adjustments (cf. Videos 2 and 7) are perceived as more suitable and usable compared to progress jumps (cf. Video 1 and 3). Concerning the latter, forward progress jumps (cf. Video 1) are considered being more intuitive compared to backward progress jumps (cf. Video 3). The non-experts and experts rate slow and fast progress speed adjustments differently. Non-experts prefer progress that, from a certain point during process execution, is growing faster (cf. Video 2) as more intuitive than progress growing slower (cf. Video 7). The experts evaluate this the other way round. Note that the same observations can be made for the permanently increased (Video 6) and decreased (Video 4) progress. In practice, an unexpected more workload is more common, e.g., due to staff absence (illness) or defect tools. This might be one reason why experts perceive this as more intuitive, as they are used to such scenarios. Unexpectedly, the 90% problem is rated slightly above neutral (cf. Video 8). With the experts rating it even better than the non-experts. This might be related to the aforementioned habitual practice. Finally, the progress variant considered being the least usable by both groups is illustrated by Video 5, where the progress finished at 80%. Consequently, progress representations should utilise the full progress spectrum between 0% and 100%.

Scenario View of Progress: Due to the additional information provided (cf. Sect. 3.3), participants get a better understanding of when and for what reason process changes become necessary. In the following analysis, first of all, Scenarios 1 to 3 are considered together, followed by an individual analysis of Scenarios 4 and 5.

Table 1. Average evaluation of intuitive progress view for each video (#) by non-experts (NE) and experts (E).

#	Description	$\varnothing_{NE}$	$\varnothing_E$
1	Recalculation of progress at 60% and forward jump in progress	3.02	3.04
2	Adjustment of progress speed at 60% and faster growing progress	4.05	3.79
3	Recalculation of progress at 60% and backward jump in progress	2.14	2.21
4	Progress becoming slower and slower over the entire process duration	3.21	3.71
5	Progress growing equally, but finishing at 80%	1.92	2.30
6	Progress becomes faster and faster over the entire duration	3.52	3.42
7	Adjustment of progress speed at 60% and slower growing progress	3.53	3.88
8	Progress growing uniformly up to 90% and then becoming slower and slower	3.13	3.33

Scenarios 1 to 3. In a nutshell, in the context of Scenario 1 (creation of instances), Scenario 2 (deletion of instances), and Scenario 3 (replacement of instances) we investigated whether progress speed adjustments (Option A) or progress jumps (Option B) are preferred by users. In all three scenarios progress speed adjustments without jumps are preferred more than progress jumps:
Scenario 1: overall (140 | 72.2%), experts (17 | 70.8%), non-experts (123 | 74.4%);
Scenario 2: overall (124 | 63.9%), experts (14 | 58.3%), non-experts (110 | 64.7%);
Scenario 3: overall (158 | 81.4%), experts (20 | 83.3%), non-experts (138 | 81.2%).
Moreover, for all three scenarios the answers are analysed in detail. Table 2 gives an overview of the eight possible answer sequences for Scenarios 1 to 3. The sequences constitute all possible combinations of preferences regarding progress jumps and are grouped by their number: none, 1, 2, and 3. Half of the experts (12 | 50.0%) and almost half of the non-experts (78 | 45.9%) vote for Option A in all three scenarios, i.e., they prefer progress speed adjustments (without jumps). Furthermore, one third of the non-experts (57 | 33.5%) preferred Option B (i.e., progress jumps) in one of the three scenarios as opposed to only 4 experts (16.7%). Additionally, only 23 non-experts (13.5%) and 7 experts (29.1%) choose Option B in two of the three scenarios. Finally, 12 non-experts (7.1%) and 1 expert (4.2%) prefer progress jumps instead of progress speed adjustments for all scenarios. Figure 6 depicts the distribution of the preferred options for both non-experts and experts. In summary, progress speed adjustments (without jumps) are significantly more preferred than progress jumps even thought there are some opposing opinions.

Moreover, Scenario 3 reflects the real world most accurately. Usually, more than one unexpected event happens during run-time. In this case, Option A (no progress jumps) is preferred by 138 (81.8%) non-experts. The experts, in turn, show similar results (20 | 83.3%). The latter can be further divided into two subgroups: computer scientists who are developing process monitoring tools and professionals from economy, administration, transport, or logistics who use these tools in their daily business. Of the latter, 100% agreed that avoiding progress jumps is the better option for Scenario 3. This confirms the previous result.

Table 2. Percentage of possible answer sequences for Scenario 1–3 (S1–S3) are given by non-experts and experts.

S 1	S 2	S 3	$\%_{NE}$	$\%_E$
A	A	A	45.9	50
A	A	B	5.3	0
A	B	A	17.0	16.7
B	A	A	11.2	0
A	B	B	4.1	4.2
B	A	B	2.3	8.3
B	B	A	7.1	16.6
B	B	B	7.1	4.2

Non-Expert   Expert

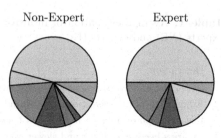

Fig. 6. Associated pie diagram for percentage distribution of Table 2.

Scenario 4. Previously entered data are still available when jumping back to an already completed form. For Option *A*, progress is determined on the basis of the already entered data. Therefore, a backward jump to a previous form sheet does not affect the progress of the process. For Option *B*, however, the progress is correlated with the currently edited form. Consequently, progress drops when jumping back to an already completed form. The non-experts (102 | 60.0%) and experts (14 | 58.3%) rate this scenario similarly and prefer the first progress determination without dropping progress.

Scenario 5. Progress is determined based on the calculated time (Option *A*) or work-based (Option *B*). Non-experts (101 | 59.4%) and experts (13 | 54.2%) preferred the work-based approach for this scenario. Especially the results of the experts, however, have revealed no clear preference. This scenario will therefore be re-considered in a more detailed study in future.

Outlook Section of the Questionnaire: This part of the study helps us to define the accepted delay due to the progress determination of process monitoring tool and gives an overview of other possible monitoring components. In the following, the maximum accepted delay due to progress determination is investigated. In the **first step**, outlier detection is conducted. When a participant specifies a maximum accepted delay of the progress determination, which is larger than total the process duration, all for possible answers of this participant related to delay of the monitoring tool (i.e., hour, day week and month accepted delays) are omitted. Additionally, if the accepted delay for a process lasting an hour is bigger than for a process lasting a day (the same applies to: hour > day > week > month) from one participant, it can be assumed that this participant has misunderstood the given format for this question. For this reason, answers from such participants are omitted as well. This results in a total of 186 participants (164 non-experts and 22 experts) for the following analysis.

In the **second step**, a `Pearson Correlation` is performed, due to the expected linear dependency. Table 3 shows the results of this analysis. Note that the result correlation may be between -1 and $+1$. The larger the absolute value of the coefficient is, the stronger the relationships between the variables are. However, in all relations there is only a `moderate correlation` $(0.40 - 0.69)$ or `strong correlation` $(0.70 - 0.89)$ and not, as expected, a `very strong correlation` $(0.90 - 1.00)$ [10].

Table 3. Analysis of Pearson and Spearman correlation with $N = 186$. Moderate, Strong, and Very strong.

Relation	Pearson		Spearman	
	NE	E	NE	E
Hour & Day	0.748 S	0.637 M	0.816 S	0.835 S
Hour & Week	0.631 M	0.621 M	0.708 S	0.644 M
Hour & Month	0.646 M	0.499 M	0.687 M	0.520 M
Day & Week	0.807 S	0.622 M	0.859 S	0.846 S
Day & Month	0.729 S	0.429 M	0.835 S	0.729 S
Week & Month	0.809 S	0.849 S	0.891 S	0.890 S

Using the results of the `Pearson Correlation`, the following hypothesis can be established and investigated in a **third step**: *The longer the total duration of a process is, the smaller the percentage share of accepted delay from the total process duration is.* Having a closer look at the average accepted delay due to progress determination we perceive the hypothesis backed up with 9.1% for the one-hour process, about 5% for both a one-day-process and a one-week-process, and 0.6% for a one-month-process. Furthermore, for some participants it can be observed that the longer the process duration is, the more similar the accepted delays become. These results indicate the existence of an absolute upper limit for the accepted delay caused by the determination of the progress. This assumption is further supported by the different delays accepted by the experts and non-experts. For this reason, an additional `Spearman Correlation` is performed (cf. Table 3), which is suitable for monotonic functions. As opposed to the `Pearson Correlation`, the `Spearman Correlation` results in significantly more strong correlations as expected.

In the **fourth step**, the maximum accepted delay caused by progress determination is visualised with the boxplot diagrams as shown in Fig. 7. For each of the four given total process duration's (hour, day, week, and month) a boxplot, which represents the accepted delay by the quartiles and the average of the experts and non-experts are depicted. All values are given in seconds (for better visibility, the scale of time (y-axis) is individual for each process duration). In summary, the smaller quartiles of the experts indicate that they have a very similar understanding of progress delays. In contrast, the results of the non-experts

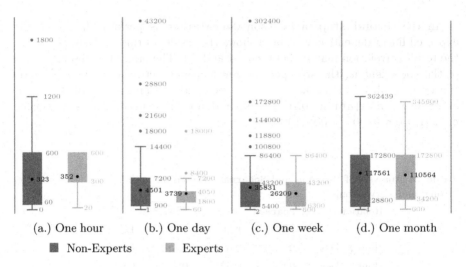

(a.) One hour (b.) One day (c.) One week (d.) One month

■ Non-Experts ▨ Experts

Fig. 7. Boxplot diagram showing the accepted delay caused by progress calculation for the different process duration's in *seconds* for non-experts and experts.

show significantly more outlier points and a larger span of the boxplot. Except for the one-hour process, the average of the accepted delays are smaller for the experts in comparison to the non-experts.

The last question investigates which monitoring component should be covered by a process monitoring tool (Question A) and which should be implemented next (Question B). No distinction is made between experts and non-experts. For Question A, five possible additional components are given and multiple answers are allowed. Figure 8(a) shows the answers to this question. Note that all five options are equally distributed (about 20%). To conclude all components seem to be relevant. Moreover, participants may name further components of a process monitoring tool. Proposed components include *Completed sub-processes*, *Early warning through pattern recognition based on machine learning*, and *Heat and failed tasks*.

For Question B, the most important monitoring component besides progress determination should be chosen. The distribution is shown in Fig. 8(b). This indicates, that *Alarm Trigger and Error Warnings* is preferred by the majority (38,7%) of the participants.

5   Related Work

To the best of our knowledge there exist no works dealing with progress determination in process monitoring tools. In participant, this applies to *object-aware BPM*, as there exists no comparable approach that offers the same flexibility as PHILharmonicFlows does (e.g. ad-hoc changes of running process instances) [2]. Consequently, no monitoring rules for processes being subject to ad-hoc changes in *object-aware BPM* approaches exist.

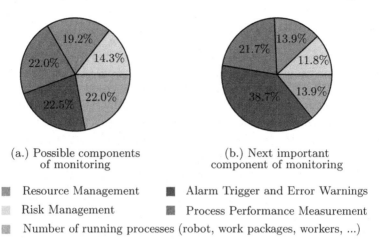

(a.) Possible components
of monitoring

(b.) Next important
component of monitoring

- Resource Management
- Risk Management
- Number of running processes (robot, work packages, workers, ...)

- Alarm Trigger and Error Warnings
- Process Performance Measurement

Fig. 8. Evaluation of which components of a monitoring shall be part of a process monitoring tool (a) and realised next (b).

Concerning *data-centric BPM*, only few approaches offer this flexibility during process execution. In *artifact-centric BPM*, for example, process executions are driven by business data [11]. Moreover, a tool for designing and modelling *artifact-centric* processes exists. There is no intuitive tool support for executing and monitoring corresponding instances in the large scale (as in the case of our PHILharmonicFlows engine) [12].

A second approach for *data-centric BPM* is *case handling*, which is based on the idea what can be done to achieve a business goal [13]. With Flowers a tool exists, which supports the design, implementation, and execution of data-centric processes, but does not provide comprehensive monitoring support [12]. Note that other data-centric approaches to BPM are available. However, all of them are limited compared to the *artifact-centric BPM* and *case handling* paradigms [12]. In summary, no related works of dynamic process changes exist.

Furthermore, there exists research emphasising the importance of progress indicators. The study presented in [14] shows that progress representation can influence perception of a user concerning the duration of a task. This progress representation influences the decision whether a task (e.g., in form of a online questionnaire) is abandoned or continued. Thereby, progress representation is displayed with different progress speed throughout the total questionnaire. As a result, this study shows that the slower the progress is in an early stage of the questionnaire the higher abandonment rates are [14]. Furthermore, the experiment presented in [15] shows that progress representations are important and useful for user-interaction tools. Additionally, [15] underlines that users prefer tools with progress visualisation.

6   Summary and Outlook

All the defined research questions could be addressed. Progress determination and visualisation should always end up with 100% progress upon process termination (SR1). When ad-hoc changes occur, progress speed adjustments without jumps are considered being more suitable and usable by the majority of the users (SR2). 100% of the experts with a profession in economy, administration, transport, or logistics agreed that no progress jumps are the best option, particularity when facing more than one ad-hoc change (SR3). The well-known 90%-problem (i.e., progress needs significant more time for the last 10% of the work than for the first 90%) is evaluated slightly more suitable than neutral. Previously entered data are still available when jumping back to an already completed form the majority of the participants preferred progress speed adjustments without jumps (SR2). Additionally, work-based progress calculation is favoured compared to time-based one by a small majority (SR3). Furthermore, the study evaluation indicates, which components are essential for monitoring tools and should therefore be considered in further tool releases.

The conducted study confirms a generally accepted understanding of process progress (Main Research Question). However, further research is needed to gain a deeper understanding of progress in data-driven processes aware information systems and its perception by end-users. For example, a *Delphie* study with a focus on experts might provide additional insight into prediction needs.

Acknowledgement. This work is part of the ZAFH intralogistic project, funded by the European Regional Development Fund and the Ministry of Science, Research and Arts of Baden-Württemberg, Germany (F.No. 32-7545.24-17/12/1).

References

1. Andrews, K., Steinau, S., Reichert, M.: Enabling runtime flexibility in data-centric and data-driven process execution engines. Inf. Syst. **101**, 101447 (2021)
2. Andrews, K., Steinau, S., Reichert, M.: Enabling ad-hoc changes to object-aware processes. In: 2018 IEEE 22nd International Enterprise Distributed Object Computing Conference (EDOC), pp. 85–94 (2018)
3. Steinau, S., Andrews, K., Reichert, M.: Executing lifecycle processes in object-aware process management. In: International Symposium on Data-Driven Process Discovery and Analysis, pp. 25–44 (2017)
4. Steinau, S., Andrews, K., Reichert, M.: The relational process structure. In: International Conference on Advanced Information Systems Engineering, pp. 53–67 (2018)
5. Steinau, S., Künzle, V., Andrews, K., Reichert, M.: Coordinating business processes using semantic relationships. In: Conference on Business Informatics, pp. 33–42 (2017)
6. Arnold, L., Breitmayer, M., Reichert, M.: Towards real-time progress determination of object-aware business processes. In: Proceedings of the 13th European Workshop on Services and their Composition, vol. 2839, pp. 14–18 (2021)

7. Arnold, L., Breitmayer, M., Reichert, M.: A one-dimensional Kalman filter for real-time progress prediction in object lifecycle processes. In: 2021 IEEE 25th International Enterprise Distributed Object Computing Workshop (EDOCW), pp. 176–185 (2021)
8. Wieringa, R.: Design Science Methodology for Information Systems and Software Engineering. Springer, Heidelberg (2014). https://doi.org/10.1007/978-3-662-43839-8
9. Yin, R.K.: Case Study Research: Design and Methods. Sage, Thousand Oaks (2009)
10. Schober, P., Boer, C., Schwarte, L.A.: Correlation coefficients: appropriate use and interpretation. Anesth. Analg. **126**, 1763–1768 (2018)
11. Cohn, D., Hull, R.: Business artifacts: a data-centric approach to modeling business operations and processes. IEEE Data Eng. Bull. **32**, 3–9 (2009)
12. Steinau, S., Marrella, A., Andrews, K., Leotta, F., Mecella, M., Reichert, M.: DALEC: a framework for the systematic evaluation of data-centric approaches to process management software. Softw. Syst. Model. **18**(4), 2679–2716 (2018). https://doi.org/10.1007/s10270-018-0695-0
13. Van der Aalst, W., Weske, M., Grünbauer, D.: Case handling: a new paradigm for business process support. Data Knowl. Eng. **53**, 129–162 (2005)
14. Conrad, F., Couper, M., Tourangeau, R., Peytchev, A.: The impact of progress indicators on task completion. Interact. Comput. **22**, 417–427 (2010)
15. Myers, B.: The importance of percent-done progress indicators for computer-human interfaces. ACM SIGCHI Bull. **16**, 11–17 (1985)

Enabling Conformance Checking for Object Lifecycle Processes

Marius Breitmayer[(✉)][iD], Lisa Arnold[iD], and Manfred Reichert[iD]

Institute of Databases and Information Systems, Ulm University, Ulm, Germany
{marius.breitmayer,lisa.arnold,manfred.reichert}@uni-ulm.de

Abstract. In object-aware process management, processes are represented as multiple interacting objects rather than a sequence of activities, enabling data-driven and highly flexible processes. In such flexible scenarios, however, it is crucial to be able to check to what degree the process is executed according to the model (i.e., guided behavior). Conformance checking algorithms (e.g., Token Replay or Alignments) deal with this issue for activity-centric processes based on a process model (e.g., specified as a petri net) and a given event log that reflects how the process instances were actually executed. This paper applies conformance checking algorithms to the behavior of objects. In object-aware process management, object lifecycle processes specify the various states into which corresponding objects may transition as well as the object attribute values required to complete these states. The approach accounts for flexible lifecycle executions using multiple workflow nets and conformance categories, therefore facilitating process analysis for engineers.

Keywords: Data-centric process management · Conformance checking · Process analysis · Object lifecycle processes

1 Introduction

Activity-centric approaches to business process management focus on the control-flow perspective of business processes, i.e., the order in which individual activities shall be executed. Consequently, activity-centric processes consist of activities that must be executed in a pre-specified order. While activities may require data during their execution, their actual specification (i.e., the data provided during activity execution) is considered as a black-box. Alternative paradigms such as data-centric and -driven process management [21] represent a process in terms of multiple interacting objects, allowing for greater flexibility through the use of declarative rules and generated forms. The individual behavior of an object is usually data-driven and described in terms of lifecycle processes. A lifecycle process specifies the states an object may transition during its lifecycle and the data required to complete each state. It, therefore, enables a white-box approach regarding process data. Examples of object-centric and data-driven process management approaches include artifact-centric processes [11], case handling [5], and object-aware process management [16]. Despite the

© The Author(s), under exclusive license to Springer Nature Switzerland AG 2022
R. Guizzardi et al. (Eds.): RCIS 2022, LNBIP 446, pp. 124–141, 2022.
https://doi.org/10.1007/978-3-031-05760-1_8

inherent flexibility of object-centric and data-driven approaches, the problem of not always executing a lifecycle process according to its pre-specified behavior applies to this paradigm as well. Deviations may be caused by users behaving differently than expected, ad-hoc behavioral changes [7], or errors introduced during the modeling or deployment of the lifecycle processes. Dynamic behavioral changes, for example, enable a variety of runtime adaptations of the lifecycle process model of a particular object. Examples include the insertion, reordering and deletion of lifecycle states, the insertion or deletion of object attributes, or objects in general. Furthermore, dynamic changes may be applied to individual objects and lifecycle instances (i.e., ad-hoc changes), respectively, as well as to lifecycle models in general (i.e., lifecycle evolution).

Another layer of complexity for checking conformance of object-centric and data-driven processes is their inherent flexibility. In a nutshell, the lifecycle process modeled for each business object describes its *guided* behavior, while accounting for *tolerated* state transitions. Nevertheless, there exist additional executions, which deviate from the modeled behavior, but correspond to correct executions of a lifecycle process as well. Moreover, the latter may occur within individual lifecycle states (i.e., when setting the attributes by filling corresponding form fields) as well as at transitions between them.

Assume a *Student* submits a solution to an exercise using a form. As long as all required form fields are set, the submission may be handed in. The order in which the form is filled, however, is arbitrary allowing for deviations from the underlying lifecycle model that was used to generate the form and its logic (i.e., its *guided* behavior). In a nutshell, the execution of object lifecycle processes operates within implicitly defined boundaries, and, therefore, tolerates certain deviations during lifecycle execution.

The approach presented in this paper is capable of identifying which object lifecycle process executions conform with the guided behavior of lifecycle processes, which executions are tolerated due to the built-in flexibility of lifecycle processes, and which executions constitute deviations from the lifecycle process.

The paper is structured as follows: Sect. 2 introduces PHILharmonicFlows, our approach to object-centric and data-driven process management. Section 3 describes the problem addressed by the paper. Section 4 describes the granularity and flexibility of object lifecycle processes and discusses how we can formally represent the latter through various workflow nets. In Sect. 5, we introduce conformance categories derived from conformance checking results in the context of object lifecycle processes. Section 6 evaluates our approach using multiple event logs. In Sect. 7, we relate our work to existing approaches for conformance checking. Section 8 summarizes the paper and provides an outlook.

2   Fundamentals

PHILharmonicFlows enhances the concept of object-centric and data-driven process management with the concept of *objects*. Each real-world business object is represented as one such object. The latter comprises data, represented in terms of *attributes,* and a state-based process model describing object behavior in terms of an *object lifecycle model* (cf. Fig. 1).

The data- and process-aware e-learning system PHoodle, a sophisticated application implemented with PHILharmonicFlows, for example, includes objects such as *Lecture, Exercise,* and *Submission.* For the *Submission* object (cf. Fig. 1), attributes include *Exercise, E-Mail, Files,* and *Points.* The corresponding object lifecycle process is shown in Fig. 1. It describes the object behavior in terms of *states* (*e.g., Edit, Submit, Rate,* and *Rated*) as well as *state transitions.* Furthermore, each state may comprise several *steps* (e.g., steps *Exercise, E-Mail,* and *Files* in state *Edit*), with each step referring to exactly one object attribute. In other words, the steps of a lifecycle process define which attributes need to be written before completing the state and transitioning to the next one.

At runtime, a lifecycle allows for the dynamic and automated generation of forms (cf. Fig. 1). Accordingly, data acquisition in PHILharmonicFlows is based on the information modeled in both states and steps.

The lifecycle of a *Submission* object (cf. Fig. 1) can be interpreted as follows:

Edit is the initial state of the *Submission* object as it has no incoming transitions. After a student has provided data for steps *Exercise, E-Mail* and *Files,* the *Submission* may transition to state *Submit.*

State *Submit,* in turn, shall enable students to alter their submission prior to the exercise deadline by following the backwards transition. This allows returning back to state *Edit,* hence enabling changes to attributes *Exercise, E-Mail* and *Files.* A *Submission* automatically transitions from state *Submit* to *Rate* once the exercise deadline is reached, and tutors may then rate the final submission. In state *Rate,* a *Tutor* may read the provided attribute values from previous steps, provide data for step *Points,* and transition the submission to state *Rated.* Rated is the end state in which students may check their points.

This simple example emphasizes the importance of data executing an object lifecycle process. While an object instance may only be in one active state at a time, we also support choices [7, 20].

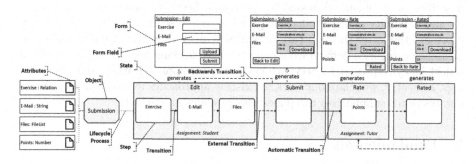

Fig. 1. Example object lifecycle process of object *Submission*

Generally, a business process not only comprises one single object, but involves multiple objects (e.g., *Submissions*, *Lectures* or *Exercises*) and their corresponding lifecycles. In PHILharmonicFlows, these objects are captured in a conceptual *data model* together with their semantic relations (including cardinality constraints) [16]. A semantic relation denotes a logical association between two objects (e.g., a relation between a *Lecture* and an *Exercise* implies that multiple exercises may be related to a single lecture). At runtime, each object may be instantiated multiple times, each representing an individual *object instance* [7]. The lifecycle processes of different object instances are then executed concurrently. Additionally, relations (e.g., between a *Lecture* instance and an *Exercise* instance) are instantiated enabling associations between two object instances. This results in novel information (e.g., one *Exercise* instance belongs to another *Lecture* instance), and intertwines the executed instances [16].

3   Problem Statement

Conformance checking leverages information from process event logs to correlate a process model with reality in order to assess its quality with respect to the behavior documented in the event log [2]. Thus, conformance checking measures how the recorded behavior of a process fits to its modeled representation by, for example, calculating a corresponding fitness value. Fitness measures to which extend a model can represent the behavior documented in an event log. This requires different representations of a process with respect to the recorded (e.g., an event log) and modeled behavior (e.g., a Petri net of a lifecycle process).

For activity-centric processes, where activities are mostly considered as black-boxes, existing conformance checking algorithms relate a process model to an event log. Deviations from the process model are identified reducing the calculated fitness value between event log and process model. In the context of object-aware process management, where the behavior of objects is modeled using a white-box approach (i.e., object behavior is explicitly modeled) and the execution of object lifecycle processes is data-driven (i.e., based on the availability of data) this problem is of great importance as well. However, due to the object-centric and data-driven processing of object lifecycles and the flexibility offered in this context [7], conformance issues are more challenging to address.

When processing object lifecycles, executions may deviate from the modeled lifecycle process, but still remain correct executions, due to the built-in flexibility of lifecycle processes. For example, the attributes of the form generated for state *Edit* in Fig. 1 may be filled in any order to complete the state (though there is a guidance in which order the form fields shall be filled according to the pre-specified sequence of steps within a state) or attribute values may be changed after having been set before. Additionally, *Submission* objects may return to previous states using backwards transitions (cf. Fig. 1). Both scenarios reflect tolerated execution behavior, but also deviations from the guided lifecycle behavior (cf. Tables 1 and 2).

Consequently, conformance checking of object lifecycle processes must account for both the guided and the tolerated lifecycle executions for individual states as well as the transitions between states. Tolerated executions are specified with respect to the order of states (i.e., the order in which single forms shall be processed) as well as the order of the steps within a state (i.e., the order in which fields within a form are organized). Tables 1 and 2 illustrate the different behavioral categories on two granularity levels for object lifecycle processes. Process engines capable of executing object-centric processes such as PHILharmonicFlows [16] or FLOWer [5] may generate all three behavior categories at runtime through the use of advanced concepts such as dynamic changes [6].

Conformance checking for object-aware lifecycle processes is multidimensional. On one hand, several levels of granularity exist at which fitness needs to be measured (state- and step-level, cf. Section 4). On the other, for each granularity level, it needs to be distinguished between guided and tolerated behavior to identify actual deviations. Consequently, a single fitness metric might not be sufficient to cover both dimensions.

Table 1. State level behavior

Behavior	Description	Example (cf. Fig. 1)
Guided	The lifecycle reaches its end state without following any backwards transitions during lifecycle execution.	<Edit, Submit, Rate, Rated>
Tolerated	The lifecycle reaches its end state, but backwards transitions were chosen during lifecycle execution.	<Edit, Submit, Edit, Submit, Rate, Rated, Rate, Rated>
Deviating	The lifecycle transitions to a non-reachable or unspecified state during its execution.	<Edit, Submit, Edit, Rated> <Edit, Submit, NewState, Rate, Rated>

Table 2. Step level behavior

Behavior	Description	Example (State Edit of Fig. 1)
Guided	All fields of the form of a lifecycle state are filled according to the pre-specified order of steps.	<Exercise, E-Mail, Files>
Tolerated	All mandatory fields of the form have been filled prior to state completion.	<E-Mail, Exercise, Files> <Files, E-Mail, Exercise>
Deviating	Required steps have been skipped or additional steps (i.e., attributes) have been added.	<Files> <Files, Exercise, Points>

4 Granularity and Flexibility of Lifecycle Processes

When analyzing object lifecycle executions, we need to account for the granularity (i.e., state and step level), while distinguishing between guided, tolerated and deviating behavior (cf. Tables 1 and 2). We, therefore, transform an object lifecycle process into a set of workflow nets, while accounting for granularity as well as the various degrees of flexible execution.

4.1 Granularity of Object Lifecycle Processes

To account for the granularity of object lifecycle processes, we analyze the behavior of each lifecycle process on two granularity levels, i.e., state and step level (cf. Fig. 2). Concerning the state level, we focus on the transitions between states (e.g., state *Edit* must be completed before state *Submit* may be activated). On step level, we analyze the logic of the steps within a state. Note that this logic is used to guide users through a form. Figure 2 depicts the two granularity levels. The transformation of an object lifecycle process into a set of workflow nets of different granularity allows checking the conformance with respect to different levels of an object lifecycle (i.e., state and step level) separately. Furthermore, we are able to categorize deviations with respect to their origin, i.e., we can analyze whether a deviation results from unplanned state changes or from the flexible processing of a single state (i.e., the processing of its form). In turn, this allows for a more fine-grained conformance checking enabling data-driven process improvement. By solely considering the granularity, we are able to identify the origin of a deviation. However, we are unable to distinguish whether or not the latter are *tolerated* due to built-in flexibility. We therefore need to consider flexibility on both levels as well.

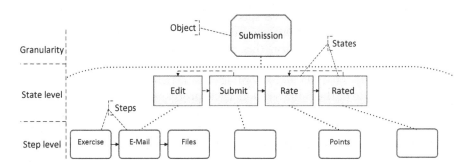

Fig. 2. Granularity of Object Lifecycle Processes (derived from Fig. 1)

4.2 Flexibility in Object Lifecycle Processes

In general, object lifecycle processes define the *guided behavior* of an object and provide corresponding user guidance during lifecycle execution. However, due to their flexible execution nature [20], object lifecycle processes allow for *deviating behavior* at both granularity levels (cf. Tables 1 and 2). When checking the conformance of an object lifecycle process with an event log, differentiating between these categories offers promising perspectives for lifecycle improvement on one hand, but introduces additional challenges on the other. Note that conformance checking of object lifecycle processes must therefore account for both the granularity levels and the flexible execution behavior.

We address this challenge by distinguishing between *guided* and *tolerated behavior*. Accordingly, we use multiple workflow nets that can be derived from lifecycle processes on both granularity levels. This allows distinguishing between guided, tolerated, and deviating behavior on both granularity levels.

Definition 1. *Workflow Net [1]*
A Workflow Net is a Petri net $N = (P, T, F, i, o)$ where:
P constitutes a finite set of places and T a finite set of transitions, $P \cap T \neq \emptyset$,
$F \subseteq (P \times T) \cup (T \times P)$ represents a set of directed arcs, called flow relation.
i is the source place ($\bullet i = \emptyset$) and o constitutes the sink place ($o \bullet = \emptyset$)
All other nodes are on a path from i to o.

Note that the workflow net depicted in Fig. 3 contains 6 places, 5 transitions, and 10 arcs. We use multiple workflow nets as representations of the lifecycle process behaviors described in Tables 1 and 2. This, in turn, allows us to distinguish between guided, tolerated, and deviating behavior.

4.3   Flexibility on State Level

On the state level, guided behavior is affected when backwards transitions are followed during lifecycle execution (cf. Table 1).

Guided state level behavior corresponds to activate the states of the lifecycle process exactly according to its specified order (cf. Fig. 1 and Table 1). By following a backwards transition, one may return to a preceding state, which has already been passed. In turn, this indicates that this state has not been properly processed such that object attribute changes become necessary (e.g., by uploading an updated file and assigning it to attribute *Files* in state *Edit*). In one such scenario of the submission lifecycle process (cf. Fig. 1), students may alter their submission by jumping back to state *Edit* following the corresponding backwards transition. When checking conformance, such backwards transitions still correspond to tolerated, but not to guided behavior (cf. Table 1). Regarding the lifecycle process from Fig. 1, the guided behavior on the state level translates to the workflow net displayed in Fig. 3. Each state is translated to a place in the workflow net, external and automatic state transitions are translated to transitions between these places, whereas backwards transitions are neglected (i.e., they correspond to tolerated behavior). Note that choices between states may be represented as well.

Tolerated state level behavior covers backwards transitions, which allow returning to a previous state to account for foreseeable exceptions during lifecycle process execution as well. While these (still planned) deviations are not

Fig. 3. Guided state level Workflow Net for Object Submission

covered by the *guided* state behavior (cf. Fig. 3), their execution at runtime is still *tolerated*. Considering the Submission lifecycle process (cf. Fig. 1), for example, tolerated behavior allows returning from state *Submit* to state *Edit* as well as from state *Rated* to *Rate*. This enables two additional scenarios. First, students may return their submission to state *Edit*, which allows them to change their previous submissions (e.g., update uploaded files). Second, tutors may return submissions to state *Rate,* which allows changing the value of the lifecycle process step *Points,* e.g., if the tutor overlooked mistakes in a previously rated submission. While the lifecycle model from Fig. 1 accounts for such scenarios, the latter indicate that previous state completions were incorrect (e.g., upload of a wrong file). We generate the "tolerated net" (cf. Fig. 4) similar to the guided behavior, but do not neglect backwards transitions. Algorithm 1 describes the generation of both guided and tolerated nets on state granularity in pseudo code.

Fig. 4. Tolerated State Level Workflow Net for Object Submission

---

Algorithm 1. Workflow Net Generation Algorithm on State Level Granularity

---

Require: OLP, NetType   ▷ Object lifecycle process, guided or tolerated behavior
  $PetriNet \leftarrow$ **new**
  $PetriNet.addPlace(source)$
  $PetriNet.addPlace(sink)$
  for all *states* in OLP **do**
    $PetriNet.addPlace(state)$
  end for
  for all t in OLP **do**   ▷ Transitions of OLP
    if $t.source.state \neq t.target.state$ **then**   ▷ Transition is external or backwards
      if $NetType = guided$ AND $t.type = backwards$ **then**
        //Do Nothing
      else
        $PetriNet.addTransition(t)$
        $PetriNet.addArcs(t.source, t.target)$ ▷ Arcs to connect t with in- and output places
      end if
    end if
  end for
  $PetriNet.makeWFN()$ ▷ Connect sink and source to places of first and last states

---

4.4   Flexibility on Step Level

As opposed to the state level, step level behavior covers intra-state behavior, i.e., the steps of a specific state and the order and constraints for their execution.

In PHILharmonic Flows, the behavior on step level is reflected by the control flow logic for processing the corresponding form. The behavior on step level is therefore directly connected to the actual data acquisition, i.e., steps of a state and their order are used to automatically generate role-specific forms at runtime.

Each step of a state corresponds to a single attribute that may be set while the state is active. A state may only be completed once all mandatory attributes have been set, i.e., its corresponding form has been properly filled. Note that the transitions between the steps of a given state are used to organize fields in the generated forms (and cursor control). For the step level, guided behavior means obeying the pre-specified execution logic of the steps when setting the attributes (cf. Table 2). Each attribute may be changed any number of times when processing the respective form field, and conditional attributes are possible as well. Considering the generated forms, guided behavior corresponds to the form being filled according to the pre-specified order of steps (cf. Table 2). Figure 5 depicts the guided net derived for state *Edit* of the *Submission* object.

Fig. 5. Guided step level Workflow Net for State Edit

For the tolerated state level behavior there only exist two constraints. First, an attribute (i.e., a step) needs to be set before it may be changed. Second, a state may only be completed once all mandatory attributes have been set. The workflow net depicted in Fig. 6 represents this behavior for state *Edit* of the *Submission* object (cf. Fig. 1).

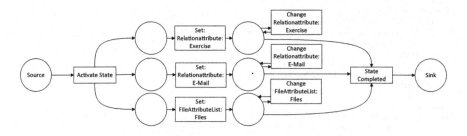

Fig. 6. Tolerated step level Workflow Net for State Edit

On step level, conditional steps (e.g., suppose a statement would be required if step *Points* in state *Rate* is below a certain threshold) may be represented through choice constructs and corresponding, additional state completion transitions. Algorithm 2 generates both the guided and tolerated workflow net (cf. Fig. 6) on step level in pseudo code.

Algorithm 2. Workflow Net Generation Algorithm on Step Level Granularity

---

Require: OLP, NetType, OLPstate   ▷ object lifecycle process, guided/tolerated, selected state
    PetriNet ← **new**
    PetriNet.addPlace(source)
    PetriNet.addPlace(sink)
    for all *step* in *OLP.getSteps(OLPstate)* **do**
        PetriNet.addBehavior(step, NetType)   ▷ one or two places & transitions (Set & Change)
    end for
    if NetType = tolerated **then**
        PetriNet.addAndConnect()   ▷ Use tolerated syntax to connect places with Sink and Source
    else
        for all *t* in *OLP* **do**   ▷ Transitions of OLP
            if *t.source.state = t.target.state* AND *t.source = OLPState* **then**
                PetriNet.connectSteps(t)   ▷ Connect steps according to transition of OLP
            end if
        end for
    end if
    PetriNet.makeWFN()   ▷ Add & connect one place (guided), connect sink and source

---

5 Conformance Checking of Lifecycle Processes

After having shown how to generate the workflow nets for each granularity level while at the same time accounting for lifecycle flexibility, we utilize these nets to check conformance for each level individually. On both granularity levels, however, using workflow nets that reflect guided behavior (e.g., Figs. 3 or 5) only enables us to check whether an object lifecycle was executed according to the guided behavior (i.e., both states and steps are executed based on the order described by the guided behavior). Deviations from this behavior may, in turn, be tolerated by the object lifecycle process due to its built-in flexibility.

A similar scenario arises when only using workflow nets representing tolerated behavior (e.g., Figs. 4 or 6), which represents all lifecycle executions that may be realized without any interventions from process supervisors (i.e., all correct executions). Based on the fitness value between the workflow nets representing tolerated behavior on one hand and an event log on the other, we are only able to check whether an object lifecycle process instance changed states, or a state was processed according to the tolerated behavior. We are unable to figure out which object lifecycle process instances were executed according to guided behavior (i.e., whether states were transitioned, or steps were processed according to the guided behavior).

When considering both nets of a granularity level in combination (e.g., Figs. 3 and 4 or 5 and 6), we can calculate two fitness values for both the state and the step level (i.e., for each state). One fitness value corresponds to the fitness regarding guided behavior, the other to tolerated behavior for a selected level of granularity (i.e., state or step level). In combination, the two fitness values allow categorizing each lifecycle process instance according to the behavior and granularity levels described in Tables 1 and 2. Furthermore, we can distinguish whether deviations from the guided behavior are due to exceptions or due to the flexibility inherent to object-aware processes.

Note that we use Alignments [3] to calculate the resulting fitness values - alignments are the de-facto standard approach to evaluate fitness [12] and are able to cope with log entries unrelated to the lifecycle process model (e.g., executing an unmodeled step or state). Alignments connect execution traces with a valid execution sequence from a process model through log, model or synchronous moves [3,10]. Fitness is then calculated using costs of identified log and model moves. Other algorithms (e.g., Token Replay [8]) may be used as well.

5.1 Conformance Categories

When categorizing lifecycle process executions, we either focus on the state changes (i.e., state level) or individual states (i.e., step level) recorded in the event log. Usually, event data are recorded during the execution of a process and consist of cases and activities [2]. In the context of object lifecycle processes, we use cases to identify individual object instances. A case comprises information on object lifecycle process states and steps (i.e., on whether the form field corresponding to a step has been set, changed or an object instance has transitioned to another state) in terms of activities. Note that this allows for the use of existing conformance checking algorithms (e.g., alignments) to lifecycle processes. During conformance checking, we consider event log subsets based on the granularity levels of each lifecycle process. Figure 7 depicts an example event log from the *Phoodle* scenario.

User	Case		Filter	Activity		Attribute Value	Timestamp
	User	Object Instance	Object State	Object Type	Method	Parameter 1	Parameter 2	Timestamp
1								
2	person78@uni-ulm.de	person78@uni-ulm.de_Di 12:00 gerade	Edit	Tutorial	Activate State			29.04.2019 23:56:17
3	person78@uni-ulm.de	person78@uni-ulm.de_Di 12:00 gerade	Edit	Tutorial	Set: RelationAttribute	Lecture	Datenbanken	29.04.2019 23:56:20
4	person78@uni-ulm.de	person78@uni-ulm.de_Di 12:00 gerade	Edit	Tutorial	Set: StringAttribute	Slot	Diensta	29.04.2019 23:56:24
5	person78@uni-ulm.de	person78@uni-ulm.de_Di 12:00 gerade	Edit	Tutorial	Change StringAttribute	Slot	Dienstag 1	29.04.2019 23:56:32
6	person78@uni-ulm.de	person78@uni-ulm.de_Di 12:00 gerade	Edit	Tutorial	Change StringAttribute	Slot	Dienstag 12	29.04.2019 23:56:34
7	person78@uni-ulm.de	person78@uni-ulm.de_Di 12:00 gerade	Edit	Tutorial	Change StringAttribute	Slot	Dienstag 12.00	29.04.2019 23:56:43
8	person78@uni-ulm.de	person78@uni-ulm.de_Di 12:00 gerade	Edit	Tutorial	Change StringAttribute	Slot	Dienstag 12.00 - 12.30	29.04.2019 23:56:49
9	person78@uni-ulm.de	person78@uni-ulm.de_Di 12:00 gerade	Edit	Tutorial	Change StringAttribute	Slot	Dienstag 12.00 - 12.30 herade	29.04.2019 23:57:01
10	person78@uni-ulm.de	person78@uni-ulm.de_Di 12:00 gerade	Edit	Tutorial	Change StringAttribute	Slot	Dienstag 12.00 - 12.30 gerade K	29.04.2019 23:57:11
11	person78@uni-ulm.de	person78@uni-ulm.de_Di 12:00 gerade	Edit	Tutorial	Change StringAttribute	Slot	Dienstag 12.00 - 12.30 gerade KW	29.04.2019 23:57:15
12	person78@uni-ulm.de	person78@uni-ulm.de_Di 12:00 gerade	Edit	Tutorial	State completed			29.04.2019 23:57:16
13	person78@uni-ulm.de	person78@uni-ulm.de_Di 12:00 ungerade	Edit	Tutorial	Activate State			29.04.2019 23:57:21
14	person78@uni-ulm.de	person78@uni-ulm.de_Di 12:00 ungerade	Edit	Tutorial	Set: RelationAttribute	Lecture	Datenbanken	29.04.2019 23:57:24
15	person78@uni-ulm.de	person78@uni-ulm.de_Di 12:00 ungerade	Edit	Tutorial	Set: StringAttribute	Slot	Dienstag	29.04.2019 23:57:26

Fig. 7. Event Log Example for State Edit of Object Tutorial (anonymized due to GDPR)

Concerning the state level, we consider those event log entries that are related to state changes (i.e., the transitions in Figs. 3 + 4). In contrast, the step level conformance checking considers events related to the writing of object attributes (i.e., the transitions in Figs. 5 + 6). Consequently, an event log subset either documents how object instances transitioned between object states or how individual states were processed (i.e., in which order the fields of its corresponding form, i.e., object attributes, were set). Aligning a log subset with the two corresponding workflow nets (cf. Figs. 3 + 4 for the state level of object lifecycle *Submission* or Figs. 5 + 6 for state *Edit*), we obtain two fitness values. Based on the latter, we can categorize each lifecycle process instance into one of the following categories, depending on the behavior captured in the event log:

- **Guided behavior:** The fitness value obtained for the guided net equals to 1. The tolerated fitness also equals to 1 as it generalizes the guided behavior.
  - **State level:** the state changes of the object lifecycle process instance fully comply with the guided behavior.
  - **Step level:** the steps of a state were executed according to the guided behavior (i.e., the generated form was filled in from top to bottom).
- **Tolerated behavior:** The fitness from the guided net is below 1, but the fitness from the tolerated net still equals 1. Deviations from the guided behavior model captured in this category are *correct* executions due to the built-in flexibility. However, process improvements might be possible.
  - **State level:** the object lifecycle process instance changed states correctly, though its execution utilizes the built-in flexibility of lifecycle processes (e.g., by following backwards transitions)
  - **Step level:** the steps were executed correctly with respect to the built-in flexibility of the generated forms (e.g., the form was filled in any order).
- **Deviating behavior:** Both fitness values are below 1. Deviations occurred that are not tolerated by the built-in flexibility of object lifecycle processes.
  - **State level:** deviations include, but are not limited to states not being reachable from the currently active state, to ad-hoc (i.e., not pre-specified) backwards transitions, or to state changes not allowed by the object lifecycle process.
  - **Step level:** deviations may result from states that are completed while not all required attributes (i.e., steps) are set, steps not being part of the lifecycle process state are set, steps are changed after the state is completed or steps are changed before being set.

5.2   Leveraging Conformance Categories for Process Analysis

When analyzing object-centric and data-driven processes, the introduced conformance categories provide useful insights for process engineers into potential model improvements. In turn, this facilitates problem detection through the discovery of actual deviations and tolerated behavior. To identify whether object lifecycle instances deviate from the lifecycle model (e.g., through ad-hoc changes executed by process supervisors) or are executed according to the tolerated behavior yields useful information for improving and evolving lifecycle processes.

Deviating behavior on the state level, for example, indicates that ad-hoc changes were required during lifecycle process execution. In this scenario, the lifecycle model does not allow for all the behavior required in practice, i.e., is has turned out to be too restrictive during the processing of certain state transitions.

Furthermore, tolerated behavior in the processing of a certain state may indicate that the ordering of the steps within this state might not be optimal.

Conformance categories are capable of prioritizing improvement efforts. On both granularity levels, tolerated behavior captures behavior inherently supported by the built-in flexibility and, thus, not requiring any interventions by process supervisors. This indicates improvement potential with respect to usability (i.e., forms may be optimized by reordering steps of a state with highly tolerated behavior). Deviating behavior, in turn, covers behavior due to either implementation mistakes or explicit interventions by process supervisors (e.g., through ad-hoc changes at runtime). As a result, when analyzing object lifecycle processes, deviating behavior should be investigated with higher priority. Our approach provides guidance for process engineers in analyzing and improving lifecycle processes.

6   Experimental Evaluation

To demonstrate the applicability of our approach for checking the conformance of lifecycle processes, we implemented a proof-of-concept prototype[1]. The latter includes a translator that enables the generation of the different workflow nets based of an object-aware process [16]. The implementation of this translator uses *python* and the *pm4py* framework [9]. The implemented algorithms are illustrated in terms of pseudo-code in Algs. 1 and 2. To evaluate the conformance checking approach described in Sect. 5, we used multiple event logs to check their conformance with each of the derived workflow nets. First, we generated event logs using the extended Petri net playout feature of *pm4py* for the derived workflow nets to simulate all allowed process executions. For this purpose, we generated all traces that are allowed according to each workflow net, up to a trace length of 10. However, any other trace length would be possible as well. The resulting event logs contain 6 (tolerated state level), 1 (guided state level), 8334 (tolerated step level) and 56 (guided step level) traces respectively.

Table 3. Object submission - playout

% of traces	State Level		Step Level (State *Edit*)	
	Guided log	Tolerated log	Guided log	Tolerated log
Guided behavior	100 %	16.66 %	100 %	0.6719
Tolerated behavior	0 %	83.34 %	0 %	99.3281

---

[1] All event logs are provided at https://www.researchgate.net/project/Lifecycle-Conformance-Checking-RCIS.

The results from Table 3 show that we are able to distinguish between tolerated and guided behavior using the described workflow nets for conformance checking. The generated event logs, however, do not contain actual deviations (cf. Section 5) as they represent all allowed playouts of the derived workflow nets. All simulated traces, therefore, belong to either guided or tolerated behavior. On state level (cf. Table 3), 16.66% of the tolerated traces fit the guided behavior (i.e., Category Guided Behavior State Level for Tolerated Log). Concerning the step level granularity of state *Edit* (cf. Table 3), only 0.6719% of the tolerated traces fit to the actual guided behavior (i.e., Category Guided Behavior Step Level for Tolerated Log). This indicates the high degree of flexibility an object lifecycle process allows for a state with only 3 steps and a trace length up to 10.

To evaluate whether the approach is able to identify deviating behavior in an event log, we generated an additional event log that contains deviating behavior as well. For this purpose, we randomly simulate behavior within an event log that may not only represent tolerated and guided, but also deviating behavior, by randomly picking from the set of transitions. In practice, such behavior can be observed in the context of ad-hoc changes or implementation mistakes (e.g., while collecting event logs). Algorithm 3 indicates how we generated the event logs used for checking conformance. We generated event logs with 1000 traces of random length between 5 and 8 in order to group traces according to the categories presented in Sect. 5.

Algorithm 3. Algorithm to Generate Event Logs with Deviations

---

Require: PetriNet, TraceNumber, TraceLength   ▷ Petri net, number of log traces and length
  OriginalTransitions = PetriNet.transitions.copy()
  for trace = 0 to TraceNumber **do**
    Transitions = OriginalTransitions
    Eventlog.add(Initial State)   ▷ Activate State or Source to first state
    for i = 0 to TraceLength **do**
      Transition = random.choice(Transitions)   ▷ Pick random transition from net
      if Transition = "Set:" **then**   ▷ Probibit, that one step is set multiple times in a trace
        Eventlog.add(Transition)
        Transition.remove(Transition)
      else
        Eventlog.add(Transition)
      end if
    end for
    Eventlog.add(FinalState)   ▷ State completed or last state to sink
  end for

Table 4 shows the categories into which the randomly generated traces are assigned according to their behavior documented in the event log. We are not only able to differentiate between guided and tolerated behavior but can also identify deviating behavior with the approach. However, note that the event logs used in the sketched evaluation constitute two edge cases of object lifecycle process executions, as they either contain no deviations (cf. Table 3) or a high ratio of deviations (cf. Table 4).

We further evaluated the approach using an event log we collected from a real-world deployment of *Phoodle* (cf. Section 2) in which 133 students used the

Table 4. Categories for object submission - random

% of traces (#)	State level	Step level (state Edit)
Guided behavior	0.1 % (1)	0.4 % (4)
Tolerated behavior	0.2 % (2)	7.0 % (70)
Deviating behavior	99.7 % (997)	92.6 % (926)

system during a university course over a period of 4 months (cf. Fig. 7). When applying the approach, all 51 lifecycle process instances of object *Tutorial* showed deviating step level behavior for state *Edit* (cf. Table 5). Upon closer inspection, according to the event log, attribute *Lecture* was set in state *Edit* (cf. lines 3 and 14 in Fig. 7), while the lifecycle process required attribute *Tutor*. In the next step, we repaired the event log to set the correct attribute, and thus 28 of 51 tutorial lifecycle process instances corresponded to guided and 4 to tolerated behavior. Note that the remaining 19 instances had additional deviations not related to the repaired deviation (e.g., attributes were changed after the state had been completed using ad-hoc changes [7]).

Table 5. Phoodle Log Tutorial - State Edit

% of traces (#)	Initial	Repaired
Guided step level behavior	0 % (0)	54.90 % (28)
Tolerated step level behavior	0 % (0)	7.85 % (4)
Deviating step level behavior	100 % (51)	37.25% (19)

Overall, the evaluation has shown that we are able to pinpoint which granularity level of an object lifecycle process is non-conforming (i.e., deviations regarding state transitions or individual states). Additionally, we can account for the flexible (i.e., tolerated) execution of object lifecycle processes through the use of multiple workflow nets, therefore enabling sophisticated and holistic deviation detection.

7   Related Work

This work is related to two research areas: conformance checking and object-/data-centric process management. In [19], conformance checking is presented as multidimensional quality metrics for processes and their corresponding event logs. Furthermore, best-effort metrics to assess the different quality dimensions are introduced. One algorithm to check conformance is Token Replay [19], which can identify those parts of the process model and event log that fit together. Furthermore, it enables diagnostics related to deviations by replaying the event log on a Petri net covering the execution behavior of the process model. Additional algorithms have emerged that enable conformance assessment by aligning process model and event log [3]. Alignments have already been adapted to various scenarios, e.g., large processes [18] or declarative processes [17]. Recently, another token replay approach emerged [8]. Finally, first approaches for discovering object-centric Petri nets have been proposed [4].

Some conformance checking approaches exist for artifact-centric conformance checking [13–15] as well. To some degree these approaches are similar to ours. However, differences arise due to the fact that we focus on business objects instead of proclets, artifacts, or UML diagrams. Compared to [13], our approach does not use UML state and activity diagrams to generate the Petri net. While [14] uses conformance checking, the presented approach is able to identify behavioral and interaction conformance with respect to proclets (Petri nets, including communication ports). As a result, no translation from proclets to Petri nets is required. The work presented in [15] focuses on the interaction between multiple artifacts rather than the behavior of object lifecycle processes in isolation. Furthermore, to the best of our knowledge, none of the existing approaches accounts for flexibility during conformance checking of data-centric and -driven processes.

8   Summary and Outlook

This paper presented an approach for checking the conformance of single object lifecycle processes. We introduced two granularity levels for enacting lifecycle processes granularity levels as well as built-in flexibility concepts. We then incorporate them during conformance checking to differentiate between guided, tolerated, and deviating behavior. Checking conformance with multiple nets allows categorizing each lifecycle execution based on the behavior captured in the event log. Furthermore, we are able to account for the flexible nature of object lifecycles through conformance categories that allow us to distinguish between deviations tolerated due to the flexibility of object lifecycles and actual deviations. Additionally, we can account for flexibility regarding transitions between states, and the behavior of individual states. When analyzing data-centric and -driven processes, conformance categories provide guidance for process engineers with respect to which parts of object lifecycle processes are of particular interest.

In future work, we plan to extend the presented approach in a two-fold manner: First, we plan to incorporate constraints between object lifecycle processes. This will allow us to further improve conformance checking of object-centric processes regarding the inter-object granularity level. The latter considers constraints between different object lifecycle processes, rather than lifecycle processes in isolation. Second, we want to provide detailed information on the origin of the deviation to further facilitate process improvement.

Acknowledgments. This work is part of the SoftProc project, funded by the KMU Innovativ Program of the Federal Ministry of Education and Research, Germany (F.No. 01IS20027A).

References

1. van der Aalst, W.M.P.: Verification of workflow nets. In: Azéma, P., Balbo, G. (eds.) ICATPN 1997. LNCS, vol. 1248, pp. 407–426. Springer, Heidelberg (1997). https://doi.org/10.1007/3-540-63139-9_48
2. van der Aalst, W.M.P., et al.: Process mining manifesto. In: Daniel, F., Barkaoui, K., Dustdar, S. (eds.) BPM 2011. LNBIP, vol. 99, pp. 169–194. Springer, Heidelberg (2012). https://doi.org/10.1007/978-3-642-28108-2_19
3. van der Aalst, W.M.P., Adriansyah, A., van Dongen, B.: Replaying history on process models for conformance checking and performance analysis. WIREs Data Min. Knowl. Discovery 2(2), 182–192 (2012)
4. van der Aalst, W.M.P., Berti, A.: Discovering object-centric Petri Nets. Fundamenta informaticae 175(1/4), 1–40 (2020)
5. van der Aalst, W.M.P., Weske, M., Grünbauer, D.: Case handling: a new paradigm for business process support. DKE 53(2), 129–162 (2005)
6. Andrews, K., Steinau, S., Reichert, M.: Enabling ad-hoc changes to object-aware processes. In: 22nd International Enterprise Distributed Object Computing Conference (EDOC 2018), pp. 85–94. IEEE Computer Society Press, October 2018
7. Andrews, K., Steinau, S., Reichert, M.: Enabling runtime flexibility in data-centric and data-driven process execution engines. Inf. Syst. 101, 101447 (2021)
8. Berti, A., van der Aalst, W.M.P.: A novel token-based replay technique to speed up conformance checking and process enhancement. In: Koutny, M., Kordon, F., Pomello, L. (eds.) Transactions on Petri Nets and Other Models of Concurrency XV. LNCS, vol. 12530, pp. 1–26. Springer, Heidelberg (2021). https://doi.org/10.1007/978-3-662-63079-2_1
9. Berti, A., van Zelst, S.J., van der Aalst, W.M.P.: Process mining for Python (PM4Py): bridging the gap between process- and data science. CoRR abs/1905.06169 (2019)
10. Carmona, J., Dongen, B., Solti, A., Weidlich, M.: Conformance Checking: Relating Processes and Models, January 2018. https://doi.org/10.1007/978-3-319-99414-7
11. Cohn, D., Hull, R.: Business artifacts: a data-centric approach to modeling business operations and processes. IEEE TCDE 32(3), 3–9 (2009)
12. Dunzer, S., Stierle, M., Matzner, M., Baier, S.: Conformance checking: a state-of-the-art literature review. CoRR abs/2007.10903 (2020)
13. Estañol, M., Munoz-Gama, J., Carmona, J., Teniente, E.: Conformance checking in UML artifact-centric business process models. Softw. Syst. Model. 18(4), 2531–2555 (2018). https://doi.org/10.1007/s10270-018-0681-6
14. Fahland, D., de Leoni, M., van Dongen, B.F., van der Aalst, W.M.P.: Behavioral conformance of artifact-centric process models. In: Abramowicz, W. (ed.) BIS 2011. LNBIP, vol. 87, pp. 37–49. Springer, Heidelberg (2011). https://doi.org/10.1007/978-3-642-21863-7_4
15. Fahland, D., de Leoni, M., van Dongen, B.F., van der Aalst, W.M.P.: Conformance checking of interacting processes with overlapping instances. In: Rinderle-Ma, S., Toumani, F., Wolf, K. (eds.) BPM 2011. LNCS, vol. 6896, pp. 345–361. Springer, Heidelberg (2011). https://doi.org/10.1007/978-3-642-23059-2_26
16. Künzle, V., Reichert, M.: PHILharmonicFlows: towards a framework for object-aware process management. JSME 23(4), 205–244 (2011)
17. de Leoni, M., Maggi, F.M., van der Aalst, W.M.P.: Aligning event logs and declarative process models for conformance checking. In: Barros, A., Gal, A., Kindler, E. (eds.) BPM 2012. LNCS, vol. 7481, pp. 82–97. Springer, Heidelberg (2012). https://doi.org/10.1007/978-3-642-32885-5_6

18. Munoz-Gama, J., Carmona, J., van der Aalst, W.M.P.: Single-entry single-exit decomposed conformance checking. Inf. Syst. **46**, 102–122 (2014)
19. Rozinat, A., van der Aalst, W.M.P.: Conformance checking of processes based on monitoring real behavior. Inf. Syst. **33**(1), 64–95 (2008)
20. Steinau, S., Andrews, K., Reichert, M.: Executing lifecycle processes in object-aware process management. In: Ceravolo, P., van Keulen, M., Stoffel, K. (eds.) SIMPDA 2017. LNBIP, vol. 340, pp. 25–44. Springer, Cham (2019). https://doi.org/10.1007/978-3-030-11638-5_2
21. Steinau, S., Marrella, A., Andrews, K., Leotta, F., Mecella, M., Reichert, M.: DALEC: a framework for the systematic evaluation of data-centric approaches to process management software. Softw. Syst. Model. **18**(1), 2679–2716 (2019)

A Framework to Improve the Accuracy of Process Simulation Models

Francesca Meneghello[1(✉)], Claudia Fracca[1,2], Massimiliano de Leoni[2], Fabio Asnicar[1], and Alessandro Turco[1]

[1] ESTECO SpA, Trieste, Italy
{meneghello,asnicar,turco}@esteco.com
[2] University of Padua, Padua, Italy
claudia.fracca@phd.unipd.it, deleoni@math.unipd.it

Abstract. Business process simulation is a methodology that enables analysts to run the process in different scenarios, compare the performances and consequently provide indications into how to improve a business process. Process simulation requires one to provide a simulation model, which should accurately reflect reality to ensure the reliability of the simulation findings. This paper proposes a framework to assess the extent to which a simulation model reflects reality and to pinpoint how to reduce the distance. The starting point is a business simulation model, along with a real event log that records actual executions of the business process being simulated and analyzed. In a nutshell, the idea is to simulate the process, thus obtaining a simulation log, which is subsequently compared with the real event log. A decision tree is built, using the vector of features that represent the behavioral characteristics of log traces. The tree aims to classify traces as belonging to the real and simulated event logs, and the discriminating features encode the difference between reality, represented in the real event log, and the simulation model, represented in the simulated event logs. These features provide actionable insights into how to repair simulation models to become closer to reality. The technique has been assessed on a real-life process for which the literature provides a real event log and a simulation model. The results of the evaluation show that our framework increases the accuracy of the given initial simulation model to better reflect reality.

Keywords: Business process simulation · BPMN model · Decision tree · Declarative language · Event log comparison

1 Introduction

Business process simulation refers to techniques for the simulation of business process behavior on the basis of a process simulation model, a process model extended with additional information for a probabilistic characterization of the different run-time aspects (case arrival rate, task durations, routing probabilities, resource utilization, etc.). Simulation provides a flexible approach to analyse and improve business processes. Through simulation experiments, various 'what if' questions can be answered, and redesigning alternatives can be compared with respect to some key performance indicators. The main idea of business process simulation is to carry out a significantly large

© The Author(s), under exclusive license to Springer Nature Switzerland AG 2022
R. Guizzardi et al. (Eds.): RCIS 2022, LNBIP 446, pp. 142–158, 2022.
https://doi.org/10.1007/978-3-031-05760-1_9

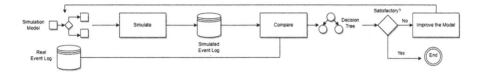

Fig. 1. The framework of the proposed approach.

number of runs, in accordance with a simulation model. Statistics over these runs are collected to gain insight into the processes, and to determine the possible issues (bottlenecks, wastes, costs, etc.). By applying different changes to the simulation model, one can assess the consequences of these changes without putting them in production, and consequently can explore dimensions to possible process improvements.

A successful application of business process simulation for process improvement relies on a simulation model that reflects the real process behavior; conclusions are drawn on an unrealistic simulation model lead to process redesigns that may not yield improvements, or even may worsen the performances.

This paper puts forward a framework to assess and improve the accuracy (i.e., the realism) of a simulation model M starting from the differences figured out by our framework.

Along with M, the framework requires an event log $\mathcal{L}_r$ that records executions of the process modelled by M. The framework is based on the idea that many process runs are carried out using M, thus generating a simulated event log $\mathcal{L}_s$. If $\mathcal{L}_s$ is similar to $\mathcal{L}_r$, then M is accurate. To compute the similarity, the framework builds a decision tree that classifies the traces of $\mathcal{L}_r$ or $\mathcal{L}_s$. If the two logs are similar, the decision tree is unable to discriminate and, hence, correctly classify. The decision tree features encode the behavior observed in traces, such as whether two casually dependent activities follow in traces, activity durations, or certain declarative rules based on Linear Temporal Logic (LTL) formula.

Section 2 further introduces the framework, using an intuitive example. Section 3 introduces the basic concepts that are used throughout the paper. Section 4 reports on the discriminating features used to create the decision tree model from the two logs, and on a final data preprocessing step before training the model. Section 5 discusses on the use of decision trees for event logs discrimination. Section 6 illustrates the results of our evaluation, while Sect. 7 discusses related work. Section 8 summarizes the contributions and outlines future work directions.

2   Overall Idea and Motivation

The framework proposed in this paper can be summarized as in Fig. 1. The starting point is an initial simulation model and a real event log. The simulation model can be drawn by process analysts on the basis of insights from stakeholders, or it can be discovered using combinations of Process Mining techniques [9,11]. The simulation model is used to generate the traces composing a simulated log. After generating the simulating model, the framework aims to compare the two logs for differences; to do

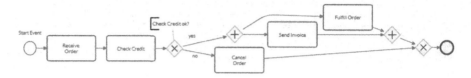

Fig. 2. BPMN model of process order.

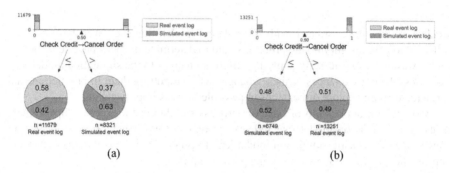

Fig. 3. Decision trees obtained from the comparison between the real log and the simulation log originated from the simulation model over the BPMN in Fig. 2.

so, a decision tree is built that highlights the differences and indicates how close our simulation model represents reality. If the decision tree indicates that the simulation model accurately reflects reality, the application of the framework concludes, and the model is deemed as appropriate. If the decision tree highlights significant differences, the model is modified, using the differences as guidelines for improvement. The new simulation model can be used to generate a new simulation log and compared with the real log for the difference. It follows that the simulation model is improved iteratively through a sequence of improvement steps: the framework more and more improves the accuracy of the simulation model, until the differences are considered negligible.

As an example, let us consider the BPMN model in Fig. 2, which refers to a purchase-order process composed of five activities. In particular, the exclusive gateway routes the control flow of the process based on the result of the previous activity *Check Credit*. If the *Check Credit* is successful, the order is carried; otherwise, the order is canceled. Let us suppose to have a real event log $\mathcal{L}_r$. The simulation model, composed by the BPMN in Fig. 2 and the simulation parameters (e.g., case arrival rate, branching probabilities), is simulated as many times as the number of traces in $\mathcal{L}_r$ in order to generate a simulated event log $\mathcal{L}_s$. The framework builds a decision tree that is trained via a multiset of feature vectors, one vector for each trace in $\mathcal{L}_r$ and $\mathcal{L}_s$. Let us suppose to obtain the decision tree in Fig. 3a. The root of the tree contains the feature CheckCredit→CancelOrder. For a given trace σ, this feature value is equal to the number of times that the activity *Check Credit* is followed by the activity *Cancel Order* and the opposite never happens in σ. Traces of $\mathcal{L}_r$ and $\mathcal{L}_s$ are then associated to decision tree leaves, depending on the values of the features that appear in the traces. Each leaf of the tree is represented as a pie chart, which describes the fractions of the traces inside the leaf that belong to $\mathcal{L}_r$ and $\mathcal{L}_s$. The right leaf in Fig. 3a contains the traces with the root feature equal to 1, i.e., the traces containing the activity *Check Credit* followed by *Cancel Order*. This leaf contains the traces for orders that are canceled, while the left for completed orders. The pie chart of the right leaf contains more traces from $\mathcal{L}_s$ log with respect to $\mathcal{L}_r$ log, which indicates that the simulation model sets a too high probability to cancel more orders than the real process. Starting from this rule, we can improve and fix the simulation model in order to have the same percentage of completed orders. In this case, the mistake is very likely associated with the branching probability of the exclusive gateway *Check Credit*. We can hence improve the model by better tuning the probability at the gateway. The new simulation model is run to obtain a simulated log to repeat the comparison based on the decision tree construction. The new decision tree is, e.g., as in Fig. 3b: now the leaves almost contain the same numbers of traces from $\mathcal{L}_r$ and $\mathcal{L}_s$. It follows that the decision tree model is not able to well discriminate whether a trace belongs to the real or simulated event log, thus positively confirming the accuracy of the simulated model.

3   Preliminary

This section introduces the preliminary concepts to later illustrate the technique's details. First, we present the concepts of events, traces and event logs, and some related notations.

Definition 1 (Events). *Let $\mathcal{A}$ be a set of activity labels. Let $\mathcal{T}$ be the universe of timestamps. Let $\mathcal{I} = \{\texttt{start}, \texttt{complete}\}$ be the life-cycle information. An event $e \in \mathcal{A} \times \mathcal{T} \times \mathcal{I}$ is a tuple consisting of an activity label, a timestamp of occurrence, and the life-cycle information.*

In the remainder, given an event $e = (a, t, i)$, $act(e) = a$ returns the activity label, $time(e) = t$ returns the timestamp, and $life(e) = i$ is the information whether e refers to the starting or completion of an activity. In practice, several event logs are composed of events where the life-cycle information is not present. In this case, we assume that those events refer to the completion. There might be events other than related to the starting or completion of activities: those events are simply ignored. Events also carry a payload consisting of attributes taking on values: they are also ignored.

Definition 2 (Traces and Event Logs). *Let $\mathcal{E}_\mathcal{A}$ the universe of the events defined over a set $\mathcal{A}$ of activity labels. A trace $\sigma = \langle e_1, \ldots, e_m \rangle \in \mathcal{E}_\mathcal{A}^*$ is a sequence of events, with the constraint that, for all $0 < i < j \leq m$, $time(e_i) \leq time(e_j)$. An event log $\mathcal{L}_\mathcal{A}$ is a set of traces, namely $\mathcal{L}_\mathcal{A} \subset \mathcal{E}_\mathcal{A}^*$.*

In the remainder, we use the shortcut $e \in \mathcal{L}_\mathcal{A}$ to indicate that there is a trace $\sigma \in \mathcal{L}_\mathcal{A}$ such that $e \in \sigma$. Also, we drop the subscript $\mathcal{A}$ when it is clear from the context. Finally, given a trace σ, the notation $complete(\sigma)$ refers to the sequence of events in σ after removing the events referring to the starting of activities, and retaining the same order, i.e. $complete(\sigma) = \{e \in \sigma \mid life(e) = \texttt{complete}\}$. Table 1 shows an example of an event log related to the management of order requests.

Table 1. A fragment of an event log of a process about dealing with orders.

Case ID	Activity	Timestamp	Life-cycle	Case ID	Activity	Timestamp	Life-cycle
12	Receive order	16-08-20 08:30	Start	14	Receive order	17-08-20 18:00	Start
12	Receive order	16-08-20 08:45	Complete	14	Receive order	17-08-20 18:45	Complete
13	Receive order	16-08-20 10:30	Start	13	Cancel order	17-08-20 18:56	Complete
13	Receive order	16-08-20 10:45	Complete	12	Send invoice	18-08-20 10:40	Start
12	Check credit	17-08-20 12:27	Start	12	Send invoice	18-08-20 10:42	Complete
12	Check credit	17-08-20 12:32	Complete	14	Check credit	18-08-20 13:11	Start
12	Fulfill order	17-08-20 14:40	Start	14	Check credit	18-08-20 13:32	Complete
12	Fulfill order	17-08-20 14:50	Complete	14	Fulfill order	18-08-20 14:25	Start
13	Check credit	17-08-20 16:40	Start	14	Send invoice	18-08-20 14:27	Complete
13	Check credit	17-08-20 17:40	Complete	14	Send invoice	18-08-20 14:40	Complete
13	Cancel order	17-08-20 17:56	Start	14	Fulfill order	20-08-20 10:50	Complete

Table 2. List of DECLARE constraints used in our techniques.

Constraints	Description
$Init(a)$	a should be the first activity in a trace
$End(a)$	a should be the last activity in a trace
$CoExistence(a, b)$	If one of the activities a or b is executed, the other one also has to be executed
$Response(a, b)$	When a is executed, b has to be executed after a
$AlternateResponse(a, b)$	When a is executed, b has to be executed after a and no other a can be executed in between
$Precedence(a, b)$	b has to be preceded by a
$AlternatePrecedence(a, b)$	b has to be preceded by a and another b cannot be executed between a and b
$Succession(a, b)$	a occurs if and only if it is followed by b

Some of the differences can be given as constraints of DECLARE, i.e., a declarative process modeling language [1]. This language indeed defines the behavior of the business process as a set of constraints. An example of DECLARE constraint is $Init(a)$ and it states that every instance must start with the execution of activity a. Another example can be $Precedence(a, b)$, and it imposes that the activity b occurs only if preceded by the activity a. The full list of DECLARE constraints can be found in [1]. The list of constraints and the related descriptions that are used in this paper are listed in Table 2. Given this framework, we can notice that each DECLARE constraint can be formally defined as a LTL formula, which can ultimately be represented as a final state automaton over the alphabet of activities.

Definition 3 (DECLARE Constraint as Final State Automaton). *Let A be a set of activity labels. The DECLARE constraint can be formulated as a final state automaton $C = (A, Q, q_0, \delta, E)$ over a set A of activities, where: (i) A is the activity labels; (ii) Q is a finite, non-empty set of states; (iii) $q_0 \in Q$ is an initial state; (iv) $\delta \in Q \times A \rightarrow Q$ is the state-transition function; (v) $E \subseteq Q$ is the set of final states.*

The automaton that represents a DECLARE constraint accepts all and only those log traces that satisfy the constraint. Note how the state-transition function is total: log traces can always be replied on the automaton.

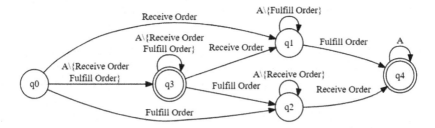

Fig. 4. This figure shows the automaton for the DECLARE constraint *CoExistence(Receive Order, Fulfill Order)*, where $\mathcal{A}$ is the who set of activity labels.

Example 1. Given the set of activity labels related to the event log in the Table 1. Figure 4 shows the automaton $\mathcal{C} = (\mathcal{A}, Q, q_0, \delta, E)$ for the DECLARE constraint *CoExistence (Receive Order, Fulfill Order)*. This DECLARE constraint states that if one of the two activities, *Receive Order* or *Fulfill Order*, is executed, then the other also has to be executed. As an example, the trace with the CASE ID = 12 in the Table 1 is accepted by the automaton, while the trace with CASE ID = 13 is not accepted.

4   Our Framework

Our framework requires a simulation model and a real event log $\mathcal{L}_r$ as input. The basic idea is to verify the quality of the simulation model by generating an event log $\mathcal{L}_s$ from the latter and find the event log features that discriminate $\mathcal{L}_r$ and $\mathcal{L}_s$. These features can range from the occurrences of activities and their sequencing to DECLARE constraints, and temporal features i.e. the activity durations. The discriminating features pinpoint the differences between the two event logs and provide valuable insights into repairing the simulation models.

In this structure, we employ decision tree learning to find the discriminating features. Given a set of features $\mathcal{F}$, and their potential values $\mathcal{V}$. Decision trees are learned from a multiset of pairs (x, y) where $x \in \mathcal{F} \rightarrow \mathcal{V}$ is a function that assigns values to (a subset of) the features, and y is the class value. The label y is whether the trace belongs to the real or simulated event log. In our framework, the feature function is extracted from an event log via a customizable mapping function:

Definition 4 (Trace-to-Feature Mapping Function). *Let $\mathcal{L}$ be an event log. Let $\mathcal{F}$ be a set of features. Let $\mathcal{V}$ be the set of potential values. A trace-to-features mapping function is a function $\rho^{\mathcal{L}} : \mathcal{L} \rightarrow (\mathcal{F} \rightarrow \mathcal{V})$ such that, for each trace $\sigma \in \mathcal{L}$, it returns a feature-to-value function $z = \rho^{\mathcal{L}}(\sigma)$ that assigns a value $z(f) \in \mathcal{V}$ to each feature $f \in \mathcal{F}$.*

With these concepts at hand, given a real event log $\mathcal{L}_r$ and a simulated event log $\mathcal{L}_s$ the training set $T_{\mathcal{L}_r, \mathcal{L}_s}$ of the decision tree is constructed as follows where real and sim denote the two possible class values for respectively real and simulated log trace:[1]

---

[1] Symbol ⊎ indicates the union of multisets where duplicates are retained.

$$T_{\mathcal{L}_r, \mathcal{L}_s} = \biguplus_{\sigma_r \in \mathcal{L}_r} (\rho^{\mathcal{L}}(\sigma_r), \texttt{real}) \uplus \biguplus_{\sigma_s \in \mathcal{L}_s} (\rho^{\mathcal{L}}(\sigma_s), \texttt{sim}) \tag{1}$$

In the remainder, from Sects. 4.1 to 4.3, we introduce several trace-to-features mapping functions supported in our implementation to construct the training sets for the decision tree algorithm, i.e., *Basic Features, Extended Features for Declare Rules and Temporal Features*.

4.1 Basic Features

In this section, we describe the features related to the control-flow perspective. First of all, we define the features related to the activities occurrences, i.e., if an activity is performed in the given trace or not.

Definition 5 (Activity Function). *Let $\mathcal{L}$ be an event log over the set $\mathcal{A}$ of activities. Let define a trace-to-function mapping function $\rho^{\mathcal{L}}_{activity} : \mathcal{L} \to (\mathcal{F} \to \mathcal{V})$ in which the set of features $\mathcal{F}$ is $\mathcal{A}$. For each trace $\sigma \in \mathcal{L}$, it returns a function $z = \rho^{\mathcal{L}}_{activity}(\sigma)$ that assigns a value $z(f) \in \mathcal{V}$ for each activity feature $f \in \mathcal{F}$. The value $z(f)$ is the number of times an activity a is executed within the trace, i.e. $|e \in complete(\sigma) : act(e) = a|$.*

Example 2. Let σ the trace with the CASE ID $= 12$ in the Table 1. Let fix the set of features $\mathcal{F}$ equal to the set $\mathcal{A}$, i.e. $\mathcal{F} = \{\texttt{ReceiveOrder}, \texttt{CheckCredit}, \texttt{FulfillOrder}, \texttt{SendInvoice}, \texttt{CancelOrder}\}$. Given the trace σ, the function $\rho^{\mathcal{L}}_{activity}$ maps the trace into a function z that assigns for each $f \in \mathcal{F}$ a value in $\{0, \ldots, n\}$. The trace contains the activities: *Receive Order, Check Credit, Fulfill Order*, and *Send Invoice*. Therefore we have that $z(\texttt{ReceiveOrder}) = 1$, $z(\texttt{CheckCredit}) = 1, z(\texttt{FulfillOrder}) = 1, z(\texttt{SendInvoice}) = 1, and$ $z(\texttt{CancelOrder}) = 0$.

In the following, we define the second type of features related to the control-flow perspective, i.e., these features encoded the causality relation between activities:[2]

Definition 6 (Causality Relation). *Given an event log $\mathcal{L}$ defined over a set $\mathcal{A}$ of activities. $a \to_{\mathcal{L}} b$ is a causality relation in $\mathcal{L}$ iff there is a trace $\sigma = \langle e_1, \ldots, e_m \rangle \in \mathcal{L}$ s.t $\langle e_i, e_{i+1} \rangle \subseteq complete(\sigma) \mid act(e_i) = a \wedge act(e_{i+1}) = b$ and $\nexists \sigma' = \langle e'_1, \ldots, e'_m \rangle \in \mathcal{L}$ s.t $\langle e'_i, e'_{i+1} \rangle \subseteq complete(\sigma') \mid act(e'_i) = b \wedge act(e'_{i+1}) = a$.*

Definition 7 (Causality Relation Function). *Let $\mathcal{L}$ be an event log defined over a set $\mathcal{A}$ of activities. We introduce a trace-to-feature mapping function $\rho^{\mathcal{L}}_{\to} : \mathcal{L} \to (\mathcal{F} \to \mathcal{V})$ in which the set of features $\mathcal{F}$ coincides with the set of causality relation in $\mathcal{L}$. The causality relation function $\rho^{\mathcal{L}}_{\to}(\mathcal{L})$ returns a function $z(f)$ that, to each causality relation $f = (a \to_{\mathcal{L}} b) \in \mathcal{F}$, assigns the numbers of times an event for activity a is followed by an event for activity b in $\mathcal{L}$.*

Example 3. Let σ the trace with the CASE ID $= 12$ in the Table 1. Let fix the set of features $\mathcal{F}$ equal to the set of all possible causality relation in $\mathcal{L}$, i.e. $\mathcal{F} = \{\texttt{ReceiveOrder} \to \texttt{CheckCredit}, \ldots, CheckCredit \to CancelOrder\}$. In this case we have for instance that $z(\texttt{CheckCredit} \to \texttt{FulfillOrder}) = 1$, and $z(\texttt{CheckCredit} \to \texttt{CancelOrder}) = 0$.

---

[2] Given two sequences s and s', $s' \subseteq s$ indicates that s' is a sub-sequence of s.

4.2 Extended Features for Declare Rules

We also want to support more complex features related to the control-flow perspective using the DECLARE constraints. Given a real event log $\mathcal{L}_r$ and the simulated event log $\mathcal{L}_s$ as in our framework, we compute the sets of DECLARE constraints over these two event logs via MinerFul Miner [8]. We denote these constraint sets are D_r and D_s respectively. As mentioned in Sect. 3, we only support the constraints in Table 2, excluding the constraints already covered or extended by the *Basic Features* discussed in Sect. 4.1. For instance, the *Activity Features* already cover the constraint *Participation(a)* and *AtMostOne(a)*. The first constraint requires that the activity a occurs at least once and the other that the activity a occurs at most once. However, the *Activity Feature* related to the number of occurrences of the activity a is certainly more detailed. Also, we decide to remove the negative DECLARE constraints because they might be overly complex and not give useful insight. After removing these constraints, we perform the symmetrical difference between the two sets of constraints, i.e., $(D_r \cup D_s) \setminus (D_r \cap D_s) = D_t$, obtaining the final set of constraints to consider. In fact, the constraints in common cannot pinpoint differences. Note also that for each $d_j \in D_t$ we have the corresponding constraint automaton $\mathcal{C}_j = (\mathcal{A}^j, Q^j, q_0^j, \delta^j, E^j)$.

Definition 8 (DECLARE Function). *Let $\mathcal{L}$ be an event log over a set $\mathcal{A}$ of activities. Let define a trace-to-function mapping function $\rho_{declare}^{\mathcal{L}} : \mathcal{L} \to (\mathcal{F} \to \mathcal{V})$ in which the set of feature $\mathcal{F}$ is the set of DECLARE constraints D_t. For each trace $\sigma \in \mathcal{L}$, it returns a function $z = \rho_{declare}^{\mathcal{L}}(\sigma)$ that assigns a value $z(f) \in \mathcal{V}$ to each DECLARE constraint $f \in \mathcal{F}$. The value $z(f) \in \{\texttt{True}, \texttt{False}\}$ corresponds to the evaluation of the trace complete(σ) on the respective automaton, i.e. set to $\texttt{True}$ if the automaton of the DECLARE constraint accepts the trace, and $\texttt{False}$ otherwise.*

Example 4. Let σ the trace with the CASE ID $= 12$ in the Table 1. Let fix the set of features $\mathcal{F}$ equal to the set of the DECLARE constraints D_t. For instance, let take the DECLARE constraint *CoExistence(Receive Order, Fulfill Order)* represented in the automaton in the Fig. 4. In this case we have that $z(\texttt{CoExistence(ReceiveOrder, FulfillOrder)})$ is equal to $\texttt{True}$.

4.3 Temporal Features

In this section, we describe the features related to the time perspective. For each $a \in \mathcal{A}$, we define the activity duration feature p_a that represents the duration of the activity a in the trace. Let define with P the set of all the activity duration features related to $\mathcal{A}$, i.e., $P = \bigcup_{a \in \mathcal{A}} p_a$.

Definition 9 (Activity Duration Function). *Let σ a trace in $\mathcal{L}$. Let define a trace-to-function mapping function $\rho_{duration}^{\mathcal{L}} : \mathcal{L} \to (\mathcal{F} \to \mathcal{V})$ in which the set of features $\mathcal{F}$ is equal to the set of activity duration P. For each trace $\sigma \in \mathcal{L}$, it returns a function $z := \rho_{duration}^{\mathcal{L}}(\sigma)$ that assigns a value $z(f) \in \mathcal{V}$ for each feature $f \in \mathcal{F}$. The value $z(f) \in [-1, \infty)$ corresponds to the activity duration.*

Table 3. The corresponding training set of the fragment event log in Table 1.

time:Fulfill Order		Check Credit→Cancel Order		CoExistence(Fulfill Order, Send Invoice)		Class Label
10 min	...	0	...	true	...	real
−1	...	1	...	false	...	sim
25 min		0		true		real

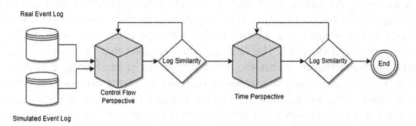

Fig. 5. The framework of the proposed iterative approach. The inputs are the real and the simulated event logs. We have two phases, related to the control-flow and the time perspectives respectively. The blacks-box refer to the framework in Fig. 1.

Example 5. Let σ the trace with the CASE ID $= 12$ in the Table 1. Let fix the set of features $\mathcal{F}$ equal to the set P, i.e. $\mathcal{F} = \{$time:ReceiveOrder,time:Check Credt,time:FulfillOrder,time:SendInvoice,time:CancelledOrder$\}$. Therefore, for example, we have that $z(time : ReceiveOrder) = 15\ minutes$, and that $z(time : CancelledOrder) = -1$ due to the absence of the activity *Cancel Order* in the trace.

Using these trace-to-features mapping functions, we can construct the multiset $T_{\mathcal{L}_r,\mathcal{L}_s}$ as in Eq. 1. Table 3 represents an example of the resulted multiset for the traces contained in the Table 1.

4.4   Application of the Framework Using Different Features

As mentioned in Sect. 2, our framework improves the simulation model by repeatedly comparing the same real event log with different simulated event logs obtained via more and more accurate simulation models. Since the control-flow of the simulation models (the activities, events, gateways, etc.) is tightly correlated to the temporal features (e.g., the time elapsed between the execution of two non-consecutive activities), we separately employ the framework using control-flow and temporal features.

The framework is firstly applied to consider the *Basic* and *Extended features for Declare rules* (cf. Sects. 4.1 and 4.2). Subsequent iterations are carried out to improve the simulation model, until a decision tree is constructed that shows an accurate model (namely the tree is unable to distinguish the traces of the simulation log from those of the real event log), as in Fig. 5. Then, we reapply the framework, focusing on the temporal features discussed in Sect. 4.3. Section 6 discusses a case-study assessment where we indeed employ the framework where we first focus on the control flow of the simulation model, and then on the temporal features.

4.5   Feature Selection and Normalization

The multiset $T_{\mathcal{L}_r, \mathcal{L}_s}$ obtained from the Eq. 1 encoding basic, extended and/or temporal features is used to train a decision tree model. As last step beforehand, we preprocess $T_{\mathcal{L}_r, \mathcal{L}_s}$ to improve the quality of the trained decision tree. Initially, we perform feature selection and remove those features do not show sufficient variability. Features over discrete and numerical domains (e.g., the number of occurrence of activities in a trace) are filtered out if more than $X\%$ of the set elements take on the same value, where X is customizable and typically is around 90%. Any feature defined over continuous domains (e.g., temporal features) is removed if the mean and standard deviation considered for the multiset elements related to the real event-log traces are distant less than $(1 - X)\%$ from the mean and standard deviations for the elements related to the simulated event-log traces, with X customizable. Then, we normalize the values of the features of the elements of the multiset, using a traditional z-score normalization [2]: the values are transformed into distribution with a mean of 0 and a standard deviation of 1, i.e., a common scale. For each value, we subtract the mean value and divide it by the standard deviation of the respective feature. Normalization may be useful when learning a decision tree because the learning algorithms tend to give more weight and importance to features characterized by larger values.

5   Discussion

Our technique aims at a business simulation model that is able to generate traces that exhibit all and only the behavioral characteristics of the traces of the real event log. The behavioral characteristics of interest can be customized, such as with those in Sect. 4. Note that, however, the real event log does not need to contain every potential trace, but it is enough to just contain traces that exhibit these behavioral characteristics. It follows that the simulation model does not need to generalize beyond the real event log. This explains why we use the entire feature vector multiset for training: the tree needs to just classify the traces in the real and simulated event log, and can ignore potential future traces that exhibit characteristics not observed in the real and simulated event log. We however acknowledge the importance to have simulated models that generalize beyond the real event log, and we aim to investigate the generalization question as future work.

Training on the entire dataset might lead to an overfitting decision-tree model, namely with an excessive number of nodes and with leaves associated to single traces. We implemented decision-tree pruning by limiting the maximum depth of the tree and the maximum number of leaves, in order to mitigate overfitting. Pruning also improves the decision-tree clarity and readability.

Recall that we aim at a decision tree that cannot distinguish traces of the two event logs. Considering the real and simulated event log have the same number of traces, the most favourable tree is such that each leaf is associated to the same number of real and simulated log traces. This leads to a metrics for log similarity that considers the weighted average of the distribution of the classes in the leaves. The metric can take on values between 0 and 1. Value 0 means that the decision tree is able to distinguish each trace of the real-life event log from each trace of the simulated one, i.e., the simulation

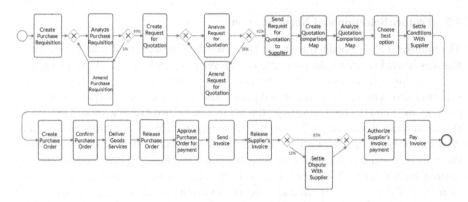

Fig. 6. The initial BPMN model for the simulation of an order purchase process [5], used in our case study.

model does not reflect the real process behavior. In the opposite case, Value 1 means that the two event logs are indistinguishable.

Definition 10 (Log Similarity). *Let* $B = \{\beta_1, \ldots, \beta_n\}$ *the set of all leaves in the decision tree model* $\mathcal{DT}$. *Let be* $\eta_r \in T_{\mathcal{L}_r, \mathcal{L}_s}$ *instances of multiset with the class label equal to* real *and* $\eta_s \in T_{\mathcal{L}_r, \mathcal{L}_s}$ *instances with* sim *as class label. For each* $\beta_i \in B$, m_{β_i} *is the total number of instances in* β_i, *of which* η_{r_i} *and* η_{s_i} *are the percentage of instances with* real *and* sim *label, respectively. The similarity of* $\mathcal{L}_r, \mathcal{L}_s$ *with respect the decision tree* $\mathcal{DT}$ *is the follow:*

$$Log\ Similarity(\mathcal{DT}^{\mathcal{L}_r, \mathcal{L}_s}) = \frac{(1-|\eta_{r_1}-\eta_{s_1}|)\cdot m_{\beta_1}+\ldots+(1-|\eta_{r_n}-\eta_{s_n}|)\cdot m_{\beta_n}}{m_{\beta_1}+\cdots+m_{\beta_n}}$$

Example 6. The log similarity related to the decision tree model in Fig. 3a is:

$$Log\ Similarity(\mathcal{DT}^{\mathcal{L}_r, \mathcal{L}_s}) = \frac{(1-|0.58-0.42|)\cdot 11679+(1-|0.37-0.63|)\cdot 8321}{11679+8321} = 0.80$$

Note how the above-defined metrics is tightly coupled with the typical decision-tree accuracy metrics: the higher is the log-similarity metrics, the lower is the accuracy. Indeed, less accuracy means that the decision tree has lower capabilities to classify traces. This implies that simulated and real logs are similar, and consequently the simulation model is capable to generate all and only the traces in the real event log.

6   Implementation and Experiments

Our approach has been implemented as a Python command-line tool.[3] Python language provides the main libraries and frameworks for process mining and machine learning. In particular, the libraries scikit-learn and PM4py are used to implement our approach.[4]

---

[3] https://github.com/francescameneghello/A-Framework-to-Improve-the-Accuracy-of-Process-Simulation-Models.git.

[4] scikit-learn: https://scikit-learn.org/stable/, PM4py: https://pm4py.fit.fraunhofer.de/.

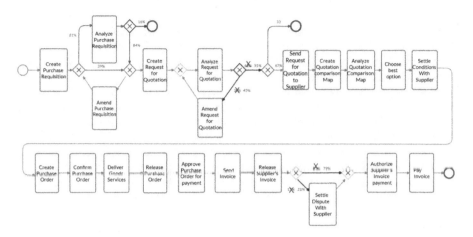

Fig. 7. The initial BPMN model for the simulation. The elements colored with green represent the changes derived from the first iteration, while those colored with purple are derived from the second iteration. The last, third iteration produced no change.

As input, the tool takes the real and the simulated event logs in XES format, together with the respective declare-constraints set in CSV format. These sets were discovered via MinerFul Miner [8]. The output of our implementation is a decision tree model such as those in Fig. 8. In order to test the applicability of our approach, we conducted a case study related to an order purchase process. We used a real event log and an existing accordant simulation model coming from literature by Camargo et al. [5].[5] The latter is composed of the BPMN model in Fig. 6 and several simulation parameters. Hence, we run this simulation model to obtain the simulated event log. In the remainder, we iteratively apply our technique to further improve the existing simulation model.

First Step: Control Flow Perspective. The first step of the improving iterative process is the analysis of the control-flow perspective using the *Basic Features* and the *Extended Features for Declare Rules*. Figure 8a shows the first comparison result a Log Similarity equal to 0.38, which illustrates the need to improve the simulation model. We leveraged on the rules discovered by the decision tree model in Fig. 8a. The CoExistence constraint in the root highlights that for 374 traces in the real log, the activities *Settle Conditions With Supplier* and *Analyze Purchase Requisition* do not coexist while the model requires both activities to occur. In fact, by analyzing the real log, we observed that the *Analyze Purchase Requisition* activity is not always executed. Hence, an exclusive gateway is introduced before this activity to make it optional (see Fig. 7). When both activities are performed, the decision tree pinpoints via the node of the AlternatePrecedence constraint that the activity *Release Supplier's Invoice* is more often preceded by *Analyze Request For Quotation* in the simulated log, compared with the real event log. This is actually caused by having more traces in the real event log where *Analyze*

---

[5] The event log is available at http://fluxicon.com/academic/material/ while the accordant simulation model is available at https://github.com/AdaptiveBProcess/Simod.

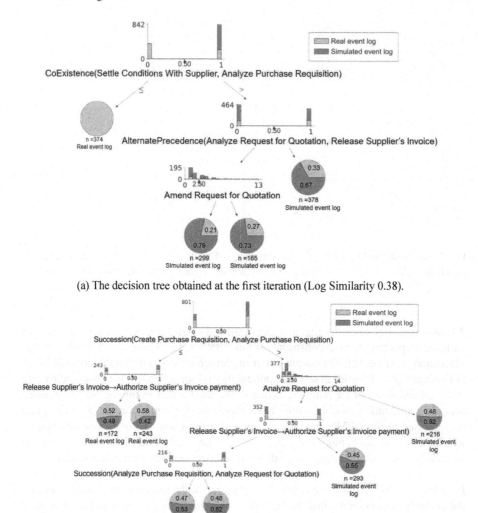

(a) The decision tree obtained at the first iteration (Log Similarity 0.38).

(b) The final decision tree obtained at the third and last iteration (Log Similarity 0.91).

Fig. 8. Decision trees generated at the first and last iteration of the framework, using the *Basic Features* and the DECLARE features.

Request For Quotation is not followed by the other activity. This might trigger at a first glance to make *Release Supplier's Invoice* optional via an exclusive gateway. We tried to do so, and we saw that an alternate-precedence constraint remained in the decision tree between *Analyze Request For Quotation* and any of the other activities following it (e.g., *Send Invoice*). We thus concluded that the real process allows for termination after *Analyze Request For Quotation*: therefore, we added an exclusive gateway before

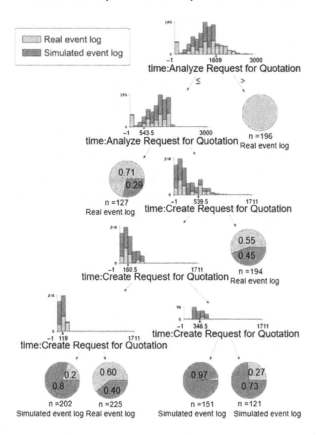

Fig. 9. Decision trees are generated at the first iteration of the second step of our iterative framework, using the *Activity Duration Features.*

Send Request for Quotation to Supplier to the model, along with a BPMN's end event (see Fig. 7).

We iterated once more the procedure. Space limitation prevents us from showing the new decision tree and discussing it. The new iteration led to some changes in the routing probabilities, as shown with purple in Fig. 7. This second iteration was followed by a third associated with the decision tree in Fig. 8b: even if the decision tree contains a few branches, one can see that the distribution of traces at leaves is well balanced. Indeed, the Log Similarity is 0.91. So, we proceeded with the second phase regarding the analysis of the time perspective through the *Activity Duration Features.*

Second Step: Time Perspective. The first iteration of time analysis produced the decision tree in Fig. 9 with Log Similarity equal to 0.42, so we tried to improve the simulation model. The root's feature `time:AnalyzeRequestForQuotation` evidences that only in the real log the activity *Analyze Request For Quotation* takes more than 26 min to complete. The remaining nodes pinpoint several deviations also about the processing time of the activity *Create Request for Quotation*. Moreover, each node presents

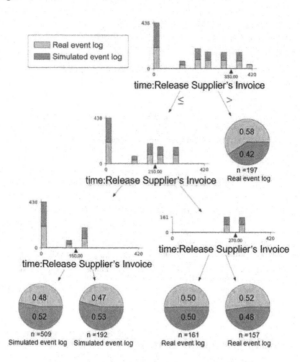

Fig. 10. Decision trees are generated at the last iteration of the second step of our iterative framework, using the *Activity Duration Features*.

a histogram that underlines the variations in the distribution of processing times for both logs. For the first activity, we changed the distribution of activity duration from a uniform distribution to a custom discrete distribution, that assigns a sample of points to the corresponding probability. For the second one, we retained the same normal distribution and we only adjusted the mean and standard deviation. After two iterations, we obtained the decision tree in Fig. 10 with Log Similarity equal to 0.94.

7 Related Works

Our technique is centered around comparing a real and a simulated event log to pinpoint common, behavioural differences. Several approaches exist for the comparison of two event logs from the same process [3,4,6,10,12]. Bolt et al. propose an approach where pairs of event logs are shown as automata and statistically-significant differences are highlighted through different colors [4]. Since transitions systems explicitly represent all the interleavings of execution of activities, differences are usually captured at low level. Low-level differences are also returned by the technique Nguyen et al. [10], where a differential graph of the differences between two event logs is constructed. As also highlighted in the evaluation presented in Taymouri's work [12], the two aforementioned research works yield an explosion of differences, which provide few actionable insights into how to improve a simulation model. In contrast, a decision tree model is able to detect the main differences between the logs, i.e., differences affecting a significant number of traces, and to return them in form of compact and high-level behavioral rules, which help process analysts to improve a simulation model. Beest et al. [3] present a method for diagnosing the differences between two event logs via natural language statements capturing behavior present in one log but not in other, but they only consider differences related to the ordering of activities and to branching probabilities. Taymouri et al. [12] proposed a hybrid machine learning approach for process variant analysis based on Discrete Wavelet Transformation (DWT). This approach uses a Support Vector Machine (SVM) to extract the features that provide enough information to discriminate the two process variations, and they are plotted using directly-follows graphs. The above-mentioned research stop at considering the *Basic Features* as introduced in Sect. 4.1, thereby ignoring DECLARE features, which allow one to capture and compare more complex behavioral patterns. Cecconi et al. [6] is the only work that consider DECLARE features when comparing two event logs. However, it does not consider the *Temporal Features*, nor does it attempt to filter out non-discriminating DECLARE rules. The latter also implies that a DECLARE miner may potentially mine redundant and inconsistent rules, which are subsequently returned (see [7]).

8 Conclusion

A successful application of process simulation to analyze and improve a process passes through a realistic simulation model, namely which accurately represents the potential real process executions. This enables analysts to improve the real process and not the supposed one. This paper has proposed a framework to assess and improve the accuracy of a simulated model to reflect the real behavior. The input is the simulation model and an event log that records real process executions. The simulation model is run to generate simulated event logs that are compared with the real log for differences. Differences are shown in form of a decision tree that classifies traces from the real event log and those from the simulated log. The tree provides actionable insights into how to modify the model to be more accurate. By repeatedly comparing with more and more accurate models and by using different behavioral dimensions of comparison, the framework aims to obtain an accurate simulation model which analysts can rely on.

The framework has been applied on a purchasing process, for which a simulation model and event log were available from literature. Our framework was able to further improve the accuracy of the simulation model, thus illustrating the benefits of the framework proposed.

There are multiple directions of future work. First, we want to investigate the question of simulation-model generalizability (cf. Sect. 5). Second, we aim to extend the set of decision tree features available in our operationalization of the framework, which now refers to control-flow and time: we also want to explicitly consider the resource, cost, and data perspectives. Second, we want to investigate how statistical approaches determine the number of traces in the event log: we presently simulate as many traces as the real event log, but it is possible to a small number would be statistically sufficient. Fourth, we want to extend the framework to provide concrete recommendations on how to modify the simulation model: the current framework focuses on providing insights, but the effort of transforming those insights into actual modifications is left to the process analysts.

References

1. van der Aalst, W.M.P., Pesic, M.: DecSerFlow: towards a truly declarative service flow language. In: WS-FM (2006)
2. Al Shalabi, L., Shaaban, Z., Kasasbeh, B.: Data mining: a preprocessing engine. J. Comput. Sci. **2**, 735–739 (2006)
3. van Beest, N.R.T.P., Dumas, M., García-Bañuelos, L., La Rosa, M.: Log delta analysis: interpretable differencing of business process event logs. In: Motahari-Nezhad, H.R., Recker, J., Weidlich, M. (eds.) BPM 2015. LNCS, vol. 9253, pp. 386–405. Springer, Cham (2015). https://doi.org/10.1007/978-3-319-23063-4_26
4. Bolt, A., de Leoni, M., van der Aalst, W.M.: Process variant comparison: using event logs to detect differences in behavior and business rules. Inf. Syst. **74**, 53–66 (2018)
5. Camargo, M., Dumas, M., González-Rojas, O.: Automated discovery of business process simulation models from event logs. Decis. Supp. Syst. **134**, 113284 (2020)
6. Cecconi, A., Augusto, A., Di Ciccio, C.: Detection of statistically significant differences between process variants through declarative rules. In: Polyvyanyy, A., Wynn, M.T., Van Looy, A., Reichert, M. (eds.) BPM 2021. LNBIP, vol. 427, pp. 73–91. Springer, Cham (2021). https://doi.org/10.1007/978-3-030-85440-9_5
7. Di Ciccio, C., Maggi, F.M., Montali, M., Mendling, J.: Resolving inconsistencies and redundancies in declarative process models. Inf. Syst. **64**, 425–446 (2017)
8. Di Ciccio, C., Mecella, M.: On the discovery of declarative control flows for artful processes. ACM Trans. Manag. Inf. Syst. (TMIS) **5**, 1–37 (2015)
9. Martin, N., Depaire, B., Caris, A.: The use of process mining in a business process simulation context: Overview and challenges. In: 2014 IEEE Symposium on Computational Intelligence and Data Mining (CIDM), pp. 381–388. IEEE (2014)
10. Nguyen, H., Dumas, M., La Rosa, M., ter Hofstede, A.H.M.: Multi-perspective comparison of business process variants based on event logs. In: Trujillo, J.C., et al. (eds.) ER 2018. LNCS, vol. 11157, pp. 449–459. Springer, Cham (2018). https://doi.org/10.1007/978-3-030-00847-5_32
11. Rozinat, A., Mans, R.S., Song, M., van der Aalst, W.M.: Discovering simulation models. Inf. Syst. **34**, 305–327 (2009)
12. Taymouri, F., La Rosa, M., Carmona, J.: Business process variant analysis based on mutual fingerprints of event logs. In: Dustdar, S., Yu, E., Salinesi, C., Rieu, D., Pant, V. (eds.) CAiSE 2020. LNCS, vol. 12127, pp. 299–318. Springer, Cham (2020). https://doi.org/10.1007/978-3-030-49435-3_19

Analyzing Process-Aware Information System Updates Using Digital Twins of Organizations

Gyunam Park[1]([✉]) [ID], Marco Comuzzi[2] [ID], and Wil M. P. van der Aalst[1] [ID]

[1] Process and Data Science Group (PADS), RWTH Aachen University,
Aachen, Germany
{gnpark,wvdaalst}@pads.rwth-aachen.de
[2] Department of Industrial Engineering, UNIST, Ulsan, Korea
mcomuzzi@unist.ac.kr

Abstract. Digital transformation often entails small-scale changes to information systems supporting the execution of business processes. These changes may increase the operational frictions in process execution, which decreases the process performance. The contributions in the literature providing support to the tracking and impact analysis of small-scale changes are limited in scope and functionality. In this paper, we use the recently developed Digital Twins of Organizations (DTOs) to assess the impact of (process-aware) information systems updates. More in detail, we model the updates using the configuration of DTOs and quantitatively assess different types of impacts of information system updates (structural, operational, and performance-related). We implemented a prototype of the proposed approach. Moreover, we discuss a case study involving a standard ERP procure-to-pay business process.

Keywords: Business process · System update · Impact · Digital twin

1 Introduction

Process-Aware Information Systems (PAIS), such as ERP and CRM, play a key role in modern organizations, underpinning the execution of business processes [8]. As the environment surrounding an organization dynamically changes and the competition becomes more intensive, a demand to accordingly change the implementation of PAIS may arise.

Modern digital transformation often involves frequent and small-scale changes, or *updates*, to information systems, which are vital to enable continuous adaptation to dynamic business environments [16]. The need for such capabilities is illustrated by the measures needed to handle the Covid-19 pandemic. For instance, organizations had to devise measures to adapt to varying degrees

This work is supported by the Alexander von Humboldt (AvH) Stiftung and the 0000 Project Fund (Project n. 1.220047.01) of UNIST (Ulsan National Institute of Science & Technology).

R. Guizzardi et al. (Eds.): RCIS 2022, LNBIP 446, pp. 159–176, 2022.
https://doi.org/10.1007/978-3-031-05760-1_10

of home-working workforce, or to variations of sales and orders that deviate extensively and unexpectedly from the usual seasonality.

PAIS updates affect the operations of an organization. However, due to the complex design and deep operational pervasiveness of modern PAIS, it is challenging to assess the impact of PAIS updates. Contributions in the PAIS literature in this direction are limited. Existing approaches mainly take a model-driven view on impact analysis [6,17,20], and concrete implementations are missing. Taking a design science standpoint [22], we address this research gap by developing (software) artifacts supporting the impact analysis of PAIS updates.

In this work, we distinguish specifically among three types of impacts: *structural*, *operational*, and *performance* impacts. At the structural level, it is important to assess the magnitude of the scope of a proposed update, such as the number of business objects and business functions that it affects. At the operational level, the impact of a PAIS update concerns the running instances of business processes involving the business objects and functions subject to change. For instance, removing the need to collect some customer information for privacy reasons may affect those customers for whom the information has been collected already. Most importantly, PAIS updates may impact on diagnostic measures of business processes: for instance, decreasing the level of tolerated mismatch between invoices and payments may slow down the processing of procurement cases, which in turn decreases the throughput of the procurement process.

To analyze the structural/operational/performance impact, we rely on Digital Twins of Organizations (DTOs). A DTO is a digital replication of an organization's operations, providing a transparent view to the business processes of the organization and enabling process analysts to analyze existing operational frictions and identify improvement opportunities. Moreover, DTOs allow to monitor business processes and trigger management actions, e.g., adding more resources and configuring business rules if improvements are available.

Specifically, we use a *Digital-Twin Interface Model* (DT-IM) [18] that (i) represents PAISs based on Object-Centric Petri Nets (OCPNs) [1] and (ii) models PAIS updates with the notion of *actions*. We compute the impact of a PAIS update with the DT-IM by analyzing the elements of the OCPN that it modifies (structural impacts), current states affected by the update (operational impacts), and the new value of diagnostic measures after the update (performance impacts).

From a design science standpoint, the practical relevance of the problem that we address lies with the increasing relevance of digital transformation for modern enterprises [16]. The rigor of the design process is ensured by extending a sound conceptual model of DTOs already published in the literature, i.e., DT-IMs. The design search process is inspired by the existing literature on ERP post-implementation change management [17], from which we draw ideas to model system updates and evaluate their impact. In this work, we limit the evaluation of the artifact to the feasibility of the proposed approach. To this aim, we have built a Web-based impact analysis software artifact and conducted a case study based on a procure-to-pay process loosely modeled following the standard one of the SAP ERP system.

The remainder is organized as follows. We discuss the related work in Sect. 2. Then, we present the preliminaries in Sect. 3. Next, we introduce the DT-IM in Sect. 4 and an approach to impact analysis based on the DT-IM in Sect. 5. Afterward, Sect. 6 introduces the implementation of a web application and a case study using the web application. Finally, Sect. 7 concludes the paper.

2 Related Work

Digital twins enabled by increasingly powerful data modeling and analysis capabilities have been envisioned by Gelernter in the 1990s [10] and have found widespread adoption in engineering in the last few years [9]. The idea of digital twins of organizations has emerged recently [4] as a means to address the challenges of information processing in modern digital transformation.

Mendling et al. [16] recently have highlighted an ongoing tension between business process management and digital transformation (DT): processes must be flexible to adapt to the continuously changing requirements of DT, while DT must rely on some process compliance to avoid a continuous disruption of business operations.

Business processes and the information systems supporting them can evolve to support changing business requirements through configuration and adaptation [7,13]. The former entails that all the possible evolution options are known a priori and can be captured into configuration tables, whereas adaptation involves ad-hoc modification of the process models or systems. Configuration is obviously more agile, but often it cannot address unexpected situations.

Business process flexibility can also be achieved through run-time adaptation. Along this direction, van Beest et al. [2] have proposed an approach for repairing processes at run-time when they interfere, e.g., when inconsistent writing operations occur. Marrella [15] have proposed to use automated planning techniques to support process adaptation to changing environments.

An issue closely related to process and system adaptability is the tracking of their evolution over time, understanding, in particular, the impact of proposed updates. In the context of cross-organizational information systems [6] or multi-tenant cloud systems [12], the evolution of systems and processes can be tracked by the evolution of Service Level Agreements (SLAs), which are updated by changing business requirements.

In the context of ERP systems, Soffer et al. [20,21] have proposed an ERP modeling language and a related methodology to align the ERP system capabilities with the enterprise requirements. The modeling language allows to express dependencies between business processes and objects, but it does not allow to express changes of them and their impact. Parhizkar and Comuzzi [5,17] have proposed a methodology and initial tool support to characterize the impact of ERP post-implementation changes inspired by the engineering change management literature. Lin et al. [14] have proposed a method that supports users during the execution of ERP post-implementation changes but which does not support the evaluation of the impact of such changes.

In summary, methods and tools to holistically support the tracking of information systems updates, understanding, in particular, their impact, remain
limited in scope, functionality, and degree of automation. This paper tackles
this research gap by proposing to use DTOs to develop tools that can (semi-
)automatically support the assessment of PAIS updates.

3   Background and Preliminaries

This section introduces a conceptual model of PAISs (Subsect. 3.1) and the
OCPNs, which we use later to formally model PAISs (Subsect. 3.2).

3.1   Process-Aware Information Systems (PAISs)

Figure 1 introduces a meta-model of PAIS entities, updates, and impacts of PAIS
updates considered in this work. This is inspired mainly by the work of Parhizkar
and Comuzzi [17] in the context of ERP systems.

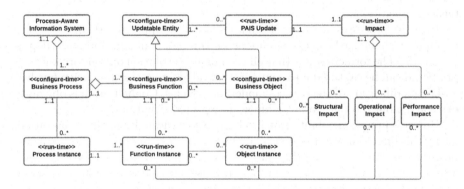

Fig. 1. A meta model of PAISs as a *UML 2.0 class diagram*

PAIS Entities. The meta model separates entities concerned with the
configure-time from entities related with the run-time. At configure-time, a PAIS
is constituted by a set of *business processes*, which in turn orchestrate a set of
business functions that manipulate, using CRUD operations, *business objects*.
At run-time, a PAIS instantiates instances of processes, functions, and objects.

PAIS Updates. A PAIS may be modified by creating, updating, or deleting
an existing entity, i.e., an object, function or process. In this work, we focus
on updates to business functions and objects. Creating a new entity has no
impact on the business operations since it extends the functionality of a system
and it only applies to new instances of business processes, functions, or objects.
Deleting an entity can be seen as a special case of updating it. Updates to
business processes, such as modifications of their control flow, are left out of
scope.

A PAIS update is associated with an updatable entity, i.e., business functions and objects. First, an *update of a business function* is achieved by changing business rules and functionalities defining it. For instance, a *place order* function in an ERP system can be updated by changing the minimum price to create orders (i.e., a business rule) or changing a set of attributes that the function will update in a business object (i.e., a functionality). Such updates not only affect the business function, but also instances of the business function.

Second, an *update of a business object* is achieved by adding or removing attributes to it. For instance, the *order* business object can be updated by adding *payment terms* to specify the conditions of the payment, e.g., received by the end of the month, within seven days or through monthly installments. The update affects both business objects and their instances.

Impacts of PAIS Updates. A PAIS update results in an impact to the system. Below are the three types of impacts that we consider.

- *Structural impacts* concern configure-time entities, i.e., business objects and functions. They include:
    - *structural object impact*: the impact on business objects, e.g., the number of impacted business objects, and
    - *structural function impact*: the impact on business functions, e.g., the number of impacted business functions.
- *Operational impacts* concern run-time entities, i.e., the instances of business objects and functions. They include:
    - *operational object impact*: the impact on object instances of business objects, e.g., the number of object instances of impacted business objects, and
    - *operational function impact*: the impact on object instances of business functions, e.g., the number of objects of impacted business functions.
- *Performance impacts* concern changes in diagnostic measures of business objects and functions. They include:
    - *object performance impact*: the performance impact on business objects, e.g., difference in the avg. service time for a business object before/after updates, and
    - *function performance impact*: the performance impact on business functions, e.g., difference in the avg. waiting time for a business function before/after updates.

3.2   Object-Centric Petri Nets (OCPNs)

In this work, we model a PAIS introduced in Subsect. 3.1 with a DT-IM that uses an OCPN as its core formalism. This subsection introduces OCPNs. First, a Petri net is a directed bipartite graph of places and transitions. A labeled Petri net is a Petri net with the transitions labeled.

Definition 1 (Labeled Petri Net).   *Let* $\mathbb{U}_{act}$ *be the universe of activity names. A labeled Petri net is a tuple* $N=(P,T,F,l)$ *with* P *the set of places,* T *the set of transitions,* $P \cap T = \emptyset$, $F \subseteq (P \times T) \cup (T \times P)$ *the flow relation, and* $l \in T \nrightarrow \mathbb{U}_{act}$ *a labeling function.*

A marking $M_N \in \mathcal{B}(P)$ is a multiset of places. A transition $tr \in T$ is *enabled* in marking M_N if its input places contain at least one token. The enabled transition may *fire* by removing one token from each of the input places and producing one token in each of the output places.

Each place in an OCPN is associated with an object type to represent interactions among different object types. The variable arcs represent the consumption/production of a variable amount of tokens in one step.

Definition 2 (Object-Centric Petri Net). *Let $\mathbb{U}_{ot}$ be the universe of object types. An object-centric Petri net is a tuple $ON=(N, pt, F_{var})$ where $N=(P, T, F, l)$ is a labeled Petri net, $pt \in P \to \mathbb{U}_{ot}$ maps places to object types, and $F_{var} \subseteq F$ is the subset of variable arcs.*

Figure 2 shows an OCPN describing a part of the peer review process for an academic conference. There are two types of places associated with two object types, i.e., *conf* (denoted in red) and *subm* (cyan). For instance, $pt(p1)=pt(p2)=conf$, $pt(p3)=pt(p5) = subm$, and $(p3, t2)$, $(t2, p5)$, and $(p7, t5)$ are variable arcs.

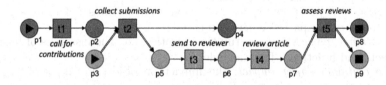

Fig. 2. An example of object-centric Petri nets

Definition 3 (Marking). *Let $\mathbb{U}_{oi}$ be the universe of object identifiers. Let $ON=(N, pt, F_{var})$ be an object-centric Petri net, where $N=(P, T, F, l)$. $Q_{ON}=\{(p, oi) \in P \times \mathbb{U}_{oi} \mid type(oi)=pt(p)\}$ is the set of possible tokens. A marking M of ON is a multiset of tokens, i.e., $M \in \mathcal{B}(Q_{ON})$.*

For instance, marking $M_1=[(p4, c1), (p7, s1), (p7, s2)]$ denotes three tokens where place $p4$ has one token referring to object $c1$ of type *conf* and $p7$ has two tokens referring to objects $s1$ and $s2$ of type *subm*.

A binding describes the execution of a transition consuming objects from its input places and producing objects for its output places. A binding (tr, b) is a tuple containing a transition tr and a function b mapping the object types of the input/output places to sets of object identifiers. For instance, $(t5, b1)$ describes the execution of transition $t5$ with $b1$ where $b1(conf)=\{c1\}$ and $b1(subm)=\{s1, s2\}$.

A binding (tr, b) is *enabled* in marking M if all the objects specified by b exist in the input places of tr. For instance, $(t5, b1)$ is enabled in marking M_1 since $c1$, $s1$, and $s2$ exist in its input places, i.e., $p4$ and $p7$.

A new marking M' is reached by executing an enabled binding (tr, b) at marking M, denoted by $M \xrightarrow{(tr,b)} M'$. For instance, by executing $(t5, b1)$, $c1$ is removed from $p4$ and added to $p8$, while $s1$ and $s2$ are removed from $p7$ and added to $p9$, leading to the new marking $M'=[(p8,c1),(p9,s1),(p9,s2)]$.

Finally, a relation function $rel \in T \to \mathcal{P}(P)$ maps a transition to a set of places associated with the transition. It is defined recursively: for any $tr \in T$, (1) $rel(tr)=\emptyset$ if $\bullet tr=\emptyset$ and (2) $rel(tr)= \bullet tr \cup \bigcup_{p \in \bullet tr, tr' \in \bullet p} rel(tr')$, where $\bullet tr$ is a set of input places of tr, i.e., $\bullet tr=\{p \in P \mid (p,tr) \in F\}$. For instance, $rel(t2)=\{p1,p2,p3\}$ and $rel(t5)=\{p1,\ldots,p7\}$.

4   Modeling PAISs: Digital Twin Interface Model

In this work, we use a *digital twin interface model* to model PAIS entities along with PAIS updates introduced in Subsect. 3.1.

4.1   Modeling PAIS Entities

A digital twin interface model consists of 1) an OCPN, 2) valves, 3) guards, and 4) operations. A *valve* is a system configuration, e.g., minimum required quantity to place orders. A *guard* is a formula composed of attributes, including valves, with relational operators (e.g., $\leq, \geq, =$) and logical operators (e.g., conjunction $\wedge$, disjunction $\vee$, and negation $\neg$). $F(X)$ denotes the set of such formulas defined over a set of attributes X. An *operation* describes a business operation, e.g., updating the quantity and price of an order.

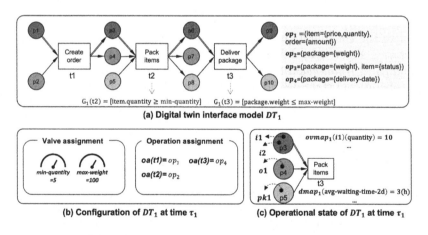

Fig. 3. An example of digital twin interface models

Definition 4 (Digital Twin Interface Model (DT-IM)). *Let $\mathbb{U}_{valve}$ be the universe of valve names. Let $\mathbb{U}_{attr}$ be the universe of attribute names. A digital twin interface model, denoted as DT, is a tuple (ON, V, A, G, O) where*

- *$ON=(N, pt, F_{var})$ is an object-centric Petri net, where $N=(P,T,F,l)$,*
- *$V \subseteq \mathbb{U}_{valve}$ is a set of valve names,*

- $A \subseteq \mathbb{U}_{attr}$ is a set of attribute names,
- $G \in T \to (F(V \cup A) \cup \{true\})$ associates transitions with guards, and
- $O \subseteq OT \nrightarrow \mathcal{P}(A)$ is a set of operations associating an object type to a set of attributes to be updated, where $OT=\{pt(p) \mid p \in P\}$.

The transitions in the OCPN of a DT-IM represent *business functions*, whereas the object types associated with places indicate *business objects*. Guards and operations represent the business rule(s) and functionality (writing operations) of *business functions*, respectively. Figure 3(a) shows an DT-IM, $DT_1=(ON_1, V_1, A_1, G_1, O_1)$, representing a PAIS supporting a simple order management process. It involves three object types: *item*, *order*, and *package*. First, an order is created with multiple items. Next, the item is packed and the packaged is delivered to the customer. Valves are depicted using red italic fonts, and guards are described in squared brackets. Operations are described in the gray box. For instance, $op_2 \in O_1$ describes a business operation writing *weight* of *package*, i.e., $op_2(package)=\{weight\}$, $op_2(item)= \perp$, and $op_2(order) = \perp$.

A configuration defines the semantics of a DT-IM by determining the value of valves and the assignment of operations to transitions.

Definition 5 (Configuration). *Let $\mathbb{U}_{val}$ be the universe of attribute values. Let $DT=(ON, V, A, G, O)$ be a DT-IM with $ON=(N, pt, F_{var})$ and $N=(P, T, F, l)$. A configuration $conf_{DT} \in (V \to \mathbb{U}_{val}) \times (T \to O)$ is a tuple of a valve assignment va and an operation assignment oa. We denote $\Sigma_{DT}=(V \to \mathbb{U}_{val}) \times (T \to O)$ to be the set of all possible configurations of DT.*

Given $conf_{DT}=(va, oa) \in \Sigma_{DT}$, $\pi_{va}(conf)=va$ and $\pi_{oa}(conf)=oa$. Moreover, we denote the configuration of digital twin interface model DT at $\tau \in \mathbb{U}_{time}$ as $conf_{DT,\tau}$. Figure 3(b) describes a configuration of DT_1 at $\tau_1 \in \mathbb{U}_{time}$, where $\pi_{va}(conf_{DT_1,\tau_1})(min\text{-}quantity)=5$, $\pi_{oa}(conf_{DT_1,\tau_1})(t2)=op_2$, etc.

An operational state describes the current status of a business process, i.e., which objects reside in which parts of the process, using the marking of OCPNs. Moreover, it represents the various diagnostics about the performance of the process, e.g., the average waiting time of activity in the last seven days. We denote Δ_{DT} to be the set of all possible diagnostics of the digital twin interface model DT.

Definition 6 (Operational State of A DT-IM). *Let $\mathbb{U}_{vmap}=\mathbb{U}_{attr} \nrightarrow \mathbb{U}_{val}$ be the set of all partial functions mapping a subset of attribute names to values. Let $DT=(ON, V, A, G, O)$ be a DT-IM with $ON=(N, pt, F_{var})$ and $N=(P, T, F, l)$. An operational state of DT is a tuple $os_{DT}=(M, ovmap, dmap)$ where*

- $M \in \mathcal{B}(Q_{ON})$ *is a marking of ON,*
- $ovmap \in OI \to \mathbb{U}_{vmap}$ *is an object value assignment where $OI=\{oi \in \mathbb{U}_{oi} \mid \exists_{p \in P} (p, oi) \in M\}$, and*
- $dmap \in \Delta_{DT} \nrightarrow \mathbb{R}$ *is a diagnostics assignment such that, for any $diag \in \Delta_{DT}$, $dmap(diag) = \perp$ if $diag \notin dom(dmap)$.*

Given $os_{DT}=(M, ovmap, dmap)$, $\pi_M(os_{DT})=M$, $\pi_{ovmap}(os_{DT})=ovmap$, and $\pi_{dmap}(os_{DT})=dmap$. Figure 3(c) describes an operational state of DT_1 at $\tau_1 \in \mathbb{U}_{time}$, i.e., $os^1_{DT_1}=(M_1, ovmap_1, dmap_1)$. In the remainder, we denote the operational state of the digital twin interface model DT at time τ as $os_{DT,\tau}$.

We define the semantics of DT-IM by extending the semantics of OCPNs with configurations and operational states. To this end, we use the notion of digital twin bindings.

Definition 7 (Digital Twin Binding). *Let $DT=(ON, V, A, G, O)$ be a DT-IM with $ON=(N, pt, F_{var})$. A digital twin binding of DT is a tuple $((tr, b), w, \tau)$ where (tr, b) is a binding of ON, $w \in \mathbb{U}_{ot} \to \mathcal{P}(\mathbb{U}_{attr})$ is a write function, and $\tau \in \mathbb{U}_{time}$ is a timestamp. A digital twin binding $((tr, b), w, \tau)$ is enabled with $conf_{DT,\tau}$ and $os_{DT,\tau}$ if the following conditions are satisfied:*

- *(tr, b) is enabled at $\pi_M(os_{DT,\tau})$,*
- *guard $G(tr)$ evaluates to true w.r.t. valve assignment $\pi_{va}(conf_{DT,\tau})$ and object value assignment $\pi_{ovmap}(os_{DT,\tau})$, and*
- *w corresponds to the assigned operation of tr, i.e., $w=\pi_{oa}(conf_{DT,\tau})(tr)$.*

Digital twin binding $((t2, b), w, \tau_1)$, where $b(item)=\{i1, i2\}$, $b(order)=\{o1\}$, $b(package)=\{pk1\}$, and $w(package)=\{delivery\text{-}date\}$, is enabled with the configuration of Fig. 3(b) and the operational state of Fig. 3(c) since $(t2, b)$ is enabled at $\pi_M(os_{DT_1,\tau_1})$, $G_1(t2)$ evaluates to true, and w corresponds to $\pi_{oa}(conf_{DT_1,\tau_1})(t2)$.

4.2   Modeling PAIS Updates

Next, we model the PAIS updates introduced in Subsect. 3.1 using the notion of *actions* in DT-IMs. An action updates the configuration of a DT-IM. First, updating valve assignments of the configuration corresponds to the update of business functions, e.g., updating the value of *min-quantity* changes the business rule of the function *pack items*. Second, updating operation assignments of the configuration corresponds to both 1) the update of business functions, e.g., updating the operation of *pack items* from op_2 to op_3 changes the functionality of it (by updating the attribute *status* of items in addition to *weight* of packages), and 2) the update of business objects (by modifying items to have a new attribute *status*).

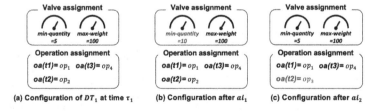

(a) Configuration of DT_1 at time τ_1 (b) Configuration after ai_1 (c) Configuration after ai_2

Fig. 4. A configuration of DT_1 at τ_1 and new configurations after applying ai_1 and ai_2

Definition 8 (Action). *Let DT be a DT-IM. An action $act \in \Sigma_{DT} \to \Sigma_{DT}$ updates the configuration. A_{DT} is the set of all possible actions defined over DT.*

An action instance describes the application of an action. An action is applied at a certain start time and, in principle, the configuration change that it entails can remain in place for the foreseeable future until a condition, e.g., on performance metrics, is met or until a specific end time. For simplicity, in this work, we consider only the latter case.

Definition 9 (Action Instance). *Let DT be a DT-IM. An action instance $ai \in A_{DT} \times \mathbb{U}_{time} \times \mathbb{U}_{time}$ is a tuple of an action, start timestamp, and end timestamp. AI_{DT} is the set of all possible action instances defined over DT.*

For instance, $ai_1 = (act_1, \tau_1, \tau_2) \in AI_{DT_1}$ describes the application of $act_1 \in A_{DT_1}$ to the configuration of DT_1 at τ_1 (Fig. 4(a)) leading to the configuration depicted in Fig. 4(b) until τ_2. $ai_2 = (act_2, \tau_1, \tau_3) \in AI_{DT_1}$ describes the application of $act_2 \in A_{DT_1}$ to to the configuration of DT_1 at τ_1 (Fig. 4(a)), producing the configuration shown in Fig. 4(c) until τ_3.

The application of actions results in different *effective changes*, depending on the configuration at the start of the action instance. An effective change of an action instance denotes the valves and transitions whose values and activity assignments are changed due to the action.

Definition 10 (Effective Change). *Let DT be a DT-IM and $ai = (act, st, ct) \in AI_{DT}$ an action instance. An effective change of ai is a tuple of a set of valves and a set of transitions, i.e., $\delta_{ai} = (V_c, T_c)$ with $V_c = \{v \in V \mid \pi_{va}(conf_{DT,st})(v) \neq \pi_{va}(act(conf_{DT,st}))(v)\}$ and $T_c = \{tr \in T \mid \pi_{oa}(conf_{DT,st})(tr) \neq \pi_{oa}(act(conf_{DT,st}))(tr)\}$).*

As noted in Fig. 4(b) and (c) with red fonts, the effective change by ai_1 is valve *min-quantity*, i.e., $\delta_{ai_1} = (\{min\text{-}quantity\}, \emptyset)$, and the effective change by ai_2 is the operation assignment of $t2$, i.e., $\delta_{ai_2} = (\emptyset, \{t2\})$.

5 Impact Analysis

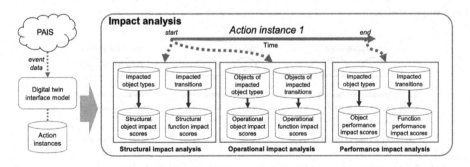

Fig. 5. An overview of the proposed impact analysis using digital twin interface models

This section introduces an approach to impact analysis of PAIS updates based on digital twin interface models. Figure 5 shows an overview of the proposed approach consisting of three components: structural/operational/performance impact analysis. Using a DT-IM representing a target PAIS, we analyze structural and operational impacts of an action instance, i.e., PAIS updates, at its execution and performance impacts at its completion.

5.1 Structural Impact Analysis

Structural impact analysis identifies the *structural object/function impacts* of an action instance. To this end, we identify the object types, i.e., business objects, and transitions, i.e., business functions, affected by an action instance. First, an object type is considered to be impacted if the operations newly assigned by an action instance introduce new attributes to update or remove existing attributes for the object type.

Definition 11 (Impacted Object Types). *Let DT be a DT-IM and $ai=(act, st, ct) \in AI_{DT}$ an action instance. $IOT_{DT}(ai) \subseteq \mathbb{U}_{ot}$ denotes the set of the object types impacted by ai, i.e., $IOT_{DT}(ai)=\{ot \in \mathbb{U}_{ot} \mid \delta_{ai}=(V_c, T_c) \wedge tr \in T_c \wedge \pi_{oa}(conf_{DT,st})(tr)(ot) \triangle \pi_{oa}(act(conf_{DT,st}))(tr)(ot) \neq \emptyset\}\}$, where $\triangle$ denotes a symmetric difference of sets.*

For instance, $IOT_{DT_1}(ai_2)=\{item\}$ since the effective change of ai_2 (i.e., $(\emptyset, \{t2\})$) in the operation assignment of $t2$ introduces the new attribute *status* of *item*, i.e., $\pi_{oa}(conf_{DT_1,\tau_1})(t2)(item) \triangle \pi_{oa}(act_2(conf_{DT_1,\tau_1}))(t2)(item) \neq \emptyset$ (i.e., $\{status\} \triangle \emptyset \neq \emptyset$).

Next, the transitions of a DT-IM are considered to be impacted if they are associated with the effective change in valve assignments and operation assignments. First, changes in the valve assignment influence transitions by changing the meaning of the guard associated with them, e.g., changing the valve *min-quality* affects the guard of *pack items*. Second, changes in the operation assignment affect transitions by changing their functionality.

Definition 12 (Impacted Transitions). *Let $DT=(ON, V, A, G, O)$ be a DT-IM with $ON=(N, pt, F_{var})$. Let $ai \in AI_{DT}$ be an action instance. $IT_{DT}(ai) \subseteq T$ denotes the set of the transitions impacted by ai, i.e., $IT_{DT}(ai)=\{tr \in T \mid \delta_{ai}=(V_c, T_c) \wedge ((\exists_{v \in V_c} v \in G(tr)) \vee tr \in T_c)\}$.*

For instance, $IT_{DT_1}(ai_1)=\{t2\}$ since the effective change of ai_1 (i.e., $(\{mi n\text{-}quantity\}, \emptyset)$) in valve *min-quantity* affects the guard associated to $t2$. Moreover, $IT_{DT_1}(ai_2)=\{t2\}$ since the effective change by ai_2 (i.e., $\delta_{ai_2}=(\emptyset, \{t2\})$) change the functionality of $t2$.

Once the impacted object types/transitions are identified, various structural object/function impact scores can be measured. In this work, we focus on basic count-based measures, e.g., how many object types and transitions are impacted. These measures can be absolute or relative, i.e., normalized by the total number of respective entities.

Additional measures can be obtained by applying filtering or prioritizing to count-based measures. For instance, by filtering *financial* objects, such as *invoice*, we can measure the absolute/relative impact on objects that are relevant for the finance department. Prioritizing refers to weighting differently the impact on different types of entities. For instance, a higher weight can be given to the impacts on *verification*-related activities in a process.

5.2 Operational Impact Analysis

Operational impact analysis aims to analyze *operational object/function impacts* of an action instance. To that end, we identify the existing objects of the impacted object types (for the former) and the objects related to the impacted transitions (for the latter) in a DT-IM. First, to identify the existing objects of the impacted object types, we use markings from the operational states of the DT-IM.

Definition 13 (Objects of Impacted Object Types). *Let* $DT=(ON,V, A,G,O)$ *be a DT-IM with* $ON=(N,pt,F_{var})$. *Let* $ai=(act,st,ct) \in AI_{DT}$ *be an action instance and* $IOT_{DT}(ai)$ *the impacted object types by* ai. $\widehat{IOT}_{DT}(ai) \subseteq \mathbb{U}_{oi}$ *denotes the set of objects of* $IOT_{DT}(ai)$, *i.e.,* $\widehat{IOT}_{DT}(ai)=\{oi \in \mathbb{U}_{oi} \mid p \in dom(pt) \land pt(p) \in IOT_{DT}(ai) \land (p,oi) \in \pi_M(os_{DT,st})\}$.

For instance, $\widehat{IOT}_{DT_1}(ai_2)$ is a set of objects associated with all tokens in *item* places, i.e., $i1, i2, \dots$ of marking $\pi_M(os_{DT_1,\tau_1})$.

Next, we identify objects related to impacted transitions. An object is related to a transition if it may perform the transition in the future.

Definition 14 (Objects of Impacted Transitions). *Let* DT *be a DT-IM. Let* $ai=(act,st,ct) \in AI_{DT}$ *be an action instance and* $IT_{DT}(ai)$ *the impacted transitions by* ai. $\widehat{IT}_{DT}(ai) \subseteq \mathbb{U}_{oi}$ *denotes the set of objects of* $IT_{DT}(ai)$, *i.e.,* $\widehat{IT}_{DT}(ai)=\{oi \in \mathbb{U}_{oi} \mid tr \in IT_{DT}(ai) \land p \in rel(tr) \land (p,oi) \in \pi_M(os_{DT,st})\}$.

For instance, $\widehat{IT}_{DT_1}(ai_1)$ is a set of objects associated with all tokens in $rel(t2)=\{i1, i2, o1, pk1\}$ of marking $\pi_M(os_{DT_1,\tau_1})$.

Based on the objects of impacted object types/transitions, we measure operational object/function impact scores. As for the operational impact analysis, in this work, we focus on basic count-based measures, e.g., how many objects of impacted object types/transitions are impacted by an update.

Also in this case, we can apply filtering or prioritizing to define new measures. For instance, objects can be filtered based on the value of specific attributes or the stage of their lifecycle, e.g., orders higher than a certain amount or from premium customers, or objects for which payments have been cleared. Regarding objects of impacted transitions, we can filter or prioritize objects that lie directly in the queue for the impacted transition, e.g., giving a higher weight to objects currently waiting for the impacted transition.

5.3   Performance Impact Analysis

Performance impact analysis aims at analyzing *object/function performance impacts*. First, to analyze the former, we compare diagnostics related to impacted object types before and after applying an action instance.

We define the object performance impact analysis as follows. Let $ai=(act, st, ct)$ be an action instance and $iot \in IOT_{DT}(ai)$ an impacted object type. Let $diag_{iot} \in \Delta_{DT}$ be a diagnostics relevant to iot, e.g., the average total service time of iot. We measure the performance impact of ai on iot w.r.t. $diag_{iot}$ as follows: $\pi_{dmap}(os_{DT,ct})(diag_{iot}) - \pi_{dmap}(os_{DT,st})(diag_{iot})$.

Next, we analyze function performance impacts by comparing diagnostics associated with impacted transitions before and after applying an action instance. Examples of relevant diagnostics are the average service time of the transition, or the average waiting time of the transition.

We formally define the function performance impact analysis as follows. Let $ai=(act, st, ct)$ be an action instance and $it \in IT_{DT}(ai)$ an impacted transition. Let $diag_{it} \in \Delta_{DT}$ be a diagnostics relevant to it, e.g., the average service time of it. We measure the performance impact of ai on it w.r.t. $diag_{it}$ as follows: $\pi_{dmap}(os_{DT,ct})(diag_{it}) - \pi_{dmap}(os_{DT,st})(diag_{it})$.

In this work we consider general purpose diagnostics, such as the average total service time of impacted object types, or the average total waiting time of impacted object types [19]. Other diagnostics can be defined applying the filtering and prioritizing principles introduced earlier, e.g., considering the average total waiting time of objects of a certain type, or giving more weight to the waiting time of certain types of objects. Diagnostics can also be defined on a domain-specific basis, e.g., process-specific KPIs.

6   Evaluation

This section presents the implementation of the approach presented in this paper and evaluates its feasibility by applying it to a simulated PAIS.

6.1   Implementation

We have implemented a cloud-based Web service to support the impact analysis with a dedicated user interface. Sources, manuals, and a demo video are available at https://github.com/gyunamister/impacta. The service comprises the following four functional components:

Designing DT-IMs. This component supports the design of DT-IMs based on event data and domain knowledge. The input is event data of the standard OCEL [11], valves, guards, and operations in a JSON-based format. The event data are used to discover an OCPN using the technique introduced in [1], and valves, guards, and operations enhance the discovered OCPN, completing the design of a DT-IM.

Updating Configurations and Operational States. This component updates the configuration and operational state of a DT-IM in sync with the updates in a target PAIS. To this end, it is connected with the PAIS, specifying: 1) the source of the current setting of the PAIS and 2) the source of the streaming event data from the PAIS. Using the current setting, the configuration of the DT-IM is updated. Then, the operational state is updated by replaying the streaming event data using the token-based replay technique described in [3].

Defining and Executing Actions. The goal of this component is to 1) define actions based on the available valves and operations and 2) instantiate them as action instances by specifying start and completion times. To this end, the service provides visual information to support the definition of actions and action instances. Once executed, an action instance changes the configuration setting of the system and, accordingly, the configuration of the DT-IM.

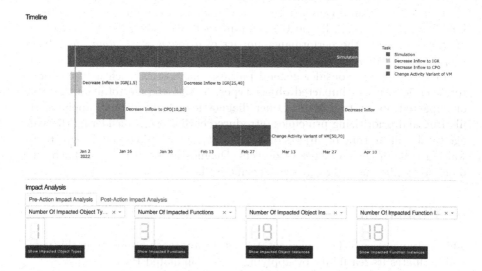

Fig. 6. Screenshot of the implementation (Color figure online)

Analyzing Structural/Operational/Performance Impacts. This component evaluates the impact of action instances. Figure 6 shows a screenshot of the Web service's interface. The timeline in the upper part shows the current status, including the current timestamp (yellow vertical line) and the overview of the action instances. Specifically, five action instances (horizontal bars) of three different actions (distinguished by colors) are scheduled to be executed. For instance, the first action instance named *Decrease Inflow to IGR* is effective from 26-12-2021 to 31-12-2021. Note that we compute structural/operational impacts at the start of action instances and performance impacts at the end of them.

6.2   Case Study

Using the implementation, we have conducted a case study on an artificial PAIS that supports a procure-to-pay process. The system is developed to reflect a real-life SAP ERP system supporting a procure-to-pay processes by using the same business objects, functions, and rules found in SAP. Using the artificial PAIS, we simulate the procure-to-pay process with 24 resources with different capacities and performance. Purchase orders are created by following the exponential distribution; the business hours are set as 9–17 from Monday to Friday; work assignments are scheduled using the *First-in-First-out* rule.

Figure 7(a) shows the DT-IM representing the PAIS (DT_{p2p}). The process involves five object types. First, a purchase requisition is created with multiple materials. Next, a purchase order is created based on the purchase requisition and material. A goods receipt is produced after receiving the materials of the purchase order. Afterward, the material is verified and issued for various purposes, and concurrently the invoice for the purchase order is received. Finally, the invoice is cleared. Figure 7(b) shows the default configuration of the system.

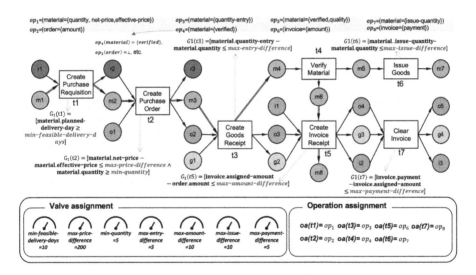

Fig. 7. A digital twin interface model of the PAIS supporting a procure-to-payment process and its configuration

Using the DT-IM and configuration, we define actions, A_1 and A_2, as follows:

- A_1 increases valve *min-quantity* to 10 to reduce the inflow to *create purchase order*, i.e., for any $conf_{DT_{p2p}} \in \Sigma_{DT_{p2p}}$, $A_1(conf_{DT_{p2p}})=conf_{DT_{p2p}}^{A_1}$ such that $\pi_{va}(conf_{DT_{p2p}}^{A_1})(min\text{-}quantity)=10$, and
- A_2 changes the operation of *verify materials* to op_5 to additionally update *quality* of materials, i.e., for any $conf_{DT_{p2p}} \in \Sigma_{DT_{p2p}}$, $A_2(conf_{DT_{p2p}})=conf_{DT_{p2p}}^{A_2}$ such that $\pi_{oa}(conf_{DT_{p2p}}^{A_2})(t4)=op_5$, where $op_5(material)=\{verified, quality\}$.

Using the actions, we define the following action instances: $AI_1=(A_1,1,5)$ and $AI_2=(A_2,10,15)$. Note that, for the ease of the simulation, we abstract the timestamp to time steps each of which has the scale of 24 h, starting from 09:00 20-12-2021. For instance, AI_1 is effective from 09:00 21-12-2021 to 09:00 25-12-2021.

Table 1. Results of the impact analysis on AI_1 and AI_2

Impact	Metric	AI 1	AI 2
Structural object impact	Total number of impacted business objects	0	1
Structural function impact	Total number of impacted business functions	1	1
Operational object impact	Total number of object instances of the impacted business objects	0	254
Operational function impact	Total number of object instances of impacted business functions	51	153
Object performance impact	Difference in avg. total service time for purchase orders	−4 m	
Object performance impact	Difference in avg. total service time for materials		1.4 h
Function performance impact	Difference in avg. sojourn time of create purchase order	−9 m	
Function performance impact	Difference in total number of purchase orders	−13	
Function performance impact	Difference in avg. sojourn time of verify material		1.6 h

Table 1 shows the result of the impact analysis on AI_1 and AI_2. The action instance AI_1 affects one transition, i.e., *create purchase order* and 51 running objects related to it. As a result of the action, the average total service time of purchase orders has been improved by 4 min, while the average sojourn time of *create purchase order* has been reduced by 9 min. This is due to the decrease in the queue for the activity, resulting from the new business rule setting the higher minimum quantity for creating purchase orders. Moreover, the number of purchases has been reduced by 13 during the execution of the action.

The action instance AI_2 affects one object type and one transition, i.e., *material* and *verify material*. Besides, 254 running objects of *material* and 153 objects of *verify material* have been affected by the action. As a result of the action, the average total service time of purchase orders has increased of 1.4 h, while the average sojourn time of *verify material* has increased of 1.6 h.

7   Conclusions

In this paper, we proposed an approach to impact analysis of PAIS updates based on a DT-IM. PAIS updates are modeled as updates of the configuration in a DT-IM. Next, we identify PAIS entities impacted by PAIS updates and measure the structural/operational/performance impacts based on such entities. We have implemented the approach as a Web application and discussed a case study on a standard Procure-to-Pay process.

The proposed approach has several limitations. First, the identification of objects related to impacted business object types and transitions is limited to existing objects in the process and the future objects entering the process are not considered. Second, we identify objects related to impacted transitions, including all objects that potentially execute the transition. However, some objects may bypass the transition, e.g., a patient expected to perform surgery may die before it, or a doctor may decide for an emergency treatment at the last moment. Finally, the performance impact analysis does not isolate the objects subject to action instances to evaluate the performance impact, instead indirectly evaluating changes in diagnostics over all existing objects in the process.

Besides addressing the above limitations, as future work, we plan to extend the approach to predict the performance impact to provide timely and accurate information before the execution of any update. Another direction of future work is to improve the performance impact analysis such that it completely isolates the instances affected by changes to provide more realistic performance impact measures. Finally, we also plan to evaluate the proposed approach's ease of use and usefulness with business analysts in real-world situations.

References

1. van der Aalst, W.M.P., Berti, A.: Discovering object-centric Petri Nets. Fundam. Informaticae **175**(1–4), 1–40 (2020)
2. van Beest, N.R., Kaldeli, E., Bulanov, P., Wortmann, J.C., Lazovik, A.: Automated runtime repair of business processes. Inf. Syst. **39**, 45–79 (2014)
3. Berti, A., van der Aalst, W.M.P.: A novel token-based replay technique to speed up conformance checking and process enhancement. Trans. Petri Nets Other Model. Concurr. **15**, 1–26 (2021)
4. Caporuscio, M., Edrisi, F., Hallberg, M., Johannesson, A., Kopf, C., Perez-Palacin, D.: Architectural concerns for digital twin of the organization. In: Jansen, A., Malavolta, I., Muccini, H., Ozkaya, I., Zimmermann, O. (eds.) ECSA 2020. LNCS, vol. 12292, pp. 265–280. Springer, Cham (2020). https://doi.org/10.1007/978-3-030-58923-3_18
5. Comuzzi, M., Parhizkar, M.: A methodology for enterprise systems post-implementation change management. Ind. Manag. Data Syst. (2017)
6. Comuzzi, M., Vonk, J., Grefen, P.: Measures and mechanisms for process monitoring in evolving business networks. Data Knowl. Eng. **71**(1), 1–28 (2012)
7. Döhring, M., Reijers, H.A., Smirnov, S.: Configuration vs. adaptation for business process variant maintenance: an empirical study. Inf. Syst. **39**, 108–133 (2014)
8. Dumas, M., van der Aalst, W.M.P., Hofstede, A.H.T.: Process-Aware Information Systems: Bridging People and Software Through Process Technology. Wiley, Hoboken (2005)
9. Eramo, R., Bordeleau, F., Combemale, B., van Den Brand, M., Wimmer, M., Wortmann, A.: Conceptualizing digital twins. IEEE Software (2021)
10. Gelernter, D.: Mirror Worlds: Or the Day Software Puts the Universe in a Shoebox... How It Will Happen and What It Will Mean. Oxford University Press (1993)
11. Ghahfarokhi, A.F., Park, G., Berti, A., van der Aalst, W.M.P.: OCEL: a standard for object-centric event logs. In: Bellatreche, L., et al. (eds.) ADBIS 2021. CCIS, vol. 1450, pp. 169–175. Springer, Cham (2021). https://doi.org/10.1007/978-3-030-85082-1_16

12. Kumara, I., Han, J., Colman, A., van den Heuvel, W.-J., Tamburri, D.A.: Runtime evolution of multi-tenant service networks. In: Kritikos, K., Plebani, P., de Paoli, F. (eds.) ESOCC 2018. LNCS, vol. 11116, pp. 33–48. Springer, Cham (2018). https://doi.org/10.1007/978-3-319-99819-0_3
13. La Rosa, M., Dumas, M., Ter Hofstede, A.H., Mendling, J.: Configurable multi-perspective business process models. Inf. Syst. **36**(2), 313–340 (2011)
14. Lin, Y.Y., Nagai, Y., Chiang, T.H., Chiang, H.K.: SUCCERP: the design science based integration of ECS and ERP in post-implementation stage. Int. J. Eng. Bus. Manag. **13**, 18479790211008812 (2021)
15. Marrella, A.: Automated planning for business process management. J. Data Semant. **8**(2), 79–98 (2019)
16. Mendling, J., Pentland, B.T., Recker, J.: Building a complementary agenda for business process management and digital innovation (2020)
17. Parhizkar, M., Comuzzi, M.: Impact analysis of ERP post-implementation modifications: design, tool support and evaluation. Comput. Ind. **84**, 25–38 (2017)
18. Park, G., van der Aalst, W.M.P.: Realizing a digital twin of an organization using action-oriented process mining. In: ICPM 2021, pp. 104–111 (2021)
19. Park, G., Adams, J.N., van der Aalst, W.M.P.: OPerA: object-centric performance analysis (2022). https://arxiv.org/abs/2204.10662. https://doi.org/10.48550/ARXIV.2204.10662
20. Soffer, P., Golany, B., Dori, D.: ERP modeling: a comprehensive approach. Inf. Syst. **28**(6), 673–690 (2003)
21. Soffer, P., Golany, B., Dori, D.: Aligning an ERP system with enterprise requirements: an object-process based approach. Comput. Ind. **56**(6), 639–662 (2005)
22. Wieringa, R.J.: Design Science Methodology for Information Systems and Software Engineering. Springer, Heidelberg (2014). https://doi.org/10.1007/978-3-662-43839-8

Hybrid Business Process Simulation: Updating Detailed Process Simulation Models Using High-Level Simulations

Mahsa Pourbafrani[✉] and Wil M. P. van der Aalst

Chair of Process and Data Science, RWTH Aachen University, Aachen, Germany
{mahsa.bafrani,wvdaalst}@pads.rwth-aachen.de

Abstract. Process mining techniques transfer historical data of organizations into knowledge for the purpose of process improvement. Most of the existing process mining techniques are "backward-looking" and provide insights w.r.t. historical event data. Foreseeing the future of processes and capturing the effects of changes without applying them to the real processes are of high importance. Current simulation techniques that benefit from process mining insights are either at detailed levels, e.g., *Discrete Event Simulation* (DES), or at aggregated levels, e.g., *System Dynamics* (SD). System dynamics represents processes at a higher degree of aggregation and accounts for the influence of external factors on the process. In this paper, we propose an approach for simulating business processes that combines both types of data-driven simulation techniques to generate holistic simulation models of processes. These techniques replicate processes at various levels and for different purposes, yet they both present the same process. SD models are used for strategical what-if analysis, whereas DES models are used for operational what-if analysis. It is critical to consider the effects of strategical decisions on detailed processes. We introduce a framework integrating these two simulation models, as well as a proof of concept to demonstrate the approach in practice.

Keywords: Process mining · Discrete event simulation · Hybrid process simulation · Scenario-based predictions · System dynamics

1 Introduction

After bringing transparency into processes, the process mining mission is to find data-supported ways to improve the processes in different aspects, e.g., performance metrics. In [2], the capability of process mining techniques to design realistic simulation models is discussed. Process mining supports designing the simulation models by capturing all the aspects of the process in detail. However,

Funded by the Deutsche Forschungsgemeinschaft (DFG, German Research Foundation) under Germany's Excellence Strategy-EXC-2023 Internet of Production - 390621612. We also thank the Alexander von Humboldt (AvH) Stiftung for supporting our research.

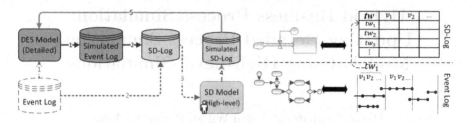

Fig. 1. General idea of designing comprehensive simulation models for business processes. We update the detailed simulations (DES) by the aggregated simulation (SD) results to take the high-level/strategical effects into account. The dotted arrows represent the optional steps in the approach, e.g., SD models can be given as inputs or can be derived from SD-Logs (left). The different aggregated levels of data used in or generated by DES and SD simulations (right).

some influential factors remain undiscovered. These undiscovered factors eventually affect the simulation results. For instance, the efficiency of resources or the effect of workload on the speed of resources is not taken into account when a discrete event simulation model for a process is designed.

The presented approach in [12] is based on using *process mining* and *System Dynamics* (SD) to tackle these types of problems. System dynamics techniques model a system and its boundary, i.e., environmental variables which influence the system and capture these influences over steps of time. The advantage of this approach is that the process variables are designed based on event logs at higher levels, i.e., not a matter of single instances. For instance, the average waiting time of customers in the process per day has a more significant influence on the number of allocated resources per day than a long waiting time of a single customer. It should be noted that high-level simulation techniques such as SD ignore the provided details, which improves the accuracy of the simulation results to some extent. On the contrary, detailed simulations in process mining, i.e., *Discrete Event Simulation* (DES), lack the high-level effects of process variables on each other as well as quality-based variables, e.g., the effect of tiredness

Table 1. The general comparison of Discrete Event Simulation (DES) and System Dynamics (SD) techniques in process mining.

	DES in PM	SD in PM
Goal	– Detailed simulation of processes – Mimicking processes – Operational	– High-level simulation of processes – Policy and decision-making – Strategical
Usage	Operational	Strategical
Data	Detailed event logs	– Coarse-grained process logs: – aggregated process variables over time
Simulation step	Events	Time steps, e.g., day
Weakness	Not capturing external factors	Evaluation of results

of resources on the execution time of cases. The overview of the comparison of DES and SD techniques in process mining simulation is shown in Table 1. Note that standard event logs and aggregated process variables (higher level logs) are referred to as detailed event logs and coarse-grained process logs, respectively. Figure 1 (right) depicts the data granularity level in process mining for both designing and re-generating simulation results, i.e., events in DESs are transformed into aggregated process variables such as v_1 and v_2 at each time step tw in SDs.

The raised concerns for simulation in process mining as mentioned specifically for discrete event simulation in [1] show that only DES models in process mining for simulating processes are not sufficient. A comprehensive business process simulation should be able to exploit the detailed process steps, i.e., workflow and resources for every single case, and the strategical perspective, and external factors at the same time. DES and SD are at different levels and for different purposes, yet, represent the same process with two views. Therefore, exploiting these two techniques in process mining makes designing a comprehensive simulation of a business process possible. The direction of interaction between these two techniques is based on business processes and scenarios [5]. The direction can be one model updating the second model or both models updating each other, resulting in bi-directional interaction.

In this paper, we propose a framework to generate comprehensive simulation models for business processes. The framework aims to combine the advantages of both simulation techniques as shown in Fig. 1. Using an event log of a process that can also be generated by the process DES model, we extract possible aggregated process variables, e.g., average arrival rate per day. The event log can be achieved by a DES model of the business process using approaches such as [17]. SD models are designed based on the generated coarse-grained process logs, i.e., SD-Logs, out of event logs [14]. The preprocessing step of our framework is generating an event log, the corresponding SD-Log, and two models at different levels. Using the provided input from the preprocessing step, we design a method to define and discover the overlapping variables between two process simulation models since they are at different levels. The transformation phase to update the DES model using the updated variables from the SD model is the critical step of our approach, in which we use the designed detailed simulation models in process mining and a list of possible variables in an SD-Log. For instance, the efficiency of resources, in reality, is not 100%, therefore, using an SD model, we can incorporate the effect of tiredness, workload per day on the resources' efficiency, or the effect of their expertise and then update the resource service time in the DES model. With the continuation of DES execution with the updated variables, a new event log is generated, which includes the simulated effect from the SD model.

The remainder of this paper is organized as follows. In Sect. 2, we present related work. In Sect. 3, we introduce background concepts and notation. In Sect. 4, we present the proposed approach, which we evaluate in Sect. 5. Section 7 concludes this work.

2   Related Work

Employing *Discrete Event Simulation* (DES) in process mining is a common approach to simulate business processes. The provided insights by process mining help the traditional business process simulation techniques to generate more accurate results based on the history of the business processes. In [17], all the aspects of a process from control-flow, organizational, performance status, and decision points are discovered from process mining techniques and considered in designing the simulation model. *Colored Petri Net* (CPN) models help in capturing both the activity-flow of processes as well as other aspects. CPN Tools [15] offers a platform for designing and simulating the CPN models. In [16], an approach for the automatic generation of CPN models for running on the CPN Tools is presented. Other approaches in the area of business process simulation exploit the process models based on the BPMN notation and improve the quality of the models using the provided information on the corresponding event logs [4].

Conducting high-level simulation models for processes is proposed in [12]. The feature extraction from event logs with a given period of time, e.g., one day, is performed before designing a model and the exploited modeling technique is System Dynamics (SD). The variables are captured with their relations and the generated model is used for simulating the process on the given time window.

These two approaches have weaknesses that can be covered by each other [6]. Using only aggregated level modeling, the details of process instances are neglected. At the same time, in detailed simulation techniques, external factors, as well as aggregated influences are ignored. In different simulation areas, the combination of DES and SD are exploited [7]. To connect two simulation models, three directions are specified [5]: (1) DES results update SD model, (2) SD results update the DES before simulation, and (3) both update each other in different phases. According to [3], the first sort of interaction is more common. It is usually more important to capture the influence of high-level decisions on detailed systems. In [19], the combination is used to perform simulation for a case study in the healthcare area where the number of newly infected patients is predicted using SD and inserted into the DES model of the serving patients in the hospital.

In process mining, both simulation techniques are proposed to be used individually on processes based on event logs. In our approach, we propose to exploit high-level simulation results for strategical scenarios and consider detailed simulation modeling of processes. These two types of modeling are supported by event data and the generated models are valid due to the existence of the previous executions of processes, event logs. To the best of our knowledge, this is the first time that both high-level and detailed simulation techniques for business processes are taken into account together.

3   Preliminaries

In this section, we define the concepts and notations used in process mining and system dynamics simulation including coarse-grained process logs, i.e.,

Table 2. A part of an event log for a sample process inside a hospital. Each row represents a unique event indicating a specific case ID, activity, resources, and timestamps.

Case ID	Activity	Start timestamp	Complete timestamp	Resource
154	Registration	01.01.2021 11:45:00	01.01.2021 11:57:00	Resource 1
155	Admission to the ward	01.01.2021 11:57:10	01.01.2021 12:40:52	Resource 1
154	Registration	01.01.2021 11:47:17	01.01.2021 12:05:01	Resource 2
156	Registration	01.01.2021 12:51:23	01.01.2021 13:02:47	Resource 1
⋮	⋮	⋮	⋮	⋮

aggregated process variables over time. Since these coarse-grained logs are utilized for SD simulation, they are referred to as SD-Logs.

3.1 Process Mining

The stored event data of processes in the form of event logs are used for process mining techniques. The form of logs which we use in our approach is defined in Definition 1.

Definition 1 (Event Log). *Let C be the universe of cases, A be the universe of activities, R be the universe of resources and T be the universe of timestamps. We call $\xi = C \times A \times R \times T \times T$ the universe of events. The event e is a tuple $e = (c, a, r, t_s, t_c)$, where $c \in C$ is the case identifier, $a \in A$ is the corresponding activity for the event e, $r \in R$ is the resource, $t_s \in T$ is the start time, and $t_c \in T$ is the complete time of the event e, where $t_s \leq t_c$. We assume that events are unique and an event log L is a set of events, i.e., $L \subseteq \xi$.*

We also define projection functions, $\pi_C : \xi \to C$, $\pi_A : \xi \to A$, $\pi_R : \xi \to R$, $\pi_{T_S} : \xi \to T$ and $\pi_{T_C} : \xi \to T$ for attributes of events. A sequence of events w.r.t. timestamp with the same case identifier represents a process instance (trace). Consider Table 2 where the first row is the event $e = (c, a, r, t_s, t_c)$ for the patients with case ID "154" as c, the activity "admission to the ward" as a which was started at timestamp "01.01.2021 11:45:00" as t_s by resource "resource 1" as r and was completed at timestamp "01.01.2021 11:57:00" as t_c.

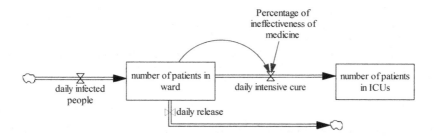

Fig. 2. A simple example stock-flow diagram.

Table 3. A part of an SD-Log for the sample hospital event log. The columns represent the process variables and the rows represent the steps.

Time window daily	Arrival rate	Finish rate (release rate)	Avg service time (time in the hospital)	Avg waiting time in process
1	180	180	0.3590	0.6099
2	147	140	0.4156	0.5409
3	160	162	0.4011	0.5961
4	116	119	0.4455	0.4908
⋮	⋮	⋮	⋮	⋮

3.2   System Dynamics

System dynamics techniques model complex systems and their environment at a higher level of aggregation over time [18]. The *causal-loop diagram* and the *stock-flow diagram* are the essential diagrams in system dynamics by which the relations between all the influential factors in/outside a system in both conceptual and mathematical ways are represented, respectively [18]. Figure 2 shows a simple stock-flow diagram where the *number of patients in the ward* in each day is calculated as follows: *number of patients in the ward = number of patients alreay in the ward + today infected people− todaty intensive cure − today release.* The values of stock-flow elements get updated in each step based on the current/previous values of the other elements that influence them.

SD-Logs In order to design system dynamics models for processes, event logs should be transformed into the SD-Logs. SD-Logs are required for generating simulation models as well as populating them using the values of variables with the purpose of validation [9].

Definition 2 (SD-Log). *Let $L \subseteq \xi$ be an event log, let V be a set of process variables, and let $\delta \in \mathbb{N}$ be the selected time window. An SD-Log of L, given δ, $sd_{L,\delta}$ is $sd_{L,\delta} \in \{1, \ldots, k\} \times V \to \mathbb{R}$, s.t., $sd_{L,\delta}(i, v)$ represents the value of process variable $v \in V$ in the ith-time window ($1 \leq i \leq k$) where $k = \lceil \frac{(p_c(L) - p_s(L))}{\delta} \rceil$.*

If L and δ are clear from the context, we exclude them from the generated SD-Logs and write sd instead. Definition 2 also indicates the format of generated outcomes as an SD model simulation in which the values of the variables in the simulation are generated. Table 3 is a part of a sample SD-Log which shows the generated SD-Log with $\delta = 1$ *day* that includes different process variables, e.g., in the first time window (day) 180 cases arrived at the process.

4   Approach

Figure 3 represents the framework including three main components: DES, SD simulation, and developing and updating interface variables. Components (1) and (2) each have two steps: discovering/designing simulation models and executing the discovered simulation models. Given an event log, and the discovered DES simulation model, the process can be simulated (1). The generated event log is inserted into the SD-Log generator and the output is used to populate the SD model (2). Having both models populated with the data and ready to run, it is time to design the connection to update DES based on SD results (3). We use DES models in the form of *Colored Petri Net* (CPN) models. In the second component, event logs are transformed automatically into multiple variables describing the process (SD-Logs) over a specific period of time, e.g., per day, as introduced in [9].

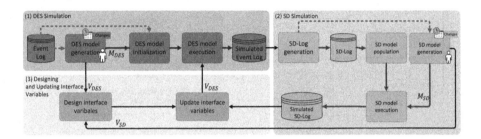

Fig. 3. The framework starts with the design and simulation of DES models (M_{DES}) and generates an event log (1). The event log is transformed into SD-Logs for generating/populating the SD model (M_{SD}) (2). Possible interaction interfaces between two models are discovered (3), e.g., simulation parameters in both models ($V_{DES} \cap V_{SD}$). Then, the detailed simulation model parameters for execution (V_{DES}) get updated by the results of the high-level simulation model. Dashed lines indicate the optional steps in designing DES and SD models.

To systematically address the connection between two models, we consider designing detailed simulation models based on the process mining insights. Furthermore, we define and extract a collection of possible variables for designing high-level simulation models from an event log [9]. The next step is to use the provided framework to update the interface variables, i.e., variables that exist in both detailed and high-level models, see Fig. 3 (part 3). Consider that the target scenario is to measure the influence of advertising investment on the acquisition rate of new customers (cases) in the process in two months. The DES model used to generate the event log is designed to simulate a specific number of cases per day. The corresponding SD model is developed based on the event log and the relevant scenario, and the new arrival rate value is predicted, e.g., as a result of viral marketing or the effects of billboard sites. When the new DES model is run with the new arrival rate, the event log is updated. The updated event log clearly reveals whether the process is capable of handling the additional cases in terms of resources.

4.1 DES Simulation

Simulation parameters such as arrival rate and average service duration of various activities for regenerating the process should be initialized before running a DES model. As a result, the first component considers the DES model discovery and execution steps separately.

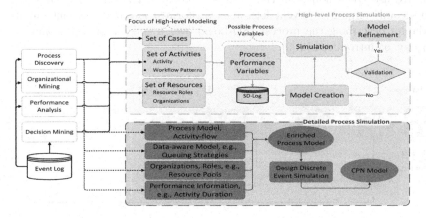

Fig. 4. Using process mining to generate simulation models of processes at different levels. Detailed process simulation models (bottom) include process activities, their performance metrics, resources, and all the possible design choices for handling the cases in the simulation, e.g., queue strategies. Our proposed framework for using process mining and system dynamics together in order to design valid models to support the scenario-based prediction of business processes at higher levels (up). Discovery and conformance techniques in process mining provide insights into the processes in different aspects, e.g., a set of activities, which are the potential design choices for aggregated simulation models.

4.1.1 Designing DES Model

The process of designing a detailed simulation model of a process using process mining is started with discovering a process model and enriching that with other aspects. For instance, for the simulation environment, often the arrival process is sampled from a *negative-exponential* distribution. To capture possible executable aspects of processes, we design process simulations based on the process mining insights as shown in Fig. 4 (bottom). In this work, we consider the designing process of the DES simulation model starting from an event log or designed based on the highlighted parameters in Fig. 4 by domain knowledge.

4.1.2 Executing DES Model

While simulating a process, all the mentioned aspects in the process simulation model can be changed for simulating different scenarios. Parameters such as the arrival rate function and performance parameters of activities in the process such as duration, number of resources can be changed as well as changing the serving queuing strategy and the flow of activities for applying different scenarios to the process. The designed DES simulation using process mining enables us to discover the change points in the process which can be updated by high-level simulation models of the process. In the DES execution step, the parameters require to be initialized. We refer to the set of simulation parameters in a DES model as V_{DES}.

Definition 3 (DES Simulate). *Let ξ be the universe of events, $n \in \mathbb{N}$, and $\mathcal{M}_{DES}$ be the universe of discrete event simulation models. Sim_{DES} : $\mathcal{M}_{DES} \times \mathbb{N} \to 2^{\xi}$. Given simulation model m_{DES}, its set of initial values of paramters V_{DES}, the specified period of time n and the start time of the simulation, $Sim_{DES}(m_{DES}^{V_{DES}}, n)=L \subseteq \xi$ simulates the process.*

Table 4. List of possible process variables generated from coarse-grained event logs [9]. The variables can be generated at different levels, e.g., the whole process or single activity level. The table shows the possibility of applying different aggregation functions (AF) on top of the performance indicators (IN) for different aspects (AS). The valid combinations provide process features, which along with the selected design choices form process variables [9].

Validator	IN												
	Value	Count					Service time			Waiting time			Time in process
AS													
AF	Numerical variable	Categorical variable	Numerical variable	Case	Resource	Activity	Case	Resource	Activity	Case	Resource	Activity	Case
Sum	True	False	True	False	False	False	True	True	True	True	True	True	True
Average	True	False	True	False	False	False	True	True	True	True	True	True	True
Median	True	False	True	False	False	False	True	True	True	True	True	True	True
Null	False	True	False	True	True	True	False	False	False	False	False	False	False

Function Sim_{DES} illustrates the simulation process for a given DES model of a process at an abstract level. Note that, in the approach, the KPIs are measured over simulated event logs in a specific period of time. Therefore, in Sim_{DES}, we consider the simulation duration, e.g., one day, to be a given input by the user.

4.2   SD Simulation

The second component aims to deliver data-driven SD simulations of processes in order to integrate detailed and high-level simulations. To accomplish this, we use event logs of processes to extract a number of performance parameters from the current state of the process and provide an interactive platform for modeling the performance metrics as system dynamics models. The models that are built can address *what-if* queries.

4.2.1   Designing SD Model

The advantage of the introduced approach in [12] for generating high-level simulation models is that the variables, i.e., simulation elements, are directly generated based on real values and can be validated. The relations that form the models are also supported by the detected behavior in event logs as shown in Fig. 4 (top). To define aggregated process variables over steps of time on the specified part of the process for simulation, the performance indicators (IN), process aspects (AS), and aggregation functions (AF) are required. The list of possible process variables given the three criteria for the selected focus of simulation is determined using the valid combinations in Table 4. For instance, the average (AF) number (IN) of resources (AS) is a process variable that can be measured over steps of time, e.g., daily. The provided list will eventually be used to facilitate the integration phase of two simulation models for determining interface variables for updates. We refer to the generated values of the possible process variables ($\mathcal{V}$) over time steps as SD-Logs (Definition 2).

Definition 4 (SD-Log Generation). *Let $L \subseteq \xi$ be an event log, $\delta \in \mathbb{N}$ be the selected time window, and $\mathcal{L}_{SD}$ be the universe of SD-Logs defined in Definition 2. $sdGen : 2^{\xi} \times \mathbb{N} \rightarrow \mathcal{L}_{SD}$, such that, for the given L and δ, $sdGen(L, \delta) = sd_{L,\delta}$ generates the corresponding $sd_{L,\delta} \in \mathcal{L}_{SD}$.*

In Definition 4, we define a function to generate SD-Logs based on event logs. Given event log L and time window δ, function $sdGen$ generates the corresponding SD-Log $sd_{L,\delta}$. The size of the time window used to generate the SD-Logs is critical. In [13], multiple time series models are trained, and the SD-Logs are generated using the one with the smallest error.

4.2.2   Executing SD Model

The SD models are designed with the help of extracted SD-Logs along with users and high-level target scenarios. For a process, SD simulation is performed for the given time step (δ) using function Sim_{SD}. The generated SD-Log ($sd_{L,\delta}$) of the process and the designed SD model (m_{SD}) are the inputs. The set of simulation variables in an SD model is referred to as V_{SD}. The future values of variables are produced in the form of SD-Logs as a result of SD simulations.

The set of SD simulation variables (V_{SD}) can include all or some of the process variables ($\mathcal{V}$) in the generated SD-Log from an event log, as well as a set of external variables (V_{EX}), i.e., $V_{SD} \subseteq \mathcal{V} \cup V_{EX}$. The SD models should be populated with the values and equations to be executable and generate simulation results in the form of SD-Logs. Consider the model in Fig. 2, where variable *daily infected people* should be populated, e.g., from the SD-Log, in order to create future values of variable the *number of patients in the ward* over time. It should be noted that the *number of patients in the ward* in the model is derived using an equation and does not need to be directly supplied. For variable $v \in V_{SD} \cap \mathcal{V}$, there are multiple possibilities of initializing and populating the SD model. For instance, for every simulation step, the value of v is taken from the corresponding row in the SD-Log, or generated by a random generator function. The random generator function is based on the distribution of values of variable v over time.

Definition 5 (SD Simulate). *Let $\mathcal{L}_{SD}$ be the universe of SD-Logs, $\mathcal{M}_{SD}$ be the universe of system dynamics models, and $j \in \mathbb{N}$ be the number of time steps. $Sim_{SD} : \mathcal{L}_{SD} \times \mathcal{M}_{SD} \times \mathbb{N} \to \mathcal{L}_{SD}$. For instance, $Sim_{SD}(sd, m_{SD}, j) = sd' \in \mathcal{L}_{SD}$ simulates the given $m_{SD} \in \mathcal{M}_{SD}$ over j time steps using the provided values in the sd and the simulation result is represented as an SD-Log (sd').*

The defined *sdGen* and Sim_{SD} enable the main steps in simulating the process models at higher levels for the focused parts of the processes as the targets of high-level simulation. These functions are used later in our framework for integrating high-level simulations and detailed simulations for business processes.

4.3   Designing and Updating the Interface Variables

The activity flow of a process, duration of each activity, batching, or queuing strategies can be updated based on high-level decisions derived from simulation models. The provided list of the changeable parameters in the detailed simulation models and the presented process variables in SD models are the baseline of finding the interfaces between these two types of models for interactions. Exploiting the simulation parameters in the detailed simulation models, we discover the ones which can be changed or get influenced by external factors or high-level decisions. In DES models of processes, all the shown process mining insights in Fig. 4 are considered to be simulation parameters that can be changed in order to perform different simulation scenarios of the processes. The changes of the DES simulation parameters can be driven from the high-level simulation model of the process, e.g., the flow of activities, the policy of handling the queues in activities, or resources based on the designed SD models. However, our goal is to automate the interaction between the two simulation models. Therefore, we focus on the parameters that can be found directly in the SD-Logs and not rely on the design choices of SD models.

Table 5. The sample mapping table for finding the interface variables in SD-Logs and DES parameters, which enables interaction of the two models of processes possible. The table is generalized, e.g., type of cases (categorical and numerical attributes), organizations and type of resources, and activities follow the same mapping table.

DES process insights			SD Process Variables	
Simulation parameters execution configuration			Simulation aspects	
Case	Number of cases (Case intervals)	=	Number of cases	Case
Activity	Processing time	=	Service Time	Activity
Resource	Processing time	=	Service Time	Resource
	Number of resources	=	Number of resources	

Table 5 shows the overview of a sample interface variables that can be found in both the DES model and the SD model of a process. These parameters and aspects enable the automatic updating of their values in one of the models based on the other. They can get extended w.r.t. designed models and used parameters. To eliminate the development details of the interaction from SD results to the DES model, the process is considered as a general method. The method looks for all the existing variables in the SD simulation results (V_{SD}) which are in the form of SD-Logs. Afterward, it updates the values of the corresponding simulation parameters in the DES model (V_{DES}) with the last values in the simulated SD-Log, i.e., the predicted values of variables in the last simulated steps. For instance, for the variable average service time of resources from *sells department*, i.e., v, a new DES model is generated in which the value of v is replaced by the last values of v in the simulated SD-Log. As a result, if the time window is one day and the SD model is run for 30 days, the value of v is taken from $sd(30, v)$.

Algorithm 1 presents the interaction between two simulation models as described. The algorithm starts with simulating the DES model for a specific period of time with the real values of the simulation parameters from process mining insights. Considering δ as the time window, the simulation duration is derived from $k * \delta$ where k is the number of steps (window of time) for simulation, e.g., $k = 20$ and $\delta=1$ *day*, the simulation duration is 20 days. The simulated event log with the same time window δ is used to generate an SD-Log, which is used to populate the SD model. Note that δ for generating SD-Logs can be different. After the SD model refinement, i.e., adding external factors, the values of interface variables in the DES model are updated by their new values as SD model simulating results. Running the DES model, a new event log is generated in which the KPIs can be measured and the effects of high-level changes can be tracked.

Algorithm 1: General algorithm of updating process mining detailed simulation based on the changed in the gateway variables.

Input: Detailed process simulation model $m_{DES}^{V_{DES}}$ initialized withe a set of parameters V_{DES}, High-level simulation model m_{SD} and its variables V_{SD}, time window δ, and k the number of time steps

Output: updated event log using the SD results L'

1 $L = Sim_{DES}(m_{DES}^{V_{DES}}, k * \delta)$;

2 $sd = sdGen(L, \delta)$;

3 $sd' = Sim_{SD}(m_{SD}, sd, 1 * \delta)$;

4 **foreach** v *in* $V_{DES} \cap V_{SD}$ **do**

5 $\quad$ Set value of v in m_{DES} to be $sd'(k, v))$;

6 **end**

7 Return updated m_{DES} as m'_{DES};

8 $L' = Sim_{DES}(m_{DES}^{'V_{DES}}, k * \delta)$;

9 return L';

---

5   Proof of Concept

We designed a scenario to demonstrate the need for hybrid simulation of processes and how to address that using the proposed data-driven approach. A process is designed using CPN Tools in the form of a Colored Petri Net. We simulate the process and generate simulated event logs. The corresponding SD-Logs are extracted from the simulated event logs, and finally, both CPN models and SD models are considered as inputs of the approach. The SD models are used to update the CPN models for the next simulation step. In each step of simulating the process, specific process KPIs are calculated. These KPIs represent the effect of high-level simulation models on the detailed process.

5.1   Implementation

As a proof of concept, the platform for running and updating a DES model of a process (CPN model) based on the results of simulating the corresponding SD model is developed. The platform is in the form of a Jupyter notebook, which includes the instructions for re-running the experiments and performing additional analysis. The designed CPN model, the SML[1] file, the SD-Log, the SD-model, and the python platform are publicly accessible.[2] The supplied tool in [10] enables producing a ready-to-execute CPN model in the CPN Tools from an event log[3] for the purpose of defining different processes and scenarios. The presented tool and approach in [11] can also be used to evaluate the quality of the generated simulation model. The automatically generated models are able to generate event logs following various changes, e.g., incorporating the effects of high-level simulation models as presented in this paper. In addition, given an event log, the presented tool in [8] supports the data-driven SD model generation.

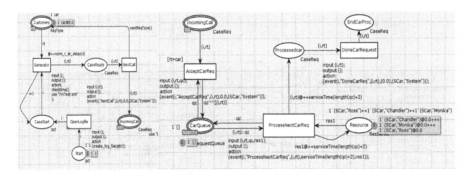

Fig. 5. The designed CPN simulation model using CPN Tools for handling the requests in one of the departments in the sample company.

---

[1] Standard ML.
[2] https://github.com/mbafrani/PMSD.
[3] https://cpn-model-process-discovery-1.herokuapp.com/generate-cpn-model/.

5.2   Designed Process and Scenario

We created a process within a process mining firm that offers customer support by handling two types of requests in two departments, namely new client inquiries and current customer support. Working days are from Monday to Friday and the working hours are from 9:00 am to 5:00 pm (including 1-hour lunchtime). The process is modeled using the CPN Tools, as partly shown in Fig. 5. On average, every 5 min, one new request is received by the company. The request arrival is modeled as a negative exponential distribution. The number of requests in the queue is limited to 20 and more upcoming requests will be automatically rejected. The service time spent by the resources in each department for executing requests is derived from a normal distribution. We designed the process model, such that resources perform the process of the request faster if the number of requests in the line is higher. This effect, i.e., the queue length on the processing time of requests, is modeled as an exponential nonlinear relation between the number of people in the queue and the service time.

In the current scenario, the company is looking to increase the number of handled requests, decrease the rejected requests, and have a more realistic simulation of their process. In the detailed simulation model, the resources considered working with full efficiency, e.g., %100 of the available time during the day. It is also considered that all the resources have the same level of expertise, e.g., the same speed in handling requests. One of the potential actions is to increase the expertise level of resources by means of training. To capture the effects of training and resource efficiency on the process KPIs, a higher-level simulation model is required. This model should have aggregated time since training requires time to eventually appear in the service time of handling cases, e.g., not all the resources get trained at once.

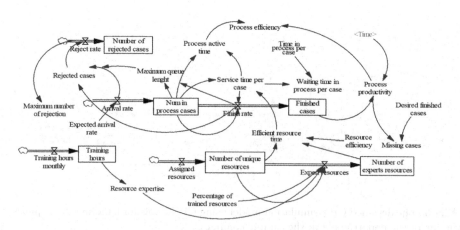

Fig. 6. The designed stock-flow diagram based on the generated event log using the CPN model in Fig. 5. The model extended by capturing the effect of training over three months on the resource efficiency, which shows the actual service time considering their efficiency per day.

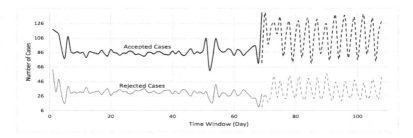

Fig. 7. A part of the process KPIs in 4 months. The number of successfully handled requests, rejected request and average service time of handling the requests by resources for the base CPN model and the updated ones with SD model results, i.e., the dashed lines.

5.3 Hybrid Simulation of the Sample Process

Running the detailed simulation model for 30 days, we calculate a set of important KPIs, such as the number of addressed requests and rejected requests. Using the generated event log of the CPN model, we generate SD-Log for daily process variables and generate the SD model using the tool presented in [8]. As shown in Fig. 6, the SD model includes the external factors of efficiency and training into account. In this scenario, after a specific amount of training, e.g., 300 h, the resources become more experts, and they are able to handle more requests in less time, e.g., %20 faster on average. The training hours per month and the percentage of resources that receive the training affect the resources' efficiency, the number of finished and rejected cases over time. To capture these effects on the DES model of the company, we run the SD model for a couple of months, e.g., 4 months. The result of changing the efficiency of resources on the available time and number of handled requests is derived from running the SD model. The SML function, in the CPN model, checks for new updates on the average service time, i.e., the interface variable which is common in two simulation models, before execution. For instance, Eq. 1 presents the SML function that reads the values of variables and their function from "CurrentValues.sml" and check if the new updates are available in the updated "UpdatedValues.sml" for the execution. The average service time gets overwritten by the execution of SD models in the corresponding SD-Logs.

$$
\begin{aligned}
&fun\ checktime()\ = \\
&if\ OS.FileSys.compare(OS.FileSys.fileId("CurrentValues.sml"), \\
&OS.FileSys.fileId("UpdatedValues.sml"))\ =\ EQUAL \\
&then\ use\ "CurrentValues.sml" \\
&else\ use\ "UpdatedValues.sml";
\end{aligned}
\tag{1}
$$

Using our framework, we update the average duration of handling requests by resources with the new value of service time from the SD model. Variables such as training and expertise of the resources are not easy to be captured and included in the discrete event simulation, i.e., the aggregated timing simulation is required to reflect the effect of changes such as training as well as defining the effect quantitatively is not a straight-forward step. System dynamics modeling enables us to handle such effects in detailed simulation models of processes. As illustrated in Fig. 7, the impact of training on average service time and, eventually, the number of handled cases are obvious after about 2 months. The impact of efficiency and training is considered, resulting in more accurate KPIs.

6   Discussion

The primary goal of this paper is to demonstrate the importance of comprehensive data-driven process simulation modeling at various levels and their interaction for businesses. Furthermore, the potential of creating various simulation models from event logs and applying results of higher-level and what-if analyses on the specific operational process is demonstrated by means of a sample scenario. It should be noted that providing a general framework for automatically combining and executing both DES and SD models at the same time is a challenging task. Moreover, even if the bases of models are built automatically using process mining insights, human domain knowledge is still crucial. We do not focus on eliminating the role of the user in the modeling phase, as it is an essential component of any effective simulation model design, specifically in strategical simulation models, e.g., SD. There are further considerations such as defining and locating interface variables. This issue can also be mitigated using the simulation parameters in DES and the provided set of process variables in SD-Logs. There are a few steps that will be needed in the future to make the technique more effective. For instance, (1) predefined scenarios that are simple and restricted to the variables in the extracted SD-Logs, or (2) substituting the DES Engine with a simple yet powerful engine that requires less expertise of modeling, e.g., SML programming in this work, will improve the framework. The assessment is a simple version of a real-world scenario with a synthetically constructed process, intended primarily to highlight the necessity and practicality of integrating two simulation-driven processes based on event data.

7   Conclusion

Simulating business processes enables organizations to examine the consequences of changes on their processes without implementing them directly. However, most simulation models are unable to capture reality. Although forward-thinking process mining techniques such as discrete event simulations attempt to address the accuracy issue of simulation models by leveraging process mining insight, there are a few unexpected issues such as the effects of external factors in the process or the role of quality-based variables. In this research, we suggested a strategy that takes advantage of both discrete event simulation approaches in process mining and high-level simulation techniques such as system dynamics simulation to mimic processes at a detailed level while applying high-level decisions. We simulate the processes with both models, and the interplay of the models results in simulation models including detailed and high-level aspects. Using common scenarios in businesses, we demonstrated the use of our technique, including the evaluation of simulation findings.

References

1. van der Aalst, W.M.P.: Business process simulation survival guide. In: vom Brocke, J., Rosemann, M. (eds.) Handbook on Business Process Management 1. IHIS, pp. 337–370. Springer, Heidelberg (2015). https://doi.org/10.1007/978-3-642-45100-3_15
2. van der Aalst, W.M.P.: Process mining and simulation: a match made in heaven! In: Computer Simulation Conference, pp. 1–12. ACM Press (2018)
3. Brailsford, S.C., Desai, S.M., Viana, J.: Towards the holy grail: combining system dynamics and discrete-event simulation in healthcare. In: Proceedings of the 2010 Winter Simulation Conference, pp. 2293–2303. IEEE (2010)
4. Camargo, M., Dumas, M., González, O.: Automated discovery of business process simulation models from event logs. Decis. Supp. Syst. **134**, 113284 (2020)
5. Jovanoski, B., Minovski, R., Voessner, S., Lichtenegger, G.: Combining system dynamics and discrete event simulations - overview of hybrid simulation models. J. Appl. Eng. Sci. **10**, 135–142 (2012)
6. Morecroft, J., Robinson, S., et al.: Explaining puzzling dynamics: comparing the use of system dynamics and discrete-event simulation. In: Proceedings of the 23rd International Conference of the System Dynamics Society, pp. 17–21. System Dynamics Society Boston, MA (2005)
7. Morgan, J., Belton, V., Howick, S.: Lessons from mixing OR methods in practice: using DES and SD to explore a radiotherapy treatment planning process. Health Syst. **5** (2016)
8. Pourbafrani, M., van der Aalst, W.M.P.: PMSD: data-driven simulation using system dynamics and process mining. In: Demonstration and Resources Track at BPM 2020 Co-located with the 18th International Conference on Business Process Management (BPM 2020), pp. 77–81 (2020). http://ceur-ws.org/Vol-2673/paperDR03.pdf
9. Pourbafrani, M., van der Aalst, W.M.P.: Extracting process features from event logs to learn coarse-grained simulation models. In: La Rosa, M., Sadiq, S., Teniente, E. (eds.) CAiSE 2021. LNCS, vol. 12751, pp. 125–140. Springer, Cham (2021). https://doi.org/10.1007/978-3-030-79382-1_8
10. Pourbafrani, M., Balyan, S., Ahmed, M., Chugh, S., van der Aalst, W.M.P.: GenCPN: automatic generation of CPN models for processes. In: Proceedings of Demonstration Track at ICPM 2021 Co-located with 3rd International Conference on Process Mining (2021)
11. Pourbafrani, M., van der Aalst, W.M.P.: Interactive process improvement using simulation of enriched process trees. arXiv preprint arXiv:2201.07755 (2022)
12. Pourbafrani, M., van Zelst, S.J., van der Aalst, W.M.P.: Scenario-based prediction of business processes using system dynamics. In: OTM 2019 Conferences, 2019, pp. 422–439 (2019). https://doi.org/10.1007/978-3-030-33246-4_27
13. Pourbafrani, M., van Zelst, S.J., van der Aalst, W.M.P.: Semi-automated time-granularity detection for data-driven simulation using process mining and system dynamics. In: Dobbie, G., Frank, U., Kappel, G., Liddle, S.W., Mayr, H.C. (eds.) ER 2020. LNCS, vol. 12400, pp. 77–91. Springer, Cham (2020). https://doi.org/10.1007/978-3-030-62522-1_6

14. Pourbafrani, M., van Zelst, S.J., van der Aalst, W.M.P.: Supporting automatic system dynamics model generation for simulation in the context of process mining. In: Abramowicz, W., Klein, G. (eds.) BIS 2020. LNBIP, vol. 389, pp. 249–263. Springer, Cham (2020). https://doi.org/10.1007/978-3-030-53337-3_19

15. Ratzer, A.V., et al.: CPN tools for editing, simulating, and analysing coloured Petri Nets. In: van der Aalst, W.M.P., Best, E. (eds.) ICATPN 2003. LNCS, vol. 2679, pp. 450–462. Springer, Heidelberg (2003). https://doi.org/10.1007/3-540-44919-1_28

16. Rozinat, A., Mans, R.S., Song, M., van der Aalst, W.M.P.: Discovering colored Petri Nets from event logs. STTT **10**(1), 57–74 (2008)

17. Rozinat, A., Mans, R.S., Song, M., van der Aalst, W.M.P.: Discovering simulation models. Inf. Syst. **34**(3), 305–327 (2009)

18. Sterman, J.D.: Business Dynamics: Systems Thinking and Modeling for a Complex World. McGraw-Hill (2000)

19. Viana, J., Brailsford, S., Harindra, V., Harper, P.: Combining discrete-event simulation and system dynamics in a healthcare setting: a composite model for chlamydia infection. Eur. J. Oper. Res. **237**(1), 196–206 (2014)

Business Process Mining

Towards Event Log Management for Process Mining - Vision and Research Challenges

Ruud van Cruchten[(✉)] and Hans Weigand

Tilburg School of Economics and Management, Tilburg University, Warandelaan 2, 5037 AB Tilburg, The Netherlands
{r.m.e.vancruchten,h.weigand}@tilburguniversity.edu

Abstract. Organizations act in dynamic and constantly changing business environments, as the current times unfortunately illustrate. As a consequence, business processes need to be able to constantly adapt to new realities. While the dynamic nature of business processes is hardly ever challenged, the complexity of processes and the information systems (IS) supporting them make effective business process management (BPM) a challenging task. Process mining (PM) is a maturing field of data-driven process analysis techniques that addresses this challenge. PM techniques take event logs as input to extract process-related knowledge, such as automatically discovering and visualizing process models. The popularity of PM applications is growing in both industry and academia and the integration of PM with machine learning, simulation and other complementary trends, such as Digital Twins of an Organization, is gaining significant attention. However, the success of PM is directly related to the quality of the input event logs, thus the need for high-quality event logs is evident. While a decade ago the PM manifesto already stressed the importance of high-quality event logs, stating that event data should be treated as first-class citizens, event logs are often still considered as "by-products" of an IS. Even within the PM research domain, research on event logs is mostly focused on ad-hoc preparation techniques and research on event log management is critically lacking. This paper addresses this research gap by positioning event logs as first-class citizens through the lens of an event log management framework, presenting current challenges and areas for future research.

Keywords: Event log management · Process Mining · Requirements engineering · Artifact network · Data quality

1 Introduction

The digitization of business processes and the subsequent explosion of available event data to organizations is rapidly changing the business process management (BPM) discipline. Nowadays, BPM researchers and practitioners face the challenge of extracting valuable insights from large volumes of data regarding business process execution and their environments. Process mining (PM) emerged as a research field to address this challenge, bridging the gap between data science, which is often process agnostic, and process science, often more focused on modelling rather than data analysis [3].

© The Author(s), under exclusive license to Springer Nature Switzerland AG 2022
R. Guizzardi et al. (Eds.): RCIS 2022, LNBIP 446, pp. 197–213, 2022.
https://doi.org/10.1007/978-3-031-05760-1_12

PM techniques enable the (semi-) automatic extraction of process knowledge, such as the discovery of process models, from process execution data, called event logs, generated by an information system (IS). PM is able to answer crucial BPM question and can be applied to manage and improve processes based on data-driven insights rather than stakeholder's idealized view on reality [3]. However, other applications of PM have been reported extensively as well [21] and the integration of PM techniques with trends such as Digital Twins (DT) [37], process simulation and prediction [9, 34] and Robotic Process Automation (RPA) [28] is gaining significant attention. Strikingly, current methodologies for PM are primarily focused on project-based applications of PM [4, 6, 15] providing little to no guidance on the recurrent or *continuous* application of PM in organizations.

The extent to which PM techniques can be applied and relied upon is directly related to the quality of the input event log: i.e. *garbage in-garbage out*. However, a general definition of what constitutes event log quality is lacking [46]. In this paper we adopt a semiotic view on event log quality that next to a *syntactic* and *semantic* perspective includes a *pragmatic* perspective on quality as well, i.e. the applicability of the event log for PM techniques and the usefulness of the results constitute its quality [38].

A decade ago, the PM manifesto already stressed the importance of high-quality event logs, stating that event data should be treated as first-class citizens [4]. Since then, the field of PM has matured but research on event log preparation and quality has not gained significant attention [21]. With the increasing popularity of PM in both research and practice, the need to continuously produce high-quality event logs will increase as well [32, 46]. Therefore, the goal of this paper is to position event logs as first-class citizens and forward research on event logs by answering the research question: *What are the requirements for Event Log Management, from a Process Mining perspective?*

In order to answer this question, we present an event log management (ELM) framework by adapting a data management (DM) framework to the context of PM. We discuss the ELM framework in light of current PM research literature to identify research opportunities and challenges. In doing so, we elaborate on how managing event logs differs from managing (business) data in general. We adopt a scenario-based requirements engineering approach as this provides an effective method for eliciting initial design requirements for the framework as well as a means of communicating the design rationale and goal [41].

The remainder of this paper is structured as follows: Sect. 2 provides some preliminaries on process mining and event logs. Section 3 introduces a general PM application scenario and corresponding artifact network. Section 4 presents the event log management framework. Section 5 discusses the framework and presents some limitations. Section 6 concludes this paper and presents future research challenges.

2 Preliminaries

2.1 Process Mining

PM is a maturing research field in which many algorithms aimed at various analysis goals have been proposed to extract process knowledge from event logs. Traditionally, three types of PM can be distinguished: *process discovery*, aimed at discovering and visualizing process models from event logs without prior information, *compliance checking*, aimed at comparing a normative process model with the behavior seen an event log, and *process enhancement*, aimed at repairing or improving a process to better achieve its goal by means of data-driven performance analysis [4].

However, as PM is gaining significant attention in both research and practice, other applications of PM techniques have been reported as well. For example, PM techniques aimed at discovering organizational structures, detecting concept drift (i.e. changes in the process over time), discovering process decision points, optimizing resource allocation and predicting the duration of-, or next activity in, a process instance [21]. In fact, PM is not limited to the analysis of business processes only. In many other application domains the added value of PM has been demonstrated, such as clinical path- or patient treatment analysis in healthcare, incident and change management in ICT, analysis of production processes and industrial machines in manufacturing, behavioral analysis of student's learning processes in education, risk analysis in finance and many more [21, 30]. In fact, activities within any context that can be defined as a process and of which event data is recorded can be analyzed using PM techniques.

2.2 Event Logs

An *event log* consists of data regarding the execution of activities in a business process (i.e., *events*) and can be seen as a multiset of *traces*. Each trace in an event log describes the sequence events within the lifecycle of a particular process instance, called a *case* (e.g., an order). Each unique sequence is called a *variant*. The events within a case are usually ordered by a *timestamp*, which is required to discover causal dependencies within a process model [3]. Thus, the minimum data required to construct an event log are: a unique identifier for the process instance (i.e., case id), activity executions (i.e., events) and their timestamps. Additional information such as the resource that performed the activity or other attributes associated with the case or event, e.g., order size, can be added as well. These attributes allow for additional insights or perspectives on the process execution.

Event data constituting an event log is often generated as a by-product of an IS for purposes such as recovery or debugging [4]. While there are ISs, such as BPM or workflow systems, that (can) intentionally generate process logging, in most cases an event log is not readily extracted from an IS but created through manual or (semi-) automatic preparation of event data from one or multiple sources [2].

3 Method

In systems development, scenario-based requirement engineering (SBRE) is a technique that is used to analyze system requirements or for user-designer communication, i.e., to present a vision or to motivate a design rationale. Scenarios can range from informal narratives and use cases, e.g., descriptions of a system's use including its technical and social environment, to more formal models [41]. In this paper we use both ends of the spectrum as we present a descriptive narrative on the application of PM and derive a formal artifact network from this narrative. The scenario serves both as a design inspiration to elicit requirements, as well as highlight the need for an event log management system.

3.1 Narrative

As a scenario, consider the application of PM in an organization to monitor and manage a business process. The organization has implemented a PM tool that supports common PM activities including discovery and conformance checking, as well as the creation of process monitoring dashboards. John is a process analyst and has been trained to properly the use the PM tool and interpret its results. John reports directly to the process owner Jane, who is responsible for the proper execution of the process and adherence business rules. While John is able to use the PM tool and analyze its results, he has no experience in data preparation. He therefore goes to the IT department to ask for help in creating an event log. Robert, a data engineer, works together with John to identify the required event data and desired abstraction level of the process activities they represent. Robert then designs and creates a data pipeline, in which the required event- and business data are extracted, transformed (e.g., clustering events) and finally loaded into a data structure suited for the PM tool. The data pipeline runs at regular intervals so that John is able to report regularly on the process execution performance to Jane.

At some point in time, John discovers that an activity in the process is not being executed that is required to adhere to new law and regulations. He presents his findings to Jane, who then decides to add the activity to the process design and derive a new business rule. Jane then asks the system administrator to implement the new business rule in the system to enforce the execution of the mandatory activity. However, John complains to Robert that the activity is still not present in the PM results, despite the rule being implemented and enforced in the system. Robert checks the system's logging and confirms that events related to the new activity are being recorded. He then checks the data pipeline that creates the event log and realizes that his initial code to extract and transform events has to be updated to include the new activity's events as well. He updates the event log's design and the code of the data pipeline. With the updated event log John is now able to check the conformance on the newly added activity. In doing so, he discover that the addition of the new activity has introduced some inefficient process behavior. He reports these results to Jane who addresses the involved employees on their behavior and uses the PM results to monitor the process performance.

Having heard about the promising results of PM Jennifer, the owner of different process, enthusiastically reaches out to John to ask for a process monitoring solution as well. John discusses the new event log requirements with Robert, who then creates a new data pipeline to continuously produce the second event log, which John again uses to create the process monitoring dashboard for Jennifer. As word keeps spreading on the benefits of PM, the number of requests for event logs increases, and as the adoption of PM in the organization grows, so does the dependence on high-quality event logs. John and Robert quickly realize that they need a system to manage the increasing number of event logs, as well as define policies and rules on who is in charge of guarding the quality and proper use of event logs. They already heard a colleague with a data science degree talk about the exciting possibility of event logs for process prediction or process for simulation…

3.2 Artifact Network

As the narrative shows, event logs often do not serve a direct practical goal for end-users (e.g., process owners) but are intended to be components of a larger whole, i.e., the event log is part of a PM application. This implicates that creating an event log always involves the integration of the event log into a network of artifacts, which is an ongoing process as artifacts continuously evolve over time. An *artifact network* specifies the dependencies and relationships in a network of imbricating artifacts [45]. Figure 1 presents an artifact network for the context of a PM application and positions event logs as design artifacts into their social and technical context. The artifact network highlights that managing an event log is no trivial task and that event log quality is dependent on the network as a whole. OntoUML is used to model the network in Fig. 1, as it is a well-founded language for ontology-driven conceptual modeling.

The scope of the network is limited to technological objects and related artifacts required for the application of PM in an organizational context. While it is a good practice to specify a provider and customer for each activity, we have only modeled the most important ones to limit the size and complexity of the network for the purpose of this paper. It is assumed that an event log is sourced from one or more ISs that are implemented in an organizational context to support the execution of one or more (business) processes. The business process execution creates business data and event data. The event log serves a particular PM analysis goal and is created via a data pipeline performing certain preparation tasks, e.g., data integration, transformation and/or enrichment. Furthermore, the event log is used in a PM tool, in which a PM algorithm is implemented that produces a certain result type (e.g., process model or rules) to fulfill the analysis goal of the agent(s). The PM tool is assumed to be "naïve" in a sense that the underlying PM algorithm does not take the quality of the input into account. Finally, it is assumed that business rules are used for compliance checking and that these business rules are built and maintained separately. Note however that sometimes business rules are used in the event log transformation, e.g., to aggregate events, so changes in one can have an impact on the other. This is one example of the artifact network being an imbricating and adaptive system. The artifact network is intended to serve as reader companion for Sect. 4, to specify and highlight the artifacts involved in managing event logs.

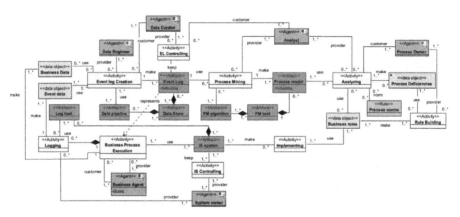

Fig. 1. General event log artifact network in the context of PM

4 Event Log Management Framework

In order to discuss the functionalities of event log management we adopt the DAMA-DMBOK data management functions framework, as depicted in Fig. 2 [13]. Data management focusses on managing the production and use of data as a valuable asset for organizations. In the context of event logs, we define event log management as the development, execution, and supervision of plans, policies, programs, and practices to deliver-, control-, protect- and enhance the value, and thus quality, of event logs throughout their lifecycle.

Event log management activities can be divided into three main categories, namely the event log lifecycle, event log foundation, and event log governance activities. The *event log lifecycle* involves all the activities and processes related to the design, creation, storage, usage, maintenance and/or enhancement, archival and finally deletion of event logs. Throughout the event log lifecycle, *foundational activities* are performed to ensure the security, privacy, compliance and quality of event logs. *Event log governance* focusses on how decisions are made about event logs and how people and processes are expected to behave in relation to event logs. The governance activities guide both the lifecycle and foundational management activities. The next sections discuss each category in more detail, relating to current PM research literature and the artifact network depicted in Fig. 1.

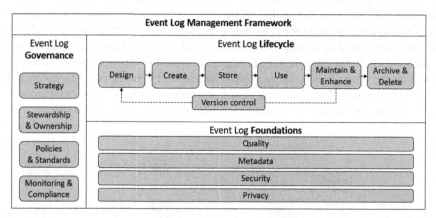

Fig. 2. Event log management framework

4.1 Event Log Lifecycle

A recent systematic literature review within the PM domain shows that research is mainly focused at the development, enhancement and application of PM techniques [30]. Research on event log preparation is scarce and mainly addresses the gathering and cleaning of event logs [21]. However, from a management perspective, the lifecycle of an event log is broader than "preparing" as it involves all the activities to manage an event log from the initial design to final deletion. The following subsections discuss each lifecycle activity and present the state of the art in PM research, from which the functionalities and future research challenges to perform event log management are derived.

Design

Existing methodologies for applying PM, such as the PM^2 methodology [15], do not explicitly design an event log, rather they consist of activities to plan, i.e., determine the scope and goal, extract and prepare the data required to construct an event log. However, considering event logs as first-class citizens, entails that they should be treated as artifacts and objects of design. This implicates that the first activity in the lifecycle of an event log is to create an explicit design for an event log.

The artifact ontology point of view [45] makes a distinction between technical objects and artifacts. Humans create many technical objects every day, but not all of them instantiate an explicitly designed artifact. Event logs that are a by-product of an IS without a PM use-context in mind are technical objects. An artifact universal has at least a use plan (e.g., PM for auditing), and a consistent make plan. The make plan describes the structure of the artifact (e.g., the XES data structure standard) and its components [45], for instance, that each record must contain a case id. Only when a technical object (i.e., dataset) is realized through the make plan specifications, it can produce the intended use effect and be considered an instance of the artifact (i.e., event log).

Considering an event log as a design artifact implicates that its *quality* can be measured in terms of adherence to the design requirements. For example, the desired level of abstraction of events should be specified in the event log's use plan, from which the required clustering or transformation steps can be derived in the make plan [12]. Only when a dataset is created via the transformation steps of the make plan, it adheres to the event log's design and can realize its intended use effect and thus be considered of good quality.

The advantage of making an explicit design is that it inherently abstracts from the implementation, so independent of a specific coding language or tool. From a management perspective, this is beneficial as the artifacts affecting the event log, e.g., IS, can change over time and can cause the implementation of the design to be updated or recreated. Furthermore, the underlying design assumptions of the event log itself may evolve over time, as highlighted by the current trend towards an object-centric event log (OCEL) in which the set of objects related to the event execution are taken into account as well [20]. Consequently, the OCEL design rationale significantly affects the downstream lifecycle activities, i.e., creation, storage and use, and artifacts as well.

Having an explicit design is also necessary in order to effectively communicate and manage all the involved stakeholders and their goals, i.e., customers of the event log. The event log design creates transparency on what data is used, for what purpose and what logic or manipulation takes place. Thus, making an explicit design also contributes to responsible and FAIR data science.

Create

Current research on event log preparation mostly focusses on a specific PM technique, application domain, data source or quality issue(s) [18, 32]. A recent Delphi study reports that complex data preparation and poor data quality are still considered as extremely relevant challenges for PM applications [33]. However, in existing PM methodologies, present event log preparation as a manual, ad-hoc activity [4, 6, 15], providing little guidance on how to continuously create an (high-quality) event log.

From a management perspective, having a repeatable and transparent event log preparation method is imperative. This could be realized by operationalizing the event log's design requirements into a systematic method to create the event log, e.g., a data pipeline. We argue that data pipelines for event logs will become important in the future as organizations applying PM require a reliable and automated system that continuously provides high-quality event logs.

An important development would be a tool, as part of an ELM system, whose function is to manage the data pipeline on the basis of a control cycle that includes continuous process monitoring from a data quality perspective. Such a tool could be achieved by separating the rules and logic into a separate repository from the execution code. Having a separate logic repository creates a transparent, auditable and manageable data pipeline.

As depicted in Fig. 1, several artifacts and their stakeholders, are involved in the creation of an event log, namely an IS, a logging system and a data pipeline artifact, for which we envision a repository with data preparation and transformation rules that are executed by a separate script [12]. And finally an artifact to store the event log. Furthermore, these artifacts are connected to other artifacts that indirectly influence the event log creation, namely the process design (i.e., process model and/or business rules) and PM application (i.e., the intended use of event log). This implicates that designing and creating a data pipeline for an event log creation is a multidisciplinary activity that requires close collaboration between a several stakeholders, e.g., a data engineer, a data curator, a PM analyst and the IS- and process owner(s).

Store
The XML-based eXtensible Event Stream (XES) log format is the current de-facto standard data structure for storing event logs, requiring an explicit case-notion to correlate events to provide a single view on the process [29, 43]. However, in reality a single unifying case notion often cannot be defined in a process, as it involves various objects that interact which each other, e.g., orders, invoices and packages. Selecting one of the objects as a case notion forces a specific view on the process which "flattens" the event log and can lead to divergence and convergence problems and even incomplete process models [23]. Thus, transforming multi-dimensional data into a flat event log introduces redundancy and a variety of data quality issues [27, 29]. Recent developments have been made to address this challenge, such as the XOC format [29] or the OCEL standard [23]. These data structures sit between object-centric databases and flat event logs, allowing for flexibility and multiple perspectives on the process based on which object is selected as a case notion [29].

While standardization of data formats improves the interoperability of event logs between various PM techniques, it does not address the challenge of extracting and creating an event log. The misalignment between data sources and event log formats highlights the fact that a data pipeline to provide high-quality event logs for PM is to be treated as a design task in its own. As data structures for event logs have been shown to evolve over time, an effective data storage artifact should be data structure agnostic, meaning that it can support multiple formats for event logs and multiple views. The concept of data independence for event logs could be investigated, as this addresses the abstraction of data on a physical, logical and view level. Alternative storage artifacts such as graph databases [19], enabling direct PM on databases rather than materializing event logs [14], or to simplify event log creation for the end-user by means of ontology-based data access [10] are all interesting areas that are being actively researched to address the challenge of data creation and storage for PM.

Use

Since its inception in the early 2000s, research on PM has mainly focused on the development of technical - rather than organizational aspects of PM [25]. Since event data and process are not limited to a specific application domain these techniques have been adopted in a variety of organizations across domains. Furthermore, PM is increasingly being adopted in other applications as well, e.g., RPA, digital transformation or Digital Twins [22, 28, 37]. However, research on organizational adoption or maturity models for PM is very scarce [25, 26, 42]. It must be noted that in several cases, such as Digital Twins, adoption of PM mainly amounts to the use of event logs.

The increasing popularity of PM outside of academia is illustrated by growing number of PM tool vendors [1]. A current trend is the shift *from* PM being a tool for a data scientist in a process improvement project *to* continuous process monitoring being a service to various enterprise-wide applications. Event log management and monitoring will therefore become a goal on its own as PM use matures in organizations. PM will be one of the event log's users, but so may be automated process execution, or continuous auditing. We therefore urge for more research dedicated to event logs as these will become first-class citizens for organizations using them for multiple purposes.

Maintain, Enhance and Version Control

Since business processes, the artifacts involved, as well as the goals of stakeholder change over time, so should the event log. Since PM currently is treated as a project, not as a continuous application, the maintenance and enhancement (M&E) of event logs is often not considered or seen as a part of the iterative nature of conducting PM projects. From an event log management perspective, we define M&E as a distinct activity that manages the adaptability of the event log and monitors changes in the artifact network and the stakeholder goals that influence the event log design. An implementation of a version control system would enable organizations to keep track of current and previous versions of the event log design, and even roll-back when necessary.

In order to perform this monitoring task, the metadata of the event log itself needs to be recorded, as well as metadata of related artifacts such as the process design and IS. A minor change in an artifact (e.g., a IS version update) could lead to some minor maintenance on the event log design and data pipeline. A major change, e.g., the new legislation as presented in the scenario, leads to a new iteration of the lifecycle as this severely impacts the event log design and hence all downstream lifecycle activities. Thus, in order to maintain event logs a version control system is required.

Archive and Delete

The final activity in the event log lifecycle is the archival and deletion of the event log, which has been neglected in current research. An event log archive has the function to store obsolete event logs, on which no maintenance or general usage occurs. While organizations would like to save all of their data, in practice this is often not feasible as the volume of archived event logs inevitably grows, or it is legally not allowed. Thus, storage costs and regulatory retention periods require organizations to properly destroy their event logs at some point. However, we argue that an event log archive should not be considered solely as an expense to organizations. Historical event logs can form a rich source of information for the analysis of business process evolution, e.g., concept drift, and can be used for (internal) auditing applications.

4.2 Event Log Foundation

Event log foundational activities include, but are not limited to, the management of event log quality, metadata, security and privacy. These activities are performed throughout all the lifecycle management activities to ensure the trustworthiness, reliability and usefulness of the event log.

Quality

Since many PM techniques do not take the quality of an event log into account, determining the reliability of the PM results highly depends on transparency of the data preparation and the ability to assess the quality of the event log. However, in order to measure event log quality (ELQ), one must define it first. While ELQ has been widely recognized as a major challenge in the application of PM, extensive research on ELQ and its impact on PM results is scarce. A recent systematic mapping study of PM research classifies only 2.56% of the publications as covering broader data preprocessing activities such as gathering, transforming and cleaning event logs [21].

As in the case of general research on DQ, no unified definition of ELQ exists within the PM domain [46]. The distinct characteristics of event logs, such as temporal constraints and correlation among events, give rise to unique ELQ issues [40]. Furthermore, PM techniques assume that the behavior recorded in an event log reflects reality and are unable to distinguish between traces showing infrequent behavior or incorrect logging, such as traces missing events or incorrect timestamps, i.e., noise [11, 44]. While some approaches to define ELQ have been presented, such as the framework in [8] or imperfection patterns in [40], a solid theoretical foundation for ELQ is critically lacking [46] as these approaches are based on the author's experience and intuitive understanding on what constitutes quality.

From an ELM perspective, ELQ is dependent on the adherence to the design requirements. This implicates that explicit quality criteria can be derived or included in the event log's design. We argue that the design specification of an event log resembles the semiotic view on DQ, which specifies three levels quality, namely syntactic, semantic and pragmatic quality. *Syntactic quality* defines quality as the degree to which data conforms to integrity rules based on a data-, system- or domain model of an application, e.g., adherence to the XES or OCEL standard. *Semantic quality* describes the degree to which data objectively corresponds to represented external phenomena, e.g., the real world event that took place. *Pragmatic quality* describes the degree to which data is suitable to perform a certain task, e.g., interpretable process discovery results. Furthermore, in order to meaningfully measure ELQ, the hierarchy of data should be taken into account when operationalizing quality criteria, e.g., Completeness can be interpreted differently on a log, case or event (record) level, requiring different metrics at each level.

Metadata

An important prerequisite to manage and hence measure the lifecycle and foundational ELM activities is the recording of *metadata* [16]. For example, in order to measure an event log's quality dimension *Completeness*, metadata of the IS- and process design, the logging system as well as the data pipeline execution is required as these artifacts (in)directly affect the event log design and creation as shown in Fig. 1.

Metadata from the IS can be used to check whether all business objects, e.g., *order*, that are stored in the database are represented in the event log by a case identifier. Process design metadata can be used to check whether all activities are represented in the event log, as the narrative example illustrates via missing events of the newly added process activity. Metadata of the logging system itself can provide information on the trustworthiness of the event log, i.e., has the logging of events been executed without errors. Finally, metadata on the data pipeline enables a quality assessment of the data preparation, e.g., the number of cases before and after transformation remains the same.

Security and Privacy

The exploitation of event logs for security audits [5] and other security challenges, such as anomaly or intrusion detection [35], have been demonstrated before. However, research on methods to secure or encrypt the event log itself is scarce [39]. Since business processes often involve multiple independent organizations, analyzing event data from inter-organizational business processes is very interesting for improving operational performance. However, sharing inter-organizational event logs gives rise to confidentiality and privacy concerns which have been largely neglected by current research [17]. We argue that sharing event logs with third parties not only requires proper security techniques and methods, but also requires dedicated management attention to ensure compliance to law and regulations, such as the GDPR and HIPAA. Preferably, sensitive event logs are not copied, but instead the interorganizational analysis is performed with a decentralized algorithm.

Privacy concerns of the information captured by event logs do not apply to an inter-organizational context only. Within an organization, analyzing event logs can expose sensitive information that may reveal personalized information too. However, the impact of privacy regulations, such as the GDPR, on the technical design and application of PM techniques in organizations requires more research attention [31]. Only as of recent, privacy preserving event log preparation and PM techniques are being researched [20, 31]. Having an explicit event log design and separating the data pipeline logic from the execution code will increase the transparency and auditability of these artifacts, enabling effective compliance monitoring to privacy law and regulation.

4.3 Event Log Governance

As organizations increasingly adopt PM for BPM and other applications, their dependency on event logs increases. Hence, organizations applying PM continuously not only face a technical- but also managerial challenge. Since business processes cut across functional siloes, systems, and even organizations, managing an event log can prove a challenging task. In order to address these managerial challenges, organizations have to govern their event logs. Adopting the DAMA-DMBOK definition of data governance, we define event log governance as the exercise of authority and control (e.g., planning, monitoring, and enforcement) over the management of event log assets [13].

Event log governance guides all other event log activities though the planning, monitoring and enforcement of both the event log lifecycle- and foundational activities. It focusses on how decisions are made about event logs and how people and processes are expected to behave in relation to event logs. This is realized through the creation of a strategy, policies, standards and cultivating stewardship and ownership of event logs. Effective governance requires an ongoing effort to ensure that organizations get value from event logs while reducing related risks. While established frameworks and mechanisms for corporate, IT and data governance exist, research on event log- or PM governance is very scarce [7, 24].

Strategy

Having a project perspective on PM implicates that the use of event logs and application of PM lacks a clear organizational goal and direction for the long term. Developing a strategy reduces the risk that organizations apply PM fragmented and hence do not use PM and event logs to their full potential. Furthermore, a strategy should exist to ensure that PM is applied to help achieve the goals of the organization. As event logs can be used for a diverse set of PM applications and analysis goals, a central vision is required on how to reach the organizational goals and improve upon them. A maturity model for PM adoption could aid organizations in creating a strategy to develop their PM capabilities. However, in current PM literature research on governance or organizational adoption is critically lacking.

A recent work presents a first step towards event log governance by establishing a theoretical data quality governance framework for event logs [24]. Consistent with Fig. 1, this framework identifies IT and Process management as two key capabilities for ELQ management. However, the framework still considers PM application from a project, rather than continuous perspective. Furthermore, the framework focusses on quality while the scope of event log governance is much broader.

Since PM sits between data- and process science we theorize that the maturity of an organizations in these areas influences the maturity of PM adoption strategies as well. Future research could look into combing existing maturity models for IT, DM and BPM in order to develop a maturity model specific for PM adoption.

Stewardship and Ownership

From an ELM perspective, the role of a data steward or curator needs to be implemented in organizations, who is charged with the responsibility of preserving the quality, privacy, confidentiality and adherence to law and regulations of the event log and its use. As shown in Fig. 1, the data curator has to manage all stakeholder goals, such as the process and system owner, the PM analyst and data engineer, as well as be able to check the compliance of the event log and data pipeline artifact to security-, privacy requirements and other regulations. That this is no trivial task is highlighted by the Siemens case reported in [36], in which the project decided to *exclude* any event data transformations or filtering in order to urge "front-end users", i.e., PM users, to improve the master data of the IS to fully benefit of the PM results.

Next to event log stewardship the ownership, i.e., accountability on the effective and efficient use of event logs, should be embedded in organizations. Depending on the organization's data- or process management maturity, event log ownership could be assigned to either an existing data (domain), process- or analytics solution owner. However, we theorize that because event logs bridge the gap between data-, process- and analytics, assigning effective ownership will prove a challenging task to organizations who have not yet implemented effective governance on any of these topics.

Policies, Standards, Monitoring and Enforcement
Similar to data- and process management, effective ELM requires organizations to define policies and standards for the creation, access and use of event logs. However, crucial prerequisites to ensure, or sometimes enforce, adherence to these policies and standards are proper monitoring, e.g., through metadata, and implemented stewardship and ownership.

5 Discussion and Limitations

The aim of this article is to present a vision on event log management and an initial design of the functionalities of ELM. The goal of the framework design is to provide a useful starting point for further research, not necessarily to provide the 'best' or 'most exhaustive' framework, as this cannot be defined and, in fact, may be a moving target. In design science literature, the problem of not being able to find an optimal solution is described as 'design being a search process'. We hope that by highlighting current research challenges on ELM, the search process for effective and useful ELM solutions continues beyond this work as it opens various avenues for further research.

Furthermore, the realization of these functionalities calls for further research, discussing the implementation choices and technical design(s). For example, is it better to create a single ELM system that supports all the functionalities or is a network of applications more realistic? We posit that existing solutions for data management should be investigated on the suitability for event log management, as from data perspective, event logs can be considered as a dataset. However, event data poses some unique challenges as compared to "traditional" DM that is focused on managing transactional and business data. Therefore, we argue that, similar to PM, event log management should bridge the gap between data and process management, as event data is highly dependent on the system and process design and thus requires close collaboration between these disciplines.

The scope of this study is limited to a systematic presentation of event log management, embedded in the PM literature, but without empirical evaluation. The narrative scenario and artifact network presented are a realistic, but single view on the application and adoption of PM in organizations. This poses the threat of confirmation bias, i.e., the scenario reflects the authors preconception of event log management requirements. However, we argue that reasoning with both an abstract example as well as concrete model addresses this threat as it links the specific (narrative) with the general (model).

In light of the goal of this paper, the scenario-based approach provides an effective mean to communicate the need for, and vision on event log management. Even our in relatively simple scenario where an event log is sourced from a single IS, ELM quickly becomes a complex and challenging task that involves multiple artifacts and stakeholders. We therefore argue that in realistic real-world scenarios where multiple systems and more stakeholders are involved, ELM only becomes more complex and hence requires dedicated attention in order for organizations to effectively and continuously apply PM.

6 Conclusion and Future Research

This article positions event logs as first-class citizens and discusses the what activities constitute event log management. The results can be read as both a research agenda, as well as a guideline for organizations to identify and implement event log management activities. First of all, event logs should be considered as design artifacts with an explicit design. This design can be used to communicate and manage stakeholder goals, as well as measure the quality of the event log as adherence to the design's requirements. Secondly, in order to consistently and continuously create event logs a data pipeline is required. From a management perspective, it is desired to separate the transformation logic and rules from the execution code to make this data pipeline transparent and auditable. Third, data independence for event log storage is required to separate the physical storage from the logical structure and views. This will ensure flexibility and adaptability of the event log data structure for future applications. Fourth, version control for event logs is required to maintain event logs in constantly changing business environments. Finally, archival of event logs is required to store event logs for historical analysis and compliance to regulatory retention periods.

Several avenues for future research on foundations for event log management are identified, such as the security and privacy of event logs, the quality of event logs as well as an event log metadata system to measure and manage these foundational activities. Finally, research on event log governance aspects is required, such as an adoption strategy and maturity model for PM and future application of event logs, as well as research on event log policies, procedures and the cultivation of stewardship and event log ownership in organizations adopting PM.

References

1. Aalst, W.: Academic view: development of the process mining discipline. In: Reinkemeyer, L. (eds.) Process Mining in Action, pp. 181–196. Springer, Cham (2020). https://doi.org/10.1007/978-3-030-40172-6_21
2. Aalst, W.M.P.: Extracting event data from databases to unleash process mining. In: vom Brocke, J., Schmiedel, T. (eds.) BPM – Driving Innovation in a Digital World. MP, pp. 105–128. Springer, Cham (2015). https://doi.org/10.1007/978-3-319-14430-6_8
3. van der Aalst, W.M.P.: Process Mining: Data Science in Action. Springer, Cham (2016). https://doi.org/10.1007/978-3-662-49851-4
4. van der Aalst, W., et al.: Process mining manifesto. In: Daniel, F., Barkaoui, K., Dustdar, S. (eds.) BPM 2011. LNBIP, vol. 99, pp. 169–194. Springer, Heidelberg (2012). https://doi.org/10.1007/978-3-642-28108-2_19

5. Accorsi, R., et al.: On the exploitation of process mining for security audits: the conformance checking case. In: Proceedings of the 27th Annual ACM Symposium on Applied Computing, pp. 1709–1716 (2012)

6. Aguirre, S., et al.: Methodological proposal for process mining projects. Int. J. Bus. Process Integr. Manag. **8**(2), 102–113 (2017)

7. Baijens, J., et al.: Establishing and theorising data analytics governance: a descriptive framework and a VSM-based view. J. Bus. Anal. 1–22 (2021)

8. Bose, R.P.J.C., et al.: Wanna improve process mining results ? It's high time we consider data quality issues seriously. In: IEEE Symposium on Computational Intelligence and Data Mining (CIDM 2013), pp. 127–134 (2013)

9. Brockhoff, T., et al.: Process prediction with digital twins, pp. 182–187 (2021)

10. Calvanese, D., Kalayci, T.E., Montali, M., Tinella, S.: Ontology-based data access for extracting event logs from legacy data: the onprom tool and methodology. In: Abramowicz, W. (eds.) Business Information Systems. BIS 2017. LNBIP, vol. 288, pp. 220–236. Springer, Cham (2017). https://doi.org/10.1007/978-3-319-59336-4_16

11. Cheng, H.J., et al.: Process mining on noisy logs - can log sanitization help to improve performance? Decis. Support Syst. **79**, 138–149 (2015)

12. van Cruchten, R.M.E., et al.: Process mining in logistics: the need for rule-based data abstraction. In: 2018 12th International Conference on Research Challenges in Information Science (RCIS), pp. 1–9 (2018)

13. DAMA International: Data Management. In: DAMA-DMBOK Data Management Body of Knowledge, 2nd edn. Technics Publications (2017)

14. Dijkman, R., Gao, J., Syamsiyah, A., van Dongen, B., Grefen, P., ter Hofstede, A.: Enabling efficient process mining on large data sets: realizing an in-database process mining operator. Distrib. Parallel Databases **38**(1), 227–253 (2019). https://doi.org/10.1007/s10619-019-072 70-1

15. van Eck, M.L., et al.: PM2: a process mining project methodology. In: Zdravkovic, J., et al. (eds.) Advanced Information Systems Engineering, pp. 297–313. Springer International Publishing, Cham (2015)

16. Eichler, R., et al.: Modeling metadata in data lakes—a generic model. Data Knowl. Eng. **136**, 101931 (2021)

17. Elkoumy, G., Fahrenkrog-Petersen, S.A., Dumas, M., Laud, P., Pankova, A., Weidlich, M.: Secure multi-party computation for inter-organizational process mining. In: Nurcan, S., Reinhartz-Berger, I., Soffer, P., Zdravkovic, J. (eds.) BPMDS/EMMSAD -2020. LNBIP, vol. 387, pp. 166–181. Springer, Cham (2020). https://doi.org/10.1007/978-3-030-49418-6_11

18. Emamjome, F., et al.: Alohomora: unlocking data quality causes through event log context. In: Proceedings of the 28th European Conference on Information Systems (ECIS2020), pp. 1–16 (2020)

19. Esser, S., et al.: Multi-dimensional Event Data in Graph Databases. Springer, Heidelberg (2021)

20. Fahrenkrog-Petersen, S.A., et al.: PRETSA: event log sanitization for privacy-aware process discovery. In: Proceedings of 2019 International Conference on Process Mining, ICPM 2019, pp. 1–8 (2019)

21. dos Santos Garcia, C., et al.: Process mining techniques and applications – a systematic mapping study. Expert Syst. Appl. **133**, 260–295 (2019)

22. Geyer-Klingeberg, J., et al.: Process mining and robotic process automation: a perfect match. In: 16th International Conference on Business Process Management, July 2018

23. Ghahfarokhi, A.F., et al.: OCEL Standard

24. Goel, K., et al.: Data governance for managing data quality in process mining. In: Proceedings of the 42nd International Conference on Information Systems (ICIS 2021) (2021)

25. Grisold, T., et al.: Adoption, use and management of process mining in practice. Bus. Process Manag. J. **27**(2), 369–387 (2021)
26. Jacobi, C., et al.: Maturity model for applying process mining in supply chains: literature overview and practical implications. Logist. J. **2020**, 9–14 (2020)
27. Jans, M., et al.: From relational database to event log: decisions with quality impact. In: Teniente, E., Weidlich, M. (eds.) Business Process Management Workshops. BPM 2017. LNBIP, vol. 308, pp. 588–599. Springer, Cham (2017). https://doi.org/10.1007/978-3-319-74030-0_46
28. Leno, V., Polyvyanyy, A., Dumas, M., La Rosa, M., Maggi, F.M.: Robotic process mining: vision and challenges. Bus. Inf. Syst. Eng. **63**(3), 301–314 (2020). https://doi.org/10.1007/s12599-020-00641-4
29. Li, G., de Murillas, E.G.L., de Carvalho, R.M., van der Aalst, W.M.P.: Extracting object-centric event logs to support process mining on databases. In: Mendling, J., Mouratidis, H. (eds.) CAiSE 2018. LNBIP, vol. 317, pp. 182–199. Springer, Cham (2018). https://doi.org/10.1007/978-3-319-92901-9_16
30. Maita, A.R.C., et al.: A systematic mapping study of process mining. Enterp. Inf. Syst. **12**(5), 505–549 (2018)
31. Mannhardt, F., et al.: Privacy-preserving process mining: differential privacy for event logs. Bus. Inf. Syst. Eng. **61**(5), 595–614 (2019)
32. Marin-Castro, H.M., et al.: Event log preprocessing for process mining: a review. Appl. Sci. **11**(22), 1–29 (2021)
33. Martin, N., et al.: Opportunities and challenges for process mining in organizations: results of a Delphi study. Bus. Inf. Syst. Eng. **63**(5), 511–527 (2021). https://doi.org/10.1007/s12599-021-00720-0
34. Martin, N., et al.: The use of process mining in business process simulation model construction structuring the field. Bus. Inf. Syst. Eng. **58**(1), 73–87 (2015)
35. Mishra, V.P., et al.: Process mining in intrusion detection-the need of current digital world. In: Singh, D. et al. (eds.) Advanced Informatics for Computing Research, pp. 238–246. Springer, Singapore (2017). https://doi.org/10.1007/978-981-10-5780-9_22
36. Nguyen, G.-T.: Siemens: driving global change with the digital fit rate in Order2Cash. In: Reinkemeyer, L. (eds.) Process Mining in Action, pp. 49–57. Springer, Cham (2020). https://doi.org/10.1007/978-3-030-40172-6_9
37. Park, G., et al.: Realizing a digital twin of an organization using action-oriented process mining. In: Proceedings of 2021 3rd International Conference on Process Mining, ICPM 2021, pp. 104–111 (2021)
38. Price, R.J., et al.: Empirical refinement of a semiotic information quality framework. In: Proceedings of Annual Hawaii International Conference on System Sciences, p. 216 (2005)
39. Rafiei, M., von Waldthausen, L., van der Aalst, W.M.P.: Supporting confidentiality in process mining using abstraction and encryption. In: Ceravolo, P., van Keulen, M., Gómez-López, M.T. (eds.) SIMPDA 2018-2019. LNBIP, vol. 379, pp. 101–123. Springer, Cham (2020). https://doi.org/10.1007/978-3-030-46633-6_6
40. Suriadi, S., et al.: Event log imperfection patterns for process mining: towards a systematic approach to cleaning event logs. Inf. Syst. **64**, 132–150 (2017)
41. Sutcliffe, A.: Scenario-based requirements engineering. In: Proceedings of 11th IEEE International Requirements Engineering Conference, 2003, pp. 320–329 (2003)
42. Syed, R., Leemans, S.J.J., Eden, R., Buijs, J.A.C.M.: Process mining adoption. In: Fahland, D., Ghidini, C., Becker, J., Dumas, M. (eds.) BPM 2020. LNBIP, vol. 392, pp. 229–245. Springer, Cham (2020). https://doi.org/10.1007/978-3-030-58638-6_14
43. Verbeek, H.M.W., Buijs, J.C.A.M., van Dongen, B.F., van der Aalst, W.M.P.: XES, XESame, and ProM 6. In: Soffer, P., Proper, E. (eds.) CAiSE Forum 2010. LNBIP, vol. 72, pp. 60–75. Springer, Heidelberg (2011). https://doi.org/10.1007/978-3-642-17722-4_5

44. Weber, P., et al.: A principled approach to mining from noisy logs using heuristics miner. In: Proceedings of 2013 IEEE Symposium Computational Intelligence Data Mining, CIDM 2013 - 2013 IEEE Symposium Series on Computational Intelligence, SSCI 2013, pp. 119–126 (2013)
45. Weigand, H., et al.: An artifact ontology for design science research. Data Knowl. Eng. **133**, 101878 (2021)
46. Wynn, M.T., Sadiq, S.: Responsible process mining - a data quality perspective. In: Hildebrandt, T., van Dongen, B.F., Röglinger, M., Mendling, J. (eds.) BPM 2019. LNCS, vol. 11675, pp. 10–15. Springer, Cham (2019). https://doi.org/10.1007/978-3-030-26619-6_2

Process Mining for Process Improvement
- An Evaluation of Analysis Practices

Kateryna Kubrak[1]([✉])(iD), Fredrik Milani[1], and Alexander Nolte[1,2]

[1] University of Tartu, Tartu, Estonia
{kateryna.kubrak,fredrik.milani,alexander.nolte}@ut.ee
[2] Carnegie Mellon University, Pittsburgh, PA, USA

Abstract. Organizations have a vital interest in continuously improving their business processes. Process analysts can use process mining tools that provide data-driven discovery and analysis of business processes to achieve this. Current research has mainly focused on creating and evaluating new tools or reporting process mining case studies from different domains. Although usage of process mining has increased in industry, insights into how analysts work with such methods to identify improvement opportunities have consequently been limited. To reduce this gap, we conducted an exploratory interview study of seven process analysts from different domains. Our findings indicate that process analysts assess improvement opportunities by their impact, the feasibility of required implementation, and stakeholders' input. Furthermore, our results indicate that process mining tools, when used to identify improvement opportunities, do not provide sufficient support for analysis, requiring process analysts to use additional tools. Lastly, analysts use storytelling to frame and communicate their findings to various stakeholders.

Keywords: Process mining · Business process analysis · Business process improvement

1 Introduction

Organizations engage in business process management (BPM) to continuously improve their business processes. In doing so, process analysts model the business processes, use a variety of methods to analyze them, and then, based on the results of the analysis, propose and implement changes to the processes [10]. In recent years, process analysts have begun using data-driven methods, such as process mining, to improve processes [8,24]. Therefore, process analysts have begun to incorporate commercial process mining tools, such as Disco[1], Celonis[2], and Apromore[3] in their continuous BPM work [24]. Process mining tools use

---

[1] https://www.fluxicon.com/disco/.
[2] https://www.celonis.com.
[3] https://apromore.org.

© The Author(s) 2022
R. Guizzardi et al. (Eds.): RCIS 2022, LNBIP 446, pp. 214–230, 2022.
https://doi.org/10.1007/978-3-031-05760-1_13

event logs, i.e., data recorded from process executions, to enable automated discovery of business process models and process analysis [1]. With such data-driven tools, process analysts gain a more complete and accurate understanding of the process execution and save time when discovering and analyzing the business processes [1].

The benefits of process mining for improving business processes have been demonstrated in different industries [24,29], such as logistics [16], manufacturing [27], telecommunication services [20], and auditing [15]. To this end, methodologies for applying process mining, such as PM^2 framework [30], have been proposed. Similar methodologies have also been proposed for specific industries, such as for healthcare [23]. While such methodologies can help process analysts, the analysis conducted to identify improvement opportunities is still manual. Furthermore, methods for applying process mining tools stipulate steps to take and *what* to analyze, but not *how* to analyze. Some studies explore practical aspects of process mining, such as process managers' perception of adopting, using, and managing process mining [13] and how process mining is used by organizations [29]. However, the majority of works mainly consider technical aspects, i.e., development and improvements of process mining techniques [6]. Few studies explore how process analysts use and work with process mining in process improvement initiatives, although the need to research teams and skills needed for successful process mining projects has been highlighted [22]. Thus, there is a gap in how analysts use process mining to identify improvement opportunities, assess which improvement opportunities to pursue, and communicate analysis results to relevant stakeholders.

This paper explores how process analysts working with business process improvement, incorporate and use process mining solutions to discover, analyze, and communicate improvement opportunities. Therefore, this paper's research objective is to explore "*how process analysts work with process mining when engaged in process improvement initiatives?*" In addressing this research objective, we specifically explore three research questions. The first relates to how process analysts use process mining to identify improvement opportunities and which improvement opportunities to address. Process analysts also use process mining to present the findings of their analysis to stakeholders. Therefore, the third research question concerns *how* process analysts use process mining to communicate their findings to stakeholders. To this end, we explore the following research questions.

RQ$_1$. *How do process analysts use process mining to identify improvement opportunities?*

RQ$_2$. *How do process analysts use process mining to select improvement opportunities to address?*

RQ$_3$. *How do process analysts use process mining to communicate their findings?*

To address these research questions, we conducted an exploratory interview study. We interviewed process analysts who use process mining to discover, analyze, and identify improvement opportunities in their daily work. We present findings on how analysts use process mining when working with business process improvement. More specifically, we describe the strategies analysts employ when using process mining to identify improvement opportunities, compare and assess such opportunities, and communicate their findings. Thus, our findings add to our understanding of process mining solutions use and utility in practice.

The derived insights can be useful for process analysts and researchers in the field of process mining. Process analysts can gain a broader understanding of how to use process mining tools to identify improvement opportunities. Insights on how process analysts use process mining tools to identify improvement opportunities can help researchers develop data-driven discovery of such opportunities, especially for efficiency gains. Finally, our findings can possibly be insightful for developers of process mining tools since they can improve their tools to accommodate process analysts' needs better.

The remainder of this paper is structured as follows. Section 2 presents the background and related work. Then, in Sect. 3, we present the research method, while Sect. 4 presents the results. The findings are discussed in Sect. 5, and finally, we conclude the paper in Sect. 6.

2   Background and Related Work

Process analysts benefit from using process mining to analyze event logs, i.e., data recorded from the process executions [1]. To this end, a set of methodologies have been proposed that aid the implementation of process mining in process improvement initiatives. For instance, Bozkaya et al. [4] and Rojas et al. [25] propose process mining methodologies that address the specific needs of business process analysis in healthcare. The PM^2 framework [30], on the other hand, is industry agnostic and includes planning, extraction, and processing of event log data, mining and analysis of data, and evaluation of results. Other frameworks, such as *Process Diagnostics Method* [4] and the *L* life-cycle model* [1] support analysts with structuring their work when employing data-driven methods for process discovery and analysis. Thus, the common denominator of such work is that they present an overall methodology for process mining projects. Our work is complementary as we focus on a specific step of such methodologies by exploring how process analysts use process mining to identify and assess improvement opportunities.

Process mining has been applied to real-life event logs for discovery and analysis. Such case studies have been conducted in different domains, such as the customer fulfillment process of a telecommunication company [20], IT management services [31], library information systems [19], agile software development [21], and the cargo release process of a logistic company [16]. Such studies focus on reporting the results obtained by applying process mining and illustrate the value of process mining in industry [7]. In this paper, we provide insight into how such results were obtained by exploring how analysts use process mining to identify improvement opportunities.

Grisold et al. [13] studied organizational and managerial aspects of process mining while Thiede et al. [29] reviewed 144 research papers to understand the use of process mining by organizations. In a similar vein, in Emamjome et al. [12], 152 case study papers were reviewed to assess the maturity of process mining in practice by using diffusion of process mining and thoroughness of their application as criteria. Syed et al. [28] focus on identifying challenges and enablers of process mining by interviewing stakeholders of one particular organization. Similarly, Klinkmüller et al. [17] reviewed 71 process mining analysis reports to examine the information needed to solve domain-specific problems with process mining tools. These works primarily focus on the organizational perspective for usage of process mining. Our contribution, however, focuses on how process analysts came to the results summarized in the above mentioned case studies and reports. Furthermore, in our work we consider additional aspects besides information needs of process analysts.

3   Empirical Method

To understand how process analysts identify ($\mathbf{RQ}_1$) and select improvement opportunities ($\mathbf{RQ}_2$), and how they communicate their findings to different stakeholders ($\mathbf{RQ}_3$), we conducted an exploratory interview study. This approach is suitable because our aim was to explore a phenomenon – how analysts utilize process mining to improve processes – and gain insights into "how" it takes place from the perspective of the individuals that are involved. We chose semi-structured interviews because they enable a more open conversation between interviewer and interviewees which allows for novel topics to emerge while at the same time providing sufficient structure for a focused conversation on specific topics related to our research focus [11]. In the following, we will elaborate on the specifics of our study setup (Sect. 3.1), data collection, and analysis procedure (Sect. 3.2).

3.1   Study Setup

We recruited a total of seven participants for our study (see Table 1). We selected them across two main dimensions: (1) internal process analysts and consultants, and (2) experience as a process analyst. We chose this differentiation as it can be expected that approaches to identify improvement opportunities vary among individuals familiar with the processes they are tasked to improve (internal process analysts) and those brought in as external experts (consultants) as well as their job experience. Moreover, we also selected our participants from different domains and companies to cover a variety of contexts and use cases. Having conducted six interviews, we noted data saturation, i.e., no new information being provided by additional interviews. We, however, sought and conducted one more interview to ensure we had enough interviews [14]. We conducted individual online interviews with each of the seven selected participants. The interviews lasted between 29 and 46 min each.

Table 1. Study participants

Code	Domain	Project	Study role (experience)
I-01	Electrical engineering	Improving order-to-fulfillment process across multiple countries	Internal process analyst (2 years)
I-02	Insurance	Improving claim-to-resolution process (esp. customer notification)	Internal process analyst (1 year)
I-03	Public services	Improving application-to-approval process for immigration (esp. waiting times)	Internal process analyst (1 year)
I-04	Data science	Improving application-to-approval process (esp. reworks)	Consultant (4 years)
I-05	Auditing	Analyzing claim-to-resolution process at a regional paying agency	Consultant (2 years)
I-06	Process mining	Analyzing standardization and harmonization of processes	Consultant (5 years)
I-07	E-commerce	Improving order-to-cash process (esp. manual tasks)	Internal process analyst (1 year)

3.2 Data Collection and Analysis

Prior to conducting the interviews, we developed an interview guide[4] based on our three main research questions. Thus, the guide included questions related to how interviewees identified process improvement opportunities (e.g., *What was the specific improvement opportunity identified?*, *What were the criteria/measures to identify the improvement opportunities/bottlenecks?*, c.f. RQ_1), and how they decided which opportunity to proceed with (e.g., *How was it decided which one to select?*, *Who made this decision?*, c.f. RQ_2). We also specifically asked each interviewee how they communicated those opportunities they deemed reasonable to implement (e.g., *Who were the results presented to?*, *How did you present your results?*, c.f. RQ_3).

During the interviews we also asked interviewees to provide us with documents, such as frameworks, screenshots of process models, and data tables pertinent to their projects. The interviewees conditioned the interviews on the materials not being made publicly available due to the sensitivity of the contents though. We started each interview by asking the interviewee to think about a recent process improvement initiative they conducted. We proceeded to ask the questions included in the interview guide in the context of this particular project. During the interviews, we did not always stick to the sequence of questions as included in the interview guide, but rather followed the flow of the interview while making sure to cover all aforementioned topics.

---

[4] Link to the full interview guide and the coding scheme: https://doi.org/10.6084/m9.figshare.19071206.v1.

To analyze the interviews, we first transcribed them by using the transcription tool Otter.ai[5]. After manually reviewing and correcting the transcriptions, we conducted a thematic analysis [5] of interview transcripts and documents. Thus, we first familiarized ourselves with the data and created an initial set codes based on our research questions which included codes such as *"improvement opportunity"* and *"analysis methods"* to discover how the interviewees identified improvement opportunities (RQ_1). Moreover, we utilized codes such as *"improvement opportunity impact"* to identify how improvement opportunities were selected (RQ_2) and *"communication media"* to identify how improvement opportunities were reported (RQ_3). One researcher then applied these codes to the interview transcripts before we discussed the coding results in the research team. The discussion subsequently yielded additional codes such as *"context"*, *"process data"*, *"communication strategies"*. The updated coding scheme was then applied to the transcripts by the same researcher before we again discussed the coding results in the team. We iterated this procedure three times until we did not discover any new codes. The final coding scheme included 14 distinct codes (See footnote 4). During the previously described procedure, we used the documents provided by the participants as additional context information to aid our understanding of the responses.

4   Findings

Here, we present the results of our study. We begin with RQ_1 on how improvement opportunities are identified, followed by RQ_2 on how analysts determine which opportunities to address. Finally, we present the results on how process analysts communicate their findings (RQ_3).

4.1   Identifying Improvement Opportunities

The first research question considers how process analysts use process mining to identify improvement opportunities. Our study shows that analysts take a structured approach when using process mining tools. Analysts identify improvement opportunities by visually analyzing discovered process models, use process mining tools to filter event-logs, and produce process variants for analysis and comparison. In so doing, analysts rely on their combined domain knowledge and process mining skills.

F1. Study participants identify improvement opportunities in a structured manner. For instance, one consultant stated that s/he *"follows the Celonis approach mostly"* (I-04)[6]. An internal analyst (I-01) shared that they follow a high-level framework (doc.[7]) with four steps (select, mine, implement, confirm). Another internal analyst, however, stated that *"I don't use a framework but simply try to find out what each project needs"* (I-02). S/he also confirmed that

---

[5] https://otter.ai/.

[6] https://www.celonis.com/ultimate-guide.

[7] We use "doc." to mark findings that are based on documents shared.

"obviously, there is always a part of extracting, cleaning, and analyzing data, interviewing process members, and concluding insights" (I-02). Similarly, one consultant expressed that *"we don't have a formalized methodology"* (I-05), but, at the same time, *"the way we work is very much like the PM^2 methodology"* (I-05). Thus, process analysts we interviewed follow a structured approach, either iterative or sequential, even if the approach is not explicitly stated.

Table 2. Summary of the findings

#	Finding
RQ$_1$. *How do process analysts use process mining to identify improvement opportunities?* (Sect. 4.1)	
F1	Approaching improvement opportunities identification using structured methods (e.g., process mining frameworks, guidelines)
F2	Dividing the big problem into sub-problems to investigate them separately
F3	Visually analyzing discovered process models, particularly, through filtering and variants analysis and comparison
F4	Finding a compromise between the domain knowledge and process mining outlook of the problem
RQ$_2$. *How do process analysts select improvement opportunities to address?* (Sect. 4.2)	
F5	Assessing the impact of the finding on the process in terms of its location and number of cases and variants involved
F6	Analyzing the dependency on entities outside of the process or the organization
F7	Assessing the financial gain of the finding
F8	Using other tools rather than process mining tools for advanced view into the analyzed data and visualizations
RQ$_3$. *How do process analysts communicate their findings?* (Sect. 4.3)	
F9	Using storytelling to present the finding(s) and selecting visual representations according to the story
F10	Adjusting the communication to the client needs' (i.e., including more technical or business details)
F11	Relieving the findings of process mining details and communicating the results and implications

F2. The analysts interviewed, be it explicitly or implicitly, use an overall framework for applying process mining to identify improvement opportunities. However, identifying improvements requires an in-depth analysis of the processes. To this end, analysts employ different tactics. We observed that analysts decompose the problem into smaller and less complicated parts. For instance, one consultant found *"simplification of the problem"* (I-06) to work. S/he stated that *"dividing the problem into smaller chunks to make it manageable is something which I believe works"* because *"you [...] do not want to start with a very* *complex process, so we start [with thinking] what is the best candidate to start with?"* (I-06). Similarly, an internal analyst expressed that *"I think it's better to define one or two improvement areas and just help to improve there"* (I-01).

F3. Study participants also identify improvement opportunities from the discovered process models. As one interviewee confirmed, *"I use like the actual process visualizations to find bottlenecks"* (I-07). More specifically, the analysts interviewed examine the process performance. For instance, an internal analyst focused on waiting times by visually examining the process model to understand *"what kind of steps are involved [...], how long time between steps [...], how many times do the users call"* (I-03). One consultant uses the models to analyze whether *"there are any bottlenecks around the approvals"* (I-05). The analysts studied also combine usage of process mining tools with other analytical tools. For instance, one interviewee conveyed that process maps are useful but *"much more valuable when you have business intelligence capabilities"* (I-03). Similarly, a consultant expressed that for complex things, it is easier to *"move the data out"* and use Python scripts because *"you can do it better with the script"* (I-05). Thus, interviewed process analysts use process models to identify improvement opportunities, such as bottlenecks.

The interviewed analysts filter and compare process models by attributes to identify improvement opportunities. For instance, one interviewee *"showed the differences in process performance for different types of claims"* (I-05). Other internal analysts *"compared it across countries"* (I-01) or considered different years (I-02). What to filter for depends on the business input. According to one consultant, *"before, we used to show the simple variant explorer, but they didn't really work [...] therefore, we are not using the algorithmic grouping of the different variants, but from the business domain, what does make sense to make this grouping"* (I-06). When filtering and comparing, the participants specifically use process mining tools. For instance, one interviewee expressed that *"I mostly used like all of the different types of filterings and, and graphs that you can make in Apromore"* (I-07). Therefore, process analysts seem to discover variants that are compared to identify improvement opportunities. The variants, however, should be defined according to what is sensible from the business perspective.

F4. The visualized process models that analysts use can be misunderstood if one does not understand the data or the business underlying the models. For example, one interviewee said: *"if you don't understand how the data is generated and where the data comes from, you might misinterpret the visualization of the process mining software and, oftentimes, if you don't understand the business process, you might overreact to exceptions that are shown on the process map"* (I-05). Another interviewee, when asked about the same issue, emphasized the importance of domain knowledge. As s/he expressed it, *"I think the most important thing in doing projects like this would be to really understand the process from the people who work with it every day; so like talking to the domain experts, because that's where I think I got the most valuable insights."* (I-07). The reason is that the process model does not provide all the information needed to identify improvement opportunities.

At the same time, knowledge about process mining is essential for identifying improvement opportunities. The implications of lacking process mining knowledge is that it makes it difficult to evaluate the credibility of the finding as *"okay, you have a finding, but is it a false positive, is it there, do you see it, how do you see it, is it important from the process mining view?"* (I-06). These results suggest that understanding of process mining and the domain are necessary for identifying improvement opportunities. One consultant expressed that they ensure that the process mining team has the necessary domain knowledge. Likewise, one consultant emphasizes the necessity to understand both the data underlying the process visualization and the business processes being analyzed.

Overall, results indicate that process analysts we interviewed follow a structured approach when using process mining to identify improvement opportunities. Such improvement opportunities are identified by manually examining the process models. Analysts look for specific improvement opportunities, such as waiting times, and explore the process models to identify previously unknown connections. Analysts also filter the event log to produce variants that are compared, from which improvement opportunities are detected. Finally, our findings show that both domain knowledge and process mining skills are required to identify relevant and credible improvement opportunities.

4.2 Selecting Improvement Opportunities

Here, we present the results concerning the second research question of how process analysts prioritize improvement opportunities to select the one(s) address. Our results show that process analysts consider the impact the improvement opportunity has on the process, the feasibility of the changes, and the potential savings that can be achieved if the improvement opportunity is addressed.

F5. The interviewees assess the impact the identified improvement opportunities have on the overall process. The impact can be assessed by considering the location of the improvement opportunity in the process. For instance, an internal analyst said that it is vital to *"see in the process where this would actually have an impact on"* (I-02). The impact can also be considered by considering if the improvement opportunity *"involves a large population of the process"* (I-06). For instance, in one case involving variants, a consultant expressed that the impact is considered *"with the total number of variance that we see, what is the ratio between the total number of variances and ratio of the process population"* (I-06). Thus, studied process analysts assess improvement opportunities by their impact on the business process.

F6. Another parameter used to determine which improvement opportunities to address is the dependency of the required changes. The dependency is assessed by considering whether the changes require involving external processes or other departments. As one consultant, when discussing a particular improvement opportunity, put it, *"but that's external to the process, so you can't do anything with that."* (I-05). Another aspect that impacts the dependency is the input of process experts, subject-matter experts, and end-users. For example, an internal analyst said that *"I presented to them everything that I found based on* *the data, [...] and then they gave me feedback about what they would implement and would not implement and why."* (I-07).

F7. The main measure used to assess and prioritize improvement opportunities is its financial gain. In the end, *"it's always about the money"* (I-01). The savings that can be realized must be estimated. As one internal analyst put it, *"we build some sort of a business case on how much we can save"* (I-01). The improvements that produce the greatest gains are prioritized *"based on where we can gain the most, and where the biggest problems are"* (I-01). If the financial gain is not sufficient, the improvement opportunity is rejected. One internal analyst stated that *"90% of them are rejected and we [...] want to focus on the ones that will bring us a lot of savings. For example, [...] working on 100 orders is not really saving, it's like someone spends five minutes on this, we're not going to spend all our efforts and time to tackle this problem"* (I-01).

F8. A standard view among the participants was that additional tools, such as Python or Tableau, are needed for visualization. For instance, one consultant shared that s/he uses both Python and Tableau (doc.) because the visualization of process mining software provides *"general statistics, which is really, really, really cool, but oftentimes, if you want to go into more details, you need something more, and that is why I use Tableau"* (I-05). The same reason was given by an analyst expressing that *"[I] use Disco for process mining and write Python scripts when [I] need additional visualizations"* (I-02). Another analyst applied the same tactics but with Excel (doc.) For instance, regarding the process mining tool limitation, *"I also used just Excel for kind of correlation visualization and percentage analysis, because that's not what you can do [with process mining tool]."* (I-07).

In summary, these results show that improvement opportunities are assessed by their impact on the process and their ratio of other cases. Besides, the feasibility of the required changes, input from process and subject-matter experts are also considered. Process analysts also require visualization and advanced data analysis which is not provided by existing process mining tools. Therefore, analysts rely on additional tools, such as Python and Tableau, to analyze and visualize process data.

4.3 Communicating Results

The final research question concerns how process analysts use process mining to communicate their findings. Interviewed process analysts communicate their findings by developing a story that is supported by data and visualizations. Furthermore, analysts simplify the results to suit the audience.

F9. The interviewed process analysts communicate their findings by framing them as a story. In the words of an internal analyst, communicating the findings is *"like storytelling for managers with process mining"* (I-02). For instance, one consultant explained that *"when presenting something to management, you don't have time, and they don't have the attention span to listen to the whole thing and to understand all the details"* (I-05). To this end, to make the findings digestible and relatable, *"you take that piece of information or data and try* *to put it in a simplified context that works as a narrative and easy enough to understand"* (I-05). Given that all the information cannot be shared, a story facilitates putting *"away your technical geeky part and think with a business mind"* (I-05) to consider if *"from a business point of view, this one might have an impact, this one makes sense, and this is strong enough to make a change"* (I-05).

Similarly, a story guides the analysts in what visualizations to use. For instance, one interviewee said that *"when it comes to additional visualizations for reporting, I use outputs from Disco thinking what screenshots exactly would fit what I want to tell"* (I-02). The types of visualizations used vary (doc.), but often, they are *"basically just a screenshot from the system"* (I-01). However, analysts also use process mining to communicate their findings. One internal analyst reported that *"typically, I show them [end-users] things interactively while explaining in parallel"* (I-03).

F10. The communication of findings should be adjusted to the client. For instance, one consultant said that *"different clients have different modes of communication. Some clients require very formal approaches"* (I-04). Regardless of audiences' preferences, process analysts modify the contents by simplifying and adding clarifications. For instance, one consultant creates new visualizations when communicating the findings. *"I tend to use overly complex visuals because I understand them because I created them. But since I'm not sure anybody else would understand it, [...] then I create another one that is very specifically targeted to communicating a message"* (I-05). The same interviewee expressed that *"the ideal process mining software would allow you to be focused on the analytic visuals, but also would let you make or parameterize simplified visuals for communications"* (I-05).

F11. The importance of simplifying is because if the target audience *"would see a process and a very complex process, but then the next question for them would be, So, what shall I do with this?"* (I-06). To simplify, an internal analyst frequently changes the names of the activities when communicating the findings *"when I need to report to some managers who do not understand the names of the activities"* (I-02). The layout of the process diagrams can also matter. The internal analyst who compared the process model for two different years noted that *"the process maps for two different years had different layouts which caused the management to think that one step disappeared, whereas it was just arranged differently"* (I-02).

In summary, analysts we interviewed frame their findings as a story and select data and visualizations according to the storyline. In selecting visualizations, analysts use screenshots and use the process mining tools interactively. Furthermore, the findings' contents are often simplified, and, if needed, custom visualizations are used.

5   Discussion

Our findings provide indications on how analysts identify improvement opportunities with process mining ($\mathbf{RQ}_1$), what aspects influence the assessment of such opportunities ($\mathbf{RQ}_2$), and finally, how the findings are communicated ($\mathbf{RQ}_3$).

With regards to $\mathbf{RQ}_1$, our findings suggest that process analysts take a structured approach when using process mining for improving processes. This is according to the previous results that confirm the need for a structured approach to process mining projects [1,4,23,30]. Process analysts seem to develop methods based on their experiences. This might be due to standard methodologies describing general activities rather than defining more detailed guidelines and specific steps for process analysts to follow [2]. However, we note that such methods are similar to standards ones, such as the PM^2 [30].

Our findings also suggest that process analysts use process mining to find specific weaknesses, such as waiting times, in business processes by, for instance, filtering the event-logs along various dimensions. Also, analysts identify improvement opportunities by exploring the processes with process mining. When exploring, analysts use filtering and variant analysis to find connections between various process parameters that suggest potential improvement opportunities. These findings are aligned with experiences reported in process mining case studies, such as [16,20,31]. However, analysts do not seem to use thematic analysis templates such as those proposed in Djurica et al. [9]. This might be due to such templates not being commonly integrated with process mining tools or that analysts are unaware of them. Our findings on the tactics of process analysts to decompose process issues and apply business-driven rationale for defining variants provide insights not commonly discussed in process mining studies.

Our findings indicate that process analysts use additional tools besides process mining tools. Process mining tools are predominantly used for the discovery of process models and filtering. However, such tools seem to lack the functionalities that process analysts need for visualization. Therefore, process analysts use other tools for visualization techniques not specific to process mining tools. In this regard, our finding is consistent with that of Klinkmüller et al. [17]. In contrast, our findings provide insights as to the reasons why process analysts use other tools.

As to $\mathbf{RQ}_2$, expectedly, process analysts use the relative financial gain as the main criterion for assessing which improvement opportunities to address. However, it is interesting to note that process analysts consider process impact, the feasibility of implementation, and input from other stakeholders, such as process experts, subject-matter experts, and end-users, when prioritizing and determining which opportunities to address. These findings, besides the financial gain as primary criterion, have not previously been extensively discussed.

Furthermore, we note that dependency on entities outside of the process or the organization is also considered. This finding is aligned with that of Thiede et al. [29] who found that process mining is mainly concerned with a single process in a single organization. Process managers find it challenging to select which process to analyze [13]. Restrictions on access and use of relevant and required data [13] might constrain process selection to a single process whereas the analysis could provide more value if cross-system and cross-organizational processes are analyzed with process mining. Therefore, we found that expanding the scope beyond the process being analyzed with process mining is relevant, but process mining tools do not support it sufficiently.

As to (**RQ₃**), our findings indicate that analysts frame their findings as a narrative when communicating their results. Such narratives determine what data and aspects to emphasize. Furthermore, process analysts simplify the analysis and the visualizations when presenting them. Analysts could use process patterns [3] and anti-patterns [18] to facilitate the communication of the improvement opportunities. However, analysts do not seem to use them. Similarly to redesign patterns [9] discussed earlier, this might be due to analysts not being aware of them or that they are not integrated as visualization aids in the process mining tools and, therefore, not feasible to use.

Finally, process analysts use process mining tools to communicate their findings, either by using such tools interactively or by taking screenshots. Our findings shed additional insights on the process behind the results presented in case studies of process mining, such as [16,19,21,31]. However, such studies do not use a story to select and frame the data when presenting their results.

5.1 Implications

Our research has implications for process analysts and developers of process mining tools. More specifically, our findings can be useful for practitioners by providing them with insights on how process analysts work with process mining when engaged in business process improvement initiatives. In addition, providers of process mining solutions can improve their solutions by considering and incorporating visualizations that better cater to the needs of practitioners.

Out findings can be helpful for practitioners. Practitioners manually analyze the output of process mining solutions and, therefore, improvement opportunities can remain undetected. Analysts might overlook an opportunity, or detection might require analysis that is not feasible or possible with existing process mining tools. As a first step, a set of process mining analysis templates can be developed that help analysts to identify common improvement opportunities. While there are a few available, such as discovering rework[8], there are no validated collection of such templates. However, as the next step, insights into how analysts use process mining tools to identify improvement opportunities can help researchers develop algorithms for data-driven discovery of improvement opportunities.

Our findings also have implications for developers of process mining tools. Understanding how analysts use process mining tools to analyze and identify improvement opportunities can help developers to develop process mining tools to serve their end-users better. For instance, such tools can be enhanced to use visualization and patterns to identify opportunities, facilitate analysis of

---

[8] https://fluxicon.com/blog/2017/03/how-to-identify-rework-in-your-process/.

dependencies of a process with other processes, improve visualizations for communication purposes, and incorporate support for financial implications of the processes. Frameworks that aid developers with choosing visualization methods for process mining outputs have been proposed [26]. However, such frameworks focus on descriptive process mining and do not extend beyond process discovery and implicit analysis. Our findings provide insight into what should be considered when developing visualizations for process improvement opportunities.

5.2 Limitations

Our study aimed to explore how analysts use process mining when working with process improvements. These aspects have received limited attention in research so far. It is, thus, appropriate to conduct an exploratory interview study for the given research context. However, there are inherent limitations related to such study designs. We studied seven process analysts who worked on a specific process improvement project within a specific company that operates within a specific domain and utilized specific process mining tools. While we made theoretically motivated selections and selected participants across different domains working in different companies, it can be expected that other study participants working on various process improvement initiatives in different companies, utilizing different process mining, might yield different results.

Moreover, our study population only contained one senior analyst. This can be expected, though, since process analyst commonly is a junior position. We also did not conduct all interviews in the same way, which could inhibit comparability. Our interest was on discovery rather than comparison or prioritization. We also further mitigated this threat by ensuring that we would cover the same topics related to our research questions in all interviews. Another limitation may be related to the number of interview participants. We noted data saturation [14] after six interviews but conducted one more to ensure we had not missed anything. However, we acknowledge this may still remain a limitation to how generalizable our findings might be.

Additionally, a single researcher's coding of the interview transcripts might induce interpreter bias. We attempted to mitigate this bias by ensuring that we followed an established analysis procedure and collaboratively discussing findings at multiple points during the analysis. We also abstain from making causal claims and prioritizing specific findings or interpretations. Instead, we provide a detailed description of how different participants utilized process mining to identify and select improvement opportunities and reported their findings.

6   Conclusion

This paper presented findings from an exploratory interview study on how analysts use process mining to improve business processes. The study specifically aimed at exploring how process analysts use process mining to discover process improvement opportunities, select which ones to address, and communicate their findings. We conducted seven interviews with process mining practitioners, namely internal process analysts and consultants.

Our findings provided tentative insights on the usage of specific methodologies in process improvement projects and the application of process mining. To this end, process analysts follow standardized methods such as PM^2 but also develop their own methods catered to specific organizational needs. When implementing the projects, they do not use process mining tools in isolation but combine them with other analytical tools, such as Tableau, for deeper insight into data. Additionally, we discuss the criteria that process analysts use when prioritizing improvement opportunities, among which are financial gain and dependency on external entities. Our findings also provide indications on how process analysts communicate their findings from process improvement projects, with storytelling being a common method.

Our findings also indicate the importance of visualization of data and process models to identify improvement opportunities. Therefore, for future work, we aim at exploring how visualization can be used to facilitate the identification of improvement opportunities.

Acknowledgments. This research is funded by the European Research Council (PIX Project).

References

1. Aalst, W.: Data Science in Action. In: Process Mining, pp. 3–23. Springer, Heidelberg (2016). https://doi.org/10.1007/978-3-662-49851-4_1
2. Aguirre, S., Parra, C., Sepúlveda, M.: Methodological proposal for process mining projects. Int. J. Bus. Process Integr. Manag. **8**(2), 102–113 (2017)
3. Becker, J., Bergener, P., Breuker, D., Räckers, M.: An empirical assessment of the usefulness of weakness patterns in business process redesign. In: 20th European Conference on Information Systems, ECIS 2012, p. 203 (2012)
4. Bozkaya, M., Gabriels, J., van der Werf, J.M.: Process diagnostics: a method based on process mining. In: 2009 International Conference on Information, Process, and Knowledge Management, pp. 22–27. IEEE (2009)
5. Braun, V., Clarke, V.: Using thematic analysis in psychology. Qual. Res. Psychol. **3**(2), 77–101 (2006)
6. vom Brocke, J., Jans, M., Mendling, J., Reijers, H.A.: Call for papers, issue 5/2021. Bus. Inf. Syst. Eng. **62**(2), 185–187 (2020)
7. Corallo, A., Lazoi, M., Striani, F.: Process mining and industrial applications: a systematic literature review. Knowl. Process. Manag. **27**(3), 225–233 (2020)
8. Dakic, D., Stefanovic, D., Lolic, T., Narandzic, D., Simeunovic, N.: Event log extraction for the purpose of process mining: a systematic literature review. In: Innovation in Sustainable Management and Entrepreneurship, pp. 299–312 (2020)
9. Djurica, D., Gross, S., Yeshchenko, A., Mendling, J.: Process mining supported process redesign: matching problems with solutions. In: Proceedings of the Forum at Practice of Enterprise Modeling 2020, PoEM-Forum 2020, pp. 24–33. CEUR (2020)

10. Dumas, M., La Rosa, M., Mendling, J., Reijers, H.A.: Fundamentals of Business Process Management. Springer, Heidelberg (2018). https://doi.org/10.1007/978-3-662-56509-4_10
11. Edwards, R., Holland, J.: What is qualitative interviewing? A&C Black (2013)
12. Emamjome, F., Andrews, R., ter Hofstede, A.H.M.: A case study lens on process mining in practice. In: Panetto, H., Debruyne, C., Hepp, M., Lewis, D., Ardagna, C.A., Meersman, R. (eds.) OTM 2019. LNCS, vol. 11877, pp. 127–145. Springer, Cham (2019). https://doi.org/10.1007/978-3-030-33246-4_8
13. Grisold, T., Mendling, J., Otto, M., vom Brocke, J.: Adoption, use and management of process mining in practice. Bus. Process. Manag. J. **27**(2), 369–387 (2021)
14. Guest, G., Bunce, A., Johnson, L.: How many interviews are enough? An experiment with data saturation and variability. Field Methods **18**(1), 59–82 (2006)
15. Jans, M., Alles, M., Vasarhelyi, M.: The case for process mining in auditing: sources of value added and areas of application. Int. J. Account. Inf. Syst. **14**(1), 1–20 (2013)
16. Kedem-Yemini, S., Mamon, N.S., Mashiah, G.: An analysis of cargo release services with process mining: a case study in a logistics company. In: Proceedings of the International Conference on Industrial Engineering and Operations Management, pp. 726–736 (2018)
17. Klinkmüller, C., Müller, R., Weber, I.: Mining process mining practices: an exploratory characterization of information needs in process analytics. In: Hildebrandt, T., van Dongen, B.F., Röglinger, M., Mendling, J. (eds.) BPM 2019. LNCS, vol. 11675, pp. 322–337. Springer, Cham (2019). https://doi.org/10.1007/978-3-030-26619-6_21
18. Koschmider, A., Laue, R., Fellmann, M.: Business process model anti-patterns: a bibliography and taxonomy of published work. In: vom Brocke, J., Gregor, S., Müller, O. (eds.) 27th European Conference on Information Systems - Information Systems for a Sharing Society, ECIS 2019 (2019)
19. Kouzari, E., Stamelos, I.: Process mining applied on library information systems: a case study. Libr. Inf. Sci. Res. **40**(3), 245–254 (2018)
20. Mahendrawathi, E., Astuti, H.M., Nastiti, A.: Analysis of customer fulfilment with process mining: a case study in a telecommunication company. Procedia Comput. Sci. **72**, 588–596 (2015)
21. Marques, R., da Silva, M.M., Ferreira, D.R.: Assessing agile software development processes with process mining: a case study. In: 2018 IEEE 20th Conference on Business Informatics (CBI), vol. 1, pp. 109–118. IEEE (2018)
22. Martin, N., et al.: Opportunities and challenges for process mining in organizations: results of a Delphi study. Bus. Inf. Syst. Eng. **63**(5), 511–527 (2021)
23. Rebuge, Á., Ferreira, D.R.: Business process analysis in healthcare environments: a methodology based on process mining. Inf. Syst. **37**(2), 99–116 (2012)
24. Reinkemeyer, L.: Process Mining in Action. Springer, Cham (2020). https://doi.org/10.1007/978-3-030-40172-6_1
25. Rojas, E., Sepúlveda, M., Munoz-Gama, J., Capurro, D., Traver, V., Fernandez-Llatas, C.: Question-driven methodology for analyzing emergency room processes using process mining. Appl. Sci. **7**(3), 302 (2017)
26. Sirgmets, M., Milani, F., Nolte, A., Pungas, T.: Designing process diagrams - a framework for making design choices when visualizing process mining outputs. In: On the Move to Meaningful Internet Systems, OTM 2018, pp. 463–480 (2018)
27. Son, S., et al.: Process mining for manufacturing process analysis: a case study. In: Proceedings of 2nd Asia Pacific Conference on BPM, Brisbane, Australia (2014)

28. Syed, R., Leemans, S.J.J., Eden, R., Buijs, J.A.C.M.: Process mining adoption. In: Fahland, D., Ghidini, C., Becker, J., Dumas, M. (eds.) BPM 2020. LNBIP, vol. 392, pp. 229–245. Springer, Cham (2020). https://doi.org/10.1007/978-3-030-58638-6_14

29. Thiede, M., Fuerstenau, D., Barquet, A.P.B.: How is process mining technology used by organizations? A systematic literature review of empirical studies. Bus. Process. Manag. J. **24**(4), 900–922 (2018)

30. van Eck, M.L., Lu, X., Leemans, S.J.J., van der Aalst, W.M.P.: PM^2: a process mining project methodology. In: Zdravkovic, J., Kirikova, M., Johannesson, P. (eds.) CAiSE 2015. LNCS, vol. 9097, pp. 297–313. Springer, Cham (2015). https://doi.org/10.1007/978-3-319-19069-3_19

31. Vázquez-Barreiros, B., Chapela, D., Mucientes, M., Lama, M., Berea, D.: Process mining in it service management: a case study. In: ATAED@ Petri Nets/ACSD, pp. 16–30 (2016)

Data-Driven Analysis of Batch Processing Inefficiencies in Business Processes

Katsiaryna Lashkevich, Fredrik Milani$^{(\boxtimes)}$, David Chapela-Campa, and Marlon Dumas

University of Tartu, Narva mnt 18, 51009 Tartu, Estonia
{katsiaryna.lashkevich,fredrik.milani,david.chapela,marlon.dumas}@ut.ee

Abstract. Batch processing reduces processing time in a business process at the expense of increasing waiting time. If this trade-off between processing and waiting time is not analyzed, batch processing can, over time, evolve into a source of waste in a business process. Therefore, it is valuable to analyze batch processing activities to identify waiting time wastes. Identifying and analyzing such wastes present the analyst with improvement opportunities that, if addressed, can improve the cycle time efficiency (CTE) of a business process. In this paper, we propose an approach that, given a process execution event log, (1) identifies batch processing activities, (2) analyzes their inefficiencies caused by different types of waiting times to provide analysts with information on how to improve batch processing activities. More specifically, we conceptualize different waiting times caused by batch processing patterns and identify improvement opportunities based on the impact of each waiting time type on the CTE. Finally, we demonstrate the applicability of our approach to a real-life event log.

Keywords: Process mining · Batch processing · Cycle time efficiency

1 Introduction

Companies seek to continuously improve their business processes by incrementally addressing process weaknesses, such as process wastes [33]. Waiting time is a waste in business processes [13]. Waiting times are inevitable in real-life processes, but they can be reduced to improve process efficiency. For instance, waiting times occur with batch processing, i.e., accumulating cases to process them collectively as a group [31]. While batch processing reduces processing time and cost by exploiting economies of scale [32], it also introduces increased waiting times due to case accumulation [7,16,19,32]. Therefore, an optimal batch processing strategy can efficiently balance the processing and waiting times associated with batching. In this regard, practitioners use Cycle Time Efficiency (CTE) to measure the ratio of processing time to the total cycle time [14]. CTE measures the value-adding part, i.e., the processing time compared to the waiting time [14]. As such, CTE-driven process improvements focus on increasing the value-adding part of the process by reducing the waiting time [14,37].

© The Author(s) 2022
R. Guizzardi et al. (Eds.): RCIS 2022, LNBIP 446, pp. 231–247, 2022.
https://doi.org/10.1007/978-3-031-05760-1_14

Modern systems log execution data that can be used in process mining techniques. Such techniques use event logs to enable discovery, analysis, and identification of improvement opportunities in business processes [1]. For instance, process mining techniques enable the discovery of wastes, such as hand-offs between process resources [6], over-processing [38], and waiting times [1]. Process mining techniques have also been used to discover batch processing [19,24,25,39,41]. However, existing techniques focus on batch processing discovery. Therefore, there is a gap in using process mining for batch processing analysis, i.e., to provide analysts with information on how batch processing, for instance, impacts the CTE. More specifically, there is a gap in how much different types of waiting times associated with batch processing contribute to process inefficiencies. To address this gap, we ask (RQ1) *What types of waiting times are associated with batch processing?*, (RQ2) *How can waiting times caused by batch processing be identified from event logs?*, and (RQ3) *How can improvement opportunities, expressed as inefficiencies due to batch processing, be identified from event logs?*

We propose a process mining-based approach for the discovery and analysis of inefficiencies caused by batch processing. In our approach, we first identify batch processing from an event log. Then, we analyze the incurred waiting times per type. For this step, we first define the types of waiting times incurred by batch processing. Finally, we identify improvement opportunities by analyzing how each type of waiting time contributes to CTE inefficiencies. As such, the contribution of this paper is an approach for the discovery and analysis of inefficiencies due to batch processing. The approach enables process analysts to identify improvement opportunities in processes that have the batch processing. We apply our approach to a real-life case using an event log of a production process of a manufacturing company to identify batch processing inefficiencies.

The rest of the paper is structured as follows. Section 2 provides the background and related work. In Sect. 3, the approach is presented. Section 4 covers the evaluation of the approach, and Sect. 5 concludes the paper.

2   Background and Related Work

Batch processing is when a resource accumulates cases to process them together as a group [31]. Batch processing is often used to solve scheduling problems [34], reduce processing times and costs [32,34] by executing multiple cases together. However, introducing batch processing requires accumulating cases, which increases waiting time [7,16,19,32]. As such, batch processing implies a trade-off between processing and waiting time [19,25]. As batch processing impacts processing and waiting times, we can use the CTE metric to measure batch processing efficiency. CTE calculates the ratio of the processing time relative to the cycle time of batch processing activities. CTE ratio close to 1 indicates that the process has comparatively low waiting to processing time and, thus, little room for improvement. However, low CTE indicates a comparatively high waiting time to processing time. Therefore, there is an improvement opportunity [14]. CTE is used to identify time-related process inefficiencies [4,18,20].

For instance, in [20], it has been applied for measuring time-related performance in factories to detect inefficiencies. Ignizio [18] reports that CTE is the most efficient metric in identifying workstation instability in production processes. Furthermore, CTE has been used to assess the efficacy of process redesigns [37]. We, therefore, use the CTE metric to analyze batch processing (in)efficiencies.

Process mining techniques enable the discovery of batch processing behavior from event logs. For instance, Nakatumba et al. [26] address the problem of accurately reproducing batch processing behavior in simulation models. Wen et al. [41] propose a process mining technique to discover batch processing from events logs. Pufahl et al. use event logs to enhance process models with batch processing [31]. Andrews et al. [2] propose an approach to identify and quantify shelf time, i.e., idle time that exceeds acceptable duration, in business processes. Pika et al. [27] propose an approach for discovering batch processing from event logs and discusses, in particular, how to identify batch processing from multiple perspectives of a business process such as activity, resource, and data perspectives. Martin et al. [24] focus on identifying rules that trigger a batch processing activity. Similarly, Martin et al. [25] use batch processing metrics, such as frequency of batch processing, batch size, duration of activity instances, and waiting time in a batch, to describe batching behavior. This paper focuses on discovering and analyzing the waiting times associated with batch processing. Thereby, we build on existing research to identify batch processing inefficiencies.

Process mining techniques have also been applied to assess batch processing performance and explore their impact on process performance. Thus, in addition to batch discovery, in [39] batch processing behavior is visualized and quantitatively analyzed to identify specific patterns such as detecting outliers. Similarly, Klijn [19] propose quantifying batch processing performance using measures such as intra-batch case inter-arrival time, case inter-arrival time, batch interval, batch size, and batch frequency. Pufal et al.[31] examine how overall process performance can be improved in terms of time and cost by simulating batch processing. While these studies use event logs to analyze various performance dimensions of batch processing, they do not consider the different types of waiting times associated with batch processing. Furthermore, their focal point is on batch-processing performance measures. We extend existing work by identifying the different types of waiting times related to batch processing and their impact on process performance to identify potential improvement opportunities.

3 Identification of Batch Processing Inefficiencies

In this section, we present the proposed approach for discovering batch processing inefficiencies. First, we present an overview of the approach and then provide a more detailed description of each step.

The approach for identifying batch processing inefficiencies consists of three steps (see Fig. 1). First, given an event log, the batch processing activities —i.e., the activities that are executed in batches— are discovered. The output of this stage is a report listing such activities, as well as the frequency of their execution being part of a batch. The second step is to analyze their batch processing

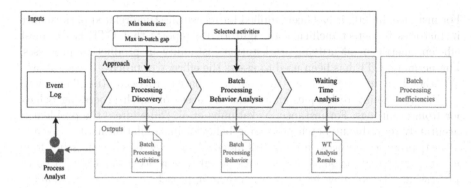

Fig. 1. Overview of the proposed approach

behavior, i.e., type of batch processing, batch size distribution, and batch activation rules. The result of the step is a report describing the batch processing behavior per activity. The final step is to analyze the time that the cases spend waiting before they are batch-processed. This includes an analysis of how batch processing and their associated waiting times impact the CTE. The final output is a report presenting results on batch-processing activities that, based on CTE, can be used to identify improvement opportunities.

3.1   Batch Processing Discovery

For the discovery of batch processing activities, we use the approach proposed by Martin et al. [23] that identifies different types of batch processing activities. They use the resource, start, and end time data to detect when an activity instance is being executed as part of a batch, i.e., by the same resource, and either in sequence, concurrently, or in parallel w.r.t. other activity instances in the batch. Thus, the necessary requirement for the event log is to have the data on the resources, start and end timestamps. In this paper, we focus on batches intentionally processed as such, i.e., the activities were not executed once they were enabled (available for processing) but accumulated and processed as a group. If a case becomes available for processing after the start of the batch, it is excluded as it was not accumulated for batch processing. Accordingly, we extend the definition of a batch with the constraint that all cases part of a batch must be enabled before the batch starts. Therefore, we filter the results obtained by Martin et al. [23] technique to remove such cases.

     To perform such filtering, we need to calculate the enabled time of each activity instance. For cases where this information is not available in the event log, we set the enabled time as the end time of the previous activity if no concurrency relation exists. To obtain the concurrency relations between the activities, we use the concurrency heuristics from the Heuristics Miner [40], as their sensibility to outliers provides more robust results than other naïve techniques.[1] As part of this filtering, we also allow the analyst to define a minimum batch size, i.e., the minimum number of cases that should be accumulated and processed as a group for it to be considered a batch (see Inputs in Fig. 1). The analyst can also define a maximum time gap between the end of processing of a particular case and the start of the next one within one batch (max in-batch gap). This input allows for the discovery of batch processing when interruptions occur. Such interruptions are common for processes executed by human resources [27], e.g., when a resource needs a break or a set-up for particular cases is required [23].

The output of this step is a report listing the activities where the batch processing occurs. For these activities, we identify their case frequency (i.e., the number of cases in which an activity is processed w.r.t. the total number of cases of the process) and batch processing frequency (i.e., the number of times an activity is processed as part of a batch w.r.t. its total number of executions). The list of batch processing activities is sorted by the batch processing frequency in descending order. An analyst can decide which batch processing activities to analyze further based on the frequencies. For instance, the list might have an activity with only 60% batch-processed cases, but the case frequency might be close to 100%. Therefore, the improvement of such batch processing would target 60% of all cases. On the other hand, the improvement of the activity with batch processing frequency close to 100% (all cases are batch-processed) and case frequency of 5% would affect a relatively small number of cases and thus, the overall impact would be negligible. Therefore, a combination of batch processing frequency and case frequency indicates the impact of the batch processing activity. Based on this information, the analyst can determine which batch processing activities to analyze further.

3.2   Batch Processing Behavior Analysis

For each identified batch processing activity, we discover its behavior, i.e., the characteristics of how the batch is processed. We describe batch processing behavior with batch processing frequency, type, size, and activation rule/-s. *Batch processing frequency* indicates the proportion of cases of an activity that are processed in a batch w.r.t. its total executions. *Batch size* illustrates the number of cases processed in a batch. The batch size can be constant (e.g., batch size = 5 cases in 100% of cases) or variable (e.g., batch size = 5 in 40% of cases, batch size = 6 in 60% of cases). *Batch processing type* describes how the cases are batch-processed. In this research, we discover five batch processing types defined by Martin et al. [23]. Parallel batch processing is when all cases are processed at the same time. In a sequential task-based batch processing, one activity is executed for all the cases in the batch, one after another. In sequential case-based batch processing, a set of activities are sequentially executed for each

---

[1]   We note that other notions of concurrency have been proposed in the field of process mining. An in-depth treatment of concurrency notions in process mining is provided in Abel Armas-Cervantes et al. [3].

case in a batch. Concurrent task-based batch processing is when one activity is executed for all cases, but with a partial overlap in case processing. In concurrent case-based batch processing, multiple activities are performed for each case with partial processing time overlap [23].

We also identify *batch activation rule*, i.e. the condition(s) that trigger the batch processing [24]. In the batch activation rule discovery, we follow the approach and batch activation rule types proposed by Martin et al. [24]. The activation rules can be volume-based (batch processing is triggered when a certain volume, weight, or number of cases are accumulated [5,11,17,22,29,30]), time-based (batch processing is time-triggered, e.g., by scheduled date and time [17,36], when a case reaches a waiting time limit [17,28]), resource-based (batch processing activation is dependent on the resource attributes, e.g., the batch is processed when the resource workload allows processing this amount of work [24]), case-based (batch processing is initiated based on the case-related attributes, e.g., arrival of the case of a particular type such as emergency or high-priority case [28]), or context-based (rules embracing other aspects out of the process scope such as meteorological conditions [24]).

Batch can be activated with a single rule, multiple rules, or on an ad-hoc basis. A single rule consists of one condition to be fulfilled to initiate batch processing, e.g., accumulated orders are shipped on scheduled date [17]. Multiple rules present a combination of different types of rules are used to activate a batch processing [28,35], e.g., the combination of volume and time-based rules [9,10, 17]: accumulated orders are shipped based either on the scheduled date or when certain weight, volume, or number of cases are collected [17]. However, when no specific pattern that confidently correlates with the start of batch processing can be identified (i.e., no activation rule is assigned [31]), it is assumed that the batch processing is determined by the resources, i.e., executed ad-hoc. In our approach, the batch activation rules are elicited using the RIPPER technique[2]. The quality of the discovered batch activation rule is measured by support and confidence parameters. The output of this step is a report detailing the behavior of each selected batch processing activity. This information serves as an input for the next step and can be used later by the analyst to identify what parameters to alter to improve batch processing.

3.3   Waiting Time Analysis of Batch Processing Activities

In this section, we describe the characteristics of the five batch processing types considered by our approach, and the waiting times they might induce. We determine three types of waiting times associated with the batch processing – waiting time for batch accumulation, waiting time of a ready batch, and waiting time to process other cases of the batch.

Parallel Batch Processing. In parallel batch processing, a resource starts and completes processing all cases of a batch at the same time [23]. First cases

---

[2] https://github.com/imoscovitz/wittgenstein.git.

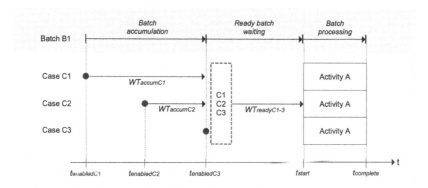

Fig. 2. Waiting times in parallel batch processing

are accumulated, and once accumulated, they can be processed. There are two waiting time types in parallel batch processing (Fig. 2). The first is *waiting time for batch accumulation*, i.e., the time it takes to build up a batch (WT_{accum} in Fig. 2). When all the cases are accumulated, a batch is ready to be processed. But a batch might not be processed as soon as it is ready. Thus, there is a second waiting time - *waiting time of a ready batch* (WT_{ready} in Fig. 2).

To identify these waiting times from the event log, we need enabled time (when a case becomes available for processing), start time (when a case starts being processed), and completion time (when a case processing is finished) [12]. The enabled time is often the same as the completion time of the preceding activity. However, they might differ where parallel activity execution occurs. Therefore, when there are two parallel activities, both of which must be completed before the processing can continue, enabled time is the completion time of the latter activity [8]. Thus, once enabled time is recorded, the case is ready to be processed ($t_{enabled}$ in Fig. 2).

Waiting time for batch accumulation is the time cases wait before a batch is accumulated and ready to be processed. A batch is accumulated when all the included cases are available for processing. Therefore, a batch is accumulated when the last-arriving case in the batch is enabled. Until the enablement of the last case, all earlier cases wait for the batch to be accumulated. Thereby, *waiting time for batch accumulation* ($WT_{accumC1}$, $WT_{accumC2}$ in Fig. 2) is calculated as the difference between the enabled time of the cases in a batch and the enabled time of the last-arrived case. In Fig. 2, a batch is activated when three cases have been accumulated. Cases (see Cases C1, C2) wait until the batch is ready to be processed (accumulated). Their waiting time is, therefore, from their enabled times (see $t_{enabledC1}$, $t_{enabledC2}$) until the enabled time of the last case included in the batch, i.e., the third case (see $t_{enabledC3}$). The earlier a case arrives, the longer it waits. The last-arriving case (see Case C3) has no waiting time for batch accumulation since its arrival marks the batch as ready to be processed.

Waiting time of a ready batch (see $WT_{readyC1-C3}$ in Fig. 2), on the other hand, is the time from when the cases are ready to be processed (batch accumulated)

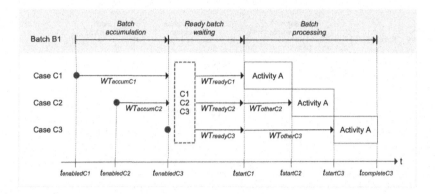

Fig. 3. Waiting times in sequential task-based batch processing

until they are processed (batch processing starts). Therefore, it is calculated as the difference between the enabled time of the last-arrived case ($t_{enabledC3}$) and the start time of the batch processing (t_{start}).

Sequential Batch Processing. In sequential batch processing, a resource accumulates a group of cases and processes them one after another. In sequential task-based batch processing (Fig. 3), the same activity is executed for all cases, one by one. However, in sequential case-based batch processing (Fig. 4), several activities are performed for each case in a batch [23].

Sequential task-based (Fig. 3) and case-based (Fig. 4) batch processing have, similar to parallel batch processing, *waiting time for batch accumulation* and *waiting time of a ready batch*. But sequential task-based and case-based batch processing also have *waiting time for other cases to be processed* within the batch processing, i.e., the time when a case waits while work is done on other cases in the same batch. This waiting time occurs when the resource is processing some case/-s in the batch (e.g., Case C1 in Fig. 3), other cases have to wait for their turn ($WT_{otherC3}$ for Case C2 and $WT_{otherC3}$ for Case C3 in Fig. 3).

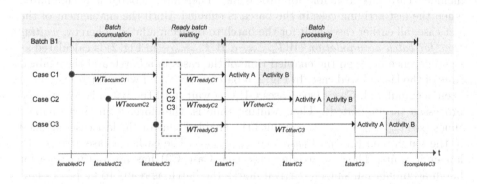

Fig. 4. Waiting times in sequential case-based batch processing

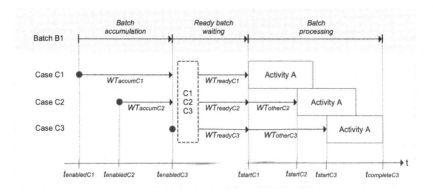

Fig. 5. Waiting times in concurrent task-based batch processing

Waiting time for other cases to be processed is calculated as the difference (interval) between the start time of the first case/-s ($t_{startC1}$ in Fig. 3) in a batch and the start time of the other case/-s ($t_{startC2}$ and $t_{startC2}$ in Fig. 3). In a sequential case-based batch processing (Fig. 4), several consecutive activities are performed for each case of a batch [23]. Waiting time types in sequential case-based batch processing correspond to those in sequential task-based batch processing (see Fig. 4).

Concurrent Batch Processing. Concurrent batch processing is when cases are processed sequentially but with partial time overlap. Concurrent task-based batch processing (Fig. 5) is to execute the same activity for all cases in the batch with an overlap in processing time. In contrast, for the case-based type (Fig. 6), multiple activities are performed for each case with an overlap in processing time [23]. In concurrent task-based (Fig. 5) and case-based (Fig. 6) batch processing, the waiting times are the same as sequential batch processing. Although there is an overlap in processing time, the cases still have to wait for their turn to be processed. Therefore, *waiting time for other cases to be processed* occurs (e.g., $WT_{otherC2}$, $WT_{otherC3}$ in Fig. 5). Given the similar activity processing times, the difference between these batch processing types lies in a shorter *waiting time for other cases to be processed* due to processing time overlaps.

Waiting Time Analysis. We analyze discovered waiting times to identify inefficiencies. We, first, determine how long cases wait on average before they are batch-processed. Then we examine the batch processing efficiency by calculating its CTE, i.e., the ratio of the processing time to the cycle time of the batch processing activity/-ies. Finally, we measure the impact of each waiting time type on batch processing efficiency. We express the impact as the potential CTE improvement, i.e., how much improvement of the CTE can be obtained if a particular waiting time type is eliminated.

The first step is to examine the average waiting time per type. We measure, for each waiting time type, the *average waiting time for batch accumulation, average waiting time of a ready batch, average waiting time for other cases to be*

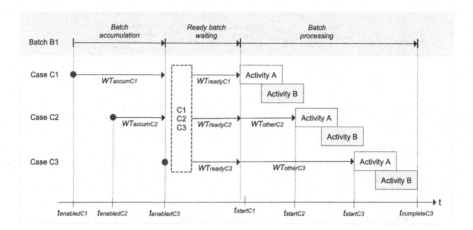

Fig. 6. Waiting times in concurrent case-based batch processing

processed, and *average total waiting time*. These metrics depict how long cases wait during batch accumulation and processing.

The analysis of waiting times provides the input required to assess the impact of discovered waiting time types on the batch processing CTE. It enables the discovery of batch processing inefficiencies from event logs. The impact of each waiting time (CTE_im) on the batch processing CTE is measured as the difference between the CTE of the batch processing activity/-ies and the CTE if that particular waiting time is eliminated. Thus, this metric illustrates the potential CTE improvement if that particular waiting time is eliminated. Thus, if the CTE of a batch processing activity is 50% and, for instance, after eliminating its *waiting time of a ready batch*, the CTE increases to 70%, the potential improvement opportunity of addressing this activity is $CTE_{im} = 40\%$.

The output of this step is a report that presents discovered waiting times per type and their impact on the batch processing activity CTE (CTE_{im}). Process analysts can use the obtained results to identify batch processing inefficiencies and from where (what particular waiting time type) the inefficiencies stem. Thus, the results can aid the analyst in identifying which batch processing activities to focus on in their improvement initiatives and how they can be improved.

4   Evaluation

In this section, we present the evaluation of our approach. First, we use a synthetic event log with artificially injected batches. With this experimentation, we validate the ability of our technique to discover batching behavior and analyze the different types of waiting times. Second, we apply our approach to a real-life event log to demonstrate its applicability in real-life scenarios. The implementation of the approach, as well as the event logs and results of the experiments, are available on GitHub[3].

---

[3] https://github.com/AutomatedProcessImprovement/batch-processing-analysis.

4.1   Experiments with Synthetic Data

For the synthetic evaluation, we artificially added batches to a simulated event log of a loan application process with the specific purpose of evaluating if our technique was able to detect them. To do this, we first grouped the activity instances of three activities of the process. Then, we assigned a new resource to each group to ensure that the activity instances of a batch did not share their resource with other activity instances. Finally, we delayed the start and end timestamps of the activity instances of each group (as well as the timestamps of succeeding activity instances) to force their execution as a batch.

To increase the variety in the synthetic evaluation, we added batches of different sizes (10, 12, and 14) and types (parallel, sequential, and concurrent) to three different activities. We left some activity instances unaltered, so not all their executions were processed in a batch. Furthermore, for some batches, we designed the batch activation to be performed based on temporal rules, e.g., at a specific time of a specific day of the week. Finally, the waiting times were also altered to create batches with different characteristics, e.g., with or without the waiting time of a ready batch.

We evaluated our approach with this event log, resulting in the discovery of all artificially added batches. Moreover, due to the displacement of the batch processing activities and their successors, batch processing was also detected in other activities. The discovered activation rules corresponded to the designed temporal constraints, indicating the day of the week and hour of activation. Finally, the performance analysis detected the correct waiting times and CTE, accurately reporting cases in which some waiting times were set to 0.

4.2   Experiments with a Real-Life Log

This section demonstrates how batch processing inefficiencies can be identified by applying our approach to a real-life event log of a manufacturing production process [21]. This event log has 225 traces, 26 activities, 4953 events, 48 resources and does not contain multitasking (i.e. the resources do not work in more than a task at the same time) [15].

Batch Processing Discovery. We applied our approach to a real-life event log. First, we discovered batch processing activities from the event log. We set the minimum batch size threshold to two cases with no in-batch time gaps (intervals between the case processing within a batch) allowed. The output of this step is a report showing each batch processing activity, their case frequency, and batch processing frequency (see Table 1). From the report, we see that, for this event log, 12 activities had batch processing. The activities are sorted in descending order based on batch processing frequency. For each activity, the report shows the case frequency and the batch processing frequency. From the report (Table 1), we note that activities "Lapping", "Packing", and "Turning Rework", have high batch processing frequencies (91.07%, 83.75%, and 66.67% respectively). These cases are predominantly processed in batches and, therefore, relevant for batch efficiency analysis. However, the "Turning Rework" activity is executed for a

Table 1. Batch processing discovery results.

Activity	Case frequency	Batch processing frequency
Lapping	58.67%	91.07%
Packing	77.78%	83.75%
Turning rework	1.33%	66.67%
Turning & Milling	72.00%	45.78%
Final Inspection Q.C.	78.22%	35.64%
Turning & Milling Q.C.	75.11%	27.78%
Grinding rework	14.67%	23.71%
Laser marking	74.22%	18.25%
Round grinding	49.33%	14.08%
Turning Q.C.	12.00%	10.91%
Flat grinding	26.22%	7.02%
Turning	10.67%	2.35%

comparatively negligent amount of cases (for 1.33% of total cases) and, therefore, has little impact on the process efficiency and is not considered for further analysis. Thus, the report shows that "Lapping" and "Packing" have high batch processing and case frequencies, and therefore, an analyst might decide to select these as an input for the next step.

Batch Processing Analysis. The next step is to analyze the behavior of selected batch processing activities. The output of this step is a report that captures batch processing type, activation rules, and batch size distribution for each batch processing activity (see Table 2). As can be seen from Table 2, the batch processing activity named "Lapping" follows parallel, sequential task-based, and concurrent task-based batch processing types, i.e., in this activity cases can be processed at the same time, sequentially one after another, with or without an overlap in processing time. It indicates that there is no specific style of activity execution. The activity "Packing", on the contrary, follows only a parallel batch processing type, i.e., cases in a batch are processed simultaneously.

Next, we discover the batch activation rules. For instance, for "Lapping", batch processing occurs from 4 to 6 h. This rule has a confidence of 0.94 and a support of 0.12. Finally, the report (Table 2) also captures the discovered batch size distribution per activity. Batch size distribution describes the number of cases that are processed in one batch. In this case, for "Lapping", the batch size distribution is 2 to 4 cases. Most commonly, in 85% of batch processing, batches include 2 cases.

Waiting time analysis. Having discovered the batch processing behavior, we focus on quantitatively analyzing the waiting times. The output for this step is a report that measures average waiting times per type and their impact on batch processing CTE (Table 3). For each activity, the average processing time (PT

Table 2. Batch processing analysis results.

Batch Processing Activity	Lapping	Packing
Batch Processing Type	Parallel, Sequential task-based, Concurrent task-based	Parallel
Activation Rules	$[hour = 4.0 - 5.0]$	$[hour \leq 5.0]$
Activation Rules Quality	$Conf. = 0.94, Supp. = 0.12$	$Conf. = 0.92, Supp. = 0.34$
Batch Size Distribution		

in Table 3) and the average waiting times per type are calculated (WT_{accum}, WT_{ready}, WT_{other} in Table 3).

The waiting times show how long, on average, cases wait for the batch to be accumulated and then processed. For instance, cases wait on an average of 4d,5h,7m while the batch is accumulated for "Lapping". Then, once the batch is accumulated, the cases wait for 2d,14h,20m before the batch processing starts. When cases are processed sequentially or concurrently, these cases wait for an additional 34m on average while other cases are processed. The existing batch processing strategy, therefore, results in a CTE of 1.10% (CTE_b). The CTE value for this activity indicates a potential improvement opportunity. Table 3 shows the impact on the CTE if the waiting times are eliminated. For instance, for "Lapping", if the waiting time for other cases to be processed is eliminated, a slight improvement will be achieved, but the CTE will remain at 1.1% ($CTE_{im3} = 0.0\%$). However, if analysts focus on reducing the waiting time of a ready batch, the CTE could be increased up to 1.78%, which corresponds to an improvement of $CTE_{im2} = 62\%$. The most significant CTE improvement can be achieved if the efforts are focused on reducing the waiting time for batch accumulation (WT_{accum}) that could improve CTE up to 2.78% ($CTE_{im1} = 153\%$).

Table 3. Waiting time analysis results.

Batch processing activity	PT	WT_{accum}	CTE_{im1}	WT_{ready}	CTE_{im2}	WT_{other}	CTE_{im3}	CTE_b
Lapping	0d,1h,50m	4d,4h,39m	153%	2d,15h,31m	62%	34m	0%	1.1%
Packing	0d,1h,0m	10d,8h,34m	267%	3d,17h,56m	33%	0d,0h,0m	0%	0.3%

The analysis demonstrates that attempts to reduce WT_{other} by changing the batch processing type (e.g., use only parallel batch processing in "Lapping") will not add much value. The analysis also shows the reason, i.e., that almost all the waiting time is related to the accumulation and waiting time of a ready batch. We also note from the analysis that the greatest improvements can be achieved by addressing the waiting times for accumulation for "Lapping" and "Packing". This is due to, as the analysis shows, cases being processed relatively fast w.r.t. the time they spend waiting in for accumulation. Based on the analysis, the analyst can consider discarding batch processing in favor of processing cases individually when they are available for these two activities.

5   Conclusion

In this paper, we propose an approach for analyzing event logs to identify improvement opportunities in processes with batch processing activities. In addressing RQ1, we define three types of waiting times associated with batch processing, namely *waiting time for batch accumulation* and *waiting time of a ready batch* in parallel, sequential, and concurrent batch processing, and *waiting time for other cases to be processed* in sequential and concurrent batch processing. We use these definitions to identify waiting times from event logs.

In our approach, we first discover different types of batch processing from an event log. For this, we extend existing research by also considering the waiting times associated with batch processing and discovering their impact on the CTE of batch processing activities (RQ2). Finally, we identify batch processing inefficiencies by measuring the impact of batch processing waiting times on the activity CTE. This enables analysts to identify where there are improvement opportunities and where to target the process changes (RQ3). We evaluated the approach with synthetic data and a real-life event log. The evaluation indicates that our approach can provide analysts with insights on potential batch processing inefficiencies. Thus, the analysts can take a data-driven approach to improve batch processing efficiency.

As a result of the experimentation, we detected potential improvements in the approach. Our approach uses all observations available per batch for the discovery of batch activation rules and reports on confidence and support. This process can be improved by establishing training/test partitions to discover and validate the rules. The approach applicability is, however, limited to the event logs that have data on the resources, start and end timestamps. If the log miss any of these data, the approach cannot be executed and the analyst is informed accordingly. In addition, the non-working periods of resources are currently part of the measured waiting times. Our approach could be extended by adding calendar information to improve the accuracy of both measures. Finally, the evaluation could be extended by including other synthetic event logs and a larger real-life event log with multitasking.

As future work, we plan to implement a what-if simulation analysis for batch processing activities to identify the impact of particular changes on the CTE.

This would allow analysts to change the parameters of batch processing activities and explore what changes would improve batch processing performance and by how much.

Acknowledgments. This research is funded by the European Research Council (PIX Project).

References

1. van der Aalst, W.M.P.: Process Mining - Data Science in Action. 2nd edn. Springer, Cham (2016). https://doi.org/10.1007/978-3-662-49851-4
2. Andrews, R., Wynn, M.: Shelf time analysis in CTP insurance claims processing. In: Kang, U., Lim, E.-P., Yu, J.X., Moon, Y.-S. (eds.) PAKDD 2017. LNCS (LNAI), vol. 10526, pp. 151–162. Springer, Cham (2017). https://doi.org/10.1007/978-3-319-67274-8_14
3. Armas-Cervantes, A., Dumas, M., Rosa, M.L., Maaradji, A.: Local concurrency detection in business process event logs. ACM Trans. Internet Tech. **19**(1), 16:1–16:23 (2019)
4. Arora, S., Rudnisky, C.J., Damji, K.F.: Improved access and cycle time with an "in-house" patient-centered teleglaucoma program versus traditional in-person assessment. Telemed. e-Health **20**(5), 439–445 (2014)
5. Baba, Y.: A bulk service gi/m/1 queue with service rates depending on service batch size. J. Operat. Res. Soc. Japan **39**(1), 25–35 (1996)
6. Burattin, A., Sperduti, A., Veluscek, M.: Business models enhancement through discovery of roles. In: IEEE Symposium on Computational Intelligence and Data Mining, CIDM, pp. 103–110. IEEE (2013)
7. Cachon, G., Terwiesch, C.: Matching Supply with Demand. McGraw-Hill Publishing, New York (2008)
8. Camargo, M., Dumas, M., González, O.: Automated discovery of business process simulation models from event logs. Decis. Support Syst. **134** (2020). Article no. 113284
9. Çetinkaya, S.: Coordination of inventory and shipment consolidation decisions: a review of premises, models, and justification. Applications of supply chain management and e-commerce research, pp. 3–51 (2005)
10. Cigolini, R., Perona, M., Portioli, A., Zambelli, T.: A new dynamic look-ahead scheduling procedure for batching machines. J. Schedul. **5**(2), 185–204 (2002)
11. Claeys, D., Walraevens, J., Laevens, K., Bruneel, H.: A queueing model for general group screening policies and dynamic item arrivals. Eur. J. Oper. Res. **207**(2), 827–835 (2010)
12. Delgado, A., Weber, B., Ruiz, F., de Guzmán, I.G.-R., Piattini, M.: Continuous improvement of business processes realized by services based on execution measurement. In: Maciaszek, L.A., Zhang, K. (eds.) ENASE 2011. CCIS, vol. 275, pp. 64–81. Springer, Heidelberg (2013). https://doi.org/10.1007/978-3-642-32341-6_5
13. Delias, P.: A positive deviance approach to eliminate wastes in business processes: the case of a public organization. Ind. Manag. Data Syst. **117**(7), 1323–1339 (2017)
14. Dumas, M., Rosa, M.L., Mendling, J., Reijers, H.A.: Fundamentals of Business Process Management. Springer, Cham (2013). https://doi.org/10.1007/978-3-642-33143-5

15. Estrada-Torres, B., et al.: Discovering business process simulation models in the presence of multitasking and availability constraints. Data Knowl. Eng. **134** (2021). Article no. 101897
16. Henn, S., Koch, S., Wäscher, G.: Order batching in order picking warehouses: a survey of solution approaches. In: Manzini, R. (eds.) Warehousing in the Global Supply Chain, pp. 105–137. Springer, Cham (2012). https://doi.org/10.1007/978-1-4471-2274-6_6
17. Higginson, J., Bookbinder, J.H.: Policy recommendations for a shipment-consolidation program. J. Busi. Logist. **15**(1) (1994)
18. Ignizio, J.P.: The impact of operation-to-tool dedications on factory stability. In: Proceedings of the 2010 Winter Simulation Conference, WSC 2010, pp. 2606–2613. IEEE (2010)
19. Klijn, E.L., Fahland, D.: Performance mining for batch processing using the performance spectrum. In: Di Francescomarino, C., Dijkman, R., Zdun, U. (eds.) BPM 2019. LNBIP, vol. 362, pp. 172–185. Springer, Cham (2019). https://doi.org/10.1007/978-3-030-37453-2_15
20. Kren, L., Tyson, T.: Using cycle time to measure performance and control costs in focused factories. J. Cost Manage. **16**(6), 18–23 (2002). Article no. 101897
21. Levy, D.: Production analysis with process mining technology (2014). https://doi.org/10.4121/uuid:68726926-5ac5-4fab-b873-ee76ea412399
22. Maity, A., Gupta, U.C.: Analysis and optimal control of a queue with infinite buffer under batch-size dependent versatile bulk-service rule. Opsearch **52**(3), 472–489 (2015). https://doi.org/10.1007/s12597-015-0197-6
23. Martin, N., Pufahl, L., Mannhardt, F.: Detection of batch activities from event logs. Inf. Syst. **95** (2021). Article no. 101642
24. Martin, N., Solti, A., Mendling, J., Depaire, B., Caris, A.: Mining batch activation rules from event logs. IEEE Trans. Serv. Comput. **14**(6), 1837–1848 (2021)
25. Martin, N., Swennen, M., Depaire, B., Jans, M., Caris, A., Vanhoof, K.: Retrieving batch organisation of work insights from event logs. Decis. Support Syst. **100**, 119–128 (2017)
26. Nakatumba, J.: Resource-aware business process management: analysis and support. Ph.D. thesis, Mathematics and Computer Science (2013)
27. Pika, A., Ouyang, C., ter Hofstede, A.: Configurable batch-processing discovery from event logs. ACM Trans. Manage. Inf. Syst. **13**(3) (2021)
28. Pufahl, L.: Modeling and executing batch activities in business processes. Ph.D. thesis, Universität Potsdam (2018)
29. Pufahl, L., Bazhenova, E., Weske, M.: Evaluating the performance of a batch activity in process models. In: Fournier, F., Mendling, J. (eds.) BPM 2014. LNBIP, vol. 202, pp. 277–290. Springer, Cham (2015). https://doi.org/10.1007/978-3-319-15895-2_24
30. Pufahl, L., Weske, M.: Batch activities in process modeling and execution. In: Basu, S., Pautasso, C., Zhang, L., Fu, X. (eds.) ICSOC 2013. LNCS, vol. 8274, pp. 283–297. Springer, Heidelberg (2013). https://doi.org/10.1007/978-3-642-45005-1_20
31. Pufahl, L., Weske, M.: Batch activity: enhancing business process modeling and enactment with batch processing. Computing **101**(12), 1909–1933 (2019). https://doi.org/10.1007/s00607-019-00717-4
32. Reijers, H.A., Mansar, S.L.: Best practices in business process redesign: an overview and qualitative evaluation of successful redesign heuristics. Omega **33**(4), 283–306 (2005)

33. Rohleder, T.R., Silver, E.A.: A tutorial on business process improvement. J. Operat. Manage. **15**(2), 139–154 (1997). Article no. 101897
34. Selvarajah, E., Steiner, G.: Approximation algorithms for the supplier's supply chain scheduling problem to minimize delivery and inventory holding costs. Oper. Res. **57**(2), 426–438 (2009)
35. Sha, D., Hsu, S.Y., Lai, X.: Design of due-date oriented look-ahead batching rule in wafer fabrication. Int. J. Adv. Manuf. Technol. **35**(5), 596–609 (2007)
36. Simons, J.V., Jr., Russell, G.R.: A case study of batching in a mass service operation. J. Operat. Manage. **20**(5), 577–592 (2002). Article no. 101897
37. Venkatraman, S., Venkatraman, R.: Process innovation and improvement using business object-oriented process modelling (BOOPM) framework. Appl. Syst. Innovat. **2**(3), 23 (2019). Article no. 101897
38. Verenich, I., Dumas, M., La Rosa, M., Maggi, F.M., Di Francescomarino, C.: Minimizing overprocessing waste in business processes via predictive activity ordering. In: Nurcan, S., Soffer, P., Bajec, M., Eder, J. (eds.) CAiSE 2016. LNCS, vol. 9694, pp. 186–202. Springer, Cham (2016). https://doi.org/10.1007/978-3-319-39696-5_12
39. Waibel, P., Novak, C., Bala, S., Revoredo, K., Mendling, J.: Analysis of business process batching using causal event models. In: Leemans, S., Leopold, H. (eds.) ICPM 2020. LNBIP, vol. 406, pp. 17–29. Springer, Cham (2021). https://doi.org/10.1007/978-3-030-72693-5_2
40. Weijters, A.J.M.M., Ribeiro, J.T.S.: Flexible heuristics miner (FHM). In: Proceedings of the IEEE Symposium on Computational Intelligence and Data Mining, CIDM 2011, pp. 310–317. IEEE (2011)
41. Wen, Y., Chen, Z., Liu, J., Chen, J.: Mining batch processing workflow models from event logs. Concurr. Comput. Pract. Exp. **25**(13), 1928–1942 (2013)

The Analysis of Online Event Streams: Predicting the Next Activity for Anomaly Detection

Suhwan Lee[✉] [ID], Xixi Lu [ID], and Hajo A. Reijers [ID]

Utrecht University, Utrecht, The Netherlands
{s.lee,x.lu,h.a.reijers}@uu.nl

Abstract. Anomaly detection in process mining focuses on identifying anomalous cases or events in process executions. The resulting diagnostics are used to provide measures to prevent fraudulent behavior, as well as to derive recommendations for improving process compliance and security. Most existing techniques focus on detecting anomalous cases in an offline setting. However, to identify potential anomalies in a timely manner and take immediate countermeasures, it is necessary to detect event-level anomalies online, in real-time. In this paper, we propose to tackle the online event anomaly detection problem using next-activity prediction methods. More specifically, we investigate the use of both ML models (such as RF and XGBoost) and deep models (such as LSTM) to *predict* the probabilities of next-activities and consider the events predicted unlikely as anomalies. We compare these predictive anomaly detection methods to four classical unsupervised anomaly detection approaches (such as Isolation forest and LOF) in the online setting. Our evaluation shows that the proposed method using ML models tends to outperform the one using a deep model, while both methods outperform the classical unsupervised approaches in detecting anomalous events.

Keywords: Process mining · Event stream · Anomaly detection

1 Introduction

Information systems, empowered by blockchain [6] and IoT systems [10], allow an enormous amount of event data to be generated and logged in real-time. The data analytic techniques, such as process mining, are developed to manage the big volume of recorded real-time data. Process mining is a technique to identify and acknowledge the recorded events and gain insights to improve process execution [1]. Recently, process mining has focused on process management and analysis on online settings including process discovery [3], conformance checking [2], and process monitoring techniques [13].

Anomaly detection in process mining aims to detect anomalous behavior in event data [5,12]. Such techniques have been used to identify potential fraudulent behavior to prevent compliance violations [2]. In addition, they are also used to

© The Author(s), under exclusive license to Springer Nature Switzerland AG 2022
R. Guizzardi et al. (Eds.): RCIS 2022, LNBIP 446, pp. 248–264, 2022.
https://doi.org/10.1007/978-3-031-05760-1_15

detect log quality issues to improve data quality [14]. Most existing unsupervised anomaly detection techniques focus on the *case-level* or are situated in *offline* settings [5], i.e., they take as input of a batch event log that contains a set of completed cases.

In practice, detecting anomalies in *online* streaming settings has many advantages, such as being able to take action and timely counter measures. Timely detection also helps dealing with concept drift. At the same time, online anomaly detection also faces many challenges. Unlike an offline setting, which only deals with completed cases, online detection should be able to continuously handle incomplete, ongoing cases. Moreover, online detection should pinpoint anomalies at *event-level* to allow timely, concrete reactions. For example, if credit card fraud is established, the techniques should immediately detect and pinpoint which purchase events are suspicious.

In this paper, we propose to tackle unsupervised anomalous event detection by predicting which activity is next, assuming an ongoing case. More specifically, we first learn a predictive model to predict next activities by preprocessing the completed cases into feature vectors and using machine learning-based classification algorithms (such as Random Forest, XGBoost, LSTM). When a new event of an ongoing case arrives, we apply that predictive model to predict the possible activities and their probabilities using the previous events. The less likely that an activity occurs, the more likely it is an anomaly.

We conduct an evaluation and compare our approach to other approaches that simply encode the events and apply unsupervised anomaly detection algorithms (including Isolation Forest, LOF, OCSVM). The results show that our approach performs better in terms of F1 scores. Therefore, this seems a promising direction for online anomaly detection, which raises many new research challenges.

The paper is organized as follows. Related work is discussed in the next section. Section 3 describes the preliminary knowledge for easy understanding on the proposed method. Section 4 introduces the proposed method and setup of evaluation is described in Sect. 5. The experimental results are reported in Sect. 6, while conclusions and challenges are drawn in Sect. 7.

2 Related Work

In this section, we discuss the event anomaly detection techniques that are related to our approach, listed in Table 1. We categorize existing approaches along two dimensions, (1) *offline* versus *online* and (2) *case-level* versus *event-level*. We discuss them accordingly. Then, our contributions in relation to related works are presented.

The classical approach to detect anomalies in process data is aimed at analyzing an offline event log, which has a fixed number of events. Regarding *offline*, *case-level* anomaly detection algorithms, existing works try to discover anomalous cases in an event log, which have infrequent patterns [5,18] or statistically deviate to an event log [8]. Sani et al. [18] suggested an outlier filtering method

Table 1. Comparison of related anomaly detection approaches

	Online/Offline	Target status	Algorithm
Ghionna et al. [5]	Offline	Case-level	Markov cluster
Sani et al. [18]			Occurance probability
Khatuya et al. [8]			Ridge regression
Nguyen et al. [14]		Event-level	Autoencoder
Savickas and Vasilecas [19]			Bayesian belief network
Nolle et al. [15]			LSTM
Tavares et al. [20]	Online	Case-level	Process model conformity check
Ko and Comuzzi [9]			Leverage score calculation
Neto et al. [22]			Autocloud
Van Zelst et al. [23]		Event-level	Automaton processor

based on observed subsequence, in which the activity below the threshold is classified as outlier according to the succeeding activity probability. Khatuya et al. [8] proposed to use ridge regression to estimate an anomaly score of the individual case by obtaining a statistical distribution of event log features. Although some events in the anomalous case may be normal, these methods are focused on the case rather than individual events.

In case of *offline, event-level* anomaly detection, a probabilistic process model has been proposed [19]. Savickas and Vasilecas used Bayesian Belief Networks to detect an anomalous event. An event with a low probability according to the obtained table is classified as an outlier. To improve detector performance, some methods use a deep neural network [14,15]. Nguyen et al. [14] have proposed anomalous event detection and reconstruction framework, which uses Autoencoder to extract normative features from an event log. Nolle et al. [15] recently introduced a framework to detect case and event anomalies by obtaining the probability of the next event with the deep neural network LSTM. The model based on deep neural networks outperforms existing anomaly detection techniques. However, the works that detect anomalous events assume a constant process distribution in an event log.

Regarding *online, case-level* anomaly detection, statistical leverage and data clustering models are used. Ko and Comuzzi [9] have proposed to use a sliding window with a recent event feature vector to calculate the statistical leverage score of the coming trace. When the sliding window is updated, a trace with a higher leverage score is classified as anomalous. Neto et al. [22] have described how to use Autocloud to detect anomalous cases from a stream of events. The model updates the data cluster and classifies anomalous data which is deviated from the existing cluster. To provide a process model, Tavares et al. [20] have proposed a framework that classifies an anomalous case by checking conformity between discovered model and the target case. The proposed methods are able to detect outliers in a streaming event log with a normative process change. Nevertheless, a streamed event is not simultaneously classified and the works are limited in taking into account case-level detection.

Specifically for *online, event-level* anomaly detection, Van Zelst et al. [23] have proposed automaton based filtering that uses a sliding bucket with a finite number of events. The model learns the probability of activity sequences. The advantage of this approach is that the model classifies a streaming event as soon as the event arrives. However, the detector performance may be limited, considering that the training data takes only activity occurrence and only one or two consecutive event sequence is used.

In summary, we see an opportunity in online, event-level anomaly detection to deal with process change and possible performance improvements. The traditional anomaly detection methods are not designed to cope with a streaming event log. The detector assumes a steady process and is not updated. In the case of anomaly detection in a streaming event log, existing works have mainly focused on case-level detection that are capable of catching anomalous data after the case is finished. In addition, the work for event-level detection uses a simple probabilistic model. In this paper, we propose an approach for online anomalous event detection. The proposed approach is based on the machine learning model for performance. The arrived event is classified by a retrained model taking into account possible process change.

3   Preliminaries

Before discussing the steps of the proposed approach in detail, let us explain some required preliminaries. This involves notations for the event log, the key components of pre-processing, and some detection mechanisms. The concepts discussed in this section are implemented to develop our anomaly detection approach, which is discussed in Sect. 4.

3.1   Event Log

An event log contains cases, which consist of a sequence of events. An event consists of multiple attributes including case id, activity, and timestamp. Let $\mathcal{G}_{att} = \{\mathcal{D}_1, \mathcal{D}_2, ..., \mathcal{D}_n\}$ be a set of all possible attributes and $\mathcal{D}_i$ be a set of all possible values for the attribute i. Attributes could be numerical or continuous values. For example, a timestamp takes a numerical value within an interval from beginning to end of an event log. The categorical attribute takes a value from a given set, e.g., an activity with a string data type is assigned within the labels $\{a_1, a_2, ..., a_n\}$. Therefore, we can express an event as a tuple $e = \langle c, act, tst \rangle$, where c, act, tst are case id, one label from set of activity, and a point of time in an event log for timestamp, respectively.

3.2   Next Activity Prediction

Next activity prediction is one of the techniques in process mining to predict a following activity of running case. Predictions are made using a *classifier* that takes a fixed number of independent features as input. The classifier learns mathematical function to estimate target variables. This means that a classifier extracts features and predicts a probability of following activity from previous events of a case. The data in an event log and previous events of a target case are used as input for training and testing of the classifier, respectively. Both input data are pre-processed and encoded to a feature vector of equal size. The output from the classifier consists of possible activity labels and probability of the candidates. The activity with the highest probability is selected as a prediction.

3.3   Unsupervised Anomaly Detection

The models used in unsupervised anomaly detection extract information from the data and map input matrix to a feature space. During a classifier training phase, the data in an event log is transformed and allocated to feature space. The classifier takes events from a running case, including target event, and calculates a distance between the running case and training data mapped into the feature space. If the target event sufficiently deviates from the training data, above the *anomaly threshold*, the event is denoted as anomalous. Otherwise, the event is normal.

4   Approach

This section presents in detail the proposed approach for detecting anomalous activities in streaming event logs: *predictive anomaly detection*. The steps of the proposed approach are shown in Fig. 1. We first explain our approach in an off-line setting in Sect. 4.1. Next, we discuss how the approach can be adapted in an online setting in Sect. 4.2.

4.1   Predictive Anomaly Detection (PAD)

The proposed approach uses machine learning classification methods to predict next activities for detecting anomalous events. We divide the approach into five steps, as shown in Fig. 1. In the first two steps, we *pre-process* the event log and *train* a model to predict next activity for an event. In the step 3–5, we classify an event e to be an anomaly or not by first *retrieving* its previous event e'. Next, using the trained model, we *predict* the next activities of e' and their probabilities. We then *detect* whether the probability of e is above a threshold. In the following, we explain the approach in depth.

Step 1 and 2: Pre-processing and Model Training. To deploy machine learning techniques, which requires feature space in regular format, we transform the finished cases into a training dataset. The dataset is suitable for model training by using *prefix-bucketing* and *feature encoding*. Since our approach relies on the output of a next activity prediction approach, pre-processing methods are adopted from existing predictive process monitoring approaches [21].

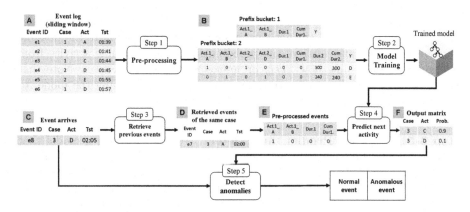

Fig. 1. Online anomalous event detection procedure: Proposed approach

Cases used for model training may have a different sequence of events. A common choice is to group cases into buckets by the same prefix length and separately pre-process the bucket for individual anomalies detector [11]. Multiple classifiers are trained to learn features from respective bucket which contains cases. Prefix bucket n contains events from first to nth event of each case.

For example, let us consider two cases in an event log, both of which have 3 events (see Fig. 1A). Two prefix length buckets are obtained, one with all prefixes of length 1, and the other with all prefixes of length 2. For each event in the buckets, its next activity is used as output label, Y, for training a classifier (Fig. 1B).

In order to train a classifier, the collected events in the buckets are required to be modified as feature vector with fixed size. For feature encoding, we transform the attributes (e.g., activity and timestamp) into suitable features with index-base encoding method to maintain the order of events. The categorical attribute, such as activity label, is encoded using a one-hot encoding scheme by considering the order of events [11]. An encoded activity label a for case c at event i is:

$$c_{i,act} = \begin{cases} 1 & \text{if } c_{i,a} = Act.i\_a \\ 0 & \text{Otherwise} \end{cases}$$

Regarding a timestamp attribute, a point of the event occurrence is transformed into duration and cumulative duration of an event and a case, respectively (see Fig. 1B). Event duration $Dur.i$ is elapsed time between a preceding event $i - 1$ and a current event i, while cumulative duration $CumDur.i$ at event i is aggregated event duration since the start of the case. Given an event timestamp, the obtained duration and cumulative duration respectively are: $Dur.i = e_{tst,i} - e_{tst,i-1}$ and $CumDur.i = \sum_{n=1}^{i} Dur.n$

After prefix-bucketing and feature encoding, the pre-processed rows and columns are concatenated together as an input feature matrix for model training.

The objective of this phase is to learn a model (i.e., a classifier) that uses previous events to predict the activity of next event. In addition to the possible next activities, we also retrieve the probabilities of each possible activity from the model to determine how *likely* or *unlikely* the next activity is. Note that we use the *completed* cases for training a model. We then apply the model to predict the next coming event of *running* cases in step 3–5.

Step 3-5: Retrieve, Predict, and Detect. We have trained a model that predicts the next possible activities. When a new event e of a running case arrives (Fig. 1C), we first retrieve its previous event e' (Fig. 1D) and the encoded event (Fig. 1E). We use this encoded event as input for the trained model to predict the possible next activities and their probability, which we wrapped into an *output matrix* (Fig. 1F).

We then use this output matrix to classify the new event e. If the activity of event e is listed on the possible activities and its probability is above a sufficient level, i.e., the *anomaly threshold*, this means that the target event occurs commonly after e' and, therefore, is classified as normal. Otherwise, we classify e as an anomaly.

As an example, let us assume the anomaly threshold is 0.15. Event e_8 in Fig. 1 arrives as the second event of Case *3* and has an activity label D. We first retrieve the previous events of Case *3*. We then apply the trained model to predict the possible next activities. According to the output matrix of the trained model, the probability of e_8 being activity D is only 0.1. Due to this probability lower than the anomaly threshold, this event classified as anomalous.

4.2 Online PAD Using Sliding Window and Retraining Interval

We have explained the proposed approach in an offline setting. In this section, we explain how the approach handles streaming events.

A *sliding window* is one approach to learn features from new observations of sequential data via updated window [7]. Before the pre-processing step, a sliding window collects a number of most recent completed cases from the streaming events and manages the data to be transformed. As a newly finished case from event streams arrives, a sliding window takes it as an input and places it at the beginning of the window. A case located at the end of the window is removed to maintain the fixed sliding window size (W). As long as W is relatively small to the total number of cases, this procedure allows the machine learning model to be agilely retrained considering possible normal behavior changes of the dataset by collecting the information from more recent observations. However, a small sliding window size may be insufficient to properly train due to the lack of information on normal behavior. As well as issues on size for model training, a small sliding window size requires frequent retraining on the model, which leads to unnecessary recalculation of the outcome. We apply multiple parameters to investigate the sensitivity of window size to detect the optimal performance of the proposed approach in our evaluation.

Besides using the sliding window, we implement a parameter called *retraining interval* R to control the retraining frequency of the model. In essence, after retraining the model at a certain point in time t, we pause to retrain the model again until R number of new cases are completed and updated in the sliding window. For example, if the absolute value of R is 1, then the model is retrained after each case is completed. If $R = W$, then the model is retrained after all the cases in the sliding window are updated. For such a retraining interval, we use a relative size to a sliding window, e.g., $R = 10\%$ of W. If the number of new cases inserted into the sliding window satisfies the retraining interval size condition, the anomaly detection model is retrained.

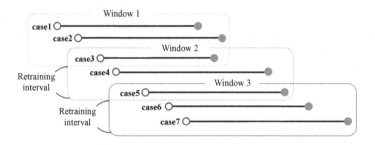

Fig. 2. The architecture of sliding window and retraining interval

Figure 2 exemplifies how the sliding windows are updated along with a retraining interval. Let us assume there are 30 cases in an event log and 10% of sliding window size with 66.7% of retraining interval size. The sliding window takes 3 cases to train a classifier. The window is updated when 2 newly finished cases are inserted which satisfies retraining interval size. Window 1 contains case1 to case3. The window is updated after two new cases, case4 and case5, are inserted into a window. After updating the window, we retrain the model and use the model for detecting anomalies in newly arrived events (e.g., case6 and case7).

5   Empirical Evaluation

The objectives of the evaluation are twofold. Firstly, we investigate how well our approach performs to detect anomalies in comparison to classical unsupervised anomaly detection as baseline. Secondly, we investigate the influence of the parameters on the detection performance, specifically the anomaly threshold, sliding window, and retraining interval. For these objectives, we implemented the proposed approach and the experiments in Python. The code to reproduce the experiments is publicly available on Github[1]. The reader can also check additional results with

---

[1] https://github.com/ghksdl6025/streaming_anomaly_detect.

different performance measurements of the proposed method that have been omitted due to the lack of space. In the following, we first explain the experiment settings, which include the dataset used, the chosen techniques, and the parameter settings. Subsequently, we discuss the results of our evaluation.

5.1   Setup of Evaluation

We have used synthetic logs with anomalous events that were used in [23]. The event logs are generated from the gathered event streams based on the 21 variations of the loan application process [4]. Then 6 anomalous event logs are created by randomly injecting the infrequently occurred events in various probability [23]. Each log follows a different probability range from 2.5% to 15% in steps of 2.5%. In this paper, the probability of generated anomalous events is denoted as *Noise level*. Each log comprises 500 cases and approximately 7600 to 8600 events depends on the noise levels. Every log has 18 different activity labels. Table 2 shows descriptive statistics of the used event logs.

Table 2. Descriptive statistics of the noise imputed event log

Noise level	Cases	Events	Activity labels
2.50%	500	7410	18
5%	500	7630	18
7.50%	500	8542	18
10%	500	7922	18
12.50%	500	7888	18
15%	500	8159	18

As discussed in Sect. 4, sliding window and retraining interval are parameters to respectively control the size of the training data and the retraining rate in the streaming setting. Figure 3 shows evaluation settings to check the influence of the sliding window and the retraining interval size. We consider 5%, 10%, and 20% as a ratio to the total number of cases in the log as sliding window size. For the retraining interval, we tested with 6 different parameters from 0% to 50% in steps of 10% as a ratio of retraining interval size to the sliding window. The influence of the sliding window is examined through changing sliding window sizes with fixed retraining interval, and vice versa. The threshold level is a proportion of anomalies in the dataset, which is used as a criteria in the anomaly detection phase. We experiment with 6 different thresholds, which have a range from 0.01 to 0.25, to investigate the influence of threshold on detection performance.

In this evaluation, we consider both a machine learning model and a deep neural network to classify anomalous events, which are typically adopted in data mining field. As a detection model for the proposed approach, we experiment with two machine learning models and one deep neural network typically

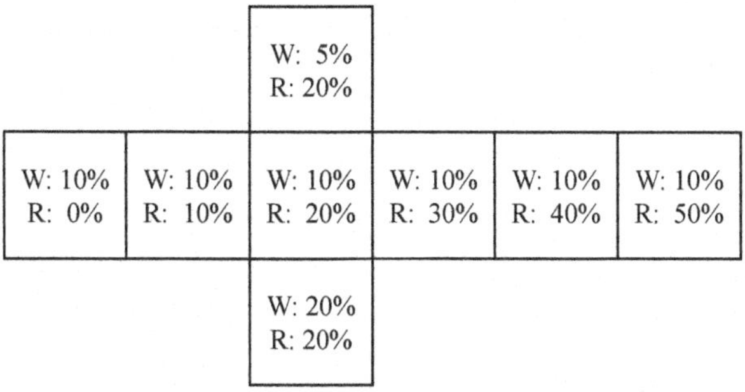

	W: 5% R: 20%				
W: 10% R: 0%	W: 10% R: 10%	W: 10% R: 20%	W: 10% R: 30%	W: 10% R: 40%	W: 10% R: 50%
	W: 20% R: 20%				

Fig. 3. Evaluation settings used in the empirical evaluation, where the window size **W** indicates the percentage of total number of cases and the retraining interval **R** the percentage of sliding window.

adopted in next activity prediction [21]: random forest (RF), extreme gradient boosting (XGB), and long short term memory (LSTM). For the baseline, we experiment with isolation forest (IForest), local outlier factor (LOF), one-class support vector machine (OCSVM), and autoencoder (AE), which are unsupervised anomalous data detection algorithms. For machine learning models, we use the classifiers provided by the Python packages Scikit-learn [17].

For the deep neural network based detection model, the experiment is conducted with the Python package Pytorch [16] with different layer sizes and a number of layers by LSTM and AE. Regarding the LSTM model for the proposed method, two LSTM layers are stacked to obtain hidden feature vectors. Multiple linear layers are followed to predict next activity labels. We implement autoencoder structure including multi-linear layers for latent variables as presented in [14].

To performance of the anomaly detection task is separately evaluated using the F-score, ranged from 0 to 1, for both normal and anomalous events. The F-score is a classification model accuracy indicator calculated from precision and recall, i.e., correct rate among positive predictions and correct decision rate among true items, respectively. The high f-score, close to 1, indicates accurate identification of the events.

5.2 Baseline - Unsupervised Anomalous Event Detection

We propose to use classical unsupervised anomaly detection as the baseline approach. The approach is shown in Fig. 4. We perform the same pre-processing step. However, the model detects the anomaly of an arrived event without predicting the next activity. This way of anomalous event detection is close to traditional outlier classification in data mining as explained in Sect. 3.3. The same pre-processing method is applied to encode input feature vector, as well as sliding window mechanism for streaming data. In the detection step, the classifier takes events of the running case with the target event as input and calculates deviated distance to the feature space. The model distinguishes the anomaly of an arrived event using an anomaly threshold.

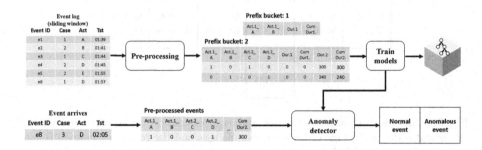

Fig. 4. Online anomalous event detection procedure: Unsupervised method

6   Result and Discussion

Detection Performance of Proposed Approach. Figure 5 shows the F-score of normal and anomalous events detection for the proposed approach and the baseline. From the results, we find that (i) the proposed approach outperforms the baseline on every noise level and (ii) the deep neural network shows a lower performance than machine learning models.

The proposed approach detects anomalous events more effectively than baseline for all every noise levels. Deep neural network based models in both approaches show lower performance than other classifiers. One possible reason for the performance gap is an issue on training data size. The sliding window size may not be big enough for proper deep learning model training.

Effects of Different Anomaly Threshold. Figure 6 shows the F-score of anomalous event detection for the proposed approach by different anomaly threshold levels. According to Fig. 6a, the F-score of both normal and anomalous event decreases with the high threshold in next activity prediction, i.e., all classifiers perform the highest capability on detection at a 0.01 threshold level, except high noise level with RF. We can observe that each classifier requires a separate threshold level for optimal performance. Moreover, RF shows the highest performance at a 0.05 threshold with high noise event log, unlike the other classifiers.

In case of the unsupervised approach, we observe two remarkable points from the analysis in Fig. 6b. These are (i) different performance patterns among implemented algorithms and (ii) the influence of high threshold on anomalous event detection. Regarding performance volatility within models for unsupervised approach, the detection ability of IForest and LOF increases with high

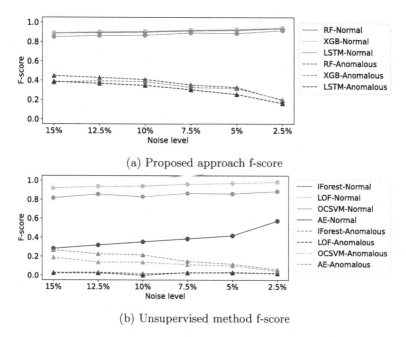

(a) Proposed approach f-score

(b) Unsupervised method f-score

Fig. 5. F-score of normal and anomalous events detection by noise level

threshold while OCSVM and AE are relatively stable at all noise levels. In contrast to the first proposed approach, low threshold to distinguish anomalies could not be applied to the unsupervised method.

Effects of Sliding Windows and Retraining Intervals. We evaluate the anomaly detection performance of the proposed approach by different training windows and retraining interval sizes.

We find that (i) the sliding window size has a positive influence on better model performance, (ii) the proposed approach is more sensitive to window size change than the unsupervised approach, and (iii) the performance change slope is getting flattened with window size. Figure 7a and 7b show the F-score of normal and anomalous event detection for the proposed approach and the unsupervised approach with sliding window change, respectively. Despite that, if both approaches were trained using the same window size, the sliding window size does not influence the performance improvement for the unsupervised approach. More specifically, the results regarding OCSVM seem to show that the increase of training window size even has a negative impact on detecting the anomalies.

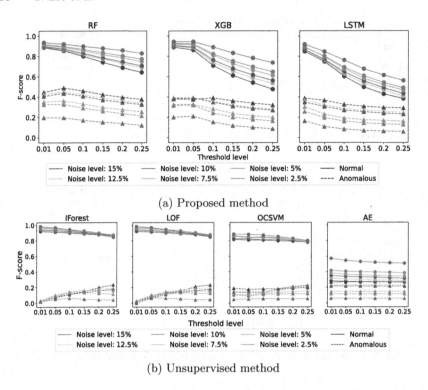

(a) Proposed method

(b) Unsupervised method

Fig. 6. Performance comparison by anomaly threshold

The performance of the proposed approach improves for both normal and anomalous events detection with a large window, i.e., more useful information is collected with a bigger window size. Along with the positive relation between model improvements and training data size, we also observe a marginal effect of increasing training window and detecting ability improvements. As in Fig. 7a, the slope between sliding window size and F-score is not linear, i.e., the obtained useful information for detecting anomalous by increasing training dataset is limited. Therefore, the efficiency of managing the training window depends on a balance between information gain and the cost of increasing the dataset.

Retraining interval size does not influence the anomaly detection ability for either the proposed approach or unsupervised method. Figure 8a and 8b show the F-score of normal and anomalous event detection for the proposed approach and the unsupervised algorithm with retraining interval changes, respectively. Generally, we can observe that the retraining interval is irrelevant to the model training phase. The performance lines across all detecting models are stable over noise level.

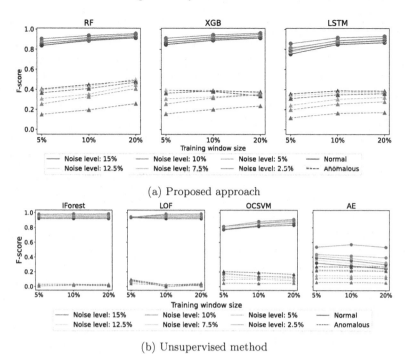

(a) Proposed approach

(b) Unsupervised method

Fig. 7. Performance comparison by sliding window size

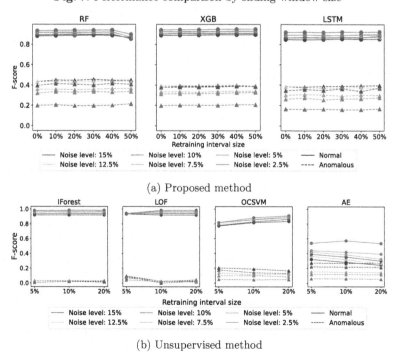

(a) Proposed method

(b) Unsupervised method

Fig. 8. Performance comparison by retraining interval size

7   Conclusion and Challenges

We have presented an approach for online event anomaly detection via next activity prediction using machine learning models. The proposed approach uses a sliding window to (re)train the model with the recently recorded cases on the event stream. The probability of possible next activities obtained from the ML models is used to classify a new event as anomalous or not.

The method uses well-known machine learning and deep learning algorithms, which can give flexibility by changing into other models later. Even though the performance of deep neural networks is lower than machine learning models, we have shown that the next activity prediction method outperforms the classical unsupervised anomaly detection method when applied to event logs.

As a relatively new research area, anomalous event detection in event streams has open challenges in several perspectives. We identify multiple challenges on online anomaly detection that still need to be addressed. Firstly, in an online setting, an event classified as (potential) anomalous may be changed into normal behavior later after the model update or after new events arrives. *(RC1) How and when can we confirm that a potential anomalous event is a definitive anomaly?* The approach proposed in this paper is designed to not updating the prediction. Nevertheless, taking such changes into account may be very informative for suggesting user actions and is a challenge itself and worthy of further investigation. Another interesting challenge related to changes is *(RC2) how can anomalies be detected while taking into account concept drift?* Finally, the online detection of potential anomalies leads to the possibility of building an online decision-making system that suggests follow-up actions for the detected anomalies. This possibility leads to the third research challenge: *(RC3) how can we build a decision-making system with the domain experts for the further investigation of the potential anomalies?*

The work presented here can be extended in several ways. In addition to solving the research challenges, we will study the post-hoc analysis on the online anomaly detecting model to provide an explanation of the cause of anomalous events. Finally, we are planning to develop the interactive online detecting model by implementing the feedback from users who select anomalous events. The collected feedback would help the model to give concrete reasoning on the predicted anomalies and improve the model performance.

References

1. van der Aalst, W.M.P.: Data science in action. In: Process Mining, pp. 3–23. Springer, Heidelberg (2016). https://doi.org/10.1007/978-3-662-49851-4_1
2. Burattin, A., Carmona, J.: A framework for online conformance checking. In: Teniente, E., Weidlich, M. (eds.) BPM 2017. LNBIP, vol. 308, pp. 165–177. Springer, Cham (2018). https://doi.org/10.1007/978-3-319-74030-0_12
3. Burattin, A., Sperduti, A., van der Aalst, W.M.P.: Heuristics miners for streaming event data. arXiv preprint arXiv:1212.6383 (2012)

4. Dumas, M., La Rosa, M., Mendling, J., Reijers, H.A., et al.: Fundamentals of Business Process Management, vol. 1. Springer, Cham (2013). https://doi.org/10. 1007/978-3-662-56509-4

5. Ghionna, L., Greco, G., Guzzo, A., Pontieri, L.: Outlier detection techniques for process mining applications. In: An, A., Matwin, S., Raś, Z.W. (eds.) ISMIS 2008. LNCS (LNAI), vol. 4994, pp. 150–159. Springer, Heidelberg (2008). https://doi. org/10.1007/978-3-540-68123-6_17

6. Guo, H., Meamari, E., Shen, C.C.: Blockchain-inspired event recording system for autonomous vehicles. In: 2018 1st IEEE international conference on hot information-centric networking (HotICN), pp. 218–222. IEEE (2018)

7. Hulten, G., Spencer, L., Domingos, P.: Mining time-changing data streams. In: Proceedings of the seventh ACM SIGKDD International Conference on Knowledge Discovery and Data Mining, pp. 97–106 (2001)

8. Khatuya, S., Ganguly, N., Basak, J., Bharde, M., Mitra, B.: Adele: anomaly detection from event log empiricism. In: IEEE INFOCOM 2018-IEEE Conference on Computer Communications, pp. 2114–2122. IEEE (2018)

9. Ko, J., Comuzzi, M.: Online anomaly detection using statistical leverage for streaming business process events. In: Leemans, S., Leopold, H. (eds.) ICPM 2020. LNBIP, vol. 406, pp. 193–205. Springer, Cham (2021). https://doi.org/10.1007/978-3-030-72693-5_15

10. Kolozali, S., Bermudez-Edo, M., Puschmann, D., Ganz, F., Barnaghi, P.: A knowledge-based approach for real-time IoT data stream annotation and processing. In: 2014 IEEE International Conference on Internet of Things (iThings), and IEEE Green Computing and Communications (GreenCom) and IEEE Cyber, Physical and Social Computing (CPSCom), pp. 215–222. IEEE (2014)

11. Leontjeva, A., Conforti, R., Di Francescomarino, C., Dumas, M., Maggi, F.M.: Complex symbolic sequence encodings for predictive monitoring of business processes. In: Motahari-Nezhad, H.R., Recker, J., Weidlich, M. (eds.) BPM 2015. LNCS, vol. 9253, pp. 297–313. Springer, Cham (2015). https://doi.org/10.1007/978-3-319-23063-4_21

12. Lu, X., Fahland, D., van den Biggelaar, F.J.H.M., van der Aalst, W.M.P.: Detecting deviating behaviors without models. In: Reichert, M., Reijers, H.A. (eds.) BPM 2015. LNBIP, vol. 256, pp. 126–139. Springer, Cham (2016). https://doi.org/10. 1007/978-3-319-42887-1_11

13. Maisenbacher, M., Weidlich, M.: Handling concept drift in predictive process monitoring. SCC **17**, 1–8 (2017)

14. Nguyen, H.T.C., Lee, S., Kim, J., Ko, J., Comuzzi, M.: Autoencoders for improving quality of process event logs. Expert Syst. Applicat. **131**, 132–147 (2019)

15. Nolle, T., Luettgen, S., Seeliger, A., Mühlhäuser, M.: Binet: multi-perspective business process anomaly classification. Inf. Syst. **103** (2022). Article no. 101458

16. Paszke, A., et al.: Pytorch: an imperative style, high-performance deep learning library. In: Advances in Neural Information Processing Systems, vol. 32, pp. 8026–8037 (2019)

17. Pedregosa, F., et al.: Scikit-learn: machine learning in python. J. Mach. Learn. Res. **12**, 2825–2830 (2011)

18. Sani, M.F., van Zelst, S.J., van der Aalst, W.M.P.: Improving process discovery results by filtering outliers using conditional behavioural probabilities. In: Teniente, E., Weidlich, M. (eds.) BPM 2017. LNBIP, vol. 308, pp. 216–229. Springer, Cham (2018). https://doi.org/10.1007/978-3-319-74030-0_16

19. Savickas, T., Vasilecas, O.: Belief network discovery from event logs for business process analysis. Comput. Ind. **100**, 258–266 (2018). Article no. 101458

20. Tavares, G.M., Ceravolo, P., Da Costa, V.G.T., Damiani, E., Junior, S.B.: Overlapping analytic stages in online process mining. In: 2019 IEEE International Conference on Services Computing (SCC), pp. 167–175. IEEE (2019)

21. Teinemaa, I., Dumas, M., Rosa, M.L., Maggi, F.M.: Outcome-oriented predictive process monitoring: review and benchmark. ACM Trans. Knowl. Discov. Data (TKDD) **13**(2), 1–57 (2019). 101458

22. Vertuam Neto, R., Tavares, G., Ceravolo, P., Barbon, S.: On the use of online clustering for anomaly detection in trace streams. In: XVII Brazilian Symposium on Information Systems, pp. 1–8 (2021)

23. van Zelst, S.J., Fani Sani, M., Ostovar, A., Conforti, R., La Rosa, M.: Filtering spurious events from event streams of business processes. In: Krogstie, J., Reijers, H.A. (eds.) CAiSE 2018. LNCS, vol. 10816, pp. 35–52. Springer, Cham (2018). https://doi.org/10.1007/978-3-319-91563-0_3

Process Mining: A Guide for Practitioners

Fredrik Milani[1]([✉]), Katsiaryna Lashkevich[1], Fabrizio Maria Maggi[2],
and Chiara Di Francescomarino[3]

[1] University of Tartu, Tartu, Estonia
{milani,katsiaryna.lashkevich}@ut.ee
[2] Free University of Bozen-Bolzano, Bolzano, Italy
maggi@inf.unibz.it
[3] FBK-IRST, Trento, Italy
dfmchiara@fbk.eu

Abstract. In the last years, process mining has significantly matured
and has increasingly been applied by companies in industrial contexts.
However, with the growing number of process mining methods, practitioners might find it difficult to identify which ones to apply in specific
contexts and to understand the specific business value of each process
mining technique. This paper's main objective is to develop a business-
oriented framework capturing the main process mining use cases and the
business-oriented questions they can answer. We conducted a Systematic Literature Review (SLR) and we used the review and the extracted
data to develop a framework that (1) classifies existing process mining
use cases connecting them to specific methods implementing them, and
(2) identifies business-oriented questions that process mining use cases
can answer. Practitioners can use the framework to navigate through
the available process mining use cases and to identify the process mining
methods suitable for their needs.

Keywords: Business process management · Process mining · Use
cases · Systematic Literature Review · Business value

1 Introduction

A business process is a set of activities executed in a given setting to achieve
predefined business objectives [25]. Since business processes constitute the operational foundation of organizations, companies seek to manage and improve them.
Business processes are commonly supported by information systems that record
data on the executions of the processes. These records are referred to as event
logs. An event log consists of traces that capture the execution of a business
process instance (a.k.a. a case). A case consists of a sequence of time-stamped
events, each representing the execution of an activity. Process mining is a family
of methods that analyze business processes based on their observed behavior
recorded in event logs.

In the past two decades, research on process mining has made advances
resulting in the generation of a large corpus of academic literature [63]. However,

© The Author(s) 2022
R. Guizzardi et al. (Eds.): RCIS 2022, LNBIP 446, pp. 265–282, 2022.
https://doi.org/10.1007/978-3-031-05760-1_16

the number of studies on process mining methods might be disconcerting when companies seek to understand how process mining can be applied for improving business processes. More specifically, companies might find it challenging to understand (1) which prominent process mining use cases are available and (2) what business-oriented questions such use cases answer.

As such, the objective of this study is to develop, based on a Systematic Literature Review (SLR), a business-oriented framework that classifies existing process mining use cases relating them with specific methods and with the business-oriented questions they can answer. Therefore, the framework can aid practitioners that seek to use data-driven approaches to manage their business processes in exploring how process mining methods can add value to their business and what process-related analysis such methods can support. Thus, the main research questions we seek to examine in this paper are *What are the main use cases for existing process mining methods?* and *What business-oriented questions do existing process mining use cases answer?*. We conducted the SLR following the guidelines proposed in [45], retrieved 2293 papers using a keyword-based search from electronic libraries, and filtered them according to predefined inclusion and exclusion criteria. We finally identified a corpus of 839 relevant papers that we reviewed. Then, we used the review and the extracted data to develop a business-oriented a framework that could represent a valid instrument to guide companies in how process mining can support their business.

The remainder of the paper is organized as follows. In Sect. 2, we position our work against related work. Section 3 presents the SLR protocol. Section 4 summarizes the results, and Sect. 5 presents and discusses the framework. Finally, we conclude the paper in Sect. 6.

2   Related Work

In this section, we position our work against existing reviews within the process mining field. More specifically, we consider systematic mapping studies, process mining reviews in specific industrial sectors, and reviews on how process mining is applied in industry.

In [63], dos Santos Garcia et al. present a systematic mapping study providing an overview of the main process mining branches, algorithms, and application domains. A similar study is discussed by Maita et al. in [54]. These studies highlight that most of the process mining publications can be associated with process model discovery. This is in line with our findings. However, while these mapping studies focus on the current state of the process mining research, we examined empirically validated process mining methods to elicit how they might deliver value to companies. Thus, we take a business-oriented perspective allowing practitioners to link the everyday issues they have to deal with in their organizations with the process mining techniques that can help them solving these issues. Furthermore, Maita et al. classified process mining techniques into 3 main branches, i.e., process model discovery, conformance checking, and process model enhancement as proposed in [1]. This classification is also applied in other studies, such as [18]. However, the application of process mining has evolved in recent years. Therefore, our framework extends this classification by incorporating more recent process mining use cases.

Several reviews have also been conducted within specific industry sectors. For instance, literature reviews have been conducted with focus on healthcare [6,29,33,36,61], educational processes [34], and supply chain [41]. While these reviews focus on a specific application domain, we consider business-oriented questions that are answered at a domain-agnostic level.

Finally, reviews have also been conducted on how organizations use process mining. For instance, Thiede et al. [70] show with their study that process mining is not sufficiently leveraged in the context of cross-system and cross-organizational processes. Corallo et al. [16] primarily provide an overview of software tools that support process mining analysis in industry. Eggers and Hein [27], instead, examine capabilities and practices required to enable the realization of value using process mining in an organization. These studies provide valuable insights on different aspects of how process mining is applied in industry, but they do not guide practitioners in selecting use cases and methods to answer common business-oriented questions.

3   Systematic Literature Review

Our research objective is to develop a framework for classifying process mining use cases and the business issues they might address. This objective is achieved by answering two research questions (RQ). The first research question (RQ1) aims at identifying and categorizing process mining use cases, such as conformance checking and predictive monitoring. Therefore, RQ1 is formulated as *What are the main use cases for existing process mining methods?* The second research question (RQ2) aims at eliciting the business-oriented questions that the outputs of process mining methods can answer. Therefore, RQ2 is formulated as *What business-oriented questions do existing process mining use cases answer?* To answer these questions, we employ the SLR method as it allows us to collect relevant studies and, based on the review of existing research, to develop a framework for classifying them [15]. We followed the guidelines proposed by Kitchenham [45] according to which an SLR has three consecutive phases: planning, execution, and reporting.

Our SLR is composed of two parts. The first one (SLR review) aims at identifying other SLR studies on specific process mining use cases (e.g., [5] for process model discovery and [26] for conformance checking). The list of final papers in each of these SLR studies was extracted. However, as not all process mining use cases have a dedicated SLR study, we conducted a second review (PM review) targeting papers that have applied process mining techniques to real-life event logs. The lists of final papers retrieved from each SLR study identified with the SLR review was combined with the hits obtained with the PM review. The merged list of candidate papers was subjected to content screening. We followed the same guidelines for both parts, i.e., we developed search strings, identified search sources, filtered the results according to predefined criteria, identified additional relevant papers through backwards referencing, and extracted data according to a predefined form. Below, we provide a summary of these steps. The review protocol[1] and the list of final papers[2] are available online.

The planning phase of our SLR includes developing search strings, identifying search sources, defining selection criteria, and defining the data extraction strategy. We derived the search strings from our research questions as suggested in [45]. For the SLR review, the aim was to capture SLR studies on process mining. Therefore, we used the search string *"process mining" AND "systematic literature review"*. The search strings for the PM review were, instead, derived from the research questions and included the terms "process mining", "workflow mining" (as this term is sometimes used interchangeably with "process mining"), "real-life", "real-world", and "case study". Then, we applied the search strings to electronic databases. We selected Scopus and Web of Science for both parts as they index the venues where most research on process mining is published.

We then defined the selection criteria to identify relevant studies. These criteria, expressed as exclusion (EC) and inclusion (IC) criteria, allowed us to filter the initial list of papers to keep only those that are relevant to answer the research questions. For the SLR review, duplicate papers (EC1), papers not written in English (EC2), and papers inaccessible via the digital libraries subscribed by the University of Tartu, or otherwise unavailable (EC3) were excluded. In addition, with the first inclusion criterion (IC1), we filtered out studies that were not specifically about process mining and the second one (IC2) served to identify studies that applied SLR to identify relevant papers for specific process mining use cases. Thus, studies focusing on, for instance, evaluating process mining algorithms, such as [57] were excluded.

We applied the three exclusion criteria above also for the PM review. In addition, papers having less than five pages were discarded (EC4). The list of candidate papers was then filtered based on the inclusion criteria. We first excluded papers not within the domain of process mining (IC1). Then, we identified papers that apply process mining to real-life event logs (IC2). This criterion was included for two reasons. The first one is to identify methods that are applicable to real-life, often challenging, business settings. The second one is to identify papers that address business process aspects existing in real business contexts. The third inclusion criterion (IC3) was aimed at excluding papers that consider process mining for applications unrelated to business, such as [59], where discovery algorithms for managing noisy event logs are compared. Finally, the fourth inclusion criterion filters out papers that do not provide sufficient information to elicit business-oriented questions (IC4).

The final step of planning an SLR is data extraction. The objective of this step is to define the data extraction form to reduce the opportunity for bias. We developed the data extraction form according to the suggestions provided in [31,60]. The data extraction form consists of two parts. The first part was used to

---

[1] https://doi.org/10.6084/m9.figshare.17099462.
[2] https://doi.org/10.6084/m9.figshare.12933239.

extract the metadata of the papers, i.e., paper id, title, authors, and publication year. In the second part, the data extracted concerned process mining use cases (such as process model discovery or performance analysis) and the questions being answered by the process mining methods.

We conducted both searches in February 2020. A summary of the detailed procedure applied is depicted in Fig. 1. For the SLR review, the application of the search string to the electronic databases returned 132 hits from Scopus and 61 from Web of Science, making it a total of 193 candidate papers. After having applied the exclusion criteria, 60 papers remained. Of these, 15 were removed as they did not meet IC1, resulting in 45 papers. The application of IC2 filtered out additional 26 papers, resulting in 19 papers left. Finally, backward and forward referencing identified two additional papers, resulting in a final list of 21 relevant studies. The data extraction for this part consisted in exporting all studies included in the final lists of all 21 relevant SLR publications. These were merged with the hits resulting from the search conducted for the PM review. From the 21 SLR studies, a total of 702 papers were extracted.

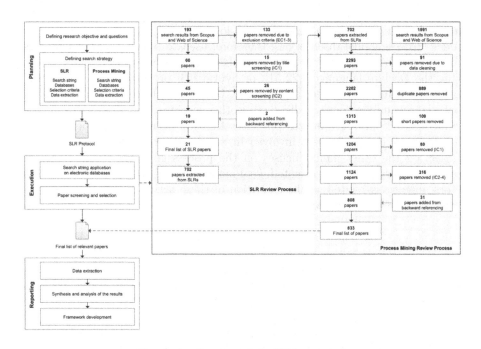

Fig. 1. Application of the SLR protocol.

For the PM review, applying the search strings resulted in 1021 hits from Scopus and 570 from Web of Science, making it 1591 hits in total. With the 702 papers added from the first part, the total number was 2293. We discarded 91 papers that were unavailable, 889 duplicates and 109 short papers. A total of 1204 papers remained. The first inclusion criterion (IC1) resulted in the removal of 80 papers. In the second filtering, where IC 2-4 were considered, 316 papers were removed, resulting in 808 papers left. Additional 31 papers were identified from the backward and forward referencing. Thus, the final list consists of 839 relevant papers.

4   Results

In this section, we first present the identified use cases of empirically validated process mining methods. Then, we relate them to the business-oriented questions they answer.

4.1   Process Mining Use Cases

The most common use case identified in the process mining literature is *process model discovery* (see Fig. 2). Business processes are commonly supported by information systems that log information on the process executions. When such logs are available, they can be used to discover process models automatically. Process model discovery takes such event logs as input and produces a process model. Process model discovery is, therefore, used to build procedural process models (e.g., using BPMN or Petri nets) [43], declarative process models (e.g., using the Declare language) [22,51], or hybrid [53] process models containing both a procedural and a declarative part. Use cases related to social network, goal, and rule mining focus on discovering other aspects of the process executions. *Social network mining* analyzes processes from an organizational perspective, i.e., discovers the performers involved in a case and their relations [2]. *Goal mining*, on the other hand, focuses on the process goals [74]. While process model discovery is activity-oriented, goal modeling seeks to discover the process actors' intentions related to the execution of these activities [20]. Finally, *rule mining* [62], also referred to as decision mining, examines the data attributes in an event log to elicit the rules behind the choices made in the process.

Event logs hold information on the actual process executions, which is not necessarily aligned with process models. Therefore, models can be enhanced using *process model enhancement* techniques. In particular, process models can be repaired to better represent the process executions [30,50] or extended with additional data recorded in the event logs [64].

Concept drift identifies changes in the process behavior. The discovery of process models from event logs implicitly assumes that the process model remains stable throughout the time period recorded in the event log. However, this is not always true, as the process might change during the recorded period. Thus, the concept drift use case focuses on detecting changes in the process behavior over time [49].

The second most common use case is *predictive monitoring*. Such use case aims at predicting the outcome of active cases, i.e., cases that are uncompleted and, therefore, still ongoing [24]. Learning from an event log of historical cases, predictive monitoring techniques are able to predict the remaining time of an

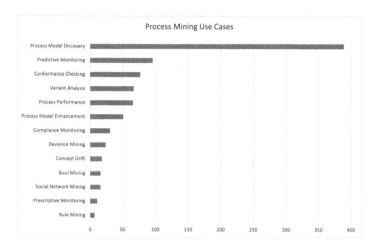

Fig. 2. Process mining use cases.

ongoing case [73], delays [65], next activities [66], waiting times [7], outcomes [68], risks [14], costs [72], or performance indicators [17]. Since hyperparameter configuration in predictive process monitoring is crucial and often difficult for users, some works provide methods to support hyperparameter optimization in predictive process monitoring [23]. *Prescriptive monitoring* can be viewed as an extension of predictive monitoring. While predictive monitoring forecasts the likelihood of a case ending up with a desirable/undesirable outcome, it does not suggest the interventions that can increase/reduce the probability of such an outcome. Prescriptive monitoring, on the other hand, seeks to identify specific interventions, such as next activities to be executed [71], resource allocation [75], resource selection [44] to improve the likelihood of a favourable outcome, or when an intervention is needed based on the trade-off cost-gain [69].

The third most common use case we identified is *conformance checking*. Conformance checking aims at examining if the behavior of a process execution, as derived from an event log, conforms with the expected behavior (represented as a process model) [13]. This can be done by simply showing to the user where the process execution deviates from the process model [12,42], or by providing a way to align the deviant process execution with the closest compliant case [3,19,46]. *Compliance monitoring* follows the same principle as conformance checking. However, while conformance checking is applied to completed process cases, compliance monitoring checks whether the behavior of active cases is compliant with predefined rules and constraints [48].

The fourth most common use case we found is *variant analysis*. Executions of a business process commonly include variants, i.e., cases that follow the same path (characterized by the same sequence of activities) [67]. Variant analysis enable identifying these variants in an event log. Variant analysis can also be applied for identifying differences and similarities between different variants [10,38].

Deviance mining, instead, aims at explaining why a certain variant deviates from the most frequently taken path [8,55].

A last use case concerns assessing *process performance* [56]. The performance measured can be the duration of a process execution [40], the resource utilization [39], or the quality of the products/services provided [4]. The performance of several connected processes can also be assessed [28] and the process performance trends over time [21].

4.2   Business-Oriented Questions

Some process mining methods answer questions that are descriptive. For instance, process model discovery answers the question *How are the cases of a procedural, a declarative, or a hybrid process executed?* The answer is expressed as a process model representing the process behavior as recorded in the event log. Process cases can commonly grouped into variants. Therefore, some process mining methods answer the question *What are the main variants of a process?* Other process mining methods answer, instead, questions to quantitatively describe processes. For instance, process mining can answer questions such as *What is the duration-, resource-, quality-related performance of a case?*

There are also methods answering comparative questions to compare two or more process cases. For instance, variant analysis methods might be used to compare different variants of a process thus answering the questions *What are the similarities between two or more variants of a process?* and *What are the differences between two or more variants of a process?* Similarly, conformance checking seeks to compare the prescribed behavior of a process with the observed behavior, i.e., comparing what a process model stipulates and how the process is executed in reality. Therefore, conformance checking answers questions such as *Where does a case differ from a process model?* Another type of comparative questions are the ones related to how the process behavior changes over time such as *How has the process behavior, or its performance changed over time?*

Process mining also answers questions that seek explanatory answers, i.e., providing information that explains relations among different entities. For instance, some process mining methods answer questions such as *How is the performance of a case affected by other factors?* Likewise, deviance mining provides explanatory information by answering the question *Why do some cases deviate from the normal flow?* Methods that compare a model with historical (conformance checking) or live (compliance monitoring) cases could provide explanatory information by answering questions such as *Given a non-compliant case, what is closest compliant case?* Predictive monitoring methods answer (forecasting) descriptive questions such as *What are the predicted remaining times, delays, next activities, waiting times, outcomes, risks, costs, or performance indicators of an ongoing case?*

Finally, there are process mining methods aiming at providing suggestions on how to redesign a process model to improve understandability, or optimize the likelihood of a favourable outcome of ongoing process executions. For instance, model repair techniques provide input for improving a process model by answering the recommendatory questions *How can a process model be repaired to better reflect the actual execution of the process?* and *How can the understandability of the mined process models be improved?* On the other hand, prescriptive monitoring methods provide recommendations on how an ongoing process case should be executed to reach a positive outcome. For instance, recommendations can be given on which variant to follow thus answering the question *What is the recommended execution path of an ongoing case?* Recommendations also extend to resources and their allocation by answering questions such as *What is the recommended resource allocation?* and *Who is the recommended process performer?*

5 Framework

This section introduces our business-oriented framework that categorizes the identified process mining use cases. The framework can be used by practitioners to be guided in the selection of the process mining methods that are the most suitable for their needs. It consists of two parts: a categorization of the main process mining use cases (RQ1) and the elicitation of the business-oriented questions that these use cases can answer (RQ2). Our categorization draws on the value-driven business process management (VBPM) proposed in [32]. According to VBPM, organizations need BPM techniques to realize at least one of the six values: efficiency, quality, compliance, agility, integration, and networking. In order to realize these values, transparency is required. Transparency is creating visibility of how processes are executed. Commonly, this is achieved with business process models. Therefore, transparency lies at the core of VBPM.

Organizations, engaging in BPM to gain in efficiency, take an internal organizational viewpoint and focus on improving the performance of their processes. Efficiency gains are achieved by, for instance, eliminating waste in the processes, reducing redundancies, and removing rework. On the other hand, organizations can focus on the outputs of the processes and on improving their quality. Organizations that hold quality as a core value, engage in BPM to explore the correlation between process characteristics and product/service quality.

Compliance as an expected value of BPM emphasizes reducing variability and increasing standardization. Financial institutions, for instance, are subjected to regulatory requirements. Therefore, such organizations gain value from designing and executing processes that comply with predefined standards. However, organizations might also gain value from having agile processes, i.e., flexible and adaptable processes. For instance, an insurance company experiencing a sharp increase in claims during severe weather conditions, would switch to a different process execution that caters to the increased volumes. Other values of BPM are integration and networking. Integration concerns the creation of business value by increasing awareness and accessibility of process models to internal stakeholders. Conversely, networking focuses on involving external stakeholders in the processes.

We use VBPM as the basis for categorizing process mining use cases for two reasons. Firstly, VBPM is derived from surveys and interviews with companies and, therefore, captures the main reasons why companies engage with BPM. Our framework is business-oriented, i.e., categorizes use cases and questions that, when answered, aim at delivering value to organizations. Therefore, using VBPM allows us to categorize process mining methods while aligning them with the business values companies seek from BPM. Secondly, process mining is applicable in different BPM lifecycle phases such as process discovery, monitoring and analysis [25]. Likewise, process mining can be applied to support the execution of different methodologies such as Six Sigma [35]. Since VBPM focuses on business value rather than on specific BPM methodologies, using VBPM as a basis, our business-oriented framework can be used to show how process mining techniques can contribute to generating value to organizations instead of framing these techniques in the context of specific BPM lifecycle phases or methodologies.

5.1   Framework Instantiation

Our framework categorizes the main process mining use cases into categories transparency, efficiency, quality, compliance, and agility. *Transparency* encompasses use cases that aim at discovering process models (process model discovery), discovering the interaction between resources in a process (social network mining), adjusting process models to capture the process executions better (process model repair), enriching the process models with additional data (process model extension), detecting the decision rules embedded in the process decision points (rule mining), and identifying the process objectives (goal mining). *Efficiency* includes process mining use cases concerning the analysis of the performance of a process (process performance), *Quality* encompasses use cases for the identification and comparison of process variants (variant analysis) and analyzing the reasons for deviations in a process case (deviance mining). *Compliance* includes use cases comparing a process model with some observed behavior (conformance checking), or predefined rules or constraints with an ongoing case (compliance monitoring). *Agility* encompasses use cases about the predictions on how ongoing process executions will unfold in the future (predictive monitoring), the description of how the process behavior changes over time (concept drift), and the prescription of actions to take, for an ongoing case to achieve a certain desired outcome (prescriptive monitoring).

We have excluded integration and networking in our framework. Integration focuses on improving the availability and accessibility of process models to internal resources, for instance, to raise their engagement in the processes. However, although process mining methods can discover process models, the distribution and accessibility of such models are beyond the scope of this field. Networking, instead, focuses on incorporating external parties within the scope of a process. Although event logs of several parties can be merged together, process mining methods treat such logs in the same way as a single internal event log. Therefore, we also excluded networking from our framework.

Table 1. Framework instantiation.

	Business-Oriented Questions Answered by Process Mining Use Cases	Sample Reference
TRANSPARENCY	*Process Model Discovery*	
	How are the cases of a procedural process executed?	[43]
	How are the cases of a declarative process executed?	[51]
	How are the cases of a hybrid process executed?	[53]
	Process Model Enhancement	
	How can a process model be repaired to better reflect the actual execution of the process?	[30]
	How can a process model be extended with additional information recorded in the event logs?	[64]
	How can the understandability of the mined process models be improved?	[11]
	Social Network Mining	
	What are the relationships among the resources involved in a process?	[2]
	Goal Mining	
	What are the process goals?	[74]
	Rule Mining	
	What are the decision rules that determine which path a case takes?	[62]
EFFICIENCY	*Process Performance*	
	What is the duration-related performance of a case?	[40]
	What is the resource-related performance of a case?	[39]
	What is the quality-related performance of a case?	[4]
	How is the performance of a case affected by other factors?	[37]
	What is the performance of multiple connected processes?	[28]
	How can the performance of a case be optimized?	[76]
	How does the performance of a case change over time?	[21]
QUALITY	*Process Variant Analysis*	
	What are the main variants of a process?	[67]
	What are the similarities between two or more variants of a process?	[38]
	What are the differences between two or more variants of a process?	[10]
	Process Deviance Mining	
	Why do some cases deviate from the normal flow?	[55]
COMPLIANCE	*Process Conformance Checking*	
	Where does a case differ from a process model?	[42]
	Given a non-compliant case, what is closest compliant case?	[3]
	Process Compliance Monitoring	
	Is an ongoing case compliant with some predefined rules and constraints?	[52]
	What are the causes of non-compliance in an ongoing case?	[9]
AGILITY	*Predictive Monitoring*	
	What is the predicted remaining time of an ongoing case?	[58]
	What are the predicted delays of an ongoing case?	[65]
	What are the next activity/activities of an ongoing case?	[66]
	What is are the predicted waiting times of an ongoing case?	[7]
	What are the predicted outcomes of an ongoing case?	[47]
	What are the predicted risks of an ongoing case?	[14]
	What are the predicted costs of an ongoing case?	[72]
	What are the predicted performance indicators of an ongoing case?	[17]
	Concept Drift	
	How has the process behavior changed over time?	[49]
	Prescriptive Monitoring	
	What is the recommended execution path of an ongoing case?	[71]
	What is the recommended resource allocation?	[75]
	Who is the recommended process performer?	[44]
	When should interventions be made to increase the probability of a positive outcome?	[69]

At the highest level, our framework categorizes process mining use cases into transparency, efficiency, quality, compliance, and agility. Each of these categories is then organized in sub-categories. For instance, transparency consists of process model discovery, repair, enhancement, social network mining, goal mining, and rule mining (see first column of Table 1). Then, we define the questions that each use case of a certain sub-category can answer, and provide a sample reference[3] (see second column of Table 1). For instance, for concept drift under agility, we have defined the question *How has the process execution changed over time?*

The questions specified in the transparency category are, as expected, descriptive, while questions specified in the efficiency category are quantitative or comparative. The questions specified in the quality and in the compliance categories are, instead, descriptive, comparative, or explanatory. Finally, the questions defined in the agility category are descriptive, recommendatory, comparative, or explanatory. Thus, we can observe that descriptive questions lie at the foundation of other questions, just as transparency constitutes the foundation of the other categories.

5.2   Limitations

The main limitations of SLR studies are selection bias and data extraction inaccuracies. These threats, although not eliminated, were reduced by adhering to the guidelines proposed by [45]. More specifically, we used well-known databases to find papers, performed backwards referencing to avoid excluding potentially relevant papers, and ensured replicability by providing access to the SLR protocol. Another limitation concerns the fact that we relied on the results reported in the literature and we did not empirically verify or assess the extent to which the use cases impact the business processes, or if they led to effective process improvements (we considered methods using real-life event logs but not necessarily tested in industrial contexts). Although this could represent a limitation for the generalizability of the results, the proposed framework still provides practitioners with valuable insights on how process mining might be applied in industry, and represents an easy-to-use instrument to understand what types of analysis can be conducted with the existing process mining methods.

6   Conclusion

Process mining methods have been growing fast in the last decades. While such methods help manage business processes, it can be challenging for practitioners to readily understand how they can deliver value, or what business-oriented questions they can answer. To fill this research gap, we propose a framework that classifies existing process mining use cases using categories transparency, efficiency,

---

[3] Due to space limitations, only sample references are included. The complete framework can be accessed at https://doi.org/10.6084/m9.figshare.17099402.

quality, compliance, and agility. Furthermore, within each of the above categories, process mining use cases can answer descriptive, comparative, explanatory, or recommendatory questions.

The SLR we conducted also highlights that several studies in the process mining literature support the discovery of process models (transparency), predictive monitoring (agility), the analysis of process performance and variants (efficiency and quality), and conformance checking (compliance). In this respect, the framework also represent an instrument allowing researchers and/or process mining companies to understand which use cases in the process mining field have already been largely explored and which ones, instead, need further investigations.

Acknowledgments. This research is funded by the European Research Council (PIX Project) and by the UNIBZ project CAT. We thank Apromore for suggesting the VBPM schema.

References

1. van der Aalst, W.M.P.: Process Mining: Discovery, Conformance and Enhancement of Business Processes. Springer, Berlin (2011). https://doi.org/10.1007/978-3-642-19345-3
2. van der Aalst, W.M.P., Reijers, H.A., Song, M.: Discovering social networks from event logs. Comput. Support. Coop. Work **14**(6), 549–593 (2005)
3. Adriansyah, A., van Dongen, B.F., van der Aalst, W.M.P.: Conformance checking using cost-based fitness analysis. In: Proceedings of the 15th IEEE International Enterprise Distributed Object Computing Conference, EDOC 2011, Helsinki, Finland, pp. 55–64. IEEE Computer Society (2011)
4. Arpasat, P., Porouhan, P., Premchaiswadi, W.: Improvement of call center customer service in a thai bank using disco fuzzy mining algorithm. In: 2015 13th International Conference on ICT and Knowledge Engineering (ICT & Knowledge Engineering 2015), pp. 90–96. IEEE (2015)
5. Augusto, A., et al.: Automated discovery of process models from event logs: review and benchmark. IEEE Trans. Knowl. Data Eng. **31**(4), 686–705 (2019)
6. Batista, E., Solanas, A.: Process mining in healthcare: a systematic review. In: 9th International Conference on Information, Intelligence, Systems and Applications, IISA 2018, Zakynthos, Greece, 23–25 July 2018, pp. 1–6. IEEE Computer Society (2018)
7. Benevento, E., Aloini, D., Squicciarini, N., Dulmin, R., Mininno, V.: Queue-based features for dynamic waiting time prediction in emergency department. Meas. Bus. Excell. **23**(4), 458–471 (2019)
8. Bergami, G., Di Francescomarino, C., Ghidini, C., Maggi, F.M., Puura, J.: Exploring business process deviance with sequential and declarative patterns. CoRR abs/2111.12454 (2021)
9. Böhmer, K., Rinderle-Ma, S.: Mining association rules for anomaly detection in dynamic process runtime behavior and explaining the root cause to users. Inf. Syst. **90**, 101438 (2020)
10. Bolt, A., de Leoni, M., van der Aalst, W.M.P.: Process variant comparison: using event logs to detect differences in behavior and business rules. Inf. Syst. **74**, 53–66 (2018)

11. Bose, R.P.J.C., van der Aalst, W.M.P.: Trace clustering based on conserved patterns: towards achieving better process models. In: Rinderle-Ma, S., Sadiq, S., Leymann, F. (eds.) BPM 2009. LNBIP, vol. 43, pp. 170–181. Springer, Heidelberg (2010). https://doi.org/10.1007/978-3-642-12186-9_16

12. Burattin, A., Maggi, F.M., Sperduti, A.: Conformance checking based on multiperspective declarative process models. Exp. Syst. Appl. **65**, 194–211 (2016)

13. Carmona, J., van Dongen, B., Solti, A., Weidlich, M.: Conformance Checking: Relating Processes and Models. Springer, Cham (2018). https://doi.org/10.1007/978-3-319-99414-7

14. Conforti, R., de Leoni, M., La Rosa, M., van der Aalst, W.M.P.: Supporting risk-informed decisions during business process execution. In: Salinesi, C., Norrie, M.C., Pastor, Ó. (eds.) CAiSE 2013. LNCS, vol. 7908, pp. 116–132. Springer, Heidelberg (2013). https://doi.org/10.1007/978-3-642-38709-8_8

15. Cooper, H.M.: Organizing knowledge syntheses: a taxonomy of literature reviews. Knowl. Soc. **1**(1), 104 (1988)

16. Corallo, A., Lazoi, M., Striani, F.: Process mining and industrial applications: a systematic literature review. Knowl. Process. Manag. **27**(3), 225–233 (2020)

17. Cuzzocrea, A., Folino, F., Guarascio, M., Pontieri, L.: A predictive learning framework for monitoring aggregated performance indicators over business process events. In: Proceedings of the 22nd International Database Engineering & Applications Symposium, IDEAS 2018, pp. 165–174. ACM (2018)

18. Dakic, D., Stefanovic, D., Cosic, I., Lolic, T., Medojevic, M., Katalinic, B.: Business process mining application: a literature review. In: Proceedings of the 29th DAAAM International Symposium, pp. 0866–0875 (2018)

19. De Giacomo, G., Maggi, F.M., Marrella, A., Patrizi, F.: On the disruptive effectiveness of automated planning for LTL$_f$-based trace alignment. In: Proceedings of the 31st AAAI Conference on Artificial Intelligence, San Francisco, California, USA, 4–9 February 2017, pp. 3555–3561 (2017)

20. Deneckere, R., Hug, C., Khodabandelou, G., Salinesi, C.: Intentional process mining: discovering and modeling the goals behind processes using supervised learning. Int. J. Inf. Syst. Model. Des. (IJISMD) **5**(4), 22–47 (2014)

21. Denisov, V., Fahland, D., van der Aalst, W.M.P.: Unbiased, fine-grained description of processes performance from event data. In: Weske, M., Montali, M., Weber, I., vom Brocke, J. (eds.) BPM 2018. LNCS, vol. 11080, pp. 139–157. Springer, Cham (2018). https://doi.org/10.1007/978-3-319-98648-7_9

22. Di Ciccio, C., Maggi, F.M., Mendling, J.: Efficient discovery of target-branched declare constraints. Inf. Syst. **56**, 258–283 (2016)

23. Di Francescomarino, C., Dumas, M., Federici, M., Ghidini, C., Maggi, F.M., Rizzi, W.: Predictive business process monitoring framework with hyperparameter optimization. In: Nurcan, S., Soffer, P., Bajec, M., Eder, J. (eds.) CAiSE 2016. LNCS, vol. 9694, pp. 361–376. Springer, Cham (2016). https://doi.org/10.1007/978-3-319-39696-5_22

24. Di Francescomarino, C., Ghidini, C., Maggi, F.M., Milani, F.: Predictive process monitoring methods: which one suits me best? In: Weske, M., Montali, M., Weber, I., vom Brocke, J. (eds.) BPM 2018. LNCS, vol. 11080, pp. 462–479. Springer, Cham (2018). https://doi.org/10.1007/978-3-319-98648-7_27

25. Dumas, M., La Rosa, M., Mendling, J., Reijers, H.A.: Fundamentals of Business Process Management, 2nd edn. Springer, Berlin (2018). https://doi.org/10.1007/978-3-662-56509-4

26. Dunzer, S., Stierle, M., Matzner, M., Baier, S.: Conformance checking: a state-of-the-art literature review. In: Proceedings of the 11th International Conference on Subject-Oriented Business Process Management, S-BPM ONE 2019, Seville, Spain, 26–28 June 2019, pp. 4:1–4:10. ACM (2019)
27. Eggers, J., Hein, A.: Turning big data into value: a literature review on business value realization from process mining. In: 28th European Conference on Information Systems, ECIS 2020 (2020)
28. Engel, R.: Analyzing inter-organizational business processes - process mining and business performance analysis using electronic data interchange messages. Inf. Syst. E Bus. Manag. **14**(3), 577–612 (2016)
29. Erdogan, T., Tarhan, A.: Systematic mapping of process mining studies in healthcare. IEEE Access **6**, 24543–24567 (2018)
30. Fahland, D., van der Aalst, W.M.P.: Model repair - aligning process models to reality. Inf. Syst. **47**, 220–243 (2015)
31. Fink, A.: Conducting Research Literature Reviews: From the Internet to Paper. Sage Publications (2019)
32. Franz, P., Kirchmer, M.: Value-Driven Business Process Management: The Value-Switch for Lasting Competitive Advantage. McGraw Hill Professional (2012)
33. Ghasemi, M., Amyot, D.: Process mining in healthcare: a systematised literature review. Int. J. Electron. Heal. **9**(1), 60–88 (2016)
34. Ghazal, M.A., Ibrahim, O., Salama, M.A.: Educational process mining: a systematic literature review. In: 2017 European Conference on Electrical Engineering and Computer Science (EECS), pp. 198–203. IEEE (2017)
35. Graafmans, T., Turetken, O., Poppelaars, H., Fahland, D.: Process mining for six sigma: a guideline and tool support. Bus. Inf. Syst. Eng. **63**(3), 277–300 (2021). https://doi.org/10.1007/s12599-020-00649-w
36. Grüger, J., Bergmann, R., Kazik, Y., Kuhn, M.: Process mining for case acquisition in oncology: a systematic literature review. In: Proceedings of the Conference on "Lernen, Wissen, Daten, Analysen", Online, 9–11 September 2020, vol. 2738, pp. 162–173. CEUR Workshop Proceedings. CEUR-WS.org (2020)
37. Hompes, B.F.A., Maaradji, A., La Rosa, M., Dumas, M., Buijs, J.C.A.M., van der Aalst, W.M.P.: Discovering causal factors explaining business process performance variation. In: Dubois, E., Pohl, K. (eds.) CAiSE 2017. LNCS, vol. 10253, pp. 177–192. Springer, Cham (2017). https://doi.org/10.1007/978-3-319-59536-8_12
38. Huang, Z., Dong, W., Duan, H., Li, H.: Similarity measure between patient traces for clinical pathway analysis: problem, method, and applications. IEEE J. Biomed. Health Inf. **18**(1), 4–14 (2014)
39. Huang, Z., Lu, X., Duan, H.: Resource behavior measure and application in business process management. Exp. Syst. Appl. **39**(7), 6458–6468 (2012)
40. Jaisook, P., Premchaiswadi, W.: Time performance analysis of medical treatment processes by using disco. In: 2015 13th International Conference on ICT and Knowledge Engineering, ICT & Knowledge Engineering 2015, pp. 110–115. IEEE (2015)
41. Jokonowo, B., Claes, J., Sarno, R., Rochimah, S.: Process mining in supply chains: a systematic literature review. Int. J. Electr. Comput. Eng. (IJECE) **8**(6), 4626–4636 (2018)
42. Kalenkova, A.A., Ageev, A.A., Lomazova, I.A., van der Aalst, W.M.P.: E-government services: comparing real and expected user behavior. In: Teniente, E., Weidlich, M. (eds.) BPM 2017. LNBIP, vol. 308, pp. 484–496. Springer, Cham (2018). https://doi.org/10.1007/978-3-319-74030-0_38

43. Kalenkova, A.A., Burattin, A., de Leoni, M., van der Aalst, W.M.P., Sperduti, A.: Discovering high-level BPMN process models from event data. Bus. Process. Manag. J. **25**(5), 995–1019 (2019)
44. Kim, A., Obregon, J., Jung, J.-Y.: Constructing decision trees from process logs for performer recommendation. In: Lohmann, N., Song, M., Wohed, P. (eds.) BPM 2013. LNBIP, vol. 171, pp. 224–236. Springer, Cham (2014). https://doi.org/10.1007/978-3-319-06257-0_18
45. Kitchenham, B.: Procedures for performing systematic reviews. Keele University, Keele, UK **33**(2004), 1–26 (2004)
46. de Leoni, M., Marrella, A.: Aligning real process executions and prescriptive process models through automated planning. Exp. Syst. Appl. **82**, 162–183 (2017)
47. Leontjeva, A., Conforti, R., Di Francescomarino, C., Dumas, M., Maggi, F.M.: Complex symbolic sequence encodings for predictive monitoring of business processes. In: Motahari-Nezhad, H.R., Recker, J., Weidlich, M. (eds.) BPM 2015. LNCS, vol. 9253, pp. 297–313. Springer, Cham (2015). https://doi.org/10.1007/978-3-319-23063-4_21
48. Ly, L.T., Maggi, F.M., Montali, M., Rinderle-Ma, S., van der Aalst, W.M.P.: Compliance monitoring in business processes: functionalities, application, and tool-support. Inf. Syst. **54**, 209–234 (2015)
49. Maaradji, A., Dumas, M., La Rosa, M., Ostovar, A.: Detecting sudden and gradual drifts in business processes from execution traces. IEEE Trans. Knowl. Data Eng. **29**(10), 2140–2154 (2017)
50. Maggi, F.M., Corapi, D., Russo, A., Lupu, E., Visaggio, G.: Revising process models through inductive learning. In: zur Muehlen, M., Su, J. (eds.) BPM 2010. LNBIP, vol. 66, pp. 182–193. Springer, Heidelberg (2011). https://doi.org/10.1007/978-3-642-20511-8_16
51. Maggi, F.M., Di Ciccio, C., Di Francescomarino, C., Kala, T.: Parallel algorithms for the automated discovery of declarative process models. Inf. Syst. **74**, 136–152 (2018)
52. Maggi, F.M., Montali, M., van der Aalst, W.M.P.: An operational decision support framework for monitoring business constraints. In: 15th International Conference on Fundamental Approaches to Software Engineering, FASE 2012, Held as Part of the European Joint Conferences on Theory and Practice of Software, ETAPS 2012, Tallinn, Estonia, pp. 146–162 (2012)
53. Maggi, F.M., Slaats, T., Reijers, H.A.: The automated discovery of hybrid processes. In: Sadiq, S., Soffer, P., Völzer, H. (eds.) BPM 2014. LNCS, vol. 8659, pp. 392–399. Springer, Cham (2014). https://doi.org/10.1007/978-3-319-10172-9_27
54. Maita, A.R.C., Martins, L.C., Paz, C.R.L., Rafferty, L., Hung, P.C.K., Peres, S.M., Fantinato, M.: A systematic mapping study of process mining. Enterp. Inf. Syst. **12**(5), 505–549 (2018)
55. Mannhardt, F., de Leoni, M., Reijers, H.A., van der Aalst, W.M.P.: Data-driven process discovery - revealing conditional infrequent behavior from event logs. In: Dubois, E., Pohl, K. (eds.) CAiSE 2017. LNCS, vol. 10253, pp. 545–560. Springer, Cham (2017). https://doi.org/10.1007/978-3-319-59536-8_34
56. Milani, F., Maggi, F.M.: A comparative evaluation of log-based process performance analysis techniques. In: Abramowicz, W., Paschke, A. (eds.) BIS 2018. LNBIP, vol. 320, pp. 371–383. Springer, Cham (2018). https://doi.org/10.1007/978-3-319-93931-5_27
57. Naderifar, V., Sahran, S., Shukur, Z.: A review on conformance checking technique for the evaluation of process mining algorithms. TEM J. **8**(4), 1232 (2019)

58. Navarin, N., Vincenzi, B., Polato, M., Sperduti, A.: LSTM networks for data-aware remaining time prediction of business process instances. In: 2017 IEEE Symposium Series on Computational Intelligence, SSCI 2017, Honolulu, HI, USA, pp. 1–7. IEEE (2017)

59. Nuritha, I., Mahendrawathi, E.: Behavioural similarity measurement of business process model to compare process discovery algorithms performance in dealing with noisy event log. Procedia Comput. Sci. **161**, 984–993 (2019)

60. Okoli, C.: A guide to conducting a standalone systematic literature review. Commun. Assoc. Inf. Syst. **37**, 43 (2015)

61. Rojas, E., Munoz-Gama, J., Sepúlveda, M., Capurro, D.: Process mining in health care: a literature review. J. Biomed. Inf. **61**, 224–236 (2016)

62. Rozinat, A., van der Aalst, W.M.P.: Decision mining in ProM. In: Dustdar, S., Fiadeiro, J.L., Sheth, A.P. (eds.) BPM 2006. LNCS, vol. 4102, pp. 420–425. Springer, Heidelberg (2006). https://doi.org/10.1007/11841760_33

63. dos Santos Garcia, C., et al.: Process mining techniques and applications - a systematic mapping study. Exp. Syst. Appl. **133**, 260–295 (2019)

64. Seeliger, A., Stein, M., Mühlhäuser, M.: Can we find better process models? Process model improvement using motif-based graph adaptation. In: Teniente, E., Weidlich, M. (eds.) BPM 2017. LNBIP, vol. 308, pp. 230–242. Springer, Cham (2018). https://doi.org/10.1007/978-3-319-74030-0_17

65. Senderovich, A., Weidlich, M., Gal, A., Mandelbaum, A.: Queue mining for delay prediction in multi-class service processes. Inf. Syst. **53**, 278–295 (2015)

66. Tax, N., Verenich, I., La Rosa, M., Dumas, M.: Predictive business process monitoring with LSTM neural networks. In: Dubois, E., Pohl, K. (eds.) CAiSE 2017. LNCS, vol. 10253, pp. 477–492. Springer, Cham (2017). https://doi.org/10.1007/978-3-319-59536-8_30

67. Taymouri, F., La Rosa, M., Dumas, M., Maggi, F.M.: Business process variant analysis: survey and classification. Knowl. Based Syst. **211**, 106557 (2021)

68. Teinemaa, I., Dumas, M., La Rosa, M., Maggi, F.M.: Outcome-oriented predictive process monitoring: review and benchmark. ACM Trans. Knowl. Discov. Data **13**(2), 17:1–17:57 (2019)

69. Teinemaa, I., Tax, N., de Leoni, M., Dumas, M., Maggi, F.M.: Alarm-based prescriptive process monitoring. In: Weske, M., Montali, M., Weber, I., vom Brocke, J. (eds.) BPM 2018. LNBIP, vol. 329, pp. 91–107. Springer, Cham (2018). https://doi.org/10.1007/978-3-319-98651-7_6

70. Thiede, M., Fuerstenau, D., Barquet, A.P.B.: How is process mining technology used by organizations? A systematic literature review of empirical studies. Bus. Process. Manag. J. **24**(4), 900–922 (2018)

71. Thomas, L., Kumar, M.M., Annappa, B.: Recommending an alternative path of execution using an online decision support system. In: Proceedings of the 2017 International Conference on Intelligent Systems, Metaheuristics & Swarm Intelligence, pp. 108–112 (2017)

72. Tu, T.B.H., Song, M.: Analysis and prediction cost of manufacturing process based on process mining. In: 2016 International Conference on Industrial Engineering, Management Science and Application (ICIMSA), pp. 1–5. IEEE (2016)

73. Verenich, I., Dumas, M., La Rosa, M., Maggi, F.M., Teinemaa, I.: Survey and cross-benchmark comparison of remaining time prediction methods in business process monitoring. ACM Trans. Intell. Syst. Technol. **10**(4), 34:1–34:34 (2019)

74. Yan, J., Hu, D., Liao, S.S.Y., Wang, H.: Mining agents' goals in agent-oriented business processes. ACM Trans. Manag. Inf. Syst. **5**(4), 20:1–20:22 (2015)

75. Zhao, W., Liu, H., Dai, W., Ma, J.: An entropy-based clustering ensemble method to support resource allocation in business process management. Knowl. Inf. Syst. **48**(2), 305–330 (2016)
76. Zhao, W., Yang, L., Liu, H., Wu, R.: The optimization of resource allocation based on process mining. In: Huang, D.-S., Han, K. (eds.) ICIC 2015. LNCS (LNAI), vol. 9227, pp. 341–353. Springer, Cham (2015). https://doi.org/10.1007/978-3-319-22053-6_38

Digital Transformation and Smart Life

How Do Startups and Corporations Engage in Open Innovation? An Exploratory Study

Maria Cecília Cavalcanti Jucá and Carina Frota Alves[(✉)]

Centro de Informática, Universidade Federal de Pernambuco, Recife, Brazil
{mccj,cfa}@cin.ufpe.br

Abstract. In recent years, the increasing market pressure and disruption have driven firms to undertake digital transformations to create value and deliver better products for customers. Large corporations frequently face difficulties to digitalize their internal processes. They have well-established and mature routines that are hard to change. By contrast, startups are recognized for their innovation capacity and agile processes. In the quest for speed and innovation, corporations are engaging with startups to match complementary goals. Corporations desire the creative potential of startups, while startups need resources that are plentiful in corporations. This paper explores how open innovation is performed from the perspective of startups and corporations. We conducted an exploratory interview study at eight startups and five corporations to understand the dynamics of their relationships during open innovation initiatives. Our results reveal the main drivers, benefits, and challenges involved in the engagement between startups and corporations. Finally, we present a set of recommendations to foster startup-corporation relationships.

Keywords: Digital transformation · Open innovation · Startup-corporation relationship · Empirical study

1 Introduction

Firms from different industries have seen dramatic and disruptive changes in recent years. To survive under such technological discontinuities and global market pressure, firms must transform their business models rapidly and frequently. Digital transformation enables firms to create value and deliver better products for customers [1]. A fundamental shift has occurred on how and where firms create value and innovate. Increasingly, value has been created from outside the firm and in collaboration with external partners [2]. This is a relevant move on how strategic information is generated and managed by several players with diverse interests and capabilities.

In line with the open innovation paradigm proposed by Chesbrough [3], large corporations are leveraging external knowledge sources and establishing a network of partners to enhance their innovation capacity. A relevant strategy to pursue open innovation initiatives is the partnership between corporations and startups. The startup-corporation engagement can be very effective by joining two, apparently, opposing forces. Startups can create new ideas and test them quickly. On the other hand, corporations have vast experience and abundant resources. Recent studies suggest the collaboration between large corporations and startups is a complementary match [4–6]. They can obtain mutual benefits from cultivating this type of relationship. Corporations desire the creative potential and agility of startups. From a startups' perspective, it is also an interesting strategic avenue. Engaging with large corporations could be a path to overcome their challenges regarding business model validation, lack of market experience, and limited financial resources.

Although previous studies aimed to understand the traditional collaborations via corporate investment funding and startups acquisition, few studies have investigated other lightweight alternatives of engagement between startups and corporations by means of open innovation [7]. This paper presents an exploratory qualitative study on how corporations engage with startups. In particular, we aim to compare and discuss the different perspectives of both players. We conducted semi-structured interviews at eight startups and five corporations to understand the dynamics of their relationships during open innovation initiatives. Our results reveal the main drivers, benefits, and challenges involved in the engagement between startups and corporations. We also present recommendations to foster startup-corporation relationships. The remainder of this paper is structured as follows. Section 2 presents a background in digital transformation and open innovation. Section 3 presents the research method. In Sect. 4 we discuss the findings of the empirical study. Section 5 proposes recommendations. Section 6 discusses related works. Finally, Sect. 7 presents conclusions, limitations, and future work.

2 Background

2.1 Digital Transformation

The acceleration in the creation of new technologies and the need to adapt to them made digital transformation increasingly important for companies of different domains and sizes to stay relevant. According to the Gartner Glossary [8]: *"digital transformation can refer to anything from IT modernization (for example, cloud computing), to digital optimization, to the invention of new digital business models"*. In a broader view, Rogers [9] points out that digital transformation is more than just the implementation of technology, it is about a holistic change in strategy towards the current market reality. Alstyne and Parker [2] agree with that view and suggest that digital transformation involves changing both the business model and changing where value is created in the organization. For Davenport and Redman [17], digital transformation requires talent in four key domains of knowledge: technology, data, process people, and organizational change capacity. The combination of adapting to new technologies, adjusting the mindset of staff to innovate, and leveraging the company's capabilities are key to digital transformation. Alstyne and Parker [2] make this even clearer when they point out that the greatest value of digital transformation comes from the inversion of value capture, moving the creation of value solely from within the firm to the outside, where it is necessary to know how to orchestrate it. In this context, open innovation is very useful mechanism to accelerate the digital transformation of traditional corporations.

2.2 Open Innovation

The openness of a corporation's vision of capturing value inside and outside is directly connected with Chesbrough's [3] definition: "*open innovation is the use of intentional knowledge inputs and processes to accelerate internal innovation and expand markets for external use of innovation, respectively*". According to [10–12], open innovation is an alternative for firms to stay tuned to new market trends, protect themselves from disruptive innovation and loss of market share. Gassmann [13] suggests that industries are more prone to engage in open innovation if they are characterized by technology intensity, technology fusion, new business models, and knowledge leveraging. Spender et al. [20] presents a literature review on the role of startups in open innovation projects. For Dodgson, et al. [14], the process of open innovation redefines the boundaries between organizations and the environment, making organizations more porous and creating a wide network of different actors, collectives, and individuals working to co-create and commercialize new knowledge. The main actors involved in these networks are corporations, medium-sized companies, startups, universities, consultants, and research centers. Relationships between corporations and startups can be an effective and collaborative way to accelerate innovation for corporations and support growth for startups [15]. There are several models on how corporate-startup relationships can be designed and executed [6, 7, 11, 15]. Chesbrough and Weiblen [6] conducted a well-known study on how corporations engage with startups to enhance corporate innovation. They identified a variety of ways that corporations can engage with startups. Based on their analysis of several open innovation initiatives, they classify the types of engagement based on two main criteria: direction of innovation flow (outside-in or inside-out), and presence or not of equity investment, as presented in Fig. 1. The four types of engagement are: corporate venturing, corporate incubation, startup program (outside-in), and startup program (platform). Corporate efforts to connect with startup ecosystems are becoming more attractive specially in periods of great uncertainty and volatility. These initiatives are

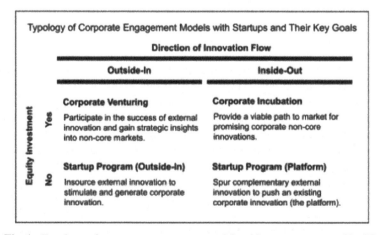

Fig. 1. Typology of corporate engagement models with startups proposed by [6].

more lightweight and fluid than traditional equity investment. Research on these types of engagement is still scarce and this paper aims to address this gap.

3 Research Method

In this paper, we aim to investigate how startups and corporations engage in open innovation initiatives. We formulated the following research questions:

RQ1 - What are the main drivers, challenges, and benefits involved in startup-corporation relationships?
RQ2 - What are the key recommendations to improve startup-corporation relationships?

Since our research questions concern a contemporary phenomenon, we conducted an empirical study by means of semi-structured interviews at startups and corporations to understand their experiences of engagement. The nature of the research is exploratory rather than confirmatory. We adopted steps proposed by Merrian and Tisdell to conduct qualitative studies [16]. To design the research protocol, we were initially inspired by the typology of engagement model presented in Fig. 1. Our main goal was to explore the whole journey of startup-corporation relationship. The questionnaire was composed of 5 parts: general context, mapping startup-corporation engagement experiences, identifying main characteristics of the engagement model, recommendations. We interviewed representants from startups and corporations to understand how their relationship began, what were the main characteristics of the relationship, what they learned during the relationship and what they would recommend to other organizations based on their experience. The mapping of the relationship aimed to explore: the **goals** of startups and corporations to start these relationships, the **approximation sources** they adopted, the main **challenges** and **benefits** from these experiences for both actors, and, finally, **recommendations** from studied startups and corporations. The complete questionnaire is available at https://tinyurl.com/m43dpvhh.

The selection of cases was based on the criteria that all organizations were required to have experience in open innovation initiatives. The first author approached contacts at accelerators and innovation institutes at the Porto Digital[1] Ecosystem to identify startups that are currently engaging with corporations in joint open innovation projects. Porto Digital is one of the largest technology hubs in Brazil. In 2020, the revenue was US$ 531 million. The innovation ecosystem has 339 companies of different sizes, with 180 startups. It has 5 innovation institutes, 7 investment institutions, 5 accelerators, and partnerships with 4 local universities. We interviewed CEOs and CTOs of 8 startups. During these interviews with startups, we asked them for contacts in corporations they are collaborating with. We were able to interview staff from 5 large corporations. In sum, the recruitment criteria were startups and corporations that have relationships with each other. The interviews were conducted by the first author via videoconference between March and November 2020. All interviews were recorded and later transcribed. The interviews lasted on average 38 min and the transcripts were consolidated in Google

---

[1] https://www.portodigital.org/home.

Docs to enable collaborative data analysis of the authors. To interpret the data from interviews, we followed Creswell's [18] approach to analyze qualitative data. As a first step, we organized and prepared the data for analysis. Then, we proceeded to reading all data to start coding. We started using open coding to analyze the transcripts. The open coding process was done by the first author, that is when themes and descriptions started emerging. The next step was interrelating themes and descriptions. Finally, we interpreted the meaning of themes/descriptions. This process was iterative, which was done multiple times to achieve final results. We associated raw data with low-level codes that were further refined into categories. Then, we used axial and selective coding to synthesize the final categories. We used MindMeister tool to visualize the categories and respective quotes from the interviews in an intuitive and collaborative manner.

4 Results

4.1 Overview of Studied Startups and Corporations

In Table 1, we present an overview of the 8 startups from the Porto Digital Ecosystem that participated in our study. The startups were chosen because of their background and notable partnerships with corporations. All startups have been in business for over 3 years. The oldest one is S6 with 15 years but the startup still explores different business models, in a context of frequent pivoting. The majority of startups have very small teams, with up to 10 employees. The startups operate in different markets, which brings a rich diversity to the research. Regarding the types of consumers, 7 out of 8 startups adopt the "B2B" Business to Business model. Regarding the revenue model, all 8 startups use the software as a service "SaaS" model. All respondents were part of the startup's "C-level",

Table 1. Overview of participant startups.

ID	Market/Domain	Type of consumers	Business model	Number of staff	Year of creation	Interviewee role
S1	Construction	B2B and B2B2C	SaaS	2–10	2018	CEO
S2	Fintech	B2B	SaaS	2–10	2016	CEO
S3	Sales and communication	B2B	SaaS	2–10	2016	CEO
S4	Social impact	B2B2C	SaaS	2–10	2016	CEO
S5	Human resources	B2B	SaaS	2–10	2017	CEO and CTO
S6	Logistics	B2B and B2B2C	SaaS	21–200	2006	CEO
S7	Analytics and big data	B2B	SaaS	11–50	2007	CTO
S8	E-commerce	B2B	SaaS	2–10	2016	CEO

which greatly benefited our study, as these people know the entire history and strategy of innovation and growth of their startups.

Table 2. Overview of participant corporations.

ID	Market/Specialty	Type of consumers	Number of staff	Year of creation	Interviewee role in the corporation
C1	Construction	B2B and B2C	51–200	1995	Commercial director
C2	Logistics and supply chain	B2B	501–1.000	1991	C2.1 Project manager C2.2 Finance manager
C3	Wholesaler, retailer, distribution and logistics	B2C and B2B	700–900	1978	CTO
C4	Cosmetics	B2C and B2B	5.000–10.000	1969	C4.1 Marketing coordinator C4.2 Innovation manager
C5	Consumer goods	B2B2C	10.001+	1872	Supply chain coordinator

The startups were selected by convenience because they are representative cases of different lifecycle stages of startups and all have experience in open innovation projects with corporations. When we interviewed participants of S1, S2, S3, S4 and S5, they mentioned their startups had relationships with corporations C1, C2, C3, C4 and C5, respectively. We were able to interview staff from these corporations. In summary, 5 out of 8 startups were portrayed with corporations they are currently engaging with. Startups S6, S7 and S8 do not have corresponding corporations because we were not able to get access to the corporations indicated by them. Therefore, these startups answered the questions taking a broad view of all corporations they had relationships with. However, we were not able to confirm the findings with the corporations. Table 2 present corporations' profiles, they were all selected due to their relationship with the interviewed startups. They all have a considerable size in terms of staff and revenue and are considered relevant players in the national market. The corporations operate in different market segments and all corporation's interviewees were directly involved in projects with startups. In particular, we were able to interview two staff in C2 and C4, which allowed us to obtain a different viewpoint in these corporations.

4.2 Engagement Models Adopted by Startups with Corporations

To understand the context of relationships between startups and corporations, it is important to analyze the engagement models used by them. The models presented in Table 3 were indicated by the participants during the interviews with startups. We classified the engagement models considering the direction of innovation flow according to proposals from [6, 11]. Using this initial categorization, we identified the following engagement models of outside-in innovation flow that studied startups adopt: *Proof of Concept, Direct Services and Co-creation, Mergers & Acquisitions.*

Table 3. Types of engagement models.

Direction of innovation flow	Engagement model	Startup ID
Outside-in	Proof of Concept	S2, S3, S4, S5, S6
	Direct Services and Co-creation	S2, S3, S6, S7
	Mergers & Acquisitions	S8
Inside-out	Venture Builders and Spin-offs	S1
	Startup Acceleration Programs	S2, S7

The *Proof of Concept* model is very popular among the startup community. It consists of a contract (paid or free) to validate the technical capacity and value of the startup solution. It is generally adopted during early relationships between startups and organizations, has a short duration (on average 3 months), and focuses on building MVPs (Minimum Viable Product) to enable the validation of the solution's value proposition. We observed that five of the studied startups adopt this model. For example, S4 described that their relation with C4 started with this model, experimenting and validating hypotheses of the product, so they ran a 3-month proof of concept to validate the idea before investing. *Direct Services and Co-creation* are a model where services are provided in a more traditional way. Startups S6 and S7 provide their standard service and get paid a regular fee, while startups S2 and S3 engage in co-creation initiatives with corporations. The last outside-in engagement model identified is *Merge & Acquisitions*. In this model, corporations go after startups, which can represent an interesting partner or a brutal competitor, making the purchase or acquisition a strategic path for the corporation. For startups, it is an interesting model when their goals change along the entrepreneurial journey, when they no longer see the possibility of growth or when the proposal is financially very attractive.

In the inside-out innovation flow, we identified two engagement models adopted by startups. *Venture Builders and Spin-offs* are a well-known model where startups are created inside corporations to serve new markets, and even define different strategies, then startups can leave the corporations to become an independent firm (spin-off). This is the case of startup S1, which was created by the founders of corporation C1 to diversify their portfolio. The *Startup Acceleration Programs* involves initiatives that corporations create a structured entrepreneurship program to share with the startup community their main challenges, looking for solutions that fit them. Startup S7 mentions how many startup programs they have participated and how this model supports their growth: *"We've already participated in some acceleration programs, where we show our product development capacity and it can end up generating new contracts, with greater scalability, security and growth potential".*

4.3 Drivers Involved in Startup-Corporation Relationships

This section presents the key drivers for startup-corporation engagement. We consider drivers as triggers or engines to create startup-corporation relationships. In this context, two categories of drivers emerged from the interviews: *(i) goals, (ii) sources of approximation.* As presented in Table 4, the startups' main goals are: *(i) Product/service development, (ii) business sustainability, and (iii) organizational growth.* Within the product/service development category, it was possible to identify the startups' goal to improve and further develop their products using corporate resources. The business's focus on sustainability was clearly the goal of maintaining the startup's financial health, since a very common situation faced by startups is the lack of financial resources. Our analysis revealed that accomplishing direct sales is a key goal for 7 of the 8 startups. As states S4: *"I had a clear focus on large corporations to make bigger sales, if not, I would take a long time to reach my breakeven (financial breakeven point)".*

Table 4. Startups' goals to engage with corporations.

Macro goals	Specific goals	ID
Product/Service development	Maturation of the startup's technological capacity	S2, S4, S6, S6
	Identify relevant opportunities to develop new features and products	S1, S2, S6
	Validate the product-market fit	S1, S4, S5, S7
Business sustainability	Accomplish a direct sale	S2, S3, S4, S5, S6, S7, S8
	Find the best revenue streams and partnership models	S4, S5
Organizational growth	Professional evolvement of the startup team to gain experience with the corporation	S1, S4, S6, S7
	Gain credibility through brand association with the corporation	S1, S2, S3, S5, S6, S8

Finally, we identified the need for organizational growth, either due to the startups' short life in the market or because of the young age and limited maturity of the team members. We also observed that several startups aim to gain credibility from associating themselves with the brand of the corporations, 6 startups mentioned the goal of marketing and brand strength.

From the perspective of corporations, their goals to engage with startups are *digital transformation acceleration, resource optimization,* and *desire to be at the forefront of industry,* as presented in Table 5. Corporations aim to learn the methodologies and mindset of startups to emulate their digital culture. Such methodologies include: agile methodologies, lean startup, design thinking, etc. Corporations aim to improve their problem-solving capacity through the discovery and development of technology-based and innovative solutions that are often skillfully mastered by startups. This goal is even more critical for traditional corporations, as C3 describes the value of startups on their digital transformation path: *"We value the relation with startups that helps us to identify and create new solutions to old problems, as well as accelerating our full digital transformation"*.

Table 5. Corporations' goals for engaging with startups.

Macro goals	Specific goals	ID
Digital transformation acceleration	Problem solving through the creation of tech-based and innovative solutions	C1, C2, C3, C4, C5
	Learn new methodologies and approaches to adopt the startups mindset (e.g., lean startup, design processes, agile methodologies)	C2, C3, C4
	Improve the team's technological and innovation capabilities	C2, C3, C4
Resource optimization	Optimize internal processes by contracting outsourcing services	C2, C3, C4
	Acquire cost-effective services	C1, C2, C3
Desire to be at the forefront of industry	Attract young and talented personnel	C3, C4, C5
	Presence in the startup and innovation ecosystems	C1, C4, C5
	Conquer or accelerate new markets through new strategic partnerships	C1, C3

Corporations focus on process improvement as part of their robust structure. Another goal of corporations is to become a reference in their industry by: attracting new talent, accelerating the entry into new markets and having presence at the innovation and startup ecosystems. By actively engaging with a startup, the corporation uses the partnership to attract new talent interested in the startup, but seeking the stability of a large corporation as C5 comments: *"through hackathons to create startups or partnerships with existing startups, we were able to map talents and attract people looking to innovate and work in our team"*.

We observed the most common **source of approximation** is through **events and networks**. Startups and corporations value attending strategic events of Porto Digital's ecosystem, these events promote the exchange of ideas between startups and corporations. Startup S2 points out how strategic it is to build strong networking and participates in events wisely: *"My strategy was to build a strong network of contacts. For that, I attended many events, knocked on many people's doors, and received many no's, but always with focus, given that this source of approximation is very important"*.

Partner scouting is another frequent source of approximation. It is a service that innovation consultancies and open innovation facilitators promote to actively identify and seek out startups in communities that are fit with the corporation's interest. This service is hired by the corporation in search of solutions that already exist on the market and works analogously to a startup "headhunting". This is a very common strategy that players of Porto Digital adopt to engage with new partners.

Finally, *corporate innovation programs* were frequently cited both by startups and corporations as mechanisms to find potential partners. We consider external programs, those that the corporation opens its problem to the market so that startups come after the corporation with proposed solutions. It is a very popular format and the operational flow only happens with proposals from participating startups. Internal programs involve the creation of an environment to enable experimentation and innovation for the corporation's own teams. They may lead to the creation of an internal startup, and eventually, the startup may leave the corporation as a spin-off.

4.4 Challenges for Startups and Corporations

In this section we discuss the challenges mentioned by startups and corporations when engaging with each other. The main challenges described by studied startups are: *(i) cultural differences, (ii) complexity of corporate processes* and *(iii) business model obstacles*.

The *cultural differences* between startups and large corporations are notable. From the perspective of startups, dealing with the lack of credibility in their delivery capacity is a great challenge, since startups may face great difficulty in transmitting self-confidence and convincing skeptical corporations, as S1 states: *"we must break this barrier that exists, often due to age prejudice or lack of professionalism, through our delivery of value"*. The basis for startups to deliver value are the innovative methodologies used in the experimentation process. An important point of resistance is the understanding of this process by corporations and the need to align their expectations. Corporations often expect a ready-made product, where in reality the startup is still developing it, sometimes even in the form of a MVP, as states S7: *"Corporations expect to buy a finished product most of the time, they rarely expect to build a product jointly. It is a great difficulty for us to manage this misalignment of expectations"*. This cultural mismatch creates problems due to the risk aversion mindset that some corporations have. By not accepting to take risks, corporations also hinder the experimentation process. Differently, startups are organizations that live in an environment of uncertainty. When dealing with corporations that do not embrace the risk inherent of innovation, a lot of friction may be generated between them.

The *complexity of corporate processes* is a matter of regulatory laws for that type and size of company. In addition to the legislation itself, the process of organizing a large corporation is naturally more complex due to the volume of people, data, and processes. Therefore, a key discord factor between startups and corporations is speed, this challenge was mentioned by all interviewed startups, as S4 explains: *"The delay in closing the contract, payment or other processes is very long, which is generally very different from the process of a startup*". S2 corroborates this view: "*Another challenge is because they are too rigid to change, to believe in innovation since the processes implemented there are already so consolidated*". Bureaucracy gains great attention from startups when several documents and internal assessments are needed to close a deal. Another perceived challenge involves the de-prioritizing innovation projects. If the project with startups does not directly provide gains to the corporate core business, it is common to see it deprioritized, with a smaller team or no corporate staff fully dedicated to it. For startups, it becomes a major obstacle to access key information for the project, in addition to slowing down the entire innovation project.

Finally, **business model obstacles** refer to limited resources and power inequality in negotiations between startups and corporations. S4 states that it is very risky for startups to start a new project without sufficient investment: "*In some cases, startups lack resources for adapting to the process of corporations. In our case, we were lucky to have our own money to invest at that point, otherwise, we wouldn't be able to run the project*". It is well-known that corporations have strong market power and a great ability to negotiate. Therefore, startups often struggle with negotiations so they can equalize their needs and prove their capabilities when partnering with corporations.

From the perspective of corporations, they reported the main challenges when interacting with startups involve: *(i) cultural differences, (ii) managerial priorities, and (iii) partnership risks.* It is interesting to note that **cultural differences** are challenges mentioned by both startups and corporations when interacting with each other. Corporations praise control, data, and predictability. In particular, corporations are not comfortable with uncertain results. Corporations C1, C3, and C4 mention how the lack of predictability on the results of the partnership with startups generate anxiety and bring frustration for the corporation's team. In the same context, low failure tolerance is also a major challenge for corporations. Startups are willing to take risks on experimentation and understand that errors are part of the innovation process. On the other hand, corporations are more concerned about the damage of failing projects, as C3 comments: "*The biggest challenge I see today is to be aware, throughout the company, that tolerance to error is necessary, tolerance to a low level of quality of delivery or performance, in order to innovate*". Another aspect related to cultural differences is that corporations have to deal with young entrepreneurs and startups that are often just starting out and are still very immature, as C4 points out: "*Many startups that we interacted with are very focused on the technical solution and often forget to evolve their management and preparation in other business issues. It is a challenge for us to deal with this lack of preparation*".

The challenge **managerial priorities** refer to the fact that corporations are already creating an awareness that the issue is internal. Studied corporations pointed out that it is hard to propose innovative projects with startups and prioritize them with corporate senior managers. Corporations mentioned the difficulties to raise internal awareness on the importance of these projects. In addition, corporations already understand the need to reduce bureaucracy in their processes to meet the speed of startups.

Finally, the challenge of **partnership risks** was often mentioned by the studied corporations. We identified that it is problematic to obtain an internal budget for innovation projects since it is quite complex to measure the ROI (Return on Investment) of initiatives with startups. The return is often obtained by employees' training on innovative approaches, publicity of engaging with the startup ecosystem, and other indirect results that are more difficult to measure. The situation is especially challenging on projects where the construction of the solution fails and, through experimentation, it is proven that it is better to pivot. In these cases, the partnership risk becomes even more critical.

4.5 Benefits for Startups and Corporations

We observed that fundamental differences between startups and corporations generate complementary strengths and opportunities during their relationships. The main benefits perceived by startups are: *(i) brand reputation and networking with corporations, (ii) accelerate product growth, (iii) business development*. The majority of startups understand the benefit obtained by the credibility of brand association with a respected corporation. Increasing the startup reputation by partnering with a corporation is an important benefit to foster the startup's image. S2 and S3 agree that having well-known corporations in their portfolio of partners brings positive financial returns. S1 states that C1's great reputation helped break down prejudice barriers that he suffered as a young entrepreneur among potential clients, especially in a more traditional market. In addition to brand association, startups also acquire a strong network and possibility to obtain mentoring from corporate executives and professionals.

Another benefit described by studied startups is *accelerate product growth*. Startups evolve and validate their business models through the experience of having as customers corporations with many demands and high expectations of quality. Engaging with corporations enables startups access of real and large data that accelerates the maturity of the solution. This benefit is perceived by startups S1, S2, S3, and S6. *Business development* is also a benefit that startups have when engaging with large corporations. The bureaucratic processes of corporations are challenging for startups, as discussed previously. On the other hand, the same bureaucratic processes can be beneficial when startups identify the need to learn how mature corporations operate, as they aim to become one in the future. S2 states: *"the processes, legal issues... all the bureaucracy in this relationship prepares us for the growth of our own business"*.

The main benefits observed by corporations during engagement with startups are: *(i) develop new solutions, (ii) stimulate innovation culture, (iii) increase optimization and efficiency*. Startups can *develop new solutions* that bring value for corporations and consequently, open the opportunity to expand corporate market position. In this context, 4 out of 5 corporations reported the speed of delivery as a great benefit of engaging with startups. C2 points out that the speed to start a pilot project with startups is quite high, and it is a differential compared to partnerships with other more mature organizations. Engaging with startups increase their productivity and delivery speed of its own team. Additionally, startups present gains for corporations by being flexible and working with ease in the customization of their products and solutions. C4 exposes this benefit and this delivery of value from startups: *"The main benefit is having the flexibility to build together a customized and specialized solution specific to our corporation's demand"*.

A benefit perceived by all corporations is the experimentation of an *innovation culture*. This search is closely linked to achieve the goal of digital transformation. In this perspective, the relationship with startups provides the benefit of rejuvenating the brand of the large corporation, which is often considered obsolete or too analogic. Interacting with startups enable corporations to solve their traditional problems through a fresh and innovative look provided by startups. Corporations find it difficult to clearly identify these problems within a traditional and routine-driven mindset. Startups are able to contribute by co-creating new solutions without internal biases and with a culture of radical experimentation deeply founded. Learning through the exchange of experiences is evidenced by C3: *"There is an intangible gain here, which is the development of the employees involved, we learn methodologies and ways of working to generate innovation"*. Finally, organizations reported the benefit of ***increase optimization and efficiency***. Startups are service providers and tend to charge very affordable price for corporations. Resource optimization is also achieved when processes that were previously done manually or by obsolete systems are updated or implemented by startups' tech-based solutions. The automation of analog processes is an important aspect to increase corporate efficiency, it is a core part of the digital transformation of traditional organizations.

5 Lessons Learned and Recommendations

During the collaboration in open innovation initiatives, startups and corporations may acquire new capabilities and obtain valuable resources that are internally scarce. In order to contribute with other startups and corporations to achieve successful open innovation projects and build win-win relationships, we summarize key recommendations provided by studied startups and corporations to foster this kind of relationship. The recommendations encapsulate the lessons learned by the participants of the study based on their own experiences and viewpoints.

5.1 Recommendations for Startups

1. Do not neglect formalities and regulations. At first, formalities and regulations are seemed as obstacles for startups, but we observed that successful startup-corporation engagement should comply with these issues. First, it is necessary to be aware of the principles and premises for the partnership and observe legal regulations. In addition, formalities help corporations to maintain their organizational structure and manage complex processes. Finally, startups should learn formal rules to support their managerial maturity.

2. Carefully negotiate and define a reasonable pricing strategy. If the relationship with corporations includes negotiation, reaching suitable agreements is often challenging for startups. We identified that it is necessary for startups to pay careful attention to their trading strategies. To gain the first contracts, it is important to keep an attractive price. Charging too high can get corporations off the table. On the other hand, if startups charge very low prices, corporations tend to think the startup won't be able to get the job done. Furthermore, startups should present their services in a trustful manner and clearly demonstrate the tangible value they will deliver to corporations.

3. Build a strong network with corporations. Strategic alliances can foster the entrepreneurial journey of startups through valuable connections, opening doors for referrals and growth support. Strengthening the network with respectable corporations is a powerful guiding process to evolve startups, enabling the connection with key actors who have valuable expertise and mentoring to share with young entrepreneurs.

4. Build an adequate delivery strategy. Satisfying the high expectations of corporations is a critical success factor. In this way, startups should build good planning and manage deliveries and deadlines. This practice will help to align expectations and strength the collaboration between corporations and startups. Carefully balancing the level of promises is also a great practice to implement, given that the boldness of startups as a surprise factor is much more alluring than promising deliveries that won't be executed.

5. Focus on customer satisfaction. Expectations are also directly linked to customer satisfaction of partnering corporations. Understanding the needs, studying the market and adapting, whenever possible, its solution to better serve corporations are essential practices to ensure satisfaction and a prosperous relationship. The flexibility of startups regarding the structure and availability of the corporation's team should also be considered.

5.2 Recommendations for Corporations

1. Align expectations with startups. The cultural differences between startups and corporations generate tensions regarding the level of delivery and commitment expected by each party. Therefore, to increase the success of the relationship it is important to seek, from the start, the alignment of delivery expectations with the startup.

2. Identify a relevant problem and seek to solve it with startups. We observed that corporations often struggle to identify relevant problems by themselves. Corporations are very immersed in their daily routines, which makes it very challenging to capture what is the root problem they aim to address. Problem framing is a premise to innovate when interacting with startups. If corporations do not understand and clarify the problems they want to solve, the relationship may lead to a loss of resources that does not generate any meaningful innovation impact.

3. Adapt bureaucratic processes to fit startup's speed. Corporations should simplify whenever possible bureaucratic routines to help the startups accelerate their processes. When the bureaucratic issue is due to a legal requirement, corporations should keep it. However, when the bureaucratic process is imposed by old fashionable organizational culture, we recommend developing new processes or making exceptions that can help to speed up the engagement with startups. It is important to have the basic infrastructure adapted as soon as possible at the beginning of the relationship. Otherwise, because of the highly uncertain environment that startups operate in, they run the risk of not surviving while waiting for corporations' approvals.

4. Embrace risk and accept the possibility of failure. Corporations should be prepared to participate in the entire innovation process, starting with the problem framing, and in an iterative manner, going through several experiments to validate the problem, explore new ideas, and collaboratively build the solution. The partnership with startups requires a tolerance to risk and awareness of eventual failures. In these situations, corporations should be prepared to learn from failed innovation experiences.

5. Learn how to identify and measure open innovation project results. It is essential to define how to measure and show the results of innovation projects carried out with startups. Defining the return on investment of innovation is not simple, as there are often intangible returns and unpredictable results involved in most open innovation projects. This measurement must be directly connected with the goals initially established. We recommend sharing the lessons learned from the project with the staff so that an innovation culture can be spread to different areas of the corporation.

6 Related Work

Differently from previous research [6, 7] that addressed only the corporations' vision, we investigated both actors' perspective so we could analyze a broad vision of their relationship. In particular, previous research of Chesbrough [6] only covered corporations from the tech industry, our study investigated corporations from different industries relating with software-based startups. Our findings show that corporations were seeking tech innovation and fostering their digital transformation throughout the relationships with startups. Prior research [19] focused on the selection of partners to conduct open innovation projects. A clear difference is that we selected the interviewees from corporations based on their relationships with the selected startups. Our results are consistent with findings from [21] confirming the growing adoption of lean startup by corporations. Another specific contribution is that we mapped three phases of the relationship: before engagement, during engagement and after engagement.

7 Conclusions, Limitations and Future Work

In the face of increasing pressure to pursue digital transformation, large corporations aim to overcome internal rigidities and leverage innovation capabilities by means of open innovation projects with startups. In this study, we investigated the dynamics of startup-corporation relationships from an exploratory interview study of eight startups and five corporations. We discussed the key drivers, challenges, and benefits involved in open innovation projects between startups and corporations. The lessons learned presented by the participants of the interviews were derived into recommendations to build win-win relationships. The practical contributions of this paper can be valuable for entrepreneurs and corporate innovators to obtain a complementary view of different types of engagement models. The results of the study may provide practical orientation, and actionable insights on how to establish and nurture startup-corporation relationships. In particular, this paper contributes to the Information Science field by exploring how startups and corporations co-create and share strategic information to improve their business models.

The following limitations were identified during the design and execution of this study. Since we use interviews as the main data collection technique, our results are centered on the personal opinions of respondents and may suffer from their prejudice and bias. The interviews were conducted and transcribed by a single researcher. To address interpretation bias, data analysis and synthesis were performed by both authors during several meetings. We were not able to conduct interviews with corporations that engaged with three of the studied startups (S6, S7, S8). Therefore, we could not confirm the views of these startups with respective partnering corporations.

In future works, we aim to carry out longitudinal case studies to follow the evolution of the relationships between startups and corporations to monitor the results, impact, and performance indicators of open innovation projects. Another future work is to deepen the analysis of what constitutes a successful and failed collaboration from the perspective of startups and corporations. Finally, we would like to understand how the flow of strategic information directs the innovation of business models of both startups and corporations.

References

1. Matt, C., Hess, T., Benlian, A.: Digital transformation strategies. Bus. Inf. Syst. Eng. **57**(5), 339–343 (2015). https://doi.org/10.1007/s12599-015-0401-5
2. Van Alstyne, M., Parker, G.: Digital transformation changes how companies create value. Harv. Bus. Rev. (2021)
3. Chesbrough, H.: Open Innovation: The New Imperative for Creating and Profiting From Technology. Harvard Business School Press (2003)
4. Mocker, V., Bielli, S., Haley, C.: Winning together: a guide to successful corporate-startup collaborations. Technical report NESTA intelligence & startup europe partnership (2015)
5. Siota, J., Prats, M., Fernandez, D., Perez, T.: Open innovation – improving your capability, deal flow, cost and speed with corporate venturing ecosystem. Technical report, IESE Business School University of Navarra (2020)
6. Weiblen, T., Chesbrough, H.: Engaging with startups to enhance corporate innovation. Calif. Manag. Rev. **57**(2), 66–90 (2015)
7. Thieme, K.: The strategic use of corporate-startup engagement. Master thesis in management of technology, Delft University of Technology (2017)
8. Gartner Glossary. https://www.gartner.com/en/information-technology/glossary/digital-tra nsformation. Accessed 22 Jan 26
9. Rogers, D.: The Digital Transformation Playbook: Rethink Your Business for the Digital Age. Columbia Business School Publishing. (2016)
10. Chesbrough, H., Vanhaverbeke, W., West, J.: Open Innovation: Researching a New Paradigm. Oxford University Press, USA (2006)
11. Bonzom, A., Netessine, S.: How do the world's biggest companies deal with the startup revolution? 500 and INSEAD Report (2016)
12. Christensen, C. M.: The Innovator's Dilemma. Harvard Business School Press, Boston (1997)
13. Gassmann, O.: Opening up the innovation process: towards an agenda. R&D Manag. **36**(3), 223–229 (2006)
14. Dodgson, M., Gann, D.M., Salter, A.: The role of technology in the shift towards open innovation: the case of procter & gamble. R& D Manag **36**(3), 333–346 (2006)

15. Alänge, S., Steiber, A.: Three operational models for ambidexterity in large corporations. Triple Helix **5**(1), 1–25 (2018)
16. Merriam, E., Tisdell, E.: Qualitative Research: A Guide to Design and Implementation, 4th edn., Jossey-Bass (2016)
17. Davenport, T., Redman, T.: Digital Transformation Comes Down to Talent in 4 Key Areas. Harv. Bus. Rev. (2020)
18. Creswell, J.W.: Research design: Qualitative, Quantitative, and Mixed Methods Approaches. Sage Publications (2009)
19. Groote, J.K., Backmann, J.: Initiating open innovation collaborations between incumbents and startups: how can David and Goliath get along? Int. J. Innov. Manag. **24**(2) (2020)
20. Spender, J.C., Corvello, V., Grimaldi, M., Rippa, P.: Startups and open innovation: a review of the literature. Eur. J. Innov. Manag. **20**(1) (2017)
21. Chesbrough, H., Tucci, C.L.: The interplay between open innovation and lean startup, or, why large companies are not large versions of startups. Strateg. Manag. Rev. **1**(2), 277–303 (2020)

Smart Life: Review of the Contemporary Smart Applications

Elena Kornyshova[1(✉)], Rebecca Deneckère[2], Kaoutar Sadouki[1],
Eric Gressier-Soudan[1], and Sjaak Brinkkemper[3]

[1] CEDRIC, Conservatoire National des Arts et Métiers, Paris, France
{elena.kornyshova,eric.gressier_soudan}@cnam.fr,
kaoutar.sadouki@lecnam.net
[2] Centre de Recherche en Informatique, Université Paris 1 Panthéon Sorbonne, Paris, France
rebecca.deneckere@univ-paris1.fr
[3] Utrecht University, Utrecht, The Netherlands
S.Brinkkemper@uu.nl

Abstract. Research efforts in various fields related to Smart life increase constantly. There are very well-established fields like Smart home, Smart city, and Smart grid, and more emergent ones, like Smart farming, Smart university, and Smart tourism. Smart life intends to enhance human life and serves as an umbrella term for all Smart topics. However, the research domain of Smart life is very diverse and multifaceted. Our main goal is to systematize the existing research work around Smart life to provide direction for the development and maintenance of Smart artefacts and applications, thus, to move towards Smart life engineering. To achieve this goal, a mandatory step is to understand and organize all the different Smart topics studied in the scientific literature by means of a systematic mapping study. We analyzed 2341 existing Smart state-of-the-art works and research agendas. We propose a taxonomy of Smart applications, study their evolution over time, analyze their venues, specific terminology, and driving factors. The resulting overview is useful to researchers and practitioners to improve the positioning of their work and to identify the research opportunities in all types of smartness for humans, organizations, and society.

Keywords: Smart life · Smart application · Systematic mapping study

1 Introduction

The number of research works in "Smart" fields grows continuously. Many related fields have appeared over the last decades. For instance, Smart home is defined in [1] as "the integration of different services within a home by using a common communication system. It assures an economic, secure, and comfortable operation of the home and includes a high degree of intelligent functionality and flexibility." In [2], Smart city is "a city that collects and utilizes data gathered from distributed sensors and video cameras that connect everything from trash bins to streetlights." Since then, appeared new concepts like Smart farming or Smart agriculture [3], Smart university [4], Smart tourism [5], and so on. Numerous state-of-the-art works exist in all these fields, as for the Smart home domain for which we identified almost 180 state-of-the-art and research agenda papers, for instance [6–8], or [9]. Despite a huge number of works in Smart fields, works aiming at classification, organization, and systematization of the different Smart fields are neglected.

We unify all these Smart fields within the concept of Smart life that we define as *"a system of integrated Smart application systems for the purpose of an enriched life experience"*. In our vision, Smart life is a set of different Smart applications (e.g., Smart home, Smart city, Smart transportation, and so on) which, in their turn, are composed of different Smart artefacts (e.g. Smart car, Smart ship, Smart railway, etc. for the Smart transportation application). We believe that optimization could and should be done to design, develop, and maintain all these various applications in a smarter way. In our vision, different methodologies, technologies, and tools could be shared and reused between the Smart fields. This leads us to promote a new field we called Smart life engineering.

To start to investigate this research direction, we should first analyze the existing works. In the already published literature on Smart life, we did not identify any detailed taxonomy on Smart applications. Most of the state-of-the-art works on Smart life deal with a particular sub-topic within this context: IoT application for energy consumption [10], Security in Internet-of-Things (IoT) [11], IoT usage in Smart cities with a classification of Smart technologies [12], computer technologies from the viewpoint of artificial intelligence [13], machine to machine usage [14], and an integrative study about IoT [15]. Only two papers are generic: one source gives a classification of Smart technologies [16]. The second one suggests a systematic literature review on Smart systems [17]. In addition, [18] reviews Information and Communication Technologies (ICT) within a "Smart Earth" concept.

In the current work, we focus on a study of Smart applications. We analyzed the titles and abstracts of 2341 State-of-the-Art- and Research Agenda-oriented papers in order to establish a taxonomy of Smart applications, analyze their venues, understand the underlying terminology, and, finally, understand what are driving factors of Smart life. We used the Systematic Mapping Study method [19] to develop our research approach.

We explain our research approach in Sect. 2. We highlight the obtained results in Sect. 3. In Sect. 4, we conclude the paper and outline our future research.

2   Research Approach

We used a systematic mapping design [19] to study the Smart life field of research. Systematic Mapping Studies (SMS) are similar to other systematic reviews (like SLR – systematic literature review synthesizing the existing research in established fields), except that they employ broader inclusion criteria to select a wider range of research papers and are intended to map out topics with a field classification rather than synthesize study results. The study presented here covers the existing work in the field of Smart life. We followed the process presented in [19] which includes five steps: definition of research questions, conducting search for primary studies, screening of papers, keywording of abstract and data extraction and mapping of studies.

Based on [19], we followed the procedure on SMS summed up in Table 1. We named differently Steps 2 and 4 as more compliant with our research process.

Table 1. Five-steps procedure of SMS on smart applications.

Number	SMS process step	Outcome
1	Definition of research questions	Review scope
2	Finding papers	All papers
3	Screening of papers	Relevant papers
4	Keywording using abstracts (classification scheme definition)	Classification scheme
5	Data extraction and mapping process	Systematic map

Step 1: Definition of research questions. The goal of this SMS is to identify and structure different Smart applications. To detail this goal, we established the following research questions:

RQ1. What are the different applications studied in the Smart life field and their evolution?
RQ2. In what venues are published research papers on Smart life applications?
RQ3. What is the specific terminology used in Smart life applications?
RQ4. What are the main driving factors of Smart life based on data about Smart applications?

Step 2: Finding Papers. This step aims at identifying a set of papers based on a relevant search string. We searched and selected papers in the SCOPUS scientific database using the SCOPUS Search API in November 2021. (This database includes all "articles being published in virtually all scholarly journals of any significance in the world"[1]). Firstly, we searched for papers containing the term "Smart" only in the title, and the number of obtained papers was more than 126 000. Thus, we limited our search to secondary research papers like state-of-the-arts and research agendas. This reduced search allowed to obtain 2410 sources with DOIs. The inclusion criterion related to the search string is given in Table 2.

Step 3: Screening of Papers. We analyzed the titles and, if needed, abstracts and papers content. We excluded 69 sources, not representing research papers or not relevant to Smart life topics papers. We obtained 2341 papers[2]. In Table 2 we summarize the exclusion criteria used to obtain the list of relevant papers.

Step 4: Classification Scheme Definition. The goal of this step of SMS is to identify the classification scheme to be applied to the obtained results. To answer the defined research questions, we classified all relevant papers accordingly to the following criteria:

---

[1] https://www.elsevier.com/__data/assets/pdf_file/0007/917179/Scopus-User-Community-Germany-API-final.pdf.

[2] The complete list of these papers is available at http://cri-dist.univ-paris1.fr/rcis22/RCIS2022_Appendix%20A%20-%20references%20list.pdf.

Table 2. Inclusion/exclusion criteria for the study on smart topics.

Selection criteria	Criteria description
Inclusion criteria (2410 sources identified)	The title includes the term «smart» and at least one of the terms "research agenda", "state-of-the-art", "review", or "survey": Search string: TITLE(smart) AND (TITLE("research agenda") OR TITLE("State-of-the-art") OR TITLE(review) OR TITLE(survey))
Exclusion criteria (2341 sources selected)	The source is not a research paper (erratum, retracted, etc.) The source is related to an abbreviation SMART, like, for instance, SMART (stroke-like migraine attacks after radiation therapy) syndrome The source mentions the term «smart», which is used in its ordinary sense, like "working smart and hard"

- Criterion related to Smart topics. Smart life is a set of Smart applications. This qualification allowed us to establish a detailed taxonomy of Smart life applications:

  - Smart Application

- Criteria related to publication venues:

  - Year
  - Venue
  - Type of Venues

In addition, we used the titles and abstracts to provide additional representations.

Step 5: Data Extraction and Mapping Process. For RQ1, the main extraction was done with relation to the Smart life topics as it required additional work to classify Smart life applications into a taxonomy. Then, we classified the identified applications into four Smart domains: Persons, Society, Environment, and Enterprises, and we analyzed the distribution of publications in these domains over time. For RQ2, the data on years and issues were already available, we completed them by qualifying the most used issues by their type (Survey-oriented, Generic, Domain-specific). For RQ3, we created keyword clouds based on papers titles and papers abstracts to have a visual representation of the most used terms in the scientific literature related to Smart life and we extracted the main used terminology. For the two clouds, we deleted the stop words, the numbers, and the special characters. Moreover, on the second one (set of abstracts), we made a pre-processing of the data to change plural nouns into singular ones if the two forms were present in the dataset.

Validity Threats. Qualitative research is based on subjective, interpretive, and contextual data. Thus, we analyzed the potential biases, which could threaten the validity of our research. [20] proposes five categories of validity. To minimize the impact of the validity threats that could affect our study, we present them with the corresponding mitigation actions in Table 3.

Table 3. Validity threats.

Validity	Actions
Descriptive validity refers to the accuracy of the data	- We unified the concepts and criteria used in the study and structured the information to be collected with a data extraction form to support a uniform recording of data
Theoretical validity depends on the ability to get the information that it is intended to capture	- We used a search string and applied it to a library including the most popular digital libraries on computer sciences and software engineering - A set of inclusion and exclusion criteria have been defined - We combined two different search methods: an automatic search and a manual search (backward and forward snowballing), to diminish the risk of not finding all the available evidence - The choice of English sources should be of minimal impact concerning the discard of other languages
Generalization validity is concerned with the ability to generalize the results	- Our set of research questions is general enough to identify and classify the findings on Smart applications
Evaluative validity is achieved when the conclusions are reasonable given the data	- Two researchers studied the papers, working independently but with an overlap of studies to identify potential analysis differences - At least two researchers validated every conclusion
Transparency validity refers to the repeatability of the research protocol	- The research process protocol is detailed enough to ensure it can be exhaustively repeated

3 Analysis of Smart Application Research

We present the obtained results in the following sub-sections organized accordingly to the defined research questions.

3.1 Taxonomy and Evolution of Smart Applications

Many applications are interconnected to each other. However, a structuration should be done. We studied the topics of the selected 2341 papers. Firstly, we followed the bottom-up approach: we studied the titles of all papers, however, when titles were not sufficient to identify the application field, we read the abstracts, and, if needed, looked inside the papers. We attributed from one to three levels of applications classification to each paper (for instance, Smart energy → Smart grid → Smart grid communication), except papers dealing with Smart artefacts, technologies, or systems without attachment to a concrete application field. The attribution of each paper to a Smart application was done by an expert. When the mapping of a topic to a Smart domain was not obvious, the paper was analyzed in more detail by two or more experts to find a consensus. Once papers qualified, we started to organize them within a taxonomy using a top-down approach: we took the major fields of Smart life like Smart home, Smart city, Smart industry, Smart healthcare, etc. and we organized the identified application fields within major fields. Considering the number of the major application fields, we organized them within four categories that we called Smart life domains: Smart applications for persons, Smart applications for environment, Smart applications for society, and Smart applications for enterprises (See Fig. 1).

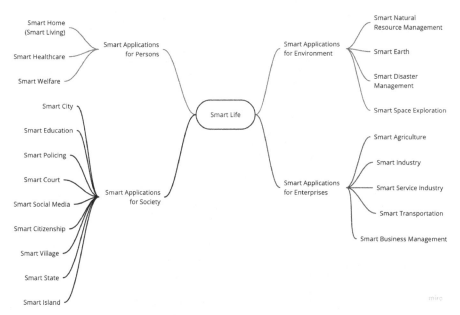

Fig. 1. Smart life domains.

- *Smart applications for persons* contain applications oriented on the personal life of humans, like Smart healthcare, Smart home (Smart living), Smart welfare, etc. (See Fig. 2).
- *Smart applications for environment* include different topics related to the environment study and preservation: Smart natural resource management, Smart earth, Smart disaster management, and Smart space exploration. (See Fig. 3).
- *Smart applications for society* deal with various not-for-profit aspects of human life in groups like Smart city and Smart village, but also applications related to the management of people living in groups: Smart citizenship, Smart state, Smart policing, and so on. (See Fig. 3).
- *Smart applications for enterprises* gather applications for-profit sectors and cover three basics sectors: Smart agriculture, Smart industry, Smart service industry, but also Smart transportation and Smart business management, the latter unifies various transversal to sector business applications (See Fig. 4).

To give some more detailed illustrations, the Smart city concept is used to solve the challenges in urban cities by providing a good livable environment for the citizens [21]. Therefore, in the taxonomy, the Smart city application domain contains sub-applications addressing security, growth, lighting, parking, and so on. With regards to Smart agriculture, new technologies are more and more used in modern agriculture and IoT farmers can remotely monitor their crop and equipment by phones and computers [22]. The Smart agriculture application domain contains usual agricultural types like Smart farming, Smart fishery, Smart crop management, and Smart livestock farming, but also some other domains influencing agriculture, like climate-smart agriculture, Smart pest management, and Smart irrigation.

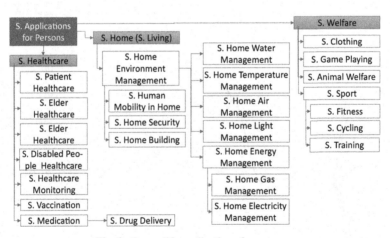

Fig. 2. Smart life applications for persons.

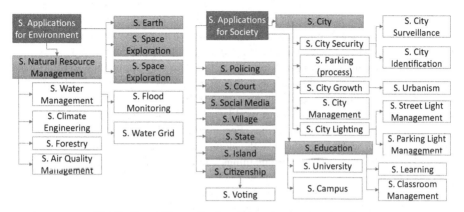

Fig. 3. Smart life applications for environment and society.

In addition to this main taxonomy, we identified the presence of "generic clusters" that can be used in different Smart applications. For instance, Smart energy is an independent application domain, within Smart industry. However, we can find it in other domains: Smart city, Smart home, etc. Another identified cluster is the Smart transportation one. This specific cluster identifies the Smart applications related to the transportation domain. These applications can be relative to several other applications, like Smart vehicle charging which can be applied in Smart cities (to identify the flow of electric vehicles in a specific area and the distribution of energy in the grid), or in a Smart home (to analyze the consumption of a specific house).

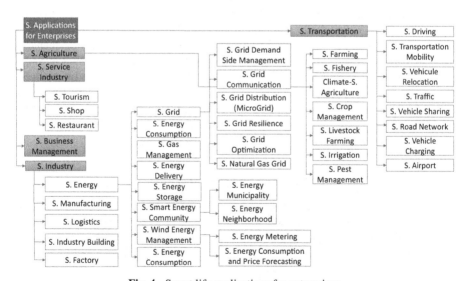

Fig. 4. Smart life applications for enterprises.

We observed that new notions are introduced in the literature, even if they are not completely implemented yet. Smart earth is defined as an innovative approach to environmental monitoring in [18]. Smart island is an attempt to classify the existing smart-related methodologies in the urban context to analyze their possible application to the islands [23]. Smart state is described in [24] as a direction of "Knowledge-Based Urban Development". In [25], a literature review is done on the usage of the blockchain technology in Smart villages. Finally, the term Smart world appeared in the literature [26, 27] as a unifying term of various Smart applications. For instance, [27] defines Smart world as "an attractive prospect with comprehensive development of ubiquitous computing involving penetrative intelligence into ubiquitous things". In our vision, this term corresponds to our vision of Smart life.

When constructing the Smart life applications taxonomy, we excluded 328 sources dealing only with Smart artefacts, Smart technologies, and Smart systems not related to specific application domains. The taxonomy does not either include Smart infrastructure-related applications as we foresee them as transversal issues, we can find infrastructure almost in all major Smart applications fields.

The evolution of Smart domains is shown in Table 4 for the period from 1986 to 2022 (publications already available in 2021). We grouped the first 21 years in the same column as the number of publications was very law for this period. We complete this chronological vision by a graphical representation illustrating the distribution of Smart domains (See Fig. 5).

We can see that the first reviews appearing in the literature concern Smart applications for enterprises (in 1986). Then literature shows interest in two other domains, namely Smart applications for persons and for society, in the 2000s. More recently, Smart applications for the environment came out more prominently.

In the Smart applications for enterprises domain (1142 papers), Smart industry dominates (891 papers), essentially with Smart energy-related topics, followed by Smart agriculture (104 papers), Smart transportation (78 papers), and Smart business management (60 papers). Smart home and Smart healthcare are equally represented in Smart applications for persons (191 and 190 papers respectively over 408), with a little part for Smart welfare. Smart applications for Environment are only represented by 16 papers, half of them focused on Smart natural resource management (9 papers). Smart city is taking the big part of the Smart applications for society (384 papers over 440), followed by Smart education (43 papers).

3.2 Sources of Papers Publishing on Smart Applications

One of our goals is to analyze the venues of papers published on Smart life. This question leads us to study the distribution of publications by type of venue: conferences, journals, or workshops, resulting in the distribution presented in Table 5. The most used type is journals (66%) with conferences at second place (27%), and a very small number of venues are presented by books and workshops.

Table 4. Smart domains evolution.

Smart Life Domains	1986-2006	2007	2008	2009	2010	2011	2012	2013	2014	2015	2016	2017	2018	2019	2020	2021	2022	Total
Persons	3	4	6	9	3	6	9	13	9	10	25	21	45	64	93	84	4	*408*
Smart healthcare	3	3	5	4	2	3	4	5	6	5	7	11	13	31	39	47	2	*190*
Smart home (Smart living)		1	1	4	1	2	5	8	3	4	18	9	27	29	45	33	1	*191*
Smart welfare			1		1					1		1	5	4	9	4	1	*27*
Society	1	1		1		1	2	4	6	13	23	37	50	107	89	100	5	*440*
Smart city	1	1		1		1	1	3	6	12	16	32	4?	96	80	87	5	*384*
Smart cityzenship											1	1		1				*3*
Smart court							1											*1*
Smart education										1	5	4	6	9	7	11		*43*
Smart island														1				*1*
Smart policing															1			*1*
Smart social media								1							1	1		*3*
Smart state										1								*1*
Smart village														2		1		*3*
Environment				1					1	3			1	2	2	6		*16*
Smart disaster management									1						1	3		*5*
Smart earth														1				*1*
Smart natural resource management				1						3				1	1	3		*9*
Smart space exploration														1				*1*
Enterprises		1		2	11	15	40	39	43	56	98	89	130	179	199	231	9	*1142*
Smart agriculture								1	2	1	2	9	8	12	26	41	2	*104*
Smart business management								1				2	5	10	15	24	3	*60*
Smart industry	7	1		2	11	14	38	37	37	50	91	72	108	137	137	145	4	*891*
Smart service industry									1		1	1		5	5	3		*16*
Smart transportation						1	2		3	5	4	5	9	15	16	18		*78*
Total by year	*1*	*6*	*6*	*13*	*14*	*22*	*51*	*56*	*59*	*82*	*146*	*147*	*226*	*352*	*383*	*421*	*18*	*2013*

We identified the principal venues publishing on Smart life (Table 6). We selected venues that published 10 and more papers on Smart life state-of-the-arts and research agendas. Three first places are taken by Renewable and Sustainable Energy Reviews (74 venues), IEEE Access (65 venues), and Energies (43 venues), thus 117 venues in the field of energy, and 65 generic venues. 262 venues have between 2 and 9 papers, and 931 venues have only one paper.

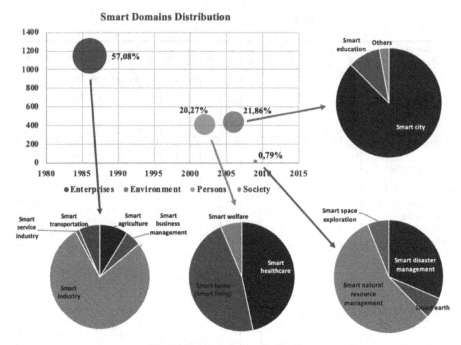

Fig. 5. Smart domains distribution.

Table 5. Papers distribution by publication type.

Publication type	Number	Percentage
Journal	1553	66%
Conference	629	27%
Book or book chapter	149	6%
Workshop	10	1%

With regards to the venues categories, we obtained 16 most publishing domain-specific venues (371 papers), 8 generic venues (205 papers), and 2 survey-oriented venues (42 papers), thus 26 venues in total. On the 619 different venues identified in our dataset, more than half can be classified as domain-specific.

Table 6. Most publishing venues on smart life and their category.

Publishing Venue on Smart Life	Number	Category
Renewable and Sustainable Energy Reviews	74	Domain-specific
IEEE Access	65	General
Energies	43	Domain-specific
Sustainability (Switzerland)	36	Domain-specific
Advances in Intelligent Systems and Computing	30	General
IEEE Communications Surveys and Tutorials	30	Survey-oriented
Applied Sciences (Switzerland)	29	General
Lecture Notes in Computer Science	29	General
Lecture Notes in Electrical Engineering	29	Domain-specific
Sensors (Switzerland)	25	Domain-specific
Sustainable Cities and Society	25	Domain-specific
Journal of Cleaner Production	23	Domain-specific
Lecture Notes in Networks and Systems	18	Domain-specific
ACM International Conference Proceeding Series	16	General
Journal of Network and Computer Applications	15	Domain-specific
Communications in Computer and Information Science	14	General
Smart Innovation, Systems and Technologies	14	Domain-specific
ACM Computing Surveys	12	Survey-oriented
Electronics (Switzerland)	12	Domain-specific
IOP Conference Series: Materials Science and Engineering	12	Domain-specific
Journal of Physics: Conference Series	12	Domain-specific
Lecture Notes of the Institute for Computer Sciences, Social-Informatics and Telecommunications Engineering	12	Domain-specific
EAI/Springer Innovations in Communication and Computing	11	General
Procedia Computer Science	11	General
Sensors	11	Domain-specific
IOP Conference Series: Earth and Environmental Science	10	Domain-specific

3.3 Smart Applications Underlying Terminology

We created two keyword clouds: one based on the title of the selected publications (Fig. 6) and the second one based on their abstract (Fig. 7). The size of the word in the cloud indicates its frequency in the dataset.

Fig. 6. Word cloud based on the papers titles.

The titles contained 2023 occurrences of the word "Smart", which put it in the first place before the others. Table 7 shows the frequency of the most popular words in this title dataset (over 200 occurrences).

Table 7. Popular words frequency in the titles.

Word	Frequency	Word	Frequency	Word	Frequency
Smart	2023	Internet	266	Things	245
Grid	476	Data	255	IoT	241
Energy	425	System	252	Systems	216
City	342	Management	249	Security	210

Some words concern the main objective of this study which is "Smart" of course. We can identify also that we are in the domain of "internet", "data", "system", "things", "IoT", and "systems" (1475 occurrences). What is more interesting however is the frequency of the set of terms "grid" and "energy" (901 occurrences) which means that this specific topic is at the heart of the selected research works in this study. The term "city" appears 342 times, "security" 210 times, and "management" 249 times which are still important frequencies.

Fig. 7. Word cloud based on the papers abstracts.

The dataset based on the selected publication abstracts contained 8098 occurrences of the word "Smart", four times more than the next frequent word. Table 8 shows the frequency of the most popular words in this dataset (over 1000 occurrences).

Table 8. Popular words frequency in the abstracts.

Word	Frequency	Word	Frequency	Word	Frequency
Smart	8098	Energy	1806	Review	1456
System	2803	Research	1795	Study	1287
Technology	2403	City	1890	Power	1252
Papers	2330	Data	1736	Challenges	1073
Grid	2326	Application	1669	Network	1044

Apart from the "Smart" word, we can also identify words concerning our research method, like "papers", "research", "study", "review" and "challenges" (7941 occurrences), as we restricted our SMS to papers which were literature studies. As in the title dataset, we identify words about our domain of "system", "technology", "data", "network", and "application" (9655 occurrences). Again, the frequency of the set of terms containing "grid", "energy", and "power" (5384 occurrences), reiterates our preceding finding that this domain is massively present in the selected publications. The term "city" appears 1890 times which is still an important frequency.

Based on the study of the keywords, we established that, except for the research study terms and the development-related ones, the most used application areas are home, agriculture, health, energy, and parking. The main purposes are metering, monitoring, manufacturing, and intelligence. The most present characteristics are mobile, electric, sustainable, and renewable. And, finally, the more frequent technologies are sensors, grids, state of control, and network.

3.4 Smart Life Driving Factors

Addressing Smart life applications raises some specific requirements. The most important ones, in our humble point of view, are related to driving perspectives. What are the main issues we should take care of to be able to successfully build this new generation of applications? There is many, but let's focus on what we expect to be the most significant ones through application domains.

The first one corresponds to the *ability to scale*. From Smart cities to Smart homes, applications and technologies have to face numerous units: sensors, computers, communication devices, people or users, buildings, data produced and consumed... Their considerable quantity implies the ability to scale, eventually to scale to the world. The panoptic model from Jeremy Bentham [28], already contradicted by Michel Foucault, is no longer a reference. This has already been considered by Cloud providers as they adopted the CAP Theorem stating that, among the 3 properties of Consistency/Adaptability/Partitioning, you can only insure two of them. This is due to the deep nature of Cloud Computing and its inherent dimensions of distribution and scaling.

Application designers will address *safety and security as key issues*. They are mandatory, we fully agree and there is no discussion about these requirements. It is very general for any application that involves human beings in someplace. From our point of view, ubiquity - sometimes also addressed as pervasivity - belongs to the deep heart of Smart life applications. Nowadays, through the pandemic crisis, most people have experienced collaborative tools such as Zoom, Teams, Google meet, or whatsoever. These tools bring more than connectivity. They took a part of ourselves and put it in some shared void enforcing belongingness. We are at the beginning of a new era where we are in the physical world and one or more virtual worlds at the same time. From Building Information Modelling (BIM) to Metaverses, the feeling of being "more than one" and being "only one" at the same time is enforced. Maybe we are moving forward to the empathy box of the mercerism described in "Do androids dream of electric sheep" from Philip K. Dick [29]?

The relationship between human beings and technology is an emerging issue. From a human perspective, *we are driving the system, not the other way round*. But this is not just black or white; there is a blurred frontier between what human needs and what Smart life brings. Sentience is the key issue dealing with requirements. Sentience has been defined a long time ago, especially in the context of animal rights. Such a concept can be derived and adapted to technology. It is easier to understand in the context of Robots, more precisely when you apply it to robots' consciousness. Recently, this concept has started to be refined in [30]. ICT should be able to sense human expectations without any explicit specification and adapt to these expectations consistently. Modern application designers will take care of sustainability and ecological footprint. We deeply believe that a sentient-ready application or technology will be able to fully take care of this requirement if properly designed and self-aware.

4 Conclusion and Future Research

Our main goal is to promote the field of Smart life and Smart life engineering to provide fundamentals for more efficient and sustainable research efforts. In this paper, we summarize the research work on "Smart" state-of-the-arts and research agendas to identify and classify Smart applications. We carried out a systematic mapping study based on 2341 publications to define a taxonomy of Smart applications, understand their distribution, analyze sources of the research works, and identify the underlying terminology. Finally, we discuss driving properties allowing us to move forward the Smart life.

Based on the obtained results, we plan to investigate how the identified driving properties could be considered to identify possible relationships between different Smart applications. We will carry out a bibliometric study for this purpose. We also intend to go further with the identification of research challenges in the field of Smart life and Smart life engineering. Our main goal is to discover if common methodological basics could be identified in the field of Smart life.

References

1. Lutolf, R.: Smart home concept and the integration of energy meters into a home based system. In: Proceedings of the 7th International Conference Metering Apparatus and Tariffs for Electricity Supply, pp. 277–278 (1992)
2. Chamee, Y.: Historicizing the smart cities: genealogy as a method of critique for smart urbanism. Telematics Inform. 55, 101438 (2020)
3. Hidayat, T., Mahardiko, R., Franky, S., Tigor, D.: Method of systematic literature review for internet of things in ZigBee smart agriculture. In: 2020 8th International Conference on Information and Communication Technology, ICoICT 2020 (2020)
4. Al-Shoqran, M., Shorman, S.: A review on smart universities and artificial intelligence. In: Hamdan, A., Hassanien, A.E., Razzaque, A., Alareeni, B. (eds.) The Fourth Industrial Revolution: Implementation of Artificial Intelligence for Growing Business Success, pp. 281–294. Springer, Cham (2021). https://doi.org/10.1007/978-3-030-62796-6_16
5. Mehraliyev, F., Chan, I., Cheng, C., Choi, Y., Koseoglu, M.A., Law, R.: A state-of-the-art review of smart tourism research. J. Travel Tour. Mark. 37, 79–91 (2020)
6. Tiwari, P., Garg, V., Agrawal, R.: Changing world: smart homes review and future. In: Moh, M., Sharma, K.P., Agrawal, R., Diaz, V.G. (eds.) Smart IoT for Research and Industry, pp. 145–160. Springer International Publishing, Cham (2022). https://doi.org/10.1007/978-3-030-714 85-7_9
7. Alam, M.R., Reaz, M.B.I., Ali, M.A.M.: A review of smart homes - past, present, and future. IEEE Trans. Syst. Man Cybern. Part C Appl. Rev. 42, 1190–1203 (2012)
8. De Silva, L.C., Morikawa, C., Petra, I.M.: State of the art of smart homes. Eng. Appl. Artif. Intell. 25(7), 1313–1321 (2012)
9. Alaa, M., Zaidan, A.A., Zaidan, B.B., Talal, M., Kiah, M.L.M.: A review of smart home applications based on Internet of Things. J. Netw. Comput. Appl. 97, 48–65 (2017)
10. Wang, D., Zhong, D., Souri, A.: Energy management solutions in the Internet of Things applications: technical analysis and new research directions. Cogn. Syst. Res. 67, 33–49 (2021)
11. Harbi, Y., Aliouat, Z., Harous, S., Bentaleb, A., Refoufi, A.: A review of security in internet of things. Wirel. Pers. Commun. 108(1), 325–344 (2019)

12. Tai-hoon, K., Ramos, C., Mohammed, S.: Smart city and IoT. Fut. Gener. Comput. Syst. **76**, 159–162 (2017)
13. Yamane, S.: Deductively verifying embedded software in the era of artificial intelligence = machine learning + software science. In: 2017 IEEE 6th Global Conference on Consumer Electronics, GCCE 2017 (2017)
14. Severi, S., Sottile, F., Abreu, G., Pastrone, C., Spirito, M., Berens, F.: M2M technologies: enablers for a pervasive internet of things. In: European Conference on Networks and Communications, EuCNC 2014 (2014)
15. Baruah, P.D., Dhir, S., Hooda, M.: Impact of IOT in current era. In: Proceedings of the International Conference on Machine Learning, Big Data, Cloud and Parallel Computing: Trends, Perspectives and Prospects, COMITCon 2019, pp. 334–339 (2019)
16. Mizintseva, M.F.: Smart technologies for smart life. In: Popkova, E.G., Sergi, B.S. (eds.) ISC 2020. LNNS, vol. 155, pp. 653–664. Springer, Cham (2021). https://doi.org/10.1007/978-3-030-59126-7_73
17. Romero, M., Guédria, W., Panetto, H., Barafort, B.: Towards a characterisation of smart systems: a systematic literature review. Comput. Ind. **120**, 103224 (2020)
18. Bouchrika, I.: A survey of using biometrics for smart visual surveillance: gait recognition. In: Karampelas, P., Bourlai, T. (eds.) Surveillance in Action, pp. 3–23. Springer, Cham (2018). https://doi.org/10.1007/978-3-319-68533-5_1
19. Petersen, K., Feldt, R., Mujtaba, S., Mattsson, M.: Systematic mapping studies in software engineering. In 12th International Conference on Evaluation and Assessment in Software Engineering, vol. 17 (2008)
20. Thomson, S.B.: Qualitative research: validity. JOAAG **6**(1), 77–82 (2011)
21. Al-Ani, K. W., Abdalkafor, A. S., Nassar, A. M.: Smart city applications: a survey. In: Pervasive Computing Technologies for Healthcare, PervasiveHealth 2019 (2019)
22. Mekala, M. S., Viswanathan, P.: A survey: smart agriculture IoT with cloud computing. In: 2017 International Conference on Microelectronic Devices, Circuits and Systems, ICMDCS 2017 (2017)
23. Desogus, G., Mistretta, P., Garau, C.: Smart islands: a systematic review on urban policies and smart governance. In: Misra, S., et al. (eds.) ICCSA 2019. LNCS, vol. 11624, pp. 137–151. Springer, Cham (2019). https://doi.org/10.1007/978-3-030-24311-1_10
24. Hortz, T.: The smart state test: a critical review of the smart state strategy 2005-2015's knowledge-based urban development. Int. J. Knowl. Based Dev. **7**(1), 75 (2016)
25. Kaur, P., Parashar, A.: A systematic literature review of blockchain technology for smart villages. Arch. Computat. Meth. Eng. (2021). https://doi.org/10.1007/s11831-021-09659-7
26. Kumari, M., Sharma, R., Sheetal, A.: A review on hybrid WSN-NGPON2 network for smart world. In: Singh, P.K., Bhargava, B.K., Paprzycki, M., Kaushal, N.C., Hong, W.-C. (eds.) Handbook of Wireless Sensor Networks: Issues and Challenges in Current Scenario's. AISC, vol. 1132, pp. 655–671. Springer, Cham (2020). https://doi.org/10.1007/978-3-030-40305-8_31
27. Liu, H., et al.: A review of the smart world. Fut. Gener. Comput. Syst. **96**, 678–691 (2019)
28. Bentham, J.: An Introduction to the Principles of Morals and Legislation. Batoche Books, Kitchener (1781)
29. Dick, P.K.: Do Androids Dream of Electric Sheep? Ballantine Books, New York (1968). Norstrilla Press. ISBN 0-345-40447-5
30. Kornyshova, E., Gressier-Soudan, E.: Introducing sentient requirements for information systems and digital technologies. In: EMCIS 2021, Dubai, United Arab Emirates (2021)

Conceptual Modelling and Ontologies

Conceptual Integration
for Social-Ecological Systems
An Ontological Approach

Greta Adamo[(✉)] [ID] and Max Willis [ID]

ITI/LARSyS, Caminho da Penteada, 9020-105 Funchal, Madeira, Portugal
{greta.adamo,max.willis}@iti.larsys.pt

Abstract. Sustainability research and policy rely on complex data that couples social and ecological systems (SESs) to draw results and make decisions, therefore understanding the dynamics between human society and natural ecosystems is crucial to tackle sustainability goals. SESs frameworks are employed to establish a common vocabulary that facilitates the identification of variables and the comparison of results. A variety of SESs approaches have been proposed and explored, however integration and interoperability between frameworks is missing, which results in a loss of relevant information. In addition, SESs frameworks often lack semantic clarity which exacerbates difficulties in developing a unified perspective. In this paper we demonstrate the use of ontological analysis to unify the main elements of two prominent SESs paradigms, the social-ecological system framework (SESF) and the Ecosystem Services (ESs) approach, to build an *integrated social-ecological perspectives* framework. The proposed conceptual framework can be adopted to combine existent and future results from the two paradigms in unified databases and to develop broader explanatory and decision-making tools for SESs and sustainability research.

Keywords: Ontological analysis · Social-ecological system framework · Ecosystem Services

1 Introduction

Analysing the relationships between the natural environment and human societies is at the core social-ecological systems (SESs) research [16]. One of the main motivation behind SESs is to build a knowledge-base useful to create a shared understanding of environmental and societal feedback and impacts [10,16]. SESs are often grounded on conceptual frameworks that support the identification of key elements and their interactions [11,34]. Two widely adopted SESs approaches are the Ecosystem Services (ESs) [6] that reflects on the natural world as support of human well-being, and the social-ecological system framework (SEFS) [39,41] that aims at specifying a common language dedicated to human-nature dynamics. Both ESs and SESF are supported by conceptual representations of

© The Author(s), under exclusive license to Springer Nature Switzerland AG 2022
R. Guizzardi et al. (Eds.): RCIS 2022, LNBIP 446, pp. 321–337, 2022.
https://doi.org/10.1007/978-3-031-05760-1_19

the system inter-linkages, the former is often associated with the *cascade model* [33] and the latter with the framework proposed by Ostrom [41].

In the context of sustainability and sustainable development [53] SESs frameworks are crucial for planning and decision-making as they create a common vocabulary, organise knowledge, define variables, and align results. For example, climate change projections and models based on environmental data are key tools for policymakers [37] and are closely related to the understanding of SESs resilience, adaptation and robustness [3,16]. Some SESs approaches adopt maps to visualise, communicate and assess relevant ESs [15] in which the identification of indicators and the selection of datasets (i.e. environmental and social) represent important methodological steps [55]. Thus defining a clear semantics for SESs components and aligning concepts among existent theories is central to create a common ground, preserve relevant knowledge and exploit information systems to maximise the production and comparison of models and results. Despite the intense development of SESs frameworks [11] and the effort to define a shared ontology that captures social-ecological interactions and pressures [39,41], SESs are still poorly defined [16]. Inconsistencies are found within specific SESs approaches, for instance there is still a lack of a standardised vocabulary and classification in ESs, which coupled with ambiguous concept semantics can affect how practitioners use and interpret ESs notions [48]. In addition, many SESs approaches are challenging to compare and integrate due to their theoretical differences [11]; this results in a disconnect between approaches, an over proliferation of concepts and variables that might explain similar phenomena, and a lack of an unified framework that hinders clear definition of indicators in the SESs community.

In this paper we provide an integration of the main SESF and ESs notions using ontological analysis as an approach for semantic clarification.[1] Although the combination and comparison of SESF and ESs is not new in the literature, see e.g. [8,45], a comprehensive semantic analysis of SESF and ESs notions and their interlinks is still missing. We propose an *integrated social-ecological perspectives* framework, that facilitates the unification of the main SESF and ESs elements. The framework can be a tool to define and integrate concepts from both paradigms, to promote unambiguous data representation, extend the reach of SESs and sustainability analysis and potentially create tools to compare results. The paper is organised as follows: Sect. 2 introduces SESs, SESF and ESs states of the art. Section 3 is dedicated to the ontological clarification of SESF and ESs components and the presentation of the integrated framework. Final considerations are presented in Sect. 4.

2   Social-Ecological Systems

SESs are complex, dynamic assemblages of social (e.g., governance and norms) and ecological (e.g., ecosystem functions and species) elements. The notion of SESs emerged in the 1970's, but over the past 20 years SESs has became a

---

[1] The images in this work can be found in high resolution at this link.

proper interdisciplinary research field that encompasses environmental and social sciences, economics, business management, engineering, computer science and humanities with approximately 12,990 publications dated in 2019 [16].

The initial focus of SESs was on resource management to understand systems' *resilience* to impacts and disturbances [10,16]; to this end Berkes and Folke developed a SESs framework [10] that explained the links between ecological, social and economical aspects by considering *ecosystem, people* and *technology, local knowledge, property right* and *institution*, and their reciprocal connections and feedbacks. More recently the SESs debate has been enriched by including the notion of systems' *robustness* [3,16], defined as the capability of a system to maintain performances under pressure. The robustness of the system may be affected by several parameters, such as institutional decisions and human behaviours, and is analysed on the basis of external and internal disturbances (e.g. natural disasters and changes in demographics vs. system reconfigurations). To capture these dynamics Anderies et al. [3] propose a SESs framework that involves *resource* used by *resource users* (e.g. fisheries-fisherpeople), the collective entity of *public infrastructure providers* (e.g. public council) and *public infrastructures*, which are differentiated between physical and social capitals (e.g. canals, ports and rules). The analysis of systems' robustness encompasses all these actors and their interactions, for example understanding the dynamics between resource users and resource extraction involves several aspects from property rights to sense of collectives, participation and policy that supports the management of common-pool resources (CPR) [42], e.g. fisheries.

Over the course of its development SESs research has proposed several conceptual frameworks that allow for the capture of human-natural ecosystems relationships by adopting different perspectives, levels of analysis and granularity [11]. In the following we review two popular frameworks, SESF [39,41] and ESs [6].

2.1 Social-Ecological System Framework

SESF stems from the field of political science [11] and evolved from different streams of research such as collective action, CPR management, governance and community self-organisation [44]. SESF is a domain ontology that aims at creating a shared ground among scholars and experts through a vocabulary that specifies complex social-ecological interactions to organise and optimise knowledge sharing and develop a diagnostic system for SESs governance [9,39,41]. SESF includes concepts and their interactions that can be used to define variables for a wide range of case studies specific to the management of CPR, for example small-scale fisheries [9] and community-oriented systems, such as irrigation [18]. A list of SESF applications can be found in [44].

SESF has expanded over the years with refinements and extensions, however for the sake of simplicity in this paper we refer to the version proposed by McGinnis and Ostrom in [39]. The framework (see Fig. 1, adapted from [39]) is organised in tiers (i.e. classes of variables) and targets a domain in which *actors* extract *resource units* that belong to a wider *resource system*. At the same time actors are also responsible for the maintenance of the resource pool based on rules defined by the *governance system*. Activities such as resource extraction and maintenance are included in *action situation* in terms of *interactions* and *outcomes* within the social-ecological system. The higher tier of the framework includes *resource systems, resource units, governance systems* and *actors*. All of these elements are involved in the *action situation* that results in reciprocal feedbacks. Finally *related ecological systems* and *social-economic-political settings* represent broader and exogenous social-ecological system settings that can pressure the system's equilibrium. The second level tiers include sub-classes of the first tier, their qualities and attributes. The full second level tiers table can be found here [39]. These variables can be adopted to assess positive and negative factors that affect self-organisation management of CPR to avoid over-exploitation [41] as well as for diagnostic processes that involve human-nature relations [9].

Despite the broad conceptual framework and range of applications, SESF presents some limitations. While data collection and analysis are becoming central to the study of SESs in conjunction with sustainable development and climate change monitoring, forecasting, environmental planning and decision-making activities, SESF remains challenging to adopt in empirical settings and in the collection of primary data. In these situations variables would need to be aligned to the data which would require a deep knowledge of the framework. This challenge is reflected also in the complexity of comparing results, data management and interoperability. Any modification of the variables list represents another issue, for although the framework is supposed to be extensible, trades-off need be considered between the introduction of new variables (e.g. bio-chemical-physical ecosystem parameters) and the maintenance of the theoretical ground of SESF. Moreover, the introduction of new domain-specific variables poses further questions, such as what precisely is a variable within SESF, how to distinguish between variables and indicators (e.g. water quality) and when to determine classification of variables and sub-variables, considering also that the definition of tier is not clearly specified [34,43,44]. These ontological challenges affect and potentially hinder the methodological setting and development of SESF and its potential applicability to sustainability studies [43,44].

Specific studies have been proposed to integrate ontological strategies and provide formal structure for SESF, manage its complexity and issues of integration and comparability [27,34]. These issues have been addressed by the Social-Ecological Systems Meta-Analysis Database (SESMAD) project [19] and the SES Library developed by the Arizona State University (ASU).[2] Despite these efforts, a clear and unified semantics for concepts and variables to facilitate comparisons among results has not been forthcoming.

---

[2] seslibrary.asu.edu.

2.2   Ecosystem Services

Nature provides humans and societies with many essential goods and benefits, such as food, water and energy [33]. The study of human dependence on the natural environment is at the core of ESs research that focuses on the role of nature in support of human life and well-being, and the effects of human-based ecosystem pressures on health, economy, politics and more [6]. In the Millennium Ecosystem Assessment (MA) [6] ESs are grouped as *provisioning* (e.g. food, water), *regulating* (e.g. climate and disease regulation), *supporting* (e.g. soil formation) and *cultural* (e.g. educational and recreational). These ESs are linked to different aspects of human well-being, such as safety and materials for life (e.g. food, shelter). Unfortunately, research outcomes from the MA reported that 60% of ESs are over-exploited and degraded, a condition that was linked for instance to poverty, loss of biodiversity and unsustainable development [6,33].

The Economics of Ecosystems and Biodiversity (TEEB) initiative [46] focussed greater attention on the valuation of ESs as a tool for decision-making [51] that allows for quantitative assessment of the importance of nature for society and welfare, and estimation of trade-offs between the presence of human activities and the preservation of natural ecosystems in a sustainability setting [25,35]. Valuations can be performed both in monetary (e.g. market value) and non-monetary terms (e.g. measures of attitudes) [35]. The valuation of ESs is often connected to spatial characteristics, and the use of data-driven maps then becomes relevant [51] for instance to visualize the geographical spread of ESs and facilitate communication among various stakeholders. Note that ESs maps can be adopted not only for economic valuations, but also for ecological and socio-cultural assessments [15,55]. The identification, mapping, assessment and valuation of ESs represent important steps to build a more sustainable and effective environmental management. Indeed, ESs and biodiversity knowledge-bases ground decisions for environmental policies, such as the EU Biodiversity Strategy [17,55].

Despite the long tradition of studies in ESs and Ecosystem Approach [33], the development of dedicated tools (see e.g. Chapter 4.4 of [15]) and their applications at sovranational and intergovernmental levels, a unified definition of ESs and relevant associated notions is missing [47]. Without a standardised conceptual ground the unification and comparison of ESs analysis outcomes is challenging [45]. In addition there remains some confusion between core ESs concepts such as *service, benefit* and *value* [47,48]. However, some scholars have recognised similarities among ESs communities in terms of production and delivery of ESs; these have been summarised and represented by the cascade conceptual framework [33,47]. The cascade (see a simplified and adapted version in Fig. 2 [47])[3] includes the main elements of ESs divided into two groups, the environmental and the social-economical systems, and the pressures that the latter exerts on the former. The ecological system focuses on the *structures, functions* and *services* of ecosystems as habitat type and composition, performed cycle, and ecosystem

---

[3] We condense the notion of ecosystem *process* with ecosystem *structure* and *function* following results reported in [48].

characteristics that can be utilised for human sustenance, health and well-being. The service element plays a role of mediator in the cascade, connecting the natural ecosystem with social-economical systems. Indeed structures and functions allow the materialisation of ecosystem services that are associated to human *values*, both monetary and non-monetary, due to the *benefits* they carry and their potential affects on well-being [47]. The cascade framework serves as a conceptual foundation for the Common International Classification of Ecosystem Services (CICES) [32], a reference framework that translates several classifications systems such as MA and TEEB and related research and provides a terminological standard for the ESs community.

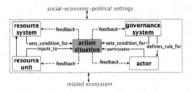

Fig. 1. SESF adapted from [39].

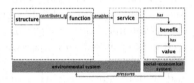

Fig. 2. ESs cascade adapted from [47].

3   Ontological Foundations for Social-Ecological Systems

We examine the ontological meanings of the main SESF and ESs elements and merge them into our proposed integrated framework. Some of the SESF concepts are complex, such as resource and governance systems, and first require the disambiguation of their "atomic" counterparts (e.g. resource and governance). In this writing we elucidate the following components: *resource, actor, governance, ecosystem structure* and *function*, ecosystem *service*, *benefit* and *value*.

The semantic clarification of the aforementioned notions follows the steps of (i) examining common-sense and literary definitions, such as consulting the Cambridge Dictionary[4] and the Lexico.com powered by Oxford[5] and field-related literature, then (ii) employing well-established foundational-, domain-, and applied ontologies research. Due to the descriptive and conceptual nature of SESs frameworks and the purpose of this paper, in the second step we mostly reference ontological studies that are applied in the domain of information systems, such as data-, information-, and conceptual modelling. For example, we reference *Unified Foundational Ontology* (UFO) [28] and applications/extensions of *Descriptive Ontology for Linguistic and Cognitive Engineering* (DOLCE) [36] (e.g. [14,50]). UFO is widely used as a grounding for conceptual modelling and DOLCE has a natural language and cognitive approach and has been widely adopted in information systems. Note that the former is based on the latter [4]. Several works

---

4 dictionary.cambridge.org.

5 www.lexico.com.

that contribute to our analysis are often interrelated, e.g. the paper of Boella et al. [12] is associated with DOLCE, Bottazzi et al. [14] propose an extension of DOLCE for organisations analysing notions such as roles and norms and Andersson at al. [4] present an ontology of value ascription useful for enterprise modelling that is based on UFO. This technique of utilising ontologies as a methodological ground for the disambiguation of concepts is described in [1].

3.1   Ontological Clarification of SESs Components

Resource. In the SESF literature the notion of resource traditionally refers to CPR, natural or human-made, which are subject to possible over-exploitation due to the challenges involved in regulating access [42]. Examples of CPR are animals, plants and artificial constructions.

The dictionaries define resources as assets that are beneficial or valued by individuals or collectives and which contribute to their functioning.[6] This condensed definition stresses the notion of resource as a valuable entity and an asset that potentially can be used, yet the definition is still unclear due to the variety of entities considered as resources. To disambiguate the semantics of resource we start from several ontological studies in the domains of enterprise modelling, manufacturing and business process modelling that define resource as the *role* that objects plays in the context of *activities* or *plans* to achieve *goals* [2,7,24,50]. Ontologically, roles are dependent upon other objects to be existent and are often realised in contexts, for instance the mud-lined trench dug x perpendicular to a stream can play the role of an irrigation canal in the context of subsistence farming. While activities are occurents performed by actors, plans are *information objects* (e.g. a document) that describe situations or a set of activities and their organisation to achieve a certain goal [12,50]. In this way, resources can be *assigned to* activities and can be *relevant for* plans [2,50].

In SESF resources are divided into *natural* and *human-made*, the latter also referred to as *artefacts* that in contrast to natural resources are typically intentionally designed by actors to have certain characteristics on the basis of plans and goals [13]. Adopting the distinction proposed in [50], resources can be played by *physical objects*, (e.g. fishes and dams) and *amount of matters*, (e.g. sands and gold.) Physical objects and amount of matters present characteristics that determine their adoption as resources in a particular plan, for example specific benthic species may play the role of resources in certain SESF studies while in others they may not. Note that resources carry *values* associated to their characteristics whether or not that resource is actually exploited [50]. Indeed these values are dependent upon the plans and goals in which the resource is allocated and potentially utilized in the future.

Actor. SESF includes the identification of relevant actors, previously named "users" [39], who are involved in the management of resources. While some

---

[6] dictionary.cambridge.org/us/dictionary/english/resource;   www.lexico.com/en/ definition/resource.

common-sense and terminological definitions of actor relate to specific stag-
ing/acting activities, a more general dictionary definition of actor is one who
takes part in an activity/process.[7] We define the ontology of actor following an
account of UFO dedicated to social entities called *UFO-C* [28] that we have
simplified and modified for the purpose of this paper and domain. The UFO
extension is based on the distinction between *agent* (e.g. persons and institutions) and *object* (e.g. rocks and norms), both of which can be physical or social.
Note that in this writing we use the terms agent and actor interchangeably.

One of the main differences between actors and objects is that the former
bears *intentional moments*, i.e. beliefs, desires and intentions, that have a *propositional content* and a directionality (e.g. "I intend x") related to a specific
situation. For example, the propositional content of intentions are goals that
can be satisfied by a situation (e.g. to catch fishes without over-exploiting the
resource). Intentional moments may trigger activities performed by actors that
are the execution of plans; these may or may not be satisfied according to the
intention-goal of the actors, and can involve the presence or use of resources.
An example of action-interaction is that of *communicative acts*, in which actors
use language for instance to share opinions, ask questions, to commit formal
acts and create *social moments* that exist due to the situation generated by the
actors. Actors may also interact with each other in complex actions (e.g. two
or more fisherpeople coordinate their fishing activities) and can use resources in
activities, the participation of resource in such activities can take several forms:
creation (i.e. the existence of the resource is the output of the activity), *termination* (i.e. the non existence of the resource is the result of an activity) and *change*
(i.e. the resource acquires or loses one or more characteristics as the output of
the activity).

Thus resources and actors can be ontologically related and this linkage can
potentially influence outcomes within and between SESs. Indeed actors can
decide over resource allocation, manipulation and consumption, thereby modifying socio-ecological balances.

Governance. Clarifying the notion of governance is not an easy task due to the
wide variety of meanings that have been attributed to it [49]. We begin with dictionary definitions of governance as the activities/actions within administrative
systems and practices for national and organisational management.[8] These definitions depict governance as a kind of action undertaken by governing-managing
states and institutions. Scoping the definitions from within the field itself and
extending into sustainability sciences, governance has been defined as a normative, rule-based and strategic process to guide behaviour in the context of policy
([52], p. 3, referencing [38]), as a social function that guides humans and societies to expected goals ([22], p. 6), as an intended result of strategic institutional

---

[7] dictionary.cambridge.org/it/dizionario/inglese/actor;
   www.lexico.com/en/definition/actor.

[8] dictionary.cambridge.org/dictionary/english/governance;
   www.lexico.com/en/definition/governance.

decisions to tackle problems [22]. Governance has also been distinguished from *government*, which is a collective entity that may perform a form of governance ([22], pp. 6–7).

Despite our efforts we were not able to find existing ontological literature on governance, however we are able to explore this through works that analyse related concepts, such as norms, organisations, roles and decisions (e.g. [12,14,20,31]). Drawing from the definitions presented above, we consider governance as a specific kind of activity performed by agents that involves norms, commitments and decisions to achieve shared goals. The activity of governance regards *policy* [52] that can be defined as an agreed plan of action formally stipulated by a group of people, e.g. organisations, institutions, governaments, or a kind of document that communicate such an agreement.[9] Thus the notion of policy encompasses both the planning and the execution of plans on the basis of a group's agreements, again based on commitments. Governance activities are typically performed by affiliated actors playing roles [14], examples are the president and the chief administrator of an organisation that share common goals described in the administration's plans. Actors involved in governance establish and recognise social objects such as *norms, social commitments* [28] and shared *decisions*; these three are parts of the plans and are directed towards specific governance goals. Norms are descriptions that can be satisfied, or not, [14]; social commitments and decisions (i.e. type of intention [30]) are social moments typically originating from actors' interaction and communication [20,28] that might be directed towards an activity [31]. Norms and social commitments are connected to the notions of *validity* and prescription, as such they guide actors' activities [14] and decisions within the scope of governance.

Structure and Function. While ecosystem structure has been defined in terms of composition, distribution and conditions that allow species to survive [23], ecosystem function carries more elaborate semantics such as specifying the operating mechanisms of an ecosystem (e.g. energy flow, nutrient cycle, regulating systems) as well as the capability of an ecosystem to deliver services useful for humans [21,23]. In order to define the semantics of structures and functions we start by briefly scoping the intuitive semantics of the former and derive implications that are useful to understand the latter.

The Cambridge dictionary defines a structure as the configuration of the parts of a system or object,[10] stressing the role of *parts* of a *whole* and their organisation. In this work we focus on a specific kind of parthood relation called *functional parthood*, in which the whole is organised in structural and functional parts [56], such as the CPU of a computer and the gills of a fish, and allows for the definition of *functional roles* [40]. These kinds of roles, which have been also formalised for UFO [29], allows for the capture of relations between systems and their components as structural and functional unities, which is useful in clarifying their semantics within SESs.

---

[9] dictionary.cambridge.org/dictionary/english/policy.

[10] dictionary.cambridge.org/dictionary/english/structure.

Functional parthood can be explained in terms of *functional dependence* [54], in which the parts play some sort of functional role in the context of the whole and vice versa. For example, certain ecosystems are functionally dependent upon specific insects to carry pollen and propagate species and certain insects are functionally dependent upon the whole ecosystem to reproduce and continue the species. The parts and the whole can be involved in an active *functioning as* role at a certain time in addition to bearing some latent functions that might be useful in the future [40,54]. Functional roles can be *social* and/or *natural* depending on the context in which the role is played.[11] For example, the president of organisation x plays a social functional role as the administrative head, the mangrove forest plays a natural functional role as an habitat for crabs, as well as a social functional role: coastal protection for human communities. While social roles are social concepts based on descriptions [2,14], natural roles are realised within specific bio-physical and chemical conditions, for instance the mangrove forest plays the natural functional role of habitat for crabs only when crabs are co-located with the mangroves.

A functional part of a whole system can be of different kinds, however in the context of ESs and SESs, three classes are most relevant, namely *replaceable*, *persistent* and *constituent* [40]. Replaceable functional parts are those that can be changed and replaced by the same kind without compromising the whole, for example individual species exemplars can be replaced by others of the same species without changing the nature of the whole ecosystem. Persistent functional parts refers to parts that exist only if the whole exists, an example is the presence of species that are dependent upon specific ecosystem dynamics to survive. Finally constituent functional parts are those that are part of the whole whether or not they are present at a certain time, such as seasonal species that contribute to an ecosystem in a certain period of the year.

Service. The core concept of ESs is represented by *service*, which is described as a benefit/outcome that natural ecosystems provide to humans, such as health and well-being, and then is useful for humans due to its value [6,47]. Ecosystem services are delivered due to the structure and functions of nature [21,47].

Dictionary definitions[12] specify a service as a kind of activity, such as the assistance provided by an organisation, business or the public sector. Service has also been defined as the correct functioning or availability of a system. In these definitions the notion of service is associated to the one of action/activity and is more often related to an intentional act performed by humans. To examine this we first focus on the analysis of the service concept, starting from the ontological, business-oriented and web-service literature to understand the differences and similarities between more classic definitions of service (e.g. agentive and intentional) and the one adopted by ESs.

---

[11] The distinction between social and natural roles can be found also in [5].

[12] dictionary.cambridge.org/dictionary/english/service;   https://www.lexico.com/en/definition/service.

Services have been described as activities, capabilities, results, changes and values [26]. A more general definition of service encompasses the notion of commitment as grounding an understanding of what a service is [26]. In this sense a party x commits to perform an action a in favor of a party y on the basis of certain conditions c. A commitment is typically an intentional act [28] that involves constraints on actions [26]. Consider that a service commitment may exist even without a service delivery; in a business example, paramedics are committed to providing aid services to their users even when nobody is calling for emergency service (i.e. a triggering event for service invocation). Services are thus a commitment-based activity (i.e. events), which we define as *commitment-based service*, that involve participants (e.g. parties). Thus services are differentiated from *goods* that are objects, transferable and owned due to their ontological structure [26]. Finally, some technological services are provided by automated systems and artificial agents, such as web-services and data queries; however these kinds of services are designed and maintained by human actors as commitment-based services.

This simplified ontological analysis of services provides an opening insight on some differences within the notion of ecosystem services. Indeed, while a service is commonly conceived as intentionally provided by someone, typically an agentive participant, in the case of ESs the role of the provider is played by nature. However, the environment does not have the same agentive characteristics of a human actor and even indulging the idea that nature has some sort of agentivity and intentionality (e.g. by being goal-oriented), it is yet not explicit nor possible to investigate if ecosystems have the intention of committing to service delivery to humans (i.e. the consumer). For these reasons we model the notion of ecosystem services twofold, from one side commitment-based services that are intentionally extracted and provided by actors (e.g. food, water, raw materials) and ecosystem services that are unintentionally provided by ecosystem structure and the involvement of functional roles, such as regulating and maintenance. Here we can see how the concept of unintentional provision of an ecosystem service may be confused with the one of natural function, as indeed ESs are not based on commitment and the ecosystem provider does not receive anything directly in exchange for the service. However ESs differ from functions due to their association with values, valuations and benefit for humans, notions that are not always associated with ecosystem function. Indeed ecosystem services, as well as commitment-based ecosystem services, are valued by actors involved in the activity of extraction or accessing and are influenced by governance decisions. Note that the notion of service in the ESs approach might be adopted beyond its classic meanings as an instrument to facilitate understanding of the value-oriented connotations of ecosystem functions and products. This consideration is not a recommendation for changing the term service from ESs, but rather an encouragement for communities of practice which employ ESs to specify the meanings of terms and adopt clear definitions to avoid ambiguity.

Benefit and Value. As introduced in Sect. 2.2, some ESs applications include economic valuations and quantitative and qualitative analysis of ecosystem services benefits [25,35]. In the cascade model goods (i.e. products) and benefits bear values, monetary or not [47]. For example, mangroves have structural and functional characteristics that provide services, such as coastal protection, that impact human well-being and therefore provide benefits associated to values. Although both values and benefits are central to the ESs approach, the former has been the subject of critical debate due to its overloaded semantics, interpretations from different communities such as economists and ecologists, and a complex literature elaborating intrinsic and instrumental values of nature [25]. We start our analysis from the notion of value, defined in dictionaries as a monetary amount, the importance or usefulness of something, a symbolic representation and a guide for behaviours and judgments,[13] to extract the semantics of benefit. This in turn is defined as something having positive characteristics or outcomes.[14] Note that we focus in particular on human attribution of monetary and non-monetary values, this is based on the interpretation of ESs as a descriptive and normative human-made instrument to assess ecosystem outcomes.

Following a simplified interpretation of the ontological study proposed in [4], which is grounded on UFO, and extending it for SESs, we continue the analysis using the notion of *value ascription* that is a contextual relation between an actor and an entity, such as service or a good. These *value objects* (i.e. object to which values are ascribed) present qualities that are central to the valuation activity either for their functional role, (e.g. insects that carry pollen could be valuated for their functional role in an ecosystem) as well as non-functional role based on actors' *preferences*, (e.g. the westerly seashore is preferred by the actor x for its aesthetic qualities). Similarly to value objects, activities (and their associated goals) can also be ascribed to values; these types of activities are called *value activities*, an example of which is a commitment-based service such as coast guard rescue and the ecosystem service of water quality provided by soil. Actors, individuals and collectives ascribe values to entities on the basis of intentional moments (e.g. desires and preferences) that are dependent on contexts, for instance coral reefs are valued as providers of recreational and/or provisioning services. Various contextual factors influence value ascription, these include *norms*, *conditions of the actors* (i.e. physical and mental) and the *environment* (e.g. temperature), *location* and *product availability*. The valuation relationship that involves both actor and value entity results in two kinds of outcomes, namely *cost specific valuation* and *benefit specific valuation* based on the desires and preferences of the actor. While cost specific valuations are "negative" and dependent on the use and access of value entities besides their economic prices, benefit specific valuations are "positive", linked to the qualities, capability and outcomes of value entities fitting the actors' desires and preferences. In the example of fishes delivering a food provisioning service, the cost associated with that service could reflect accessibility to the fish stock and the technology required for extraction,

---

[13] dictionary.cambridge.org/dictionary/english/value;
   www.lexico.com/definition/value.

[14] dictionary.cambridge.org/it/dizionario/inglese/benefit; www.lexico.com/definition/benefit.

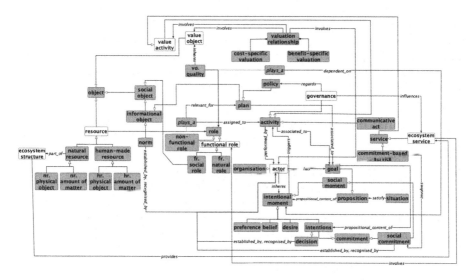

Fig. 3. Integrated framework for SESs

while the benefit of the same service is food availability and associated quality of life, health and well being in a certain context.

3.2 Integrated Social-Ecological Perspectives Framework

The section above presented a preliminary ontological analysis of the main notions of SESF and ESs to ground an integrated social-ecological perspectives framework, (see Fig. 3[15]), that provides an unambiguous semantics and conceptual organisation of the core components of SESF and ESs. Figure 3 depicts a wide spectrum of SESs concepts and their relationships as defined in the previous section. White boxes represent elements associated to SESs, and while some of them maintain the same label, such as *ecosystem service*, others are modelled following the previous ontological clarifications, for example *functional role*. The added concepts are represented using grey boxes; some of these are extracted from the ontological literature, such as *informational object*, others from the SESs literature such as *natural resource*. One of the main challenges is the handling of the concept of ecosystem service, and this has been overcome by differentiating the element of service as intentional *commitment-based* or otherwise.

While other works have concentrated on the adoption and comparison of both SESF and ESs (e.g. [8,45]) or focus on formal analysis of one of the two approaches (e.g. [34]) such an ontological analysis and integration in an unified model of the main SESF and ESs elements and relationship is unique to the literature. As a final remark, our intentions for this work is to present an approach, i.e. ontological analysis, that provides for (i) a clearer semantics of SESs

---

[15] In Fig. 3 "natural resource" "human resource", "value entity" and "functional role' are abbreviated respectively as "nr.", "hr.", "vr." and "fr.".

components and (ii) a unified framework for SESF and ESs to address the challenges related to data comparison, vocabulary disambiguation and frameworks integration. The proposed framework does not aim to replace existing conceptualisations and become yet another approach, instead the goal of this work is to refine already adopted theories and improve upon their limitations.

4   Conclusions and Future Works

This paper presents an application of ontological analysis in the context of sustainability and SESs. We introduce a framework that integrates the main components of SESF and ESs approaches with the purpose of clarifying their semantics, an issue that is still open in the SESs communities, and promote integration and comparability of studies to address sustainability challenges. We believe that the proposed framework can be the starting point to address some of the inconsistencies between SESs interpretations that are also reflected in data collection and hermeneutic activities.

As a next step we envision the extension of the integrated framework to address the complex SESF's notions of resource and governance systems as well as action situation. We also foresee the inclusion in the presented conceptualisation of the roles that technology plays in SESs and how it impacts human experience, natural ecosystems and backgrounds sustainability initiatives (Adamo and Willis, *unpublished manuscript*). Another important action will be the application and evaluation of the integrated framework, for instance in real world case studies such as in the context of marine and coastal research, and to align sustainability-relevant concepts of existing tools and methodologies, such as for ESs modelling [15].

Acknowledgements. This   research   was   funded   by   LARSyS   (Projeto   -
UIDB/50009/2020).

References

1. Adamo, G.: Investigating business process elements: a journey from the field of Business Process Management to ontological analysis, and back. Ph.D. thesis, DIBRIS, Università di Genova, Via Opera Pia, 13 16145 Genova, Italy, May 2020
2. Adamo, G., Di Francescomarino, C., Ghidini, C.: Digging into business process meta-models: a first ontological analysis. In: Dustdar, S., Yu, E., Salinesi, C., Rieu, D., Pant, V. (eds.) CAiSE 2020. LNCS, vol. 12127, pp. 384–400. Springer, Cham (2020). https://doi.org/10.1007/978-3-030-49435-3_24
3. Anderies, J.M., Janssen, M.A., Ostrom, E.: A framework to analyze the robustness of social-ecological systems from an institutional perspective. Ecol. Soc. **9**(1), 18 (2004). http://www.ecologyandsociety.org/vol9/iss1/art18/
4. Andersson, B., Guarino, N., Johannesson, P., Livieri, B.: Towards an ontology of value ascription. In: Formal Ontology in Information Systems - Proceedings of the 9th International Conference, FOIS 2016, Annecy, France, 6–9 July 2016. Frontiers in Artificial Intelligence and Applications, vol. 283, pp. 331–344. IOS Press (2016). https://doi.org/10.3233/978-1-61499-660-6-331

5. Arp, R., Smith, B.: Function, role, and disposition in basic formal ontology. Nature Precedings, p. 1 (2008)
6. Assessment, M.E., et al.: Ecosystems and Human Well-Being, vol. 5. Island press United States of America (2005)
7. Azevedo, C.L.B., et al.: Modeling resources and capabilities in enterprise architecture: a well-founded ontology-based proposal for ArchiMate. Inf. Syst. **54**, 235–262 (2015)
8. Ban, N.C., Evans, L.S., Nenadovic, M., Schoon, M.: Interplay of multiple goods, ecosystem services, and property rights in large social-ecological marine protected areas. Ecol. Soc. **20**(4), 2 (2015). https://doi.org/10.5751/ES-07857-200402
9. Basurto, X., Gelcich, S., Ostrom, F.: The social-ecological system framework as a knowledge classificatory system for benthic small-scale fisheries. Glob. Environ. Change **23**(6), 1366–1380 (2013)
10. Berkes, F., Folke, C.: Linking social and ecological systems for resilience and sustainability. Beijer International Institute of Ecological Economics. The Royal Swedish (1994)
11. Binder, C.R., Hinkel, J., Bots, P.W., Pahl-Wostl, C.: Comparison of frameworks for analyzing social-ecological systems. Ecol. Soc. **18**(4) (2013)
12. Boella, G., Lesmo, L., Damiano, R.: On the ontological status of plans and norms. Artif. Intell. Law **12**(4), 317–357 (2004)
13. Borgo, S., et al.: Technical artifacts: an integrated perspective. Appl. Ontol. **9**(3–4), 217–235 (2014). https://doi.org/10.3233/AO-140137
14. Bottazzi, E., Ferrario, R.: Preliminaries to a DOLCE ontology of organisations. Int. J. Bus. Process. Integr. Manag. **4**(4), 225–238 (2009). https://doi.org/10.1504/IJBPIM.2009.032280
15. Burkhard, B., Maes, J.: Mapping Ecosystem Services. Advanced Books **1**, e12837 (2017)
16. Colding, J., Barthel, S.: Exploring the social-ecological systems discourse 20 years later. Ecol. Soc. **24**(1) (2019)
17. Commission, E.: Eu biodiversity strategy for 2030. Bringing nature back into our lives. Communication for the Commission to the European Parliament, the Council, the European Economic and Social Committee and the Committee of the regions, p. 25 (2020)
18. Cox, M.: Applying a social-ecological system framework to the study of the Taos Valley irrigation system. Hum. Ecol. **42**(2), 311–324 (2014)
19. Cox, M.: Understanding large social-ecological systems: introducing the SESMAD project. Int. J. Commons **8**(2), 265–276 (2014)
20. Dastani, M., van der Torre, L., Yorke-Smith, N.: Commitments and interaction norms in organisations. Auton. Agents Multi-Agent Syst. **31**(2), 207–249 (2015). https://doi.org/10.1007/s10458-015-9321-5
21. De Groot, R.S., Wilson, M.A., Boumans, R.M.: A typology for the classification, description and valuation of ecosystem functions, goods and services. Ecol. Econ. **41**(3), 393–408 (2002)
22. Delmas, M.A., Young, O.R.: Governance for the Environment: New Perspectives. Cambridge University Press, Cambridge (2009)
23. Eugene, P.O.: Relationships between structure and function in the ecosystem. Jpn. J. Ecol. **12**(3), 108–118 (1962)
24. Fadel, F.G., Fox, M.S., Grüninger, M.: A generic enterprise resource ontology. In: Third Workshop on Enabling Technologies: Infrastructure for Collaborative Enterprises, WET-ICE 1994, Proceedings, pp. 117–128. IEEE (1994)

25. Farber, S.C., Costanza, R., Wilson, M.A.: Economic and ecological concepts for valuing ecosystem services. Ecol. Econ. **41**(3), 375–392 (2002)
26. Ferrario, R., Guarino, N.: Towards an ontological foundation for services science. In: Domingue, J., Fensel, D., Traverso, P. (eds.) FIS 2008. LNCS, vol. 5468, pp. 152–169. Springer, Heidelberg (2009). https://doi.org/10.1007/978-3-642-00985-3_13
27. Frey, U., Cox, M.: Building a diagnostic ontology of social-ecological systems. Int. J. Commons **9**(2) (2015)
28. Guizzardi, G., de Almeida Falbo, R., Guizzardi, R.S.: Grounding software domain ontologies in the unified foundational ontology (UFO): the case of the ode software process ontology. In: CIbSE, pp. 127–140. Citeseer (2008)
29. Guizzardi, G., et al.: UFO: unified foundational ontology. Appl. Ontol. **17**, 1–44 (2021)
30. Guizzardi, R.S.S., Carneiro, B.G., Porello, D., Guizzardi, G.: A core ontology on decision making. In: Proceedings of the XIII Seminar on Ontology Research in Brazil and IV Doctoral and Masters Consortium on Ontologies (ONTOBRAS 2020). CEUR Workshop Proceedings, vol. 2728, pp. 9–21. CEUR-WS.org (2020)
31. Guizzardi, R., Perini, A., Susi, A.: Aligning goal and decision modeling. In: Mendling, J., Mouratidis, H. (eds.) CAiSE 2018. LNBIP, vol. 317, pp. 124–132. Springer, Cham (2018). https://doi.org/10.1007/978-3-319-92901-9_12
32. Haines-Young, R., Potschin, M.: Common International Classification of Ecosystem Services (CICES) v5.1 Guidance on the Application of the Revised Structure. Fabis Consulting Ltd., The Paddocks, Chestnut Lane, Barton in Fabis, Nottingham (2017)
33. Haines-Young, R., Potschin, M., et al.: The links between biodiversity, ecosystem services and human well-being. Ecosyst. Ecol. New Synth. **1**, 110–139 (2010)
34. Hinkel, J., Bots, P.W.G., Schlüter, M.: Enhancing the Ostrom social-ecological system framework through formalization. Ecol.Soc. **19**(3), 51 (2014). https://doi.org/10.5751/ES-06475-190351
35. Liu, S., Costanza, R., Farber, S., Troy, A.: Valuing ecosystem services: theory, practice, and the need for a transdisciplinary synthesis. Ann. N. Y. Acad. Sci. **1185**(1), 54–78 (2010)
36. Masolo, C., Borgo, S., Gangemi, A., Guarino, N., Oltramari, A.: WonderWeb deliverable D18 ontology library (final). Technical report, IST Project 2001-33052 WonderWeb: Ontology Infrastructure for the Semantic Web (2003)
37. Masson-Delmotte, V. (eds.): Climate Change 2021: The Physical Science Basis. Cambridge University Press, Cambridge (2021)
38. McGinnis, M.D.: An introduction to IAD and the language of the Ostrom workshop: a simple guide to a complex framework. Policy Stud. J. **39**(1), 169–183 (2011). https://doi.org/10.1111/j.1541-0072.2010.00401.x
39. McGinnis, M.D., Ostrom, E.: Social-ecological system framework: initial changes and continuing challenges. Ecol. Soc. **19**(2), 30 (2014). https://doi.org/10.5751/ES-06387-190230
40. Mizoguchi, R., Borgo, S.: A preliminary study of functional parts as roles. In: Proceedings of the Joint Ontology Workshops 2017 Episode 3: The Tyrolean Autumn of Ontology. CEUR Workshop Proceedings, vol. 2050. CEUR-WS.org (2017)
41. Ostrom, E.: A general framework for analyzing sustainability of social-ecological systems. Science **325**(5939), 419–422 (2009)
42. Ostrom, E., Burger, J., Field, C.B., Norgaard, R.B., Policansky, D.: Revisiting the commons: local lessons, global challenges. Science **284**(5412), 278–282 (1999)

43. Partelow, S.: Coevolving Ostrom's social-ecological systems (SES) framework and sustainability science: four key co-benefits. Sustain. Sci. **11**(3), 399–410 (2016)
44. Partelow, S.: A review of the social-ecological systems framework. Ecol. Soc. **23**(4) (2018)
45. Partelow, S., Winkler, K.J.: Interlinking ecosystem services and Ostrom's framework through orientation in sustainability research. Ecol. Soc. **21**(3) (2016)
46. Pavan, S., et al.: Mainstreaming the economics of nature: a synthesis of the approach, conclusions and recommendations of TEEB. Technical report, TEEB (2010)
47. Potschin, M., Haines-Young, R., et al.: Defining and measuring ecosystem services. In: Routledge Handbook of Ecosystem Services, pp. 25–44 (2016)
48. Potschin-Young, M., et al.: Understanding the role of conceptual frameworks: reading the ecosystem service cascade. Ecosyst. Serv. **29**, 428–440 (2018)
49. Rhodes, R.A.: Governance and Public Administration. Debating Governance, vol. 5490 (2000)
50. Sanfilippo, E.M., et al.: Modeling manufacturing resources: an ontological approach. In: Chiabert, P., Bouras, A., Noël, F., Ríos, J. (eds.) PLM 2018. IAICT, vol. 540, pp. 304–313. Springer, Cham (2018). https://doi.org/10.1007/978-3-030-01614-2_28
51. Schägner, J.P., Brander, L., Maes, J., Hartje, V.: Mapping ecosystem services' values: current practice and future prospects. Ecosyst. Serv. **4**, 33–46 (2013)
52. Stephan, M., Marshall, G., McGinnis, M.: An introduction to polycentricity and governance. In: Governing Complexity: Analyzing and Applying Polycentricity, pp. 21–44. Cambridge University Press, Cambridge (2019)
53. Un, U.N.: Transforming our world: the 2030 agenda for sustainable development. Working papers, eSocialSciences (2015)
54. Vieu, L.: On the transitivity of functional parthood. Appl. Ontol. **1**(2), 147–155 (2006)
55. Vihervaara, P., et al.: Methodological interlinkages for mapping ecosystem services-from data to analysis and decision-support. One Ecosyst. **4**, e26368 (2019)
56. Winston, M.E., Chaffin, R., Herrmann, D.: A taxonomy of part-whole relations. Cognit. Sci. **11**(4), 417–444 (1987)

Contratto – A Method for Transforming Legal Contracts into Formal Specifications

Michele Soavi[1]([✉]) [ID], Nicola Zeni[1] [ID], John Mylopoulos[2] [ID], and Luisa Mich[1] [ID]

[1] University of Trento, Via Sommarive 14, 38123 Povo, TN, Italy
{michele.soavi,nicola.zeni,luisa.mich}@unitn.it
[2] University of Ottawa, 800 King Edward Ave., Ottawa, ON K1N 6N5, Canada
jmylopou@uottawa.ca

Abstract. Legal contracts have been used for millennia to conduct business trans-actions world-wide. Such contracts are expressed in natural language, and usually come in written form. We are interested in producing formal specifications from such legal text that can be used to formally analyze contracts, also serve as launching pad for generating smart contracts, information systems that partially automate, monitor and control the execution of legal contracts. We have been developing a method for transforming legal contract documents into specifications, adopting a semantic approach where transformation is treated as a text classification, rather than a natural language processing problem. The method consists of five steps that (a) Identify domain terms in the contract and manually disambiguate them when necessary, in consultation with stakeholders; (b) Semantically annotate text identifying obligations, powers, contracting parties, assets and situations; (c) Iden-tify relationships among the concepts mined in (b); (d) Generate a domain model based on the terms identified in (a), as well as parameters and local variables for the contract; (e) Generate expressions that formalize the conditions of obligations and powers using terms identified in earlier steps in a contract specification language. This paper presents the method through an illustrative example, also reports on a prototype implementation of an environment that supports the method.

Keywords: Legal contract · Smart contract · Semantic annotation · Domain model · Formal specification · Ontology

1 Introduction

Legal contracts have been used as legal basis for conducting business transactions since the late days of the Roman Empire. Today, such contracts usually come in written form and are expressed in natural language (NL). We are interested in producing formal specifications[1] [1] from such legal text that can be used to formally analyze contracts, also serve as a launching pad for generating smart contracts, information systems that partially automate, monitor and control the execution of legal contracts.

---

[1] Specifications describe *what* a contract does without describing *how*; they have been used extensively in Computer Science for software, hardware, business processes etc.

© The Author(s), under exclusive license to Springer Nature Switzerland AG 2022
R. Guizzardi et al. (Eds.): RCIS 2022, LNBIP 446, pp. 338–353, 2022.
https://doi.org/10.1007/978-3-031-05760-1_20

Towards this end, we have been developing a method for transforming legal contract documents into specifications, adopting a semantic approach where transformation is treated as a text classification, rather than a natural language processing problem. The method consists of five steps that (a) Identify domain terms in the contract and manually disambiguate them when necessary, in consultation with stakeholders; (b) Semantically annotate text identifying obligations, powers, contracting parties, assets and situations; (c) Identify relationships among the concepts mined in (b); (d) Generate a domain model based on the terms identified in (a), as well as parameters and local variables for the contract; (e) Generate expressions that formalize the conditions of obligations and powers using terms identified in earlier steps using a contract specification language.

The proposed method is intended to make the transformation systematic for users and is based on an extensive survey of legal contracts from different domains (Sales, Rentals, Transportation, Construction, Energy, etc.). The key idea behind the proposal is that we are dealing with NL text describing concepts in a rather narrow domain – legal contracts – and we can therefore assume that the text consists exclusively of defining terms, obligations and powers that determine stakeholder requirements for a contract. This insight turns the transformation into a meaning-mining problem where we keep asking "What text fragments talk about a term/obligation/power/role/asset/etc.?" rather than a vanilla NL-processing one, thereby avoiding pitfalls of NL processing of legal text that is all too often ungrammatical, ambiguous, incomplete and/or artificially structured. This insight opens the door to tool support for each step of the process, where each tool incorporates some heuristics for recognizing text that declares a term, an obligation or a power, or the constituents thereof. The target specification language for legal contracts is Symboleo [2]. The ontology offered by Symboleo for describing legal contracts is built around the concepts of obligation and power, role and asset, event and situation.

The main research contributions of this paper consist of the five-step transformation method, grounded on our past work with semantic annotation of legal text [3, 4], as well as the discovery of semantic relationships in legal text [5]. The proposed method is supported by an environment that assists a user in carrying out the steps of the method with input from lawyers and contracting parties. The workings of the method are illustrated through an example with a rental contract adopted from the Web. There is a workshop paper [6] that presents an early version of the method, using four steps rather than five and a small example and doesn't mention the Contratto environment. This work is part of a multi-year, multiply funded project that aims to develop information systems engineering methods and tools for building smart contracts.

The rest of the paper presents the research baseline for this work (Sect. 2), the five-step method (Sect. 3) through a small example, the Contratto environment (Sect. 4), related work (Sect. 5) and concludes (Sect. 6).

2   Research Baseline

Legal Contracts. Legal contracts involve promises made by contracting parties, defined in terms of obligations and powers the parties have towards each other. Every contract must have at least two parties and two assets (often one of which is money). Contracts are initially offered by a party, negotiated, signed and executed ('performed' in Law).

GaiusT. The Contratto environment was developed on GaiusT, a web-based platform intended for building annotation tools for legal documents [4]. The platform provides facilities either through a web GUI or web-services for annotating legal documents using an annotation schema. The annotation schema comprises structural and semantic tags combined with patterns described by eBNF[2] grammars. GaiusT includes modules to support the generation of the annotation schemas, using as input an annotation ontology expressed in XMI, RDF or OWL format, as well as lexical resources such as WordNet, Thesaurus, Google n-gram and Wikipedia. The generated annotation schema can be exported to a textual file.

The annotation process includes structural annotation to capture the structure of a legal document, such as title, chapters, sections, and clauses, as well as cross-references to other parts of the same document, internal references, or external ones to other documents covering applicable laws and regulations.

The annotation engine takes as input an annotation schema, lexical patterns for all concepts in the schema, a structural grammar and an input document. The annotation engine: (a) extracts plain text from files in a variety of formats (including Microsoft Word, RDF, PDF, HTML); (b) normalizes the plain text by removing unprintable characters and produces a text document where each line represents a phrase; (c) annotates text fragments with tags for structure and cross-references; and (d) annotates text fragments with semantic tags present in the annotation schema. Its output is semantically and structurally annotated text in XML format.

NomosT. The NomosT tool was developed on the GaiusT platform and is intended to semantically annotate legal text using the Nomos 2.0 modeling language ontology for law in support of compliance analysis for software requirements [5]. NomosT includes heuristic rules to identify relationships, such as Activate or Satisfy, between legal concepts.

Extensions to GaiusT and NomosT. The development of the Contratto environment adopted GaiusT and NomosT, though both tools had to be extended and revised to deal with the idiosyncrasies of contracts, rather than generic legal text. GaiusT had to be modified because the ontology it uses is for legal documents, as opposed to contracts. This resulted in a different annotation schema and different lexical markers. In addition, GaiusT was meant to support only the annotation step of the proposed Contratto process. NomosT also uses a different ontology than the Symboleo ontology adopted by Contratto and mines different relationships, see the next section.

3 The Transformation Process

We illustrate the 5-step process with an example. Consider a simple rent-to-own contract, as shown in Table 1.

---

[2] eBNF stands for 'extended Backus-Naur form' and consists of a notation for specifying programming language grammars.

Table 1. The Rent2Own contract

Whereas, John Goodman (hereafter Renter) desires to possess and have the use of the property at 35 Oxford Street, Ottawa, Canada, owned by Mark Smith (hereafter Owner) Now, therefore, the parties agree as follows: Renter shall pay Owner the sum of $300 on 20/12/2019 and the same sum on the 20th day of each month for rental of the property. If payment is late by more than thr ee days, a late fee of $30 shall be due immediately from Renter. The parties agree that the purchase price of the property is $45.000. The parties agree that $150 of each month's rent payment shall be applied towards purchase of the property. The parties a gree that if Renter fails to complete the contemplated purchase of the property for any reason, no refunds or credits shall be due to Renter. Renter shall maintain the property, at Renter's expense, in clean, good working order .

Identification of terms results in the declaration of several terms that will be used as contract parameters or local variables and are shown in Table 2.

Table 2. Identification of terms

renter means John Goodman. *owner* means Mark Smith. *parties* means the renter and the owner collectively. *property* means the single-occupancy building located at 35 Oxford Street, Ottawa, Canada, owned by the owner. *rental fee* means the monthly fee for renting the property corresponding to CAN$ 300. *rental payment days* refer to 20/12/2019 and all subsequent 20th of the month days up to 20/11/2044. *late payment fee* means a fee of CAN$30 due immediately from renter if rental payment is late by more than 3 days after rental payment day. *purchase price* means CAN$45,000. *monthly purchase contribution* represents an amount of CAN$150 of each month's rental fee that is applied towards purchase of the property. *contributed balance* means the monthly purchase contribution multiplied by the number of months from 20/12/2019. *good working order* means that the property is cleaned every week and undergoes regular maintenance.

The semantic annotation step identifies four obligations and a power each named for reference purposes. Powers are rights contracting parties have to cancel, suspend or create new obligations or powers. Moreover, this step identifies roles, namely Renter and Owner, assets, as well as situations that can trigger, activate, or terminate successfully an obligation/power. Accordingly, the output of the second step (b) would be something like what is shown in Table 3.

Table 3. Annotated text

O-rental: *<Obligation ><Role>* Renter *</Role>* shall *<Situation>* pay *<Role>* Owner *</Role>* the sum of *<Asset>* *$300 </Asset>* on 20/12/2019 the same sum on the 20th day of each month for rental of the property *</Situation> </Obligation >*   ***O-lateF:*** *<Obligation ><Situation >*If payment is late by more than three days, *</Situation> <Situation >* a late fee shall be due immediately from   *<Role >*Renter *</Role> </Situation> </Obligation >*   ***O-purchase:*** *<Obligation >* The *<Role>* Parties *</Role>* agree that *<Situation >* $150 of each month's rent payment shall be applied towards purchase of the Property *</Situation> </Obligation >*   ***P-refund:*** *<Power >*The *<Role>* Parties *</Role>* agree that   *<Situation >* if *<Role>*Renter*</Role >* fails to complete the contemplated purchase of the property for any reason *</Situation >*, *<Situation >* no refunds or credits shall be due to *<Role >*Renter*</Role > </Situation ></Power >*   ***O-care:*** *<Obligation ><Role>* Renter *</Role >* shall *<Situation >* maintain the property, at *<Role>* Renter*</Role >*'s expense, in Good Working Order.   *</Situation > </Obligation >*

However, these clauses say nothing about the Owner's obligations to rent the property to the Renter and eventually sell it, as indicated in the preamble of the contract. Accordingly, two obligations are added to Table 3 that paraphrase the preamble in terms of obligations, see Table 3a.

Table 3a. Annotated and paraphrased preamble text

O-rent *<Role>* Owner *</Role >* shall rent *<Asset>* property *</Asset>* to *<Role>* Renter *</Role>* *<Situation>* while *<Role>* Renter *<Role>* is paying the rental fee *</Situation>*   ***O-own*** *<Role>* Owner *</Role >* shall sell *<Asset>* property *</Asset>* to *<Role>* Renter *</Role>* when the *<Role>* Renter *<Role>* has completed payment of the Purchase Price *</Situation>*

A semantic annotation requires a common vocabulary (ontology) for legal contracts that defines the concepts we are looking for in the text. Sometimes the text being annotated is structured, for example, with bullets marking the clauses of the contract.

The third step (c) identifies relationships for each one of the concepts discovered in step two. For example, each obligation must have a debtor who is obliged to fulfil it, and a creditor (beneficiary). The debtors and creditors of O-rental and O-lateF are Renter and Owner respectively, while for O-purchase roles are reversed. Note that O-lateF doesn't mention a creditor, while O-purchase doesn't mention debtor and creditor, so these have to be inferred by a lawyer and/or tool addressing this step. The power P-refund has Owner as creditor and Renter as debtor. In addition, each obligation/power can have a trigger that initiates it, an antecedent that serves as precondition for it to become active, and a consequent that signals successful completion of the obligation/power. P-refund has a trigger 'If Renter fails to complete the contemplated transaction for any reason', which activates the power and enables Owner to exercise his right to withhold any funds paid by Renter towards purchase, O-lateF also has a trigger 'if payment is late ...', while other obligations take a trigger 'true', denoted by 'T', and are initiated when contract execution starts. There are no antecedents for any clauses of this contract, so they take value T, Finally, the consequents for O-rental and O-lateF are respectively 'pay Owner the sum of $300 on 20/12/2019 and the same sum on the 20th day of each month for rental of the property' and 'a late fee shall be due immediately from Renter'. The output of the third step consists of concepts and relationships, where for each obligation we list trigger, debtor, creditor, ante(cedent), cons(equent), and for each power we list trigger, creditor, debtor, ante(cedent), cons(equent). The result of this step has as shown in Table 4.

Table 4. Identified relationships for the Rent2Own contract

O-rental [trigger: 'on 20/12/2019 and on the 20th day of each month thereafter until property is purchased', debtor: Renter, creditor: Owner, ante: T, cons: 'pay Owner the sum of $300']
O-lateF [trigger: 'If payment is late by more than three days', debtor: Renter, creditor: Owner, ante: T, cons: 'a late fee of $30 shall be due immediately from Renter']
O-purchase [trigger: T, debtor: Owner, creditor: Renter, ante: T, cons: '$150 of each month's rent shall be applied towards purchase of the property']
P-refund [trigger: 'Renter fails to complete the contemplated purchase of the property for any reason', debtor: Owner, creditor: Renter, ante: T, cons: 'no refunds ... shall be due to Renter']
O-care [trigger: T, debtor: Renter, creditor: Owner, ante: T, cons: maintain the property, at Renter's expense, in Good Working Order']
O-rent [trigger: T, debtor: Owner, creditor: Renter, ante: T, cons: 'rent Property while contract is in effect']
O-own [trigger: T, debtor: Owner, creditor: Renter, ante: T, cons: 'sell Property when payment of the Purchase Price has been completed']

The reader can think of this as the backbone of the specification consisting of seven clauses, six obligations and one power, with five relationships each. In the fifth step (e) we shall embellish the backbone with expressions for each trigger, antecedent and consequent.

The fourth step (d) concerns formalizing domain terms identified in step (a) or domain terms used in the contract, such as rent and sell. These are treated as variables in the specification that take as values instances of a class that is an extension/specialization (aka isA) of the legal contract ontology used in step (b). Alternatively, variables may take values from basic programming types such as Integer, String, or enumerated types such as [1..10], or subsets thereof. In particular, property takes as values instances of the class RealEstateProperty (aka REProperty) isA Asset and has attributes ownedBy and addr. Likewise, rent is a variable of type Rent isA Situation with attribute fee, while sold is an event variable of type Sold isA Event. Note that situations occur over an interval of time while events happen instantaneously.

This step also identifies parameters for a contract, the 'givens' for each contract execution. Rent2Own doesn't have any parameters, but it could have if, for example, we wanted to define a template contract for rent-to-own with parameters renter, owner, rental fee, start date, etc. Some of the variables defined in this step are shown in Table 5.

Table 5. Some classes and local variables for Rent2Own

Domain classes
REProperty **isA** Asset **with** ownedBy: Role, addr: AddrString   /*   has values address -looking strings
Rent **isA** Situation **with** fee: CurrencyAmount
Paid **isA** Event **with** from: Role, to: Role, moyr: Month -Year: fee: CurrencyAmount
...
Variables
property: REProperty **with** ownedBy := owner, addr := '35 Oxford Street ...'
rent: Rent **with** fee := CAN$300
paid: Paid **with** from: renter, to: owner, price := CAN$45,000

The final step (e) concerns translating NL expressions such as 'on 20/12/2019 and on the 20th day of each month thereafter until property is purchased' and 'pay Owner the sum of $300' (trigger and consequent respectively of O-rental) into Symboleo expressions. For example, the trigger of O-rental is translated to 'occurs(d) **and** rentalPaymentDate(d)' where rentalPaymentDate is a 20th day of the month between 12/2019 and 11/2044 and is defined in the domain model (not shown here for lack of space). This expression says that when a 20th of the month date occurs, then O-rental is triggered and the Renter must make the consequent of O-rental true. The consequent is expressed as 'happens (paid (Renter, Owner, m-y(d), CAN$300), t) **and** t **during** d', which says that a payment event happens from Renter to Owner for the month-year of d (remember, d is date of the type dd/mm/yyyy) and the amount of CAN$300 at some time t during d. Likewise, the trigger of O-lateF can be expressed as 'occurs(d) **and** rentalPayment-Day(d) **and** happens (paid (Renter, Owner, m-y(d), CAN$300), t) **and** t **after** d+3days', which says that O-lateF is triggered if payment for a month-year happens after 3days from the due date. Finally, the consequent of O-lateF can be expressed as happens (paid (Renter, Owner, m-y(d), CAN$30), t). Note that, according to the contract, late fee is due when rent payment happens late.

More examples of formalized clauses are shown in Table 6, along with some Symboleo syntax[3]. Basically, the specification was constructed incrementally in steps (b), (c), (d) and (e) collectively using the terms identified in step (a).

Table 6. Fragments of a specification for the Rent2Buy contract

```
Domain model Rent2Buy
... ;                    /* see Table 5
end Domain;
Contract Rent2Buy( )          /* No parameters
Declarations
...                    /* see Table 5
Obligations
O-rental: occurs(d) and rentalPaymentDate(d) →
    O(Renter, Owner, T, happens(paid(Renter,Owner,m-y(d),CAN$300), t) and t
during d
O-lateF: occurs(d) and rentalPaymentDay(d) and
           happens(paid(Renter,Owner,m-y(d),CAN$300), t) and t after
d+3days →
    O(Renter, Owner, T, happens(paid(Renter,Owner,m-y(d),CAN$30), t)
O-purchase: happens(paid(Renter,Owner,moyr,CAN$300), t) →
    O(Owner, Renter, T, happens(increasedBy(contributedBal,CAN$150), t)
macro int::=[moyr,moyr+6mo]
P-refund:   ForAll   t/int   (not(happens(paid(Renter,Owner,moyr,$300),   t)   →
P(Owner,Renter,T,
           happens(cancelled(self),t') and happens(nulled(contributedBal),t')
and t' after int
...                /* includes remaining clauses expressed in Symboleo
    end Contract
```

4   The Contratto Environment

We have developed a prototype environment that supports each step of the Contratto process by (a) allowing users to carry out the transformation manually using suitable editors and tools for each step; (b) proposing to the user outputs for given inputs for all steps except step (e). For example, for the input of Table 1, Contratto (hopefully) proposes annotated text similar to that of Table 2. In our earlier work, [4] we found that such a tool can generate annotations of very good quality for legal text, as well as good

---

[3] The full specification can be found at https://legal-analysis.economia.unitn.it:7500/.

quality conceptual models [5] by discovering relationships among the entities identified by the annotation. In both cases the tool is not intended to automate the transformation, but rather to support a human by (a) Improving the quality of the output, and (b) reducing the manual effort required to carry out each step. On this point, the tool presented in [4], intended to help a human generate a conceptual model from a legal text was found to reduce manual effort by 80% while maintaining the quality of the output.

We have adopted the ideas that underlie the tool proposed in [4] to develop a tool called ContracT 1.0 that conducts structural and semantic annotation for legal contract text [7]. Moreover, we have evaluated empirically ContracT 1.0 by having human subjects that generate annotations for a given contract manually, while others simply correct the output of ContracT 1.0 to produce annotated text. The results of the experiment suggest that the use of ContracT 1.0 can improve significantly the performance of novice human annotators both with respect to precision and recall.

The Contratto environment is using a set of web APIs and a web GUI to support all steps of the process.

The process we envision begins with a user uploading a contract through the GUI and running a preliminary analysis of the contract using the Named Entity Recognizer (NER) of the Spacy[4] NLP system that exploits a pre-trained neural network combined with statistical analysis to extract contract terms in support of step (a) (Fig. 1).

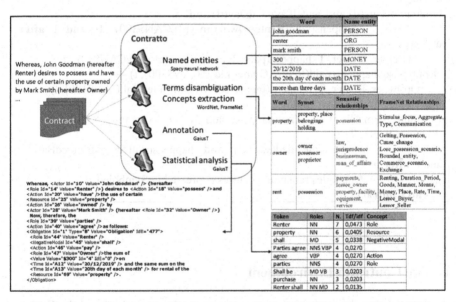

Fig. 1. Support for term identification and domain model building (steps (a) and (d))

The environment then performs annotation of the document (Fig. 3) with a refined annotation schema extracted through the GaiusT annotation, NER and statistical analysis

---

[4] Spacy is a free open-source library for Natural Language Processing implementing a neural network for pre-trained models for name entity recognition https://spacy.io/.

tools. The user can then visualize and edit the result through drag-and-drop operations (Fig. 4). For step (c) the user can rely on an updated version of NomosT to suggest relationships that the user can accept/revise.

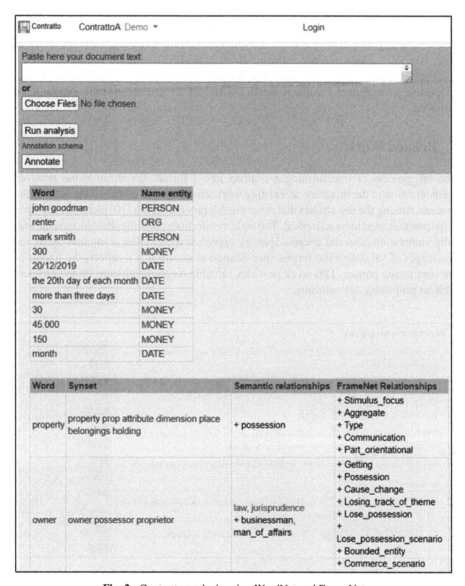

Fig. 2. Contratto analysis using WordNet and FrameNet

To support the identification of domain elements in the contract, and thus to build the domain model, the user can exploit the output of the analysis performed by two web services that use respectively WordNet [8] and FrameNet [9] to propose domain classes, their superclasses and attributes. Such web services query a local copy of WordNet ontology to retrieve, for each term of the contract, the semantic relationships – hypernymy, hyponymy, meronymy, part-whole and isA relationships – and query a local copy of FrameNet to retrieve semantic frames for verbs (Fig. 2).

The output can be used to fill the domain model. The generation of a domain model and a final specification are supported by GUI templates with placeholders that users can fill through drag-and-drop operations with elements identified through different analyses (Fig. 4).

The Contratto prototype is intended to establish the feasibility of building such a supportive environment. Further research is required on the details of the support to be provided.

5 Related Work

The full process of transforming a contract into a formal specification has received scant attention in the literature, as existing work tends to focus on one of the steps of the process. Among the few studies that cover the full process, Clack [10] identifies key open problems that need to be addressed. The author underlines the difficulties associated with fully automation, also the interdisciplinary approach required for a solution. Also, the challenges of validating the formal specification to ensure that it reflects the intents of the contracting parties. This work provides valuable recommendations on the problem without proposing any solutions.

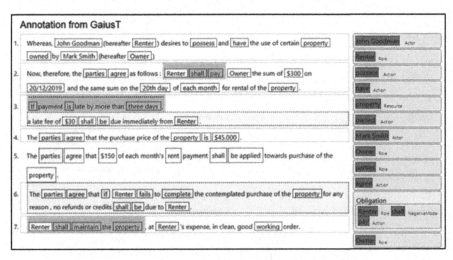

Fig. 3. Annotated document with annotated items on the side

Step (a) of the process tackles the identification of terms in a contract for disambiguation purposes. Disambiguated terms are formalized in the domain model in step (d). For contracts, ambiguity, generally defined as multiple interpretations of a text, is mainly related to vagueness, under-specification (under-determination) or generality

[11]. These are also the most difficult types of ambiguity to be solved, requiring human interpretation. Lexical, syntactic ambiguity are rare in contracts, also thanks to the use of a glossary or of a preamble. Automatic identification of these 'linguistic ambiguities' has been investigated in requirements engineering (see for example [12–14]). Tools developed in that area could be applied as a preliminary quality checking for writing contracts and are out of the scope of the paper.

Among a significant number of works concerning annotation, named entity recognition appears to be the preferred method to identify contract terms. Rule-based approaches tend to perform better when specialized for a domain by using a constrained language, such as Quaresma [15] which relies on a mixed approach based on Machine Learning (ML) and linguistic information, mostly morphological and syntactic. An NL parser is used for named-entity recognition whereas a Support Vector Machine (SVM) is used for top-level concepts such as organizations and dates. This is a promising approach although difficulties are encountered to identify cross-references that may help to better define contract terms. Term identification has also been used to generate requirements that can eventually be modelled as obligations and powers in our method. Breaux et al. [16] rely on a process called *semantic parameterization* to identify and discriminate between rights and obligations. The authors also propose strategies to identify and resolve ambiguities based on the use of a restricted NL. This proposal does not account for powers and is largely manual. Similarly, Fantoni et al. [17] perform text mining to translate contract terms into a formal specification. The approach is focused on engineering contracts but has significant elements that could be reused for every type of contract. It is supported by computational linguistic tools relying on a knowledge base consisting of keywords, concepts, relationships and formal expressions.

Semantic annotation has received much attention although the annotation of unrestricted NL still poses significant challenges. To overcome these, Libal et al. [18] normalize NL based on deontic logic and extract normative and conditional elements together, including relationships among them. Chalkidis [19] proposes that structural annotation can be learned from a benchmark, differently from our method where the structural modes are defined by a grammar. The approach combines ML and manually written post-processing rules. A popular approach, in line with our method, is represented by the use of an ontology to identify contract elements. Among them, a significant number of ontologies is based on UFO, such as UFO-S [20], grounded on Alexy's Theory of Fundamental Rights. Generally, the tools used to automatically extract ontology elements from contracts need further research.

Regarding the identification of relationships has been generally limited to syntactic relationships within a phrase (e.g., nouns or verbs) or between different sections of a contract, such as provisions and clauses. Concerning the former, an example is provided by Fischbach et al. [21] that develop a tool-supported approach named CiRA (Causality detection in Requirement Artifact) to identify causal relationships. The approach is tested with ML and Deep Learning (DL) where the best results are obtained using Bidirectional Encoder Representations from Transformers (BERT). The approach is useful to identify causality with the use of cue phrases although further ambiguity consideration would arise as several causality relationships may exist. Relationships among different contract sections have been studied to identify arguments in sections.

In Moens et al. [22] arguments are identified, annotated and classified using n-grams, adverbs, modals, text statistics and keywords. In general, such approaches are useful to identify the implications deriving from the modification of text but do not provide support or the identification of relationships among ontological elements. The use of templates, such as in [23] has been proposed as a preliminary step to explicit existing relationships, as legal texts frequently omit them. Templates work to identify three gaps: statements with no-counterpart, statements with a correlative, and statements with an implied statement.

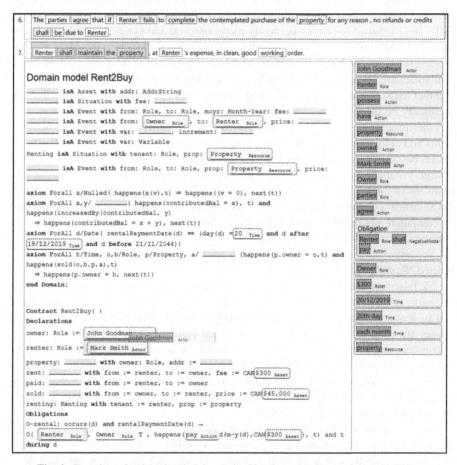

Fig. 4. Templates for domain model and specification with drag and drop facilities.

For step (d), there is significant confusion in the scientific community concerning the terms 'domain model' and 'ontology'. Frequently, 'domain model' refers to a sub-process of the general process of ontology development. Among them, Corcho et al. [24] propose an approach to support legal professionals in building ontologies with limited involvement of knowledge engineers starting from the process of building a glossary to represent concepts and relationships. An alternative approach is proposed by adopting an ontology from which a domain model is derived, such as in [25]. The approach relies on NLP techniques and focuses on the identification of the relationships between the lexical and the ontological level. Generally, the availability of tools to automatically generate a domain model from NL terms remains an open challenge.

Translating NL expressions into formal ones frequently relies on knowledge representation languages based on deontic logic. Among the first works focusing on the transformation into formal expressions, Governatori [26] proposes Business Contract Language (BCL) using Propositional Deontic Logic where a contract is composed of deontic terms for obligations, prohibitions and permissions. The approach supports as well earlier steps of the method although Propositional Logic has significant limitations on expressiveness and does not support the formalization of all contract elements. The last years saw the development of formal specifications to be implemented as smart contracts. In [27], the transformation process is presented starting with the definition of an ontology for legal contracts. The ontology is based on a Smart-Legal-Contract-Markup Language (SCML) and includes an XML schema definition to transform it into the formal specification. The approach requires significant manual effort but represents one of the few papers that covers most of our transformation process. SPESC [28] provides a framework to generate formal specifications understandable by a large audience (lawyers, programmers) although it relies on a lower level of formality. The specifications are derived manually and are expressed in a format similar to NL using eBNF and are structured around parties, assets and data type definition. Given the lower level of formality, it may be useful as an intermediate step between the NL contract and the formal specification we envision. The literature review seems to suggest using an iterative approach in generating requirements from legal NL text, as recommended in [29].

Finally, there is much recent work in the general area of ML and NL [19, 30–32], some of which relates to term identification, contract element extraction and NL inference for legal documents. Such techniques are candidates for adoption in developing Contratto tools that support individual steps of the proposed method.

6   Conclusions

We have proposed a method for transforming legal contracts into formal specifications. The method includes a systematic process, illustrated through an example, and an environment that facilitates the transformation from legal text to a specification. The environment includes several tools that automate specific tasks of the process. The proposal has only seen a preliminary evaluation by the authors using several contracts adopted from the Web.

There are two directions along which the work reported here needs to be extended and enhanced. The first dimension concerns the Contratto environment. Firstly, the tool for identifying relationships (step (c)) needs heuristic rules specifically applicable to contracts, as opposed to rules that are applicable for any legal document. Secondly, we envision a new component of the Contratto environment that includes a library of templates for different types of contracts (rental, consumer sales, real estate sales, rent-to-own, construction, transportation, etc.), along with the outputs of each of the 5 steps of the Contratto process for each template. These can serve as starting points for customizations and could significantly reduce manual effort while improving quality. Along a different dimension, the Contratto process and environment needs to be evaluated by subjects other than the authors to determine its effectiveness in comparison to ad hoc manual transformations, both in reducing required manual effort and in improving the quality of the final specification. Moreover, such evaluation needs to be performed not only with vanilla contracts, but also large ones where there are bound to be problems in scaling out methods and tools to deal with complex business transactions.

References

1. Lamport, L.: Who builds a house without drawing blueprints? Commun. ACM **58**, 38–41 (2015)
2. Sharifi, S., Parvizimosaed, A., Amyot, D., Logrippo, L., Mylopoulos, J.: Symboleo: a specification language for Smart Contracts. In: 28th IEEE Requirements Engineering Conference, RE@Next track, Zurich (2020)
3. Kiyavitskaya, N., Zeni, N., Breaux, T.D., Antón, A.I., Cordy, J.R., Mich, L., Mylopoulos, J.: Automating the extraction of rights and obligations for regulatory compliance. In: 27th International Conference on Conceptual Modelling (ER), pp. 154–168 (2008)
4. Zeni, N., Kiyavitskaya, N., Mich, L., Cordy, J.R., Mylopoulos, J.: GaiusT: supporting the extraction of rights and obligations for regulatory compliance. Requir. Eng. **20**(1), 1–22 (2013). https://doi.org/10.1007/s00766-013-0181-8
5. Zeni, N., Seid, E., Engiels, P., Ingolfo, S., Mylopoulos, J.: NomosT: building large models of law through a tool-supported process. Data Knowl. Eng. (DKE), **117**, 407–418 (2018)
6. Soavi, M., Zeni, N., Mylopoulos, J., Mich, L.: From legal contracts to formal specifications: a progress report. In: Joint Proceedings of REFSQ-2021 Workshops, OpenRE, Posters and Tools Track, and Doctoral Symposium, http://ceur-ws.org, vol. 1613, p. 0073 (2020)
7. Soavi, M., Zeni, N., Mylopoulos, J., Mich, L.: ContracT – from legal contracts to formal specifications: preliminary results. In: 13th International Working Conference on the Practice of Enterprise Modelling (PoEM) (2020)
8. Miller, G., Beckwith, R., Fellbaum, C., Gross, D., Miller, K.: Introduction to WordNet: an on-line lexical database. Int. J. Lexicogr. **3**(4), 235–244 (1990)
9. Narayana, S., Fillmore, C., Baker, C., Petruck M.: FrameNet meets the semantic web: a DAML+OIL frame representation. In: 18th National Conference on Artificial Intelligence, Edmonton, Alberta (2002)
10. Clack, C.D.: Languages for Smart and computable contracts. ArXiv preprint arXiv:2104.03764 (2021)
11. Sennet, A.: Ambiguity. The Stanford Encyclopedia of Philosophy. https://plato.stanford.edu/archives/fall2021/entries/ambiguity/. Accessed 27 Jan 2022
12. Mich, L., Garigliano, R.: Ambiguity measures in requirement engineering. In: International Conference on Software Theory and Practice (ICS) (2000)
13. Kiyavitskaya, N., Zeni, N., Mich, L., Berry, D.M.: Requirements for tools for ambiguity identification and measurement in natural language requirements specifications. Requir. Eng. **13**(3), 207–239 (2008)
14. Berry, D.M., Kamsties, E.: Ambiguity in requirements. Perspect. Softw. Requir. **753**, 7 (2012)
15. Quaresma, P., Gonçalves, T.: Using linguistic information and machine learning techniques to identify entities from juridical documents. In: Francesconi, E., Montemagni, S., Peters, W., Tiscornia, D. (eds.) Semantic Processing of Legal Texts. LNCS, vol. 6036, pp. 44–59. Springer, Heidelberg (2010). https://doi.org/10.1007/978-3-642-12837-0_3

16. Breaux, T.D., Antón, A.I.: Analyzing goal semantics for rights, permissions, and obligations. In: 13th IEEE International Conference on Requirements Engineering (RE) (2005)
17. Fantoni, G., Coli, E., Chiarello, F., Apreda, R., Dell'Orletta, F., Pratelli, G.: Text mining tool for translating terms of contract into technical specifications: development and application in the railway sector. Comput. Ind. **124**, 103357 (2021)
18. Libal, T., Pascucci, M.: Automated reasoning in normative detachment structures with ideal conditions. In: 17th International Conference on Artificial Intelligence and Law (2019)
19. Chalkidis, I., Androutsopoulos, I.: A deep learning approach to contract element extraction. JURIX (2017)
20. Griffo, C., Almeida, J.P.A., Guizzardi, G., Nardi, J.C.; From an ontology of service contracts to contract modeling in enterprise architecture. In: 21st International Enterprise Distributed Object Computing Conference (EDOC), pp. 40–49. IEEE (2017)
21. Fischbach, J., Frattini, J., Spaans, A., Kummeth, M., Vogelsang, A., Mendez, D., Unterkalmsteiner, M.: Automatic detection of causality in requirement artifacts: the CiRA approach. In: International Working Conference on Requirements Engineering: Foundation for Software Quality, pp. 19–36. Springer, Cham (2021)
22. Moens, M. F., Boiy, E., Palau, R.M., Reed, C.: Automatic detection of arguments in legal texts. In: Proceedings of the 11th International Conference on Artificial Intelligence and Law, pp. 225–230 (2007)
23. Sleimi, A., Ceci, M., Sabetzadeh, M., Briand, L. C., Dann, J.: Automated recommendation of templates for legal requirements. In: IEEE 28th International Requirements Engineering Conference (RE), pp. 158–168. IEEE (2020)
24. Corcho, O., Fernández-López, M., Gómez-Pérez, A., López-Cima, A.: Building legal ontologies with METHONTOLOGY and WebODE. In: Benjamins, V.R., Casanovas, P., Breuker, J., Gangemi, A. (eds.) Law and the Semantic Web. LNCS, vol. 3369, pp. 142–157. Springer, Heidelberg (2005). https://doi.org/10.1007/978-3-540-32253-5_9
25. Francesconi, E., Montemagni, S., Peters, W., Tiscornia, D.: Integrating a bottom–up and top–down methodology for building semantic resources for the multilingual legal domain. In: Francesconi, E., Montemagni, S., Peters, W., Tiscornia, D. (eds.) Semantic Processing of Legal Texts. LNCS, vol. 6036. Springer, Heidelberg (2010). https://doi.org/10.1007/978-3-642-12837-0_6
26. Governatori, G., Milosevic, Z.: A formal analysis of a business contract language. Int. J. Coop. Inf. Syst. **15**(04), 659–685 (2006)
27. Dwivedi, V., Norta, A., Wulf, A., Leiding, B., Saxena, S., Udokwu, C.: A formal specification smart-contract language for legally binding decentralized autonomous organizations. IEEE Access **9**, 76069–76082 (2021)
28. He, X., Qin, B., Zhu, Y., Chen, X., Liu, Y.: SPESC: a specification language for smart contracts. In: 42nd Annual Computer Software and Applications Conference (COMPSAC), vol. 1, pp. 132–137. IEEE (2018)
29. Maxwell, J.C., Antón, A.I.: Developing production rule models to aid in acquiring requirements from legal texts. In: 17th IEEE International Requirements Engineering Conference. IEEE (2009)
30. Chalkidis, I., Fergadiotis, M., Malakasiotis, P., Aletras, N., Androutsopoulos, I.: LEGAL-BERT: the muppets straight out of law school. ArXiv [cs.CL] (2020)
31. Koreeda, Y., Manning, C.D.: ContractNLI: a dataset for document-level natural language inference for contracts. ArXiv [cs.CL] (2021)
32. Devlin, J., Chang, M.-W., Lee, K., Toutanova, K.: BERT: pre-training of deep bidirectional transformers for language understanding. ArXiv [cs.CL] (2019)

Ontological Foundations for Trust Dynamics: The Case of Central Bank Digital Currency Ecosystems

Glenda Amaral[1]([✉]), Tiago Prince Sales[1], and Giancarlo Guizzardi[1,2]

[1] Conceptual and Cognitive Modeling Research Group (CORE),
Free University of Bozen-Bolzano, Bolzano, Italy
{gmouraamaral,tiago.princesales,giancarlo.guizzardi}@unibz.it
[2] Services and Cybersecurity, University of Twente, Enschede, The Netherlands

Abstract. In recent years, disruptive technologies have advanced at a rapid pace. These new developments have the power to accelerate the production and delivery, improve the quality, and reduce the costs of goods and services, as well as to contribute to individual and collective well-being. However, their adoption relies largely on user trust. And trust, due to its dynamic nature, is fragile. Therefore, just as important as to build trust is to maintain it. To build sustainable trust it is fundamental to understand the composition of trust relations and what factors can influence them. To address this issue, in this paper, we provide ontological foundations for trust dynamics. We extend our previous work, the Reference Ontology of Trust (ROT), to clarify and provide a deeper account of some building blocks of trust as well as the many factors that can influence trust relations. We illustrate the working of ROT by applying it to a real case study concerning citizens' trust in central bank digital currency ecosystems, which has been conducted in close collaboration with a national central bank.

Keywords: Trust dynamics · UFO · OntoUML · Central bank digital currency

1 Introduction

New and disruptive technologies have been developed at a rapid pace, affecting almost every area of our lives. Industrial robots, artificial intelligence algorithms, machine learning, big data, decentralized technologies, just to cite a few, have the power to accelerate the production and delivery, improve the quality, and reduce the costs of goods and services, as well as to contribute to individual and collective well-being. However, the adoption of these innovative technologies relies largely on user trust. And trust is highly dynamic. Trust is generally said to be one of the easiest things to lose and one of the most difficult things to win back. It may break in an instant or erode gradually. Therefore, it is important to build sustainable trust that is not easily lost. In the case of information

R. Guizzardi et al. (Eds.): RCIS 2022, LNBIP 446, pp. 354–371, 2022.
https://doi.org/10.1007/978-3-031-05760-1_21

technology systems and ecosystems, building sustainable trust involves addressing stakeholders' trust concerns from the system (or ecosystem) inception and their constant monitoring, as trust changes with time. But how to identify these trust concerns? What makes one trustworthy? And what factors can influence trust? In this paper we address these questions via an ontological analysis and representation of trust dynamics.

In a previous work [2], we proposed the Reference Ontology of Trust (ROT) - an ontologically well-founded reference model, grounded in the Unified Foundational Ontology (UFO) [15,16], which formally characterizes the concept of trust, as well as clarifies the relation between trust and risk, and represents how risk emerges from trust relations. This paper sheds new light on trust-related concepts and relations under the perspective of trust dynamics. We extend our previous work to clarify and provide a deeper account of (i) the different factors that can influence trust; (ii) the signals that trustees can emit to indicate their trustworthy behavior; and (iii) pieces of evidence that suggest that a trustee should be trusted. We validate and illustrate the use of ROT with a real case study on citizens' trust in central bank digital currency (CBDC) ecosystems, which was conducted in close collaboration with a national central bank.

This paper is organized as follows. In Sect. 2 we introduce the reader to the Reference Ontology of Trust. Then, in Sect. 3 we present the extensions to the ontology to provide ontological foundations for trust dynamics. We apply ROT in a real case study and discuss its results in Sect. 4. We assess related work in Sect. 5 and conclude the paper in Sect. 6 with some final considerations.

2   The Reference Ontology of Trust (ROT)

The Reference Ontology of Trust[1] (ROT) is a well-founded ontology, based on the Unified Foundational Ontology (UFO) and specified in OntoUML [15]. It formally characterizes the concept of trust, clarifies the relation between trust and risk, and represents how risk emerges from trust relations [2]. ROT makes the following ontological commitments about the nature of trust:

- **Trust is relative to an intention.** An agent, the trustor, trusts an individual, the trustee, only relative to a certain intention, on which achievement she counts on the trustee. The trustor may trust in the trustee regarding a certain intention, but not bestow such a trust regarding a different one. For example, I trust my dentist to fix a cavity in my tooth, but not to fix my computer. Furthermore, such an intention is not always atomic. For example, in the trust relation "Bob trusts a certain airline to take him on his holiday trip comfortably and safely", trust is about a complex intention, composed of (i) Bob's intention of traveling; (ii) his intention of being safe; and (iii) his intention of being comfortable.

---

[1] The complete version of ROT in OntoUML and its implementation in OWL are available at http://purl.org/krdb-core/trust-ontology.

- **A trustor is an "intentional entity".** Trustors are cognitive agents, being endowed with intentions and beliefs [9].
- **A trustee does not need agency.** A trustee is an individual capable of impacting one's intentions by the outcome of its behavior [9]. A trustee may be either an agent (e.g. a person, an organization) or an object (e.g. a car, a vaccine).
- **Trust is a complex mental state.** Trust is a mental state of a trustor regarding a trustee and its behavior. It is composed of: (i) an intention; (ii) beliefs that the trustee can perform a desired action or exhibit a desired behavior; (iii) beliefs that the trustee's vulnerabilities will not prevent it from performing the desired action or exhibiting the desired behavior; and (iv) if the trustee is an agent, beliefs that the trustee intends to exhibit that behavior. For example, a mother who trusts a babysitter to take care of her kids believes that: (i) the babysitter has experience in caring for children and is First Aid trained (a belief about the babysitter's capabilities); (ii) the babysitter is well and probably is not going to have health issues (a belief about the babysitter's vulnerabilities); and (iii) the babysitter is willing to take good care of her children (a belief about the babysitter's intention).
- **Trust is context-dependent.** A trustor may trust a trustee for a given goal in a given context, but not do so for the same goal in a different context. For example, Mary may trust her brother to drive her to the train station in a sunny day, but she does not trust him to do so when it is snowing. We assume trust relations to be highly dynamic [9].
- **Trust implies risk.** By trusting, the trustor becomes vulnerable to the trustee in terms of potential failure of the expected behavior or outcome [17, p. 21]. In trust relations, risk can emerge as consequence of either the manifestation of a trustee's vulnerability or the unsatisfactory manifestation of a trustee's capability. In the above-mentioned babysitter example, the babysitter getting sick during the term of the employment contract corresponds to the manifestation of a babysitter's vulnerability that prevents her from going to work and ultimately can hurt the mother's intention of having an adult to take care of her kids.
- **Trust can be quantified.** Our trust in a certain individual can increase or decrease in time, and we can trust certain individuals more than others. To account for these scenarios, ROT assumes that trust can be quantified, even if it does not commit to any particular scale or measurement strategy. In ROT, the quantitative perspective of trust is captured by means of (i) the trust degree—the extent to which the trustor trusts in the trustee; (ii) the belief intensity—the strength of a trustor's belief; (iii) the performance level—how well the trustor believes the trustee can perform the action; and (iv) the manifestation likelihood—how strongly the trustor believes a disposition of the trustee may be manifested through the occurrence of certain events.

The aforementioned ontological commitments are captured in the ROT diagram presented in Fig. 1. In the OntoUML diagrams depicting the ontology, we adopt the following color coding: substantials are represented in pink, modes and qualities in blue, relators in green, and classes whose instances might be of different ontological nature in gray. The reader interested in an in-depth description of the complete version of ROT is referred to [2].

3   Modeling Trust Dynamics

We previously mentioned that trust is composed of a trustor's intention and a set of her beliefs about the trustee and its behavior. However, several other factors that influence trust have been discussed in the literature [18]. For instance, Castelfranchi and Falcone [9] argue that "trust changes with experience, with the modification of the different sources it is based on, with the emotional or rational state of the trustor, with the modification of the environment in which the trustee is supposed to perform, and so on". They claim that trust is a dynamic entity because it depends on dynamic phenomena.

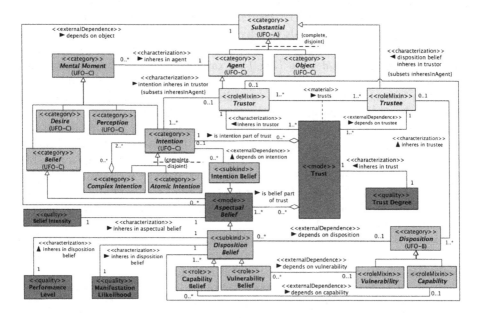

Fig. 1. Reference ontology of trust.

In this section, we extend ROT to allow it to account for trust dynamics more adequately by categorizing influence relations according to the ontological nature of the factors that explain them. These factors are: (F1) trust influencing trust (Sect. 3.1); (F2) mental biases (Sect. 3.2); (F3) trust calibration signals (Sect. 3.3); and (F4) trustworthiness evidence (Sect. 3.4). In the diagrams depicted in Figs. 2 and 3, the colored concepts represent the extensions proposed to ROT. The concepts in white belong to the original version of the ontology. In the remaining of this section, we present a detailed description of these extensions.

3.1   Trust Influencing Trust

This category represents the situation in which a trust relationship is influenced by another trust relationship. According to Castelfranchi and Falconi [9] "in the same situation trust is influenced by trust in several rather complex ways". In the same work they also discuss the phenomenon of trust creating reciprocal trust, as well as how trust relations can influence each other. In fact, countless examples can be found in real life about trust influencing trust, either positively or negatively. For instance, citizens' trust in the central bank positively influence their trust in the national currency. People's trust in the healthcare system, in the experts defining vaccination strategies, and more generally in government bodies influence their trust in vaccines.

3.2   Mental Biases

This category represents situations in which trust is influenced by *mental moments* (a concept from UFO). *Mental moments* refer to the capacity of some properties of certain individuals to refer to possible situations of reality [16]. A *mental moment* is existentially dependent on a particular agent, being an inseparable part of its mental state (Fig. 1). Examples include beliefs, desires and intentions. Beliefs have a propositional content that agents consider to be true. They can be justified by situations in reality. Examples include my belief that Rome is the Capital of Italy, and the belief that the Moon orbits the Earth. Beliefs can be formed by perceptions expressing how agents sense their environment and the things that happen around them. Desires and intentions can be fulfilled or frustrated. A desire expresses the will of an agent towards a possible situation (e.g., a desire that Italy wins the next World Cup), while an intention expresses desired states of affairs for which the agent commits to pursuing (e.g., Mary's intention of going to Paris). For an extensive discussion of mental moments, please refer to [16].

Mental moments can significantly influence trust relationships. Let us consider the example of a person who really wants to travel but cannot. One day she receives an email containing an unbelievable offer for an exotic destination that is just about to expire. Although many will immediately think it is a scam, the person's desire to travel may influence her to trust the email offer [13]. Another example would be people who are strongly committed to environmental preservation and tend to trust companies that support environmental sustainability. There is also the case of beliefs not related to specific trustees. An example discussed by [19] suggests that some religious beliefs, which prescribe honesty and mutual love, lead people to generally assume that others are usually honest, benevolent, competent, and predictable.

Another important aspect is the occurrence of events that can affect one's perception regarding a trustee. McKnight et al. [20] discuss how trust changes in response to external events and propose a model that addresses the mental mechanisms people use as they are confronted by trust-related events, which "indicates that trust may be sticky or resistant to change, but that change can and will occur" [20]. Castelfranchi and Falconi [12] claim that the success of an action performed by the trustee in order to reach a goal of the trustor depends not only on the trustee's capabilities but also on external conditions that allow or inhibit the realization of the task. To illustrate this point, the authors use the case of a violinist that will give a concert in an open environment. In general, people trust the violinist to play well. However, if it is particularly cold during the concert, their trust will decrease if they infer that the cold can hinder her ability to play. Similarly, in financial systems, the emergence of detrimental information about a financial agent can negatively affect public trust in this agent, which can lead to considerable adverse effects on one or several other financial institutions that can ultimately propagate to the entire financial system.

3.3   Trust Calibration Signals

The emission of *trust-warranting signals*, that is, signals that indicate trustworthy dispositions of a trustee, is one of the ingredients for building sustainable trust [21]. In trust relations, once the trustee's capabilities and vulnerabilities related to the beliefs of the trustor are known, it is possible to reason about the signals that the trustee should emit to indicate that it can successfully realize its capabilities and prevent its vulnerabilities from being manifested. These signals are specifically created to indicate a trustworthy behavior on the part of the trustee and therefore can influence trust. For example, information about how privacy and security measures are implemented could be provided as signals of the trustworthiness of a system. Another example is the establishment of a universal brand to create visual identity, so that users can identify the system interface elements in a clear and unambiguous way, thus facilitating the understanding and adoption of its functionalities.

Equally important are *uncertainty signals*, i.e. signals that communicate uncertainties regarding the realization of capabilities and the prevention of vulnerabilities. Some examples are the publication of uncertainties about the accuracy of scientific findings, patient communication of uncertainties on the precision of medical diagnosis, investor communication of uncertainties in forecasting financial investments returns, communication to the public about uncertainties regarding the efficacy of vaccines, among others. While trust-warranting signals contribute to trust building, uncertainty signals allow trustors to adjust their trust level appropriately to the trustee's trustworthiness, thus avoiding misplaced levels of trust. Research show that communicating uncertainty can be beneficial for maintaining trust and commitment over time [5,23]. This is because building trust that is higher than the actual trustworthiness of the trustee might set trustors' expectations too high, which may result in disappointment sooner or later.

We extend ROT to represent trust signals emitted by trustees (F3) in the following way. As illustrated in Fig. 2, the `Trustee` may emit `Trust Calibration Signals` regarding its `Dispositions` (either a `Capability` or a `Vulnerability`). `Trust Calibration Signal` is specialized into `Trust-warranting Signal` and `Uncertainty Signal`. The former represents trust-warranting signals that should be emitted by the trustee in order to ensure trustworthy behavior, while the latter represents uncertainty signals emitted by the trustee, which allow trustors to adjust their trust levels[2].

3.4   Trustworthiness Evidence

Another trust influencing factor corresponds to *trustworthiness evidence*, pieces of evidence that can make a trustor believe that the trustee should be trusted. Similarly to trust-warranting signals, they suggest that a trustee can realize its capabilities and shield its vulnerabilities. However, differently from signals, which are purposefully emitted to suggest trustworthiness, evidence result from the observation of a trustees' trustworthy actions. Examples include:

- third-party certifications and credentials (e.g. John's TOEFL certification makes me believe that he can speak English, because I trust the certificate issued by a certain authority);
- performance history (e.g. accuracy of a medical diagnosis system);
- track record (e.g. reviews from service recipients and statistics on its experience);
- recommendations (e.g. my brother trusts a car mechanic and recommends his services to me);
- reputation records (e.g. positive evaluations received by an Uber driver);
- availability (e.g. a medical doctor you rarely succeed to make an appointment with is not trustworthy);
- past successful experiences (e.g. all the products I purchased at Amazon arrived on time and in perfect condition);
- transparency (e.g. offering information on what an artificial intelligence system is doing, as well as rationale for its decisions (aka explainability));
- longevity (e.g. indication that a vendor has been in the market for a long time and that it is interested in continued business relationship with the client); and
- risk mitigation measures, which indicate that one is actively trying to prevent the manifestation of one's vulnerabilities.

---

[2] `Emits` is grounded on a communicative act of the trustee [16] and, hence, a *historical* relation in the sense of [14]. The propositional content of this act refers to a disposition, thus, grounding the (derived) `refers to` relation between the latter and `Trust Calibration Signal`.

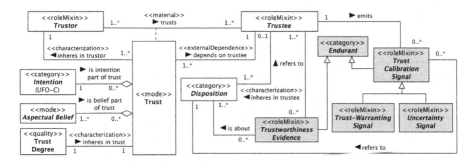

Fig. 2. ROT - Trust calibration signals and trustworthiness evidence extensions.

Ontologically speaking, a piece of `Trustworthiness Evidence` is a *social endurant*, typically a *social relator*[3] (e.g. a relator binding the certifying entity, the certified entity and referring to a capability, vulnerability, etc.), but also documents (*social objects* themselves) that represent these *social entities* (e.g., in the way a marriage certificate documents a marriage as a social relator). As illustrated in Fig. 2 we extended ROT to model `Trustworthiness Evidences` (F4) as "roles" played by *endurants* (objects, relators, etc.) related to a `Disposition` of the `Trustee`[4].

To represent the role of influences, we included the `Influence` relator, which connects the sources of influence to the aspectual beliefs of the trustor under their influence (Fig. 3). We distinguish `Influence` according to the source of influence into: (i) `Trust Influence`, associated to a `Trust` relationship (F1); (ii) `Mental Moment Influence`, associated to a `Mental Moment` (F2); (iii) `Trust Calibration Signal Influence`, associated to a `Trust Calibration Signal` (F3); and (iv) `Trustworthiness Evidence Influence`, associated to a `Trustworthiness Evidence` (F4). The property `weight` corresponds to the weight of an influence over a particular belief, as certain influences may weight more heavily than others.

---

[3] Briefly speaking, a relator (a concept from UFO) is an entity that is existentially dependent on at least two individuals, thus, mediating or binding them [14].

[4] Playing of the "role" of `Trustworthiness Evidence` for a particular focal disposition is dependent on the belief of trustors, whose propositional content makes that connection between that player and that disposition. The `is about` relation in this model is, thus, derived from the propositional content of that belief. The `refers to` relation connected to the trustee is derived from the relation between that focal disposition and its bearer.

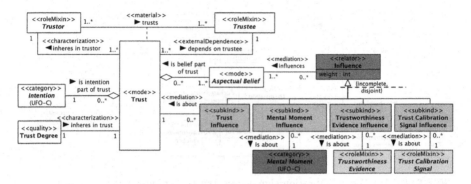

Fig. 3. ROT - Influences extensions.

4   Case Study: Citizens' Trust in CBDC Ecosystems

In this section, we present a real-world case in which we use ROT to model citizens' trust in CBDC ecosystems. We worked in close collaboration with a national central bank to analyse the case and instantiate the ontology to represent the dynamics of citizens' trust in CBDC ecosystems. Due to the sensitivity of this topic (the development of CBDC ecosystems is in full swing, and their design is not finished yet) and specific request of the central bank, the only information we can disclose is that the contributing central bank's context is a country with between 50 and 300 million citizens.

4.1   Research Method

We conducted a case study, in collaboration with a national central bank, regarding citizens' trust in CBDC ecosystems. This methodological approach is particularly appropriate when the focus is investigating a contemporary phenomenon in depth and within its real-life context, and the investigator has no control over actual behavioral events [24]. The research procedure we employed was adapted from [24].

We started by planning the case study. We defined its purpose—*to verify if the Reference Ontology of Trust can properly represent real world situations, or more specifically if it can model citizens' trust in CBDC ecosystems*—identified the areas of interest, namely, economics, financial citizenship and information technology, and selected the interviewees. We also obtained the necessary authorizations from the central bank to carry out the study.

In the collect stage, we gathered information from documentation and interviews. First, documents describing and documenting information on citizens' trust in CBDC ecosystems were collected from the literature [4,6–8,11,22] and from the central bank's website, to deepen our knowledge about the topic. Based on this documentation, we created an initial version of the ontology instantiation, to be validated and complemented at the interviews stage. Then, we conducted interviews with central bank experts in the areas of interest, namely, economics,

Table 1. Questions related to key ontology concepts.

Question	ROT concept
Citizens trust the CBDC ecosystem to···	Intention
Citizens trust the CBDC ecosystem because they believe that it can···	Belief capability
Citizens trust the CBDC ecosystem because they believe that it has mechanisms to prevent···	Belief vulnerability
How can the CBDC ecosystem indicate that it is trustworthy?	Trust calibration signal
What pieces of evidence show the CBDC ecosystem is trustworthy?	Trustworthiness evidence
What can influence citizens' trust in the CBDC ecosystem?	Influence

financial citizenship and information technology. The questions that would guide the interviews with the stakeholders (Table 1) were defined based on the main concepts of the Reference Ontology of Trust (Sects. 2 and 3). The ontology served as guidance for our work from the beginning of the case study, helping us focus on the domain being investigated and supporting the creation of the interview questions. As shown in Table 1, these questions are actually formulated based on the concepts from ROT (see column 2).

We conducted three individual interviews in the form of guided conversations, one for each expert of the areas of interest, namely, economics, financial citizenship and information technology. During the interviews we presented the initial version of the ontology instantiation to be validated and gathered information based on the aforementioned questions (Table 1). The interviews were recorded (audio) and with their feedback and validation we have improved both the ontology and the ontology instantiation, and presented them again to the central bank in a validation meeting. We had in total four sessions with the central bank, in which the modeling of citizens' trust in CBDC ecosystems were discussed in detail.

In the analyze stage, we concluded the final version of the ontology instantiation. In addition, to demonstrate the contribution and applicability of our ontology to the modeling practice, we used the ontology instantiation as a domain model to create a goal model for this case using the i^* framework [10]. Finally, we shared the results with the central bank team.

4.2 Research Context: CBDC Ecosystems

Recent innovations in the financial industry, such as cryptocurrencies, blockchains and distributed ledger technologies, smart contracts, and stablecoins have fostered the creation of financial products and services on top of decentralized technologies, giving rise to the concept of Decentralized Finance

(DeFi) [25]—the decentralized provision of financial products and services. This disruption, alongside the entry of big techs into payments and financial services, pushed central banks to investigate new forms of digital money and prepare the grounds for central bank digital currencies (CBDCs). A CDBC is a form of digital money, denominated in the national unit of account, which is a direct liability of the central bank, such as physical cash and central bank settlement accounts [1].

In general, a CBDC ecosystem would comprise elements and functions similar to traditional payment systems, with central banks facing many of the practical policy questions around access, services and structure they currently do. According to [3], at the center of any CBDC ecosystem would be a CBDC "core rulebook" outlining the legal basis, governance, risk management, access and other requirements of participants in the CBDC ecosystem. Participants in the CBDC system could include banks, payment service providers, mobile operators and fintech or big tech companies, which would act as intermediaries between the central bank and end users. This broader ecosystem would be complemented by a legal and supervisory framework and contractual arrangements between end users and their intermediaries [3]. Currently, all CBDC ecosystems are still under design. The initiatives around the world are either at the stage of experimentation, proof-of-concept, or pilot arrangements.

Consumer demand for CBDC is an important element that determines how widely a CBDC would be used. Therefore, the successful implementation of a CBDC crucially depends on citizens' motivation to adopt this new digital form of public money, which is directly related to their trust[5] in the CBDC ecosystem.

4.3   Modeling Citizen's Trust in CBDC Ecosystems

Ontology Instantiation. We adopt the following coding to refer to instances of key ROT concepts hereafter: **INT** for intention; **BEL** for disposition belief; **CAP** for capability; **VUL** for vulnerability; **TS** for trust-warranting signal; **US** for uncertainty signal; **TE** for trustworthiness evidence; and **INF** for influence.

Both the literature research and the interviews showed that *citizens trust the CBDC ecosystem to preserve their privacy* (INT1). Privacy emerges as a key feature, which can be confirmed both indirectly—by the presence of comments on the importance of privacy—and directly—by ranking privacy first, among many other features [11]. Citizens who trust the CBDC ecosystem believe that *it safeguards their privacy* (BEL1.1). This belief is related, for example, to the *CBDC ecosystem's capability to comply with the General Data Protection Regulation (GDPR)*[6] *and other privacy laws and regulations* (CAP1.1).

---

[5] Agustín Carstens, the General Manager of the Bank for International Settlements, in a recent speech at the Goethe University's ILF conference on "Data, Digitalization, the New Finance and Central Bank Digital Currencies: The Future of Banking and Money" explicitly defended that "the soul of money is trust." (https://www.bis.org/speeches/sp220118.htm).

[6] https://gdpr-info.eu/.

Interviewees also expressed that it is important that citizens feel safe to perform digital transactions in the ecosystem. *Citizens trust the ecosystem to safely make transactions using CBDCs* (INT2). They believe that *the ecosystem is safe* (BEL2.1) and that *it will be able to prevent security breaches* (BEL2.2). The former belief (BEL2.1) is related to the *ecosystem's capability to have security mechanisms* (CAP2.1), while the latter (BEL2.2) is related both to *possible security breaches* (VUL2.1)— which correspond to a vulnerability—and to the *ecosystem's capability to quickly react to risk events on security* (CAP2.2). In addition, *the existence of a cybersecurity policy* (TE2.1) is an example of trustworthiness evidence related to the capability CAP2.1.

Another aspect that emerged from the interviews and the literature is the importance of providing a simple experience for end users. *Citizens trust the CBDC ecosystem to make transactions using CBDCs easily* (INT3) and they believe both that *the ecosystem is easy to access and use* (BEL3.1) and that *it is easy to onboard the CBDC ecosystem* (BEL3.2). A possible capability of the ecosystem, related to these beliefs, is *to meet minimum usability criteria* (CAP3.1). The *existence of a manual with minimum usability requirements, which must be followed by all participants of the ecosystem* (TE3.1) is an example of trustworthiness evidence related to the capability CAP3.1. *The establishment of a universal brand to create visual identity* (TS3.1), *advertising campaigns in the media and social networks using everyday examples* (TS3.2), and *documentation available* (TS3.3) are examples of trust-warranting signals related to capability CAP3.1.

Low cost was another attribute mentioned both in the literature and by the interviewees. *Citizens trust the CBDC ecosystem to make transactions using CBDCs at a low cost* (INT4) and they believe both that *it will be offered at a low cost to its users* (BEL4.1) and that *they will not need to buy a new device to make transactions in the CBDC ecosystem* (BEL4.2). The former belief (BEL4.1) is related to the *ecosystem's capability to have lower costs for consumers and merchants* (CAP4.1), while the latter (BEL4.2) is related to the *ecosystem's capability to operate using existing, accessible technology* (CAP4.2).

An additional valuable feature identified in the collect phase is the ability to make offline payments. This feature might be particularly relevant during outages and in environments where internet availability is limited or unreliable [4]. *Citizens trust the CBDC ecosystem to make transactions wherever they need* (INT5) and they believe that *they will be able to access the system from any place* (BEL5.1). This belief is related to the *ecosystem's capability to support offline transactions* (CAP5.1). Note that intention INT5 (offline access) conflicts with intention INT4 (low cost), as technology to support offline capacity may incur additional costs.

The interviews also showed that *citizens trust the CDDC ecosystem to make transactions instantly on a 24/7 basis* (INT6). In other words, users who trust the ecosystem believe that *it is able to make instantaneous transactions* (BEL6.1) and that *it will be available when they need* (BEL6.2). These beliefs are related to the ecosystem's capability to *meet high availability parameters and processing time limits* (CAP6.1). Examples of trustworthiness evidence related to this capability are the existence of *a service level agreement that establishes high availability parameters and processing time limits* (TE6.1) and *statistics on the functioning of the ecosystem showing that this service level agreement has being fulfilled* (TE6.2). Information about *instability* (US6.1) and *low response times* (US6.2) are examples of uncertainty signals.

A further important aspect identified both in the literature and in the interviews is currency acceptance. *Citizens trust the CDBDC ecosystem to make transactions using a widely accepted currency* (INT7). And they believe that *the CBDC ecosystem operates with a digital currency widely accepted* (BEL7.1). This relates to the *capability to operate using a legal tender currency* (CAP7.1).

Equally important is the stability of the currency purchasing power. *Citizens trust the CDBDC ecosystem to make transactions using a stable currency* (INT8). And they believe that *the CBDC purchasing power has stability* (BEL8.1). This belief relates to the ecosystem's *capability to have proper mechanisms to ensure stability of CBDC purchasing power* (CAP8.1).

Finally, it was also mentioned that *citizens trust the CBDC ecosystem to have access to better financial products and services offerings* (INT9). Therefore, they believe that *they will have access to more product and service offers customized to their needs* (BEL9.1). This relates to the *ecosystem's capability to provide better customized services and products offerings* (CAP9.1). Once more, it is possible to observe the existence of conflicting intentions: the intention just mentioned (INT9) conflicts with privacy preservation (INT1), as to propose better financial services offerings, financial institutions in the ecosystem need to have access to more (private) information about the citizen.

It is important to note that in the trust relation between citizens and the CBDC ecosystem, trust is about a complex intention, composed of the aforementioned intentions (INT 1 to 9).

Furthermore, it is possible to observe the existence of trust influences. For example, *citizens' trust in a country's monetary system* (INF1) positively influences their trust on the CBDC ecosystem, just as *their trust in the central bank* (INF2) does.

Figure 4 shows a graphical representation of the ontology instantiation focusing on usability (INT3). The detailed diagrams presenting the complete case study can be found at https://purl.org/krdb-core/rot-cbdc-case-study.

Goal Model. We use the ontology instantiation as a domain model to create a goal model for this case using the *i** framework [10], presented in Fig. 5. The model shows the goals that citizens delegate to the CBDC ecosystem (through the i* dependency relation). Citizens and the CBDC ecosystem are represented as actor and agent, respectively. Citizens' intentions are represented as quality dependences. Conflicting intentions are represented in yellow, circled by a red dashed line. Entities represented in green and yellow were obtained directly by mapping elements from the ontology instantiation. For each of them, more specific goals, qualities, tasks and resources were identified and are represented in

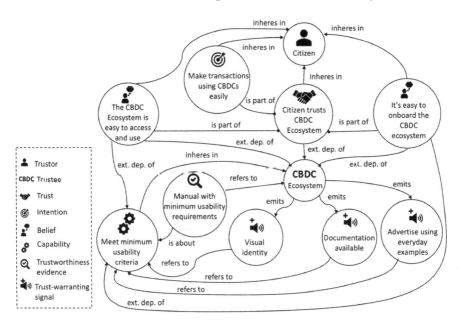

Fig. 4. Graphical representation of the ontology instantiation focusing on usability.

blue. Besides dependencies, the goal model depicts the internal perspective of the CBDC ecosystem. Beliefs are represented as goals or qualities that contribute to (help) the achievement of higher level goals. For example, beliefs such as *the ecosystem is safe* (BEL2.1), *the ecosystem is easy to use* (BEL3.1), *the CBDC is widely accepted* (BEL7.1) were represented as qualities that contribute to the ultimate goal of *being trustworthy*. Capabilities, trust calibration signals and trustworthiness evidence are represented as goals, qualities, tasks or resources that contribute to (help) the achievement of higher level goals. For example, the goal *support offline transactions* (CAP5.1) helps the achievement of *be available*. The tasks *meet minimum usability criteria* (CAP3.1) and *keep visual identity* (TS3.1) help the achievement of *be easy to use*. The resource *cybersecurity policy* (TE2.1) contributes to *comply with cybersecurity policy*. The resource *manual with minimum usability requirements* (TE3.1) contributes to the task *meet minimum usability criteria*, which in turn contributes to *be easy of use*. Conversely, vulnerabilities can be represented as goals, qualities, tasks or resources that negatively impact (hurts) the achievement of higher level goals. Finally, influences are represented as contribution links that help or hurt the achievement of higher level goals (help for positive influences and hurt for negative ones). The mapping between the ROT concepts and their representation in the i* Goal Model is presented in Table 2.

4.4   Discussion

In the validation session, the central bank experts of the aforementioned areas of interest were unanimously of the opinion that the ontology was capable of capturing all the important aspects of citizens' trust in CBDC ecosystems (perceived *usability* and *usefulness* of the approach). It was also mentioned by the interviewees that, when designing the CDBD ecosystem, it is useful to understand the intentions of the users that are related to their trust in the ecosystem, so that we can identify, at a very early stage, capabilities required to create a trustworthy and efficient environment, possible vulnerabilities that should be dealt with, as well as how to properly communicate about the ecosystem trustworthiness. Being able to identify citizens' goals provides a broad view of how CBDC can be successfully implemented, from a trustworthiness perspective. By eliciting goals, we can also identify the conflicting goals, and consequently we can be more proactive in resolving possible design issues.

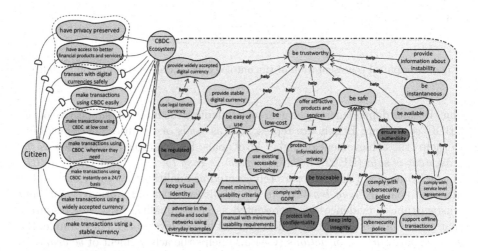

Fig. 5. A fragment of the goal model of the CBDC ecosystem.

Table 2. Representation of ROT concepts in i* Goal Model.

ROT concept	Representation in i* Goal Model
Trustor	Actor, Agent, Role
Trustee	Actor, Agent, Role
Intention	Goal dependence, Quality dependence
Belief	Goal, Quality
Capability	Goal, Quality, Task, Resource
Vulnerability	Goal, Quality, Task, Resource
Trust Calibration Signal	Goal, Quality, Task, Resource
Trustworthiness Evidence	Goal, Quality, Task, Resource
Influence	Contribution link

An interesting finding was that once we understand what composes citizens' trust and which factors may influence it, we can take these requirements into account since the ecosystem's inception, thus enabling trustworthiness by design. Furthermore, it allows the identification of potential risks in advance and the definition of risk mitigation strategies. This is because if we know which capabilities and vulnerabilities are related to the trustor beliefs, we can reason about what can go wrong with the realization of the capabilities and the manifestation of the vulnerabilities, which will hurt the intentions of the trustor. These unwanted events correspond to risk events for which mitigations strategies may be defined in advance. Another interesting finding is that trust relations require constant monitoring as the ecosystem is very dynamic and is constantly changing. And changes in the environment can influence user trust.

5 Related Work

There are some works in the literature that address the dynamic nature of trust. Riegelsberger et al. [21] propose a framework on the mechanics of trust, in which they identify contextual (temporal, social, and institutional embeddedness) and intrinsic (ability and motivation) properties that warrant trust in another actor, which they name trust-warranting properties. They also describe how the presence of these properties can be signaled. In their model, they identify two broad categories of signals: symbols and symptoms, which are analogous to the ROT concepts of trust-warranting signals and trustworthiness evidence, respectively. Despite this similarity, their work differs from what we propose here, as they do not consider uncertainty signals and other factors that may influence trust, such as other trust relations and the mental state of the trustor. Also, they do not provide an ontological account for the concepts represented in their model.

Castelfranchi and Falcone [9] made an important contribution with their theory of trust. ROT relies largely on their theory to formalize the general concept of trust and the concept of social trust. In their work, they present trust dynamics in different aspects: (i) how trust changes on the basis of the trustor's experiences, which is related to the ROT concepts of *trustworthiness evidence* and *influence*; (iii) how trust is influenced by trust; how diffuse trust diffuses trust (that is how A's trusting B can influence C trusting B or D, and so on); and (iv) how trust can change using generalization reasoning (the fact that it is possible to predict how/when an agent who trusts something/someone will therefore trust something/someone else, before and without a direct experience). These last three aspects are related to *trust influences* in ROT. Although it is rather comprehensive, their proposal does not mention the emission of signals to communicate uncertainties nor to indicate trustworthy behavior on the part of the trustee.

6 Conclusions

In this paper, we presented an ontological analysis of the factors that influence trust as well as other trust-related concepts, such as pieces of evidence that indicate a trustee's trustworthiness and the signals that the trustee may emit to indicate trustworthy behavior. To validate the ontology and demonstrate the applicability of our proposal, we conducted a real case study concerning citizens' trust in CBDC ecosystems. The case study experience confirmed that ROT can properly represent trust in this context, and suggests it could be used to represent other real cases. We acknowledge that our case study has some limitations as we only took the central bank's view regarding citizens' trust in CBDC ecosystems. Nevertheless, we rely on documentation and information from surveys conducted with citizens from the literature [4,6–8,11,22]. As future work, we plan to apply ROT to support trustworthiness by design, so that trust can be part of the design of ecosystems since their inception and be prioritized in all aspects of the ecosystem.

Acknowledgments. CAPES (PhD grant# 88881.173022/2018-01) and NeXON project (UNIBZ). The authors would like to thank the Central Bank for their time and support.

References

1. Amaral, G., Sales, T.P., Guizzardi, G.: Towards ontological foundations for central bank digital currencies. In: Proceedings of the 15th VMBO (2021)
2. Amaral, G., Sales, T.P., Guizzardi, G., Porello, D.: Towards a reference ontology of trust. In: Panetto, H., Debruyne, C., Hepp, M., Lewis, D., Ardagna, C.A., Meersman, R. (eds.) OTM 2019. LNCS, vol. 11877, pp. 3–21. Springer, Cham (2019). https://doi.org/10.1007/978-3-030-33246-4_1
3. Bank for International Settlements et al.: Central bank digital currencies: system design and interoperability. Bank for International Settlements (2021)
4. Bank for International Settlements et al.: Central bank digital currencies: users needs and adoption. Bank for International Settlements (2021)
5. Batteux, E., et al.: The negative consequences of failing to communicate uncertainties during a pandemic: The case of COVID-19 vaccines (2021)
6. Bijlsma, M., van der Cruijsen, C., Jonker, N., Reijerink, J.: What triggers consumer adoption of CBDC? De Nederlandsche Bank Working Paper (709) (2021)
7. Boar, C., Holden, H., Wadsworth, A.: Impending arrival-a sequel to the survey on central bank digital currency. Bank for International Settlements (107) (2020)
8. Boar, C., Wehrli, A.: Ready, steady, go?-Results of the third BIS survey on central bank digital currency. Bank for International Settlements (2021)
9. Castelfranchi, C., Falcone, R.: Trust Theory: A Socio-cognitive and Computational Model, vol. 18. Wiley, Hoboken (2010)
10. Dalpiaz, F., Franch, X., Horkoff, J.: iStar 2.0 Language Guide. CoRR abs/1605.07767 (2016). http://arxiv.org/abs/1605.07767
11. European Central Bank: Eurosystem report on the public consultation on a digital euro. European Central Bank (2021)
12. Falcone, R., Castelfranchi, C.: Trust dynamics: how trust is influenced by direct experiences and by trust itself. In: Proceedings of the AAMAS 2004, pp. 740–747. IEEE (2004)
13. Fischer, P., Lea, S.E., Evans, K.M.: Why do individuals respond to fraudulent scam communications and lose money? The psychological determinants of scam compliance. J. Appl. Soc. Psychol. **43**(10), 2060–2072 (2013)

14. Fonseca, C.M., Porello, D., Guizzardi, G., Almeida, J.P.A., Guarino, N.: Relations in ontology-driven conceptual modeling. In: Laender, A.H.F., Pernici, B., Lim, E.-P., de Oliveira, J.P.M. (eds.) ER 2019. LNCS, vol. 11788, pp. 28–42. Springer, Cham (2019). https://doi.org/10.1007/978-3-030-33223-5_4

15. Guizzardi, G.: Ontological foundations for structural conceptual models. Telematica Instituut, University of Twente, The Netherlands (2005)

16. Guizzardi, G. et al.: Grounding software domain ontologies in the unified foundational ontology (UFO). In: Proceedings of the 11th CIBSE, pp. 127–140 (2008)

17. Luhmann, N.: Trust and Power. Wiley, Hoboken (2018)

18. Mayer, R.C., Davis, J.H., Schoorman, F.D.: An integrative model of organizational trust. Acad. Manag. Rev. 20(3), 709–734 (1995)

19. Harrison McKnight, D., Chervany, N.L.: Trust and distrust definitions: one bite at a time. In: Falcone, R., Singh, M., Tan, Y.-H. (eds.) Trust in Cyber-societies. LNCS (LNAI), vol. 2246, pp. 27–54. Springer, Heidelberg (2001). https://doi.org/10.1007/3-540-45547-7_3

20. McKnight, D.H., Liu, P., Pentland, B.T.: How events affect trust: a baseline information processing model with three extensions. In: Dimitrakos, T., Moona, R., Patel, D., McKnight, D.H. (eds.) IFIPTM 2012. IAICT, vol. 374, pp. 217–224. Springer, Heidelberg (2012). https://doi.org/10.1007/978-3-642-29852-3_16

21. Riegelsberger, J., Sasse, M.A., McCarthy, J.D.: The mechanics of trust: a framework for research and design. Int. J. Hum. Comput. Stud. 62(3), 381–422 (2005)

22. Söilen, K.S., Benhayoun, L.: Household acceptance of central bank digital currency: the role of institutional trust. Int. J. Bank Mark. 40(1), 172–196 (2021)

23. Tomsett, Richard et al.: Rapid trust calibration through interpretable and uncertainty-aware AI. Patterns 1(4), 100049 (2020)

24. Yin, R.K.: Case Study Research: Design and Methods (Applied Social Research Methods). Sage Publications, Thousand Oaks (2008)

25. Zetzsche, D.A., et al.: Decentralized finance (DeFi). IIEL Issue Brief 2 (2020)

Abstracting Ontology-Driven Conceptual Models: Objects, Aspects, Events, and Their Parts

Elena Romanenko[1]([⊠])(ID), Diego Calvanese[1,2](ID), and Giancarlo Guizzardi[1,3](ID)

[1] Free University of Bozen-Bolzano, 39100 Bolzano, Italy
{eromanenko,giancarlo.guizzardi}@unibz.it, calvanese@inf.unibz.it
[2] Umeå University, 90187 Umeå, Sweden
[3] University of Twente, 7500 Enschede, The Netherlands

Abstract. Ontology-driven conceptual models are widely used to capture information about complex and critical domains. Therefore, it is essential for these models to be comprehensible and cognitively tractable. Over the years, different techniques for complexity management in conceptual models have been suggested. Among these, a prominent strategy is model abstraction. This work extends an existing strategy for model abstraction of OntoUML models that proposes a set of graph-rewriting rules leveraging on the ontological semantics of that language. That original work, however, only addresses a set of the ontological notions covered in that language. We review and extend that rule set to cover more generally types of objects, aspects, events, and their parts.

Keywords: Conceptual model abstraction · Complexity management of conceptual models · OntoUML

1 Introduction

The term conceptual model (CM) is heavily overloaded and covers a wide range of models, e.g., Entity-Relationship diagrams and Business Process Models. These models are used to capture information about complex and critical domains, and "play a fundamental role in different types of critical semantic interoperability tasks" [14].

Conceptual modeling is the activity of "representing aspects of the physical and social world for the purpose of understanding and communication [...] among human users" [19]. However, when ontological theories, coming from areas such as formal ontology, cognitive science, or philosophical logics, are utilized for improving the development of CMs, it is common to speak of ontology-driven conceptual modeling (see [7,22]).

In critical and complex scenarios, the number of concepts and axioms of a CM can grow significantly, leading to situations where "it is important that conceptual models are cognitively tractable" [6]. It is known that human working memory capacity in processing visual information is limited [15], and "displaying

© The Author(s), under exclusive license to Springer Nature Switzerland AG 2022
R. Guizzardi et al. (Eds.): RCIS 2022, LNBIP 446, pp. 372–388, 2022.
https://doi.org/10.1007/978-3-031-05760-1_22

a large amount of data in a single node-link diagram can be visually overwhelming and confusing" [15]. Thus, one of the most challenging aims is "to understand, comprehend, and work with very large conceptual schemas" [23].

Due to the above-mentioned reasons, complexity management of large conceptual models has been an area of intensive research. Recently published methods can be grouped into the following categories: clustering methods, relevance methods, and summarization methods [23, p.54]. The first group covers methods in which elements of the CM are divided into groups (clusters). Relevance methods rank CM elements into ordered lists according to their value, while summarization methods produce a reduced version of the original CM.

The work presented in this paper belongs to the group of summarization techniques, more specifically, to the abstraction techniques. According to Egyed [5], model abstraction is "a process that transforms lower-level elements into higher-level elements containing fewer details on a larger granularity". The main idea is to provide the user with a bird's-eye view of the model by filtering out some details. Thus, such methods by definition provide lossy transformations.

Egyed also suggested an interesting approach for model abstraction [5]. The proposed abstraction algorithm performs syntactic matching of abstraction rules on the model, and the matched pattern is replaced by the result pattern of that rule. The author claims that every application of a rule simplifies a given model. However, the suggested set of rules is complicated, and it consists of 121 patterns, 92 of which are abstraction-generating rules.

Since most of the methods for CM summarization are based on classic modeling notations (UML, ER) [23, p.44], they rely on syntactic properties of the model, such as closeness or different types of distances between model elements (see [1]). What is interesting here is that abstraction techniques even for languages with ontological semantics sometimes are based mainly on topological properties of the graphs, see [16,20].

More recently, a number of approaches for complexity management have been proposed for Ontology-Driven CM languages—most notably OntoUML [9]—by leveraging on the richer ontological semantics offered by these languages. These include [6,12,18]. The latter deals exactly with the topic of model abstraction and proposes a set of graph-rewriting rules for abstracting OntoUML patterns. However, it targets only a subset of the ontological notions present in OntoUML. For example, it did not cover any other relation but specialization, while recently OntoUML stereotypes for relations have been revised and extended (see [7]). Moreover, the technique is focused on the category of objects, leaving out categories of aspects and events and, hence, also the relations between events and endurants (objects and aspects). In this paper, we revisit and extend that work by proposing new abstraction graph-rewriting rules that more generally address types of objects, aspects, events, and their parts.

The remainder of the paper is organized as follows: Sect. 2 presents our baseline and background; Sect. 3 introduces our new abstraction rule set, which is combined with the original rule set in an unified algorithm in Sect. 4; Sect. 5 elaborates on final considerations and future work.

2   Background

2.1   A Brief Introduction to UFO and OntoUML

Unified Foundational Ontology (UFO) [9,11] is a well-grounded foundational ontology based on contributions from Formal Ontology in Philosophy, Philosophical Logic, Cognitive Psychology, and Linguistics. A quite recent study [21] has shown that UFO is the second-most utilised foundational ontology applied in conceptual modeling, and also the one with the fastest adoption rate.

OntoUML is an ontology-driven conceptual modeling language that extends UML class diagrams by defining a set of stereotypes that reflect UFO ontological distinctions into language constructs. These stereotypes extend UML's metamodel via a profile mechanism, and allow users to decorate diagram's elements. Thus, classes and associations that are decorated with OntoUML stereotypes bring precise (real-world) semantics grounded in the underlying UFO ontology. Additionally, a number of semantically motivated syntactic constraints govern OntoUML models, making them compliant with UFO [9]. Verdonck and Gailly also have shown that OntoUML is among the most used languages in ontology-driven conceptual modeling [21]. In the remainder of this section, we briefly describe a selected subset of the ontological distinctions in UFO, and how they are represented by means of OntoUML. For an in-depth discussion, philosophical justification, and formal characterization we refer to [9,13].

A first key distinction is made between *Endurants*, i.e., entities that exist in time while keeping their identity (e.g., John, Mary and their Marriage), and *Perdurants*, i.e., exist that occur in time, also generally called *Events* (e.g., John and Mary's Wedding ceremony). Endurants instantiate *Endurant Types*, which depending on their modal properties can be classified into *Ultimate Sortals*, *Subkinds*, *Roles*, *Phases*, *Categories*, *Role Mixins*, *Phase Mixins*, and *Mixins*. Perdurants instantiate *Perdurant Types*.

Ultimate Sortals are types that capture the essential properties ascribed to their instances. The term is a synonym to what is called a (natural) kind in the literature. OntoUML reserves the stereotype *Kind* for *Object* kinds (e.g., Car, Person, Planet) but also allows *Relator* kinds (e.g., Marriage, Employment, Enrollment), *Quality* kinds (e.g., Color, Weight), and *Mode* kinds (e.g., That conductor's electrical conductivity, Mary's ability to speak English, John's Dengue Fever). *Objects* are endurants that can exist independently, while *Relators*, *Modes* and *Qualities* (generally called *Aspects*) can only exist parasitic to other entities (existential dependence): *Relators* are the truthmakers of relational propositions, e.g., a Marriage is a complex relator composed of mutual commitments and claims; *Qualities* are entities that are existentially depend on their bearers and can be directly measured; *Modes*, on the contrary, represents complex intrinsic aspects that can bear their own properties.

Subkinds are subtypes specializing ultimate sortals, like Man and Woman for the kind Person, or Temporary Employment for the relator kind Employment. Ultimate sortals are static (or *Rigid*) types in the sense that they classify their instances necessarily (in the modal sense). For this reason, they are called *Rigid*

Sortals. Phases and *Roles* represent contingent (accidental, dynamic) subtypes of rigid sortals. They are hence called *Anti-Rigid Sortals*. The former class represents specializations according to intrinsic properties of instances, like being a `Child` or an `Adult` for `Person`, while the latter is the way for endurants to participate in relations, e.g., play the role of `Wife` in a `Marriage`. Rigid and anti-rigid sortals constitute the class of *Sortal Types*.

Non-Sortal Types capture properties that are common to instances of different kinds (ultimate sortals). These can be essential properties, in which case these are called *Categories*; non-essential properties, in which case these are called *Mixins*; contingent properties cross-cutting several types, in which case these are called *Role Mixins* and *Phase Mixins*. An example of a role mixin is the type `Customer` that describes properties that apply to both individuals of the kind `Person` and individuals of the kind `Organization`.

Events bear different relations to endurants. On the one hand, they can create, change, or terminate them. On the other hand, events are manifestations of aspects, like when the `unfolding of John's Dengue Fever` (an event) manifests a `particular set of pathological conditions in his body` (a mode). In this case, by being the bearer of this mode, John *participates* in that corresponding dengue fever unfolding event. *Event Types* can also specialize other event types, thus, forming their own taxonomies (e.g., `Cardiovascular Surgery` is a subtype of `Surgery` event).

2.2   Ontology-Based Model Abstraction

The approach presented in [12] is made possible by the ontological semantics of OntoUML. The approach proposed an algorithm that is based on four graph-rewriting rules, `R1`–`R4` (see Table 1, reproduced from [12]). Following these rules, fragments of the original model are abstracted into reduced counterparts, while maintaining essential information. The rationale behind these rules is that the focus should be on the representation of *Object* kinds and relations between them, so that: *(1) Relators* are to be omitted; *(2)* properties of *Non-Sortals* (e.g., *Categories*) should be moved downwards to the level of *Sortals*; and *(3)* properties of *Sortals* that are not *Object* kinds, such as *Subkinds* and *Roles*, should be pushed upwards to the level of kinds. Given this rationale, the rules are meant to be applied in a specific order.

As previously discussed, there is the need to revise and extend this approach. First, it only covers subtyping relations. Second, it neglects both taxonomies of *Aspect Types* and *Event Types*. Given that latter are not addressed, so are also the relations between *Events* and *Endurants*.

3   Abstracting Objects, Aspects, Events, and Their Parts

In the next three subsections, we introduce our new rules for abstracting types of *Objects*, *Aspects*, and *Events*, respectively. Each subsection describes first part-hood relationships and then other ontological relations involving these types of

Table 1. Original graph-rewriting rules for OntoUML model abstraction [12].

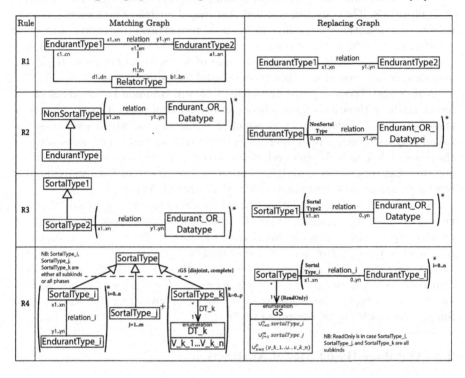

entities, the summary of which is given in Fig. 1[1]. The section ends with a list of graph-rewriting rules.

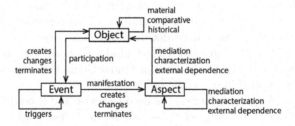

Fig. 1. *Types* and stereotyped relations between them.

---

[1] A formal definition of these stereotyped relations is given in [7] and in [2] for those relations that include *Events*.

3.1   Abstracting Objects

In the previous version of the algorithm [12], parthood relations were not considered. Despite the apparent simplicity, mereological relationships, also known as part-whole or part-of relationships, come in different forms, which behave differently in practice. An interesting research question is whether this diversity leads to different rules when abstracting such relationships and, if yes, how these rules could be formalized.

There are many mereological systems described in logic, philosophy and related fields. These systems differ depending on what axioms are included. However, the "minimal characterization of parthood relation", P, is provided by three axioms [4, Ch. 2] amounting to is termed *Minimal Mereology*:

$$\forall x P(x, x) \qquad \text{(Reflexivity)}$$

$$\forall x \forall y ((P(x, y) \wedge P(y, x)) \to x = y) \qquad \text{(Antisymmetry)}$$

$$\forall x \forall y \forall z ((P(x, y) \wedge P(y, z)) \to P(x, z)) \qquad \text{(Transitivity)}$$

However, in real-world scenarios transitivity does not always hold (see [10]). As an example, we model the case where hearts of football players can fail during a game, thus, leading to a surgery[2] (see Fig. 2)[3]. When a player as part of a team scores a goal, the whole team scores a goal. In contrast, although a football player (as a person) has a heart, it seems rather odd to speak about the heart (being part) of a football team (at least in the biological sense).

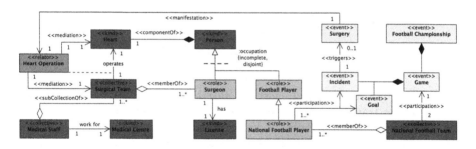

Fig. 2. Examples of different parthood relations.

---

[2] Unfortunately, incidents like this from time to time happen in real life, e.g., https://www.webmd.com/heart-disease/news/20210614/danish-soccer-player-suffered-cardiac-arrest-during-euro-match.

[3] Hereinafter, the default cardinality constraints '*' and '0..*' are not shown in the models, as well as '1' on the side of diamonds in parthood relations. For visual economy, when we have more than one part type connected to the same whole type, we join these different parthood relations in the same diamond-head (on the end connected to the whole). Parthood, nonetheless, is still defined as a binary relation.

In computer science, UML standard distinguishes between an aggregation and a composite aggregation, where the latter is a strong form that "requires a *part object* to be included in *at most one composite object* at a time. If a composite object is deleted, all of its part instances that are objects are deleted with it." [3, p. 112] Also, "compositions may be linked in a directed acyclic graph with *transitive deletion* characteristics" [3]. This standard approach does not provide enough support for the modeller to distinguish different types of parthood relationships but addresses the problem of possible independence of parts.

As for OntoUML, different types of parthood relations between *Objects* and the problem of their transitivity were considered in [9,10]. According to the specification of UFO [9,11], there are three types of *Objects* that can participate in parthood relations: *Functional Complexes*, *Collectives*, and *Quantities*.

Quantities are connected to their parts with the *SubQuantityOf* relation, which is always transitive [9, p. 184] but quite rare in practice. Since the relationship between the portion and the whole is very close, these quantities coalesce their mass without additional attributing. And because of transitivity, a relation in which a portion serves as domain can be abstracted to the whole with the proper role. As an example, we can take `Alcohol` as a sub quantity of `Wine` contained in a `Wineglass`. After abstraction we obtain that `Wineglass`, or even `Glass` when abstracting further, contains `Wine` (see Fig. 3).

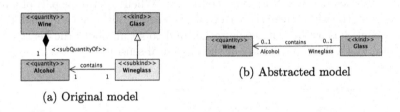

(a) Original model   (b) Abstracted model

Fig. 3. Abstracting *SubQuantityOf*.

A *Collective* can be part of another *Collective* via the *SubCollectionOf* relation, which is transitive [9, p. 186]. Thus, `Surgical Team` from Fig. 2 should receive all properties from `Medical Staff`, e.g., an employment contract with a `Medical Centre`. *Collective* also includes elements with the *MemberOf* relation, which on the contrary is intransitive [9, p. 185].

While formulating the abstraction rules for *Collectives*, we need to take into consideration the following remarks. First, members of the collection by definition of the relation have a uniform structure and are conceived as playing the same role in the collection, like a `Player` in a `Team`. If it is not the case, and we want, e.g., to distinguish between `Forward` and `Goalkeeper`, additional subcollections shall be introduced, each of which with uniform members. Second, "although member-collective is never transitive, a combination of member-collective and subcollective-collective is again *always transitive*" [10], i.e., in our example, a `Surgeon` as part of the collective also works for a `Medical Centre`.

Finally, "the member-collective relation necessarily causes the part to be seen as atomic in the context of the whole, hence, 'blocking' a possible transitive chain of part-whole relations" [9]. Thus, chains of *Collectives* could be abstracted to the most general collection together with the propagation of all relations in which these subcollections are domains. However, members of the collections must be kept because of their atomicity and relative independence. In the example from Fig. 2, this approach gives us `Medical Staff` that works for `Medical Centre`, operates `Heart` (as `Surgical Team`), and has `Surgeons` as members.

Functional Complexes include their parts with the *ComponentOf* relation, which is not transitive in general [9, p. 183]. The simplest approach is to use these parts as attributes of the whole (see `heart` of `Person` in Fig. 4).

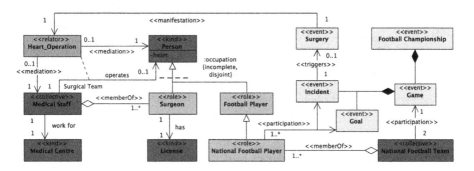

Fig. 4. Abstracting parthood relations of *Objects*.

However, there is no general rule for abstracting relations in which these parts serve as domains. In fact, the only relation that also holds for the whole object is `<<participation>>` in the *Event*, e.g., if `Heart` participated in a `Surgery`, a `Person` with that heart also participated in the same `Surgery`[4].

Abstracting from other relations that bind parts of *Objects* to other *Objects* is more challenging. As an example we can consider `Tail` as a component of `Dog`, which is used for `Greeting`. Abstracting these relations to the `Dog` 'used for' `Greeting` is definitely wrong. Instead of removing such relations, we suggest to rename them using the pattern '*Object's part RelationName*'. In the given example, that would result in `Dog` connected to `Greeting` via the relation 'Dog's tail used for'. Similar examples can be given for the `<<comparative>>` and `<<historical>>` stereotypes.

It should be noted, that so far we did not discuss situations where part of the *Object* serves as range of the relation coming from *Aspect* or *Event*. These rules with proper substantiation are considered in the corresponding sections.

After abstracting from parthood relations, further abstractions of *Objects* can be made with the help of rules from the previous version of the algorithm [12], namely, Rules R2–R4 can be applied.

---

[4] This is called the *Principle of Event Expansion* [17].

3.2 Abstracting Aspects

The first rule suggested in [12] for abstracting models was R1, the rule for eliminating *Relators*. We argue that this rule can be further extended towards other *Aspects* of relations. As an example, we use the model in Fig. 5.

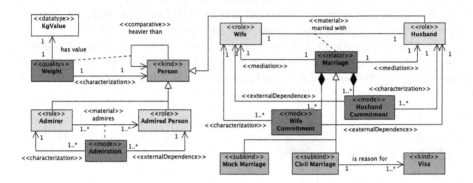

Fig. 5. Different *Aspects* of relations.

As previously discussed [8,9], *Relators* are the truthmakers of relational propositions, and their absence may lead to single-tuple/multiple-tuple cardinality ambiguity problem. Therefore, by applying R1, we have a *lossy transformation*.

What was omitted in [12] are the following features. First, *Relators* as *Endurants* could be embedded into a hierarchy and have *Subkinds* or *Phases* (an example can also be found in [8]). Second, *Relators* are complex objects in terms of how they are formalized as mereological sums of externally dependent *Modes* (for details see [7]). Third, *Relators* or their descendants may participate in other relations other than mediation relations with the relata. In the given example, Civil Marriage can serve as a reason for a spouse's Visa, while Mock Marriage cannot.

The last version of UFO distinguished *Relators*, *Qualities*, and *Modes* within the *Aspect Types* [11]. We argue that pattern for abstracting should be the same for all pre-described classes. Also, the above mentioned comments about the *Relators* can be applied to *Qualities* and *Modes* as well (see Fig. 7).

The first step when abstracting *Aspects* is addressing aspect parthood. All parts of an *Aspect* follow classical mereology axioms that were mentioned in Sec. 3.1. Also, because of transitivity, it is possible to move relations from a part of the *Aspect* to the whole. However, in most cases, like the one in Fig. 5 where Wife Commitment and Husband Commitment are parts of Marriage, a relation could already exist[5] and there is no need to create a new one.

---

[5] For details, why this is the case, see Table 1 in [7].

The second step is to abstract from aspect taxonomies by applying a modification of the original R3 rule, since we are not interested in creating an enumeration like in R4 even for disjoint and complete generalization sets because the transformation is lossy. The modification is the following. First, instead of applying R3 to *Sortal Type* (meant there as Sortal Object Type), we define an analogous rule for *Sortal Aspect Type*. Second, we need to keep the available role that is lowest in the hierarchy. In the example from Fig. 5, that leads to `Marriage` (as `Civil Marriage`) being a reason for a `Visa`.

The last step consists of abstracting from the *Aspect Type* itself. However, it can participate in the relations, decorated with <<externalDependence>>, <<mediation>> and <<characterization>> stereotypes, representing different sorts of existential dependencies (for details see [7]). Also, if the *Aspect Type* participates in the relations as the domain, those relations should be moved to the corresponding *Object Types* (i.e., those that are ranges for <<mediation>> and <<characterization>> relations only) with the corresponding modification of the relation as the '*Object's AspectRole RelationName*'. The resulting model for our example is shown in Fig. 6.

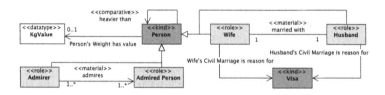

Fig. 6. Resulting model for abstracting *Aspects*.

The last rule gives us a reason to move <<externalDependence>> and <<characterization>> with <<mediation>> relations from part of an *Object* to the whole, even for intransitive parthood relations, but with some priority. So, <<mediation>> and <<characterization>> relations, being relations of *inherence* (i.e., a stronger for existential dependence), are more important to keep than <<externalDependence>>. Thus, in Fig. 4, `Heart Operation` mediates `Medical Staff` and `Person`.

Finally, *Aspects* can have relations with other *Aspects*. See, e.g., Fig. 7, where `Redirected Destination Intention` characterizes another *Mode* `Walk`. Because of that, the three above-mentioned steps, namely *(1)* abstracting from parthood, *(2)* abstracting from taxonomic relations, and *(3)* abstracting from the *Aspect Type* itself, should be applied iteratively. The result for our example is given in Fig. 8.

3.3   Abstracting Events

Event mereology is quite extensively described in [2]. Here we also take that *Event* may be composed of other *Events*. This parthood relation is the standard

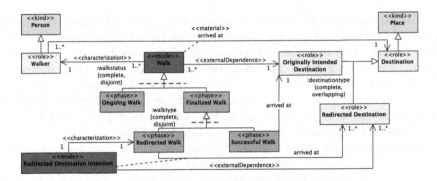

Fig. 7. Relations between *Aspects* (based on model from [11]).

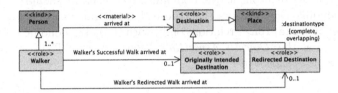

Fig. 8. Result of abstraction from *Aspects*.

one and obeys the three previously mentioned axioms. Thus, we can follow the same approach as for *Aspects* and abstract parts of the *Event* to the whole. Taking into account these arguments, in our example we can abstract the five *Events* in Fig. 4 to `Surgery` and `Football Championship` only.

The next aspect concerning the abstraction of *Events* is dealing with relations. Earlier we have mentioned that if part of an *Object* participates in an *Event*, the whole *Object* also inherits this relation. However, [2] mentioned that the opposite is also true, and when an *Object* participates in part of an *Event* it also participates in the whole *Event*. In the given example, if a `Football Team` participates in at least one `Game` of the `Football Championship`, it participates in the whole `Championship` (in the sense that Pelé played the 1962 World Cup by just playing a few games).

In some of the relations, *Events* can be associated with domains. First, we claim that a `<<triggers>>` relation, where both the domain and the range are *Events*, can be moved safely to more general *Events*.

Second, `<<creates>>` and `<<changes>>` relations from an *Event* to part of an *Object* or *Aspect* can also be abstracted to the whole type. However, the `<<terminates>>` relationship is trickier, because we can terminate the whole only if this part is an *essential* part[6]. In OntoUML notation, this is expressed as an `essential` tagged value decorating that parthood relation. Coming back to the example with `Person`, one can conclude that the termination of the `Brain`

---

[6] For the definition of essential and inseparable parthood, we refer to [9].

would result in the termination of the **Person** as a whole, but at the same time some other organs could be not strictly required.

Finally, *Events* are always manifestations of *Aspects* (or their aspect parts). Coming back to the example in Fig. 5, we can imagine the **Wedding** *Event*, which would explicitly manifest **Fiancée Commitments** and **Fiancé Commitments**. Abstracting from *Aspects* should lead to moving this manifestation to the proper *Relator*, in this case **Engagement**. As a consequence, we also have that the relata at hand participate in this **Wedding**, thus, **Fiancée** and **Fiancé** have to participate in their **Wedding**.

Events can also form taxonomies. E.g., in our first example, instead of **Surgery**, we can consider **Cardiovascular Surgery** as a *Subkind* of the more general **Surgery**. They can also be abstracted with the same rule suggested in the previous section, but now with *Event Type*. However, contrary to *Aspects*, *Events* are not completely abstracted at the end, so a modification of Rule **R4** is suggested in which all *Subkinds* of an *Event* form an enumeration.

Taking into account all these considerations, in our example, the model in Fig. 2 can be abstracted to the one in Fig. 9.

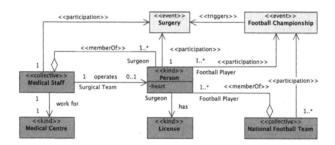

Fig. 9. After abstracting *Events*, *Aspects* and *Objects*.

3.4   Combining Abstraction Rules

Table 2 provides a summary of all previously suggested abstraction rules.

The first set of rules, **P**, is responsible for abstracting from parthood relations. The first variant of the rule is applicable to parthood relations with transitive properties (called **partOf** in general) and it can be applied to *Objects*, *Aspects*, and *Events*. The only exception there is w.r.t. the relations is termination. If we suppose that **Alcohol** is an essential part of **Wine**, its exhalation would lead to the 'end' of **Wine**. Also, if there was already a relation between a whole and another type, there is no need to create a new one.

All other rules from this set are applied to *Object Types* only and may be applied together to the same type, i.e., they are not mutually exclusive, but could work as a supplement to each other. The difference between them is in the *Type* that has a relation with the part of the *Object* at hand.

The second set of rules, **H**, is responsible for abstracting from hierarchies. The previously defined Rules **R2–R4** are still applicable to *Objects*, but two new

Table 2. Graph-rewriting rules for OntoUML model abstraction

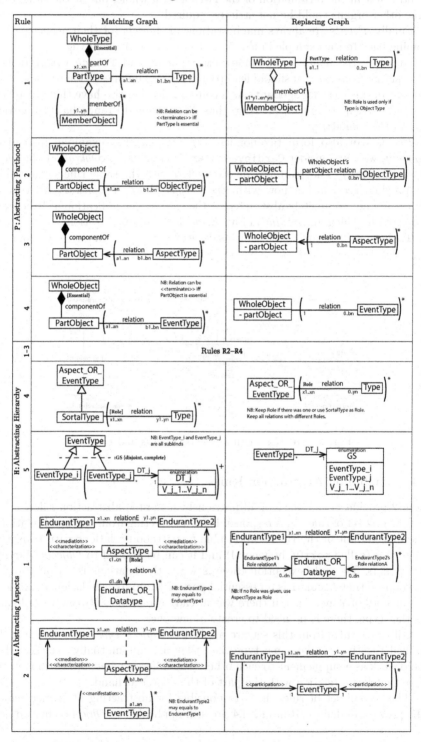

rules were introduced. Rule H4 is applicable both to *Aspect Types* and to *Event Types*. In contrast to the rules from the first set, all relations with different role names should be kept. Rule H5 keeps an enumeration of all *Subkinds* of *Events* and can be used together with Rule H4.

The last set of rules, A, is responsible for abstracting *Aspect Types* and explicitly representing the participation of *Endurant* in *Events* that manifest these corresponding aspects. Again, these two rules could work in tandem.

4 Towards Ontology-Based Model Abstraction 2.0

As we previously mentioned, the work of [12] was focused on *Object Types* and restricted to subtyping relations between these types. Our revised proposal not only takes into account different *Types*, but also considers parthood relations and some other stereotyped relationships.

Considering the original proposal, the only rule that was completely reworked is Rule R1. Rules R2–R4 are still applied to *Object Types*, although, *Aspect Types* and *Event Types* were also addressed by new rules for abstracting their taxonomies.

What is more important, in contrast to the original proposal, there isn't a single strict order in which rules must be applied.

Thus, we here suggest two variants of rule's ordering that seem meaningful, namely, a *Parallel* (Listing 1) and an *Iterative* (Listing 2) versions of our new abstraction algorithm.

Listing 1. Parallel version of the abstraction algorithm

```
// abstract from all parthood relations:
  repeat: apply P1-P4;
// abstract from hierarchies of Aspects and Events:
  repeat: apply H4-H5;
// abstract from Aspects:
  repeat: apply A1-A2;
// abstract from hierarchies of Objects:
  repeat: apply H1-H3.
```

Listing 2. Iterative version of the abstraction algorithm

```
// abstract from Aspects:
  repeat: apply P1 to Aspects;
          apply H4 to Aspects;
          apply A1-A2;
// abstract from Events:
  repeat: apply P1 to Events;
          apply H4-H5 to Events;
// abstract from Objects:
  repeat: apply P1-P4 to Objects;
          apply H1-H3.
```

The first version of the algorithm, shown in Listing 1, suggests to abstract from all *Types* almost simultaneously. Abstraction cannot be done in a completely simultaneous manner, since we are not allowed to abstract from taxonomies of *Object Types* before we abstract from *Aspect Types*. The second version, shown in Listing 2, abstracts instead in sequence, first from *Aspect Types* and then from *Event Types*, postponing as much as possible abstraction over *Object Types*.

Determining which version of the algorithm gives better results by producing more meaningful models requires further investigations, which could be aimed at answering the following questions: *(1)* Is there any difference in preferring one algorithm over the other between novel and experienced users? *(2)* Is there any difference in preferring one algorithm over the other depending on the domain and, consequently, on the characteristics of the CM? These empirical questions will be addressed in future work.

5   Final Considerations

In this paper, we have present an extended version of a tested ontology-based model abstraction for ontology-driven conceptual models. Contrary to the original algorithm, our proposal is able to deal more generally with types of Objects, Aspects, Events, and their Parts. We have suggested eight graph-rewriting rules that in compliance with the previously designed Rules R2–R4 can automatically produce an abstracted version of a complex conceptual model. These rules were designed so as to guarantee that they preserve the essential information of the original model by leveraging on the ontological semantics of OntoUML, while simplifying that model.

The research presented here is under active development[7]. As a further goal of the project, we intend to develop a tool that would allow users to interact with their models at different levels of abstraction, thus, making these models more comprehensible. The alternative versions of the algorithm proposed here assume that there is also a possibility of making the abstraction process more user-centred by taking into account the user's current goals and intentions.

Another hypothesis that should be checked is whether the proposed abstractions could help finding errors within the models, i.e., when abstracting leads to unexpected results, this could be a sign of the presence of *anti-patterns* hidden in the original model.

A repository with OntoUML models is under development[8]. We foresee to test different version of our model abstraction algorithm over the models in this repository.

---

[7] The interested reader can refer to the current version of the Visual Paradigm plugin with supported abstraction functionality on https://github.com/mozzherina/ontouml-vp-plugin.git, and the corresponding server on https://github.com/mozzherina/ontouml-server.git.

[8] https://github.com/unibz-core/ontouml-models.

References

1. Akoka, J., Comyn-Wattiau, I.: Entity-relationship and object-oriented model automatic clustering. Data Knowl. Eng. **20**(2), 87–117 (1996). https://doi.org/10.1016/S0169-023X(96)00007-9
2. Benevides, A., Bourguet, J.R., Guizzardi, G., Peñaloza, R., Almeida, J.: Representing a reference foundational ontology of events in SROIQ. Appl. Ontol. **14**, 1–42 (2019). https://doi.org/10.3233/AO-190214
3. Cook, S., et al.: Unified modeling language (UML) version 2.5.1. Standard, Object Management Group (OMG) (2017). https://www.omg.org/spec/UML/2.5.1
4. Cotnoir, A.J., Varzi, A.C.: Mereology. Oxford Scholarship Online, 1 edn. Oxford University Press, Oxford (2021). https://doi.org/10.1093/oso/9780198749004.001.0001
5. Egyed, A.: Automated abstraction of class diagrams. ACM Trans. Softw. Eng. Methodol. **11**(4), 449–491 (2002). https://doi.org/10.1145/606612.606616
6. Figueiredo, G., Duchardt, A., Hedblom, M.M., Guizzardi, G.: Breaking into pieces: an ontological approach to conceptual model complexity management. In: Proceedings of the 12th International Conference on Research Challenges in Information Science (RCIS), pp. 1–10 (2018). https://doi.org/10.1109/RCIS.2018.8406642
7. Fonseca, C.M., Porello, D., Guizzardi, G., Almeida, J.P.A., Guarino, N.: Relations in ontology-driven conceptual modeling. In: Laender, A.H.F., Pernici, B., Lim, E.-P., de Oliveira, J.P.M. (eds.) ER 2019. LNCS, vol. 11788, pp. 28–42. Springer, Cham (2019). https://doi.org/10.1007/978-3-030-33223-5_4
8. Guarino, N., Guizzardi, G.: "We need to discuss the *Relationship*": revisiting relationships as modeling constructs. In: Zdravkovic, J., Kirikova, M., Johannesson, P. (eds.) CAiSE 2015. LNCS, vol. 9097, pp. 279–294. Springer, Cham (2015). https://doi.org/10.1007/978-3-319-19069-3_18
9. Guizzardi, G.: Ontological foundations for structural conceptual models. CITIT PhD.-thesis series 05–74 Telematica Instituut fundamental research series 015, Centre for Telematics and Information Technology, Enschede (2005)
10. Guizzardi, G.: The problem of transitivity of part-whole relations in conceptual modeling revisited. In: van Eck, P., Gordijn, J., Wieringa, R. (eds.) CAiSE 2009. LNCS, vol. 5565, pp. 94–109. Springer, Heidelberg (2009). https://doi.org/10.1007/978-3-642-02144-2_12
11. Guizzardi, G., Benevides, A.B., Fonseca, C.M., Porello, D., Almeida, J.P.A., Sales, T.P.: UFO: unified foundational ontology. Appl. Ontol. **17**(1), 1–44 (2021). https://doi.org/10.3233/AO-210256
12. Guizzardi, G., Figueiredo, G., Hedblom, M.M., Poels, G.: Ontology-based model abstraction. In: Proceedings of the 13th International Conference on Research Challenges in Information Science (RCIS), pp. 1–13. IEEE (2019). https://doi.org/10.1109/RCIS.2019.8876971
13. Guizzardi, G., Fonseca, C.M., Almeida, J.P.A., Sales, T.P., Benevides, A.B., Porello, D.: Types and taxonomic structures in conceptual modeling: a novel ontological theory and engineering support. Data Knowl. Eng. **134**, 101891 (2021). https://doi.org/10.1016/j.datak.2021.101891
14. Guizzardi, G., Sales, T.P., Almeida, J.P.A., Poels, G.: Automated conceptual model clustering: a relator-centric approach. In: Software and Systems Modeling, pp. 1–25 (2021)
15. Huang, W., Luo, J., Bednarz, T., Duh, H.: Making graph visualization a user-centered process. J. Visual Lang. Comput. **48**, 1–8 (2018). https://doi.org/10.1016/j.jvlc.2018.07.001

16. Kondylakis, H., Kotzinos, D., Manolescu, I.: RDF graph summarization: principles, techniques and applications. In: Proceedings of the 22nd International Conference on Extending Database Technology (EDBT), pp. 433–436 (2019). https://doi.org/10.5441/002/edbt.2019.38
17. Lombard, L.B.: Events: A Metaphysical Study. Routledge, Abingdon (2019)
18. Lozano, J., Carbonera, J., Abel, M., Pimenta, M.: Ontology view extraction: an approach based on ontological meta-properties. In: IEEE 26th International Conference on Tools with Artificial Intelligence, pp. 122–129 (2014). https://doi.org/10.1109/ICTAI.2014.28
19. Mylopoulos, J.: Conceptual modeling and Telos. In: Conceptual Modelling, Databases and CASE: An Integrated View of Information Systems Development. Wiley (1992)
20. Pouriyeh, S., et al.: Ontology summarization: graph-based methods and beyond. Int. J. Semant. Comput. **13**(2), 259–283 (2019). https://doi.org/10.1142/S1793351X19300012
21. Verdonck, M., Gailly, F.: Insights on the use and application of ontology and conceptual modeling languages in ontology-driven conceptual modeling. In: Comyn-Wattiau, I., Tanaka, K., Song, I.-Y., Yamamoto, S., Saeki, M. (eds.) ER 2016. LNCS, vol. 9974, pp. 83–97. Springer, Cham (2016). https://doi.org/10.1007/978-3-319-46397-1_7
22. Verdonck, M., Gailly, F., Pergl, R., Guizzardi, G., Souza, B.F.M., Pastor, O.: Comparing traditional conceptual modeling with ontology-driven conceptual modeling: an empirical study. Inf. Syst. **81**, 92–103 (2019). https://doi.org/10.1016/j.is.2018.11.009
23. Villegas Niño, A.: A filtering engine for large conceptual schemas. Ph.D. thesis, Universitat Politècnica de Catalunya (2013)

Understanding and Modeling Prevention

Riccardo Baratella, Mattia Fumagalli, Ítalo Oliveira[✉],
and Giancarlo Guizzardi

Conceptual and Cognitive Modeling Research Group (CORE),
Free University of Bozen-Bolzano, Bolzano, Italy
{riccardo.baratella,mattia.fumagalli,idasilvaoliveira,
giancarlo.guizzardi}@unibz.it

Abstract. Prevention is a pervasive phenomenon. It is about blocking an effect before it happens or stopping it as it unfolds: vaccines prevent (the unfolding of) diseases; seat belts prevent events causing serious injuries; circuit breaks prevent the manifestation of overcurrents. Many disciplines in the information sciences deal with modeling and reasoning about prevention. Examples include risk and security management as well as medical and legal informatics. Having a proper conceptualization of this phenomenon is crucial for devising proper modeling mechanisms and tools to support these disciplines. Forming such a conceptualization is a matter of Formal Ontology. In fact, prevention and related notions have become a topic of interest in this area. In this paper, with the support of Unified Foundational Ontology (UFO), we conduct an ontological analysis of this and other related notions, namely, the notions of countermeasures and countermeasure mechanisms, including the notion of antidotes. As a result of this conceptual clarification process, we propose an ontology-based reusable module extending UFO and capturing the relations between these elements. Finally, we employ this module to address a few cases in risk management.

Keywords: Prevention · Ontology-driven conceptual modeling

1 Introduction

In conceptual modeling, we need to represent both structural and dynamic aspects of reality. Within the latter, we find the recurrent phenomenon of prevention. Prevention is a pervasive phenomenon. In a nutshell, it is about blocking an event before it happens or interrupting it as it unfolds, as when vaccines prevent the unfolding of diseases, when seat belts prevent happenings that would cause grave injuries, or when the activation of circuit-break blocks an overcurrent from damaging the circuit. Many disciplines in the information sciences deal with modeling and reasoning about prevention. For example, risk and security management is ultimately about prevention: once we identify the assets-at-risk, we want to prevent or mitigate the possible undesired consequences through countermeasures such as barriers and antidotes; medical informatics frequently deals with the modeling of life-threatening events and their deterrents; normative systems and contracts in legal informatics deal with the design of rules to constrain unwanted behaviors. Given the importance of the topic, having a proper conceptualization of this phenomenon is crucial for devising proper modeling mechanisms and tools to support these disciplines.

Ontology-driven conceptual modeling [30,31] is an approach that employs Formal Ontology to analyze and (re)design conceptual models and modeling methodologies, languages, and tools. This approach often relies on foundational ontologies, i.e., domain-independent philosophically-sound axiomatic theories. These ontologies provide an inventory of the most general aspects of reality, including classification and taxonomic structures, part-whole relationships, events, causality, dependence, etc. There is evidence showing that the support of foundational ontologies for building conceptual models helps to avoid bad design [29], and to improve the quality of and interoperability between these models [13,17] as well as between the systems built with their support.

One of the most used foundational ontologies in conceptual modeling is the Unified Foundational Ontology (UFO) [12]. In UFO, the unfolding of events is intimately connected to the notion of *dispositions* [8] (e.g., capacities, capabilities, powers, abilities, liabilities, vulnerabilities, intentions). In a nutshell, events are manifestations of dispositions of objects under certain situations [15]. Moreover, causality, a relation between events, is defined in terms of the activation of certain dispositions in certain situations [15].

In this paper, with the support of the Unified Foundational Ontology (UFO), we conduct an ontological analysis of prevention and related notions, namely, countermeasures and countermeasure mechanisms, including the notion of antidotes. We show how these notions can be grounded on relations between dispositions, and between dispositions and their manifestations. As a result of this conceptual clarification process, we propose an ontology-based reusable module extending UFO and capturing the relations between these elements. This module can then be reused in the design and integration of domain ontologies, conceptual domain models, as well as domain-specific modeling languages (e.g., for the domain of security modeling). Finally, we employ this module to address a few cases in the risk management domain.

The remainder of this paper is structured as follows: Sect. 2 presents a fragment of UFO used as our baseline. We also review and show the insufficiency of prominent standard analysis of dispositions in the literature (the Simple Conditional Analysis); Sect. 3 presents our ontological analysis and culminates with the proposal of a reusable model of prevention capturing the results of this analysis; Sect. 4 shows an application of this model to address some cases in risk management; Sect. 5 discusses related work and, in particular, how the ontological foundations of prevention are articulated in a competing foundational ontology, namely, BFO (Basic Formal Ontology); finally, Sect. 6 presents final considerations, including a discussion about the limitations of our approach, and future works.

2   Background

2.1   The Unified Foundational Ontology (UFO)

The Unified Foundational Ontology (UFO) is a domain-independent axiomatic theory developed to contribute to the foundations of Conceptual Modeling [14]. It is one of the most used foundational ontologies in conceptual modeling [30]. Moreover, it has been successfully employed in a number of projects in different countries, by academic, government and industrial institutions in the development of core and domain ontologies in a number of domains (e.g., Trust, legal relations and Constitutional Law, Risk and Value, Service, Software Requirements and Anomalies, Discrete Event Simulation, etc.) [14]. For these reasons, we chose UFO as a foundation for the work developed here.

UFO makes a fundamental distinction between *endurants* and *events* [12, 15]. Classically, an *endurant* is an entity that is wholly present at any moment at which it exists, being able to undergo qualitative changes while keeping its identity. A person, a car, but also qualities, such as the temperature of an object, are examples of endurants. For instance, Mary, a person, can weigh 60 Kg in some circumstances, and weigh 65 Kg in another; her weight, an individual quality, can vary in different circumstances. In both cases Mary and her weight are the same individuals before and after these changes.

The category of endurant in UFO is specialized in *objects* and (variable) *tropes*. Objects are existentially independent entities (e.g., a person, a car, a rock). Tropes (aka aspects, objectified or reified properties) are endurants that are existentially dependent on other entities (termed their bearers) in the way in which, for instance, the temperature of Mary or her capacity to play volleyball can exist only while Mary exists. A trope is said to *inhere* in an object. Inherence is a functional relation of existential dependence between a trope and its bearer. This notion of trope includes both *qualities* (e.g., the temperature of Mary, the color of a car) and *dispositions* (e.g., a capacity to play volleyball, the electrical conductivity of a material). If an object undergoes a change (e.g., Mary's loss of a limb), it might acquire, lose or have replaced some of its tropes.

Several terms are used to refer to dispositions, both in philosophy and in ordinary language: 'power', 'ability', 'function', 'potency', 'capability', 'tendency', 'potentiality', 'capacity'. There is a vast literature in philosophy on the topic of dispositions and how they are related to events and to standard (categorical) properties (e.g., height, weight, color). Authors range from those that take all properties to be dispositions (e.g., color is the disposition to refract certain light wave ranges) to those that take all properties to be categorical (e.g., water solubility is just a proxy name for a particular crystalline structure). In any case, from a cognitive point of view, it is undeniable that notions like ability, function, liability, capacity, and capability are part of our commonsensical apparatus and of our conceptual modeling toolbox. For these reasons, disposition is taken as a primitive notion in UFO, but also in other foundational ontologies such as BFO [3]. In the former, it has been shown to be an instrumental notion in providing semantics to the notion of capability in enterprise architecture [4] and defense frameworks [20].

Events can be composed of other events, forming a mereological structure: an event is *atomic* iff it has no (proper) parts; otherwise, it is a *complex event* [6,15]. A PhD defense presentation, a party, a security incident, and a natural disaster are examples of events. In contrast with endurants, which are wholly present whenever they are present, events have their parts scattered in different time points. When they are present, only some of their proper parts are present. Furthermore, events are transformations from one piece of reality to another, and this piece is called a *situation* [15]. As a particular configuration of a part of reality, situations can be factual, i.e., obtaining at a particular time point (these are called *facts*). Situations are composed of other individuals [1]. For example, the situation in which I have a fever has both me and my temperature (in a particular state) as parts; the situation in which John is married to Mary has both of them and a particular relational trope bundle (their marriage) as parts.

UFO [15], in pace with Mumford [22,23], assumes dispositions are properties that are only manifested in specific situations via the occurrence of events. For example, the disposition to attract metal of a magnet is only manifested via an event of metal attraction and movement, which is only manifested when a particular type of situation presents itself (e.g., there is a piece of metal, of a proper weight range, at a proper distance, in a surface that has the right friction).

Dispositions connect endurants and events, since the latter are manifestations of objects' dispositions. In a nutshell: a situation of a special kind *activates* a disposition; the disposition is then *manifested by* an event in which the bearer of the disposition participates; that event then *brings about* a new situation. If an event E_1 brings about a situation S that activates the dispositions that are manifested as event E_2, then we say that S *triggers* E_2, and that E_1 *directly_causes* E_2; if E_1 *directly_causes* E_2, and E_2 *directly_causes* E_3, then E_1 *causes* E_3, where *causes* is a strict partial order relation [15]. So *causes* is the transitive closure of *directly_causes* [6].

A situation that triggers an event starts when this event starts, while a situation that is brought about by an event starts when this event ends. In both cases these situations are facts. There is a unique situation that triggers a particular event occurrence, and there is a unique (maximal) situation that is brought about by an event, corresponding to the effects of the event at the moment it ends. Moreover, since it is assumed that a disposition that is being manifested must exist throughout its manifestation, this disposition is not only *present in* the situation that activated it, but it is also present in its manifestation event and in the situation brought about by this event. And, obviously, if the disposition is present in a situation or an event, then the object in which the disposition inheres is also present there (due to existential dependence) [6,15].

Figure 1 (from [2]) summarizes the individuals in UFO and their relations that are relevant for our discussion. Notice that no account for interactions between dispositions is provided, even though it is assumed, for example, that an event (manifesting a disposition) can somehow remove another disposition [6].

In the next section, we extend UFO in this respect by complementing it with a module to address the notion of prevention.

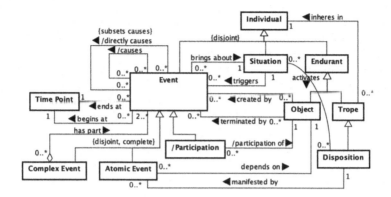

Fig. 1. Individuals in UFO (from [2])

UFO is an ontology that includes both individuals and types. So, objects, dispositions, events, situations instantiate object types, disposition types, event types and situation types, respectively.

Types can be more or less saturated, i.e., they can include in their intention *individual concepts* [12], provided that the pattern of features associated with the types is still possibly repeatable across multiple individuals. A standard (completely unsaturated) type is characterized only by general properties (e.g., the type "Physical Object" is characterized by general properties such as spatial extension, weight, color). On the other extreme, we have individual concepts, which are completely saturated, i.e., they are instantiated by exactly one individual (single reference) and always the same individual (rigid designation). A semi-saturated type is characterized by both general properties as well as individual concepts. For example, the type "President of Italy" includes the general type "President" and the individual concept "Italy", but it is still a type instantiated by many individuals (e.g., Matarella, Ciampi). In particular, semi-saturated situation types will play a role here, e.g., the situation in which I am located in a South Tyrolean City includes a reference to me as an individual[1] but it can be instantiated by situations in which I am in Bolzano, Brunico, etc.

2.2 The Simple Conditional Analysis and Its Insufficiency

Traditionally, the term *Canonical dispositions* is used to refer to disposition ascriptions that make explicit reference to their stimulus conditions and manifestations (e.g., the disposition to dissolve in water). In [8], dispositions, in general, are assumed to be reconceptualizable in terms of their canonical descriptions by

---

[1] It also includes a reference to another semi-saturated type *South-Tyrolean City*.

explicitly identifying their stimulus conditions and their manifestations. This is the first step, originally suggested by Lewis [19], to clarify what means for an object to bear a disposition. The second step would be a conceptual analysis of the resulting canonical disposition.

A intuitive analysis is the so-called *Simple Conditional Analysis*: An object o, at time t, is disposed to manifestations M when in circumstances C iff o would manifest M if it were the case that o is in C at time t.

A known problem with this analysis, however, is the following: it is not *necessarily* (in a modal sense) true that if o were to undergo to C, it would give response M. Sometimes something happens, such that, though o is in C at t, it does not respond with M accordingly [7,19]. In the alleged time gap between the activation of a disposition and its manifestation, something could happen to prevent manifestation M. The literature on dispositions suggests two ways this may occur: if the original disposition is removed from the object, or by tampering with the causal chain that is foreseen to bring about its manifestation.

In the sequel, we explore these two modes of prevention in terms of the ontological categories and relations of UFO: on the one hand, by exploring the relation between events and dispositions (events are always manifestations of dispositions), and between dispositions and situations (dispositions are only activated in certain situations); on the other hand, by exploring the notion of (indirect) causation - between the activation of a disposition and the manifestation of a distal event, there could be a chain of intermediate disposition activation, events manifestations and situations constituting a causal chain [6].

3 Unpacking the Notion of Prevention

3.1 Lifting the Discussion to the Level of Types

In order to advance our proposal for analyzing and modeling prevention, we need to extend UFO in three directions. Firstly, we have to generalize the relations between dispositions, their manifestations (events), and situations to the level of types. Secondly, we need to characterize situation types activating dispositions in terms of general requirements that make particular references to dispositions of complementary types. Thirdly, we have to define a notion of incompatibility between situations of certain types.

Let us first deal with types. One thing is the flammability of a piece of wood, another is flammability as a type of disposition. Flammability as a type can be associated with a type of event: instances of flammability are manifested via *Catching on Fire* events. Conversely, instances of the latter are always manifestations of dispositions of the flammability type. We therefore define a type-level relation of *manifestation* between event types E_T and disposition types D_T such that: $manifestation(E_T, D_T)$ implies that every instance of E_T is a manifestation of an instance of D_T.

Analogously, flammability as a type can be associated with a type of situation via a type-level *activation* relation: instances of flammability are activated by situations of a given type, i.e., they are activated by situations in which there is enough presence of oxygen, the presence of an ignition heat source, etc. This description clearly defines a type as it can be instantiated by a multitude of individual situations, each with different ignition heat sources, different portions of oxygen with different volumes, etc. The same situation type may be connected via activation to dispositions of different types: for example, a situation in which the objects are exposed to high temperature may activate flammability of woody material, but also a number of dispositions in the human body. In contrast, a disposition type is activated (on the type level) by a unique situation type. We define the activation type-level relation between situation types S_T and disposition types D_T symbolized as $activation(S_T, D_T)$.

We can also define a type-level counterpart of the UFO relation *brings about*. We call this relation bringing about of (symbolized as $bringingAboutOf(E_T, S_T)$). This relation, holding between event type E_T and situation type S_T, implies that instances of the former bring about instances of the latter.

A situation type associated via *activation* to a given disposition type must include the presence of other dispositions of a suitable kind. In this example, both oxygen and the ignition heat source have specific dispositions, which together with flammability produce a *Catch on Fire* event. In other words, particular events of catching on fire are not manifestations of flammability only but complex events composed of the interacting manifestation of all these dispositions together[2]. Here are some additional examples of this dependence between dispositions and their manifestation: a given disposition of a particular key to open a certain lock only manifests itself when it is in such a situation that the disposition to be opened is also present (inhering in that lock). The opening event is then a combined manifestation of these dispositions. Though this example involves explicitly the reciprocity of dispositions, in fact all dispositions present this partnership aspect: the disposition of a person to swim under the water can only be manifested with the presence (and manifestation) of the dispositions that inhere in the water; the disposition of an object to roll on certain surfaces can only manifest itself with the presence (and manifestation) of the dispositions of the very surfaces (e.g., the friction of a certain kind). When one disposition is manifested, the others are also manifested, each one producing its particular manifestation, which combines with each other to produce an effect.

To capture this dependence relation between a disposition d of a certain type D_T and other types of dispositions, we introduce here the *Mutual Activation Partner (MAP)* relation. $MAP(D_T, D'_T)$ implies that, in order for instances d of D_T to be activated, it needs the presence in the activating situation of an instance of D'_T so that the manifestations of d are always part of complex events that are also composed of a manifestation of instances of D'_T. As a consequence, we have that any situation type S_T bearing an activation relation to instances of D_T must have in its instances (particular situations) instances of all D'_T associated via MAP to D_T. MAP is a relation of generic dependence and, hence, asymmetric and transitive, i.e., a strict partial order relation. This proposal takes elements

---

[2] For this reason, [21, 23] call events *polygenic* manifestations.

from the notion of *mutual manifestation partnerships* put forth by many authors in the literature [5, 22–24].

Let us now make explicit an additional requirement for situations of type S_T activating dispositions of the type D_T. A situation type S_T activating a particular disposition d is semi-saturated in the following way: its instances must be situations in which d and, hence, its bearer are present. This follows directly from UFO's constraints that the manifesting disposition must be present in the situations preceding and succeeding its manifestation, and from the existential dependence between disposition and its bearer. Similarly, we can define semi-saturated event types. For example, the event of *Nina Simone's singing* is still a type that can be instantiated by multiple occurrences, all of which have Nina Simone as a participant.

Now, let's define a notion of *incompatibility* between situation types. In [16], the authors define the notion of *conflict* between situations: situation s conflicts with situation s' if they cannot obtain concurrently. We here lift this notion to the type-level by defining an incompatibility relation between situation types S_T and S'_T: $incompatible(S_T, S'_T)$ implies that there are no two instances of these two types that obtain in overlapping time intervals. Notice that these situation types are semi-saturated in a proper way, namely, they must share some relevant dispositions or objects. For instance, a situation with a damp match x is compatible with a situation that contains a different dry match x' even if the two situations temporally overlap.

Finally, lifting the discussion to the level of types is necessary for introducing the chances of that certain types of events might happen. Likelihood only inheres in types, not in individuals, i.e., a particular event either occurs or it doesn't and it cannot be repeated in different time points. In contrast, events of certain types can have instances occurring with more or less frequency and likelihood. In [28], the authors define two notions of likelihood that can be adopted here: the *Triggering Likelihood* inheres in a Situation Type, and it refers to how likely a Situation Type will trigger an Event Type once a situation of this type becomes a fact; the *Causal Likelihood* inheres in an Event Type, and it states that, given the occurrence of an event e and a certain Event Type E_T, how likely e will - directly or indirectly - cause another event of type E_T to occur.

3.2   A Model for Prevention and Related Notions

Given that events are complex bundles of dispositions' manifestations, a clear way to block the occurrence of events is somehow interfering with the interacting dispositions (i.e., the mutual activation partners). The notion of canonical dispositions, which specify the stimulus conditions and manifestations, obfuscates this interaction because it focuses on a single disposition under certain conditions. As a consequence, prevention is restricted to cases where we have the removal of dispositions from their bearers.

We argue that prevention must involve some kind of removal of dispositions from the scene (i.e., from the activating situation), but removing it from its bearer is just one way of doing so (let us call it *case a*). An obvious alternative is the removal of the bearer of that disposition from that situation (*case b*). In yet another manner, we can have the removal of a required partner disposition (*case c*) that would otherwise produce the event at hand. A special case of c (*case d*) is when a disposition *d'* of type D'_T (incompatible with a suitable mutual activation partner) is present in that situation (e.g., high humidity inhering in otherwise flammable objects, which are incompatible with a required *dryness* for those object). For example, the catching on fire of combustible object x can be prevented by: removing x's flammability, for instance, by altering the molecular structure of that object (case a), but also by removing x's from a situation where there are the right conditions (e.g., enough oxygen and ignition temperature), or by removing those conditions from that situation (e.g., producing a vacuum eliminating the presence of oxygen and its properties, humidifying the object).

All these cases can be generalized in the following rule: prevention of events of type E_T that are manifestations of dispositions of type D_T occurs when an event of type E'_T brings about a situation of a type S'_T that is incompatible with the situations required to activate instances of D_T, i.e., *manifestation(E_T, D_T)*, *activation(S_T, D_T)*, *bringingAboutOf(E'_T, S'_T)* and *incompatible(S_T, S'_T)*. In other words, an event *e* prevents the occurrence of an event *e'* iff *e* brings about a situation that is incompatible with any situation that could activate the disposition of which *e'* is a manifestation. Lifting this to the type level: *prevention(E_T, E'_T)* implies that the occurrence of events of type E_T brings about situations that are incompatible with the conditions required for the occurrence of events of type E'_T. Bear in mind that these event and situation types are semi-saturated in the proper way, i.e., guaranteeing the presence (co-reference) of the same disposition and bearer. For example, it is the event of *Humidifying object x* that prevents an event of *Catching on Fire of object x*. Obviously, humidifying flammable objects, in general, does not prevent other flammable objects from catching on fire.

This characterization of prevention follows from the MAP relation between disposition types and the consequent constraint that it imposes on the situation types S_T needed to activate a disposition of a given type D_T. As previously explained, a situation *s* of type S_T activating disposition *d* of type D_T must have present therein *d*, its bearer, and instances of dispositions of all types connected to D_T via MAP.

Intuitively, if an event *e* causes an event *e'*, and *e'* prevents events of type E_T, we would be inclined to accept that *e* prevents instances of E_T. This suggests a distinction between *direct and indirect prevention*, analogous to direct and indirect causation. Thinking about this chain of causation also allows us to deal with the case of *antidotes* [7]. As previously discussed: what makes the world tick according to UFO is a sequence of events bringing about situations that activate dispositions that bring about other situations that activate other dispositions, and so on. So, if we have that an event *e* indirectly causes an event of type E_T, we can interfere in the causal chain connecting *e* to events of type E_T. Using the mechanism we described above, we can execute an event of type E'_T that prevents one of the events of type E^i_T that would otherwise occur in that causal chain. For instance, the event in which John drinks poison would cause his death. Looking into the causal chain, we can be more precise: John's drinking of that poison causes a series of biochemical reactions in his body that eventually would cause his death. This can be avoided if John takes an *antidote* in time, i.e., if John ingests a substance that has the disposition, when manifested, of preventing an event in that causal chain.

Antidotes are a particular case of *Countermeasures*. In general, given a disposition d whose manifestations are of type E_T, countermeasures are designed interventions that endow a setting containing d with other dispositions $\{d_1, ..., d_n\}$, whose manifestations prevent any instance of E_T. More specifically, *Countermeasure Mechanisms* are designed such that: they contain dispositions of type D_T, and given the situations of type S_T that would trigger events that would (directly or indirectly) cause instances of E_T, the instances of S_T instead activate the instances D_T whose associated event type prevent E_T. For example, a circuit break contains a disposition to close the circuit in a situation where there is a current above a certain threshold. The manifestation of that disposition of the circuit breaker thus prevents the event of an overcurrent.

Our analysis makes explicit a number of ways in which countermeasures can be designed: (i) we can obviously remove the disposition d whose manifestation we want to avoid (this can be done by removing the object with that disposition from the setting at hand); (ii) we can remove from the scene required activation partners (e.g., produce a vacuum to prevent fires); (iii) we can include in that setting a disposition that is incompatible with a mutual activation partner (e.g., humidifying a flammable object, removing dryness as a required property); (iv) we can design countermeasure mechanisms surrounding the bearer of d, which have the capacity of preventing the manifestation of d.

Although our examples so far focus on disposition of physical objects, the proposed model can also be applied to social examples. For instance, if we want to prevent a theft from happening in a building, we can: remove the vulnerabilities of the access points to the building. Alternatively, we can suitably remove mutual activation partners, namely, the capacity and/or intention inhering in potential burglars. For example, by employing locking materials that are very resistant as well complex to circumvent, plus legal mechanisms that create severe liabilities for offenders, we can at the same time eliminate vulnerabilities of access points, eliminate matching capacities as well as intentions.

Figure 2 below summarizes the discussion of this session. The fragment in blue is an extension to the original UFO fragment proposed in [6, 15] (in black).

4   Prevention Applied to Risk Management

In this section, we apply our proposed model of prevention to reason about cases in risk management. Before doing that, however, let us harmonize our discussion with the vocabulary and concepts of that domain. We do that with the help of the Common Ontology of Value and Risk (COVER) [28]. In a nutshell, for COVER: *value* is a relational property emerging from the relations between certain *capacities* (dispositions) of certain objects (the *value object*) and the goals

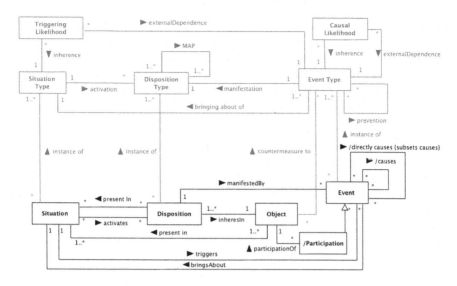

Fig. 2. Extending UFO with type-level relations to model prevention.

of a particular agent. In other words, the value of that object to that agent is the degree to which those capacities can be enacted (as manifestations) to bring about situations that satisfy the goals of the agent. *Risk*, in a sense, is anti-value, or "value with the reverse polarity" having an analogous formulation: risk is a relational property emerging from the relations between the *vulnerabilities* (dispositions) of an *object at risk*, as well as the *threatening capacities* (dispositions), and sometimes intentions (dispositions) of a *threatening entity*. The combined manifestation of these dispositions (vulnerabilities, capacities, intentions) are threat events[3] that can (directly or indirectly) cause loss events, i.e., events that hurt or break the goals of a particular agent. This last point highlights another manner in which value and risk are connected: the object at risk must be a value object to a given agent. In other words, things that have no value cannot be said to be at risk. In a sense, risk is always the risk of destruction of value.

In the sequel section, we employ our model of prevention, as well as the vocabulary put forth by COVER to address exemplars of the four *cases a–d* previously identified:

Application 1 - Therac-25 Accidents [18]: In a known tragic event, the Therac-25, a medical equipment for radiation therapy, malfunctioned and exposed several patients to an excessive and at cases lethal dose of radiation. This was caused by an anomaly in the software controlling the medical equipment. This case appears as an illustration of an UFO-based core Ontology of software risks and anomalies in [9]. Following their analysis, a preventive event of upgrading the software would have eliminated said software anomaly (a propensity to cause Race

---

[3] According to our model, vulnerabilities, capacities and intentions of a certain type are mutual activation partners!.

Conditions, i.e., a disposition) from the program copy installed in that machine. In other words, a proper event of *Software Upgrading* would have prevented threat events of the type *Megavolt X-Ray Activation*, consequent threat events of *High Dosage Radiation Exposure* and, ultimately, loss events of *Patient Death*. This exemplifies a case of prevention by removal of an undesired disposition (the vulnerability of the software copy), i.e., an exemplar of *case a*.

Application 2 - Caged Tiger in a Zoo: This example is frequently used by the community of the Bowtie methodology[4], which is a prominent professional risk assessment methodology [25]. Caging a tiger is an event that brings about a lasting situation in which the tiger, although maintaining its threatening capabilities, is separated from the public (bearers of vulnerabilities). In this case, neither the capacities of the tiger nor the vulnerabilities of members of the public w.r.t. to these capacities are present in the same situation (*case b*). Now, suppose the tiger escapes from its cage. A common countermeasure would be employing a dart projectile to sedate the animal. The correct use of this measure is an event that brings about a situation in which a number of mutual activation partners for the tiger's threatening capacities are removed, namely, the tiger's consciousness and, consequently, its intention to attack (*case c*). In both cases, there is the prevention of threatening events of the type *Attacking the Public*, which in turn prevent loss events of the type *Having members of the Public Dangerously Injured*.

Application 3 - Non-pharmaceutical Interventions: In [10], the authors conduct an ontological analysis of non-pharmaceutical interventions and, in particular, of the semantically overloaded term "Lockdown". As illustrated in Fig. 3 below, a lockdown is a complex form of non-pharmaceutical intervention that potentially aggregates different forms of *Social Distance* Measures as well as different forms of *Travel Restrictions* measures. Lockdowns are, therefore, countermeasure mechanisms (*case d*) intended to dissuade people from (i.e., ideally remove their *intention* of) placing themselves in situations where a vulnerability (to virus contamination) could be manifested - a contagion event is a manifestation of the infectability of the virus collective[5] (capacity to enter human cells, reproduce, elevate the immunological system, etc.) together with the vulnerability of human subjects. In other words, lockdowns are meant (by design) to systematically produce situations in which these dispositions (capacities, on one side, and vulnerabilities, on the other) cannot be both present (*case d*). In contrast, pharmaceutical interventions typically act by endowing human subjects with antidotes (e.g., capacity to destroy the virus before it can enter the cell - in the case of vaccines, or capacity to interrupt the causal chain between infection and the emergence of symptoms - anti-viral medicine). Notice that for all these cases we could assume a probabilistic perspective captured by the *Triggering and Causal Likelihood* properties (see Fig. 2) that can be ascribed to types of preventive events.

---

[4] E.g., https://www.bowtiepro.com/examples/htmlexport/hazardref.htm.

[5] A single virus cannot infect anyone; infectibility is a disposition of a virus collective.

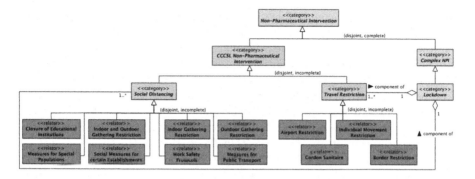

Fig. 3. A model of lockdown (from [10]).

5 Related Work

The concept of disposition as a property of objects appears in several foundational ontologies. The ISO/IEC 21838-2:2021 standard BFO is the one among these that presents the most extensive treatment of dispositions and their interactions (including different types of prevention), and this is why we focus the discussion of this section on it.

Like in UFO, in BFO, a disposition is a dependent entity that inheres in an object and that is realized by events in certain circumstances. However, unlike UFO, BFO also requires that every disposition is grounded on a physical quality of its bearer [3,11] (e.g., in the way *water solubility* is grounded on some complex crystalline structure of sugar[6]). As a consequence, in BFO, in order for an object to lose some disposition, it must lose some quality.

In [11], BFO is used in the construction of the Infectious Disease Ontology (IDO). There, three kinds of interactions between dispositions in BFO are presented: blocking dispositions, complementary dispositions, and collective dispositions.

If D_1 is a disposition and D_2 is a blocking disposition for D_1 (called blocked disposition), then it must be the case that the manifestation of D_2 prevents the manifestation of D_1. This prevention is understood in two ways: (a) *incompatible occurrences*, when the manifestation of D_1 and the manifestation of D_2 are somehow incompatible occurrences, that is, they cannot exist simultaneously or one negatively regulates the other; by definition, a process P_1 *negatively_regulates* a process P_2 when the unfolding of P_1 decreases the frequency, rate, or extent of P_2; (b) *incompatible qualities*, when the manifestation of D_2 makes an object

---

[6] UFO does not make such a commitment. The reasons are related to the fact that dispositions and qualities interact in different levels. For instance, the crystalline structure of sugar is itself grounded on the disposition of the molecules constituting sugar to bind a specific way. Moreover, as discussed before, the distinction between qualitative/categorical and disposition properties is not settled in the literature.

gain a quality that is incompatible with some quality that the same object would have acquired through the manifestation of D_1.

Now, the case of incompatible occurrences in BFO involves one key observation. On the one hand, since occurrences (events) are manifestations of dispositions and, in BFO, dispositions are grounded on physical qualities, the case of incompatible occurrences boils down to a case of *incompatible qualities*. So, (a) above is reducible to case (b). On the other hand, it might be the case that dispositions D_1 and D_2 are compatible, but their manifestations are incompatible due to the manifestation of further disposition D_3 present in the situation at stake (our case d). Consider the following example: Tom has the legal right (a disposition) to vote on candidate A; he also has the right to vote on candidate B. Now, the manifestation of the former prevents the manifestation of the latter (and vice-versa). What makes the manifestations incompatible in this case is not the incompatibility of the respective initial dispositions, but some countermeasure introduced in the situation - viz., rules of preventing double voting.

In our model, all these cases are generalized by having incompatible situations. In fact, the incompatibility of occurrences can be reduced to the incompatibility of their triggering conditions (situation types). Moreover, we advocate that in making dispositional analysis, we are seldom interested in these qualities but in the dispositions that they ground and that are removed scenes. Likewise, when analyzing tropes that must co-occur with a disposition to enable its activation, we are not interested in qualities *per se* but in the proper grounded dispositions (mutual activation partners). For example, in the flammability example, we are not interested in the volume of the oxygen mass presented by on the dispositions that can only inhere on oxygen masses of a certain volume. Further, BFO's treatment of blocking dispositions seems to omit the case where the manifestation of a disposition is prevented by the absence or removal of its bearer from the circumstances that would allow the realization of the disposition.

Complementary dispositions are somehow manifested together in the same process: for example, the functions of hammers and nails, locks and keys. To address this case, the authors proposed that different dispositions form a whole with a collective disposition that is manifested in a single event [11]. A collective disposition is defined as a "disposition inhering in an object aggregate OA in virtue of the individual dispositions of the constituents of OA and that does not itself inhere in any part of OA or in any larger aggregate in which OA is part". For example, the collective capability of two people for lifting together a heavy object; the collective capability of the crowd to do a wave due to the capability of each person to stand up at the appropriate time.

In our approach, the phenomena of blocking, complementary and collective dispositions are generalized in one single framework: complementary and collective dispositions are simply reducible to MAP relation, whereas the blocking disposition can be generalized in terms of incompatible situations.

6   Final Considerations

In this paper, we have presented an ontological analysis of the notion of prevention by relying on and extending the Unified Foundational Ontology (UFO). In particular, we relied on this ontology's notions of events, dispositions, situations, and their ties. By lifting the UFO modeling of these categories to the level of types, we managed to propose a general model of prevention between events of certain types. This model is then used to analyze the cases of: (i) antidotes, which interfere in a causal chain whose ultimate outcome we intend to prevent; (ii) other countermeasures that can (directly or indirectly) prevent events of a given type from manifesting; (iii) countermeasure mechanisms, which are designed interventions that endow a given setting containing a disposition of interest with antidotes or other countermeasures. Finally, we used this framework to interpret elements from the COVER value and risk core ontology, and applied it to analyze three cases in the literature. These cases constitute a preliminary analysis, which will be extended in future work.

COVER has been applied to analyze and extend the enterprise architecture standard Archimate in terms of its value [27] and risk [26] modeling capabilities. In future work, we intend to extend COVER with the model proposed here to arrive at a core ontology of security and safety. In the same spirit of the aforementioned works, this ontology will then serve as a basis for analyzing and extending Archimate w.r.t. to its support for modeling the latter notions.

We also intend to extend our model of prevention to generally deal with aspects of *dispositional (gradable) interference* [23]. Instead of being limited at blocking cases, we intend to deal with cases in which dispositions can either decrease or increase the degree of manifestation of other dispositions. For example, a protection against shocks does not eliminate the fragility of a glass but it reduces the manifestation of a shattering event; a catalyst is a disposition that can accelerate the manifestation of other dispositions. Currently, our model can deal with interference regarding the frequency of events (via the different likelihood properties). We believe that, in this future work, prevention can be formulated as a limited case of decreasing interference.

Finally, we intend to provide a full formalization of our model with formal validation and consistency checking.

Acknowledgement. Work supported by Accenture Israel Cybersecurity Labs.

References

1. Almeida, J.P., et al.: Towards an ontology of scenes and situations. In: Proceedings of the IEEE CogSIMA 2018, pp. 29–35. IEEE (2018)
2. Almeida, J.P.A., Falbo, R.A., Guizzardi, G.: Events as entities in ontology-driven conceptual modeling. In: Laender, A.H.F., Pernici, B., Lim, E.-P., de Oliveira, J.P.M. (eds.) ER 2019. LNCS, vol. 11788, pp. 469–483. Springer, Cham (2019). https://doi.org/10.1007/978-3-030-33223-5_39
3. Arp, R., et al.: Building Ontologies with Basic Formal Ontology. MIT Press (2015)

4. Azevedo, C., et al.: Modeling resources and capabilities in enterprise architecture: a well-founded ontology-based proposal for archimate. Inf. Syst. **54**, 235–262 (2015)
5. Baltimore, J.A.: Expanding the vector model for dispositionalist approaches to causation. Synthese **196**(12), 5083–5098 (2018). https://doi.org/10.1007/s11229-018-1695-x
6. Benevides, A.B., et al.: Representing a reference foundational ontology of events in SROIQ. Appl. Ontol. **14**(3), 293–334 (2019)
7. Bird, A.: Dispositions and antidotes. Philos. Q. **48**(191), 227–234 (1998). https://doi.org/10.1111/1467-9213.00098
8. Choi, S., Fara, M.: Dispositions. In: Zalta, E.N. (ed.) The Stanford Encyclopedia of Philosophy. Stanford University, Spring (2021)
9. Duarte, B., et al.: An ontological analysis of software system anomalies and their associated risks. Data Knowl. Eng. **134**, 101892 (2021)
10. Fabio, I., et al.: "what exactly is a lockdown?": towards an ontology-based modeling of lockdown interventions during the Covid-19 pandemic (2021)
11. Goldfain, A., et al.: Dispositions and the infectious disease ontology. In: Formal Ontology in Information Systems, pp. 400–413. IOS Press (2010)
12. Guizzardi, G.: Ontological foundations for structural conceptual models. CTIT, Centre for Telematics and Information Technology (2005)
13. Guizzardi, G.: Ontology, ontologies and the "I" of FAIR. Data Intel. **2**(1–2), 181–191 (2020)
14. Guizzardi, et al.: Towards ontological foundations for conceptual modeling: the unified foundational ontology (UFO) story. Appl. Ontol. **10**(3–4), 259–271 (2015)
15. Guizzardi, G., Wagner, G., de Almeida Falbo, R., Guizzardi, R.S.S., Almeida, J.P.A.: Towards ontological foundations for the conceptual modeling of events. In: Ng, W., Storey, V.C., Trujillo, J.C. (eds.) ER 2013. LNCS, vol. 8217, pp. 327–341. Springer, Heidelberg (2013). https://doi.org/10.1007/978-3-642-41924-9_27
16. Guizzardi, R.S.S., Franch, X., Guizzardi, G., Wieringa, R.: Ontological distinctions between means-end and contribution links in the $i*$ framework. In: Ng, W., Storey, V.C., Trujillo, J.C. (eds.) ER 2013. LNCS, vol. 8217, pp. 463–470. Springer, Heidelberg (2013). https://doi.org/10.1007/978-3-642-41924-9_39
17. Keet, C.M.: The use of foundational ontologies in ontology development: an empirical assessment. In: Antoniou, G., et al. (eds.) ESWC 2011. LNCS, vol. 6643, pp. 321–335. Springer, Heidelberg (2011). https://doi.org/10.1007/978-3-642-21034-1_22
18. Leveson, N.G., Turner, C.S.: An investigation of the Therac-25 accidents. Computer **26**(7), 18–41 (1993)
19. Lewis, D.: Finkish dispositions. Philos. Q. **47**(187), 143–158 (1997). https://doi.org/10.1111/1467-9213.00052
20. Miranda, G., et al.: Foundational choices in enterprise architecture: the case of capability in defense frameworks. In: Proceedings of IEEE EDOC 2019, pp. 31–40. IEEE (2019)
21. Molnar, G., Mumford, S.: Powers. Oxford University Press (November 2006)
22. Mumford, S.: Dispositions. Clarendon Press (2003)
23. Mumford, S., Anjum, R.L.: Getting Causes from Powers. OUP (2011)
24. Mumford, S., Anjum, R.L.: Powers and potentiality. In: Engelhard, K., Quante, M. (eds.) Handbook of Potentiality, pp. 261–278. Springer, Dordrecht (2018). https://doi.org/10.1007/978-94-024-1287-1_10
25. de Ruijter, A., Guldenmund, F.: The bowtie method: a review. Saf. Sci. **88**, 211–218 (2016)

26. Sales, T., et al.: Ontological analysis and redesign of risk modeling in archimate. In: Proceedings of the IEEE EDOC 2018, pp. 154–163. IEEE (2018)
27. Sales, T.P., Roelens, B., Poels, G., Guizzardi, G., Guarino, N., Mylopoulos, J.: A pattern language for value modeling in ArchiMate. In: Giorgini, P., Weber, B. (eds.) CAiSE 2019. LNCS, vol. 11483, pp. 230–245. Springer, Cham (2019). https://doi.org/10.1007/978-3-030-21290-2_15
28. Sales, T.P., Baião, F., Guizzardi, G., Almeida, J.P.A., Guarino, N., Mylopoulos, J.: The common ontology of value and risk. In: Trujillo, J.C., et al. (eds.) ER 2018. LNCS, vol. 11157, pp. 121–135. Springer, Cham (2018). https://doi.org/10.1007/978-3-030-00847-5_11
29. Schulz, S.: The role of foundational ontologies for preventing bad ontology design. In: 4th Joint Ontology Workshops (JOWO), vol. 2205. CEUR-WS (2018)
30. Verdonck, M., Gailly, F.: Insights on the use and application of ontology and conceptual modeling languages in ontology-driven conceptual modeling. In: Comyn-Wattiau, I., Tanaka, K., Song, I.-Y., Yamamoto, S., Saeki, M. (eds.) ER 2016. LNCS, vol. 9974, pp. 83–97. Springer, Cham (2016). https://doi.org/10.1007/978-3-319-46397-1_7
31. Verdonck, M., et al.: Ontology-driven conceptual modeling: a systematic literature mapping and review. Appl. Ontol. **10**(3–4), 197–227 (2015)

Requirements Engineering

Requirements Engineering for Collaborative Artificial Intelligence Systems: A Literature Survey

Lawrence Araa Odong[1]($\boxtimes$), Anna Perini[2], and Angelo Susi[2]

[1] University of Trento, Via Calepina, 14, 38122 Trento, TN, Italy
`lawrence.odong@unitn.it`
[2] Fondazione Bruno Kessler, Trento, Italy
`{perini,susi}@fbk.eu`

Abstract. Artificial Intelligence (AI) systems are pervasively exploited to manipulate large sets of data, support data-driven decisions, as well as to replace or collaborate with humans in performing boring tasks that require high level precision. Awareness of the need of engineering approaches that align with ethical principles is increasing and motivates attention by diverse research communities, such as AI research and Software Engineering research communities.

In our research, we focus on Requirements Engineering (RE) for Collaborative Artificial Intelligence Systems (CAIS), such as robot arms that collaborate with human operators to perform repetitive and tiring tasks. A systematic literature review was conducted to assess the state of research, which resulted in the analysis of 41 research publications. Among the main findings, a set of challenges pointed out by researchers, such as the lack of a well-structured definition for CAIS requirements and the inadequacy of current standards. A discussion of these challenges and of recommendations for addressing them is proposed, taking into account similar results from recent related work. Similarly, the requirements types mentioned in the analysed literature are analysed according to categories proposed in related work.

Keywords: Collaborative Artificial Intelligence Systems ·
Requirements Engineering · Machine learning · Collaborative robots

1 Introduction

Autonomous Systems (AS) in Artificial Intelligence have received a great deal of attention in recent years, particularly systems that operate in human-shared environments. Consider self-driving vehicles in the automotive industry, which are expected to share roads with human drivers in the coming years. With the growing user demand for smart, reliable, and efficient solutions, autonomous systems are increasingly gaining traction in other fields, such as socially-enabled domestics robots, robots in health and manufacturing industries, and self adaptive systems in smart homes [2].

© The Author(s), under exclusive license to Springer Nature Switzerland AG 2022
R. Guizzardi et al. (Eds.): RCIS 2022, LNBIP 446, pp. 409–425, 2022.
https://doi.org/10.1007/978-3-031-05760-1_24

Collaborative Artificial intelligence Systems (CAIS), is a term that has recently been coined to describe autonomous system that work together with humans to jointly analyze and execute activities required to achieve their common goal, each one playing their own roles. CAISs work together with Humans as teammates, taking different roles based on their strengths. This allows for the benefits of artificial intelligence, such as performing repetitive tasks with high levels of accuracy, while still retaining humans' flexibility and cognitive skills [48]. In this study, the term CAIS will be primarily referring to collaborating systems that share the **same physical space with humans** and also have physical components themselves.

The most common CAIS technology are industrial collaborative robots also known as cobots in the manufacturing industry. In this scenario, both the cobots and productions workers work side by side simultaneously without a safety fence. The typical industrial collaborative robot is made up of a robotic arm, a controller that controls the robot arm, and finally a camera sensor to visualise the surroundings. The robotic system also includes an ML component embedded within the controller that learns from a set of heterogeneous data sources and feedback from the environment to enable the robotic arm to interact with its human counterpart and its surroundings.

Recent studies show a growing interest in AI/ML-enabled systems by the software engineering and requirements engineering research communities with venues such as the International Workshop for Requirements Engineering for Artificial Intelligence (RE4AI)[1] and the first International Conference on AI Engineering - Software Engineering for AI (CAIN 2022)[2], which aim at bringing together leading researchers and practitioners in both software engineering and artificial intelligence to reflect on and discuss challenges and implications of designing and developing complex Artificial Intelligence systems.

The study presented in this paper concerns RE for CAIS, and has as primary objective that of assessing what challenges and types of requirements have been identified and discussed so far by researchers. Specifically, the following research questions are considered:

RQ1: What are the challenges in requirements engineering for CAIS development that are discussed in research literature?

RQ2: What types of requirements are relevant for CAIS, according to research literature?

To address them we performed a Systematic Literature Review (SLR), considering research literature in the period 2014–2020. Findings were analysed leading to the following contributions:

– Key requirements relevant to the RE process in CAIS engineering were identified and then classified according to where they are defined and measured within the system.

---

[1] https://sites.google.com/view/re4ai/home.
[2] https://conf.researchr.org/home/cain-2022.

– The study also identifies the issues and challenges encountered when performing RE for CAIS and provide recommendations about future research areas on the topic based on the study results.

We discuss these findings taking into account recent SLRs [5,28], and a qualitative study with practitioners [50]. The remainder of this paper is structured as follows. Section 2 presents related work. Section 3 describes the methodology used in the study. In Sect. 4, the results from the study are presented. Section 5 discusses the implications of the results obtained in the previous section as well as any threats to the validity to the study findings. Section 6 concludes the paper.

2   Related Work

Analyses of research literature regarding RE for AI/ML systems have been presented in a few recent studies that are summarised here below. In [28] an SLR on RE for AI is presented, which covers research published in the time period 2010–2020. The study focuses mainly on identifying available tools and techniques used, and identifying existing challenges and limitations while specifying and modelling AI requirements. Twenty-seven primary studies are considered. Differently, in our SLR, we focus on RE for a specific AI-based system type, i.e. CAIS, leading to different, but somehow related, search queries. A similar selection strategy is used, but we consider a shorter time period 2014–2020, still resulting in a higher number of primary studies to analyse.

In [5], a set of papers presenting research in software engineering are analysed, and a discussion on identified challenges posed to RE towards building AI-based complex systems is proposed in terms of a taxonomy mapping each RE activity to a set of challenges. Moreover, the identified challenges are partitioned into 3 AI-related entities, that is, challenges related to data, model, and system as a whole. Differently, in our work we perform a systematic literature review, but we consider useful to map and discuss some of our findings using the above mentioned 3 categories.

The work in [50] consists of a qualitative study with practitioners, which is based on focus group, interviews and a survey. The objective of the study is to understand the difference between the development of ML systems and development of non-machine learning systems, as perceived by practitioners. While our study was instead taking the researchers perspective, we will take into account this study's findings when discussing the results of our SLR.

3   Methodology

To address *RQ*1 and *RQ*1, a Systematic Literature Review was conducted following the guidelines of Kitchenham et al. [30]. These guidelines define a process of nine activities that can be broken down into three phases: planning, conducting, and reporting the review. The initial planning phase involved identifying why the review is needed, writing the review protocol, and identifying the research questions. The review protocol included a plan to identify which search strings to use, what search strategy to employ, and which selection criteria to enforce. To improve chances of exhausting exploration of targeted studies, we identified relevant keywords to formulate efficient search strings. The main keywords included Requirements Engineering, Artificial Intelligence, and Robotics. From these three main keywords, several alternative terms were derived such as machine learning, collaborative robots etc.

Since CAIS covers a group of interdisciplinary communities i.e., robotics, human-computer interaction, machine learning, it was challenging to formulate an effective search strategy that can efficiently extract the required papers that cover all disciplines related to CAIS. Therefore, a combination of three search strategies that involve automatic search, manual search, and snowballing was executed to realize a broad range of relevant studies related to the research questions.

An automatic search was conducted by applying a search string in Scopus. Scopus was chosen because it is comprehensive with advanced search capabilities and it contains peer-reviewed publications from computer science journals and conferences, including ScienceDirect, IEEE Xplore, ACM Digital library, and Springer research papers, enabling multiple digital repositories to be searched under one engine.

A manual search was conducted on four recent editions of the IEEE International Requirements Engineering Conference (RE), and collocated workshops[3], and on the first edition of the International Workshop on Requirements Engineering for Artificial Intelligence (RE4AI 2020).

And finally, a snowballing process was employed in this study following Wohlin's guidelines [52]. The main motive for this process was to obtain relevant papers that might not have been considered by the previous search methods. First, a set of primary papers were identified from the previous search results, then backward and forward snowballing was performed on those sets of papers. Google Scholar was used to identify the origins of the selected papers. Additionally, Google Scholar was also used to extract BibTex citations for papers whose source did not support the BibTex citation format. Since RE for CAIS is still a new and emerging field, the snowballing iteration was limited to just a single iteration due to few studies about the field.

Scoping criteria to select studies for answering the research questions are presented in Table 1. Inclusion and exclusion criteria provide a consistent way to screen selected papers from the search results. The selection process based on the criteria was realized in two iterations. The first iteration involved reading the title, abstract, and keywords of all the identified papers excluding duplicated papers. Studies that match the description of exclusion criteria were discarded in this iteration. In the second iteration, a thorough review of the remaining papers was performed, the exclusion and inclusion criteria were applied once more, this time focusing on the inclusion criterion (IC3), which ensures that the considered paper contains both RE and CAIS concepts.

---

[3] IEEE Int. RE Conference 2020, 2019, 2018, 2017.

Table 1. Inclusion and exclusion criteria

Criteria	Description
Inclusion (IC1)	The study paper be a primary study, i.e. it have original contribution
Inclusion (IC2)	The paper must be written in English
Inclusion (IC3)	The paper discusses either RE or AI/ML as its core in a CAIS domain
Exclusion (EC1)	The paper is a thesis, published at a local conference or workshop
Exclusion (EC2)	The paper is a secondary study
Exclusion (EC3)	The paper is an exact duplicate of another study
Exclusion (EC4)	The work has been published before 2014

After performing the selection process, a total of 41 papers were selected. Fourteen papers were obtained using automatic search, thirteen additional papers were obtained based on manual search. Furthermore, fourteen papers were obtained from the snowballing process.

Finally, data were extracted from each of the 41 primary studies according to a predefined extraction form. This form was used to record all relevant information about the papers under review, in particular, information addressing each research question under the review. Google sheets were used to aid in organising the extracted information. The spreadsheet used to record the relevant data can be found in the supplementary material of this study.[4]

The majority of data synthesis was performed using content analysis. According to Klaus Krippendorf [33], content analysis can be defined as a replicable research tool used to compress large amounts of data into fewer content categories based on explicit rules of code. Content analysis allows researchers to quantify and analyze the presence and relationship of specific words or concepts with relative ease [46]. The data synthesis process started by creating two classification schemas to group collected information about challenges highlighted during RE4CAIS (*RQ1*) and extracted requirements (*RQ2*). For each schema, all the extracted papers were thoroughly examined, and then an iterative coding process was performed. The coding process involved identifying sentences or paragraphs that provide information in regards to the research questions and assigning them labels. Similar labels were then clustered together into one of the categories identified in the literature or knowledge obtained from previous surveys and reviews. The classification schema of CAIS challenges resulted in a list of five categories that we use to summarise results in Table 2, while the classification schema of CAIS requirements resulted in a list of twelve categories used in Table 3.

---

[4] https://github.com/lawrencearaa/RE4CAIS-RCIS2022.

4   Results

In this section, we present the findings for each research question.

4.1   RQ1: Existing RE Challenges in CAIS

Focusing on the first research question RQ1, Table 2 summarises main challenges discussed in the considered research literature. We explain them in detail below.

Table 2. Identified challenges

ID	Challenge	Reference
CH1	Limited domain knowledge regarding CAIS	[21,24,29,38,44]
CH2	Uncertainty of CAIS environment and ML systems	[2,6,7,22,36,43,47,49]
CH3	Non-involvement of requirements engineering experts during CAIS development	[26,49]
CH4	Constraints from Standards and regulations	[5–7,17,29,49]
CH5	Difficulty in modelling data	[2,5,10,49,53]

CH1 Limited Domain Knowledge Regarding CAIS. This issue has direct implications for the definition and specification of requirements as having inadequate knowledge about a system can lead to eliciting of wrong requirements or exclusion of critical ones. Seeber et al. [44] discuss machines collaborating with humans as teammates and they highlight poorly designed tasks and insufficient information about the system as major challenges to a successful collaboration. Inadequate domain knowledge is also a problem because its implication leads to under specification of requirements due to partial knowledge about the system which ultimately lead to unclear or wrong specifications of system components. Horkoff [21] emphasized that our grasp of non-functional requirements for ML-enabled systems is lacking and that we need to establish standards for defining them. For example, how can we define trustworthiness?

CH2 Uncertainty of CAIS Environment and ML Systems. CAISs operate in human environment with a lot of unplanned scenarios and situations that can not be accounted for during the planning phase. At the same time, due to the fact that ML outputs are determined by the training data used as opposed to manually coded rules in traditional systems, the results of ML systems are always uncertain. This makes it difficult to assure a successful fulfilment of specified requirements in both scenarios due to their uncertain nature. Ishikawa et al. [22] identify the intrinsic uncertainty and unpredictability of requirements as a core reason why RE is listed as one of the most difficult activities in the development and operations of ML-based systems. Unhelkar et al. [47] introduce uncertain environment design for collaborative robots as one of the multi-disciplinary challenges that need to be resolved before introducing Cobots in human settings.

Camilli et al. [6] identified uncertain environment of CAIS as one of the major open issues that is faced while developing risk-driven compliance assurance for CAISs.

CH3 Non-involvement of Requirements Engineering Experts During CAIS Development. This challenge is concerned with the responsibilities of requirements engineers being taken up by other experts during CAIS development. Vogelsang et al. [49] stated that data scientists are responsible for writing requirements in current ML systems. Meanwhile, Kaindl et al. [26] stated that most AI systems are developed and deployed without carrying out any RE at all. Such practices result in systems that are not user-centric which are often lead to systems that do not fulfill stakeholder's needs.

CH4 Constraints from Standards and Regulations. Six studies were identified that highlight the issue of standards and compliance requirements. Belani et al. [5] pointed to a lack of mandatory standards for data exchange as one of the notable obstacles that hinder development of ML-based systems. Standards are a critical component of RE as they provide definitions and specifications for the most critical requirements used to design CAIS and other ML-based systems. Hanna et al. [17] called for the current safety standards for collaborative robots to be updated and improved. They stated that current standards are much similar to traditional robots and do not support implementation of intelligent and adaptive collaborative systems in complex applications. Four data scientists were interviewed by Vogelsang et al. [49] and also raised challenges regarding legal aspects and ethics and when asked about legal regulatory requirements. They explained that certain legal requirements constrain ability of data scientists to acquire necessary data features from datasets. Camilli et al. in [6] and [7] also state that existing CAIS standards pose challenges in realizing flexible automation that can enable Cobots to interact with humans intelligently. Results from a survey conducted on industrial experts and students by Kidal et al. [29] identified legislation as one of the main barriers to the adoption of collaborative robots, thus requiring engineers to keep regulatory requirements updated.

CH5 Difficulty in Modeling Data. Data are fundamental building blocks for CAISs with the capabilities to make or break the entire system. Five studies were identified that brought up challenges related to quality of data used in ML-based systems. Belani et al. [5] presented the following data related challenges encountered during complex AI development; (1) data volatility, meaning the behaviour of data changes over time; (2) under-utilisation of data features which can lead to under specification of requirements and lastly; (3) difficulty in performing static analysis and tracking the use of data in the system, causing traceability issues when trying to account for the actions performed by the system; and lastly difficulty in pinpointing discrimination in the data. All the above challenges greatly affect the data quality of training data which a very critical requirements for all AI systems. Vogelsang et al. [49] re-emphasised the issue of data discrimination stating that *"While discrimination is also possible in conventional rule-based algorithms, it is more critical in ML systems for two* *reasons: (1) Discrimination is more implicit in ML systems because you cannot pinpoint to the discriminating rule, (2) ML algorithms amplify discrimination bias in the data during the training process."* Aniculaesei et al. [2], pointed out that data collection and preparation as one of the major challenges being faced during the development of autonomous systems. Zhang et al. [53] state that AI models are always a form of statistical discrimination by nature and propose a framework that defines fairness metrics in an attempt to mitigate bias in data used to developed AI systems.

4.2   RQ2: Relevant Requirements Types for CAIS

RE knowledge provides definitions of different types of software systems requirements in form of taxonomies (e.g. [15]), and ontology (e.g. [1]), which attempt at capturing the different perspectives of the involved stakeholders[5]. For instance, when focusing on software systems in general, and taking the perspective of requirements engineers, guides to help differentiate between functional and non functional requirements, and their sources are important. For instance, as suggested in [1], non-functional requirements can be derived from external sources, such as regulations, specific domain standards or ethical design principles. The definitions proposed in literature for requirements for AI-based software system vary depending on the research areas where they have been proposed, and they are often complementary. So for instance *explainability* in RE context is defined in terms of a capability of the system to disclose information, which can help a user (or stakeholder) to understand why a software system produced a given results [8], while in the AI research community we can find definitions of explanations in relation to the type of data used by the ML component of the system under consideration. While analysing SLR data to address RQ2, we searched for descriptions of those requirements that can be linked to the short description given in Table 3. In the following, we comment on the requirements that recurred most in the analysed literature.

Safety. Most authors indicated safe requirements as one of the cornerstone for CAIS development. The safety requirements identified in the literature mostly introduced in the context of human safety, that is, avoiding accidents that can lead to human harm during interaction between robots and humans in close proximity. Cysneiros et al. [9] called for an interdisciplinary approach for achieving safety for autonomous systems involving collaboration among different experts in sensor technology, machine learning, control engineering, human-machine interaction, and the legal department. Aniculaesei et al. [2] identify safety as one of the core issues that need to be addressed in order to achieve a new notion called dependability which aims to improve user acceptance of autonomous systems in society.

---

[5] This knowledge is also expressed in standards used by practitioners, e.g. ISO/IEC 25000 series SQuaRE (System and software product Quality Requirements and Evaluation [31].

Table 3. Requirements relevant to CAIS

Requirement	Description	Reference
Safety	Defines safety measures when interacting with CAIS	[2,4,6,7,9,11–14,16–18,20,23–26,29,32,34–37,39,41,43,45,48,51]
Transparency	Defines how to make CAIS more understandable	[3,9,22,23,32,42,51,53]
Data	Defines how to prepare CAIS training and test data	[2,10,22,23,32,37,39]
Privacy	Defines personal information policies	[2,9,23]
Auditability	Defines accountability actions	[9,32,51]
Security	Defines protection against malware attacks or unauthorized access	[2,9,23,24,51]
Usability	Defines how easy it is to use CAIS	[29]
Reliability	Defines the ability of CAIS to operate without failure for a specified period of time	[9,11]
Performance	Defines how well CAIS must operate	[16,21]
Trustworthiness	Defines requirements that capture user trust	[6,9]
Fairness	Defines specification against bias ML data	[21,51]
Ethics	Defines human values that CAIS must abide to	[9,22]

Data. Data requirements are also prominent requirements that are relevant to CAIS development. The two main data requirements reported in the studies were data quality and data privacy. Koopman et al. [32] argued that the safety of autonomous robotic machines hinges on the accuracy of the training and validation data collection. They consider training and validation data as critical requirements for implementing safety in such systems and define the system that collects these data as safety-critical. Aniculaesei et al. [2] emphasised on the quality of collected data. They argued that data must be readable for humans so that system developers can gain new insights from the data. CAISs are ML-based systems which means that they are data-centric, therefore, all data attributes of the system must be properly defined and specified as poor data quality can lead to failure of the system as a whole.

Transparency. Transparency as a requirement is a fairly new concept that has been introduced due to the complexity of AI/ML systems. Transparency requirement facilitate in building trust between the public and robots that interact with each other. Cysneiros et al. [9] highlighted transparency as a gateway to user acceptance and as a key requirement that is a prerequisite for building robust systems and improving the adoption rate of autonomous systems that can interact with humans. Adding to that, Ishikawa et al. [22] proposed transparency as one of the new ethical requirements that should be considered when realizing machine learning functionality that affects human activities and society as a whole.

Security. Another set of requirements that are closely related to safety requirements are security requirements Cysneiros et al. [9], Aniculaesei et al. [2]. This is because a breach of security of the collaborative robots can directly lead to human harm in case the cobots are remotely controlled by a hacker. Despite the lack of attention on the need for security requirements for collaborative robots, security requirements have potentially equal ramifications as the other critical requirements to CAIS. For example, a cyber-physical attack on a collaborative robot can lead to direct physical harm of humans around in case its controlled by the attacker. Therefore, special measures need to be taken to make sure that CAISs are invulnerable to cyber-physical attacks.

Auditability. Auditability is another important requirement that was mentioned by Cysneiros et al. [9] and Koopman et al. [32]. The main aim of auditability is to tackle future unforeseeable issues such as accidents or failures so that responsible parties for the failures can be held accountable Cysneiros et al. [9]. Furthermore, Koopman et al. [32] empathized the need for auditability by highlighting the issue of legal liability in case of an accident involving autonomous vehicles.

5   Discussion

With the increase of CAIS integration within industrial solutions, new RE challenges have emerged in regards to CAIS development. The availability of data and the processing power to process the large amounts of data has led to increased adoption of CAIS in the industry, however, new issues and limitations in regards to RE techniques and methodologies have also emerged. For example, how to define and specify certain requirements in the context of RE, and how to specify details for requirements that involve uncertain actions such as human behavior towards the Cobot. Based on the challenges discovered in this study, we present some research recommendations that might overcome some of the issues presented in the study and need more attention in the RE and AI research community.

5.1   Challenges

From the first challenge (limited domain knowledge), it is observed that there is limited understanding and fragmented knowledge about CAIS and AI/ML systems in general. This can be attributed to the complex nature of such systems that involve multiple disciplinary experts from different fields. Despite recent research boost on the topic, they are being carried out by individual fields. Therefore, there is a need for more collaboration between the various fields involved in

CAIS development to better understand each others domains and how it affects the general quality of CAIS development.

A way forward to manage the challenge of uncertainty of CAIS environments is the extension of risk modelling capabilities to current requirements engineering methods and frameworks. Camilli et al. [6], proposed an adoption of risk models that can quantify risk levels of elicited CAIS requirements by integrating probability and impact functions in order to prioritize elicited requirements and related compliance requirements depending on the level of risk associated with them. A possible future research topic could be how to integrate risk modelling methods to already existing modelling languages such as i* modelling language.

Given that standards provide majority of definitions and specifications for critical CAIS requirements such as safety, privacy, security and data requirements, it is critical that requirements engineering experts participate in the development of new standards in order to provide more insight into how proposed rules affect the development of technology, backed up by information and field experience. Furthermore, organizations should include legal experts in their development teams to ensure that systems being built are compliant with regulations and standards.

As an emerging technology, CAIS started from the laboratory as any other technology where efficiency and performance are prioritised above all. But as the adoption of CAIS moves outside of research and into the real world, requirements engineering experts play a very critical role in managing the needs of every stakeholder involved in the development process. Therefore it is important for companies and organisations to consider all necessary experts (not only requirement engineering experts) involved in the development of CAIS for a successful assurance of the product.

More attention must be paid to data preparation processes of data used in developing CAIS. Hassanni et al. [19] suggest that most organisations fail to meet their data quality standards due to under-valuing work related to data preparation. This phenomenon can be attributed to a lack of incentives associated with data preparation responsibilities compared to other responsibilities such as model development. Poor data handling is especially critical when developing CAIS because it is a high stake domain that has safety impacts on humans. For example, in case that an algorithm is trained on biased data, it might fail to recognise certain groups of people that are underrepresented in the datasets hence causing accidents. Therefore, is need for data standards to set rules on how to handle data during acquisition, training, and testing of data in order to maintain data quality.

5.2 Requirements

The most discussed requirements in the context of CAIS are safety, data requirements, and transparency. Overall, a strong focus was emphasized on qualities related to safety requirements especially from robotics publications whereas publications related to software engineering emphasized more transparency and data requirements. Our observation from the SLR results also revealed reduced interest in traditional requirements such as usability, performance, and reliability according to the number of their mentions from the selected papers. However, findings from one of the related works [50] indicate that traditional requirements such as performance are still considered as critical requirements. A possible future research might be eliciting requirements and challenges from practitioners and access the research gap between researchers and practitioners.

The study was able to elicit new emerging requirements for CAIS and AI in general, such as data, transparency, trustworthiness, fairness, and ethics, which introduced new challenges that were also discussed in this study. Though mentioned by several studies, no comprehensive definitions was provided for any of the new requirements. There is need for further research on how to manage and document these new requirements.

CAIS Requirements Classification. With the aim of helping CAIS experts on how to define CAIS requirements and related technologies, a conceptual classification model for CAIS requirements has been created from the results of the SLR. The classification model was inspired by two studies. First, Horkoff et al. [27], who presented three levels of granularity over which non-functional requirements for ML-enabled systems can be defined and measured and Belani et al. [5], who used the same level of granularity to partition RE challenges for AI complex systems.

Requirements identified in the study were clustered in three categories according to their granularity level, i.e., at the system level, at the level of ML model component, and at the data level. Figure 1 displays the classification model clustering identified CAIS requirements in the study under one of the three categories: System requirements, ML model requirements, and Data requirements.

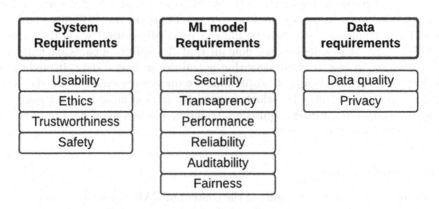

Fig. 1. Classification of CAIS requirements.

According to the latest version of SWEBOK [40], system requirements are requirements for the system as a whole, in the context of CAIS, that would be the entire collaborative robot and the human collaborator. Requirements under this category are defined and measured over the whole software system including ML and the data itself. Requirements in this category are normally satisfied by fulfilling other minor requirements in the other categories. For example, trustworthiness can only be fulfilled when fairness, reliability, and privacy are achieved.

Requirements under ML model requirements are defined and measured over the machine learning model. They represent what would have been software requirements in traditional systems which express the needs of a software product. Data requirements on the other hand are defined and measured over ML-related data. They represent data characteristics as well as constraints to the data being used in the system.

It is also possible that a single requirement can be defined and measured over multiple components of the system. Depending on the needs of the stakeholders and the software system, the granularity over which certain requirements are defined can change. Moreover different domains define requirements with different levels of granularity, therefore, certain requirements can be defined and measured over different components of the system. For example, transparency can be defined over data to ensure that data is comprehensible to data engineers, at the same time, it can also be defined over the machine learning model to help both engineers and clients understand the reasoning behind actions performed by the model.

5.3   Limitations of the Study

The main limitation of the study is the threat of subjectivity. The study was conducted using qualitative methods, therefore, there are chances of biased results. To minimise the chance of producing results with personal bias, a peer review approach was employed where co-authors that did not conduct the SLR reviewed, discussed and resolved and conflicts presented during the discussion.

The study also faced a challenge of population sample size. The original intention was to include only existing RE studies discussing CAIS topics, however, due to the limited research available, a decision was made to add non RE studies. To maintain the quality of the research, non-RE studies were parsed through SLR criteria checks to make sure that they meet the required acceptance criteria. Though not originally planned to be part of the study, the elicited non-RE studies provided great insight in regards to eliciting perspectives of different fields such as robotics and system engineers with regards to RE for CAIS systems.

6   Conclusion

The increasing adoption of AI is calling the attention of researchers on defining suitable engineering approaches. In this paper we focused on Requirements

Engineering (RE) and on a specific class of AI-based systems, namely Collaborative Artificial Intelligence System (CAIS). Our ultimate goal was to assess what challenges and types of requirements are discussed by researchers. Guided by this objective we formulated two research questions ($RQ1, RQ2$), and performed a systematic literature review, considering research published in the period 2014–2020. Forty-one primary studies were selected and analysed to answer the research questions. Findings include a set of challenges in RE for CAIS, a few suggestions for addressing them, and types of requirements that are considered relevant for CAIS. We discussed these findings, taking into account recent related work, and with the aim of pointing out open problems to be considered in future research. In our future work, we believe it will be important to complement the research perspective derived from the presented SLR with the practitioner perspective, also because in the CAIS domain we are considering standards play a key role for system certification. Moreover, we believe that it will be worth trying to integrate recently proposed methods for requirements elicitation and modelling for AI-based systems, e.g. [6,38], and to extend them with risk concepts and management techniques.

References

1. Alrumaih, H., Mirza, A., Alsalamah, H.: Domain ontology for requirements classification in requirements engineering context. IEEE Access **8**, 89899–89908 (2020)
2. Aniculaesei, A., Grieser, J., Rausch, A., Rehfeldt, K., Warnecke, T.: Toward a holistic software systems engineering approach for dependable autonomous systems. In: 2018 IEEE/ACM 1st International Workshop on Software Engineering for AI in Autonomous Systems (SEFAIAS), pp. 23–30 (2018)
3. Arrieta, A.B.: Explainable artificial intelligence (XAI): concepts, taxonomies, opportunities and challenges toward responsible AI. Inf. Fusion **58**, 82–115 (2020)
4. Askarpour, M., Lestingi, L., Longoni, S., Iannacci, N., Rossi, M., Vicentini, F.: Formally-based model-driven development of collaborative robotic applications. J. Intell. Robot. Syst. **102**(3), 1–26 (2021)
5. Belani, H., Vukovic, M., Car, Z.: Requirements engineering challenges in building AI-based complex systems. In: 2019 IEEE 27th International Requirements Engineering Conference Workshops (REW), pp. 252–255 (2019)
6. Camilli, M.: Risk-driven compliance assurance for collaborative AI systems: a vision paper. In: Dalpiaz, F., Spoletini, P. (eds.) REFSQ 2021. LNCS, vol. 12685, pp. 123–130. Springer, Cham (2021). https://doi.org/10.1007/978-3-030-73128-1_9
7. Camilli, M., et al.: Towards risk modeling for collaborative AI. arXiv preprint arXiv:2103.07460 (2021)
8. Chazette, L., Brunotte, W., Speith, T.: Exploring explainability: a definition, a model, and a knowledge catalogue. In: 29th IEEE International Requirements Engineering Conference, RE 2021, Notre Dame, IN, USA, 20–24 September 2021, pp. 197–208. IEEE (2021)
9. Cysneiros, L.M., Raffi, M., do Prado Leite, J.C.S.: Software transparency as a key requirement for self-driving cars. In: 2018 IEEE 26th International Requirements Engineering Conference (RE), pp. 382–387 (2018)
10. D'Amour, A., et al.: Underspecification presents challenges for credibility in modern machine learning. arXiv preprint arXiv:2011.03395 (2020)

11. Dede, G., Mitropoulou, P., Nikolaidou, M., Kamalakis, T., Michalakelis, C.: Safety requirements for symbiotic human-robot collaboration systems in smart factories: a pairwise comparison approach to explore requirements dependencies. Requirements Eng. **26**(1), 115–141 (2021)
12. Di Cosmo, V., Giusti, A., Vidoni, R., Riedl, M., Matt, D.T.: Collaborative robotics safety control application using dynamic safety zones based on the ISO/TS 15066:2016. In: Berns, K., Görges, D. (eds.) RAAD 2019. AISC, vol. 980, pp. 430–437. Springer, Cham (2020). https://doi.org/10.1007/978-3-030-19648-6_49
13. Franklin, C., Dominguez, E., Fryman, J., Lewandowski, M.: Collaborative robotics: new era of human-robot cooperation in the workplace. J. Saf. Res. **74**, 153–160 (2020)
14. Gleirscher, M., Calinescu, R.: Safety controller synthesis for collaborative robots. In: 2020 25th International Conference on Engineering of Complex Computer Systems (ICECCS), pp. 83–92. IEEE (2020)
15. Glinz, M.: On non-functional requirements. In: 15th IEEE International Requirements Engineering Conference, RE 2007, pp. 21–26. IEEE (2007)
16. Gualtieri, L., Rauch, E., Vidoni, R., Matt, D.T.: Safety, ergonomics and efficiency in human-robot collaborative assembly: design guidelines and requirements. Procedia CIRP **91**, 367–372 (2020)
17. Hanna, A., Bengtsson, K., Dahl, M., Eros, E., Götvall, P.-L., Ekström, M.: Industrial challenges when planning and preparing collaborative and intelligent automation systems for final assembly stations, September 2019, pp. 400–406. Institute of Electrical and Electronics Engineers Inc. (2019)
18. Hanna, A., Bengtsson, K., Götvall, P.-L., Ekström, M.: Towards safe human robot collaboration - risk assessment of intelligent automation, September 2020, pp. 424–431. Institute of Electrical and Electronics Engineers Inc. (2020)
19. Hassani, H., Silva, E.S., Unger, S., TajMazinani, M., Mac Feely, S.: Artificial intelligence (AI) or intelligence augmentation (IA): what is the future? AI **1**(2), 143–155 (2020). https://doi.org/10.3390/ai1020008
20. Hernandez, C., Fernandez-Sanchez, J.: Model-based systems engineering to design collaborative robotics applications. Institute of Electrical and Electronics Engineers Inc. (2017)
21. Horkoff, J.: Non-functional requirements for machine learning: challenges and new directions. In: 2019 IEEE 27th International Requirements Engineering Conference (RE), pp. 386–391. IEEE (2019)
22. Ishikawa, F., Matsuno, Y.: Evidence-driven requirements engineering for uncertainty of machine learning-based systems. In: 2020 IEEE 28th International Requirements Engineering Conference (RE), pp. 346–351 (2020)
23. Ishikawa, F., Yoshioka, N.: How do engineers perceive difficulties in engineering of machine-learning systems?-Questionnaire survey. In: 2019 IEEE/ACM Joint 7th International Workshop on Conducting Empirical Studies in Industry (CESI) and 6th International Workshop on Software Engineering Research and Industrial Practice (SER&IP), pp. 2–9. IEEE (2019)
24. Japs, S.: Security safety by model-based requirements engineering. In: 2020 IEEE 28th International Requirements Engineering Conference (RE), pp. 422–427 (2020)
25. Kadir, B., Broberg, O., Da Conceição, C.S.: Designing human-robot collaborations in industry 4.0: explorative case studies, vol. 2, pp. 601–610. Faculty of Mechanical Engineering and Naval Architecture (2018)
26. Kaindl, H., Ferdigg, J.: Towards an extended requirements problem formulation for superintelligence safety. In: 2020 IEEE 7th International Workshop on Artificial Intelligence for Requirements Engineering (AIRE), pp. 33–38 (2020)

27. Khan Mohammad Habibullah, J.H.: Non-functional requirements for machine learning: understanding current use and challenges in industry. In: 2021 International Requirements Engineering Conference (RE). IEEE (2021)
28. Ahmad, K., Bano, M., Abdelrazek, M., Arora, C., Grundy, J.: What's up with requirements engineering for artificial intelligence systems? In: 2021 International Requirements Engineering Conference (RE), pp. 51–54. IEEE (2021)
29. Kildal, J., Tellaeche, A., Fernández, I., Maurtua, I.: Potential users' key concerns and expectations for the adoption of cobots. Procedia CIRP **72**, 21–26 (2018)
30. Kitchenham, B., Charters, S.: Guidelines for performing systematic literature reviews in software engineering (2007)
31. Koh, S., Whang, J.: A critical review on ISO/IEC 25000 square model. In: Proceedings of the 15th International Conference on IT Applications and Management: Mobility, Culture and Tourism in the Digitalized World (ITAM15), pp. 42–52 (2016)
32. Koopman, P., Wagner, M.: Autonomous vehicle safety: an interdisciplinary challenge. IEEE Intell. Transp. Syst. Mag. **9**(1), 90–96 (2017)
33. Krippendorff, K.: Content Analysis: An Introduction to its Methodology. Sage Publications (2018)
34. Malik, A., Bilberg, A.: Complexity-based task allocation in human-robot collaborative assembly. Ind. Robot. **46**(4), 471–480 (2019)
35. Malik, A.A.: Robots and Covid-19: challenges in integrating robots for collaborative automation. arXiv preprint arXiv:2006.15975 (2020)
36. Malik, A.A., Bilberg, A.: Complexity-based task allocation in human-robot collaborative assembly. Ind. Robot Int. J. Robot. Res. Appl. **46**(4), 471–480 (2019)
37. Nakamichi, K., et al.: Requirements-driven method to determine quality characteristics and measurements for machine learning software and its evaluation. In: 2020 IEEE 28th International Requirements Engineering Conference (RE), pp. 260–270 (2020)
38. Nalchigar, S., Yu, E., Obeidi, Y., Carbajales, S., Green, J., Chan, A.: Solution patterns for machine learning. In: Giorgini, P., Weber, B. (eds.) CAiSE 2019. LNCS, vol. 11483, pp. 627–642. Springer, Cham (2019). https://doi.org/10.1007/978-3-030-21290-2_39
39. Nuzzi, C., Pasinetti, S., Lancini, M., Docchio, F., Sansoni, G.: Deep learning-based hand gesture recognition for collaborative robots. IEEE Instrum. Meas. Mag. **22**(2), 44–51 (2019)
40. Bourque, P., Fairley, R.E.(Dick): SWEBOK: Guide to the Software Engineering Body of Knowledge. IEEE (2020)
41. Pieska, S., Pitkaaho, T., Kaarlela, T.: Multilayered dynamic safety for high-payload collaborative robotic applications. Institute of Electrical and Electronics Engineers Inc. (2020)
42. Rzepka, C., Berger, B.: User interaction with AI-enabled systems: a systematic review of IS research. In: Pries-Heje, J., Ram, S., Rosemann, M. (eds.) Proceedings of the International Conference on Information Systems - Bridging the Internet of People, Data, and Things, ICIS 2018, San Francisco, CA, USA, 13–16 December 2018 (2018)
43. Saenz, J., Elkmann, N., Gibaru, O., Neto, P.: Survey of methods for design of collaborative robotics applications- why safety is a barrier to more widespread robotics uptake, vol. Part F137690, pp. 95–101. Association for Computing Machinery (2018)
44. Seeber, I., et al.: Machines as teammates: a research agenda on AI in team collaboration. Inf. Manage. **57**(2), 103174 (2020)

45. Steidel, V.: Framework for requirement analysis in the design of collaborative robots on construction sides, vol. 2348, pp. 277–282. CEUR-WS (2019)
46. Stemler, S.: An overview of content analysis. Pract. Assess. Res. Eval. **7**(1), 17 (2000)
47. Unhelkar, V., Shah, J.: Challenges in developing a collaborative robotic assistant for automotive assembly lines, 02–05 March 2015, pp. 239–240. IEEE Computer Society (2015)
48. Villani, V., Pini, F., Leali, F., Secchi, C.: Survey on human-robot collaboration in industrial settings: safety, intuitive interfaces and applications. Mechatronics **55**, 248–266 (2018)
49. Vogelsang, A., Borg, M.: Requirements engineering for machine learning: perspectives from data scientists. In: 2019 IEEE 27th International Requirements Engineering Conference Workshops (REW), pp. 245–251 (2019)
50. Wan, Z., Xia, X., Lo, D., Murphy, G.C.: How does machine learning change software development practices? IEEE Trans. Softw. Eng. **47**, 1857–1871 (2019)
51. Wickramasinghe, C.S., Marino, D.L., Grandio, J., Manic, M.: Trustworthy AI development guidelines for human system interaction. In: 2020 13th International Conference on Human System Interaction (HSI), pp. 130–136 (2020)
52. Wohlin, C.: Guidelines for snowballing in systematic literature studies and a replication in software engineering. In: Proceedings of the 18th International Conference on Evaluation and Assessment in Software Engineering, pp. 1–10 (2014)
53. Zhang, Y., Bellamy, R., Varshney, K.: Joint optimization of AI fairness and utility: a human-centered approach. In: AIES 2020, pp. 400–406 (2020)

On the Current Practices for Specifying Sustainability Requirements

Salvador Mendes, João Araujo$^{(\boxtimes)}$ (ID), and Ana Moreira (ID)

NOVA LINCS/NOVA School of Science and Technology, NOVA University Lisbon,
Lisbon, Portugal
sr.mendes@campus.fct.unl.pt, {joao.araujo,amm}@fct.unl.pt

Abstract. As sustainability becomes a fundamental concern in software development, it is important to understand how industry is addressing it. This paper discusses the results of a survey performed in industry aiming at identifying their current needs and practices to handle sustainability in agile software development. The survey includes an initial section to gather participants' information, followed by a section inquiring about the impact of sustainability on their working environment, and which methods and tools are used. The enquired population is a small subset of the IT professionals in Portugal. The main findings include lack of methods, tools, knowledge and domain experts to support elicitation and specification of sustainability requirements. Still, the participants recognise that one of the main reasons to consider sustainability is for the improvement of product quality and for creating a good reputation.

Keywords: Sustainability requirements · Agile development · Survey

1 Introduction

In recent years, the race towards sustainability has become increasingly more important and widely covered. Even so, software development methods, tools, and validation and verification techniques supporting sustainability are still lacking. Both academics and professionals are still not knowledgeable on the topic. Currently, it is agreed that sustainability implies *development that meets the needs of the present without compromising the ability of future generations to meet their own needs* [1]. This challenge calls for the integration of social equity, economic growth, and environmental preservation, considering also their effects (inter- and intra- relationships) on each other. These three dimensions—*social, economic* and *environmental*, respectively—have been integrated in a multidimensional line of thought that also encompasses an *individual* and a *technical* dimension [2]. Each dimension addresses different needs (e.g., improve employment indicators, reduce costs, reduce CO2 emissions, promote high agency, and easy system evolution) and impacts on the others and respective stakeholders

Supported by NOVA LINCS.

[3]. Therefore, sustainability-aware systems differ from other types of systems in that their functionality must explicitly balance the trade-offs between these dimensions [4].

Although there is no common definition of what sustainability in Software Engineering is, we prefer to think of sustainability as *an emergent property of a software system* [5] that should not be added to the system in later stages of the development nor looked into in isolation. We agree with the vision of sustainability as a complex composite quality attribute, formed of five complex aggregates of quality attributes, one for each of the five dimensions, which, in turn, is naturally refined into quality attributes relevant for that particular dimension [4].

Our goal is to investigate how sustainability requirements are being addressed in the agile software development industry in Portugal. It is well-known that agile development methods have become major approaches in the software industry. However, how is sustainability being addressed in agile development, in particular at requirements level?

We started by performing a systematic mapping study to identify works related to sustainability requirements specification methods and techniques in the agile context. The results indicate that very few approaches are used by agile developers. Given this, we decided to gather empirical evidence through a survey in industry. The design of this survey follows the principles proposed by Pfleeger and Kitchenham [6]. We started by defining the research question that our work aims to answer: "How are sustainability requirements addressed by professionals in their agile projects?" Therefore, we were particularly interested in understanding how IT professionals elicit and specify sustainability requirements and what their needs to addressing sustainability in software are.

Our survey enquires IT professionals, from several companies, about their methods and tools to tackle sustainability. These professionals work in several application domains. The survey is structured in 2 sections, aiming at gathering participants' information about their experience in agile projects and also their sustainability development issues (e.g., the impact of sustainability on their working environment), and the methods and tools they use to handle sustainability. The survey results revealed that there are sustainability-related shortcomings when developing software applications, such as lack of methods, tools, knowledge, and skilled domain experts. Implementation and maintenance are the main activities where sustainability is contemplated instead of requirements. Nevertheless, the participants endorse three main reasons to consider sustainability: (i) improvement of product quality, (ii) creating good reputation, for showing social responsibility, and (iii) increasing personal motivation.

The paper is organised as follows. Section 2 offers some background on agile development and sustainability. Section 3 presents the survey design while Sect. 4 discusses the results. Section 5 presents some related work, and, finally, Sect. 6 draws some conclusions and directions for future work.

2   Background

Agile Development and User Stories. Agile is known for employing continuous planning, learning and improvement, focusing on team collaboration and early delivery. This results in a process that is flexible and responsive to change. There are four core values of agile development that must always be present in mind when using or talking about agile [7]: individual and team interactions over processes and tools; working software over comprehensive documentation; customer collaboration over contract negotiation; responding to change over following a plan. One of the most popular artifacts used by agile methods is User Story. This is an informal and concise explanation of a system feature [8] or requirement, written from the perspective of the end user of that feature/requirement. A user story provides a guideline of how a piece of software delivers value to the customer, which may be external end users and also members of the organization developing the project. User stories are usually a few sentences long in simple and informal language that serves to outline the desired goal for the feature; further detail is added after with requirements once they are agreed upon.

Sustainability Engineering. As stated before, "sustainability is about finding a way to meet the needs of the present without compromising the ability of future generations to meet theirs" [9]. With this definition we can understand that sustainability is crucial for humankind's survival and longevity on this planet, and as such we must do our best efforts to achieve it.

There are three main sustainability dimensions [10]: **Environmental sustainability** related to the amounts of pollution that can be created, renewable resources that can be harvested and nonrenewable resources that can be depleted in an indefinite way without compromising the ecosystems and their natural balance; **Social sustainability** related to the capability of a social system to handle relationships between individuals and groups, including mutual trust, communication an conflicts; **Economic sustainability** related to financial aspects and business value and the ability of an company to maintain a determined amount of economic production. In Software Engineering, it is well accepted that two other dimensions can be considered when discussing sustainability [10]: the **Individual sustainability** covering topics such as individual freedom, the ability for individuals to exercise their rights, and free development; and the **Technical sustainability** that focuses on the longevity of a software system and its capability to adapt to the evolving environment conditions and requirements.

In the context of the paper, we are particularly interested in the technical dimension of sustainability, which is directly related to how well a software system is able to adapt to the ever changing environment it is inserted in, while continuing to function correctly. When talking about IT, one may refer to "sustainability by IT" (or Green by IT) and "sustainability in IT" (or Green IT) [11]. Sustainability by IT is when a system is specifically designed to educate about sustainability or aid in achieving some sustainability goals. On the other hand, sustainability in IT is when IT itself is sustainable.

Therefore, a software can only be truly green and sustainable if it comes from a green and sustainable process, therefore we can say it is critical that everyone involved in the development process has to be aware of the impacts, both positive and negative, their software can have on sustainability. Thus, we can define green and sustainable engineering as "the art of developing green and sustainable software with a green and sustainable software engineering process, i.e., the art of defining and developing software products in a way, so that the negative and positive impacts on sustainable development that result and/or are expected to result from it, over its whole life cycle, are continuously assessed, documented and used for a further optimization of the software product" [12]. The more complex the system is, the more probable it is to affect one, if not all, of the five dimensions, especially when considering that many of the dimensions are connected through common goals and requirements. Therefore, all trade-off possibilities must be discussed and analyzed before any changes to the system can be made, but with the topic of software sustainability being broad and understudied this process can become quite hard.

3 Survey Design

The main goal of this paper is to identify how agile practitioners (in Portugal) address sustainability when developing software. We carried out an online survey with companies that use agile methodologies for system development. We adopted Kitchenham and Pfleeger's guidelines for personal opinion surveys [6] and the activities to conduct online surveys in Software Engineering proposed by Punter [13]. Thus the activities managed here were: (i) define the study; (ii) design the survey; (iii) develop the questionnaire; (vi) execute the survey; (v) analyze the data and (vi) report the results.

Regarding the definition of the study, we want to know how do professionals that use agile development address sustainability requirements in their projects. Afterwards, we selected by invitation our participants, that come from different companies in Portugal. The set of participants invited play different roles in their companies (e.g., product owners, developers, testers, Scrum masters) and are experienced in different agile methods [7] (e.g. Scrum, Kanban), thus providing different viewpoints. The survey was sent to 52 professionals from several companies, from which 32 responded (a response rate of 61,5 %). Participants were informed about the purpose of the research and data confidentiality.

The survey[1] was made available through Google Forms to the participants contacted via email. We divided the questionnaire into 2 main sections: (i) General information questions to collect some indicators about their knowledge and experience in agile development, and (ii) Sustainability in software development questions. Since all the professionals were from companies that adopted agile development (a selection criteria), there was no need to explicitly ask them about the kind of software development process they used.

---

[1] https://drive.google.com/file/d/1zc6hpvq6_Nn3D-wLtkL7Mfqc5FgIDfZj/view?usp=sharing.

The first 5 questions had the purpose of gathering information about the participants. The second section consisted of 10 questions (being 8 multiple choice and 2 open-ended) to answer different aspects related to our research question: if they address sustainability during development and in which activities; in those activities how sustainability is addressed; how sustainability requirements are specified; their knowledge on sustainability in their development context; which tools and methods they used; their opinion on the impact of software on sustainability; the main challenges when addressing sustainability in their projects; and the main reasons to consider sustainability requirements in their projects. The use of spreadsheets helped with the quantitative analysis process, while an inductive approach was used for the qualitative analysis of the data.

4   A Survey on Sustainability at IT Companies

The subsequent sections introduce the participants, and the layout of the survey together with the reasoning for each of the questions. In particular, we start with a biographical note of the participants (e.g. years of experience and development methods used in their work), followed by presenting the questions and discussing the specific results for each question.

4.1   Participants

This survey was answered by 32 professionals in the Portuguese IT industry. The participants work in different companies playing different roles in software development. Table 1 summarizes our participants information.

From the total number of respondents, 32 respondents, 34.4% are professional developers (11 out of 32), 12.5% are business analysts (4 out of 32), 12.5% are Scrum masters (4 out of 32), another 12.5% are project leaders/managers (4 out of 32), 9.4% are product managers (3 out of 32), and 6.3% are testers (2 out of 32). The remaining 4 roles correspond to tech leads, operations, UX/UI and product owners. Each of these roles has a 3.1% (or 1) presence in our surveyed population. The majority of the respondents (22 out of the 32 respondents, or 68.8%) have between 1 and 5 years of experience, meaning we have a fairly new and somewhat inexperienced group of professionals. Even though this could be seen as negative, we believe younger people may be more keen to learn about and address sustainability issues in software development.

Other interesting statistics are that 93.75% of the respondents (30 out of 32) work with some variant of agile methodologies and that out of those 30 respondents, 28 work with Scrum (corresponding to 87.5% of the whole population). Finally, all the participants develop software using agile approaches.

Table 1. Respondents individual information.

ID	Preferred Agile Methods	Role in the Company	Experience with Agile (Years)
P1	Lean Development	Business Analyst	1–5
P2	Development Method (DSDM), Dynamic Systems, Feature Driven Development (FDD)	Business Analyst	1–5
P3	Development Method (DSDM), Dynamic Systems, Feature Driven Development (FDD)	Business Analyst	1–5
P4	Scrum	Business Analyst	1–5
P5	Crystal, Scrum	Developer	1 5
P6	Scrum	Developer	<1
P7	Scrum	Developer	<1
P8	Scrum	Developer	<1
P9	Kanban, Scrum, Scaled Agile Framework (SAFe)	Developer	1–5
P10	Scrum	Developer	<1
P11	Scrum	Developer	1–5
P12	Scrum	Developer	1–5
P13	Scrum	Developer	1–5
P14	Scrum	Developer	1–5
P15	Extreme Programming (XP), Scrum	Developer	1–5
P16	Extreme Programming (XP), Kanban, Lean Development, Scaled Agile Framework (SAFe), Scrum	Operations	1–5
P17	Kanban, Scrum	Product Manager	1–5
P18	Kanban, Lean Development, Scaled Agile Framework (SAFe)	Product Manager	>10
P19	Kanban, Scrum, Scaled Agile Framework (SAFe)	Product Manager	6–10
P20	Scrum	Product Owner	1–5
P21	Kanban, Scrum	Project Lead/Project Manager	1–5
P22	Scrum	Project Lead/Project Manager	1–5
P23	Kanban, Scrum	Project Lead/Project Manager	1–5
P24	Extreme Programming (XP), Lean Development, Scrum Scaled Agile Framework (SAFe)	Project Lead/Project Manager	>10
P25	Scrum	Scrum Master	1–5
P26	Kanban, Scrum, Lean Development	Scrum Master	1–5
P27	Scrum	Scrum Master	1–5
P28	Extreme Programming (XP), Feature Driven Development (FDD), Kanban, Lean Development, Scrum	Scrum Master	6–10
P29	Dynamic Systems, Kanban, Scaled Agile Framework (SAFe), Scrum	Tech Lead	6–10
P30	Scrum	Tester	<1
P31	Scrum	Tester	1–5
P32	Kanban, Scrum	UX/UI	1–5

4.2   Survey Questions and Results

Question 1. Do You Address Sustainability in Software Development?
This question aims to get an overview of how often sustainability is present in the projects developed by the participants. The options available for the participant to choose from are: never, rarely, sometimes, often, and always. Only one option

could be selected. Figure 1 synthesizes the received answers. A significant part
of the respondents 34.4% (11 out of 32) never address sustainability in their
projects, meaning that 65.6% (21 out of 32) of them do address sustainability
to different extents (even if only rarely). This is a positive aspect that leaves the
impression that developers and companies are already moving towards the right
direction. That said, it is not surprising that only 12.5% of the participants (4
out of 32) always address sustainability, since this is, after all, a new concept in
software development.

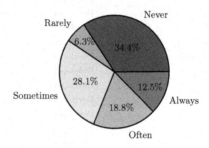

Fig. 1. Addressing of sustainability in software development

**Question 2. If You Have Answered Yes, in Which Activities Do You
Address Sustainability?** This question is intended to narrow down and identify in which activity of the software's life cycle sustainability is addressed by our
participants. In this question the respondent can choose from a number of different options namely requirements elicitation, software design, implementation,
unit testing, system/integration testing, and maintenance. Here the participant
could pick all of the options they wish. Figure 2 summarizes the results.

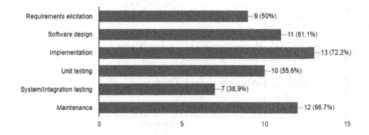

Fig. 2. Activities where sustainability is addressed

Out of the 18 participants that reported in what activities they deal with
sustainability, 86.6% of the answers do it in two or more activities. The top
three most used activities are 'Implementation' with 72.2% (13 answers), 'Maintenance' with 66.7% (12 answers) and 'Software design' with 61.1% (11 answers)
and out of the six activities used by the participants only 'System/Integration
testing' reported a use of less than 50%. There is a discrepancy in the number

of participants that answered this question (18) and the number of participants that said they address sustainability in software development in the previous question (21). This can be due to lack of a custom option where respondents can type what they want, or because 2 of these participants only have basic knowledge of sustainability and address sustainability in their projects "sometimes", and the other one only addresses it "rarely".

Question 3. If You Selected Any Option Above, How Do You Address Sustainability in Each of the Selected Activities? This was an open ended question, aiming at further deepen the previous question to be able to understand how our participants addressed sustainability in each of the activities they selected previously. By analyzing the answers to this question we concluded that simple and easy to maintain solutions that are also capable of being adapted to the businesses needs are necessary to achieve software sustainability in our participants' perspective.

Another 'popular' factor relevant to software sustainability is the reduction of technical debt, where our participants use *"a set of practices that allow us to reduce as much as possible our technical debt"*. One of the participants also said that *"ethic behaviour of the team and the software"* is a key factor. It is also mentioned that *"continuous information sharing with all stakeholders"* are important to maintain everyone up to date and achieve a continuous flow of information and so the team is able to have time to properly discuss the trade-offs between the different system qualities. Thus, adaptability, technical debt, ethics in the team and software, inclusion of all to guarantee continuous flow for information, and trade-offs among different qualities, are some of the sustainability-related aspects raised by the participants.

Question 4. How Do You Specify Sustainability Requirements? The goal of this question is to gather information on how our participants detail and specify the sustainability requirements necessary to address sustainability in their projects. Here the participant is presented with several options from which he can choose as many as needed. The options are: user stories, use case, backlog item, personas, acceptance criteria, part of the definition of done, informal text, UML models, none, and a space for free text where the participant could add a different option.

This question was analyzed in two ways: (i) regarding all of the 32 responses given and (ii) with the answers filtered to only include those who considered sustainability in their projects. The latter analysis left us with 21 answers (out of the 32). When analyzing all of the responses we can see in Fig. 3 that 75% of the respondents (24 participants) utilise user stories to specify sustainability requirements. This is to be expected since 30 participants work with a variant of agile and hence utilise 'User stories' regularly in their projects.

On the other hand, when we filter out the participants that said they do not take sustainability into account, the story repeats itself (see Fig. 4). Out of the 21 answers, 85.6% (19 participants) use "User Stories" to specify the

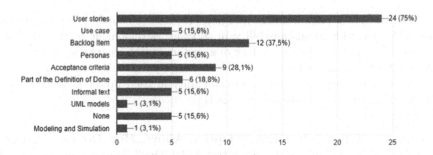

Fig. 3. Sustainability requirements specification artifacts

sustainability requirements. This was once again expected since out of those 21 participants there are also 85.6% of participants that work with a variant of agile. This suggests that developers tend to specify sustainability using formats that they already known and use regularly.

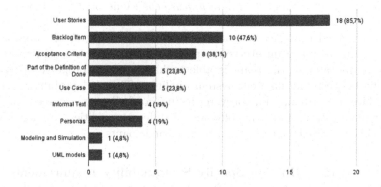

Fig. 4. Sustainability requirements specification artifacts filtered

Question 5. Which of the Following Best Describes Your Knowledge About Sustainability in Software Development? This question's objective is to ascertain the level of knowledge our participants know about sustainability in the software development realm. The available options to answer this question are: no knowledge, basic, intermediate, advanced, and expert. Only one option could be selected. The analysis of this question is summarized in Fig. 5.

It was quite surprising to see that 78% (25 participants) of the respondents say they have at least basic knowledge about sustainability in software. Even though this is positive, 13 out of those 25 only have basic knowledge on sustainability. Furthermore, only 6 participants have knowledge at an advanced (15,6%, or 5 people) or expert level (3,1%, or 1 person). This number of highly knowledgeable participants is contrasted by the number of completely uninformed ones, 7 (21,9%). This leads to the conclusion that there is an overall lack of

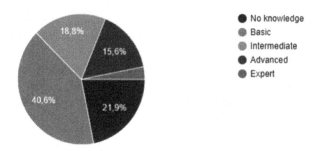

Fig. 5. Knowledge about sustainability in software development

knowledge about sustainability in software development, indicating the need for significant investment to be made in training, both by academia and industry; practitioners have to be taught and coached on the concepts of sustainability, similarly to what is happening with agile/Scrum. This lack of knowledge can also be because universities are just now starting to include sustainability in their ICT curricula. For example, some universities (e.g., from UK[2], and USA[3]) already have sustainability in their courses, but we cannot say this is common in many institutions.

Question 6. Do You Think You Have Tools or Methods to Be Able to Include Sustainability in Software Development and Maintenance? This "yes" or "no" question aims to get the participants' opinion on whether or not they have the appropriate tools to address sustainability in software development. We observed that 75% of the participants (24/32) report that they do not have tools or methods to deal with, or include, sustainability in software development. This is interesting as it supports our claim that sustainability in software development is still under researched and under developed.

Question 7. If You Answered Yes to the Previous Question, What Are the Methods You Use? This question also aims to further investigate the preceding question and as such was designed with the purpose of discovering any tools or methods our participants use, or have used in the past, to address sustainability in their projects. This is important because it might uncover methods or tools that we might have missed during our research or any sort of custom methods they might have developed.

The responses given show that participants use several types of tools that are not necessarily designed for sustainability. For example, a participant that works with the OutSystems framework[4] said *"you could use Architecture Dashboard"* to help you view any potential "design smells" and opportunities to improve

---

[2] https://www.ucl.ac.uk/sustainable/education/embed-sustainability-curriculum.

[3] https://sustainabilityinschools.edu.au/sustainability-curriculum.

[4] https://www.outsystems.com/.

their software. Others use CAST[5] (software intelligence suite that focuses on benchmarking software quality and condition) tools, KIUWAN[6] (code quality analysis SaaS focused on security) and SonarQube[7] (quality of code inspection tool). And lastly, some use *"sustainability models and articles"* while others use *"modeling and automatic code generation"* to deal with sustainability. One of our participants also said that there are tools available but the clients are not usually willing to pay for them.

Question 8. How Often Do You Think Software Has an Impact on Sustainability? This question intends on gauging the participants perceptions on whether or not software has an impact on sustainability. This question has 5 options: never, rarely, sometimes, often and always. Only one option can be selected.

The answers to this question revealed that the number of participants that thinks that software always impacts sustainability is similar to the number of participants that thinks that software has no impact on sustainability—this number being 18.8% (6 participants). Other than that, it is good to see that 81% of the participants (29 participants), think that software has some sort of impact on sustainability even if not often (see Fig. 6).

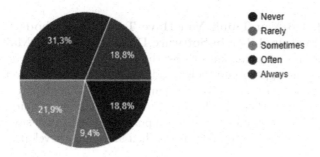

Fig. 6. How often software impacts sustainability

Question 9. In Projects Where Sustainability Was Addressed, What Were the Main Challenges Found by the Team? With this question we wanted to identify the major problems developers face when addressing sustainability in their projects. This question has multiple choices and the participant can pick as many as needed. The options are: lack of support materials, lack of experts on the domain, lack of knowledge by the team, lack of methods, lack of tools, and none.

There was also the option to choose a custom option where participants type what they want. The results from this question let us conclude that participants

---

[5] https://www.castsoftware.com/.

[6] https://www.kiuwan.com/.

[7] https://www.sonarqube.org/.

that try to implement sustainability in software encounter many types of challenges with high frequencies (see Fig. 7). As such, we can safely say this area is very undeveloped and that there are still no easy ways to address sustainability in software development.

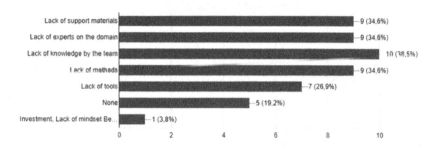

Fig. 7. Challenges found with addressing sustainability

Question 10. What Were the Main Reasons to Consider Sustainability Requirements in Your Projects? This question intends to discover what the mindset of our participants is on why they choose, or chose in the past, to include and implement sustainability requirements in their projects. The respondent can choose their answers from several options: client requirement, good reputation, legal obligations, social responsibility, organizational requirement, personal motivation, to improve the quality of the product and never was considered. Once again there was the option for the participant to add a custom option.

With this question we were able to gather the most important aspects, in our respondents opinion, to consider when addressing sustainability in a project (see Fig. 8). The most important factor was *"To improve the quality of the product"*; this factor was chosen by 53.3% of our respondents, hence picked 17 times. In second place comes *"Good reputation"* and *"Personal motivation"* tied with 34.4%, picked 11 times each.

4.3   Threats to Validity

Internal Validity. A threat to our questionnaire is that we might have asked the wrong questions, or at the very least constructed them poorly and making them ambiguous. To mitigate this, however, we structured the order of the questions and made an effort to employ clear wording, by reviewing the formulation of the questions in several meetings among the authors.

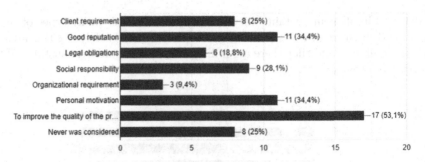

Fig. 8. Reasons to consider sustainability in projects

Construct Validity. Another threat is that we did not provide a definition for sustainability. Thus, each participant answered the questions with their own definition of sustainability in mind and as such could have failed to convey the information we needed. To mitigate this the questions were elaborated to be as objective as possible and reviewed by the senior authors.

External Validity. Also related to the participants is the fact that we did not constrain the role of the respondents; this means that we might have data that are not representative of the overall IT industry. This was, however, defined in this way by design, to allow our participants to feel more relaxed and avoid bureaucratic permissions from the managerial level that could delay the study.

Conclusion Validity. The data analysis process itself may also be a threat, due to the fact that wrong data analysis can yield wrong results and conclusions. So, to avoid this issue, we analyzed the data for each question individually through excel filters.

5   Major Insights

This section summarises some of the most relevant take away messages from the analysis of our survey. The study was limited to Portuguese professionals, where the results reveal the current situation of sustainability in agile development in Portugal, but it can serve as an indicator to be compared with companies worldwide.

Awareness About Sustainability. Although limited, there is a growing interest in considering sustainability during software development (Q1). Indeed, even if the majority of the participants realise that software has some impact on sustainability, a relevant amount of practitioners do not share this opinion. Therefore, more awareness about the importance of sustainability in IT should be raised in industry (Q8).

Advantages of Addressing Sustainability. Participants gave us many reasons to consider sustainability in their projects, where the improvement of product quality is the main one, but good reputation, social responsibility and personal motivation were also highly mentioned. This is good news, since it is an indication that sustainability is a concern worth investing in (Q10).

Techniques for Sustainability Requirements. When sustainability is addressed, implementation and maintenance are the main activities where sustainability is considered (Q2). This suggests more effort is needed to promote sustainability as part of the requirements activities. On the other hand, it was not surprising that user stories are the most popular artifact (Q4). Participants seem to adapt user stories to help specify sustainability requirements; this would help the integration of sustainability in requirements activities.

Challenges to Address Sustainability in IT. Participants pointed out to several challenges when addressing sustainability in their projects, such as lack of tools, learning materials, methods, domain experts, knowledge (Q9). Indeed, a non-surprising finding reported by our participants was the lack of methods and tools to help eliciting and specifying sustainability (Q6). When used, some practitioners rely on quality of code and code generation tools to address sustainability, but not much was used for requirements (Q7). Finally, although many respondents reported that they have basic knowledge on sustainability, more training and knowledge dissemination on sustainability is needed (Q5). This is in accordance with [14] where a clear lack of theoretical understanding and practical methods to apply sustainability to software and its development process is discussed. Indeed, the need for action is urgent and taking self-responsibility in developing critical thinking about this topic is necessary. But also there needs to be a bigger presence of sustainability discussion and goals in education to cultivate a mindset of change in future software designers. On a corporate level, there needs to be more workshops to educate developers, and clients alike about sustainability, its issues and its importance.

6   Related Work

In [15] the authors discuss the explanatory power of various sustainability indices applied in policy practice. Their results show that such indices fail to accomplish essential scientific requirements making them rather not useful if not misleading regarding policy advice.

In [16] it was identified that there was a lack of specific practices to be followed by global software development (GSD) vendors regarding the development of green and sustainable software. A study was carried on where agile practices were identified, aiming to help GSD vendors to improve their agile maturity towards green and sustainable software development.

In [17], a study shows that practitioners consider software sustainability important, but more at a technical level, where environmental considerations are missing. The study revealed that the meaning of sustainability needs to be refined for specific project and application context.

Our study reinforces the above studies by also identifying awareness of sustainability by practitioners and complements the findings by pointing out the need to provide more sustainability requirements techniques and training on the sustainability area.

7 Conclusions

This paper discusses the results of a survey we performed with a subset of the Portuguese industry aiming at identifying their current needs about sustainability as well as their practices when addressing sustainability in their agile projects. We were particularly interested in understanding how IT professionals elicit and specify sustainability requirements. To achieve this goal, we created a survey aiming at gathering participants' information, and their sustainability development issues, including methods and tools.

This survey was answered by 32 IT professionals, from several companies, working in several application domains. The analysis of the results revealed that the majority of the respondents say they address sustainability during software development even though 40% only have basic levels of knowledge on the topic. Furthermore, 75% said that they have no tools nor methods to address sustainability in software. As such, some participants have found some workarounds by using already existing pieces of software like code quality analyzers. It was interesting to note that implementation and maintenance are the main activities where sustainability is contemplated, instead of requirements. Finally, it was curious to see that participants consider sustainability to improve their product quality, and also to build good reputation in the market and to show social responsibility and increase personal motivation.

The identified shortcomings are key to decide how to improve the current state of knowledge and propose a new approach to specify sustainability requirements in an agile development context. Also, the agile processes should consider effective strategies to mitigate current difficulties (e.g., lack of knowledge, methods) in incorporating sustainability as part of their activities. Finally, we plan to replicate this study in IT companies from other countries.

Acknowledgements. We thank NOVA LINCS UID/CEC/04516/2019 for supporting this work.

References

1. Brundtland, G.H.: Our common future: development that meets the needs of the present without compromising the ability of future generations to meet their own needs. In: WCED (1987)
2. Penzenstadler, B., Femmer, H.: A generic model for sustainability with process-and product-specific instances. In: Workshop on Green In/by SE (2013)

3. Paech, B., Moreira, A., Araujo, J., Kaiser, P.: Towards a systematic process for the elicitation of sustainability requirements. In: 8th RE4SuSy Workshop, Co-located with 27th International Conference on RE (2019)
4. Albuquerque, D., Moreira, A., Araujo, J., Gralha, C., Goulão, M., Brito, I.S.: A sustainability requirements catalog for the social and technical dimensions. In: Ghose, A., Horkoff, J., Silva Souza, V.E., Parsons, J., Evermann, J. (eds.) ER 2021. LNCS, vol. 13011, pp. 381–394. Springer, Cham (2021). https://doi.org/10.1007/978-3-030-89022-3_30
5. Venters, C.C., et al.: Software sustainability: the modern tower of babel. In: CEUR WS Proceedings, vol. 1216, pp. 7–12 (2014)
6. Kitchenham, B.A., Pfleeger, S.L.: Personal opinion surveys. In: Shull, F., Singer, J., Sjøberg, D.I.K. (eds.) Guide to Advanced Empirical Software Engineering, pp. 63–92. Springer, London (2008). https://doi.org/10.1007/978-1-84800-044-5_3
7. What Is Agile? https://www.atlassian.com/agile. Accessed 15 Jan 2021
8. User Stories. https://www.atlassian.com/agile/project-management/user-stories. Accessed 15 Jan 2021
9. World Commission on Environment and Development: Our Common Future. Oxford University Press, Oxford (1987). https://EconPapers.repec.org/RePEc:oxp:obooks:9780192820808. Accessed 15 Jan 2021
10. Becker, C., et al.: Requirements: the key to sustainability. IEEE Softw. 33(1), 56–65 (2016)
11. Murugesan, S.: Harnessing green it: principles and practices. IT Prof. 10(1), 24–33 (2008)
12. Naumann, S., Dick, M., Kern, E., Johann, T.: The Greensoft model: a reference model for green and sustainable software and its engineering. Sustain. Comput. Inform. Syst. 1(4), 294–304 (2011)
13. Punter, T., Ciolkowski, M., Freimut, B.G., John, I.: Conducting on-line surveys in software engineering. In: International Symposium on Empirical Software Engineering (ISESE 2003). IEEE Computer Society (2003)
14. Chitchyan, R., et al.: Sustainability design in requirements engineering: state of practice. In: IEEE/ACM 38th ICSE, pp. 533–542 (2016)
15. Böhringer, C., Jochen, P.: Measuring the immeasurable - a survey of sustainability indices. Ecol. Econ. 63, 1–8 (2007)
16. Rashid, N., Khan, S.U.: Agile practices for global software development vendors in the development of green and sustainable software. J. Softw. Evol. Process 30(10), e1964 (2018)
17. Groher, I., Weinreich, R.: An interview study on sustainability concerns in software development projects. In: 43rd Euromicro Conference on Software Engineering and Advanced Applications (SEAA) (2017)

Defining Key Concepts in Information Science Research: The Adoption of the Definition of Feature

Sabine Molenaar, Emilie Steenvoorden, Nikita van den Berg,
Fabiano Dalpiaz[ID], and Sjaak Brinkkemper[(✉)][ID]

Department of Information and Computing Sciences, Utrecht University, Utrecht,
The Netherlands
{s.molenaar,e.r.m.steenvoorden,
i.a.n.vandenberg,f.dalpiaz,s.brinkkemper}@uu.nl

Abstract. This paper analyzes the definitions of the concept *feature* in the information science literature. The concept of feature has been defined in various ways over the last three decades. To be able to obtain a common understanding of a feature in information science, it is important to conduct a thorough analysis of the definitions that can be used in research and in practice. The main contribution of this paper is a categorization of the existing definitions, which highlights similarities and differences. By means of a *Concept Definition Review* process, we gather a total of 23 definitions from Google Scholar using five search queries complemented by backward snowballing. Our analysis organizes the definitions according to their level of abstraction and the taken viewpoint. Within the range of analyzed definitions, we do not wish to argue that one is better or worse than another. We provide, however, guidelines for the selection of a definition for a given goal. These guidelines include: popularity based on the citations count, the research field, the abstraction level, and the viewpoint.

Keywords: Feature · Requirements engineering · Information science · Definition · Literature review · Concept Definition Review

1 Introduction

The concept of *feature*, in relation to software and information systems, has been defined in many ways over the last three decades. One of the first definitions dates back to the 1990s, and it is stated in a highly influential technical report on feature-oriented domain analysis [20]. This definition seems to be adapted from the American Heritage dictionary entry for *feature*. Ever since, alternative definitions of the concept of *feature* emerged, which deviated from it.

The existence of multiple, diverging definitions has both conceptual and practical consequences. Conceptually, researchers may use the same terminology while referring to different meanings (*denotations* [28]), leading to undetected conflicts in verbal or written communication. Practically, the choice of a definition may affect the artifacts that are created based on the concept. For instance,

R. Guizzardi et al. (Eds.): RCIS 2022, LNBIP 446, pp. 442–457, 2022.
https://doi.org/10.1007/978-3-031-05760-1_26

since features are at the basis of feature diagrams [20], different definitions may lead to conflicting interpretations of a feature diagram, or to different models for the same system depending on the modeler's preferred definition.

To reach the goal of analyzing the different definition of feature, we sketch a more general literature review approach that we call *Concept Definition Review (CDR)*. Literature reviews are a widely practiced type of research method in various flavors in all scientific disciplines [12]: systematic literature review, meta-analysis, argumentative literature review, systematic mapping review. This new approach was made necessary by the need to identify the definitions of a certain term without conducting a heavyweight systematic literature review. Definitions of newly introduced concepts are usually made in an explicit statement (e.g., "We define the concept of *feature* as follows ...") at the beginning of the paper in order to establish a common understanding with the reader. For the CDR, the definition text with some context suffices and the remainder of the paper is then ignored. We envision that CDRs can be used to bring clarity regarding several concepts in the domains of information science, information systems, and software engineering, e.g., those of class, function, task, and goal.

We choose to investigate the concept of feature because of its importance both in Requirements Engineering (RE) and in Software Architecture (SA). For example, in our RE4SA framework [30], features are an elementary abstraction to define the functional architecture of a system, and atomic functional requirements are expected to justify individual features.

The rest of the paper is organized as follows. In Sect. 2, we describe the Concept Definition Review approach and its application to the concept of feature. In Sect. 3, we analyze the types of definitions, the research topics, the abstraction level, and the viewpoint on the concept of feature. Finally, we provide guidelines for researchers on the usage of the definitions of feature, and we conclude, in Sect. 4.

2 Research Method and Data Collection

The goal of this research is to recommend definitions that fit various perspectives and multiple purposes, rather than that of creating an exhaustive list of all definitions of the concept of feature. We are particularly interested in collecting and analyzing definitions from the domains of RE and SA. Therefore the main research question for this paper is formulated as follows: *"How are features defined for different purposes in the context of information science literature?"*.

2.1 Concept Definition Review

In an attempt to properly define the concept feature and categorize existing definitions in the field of information science, we devised a literature research method that focuses on analyzing and clarifying the meaning of a concept in the literature. The high-level structure of the CDR method consists of the following six steps, which are inspired by the SLR guidelines by Okoli and Schabram [25]:

1. *Purpose of the concept definition review:* the goal and research scope of the concept at hand is established by selecting the scientific sub-domains where the concept plays a critical role;
2. *Searching for papers containing definitions:* querying bibliographic indexes with the name of the concept in the identified scientific sub-domains;
3. *Relevancy screen and quality appraisal:* for each paper that fulfills the quality criteria of renowned scientific publication venues, an explicit formulation of the concept definition is to be identified;
4. *Data extraction:* various data items are collected from the literature resources, e.g. syntactical structure, research domain, and citation impact data;
5. *Synthesis of studies:* analysis of the data items provides insights on concept definition adoption, variations over abstraction levels, and evolution over time, i.e. old-fashioned definitions versus new interpretations (see, for instance, the changes of definitions by the same author group in Table 3);
6. *Documenting the concept guidelines:* based on the synthesis findings, a guideline is formulated for the most suitable concept definition usage in the research domains.

The remainder of this paper describes an instantiation of the CDR for the concept of *feature*. The six steps are illustrated by presenting our experience. A more general definition of the techniques that can be used is left to future work.

In our research, the definitions are collected by searching in Google Scholar, and by complementing those identified sources via backward snowballing. After the definitions are identified, we gathered the number of citations. The data collection is done at two points in time to observe usage evolution. The collection started in November 2018, then the research paused for two years, and additional data is collected in October 2021. The collected information is then analyzed from various perspectives. We first analyze the definition type and the research topic based on the abstract, keywords, introduction, and research topic of the venue where it has been published. Next, a categorization is made based on the level of abstraction and the viewpoint. The differentiation between abstract and technical is done through the method proposed by Apel and Kästner [4]. Based on the data collection and the analysis, we provide guidelines on how to select a suitable definition of *feature* for use in a given context.

2.2   Data Collection

Initially, relevant papers are found using the search operators in Google Scholar. Through citation tracing, other literature repositories became involved (Scopus, ACM DL, IEEE Xplore, etc.). All definitions should be related to the term *feature*; thus, this term is included in all search queries. Since that term in combination with the term 'definition' often leads to results not related to information science, more specific queries were used instead. The second term in the search query is based on other topics relevant to this paper, as explained in the introduction: RE and SA. This is aligned with the objective of our research, which aims at identifying definitions for various purposes, but within the research sub-fields we have defined.

To scale down the number of results and to assure quality, some results are excluded. The definition and the source where the definition appears should meet the following selection criteria:

- They must be written in English.
- They must be scientific literature.
- They must present a unique definition of the concept feature.

Table 1 provides an overview of the used search queries and their included results. At first, we hoped we could restrict our search to a limited number of citations. However, since the works used in the literature study range from 1990 to 2021, this would be an unfair criterion, since older works have had more time to get cited. The third criterion traces to the original first definition of the concept. Furthermore, since the aim is to provide an overview of existing definitions, less cited definitions should be featured as well for completeness. The results are presented in order of the search results (relevance in relation to the search query). It should be noted that the work by Apel et al., published in 2013, is stated as a work from 2016 by Google Scholar. However, the book itself includes a copyright text from 2013 and the foreword was also dated 2013. The identified results that did not match the selection criteria are not listed in the table.

Table 1. Overview of the search queries on Google Scholar and of the returned results.

Search query	Included results
"feature" AND "requirements engineering"	Classen et al., 2008; Kang et al., 1990
"feature" AND "software architecture"	Apel & Kästner, 2009; Kang et al., 1990; Zhang et al., 2019
"feature" AND "product lines"	Apel et al., 2013
"feature" AND "software system"	Apel et al., 2013; Apel & Kästner, 2009
"feature" AND "feature-oriented specification"	Guerra et al., 1996; Apel & Kästner, 2009
"feature" AND "source code"	Dit et al., 2013

In addition, the snowballing technique was utilized. In this case, this consisted of backwards searching. Two articles were selected as a starting point, since these two works explicitly cited various definitions of the term *feature*. Table 2 summarizes which and how many works have been found per article.

Table 2. Works found through the use of the backward snowballing technique.

Source	References	Total
Classen et al., 2008	Kang et al., 1990; Kang et al., 1998; Bosch, 2000; Czarnecki & Eisenecker, 2000; Batory, 2004; Batory et al., 2004; Pohl et al., 2005; Batory et al., 2006; Apel et al., 2007	9
Apel & Kästner, 2009	Kang et al., 1990; Kang et al., 1998; Bosch, 2000; Czarnecki & Eisenecker, 2000; Zave, 2003; Batory et al., 2004; Chen et al., 2005; Czarnecki et al., 2005; Pohl et al., 2005; Batory et al., 2006; Apel et al., 2007; Classen et al., 2008; Kästner et al., 2008	13

Table 3. Definitions of *feature* obtained in our instantiation of the CDR. The 'Year' column refers to the year of publication.

Authors	Year	Definition
Kang, Cohen, Hess, Novak & Peterson [20]	1990	*"a prominent or distinctive user-visible aspect, quality or characteristic of a software system or systems"*
Guerra, Ryan & Sernadas [17]	1996	*"is a part or aspect of a specification which a user perceives as having a self-contained functional role"*
Kang, Kim, Lee, Kim, Shin & Huh [19]	1998	*"distinctively identifiable functional abstractions that must be implemented, tested, delivered, and maintained"*
Bosch [9]	2000	*"a logical unit of behaviour specified by a set of functional and non-functional requirements"*
Czarnecki & Eisenecker [13]	2000	*"a distinguishable characteristic of a concept (e.g., system, component, and so on) that is relevant to some stakeholder of the concept"*
Zave [33]	2003	*"an optional or incremental unit of functionality"*
Batory [6]	2004	*"the primary units of software modularity"*
Batory, Sarvela & Rauschmayer [5]	2004	*"a product characteristic that is used in distinguishing programs within a family of related programs"*
Chen, Zhang, Zhao & Mei [10]	2005	*"a product characteristic from user or customer views, which essentially consists of a cohesive set of individual requirements"*
Czarnecki, Helsen & Eisenecker [14]	2005	*"a system property that is relevant to some stakeholder and is used to capture commonalities or discriminate among systems in a family"*
Pohl, Böckle & van der Linden [26]	2005	*"an end-user visible characteristic of a system"*
Batory, Benavides & Ruiz-Cortes [7]	2006	*"an increment in product functionality"*
Apel, Lengauer, Batory, Möller & Kästner [3]	2007	*"a structure that extends and modifies the structure of a given program in order to satisfy a stakeholder's requirement, to implement and encapsulate a design decision, and to offer a configuration option."*
Classen, Heymans & Schobbens [11]	2008	*"a triplet, f = (R,W,S), where R represents the requirements the feature satisfies, W the assumptions the feature takes about its environment and S its specification"*
Kästner, Apel & Kuhlemann [21]	2008	*"represents an increment in functionality relevant to stakeholders"*
Apel & Kästner [4]	2009	*"is a unit of functionality of a software system that satisfies a requirement, represents a design decision, and provides a potential configuration option"*
Apel, Batory, Kästner & Saake [2]	2013	*"is a characteristic or end-user-visible behavior of a software system"*
Dit, Revelle, Gethers & Poshyvanyk [15]	2013	*"represents a functionality that is defined by requirements and accessible to developers and users"*
Berger, Lettner, Rubin, Grünbacher, Silva, Becker, Chechik & Czarnecki [8]	2015	*"describe the functional and non-functional characteristics of a system"*
Andam, Burger, Berger & Chaudron [1]	2017	*"are high-level, domain-specific abstractions over implementation artifacts"*
Krüger, Gu, Shen, Mukelabai, Hebig & Berger [23]	2018	*"used to specify, manage, and communicate the behavior of software systems and to support developers in comprehending, reusing, or changing these systems"*
Rodríguez, Mendes & Turhan [27]	2018	*"represent needs that are gathered via meetings with customers or other stakeholders, which, once selected, are refined during a requirements elicitation process"*
Zhang, Wang & Xie [34]	2019	*"indispensably basic functional modules available to users, which can be captured by one or two words in the review"*

Overlapping references between the two works are included for both in the interest of completeness. Based on correspondence with Sven Apel, an additional three works co-written by Thorsten Berger are included (S. Apel, personal communication, February 12, 2019). More than one definition written by Apel is included and the article providing an overview of feature-oriented development is used not only as a starting point for searching for more definitions, but also because it inspired the synthesis in part. Therefore, the recommendation was gladly accepted. Moreover, these three works are published more recently than most of the other included works, providing a scientific evolution of the term *feature* over the past thirty years.

Table 3 shows the feature definitions in chronological order and, if two or more works were published in the same year, alphabetical order is applied. The references are provided via short citations to increase the readability of the table. The table highlights the high number of definitions of the term *feature* in the context of RE and SA: we identified 23 relevant ones. This leads to possible ambiguity and conflict [28] when discussing the literature in the field, and also when interpreting or creating models that refer to the concept of feature such as feature diagrams [20].

2.3 Popularity of the Definitions

After collecting the definitions, additional data about the number of citations per work was gathered. This data can suggest a first selection of a definition to use. Figure 1 shows the number of citations per article measured in 2018 (blue) and 2021 (green). Also, the percentage growth of the number of citations in this time period is visible for each work. Interestingly, from the three most cited works, Kang et al., Eisenecker and Czarnecki and Pohl et al., the latter two experienced a stronger growth: 32% and 29%, respectively, much higher than the 5% growth of Kang et al. Since more recently published works have had less time to get cited, the picture may convey a slightly skewed view. Therefore, the publication year should also be taken into account when selecting the most cited works. This can be done by looking at a trend line. From the trend line based on the citations measured in 2021, it is apparent that four works are cited significantly more often relative to the others, being the works from Kang et al. (+271 citations), Eisenecker and Czarnecki (+1,177), Bosch (+279), and Dit et al. (+245). However, using only the number of citations does not take research topics into account.

3 Analysis and Categorization of the Concept Definitions

The previous section provided a basic recommendation for selecting a definition based on the number of citations. That straightforward criterion, which is very easy to adopt, does not actually answer the question as to what definition should be used in what context. To answer this question, we propose various analyses of the definitions of Table 3, which include the categorization of the definitions

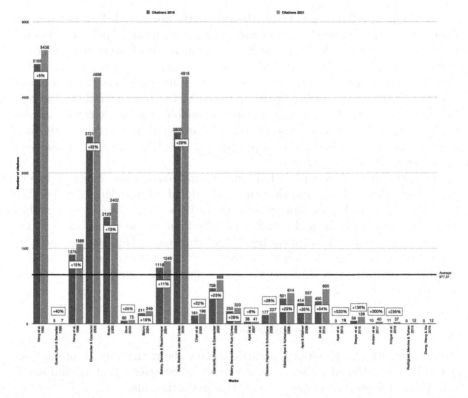

Fig. 1. Overview of number of citations on Google Scholar of works in which definitions are provided (from November 2018 to October 2021). (Color figure online)

according to various facets: definition type, research topic, level of abstraction, and viewpoint.

3.1   Definition Types

Within the broad range of types of definitions, an important distinction is that between *intensional* and *extensional*.

An *intensional* definition describes the common characteristics of the members of the category, e.g., birds have feathers, they can fly, and they have a specific shape [16]. For example, Kang et al. (1990): *"a prominent or distinctive user-visible aspect, quality or characteristic of a software system or systems"*. Here, a feature is described by means of common characteristics. Another example is that of Zhang et al. (2019): *"indispensably basic functional modules available to users, which can be captured by one or two words in the review"*. An *extensional* definition lists the members of the category, e.g. robins, eagles, nightingale, etc. Furthermore, other definition types exist. For example, *ostensive* definitions, which are like extensional definitions, but where extensional definitions call for an exhaustive list of members of the category, ostensive definitions only call for a couple of example members [32].

Besides this main distinction, other types of definitions can be identified. First off, a *stipulative* definition is used when a term is made up for the first time. Often consisting of a general category the concept belongs to, followed by a differentiator. Secondly, *lexical* definitions provide descriptions that depends on the term's use in particular communities: the definition depends on the audience. Common examples come from the legal domain. *Partitive* definitions explain concepts as being part of a greater whole. A partitive definition of feature is that of Guerra et al. (1996): *"is a part or aspect of a specification which a user perceives as having a self-contained functional role"*. Next, *functional* definitions explain actions or activities of a concept in relation to another concept. The definition of Batory et al. (2004) is functional: *"a product characteristic that is used in distinguishing programs within a family of related programs"*. Encyclopedic definitions often include additional information on the concept. Next, *theoretical* definitions attempt to add an argument for a concept and can be compared to scientific hypothesis. The last type of definition is the *synonym* definition, which describe a concept by mentioning a similar concept [31].

All definitions in this research, except for those of Guerra et al. (1996) and Batory et al. (2004), are intensional definitions, because the definitions analyze the concepts into constituent characteristics. It would be interesting to explore whether other types of definitions would be suitable for providing a clear, homogeneous characterization of the notion of feature.

3.2 Research Topics

To recommend definitions based on context, it is useful to see in what research field or sub-field a definition is proposed. This organization based on research topic is provided in Table 4.

Table 4. Feature definitions categorized by research topic.

Research topic	Related works
Feature-oriented software	Kang et al., 1990, 1998; Batory, 2004; Apel et al., 2007; Apel & Kästner, 2009; Dit et al., 2013; Zhang et al., 2019
Feature-oriented specifications	Guerra et al., 1996; Zave, 2003
Generative programming	Czarnecki & Eisenecker, 2000
Software product lines	Bosch, 2000; Batory et al., 2004; Pohl et al., 2005; Kästner et al., 2008; Apel et al., 2013; Berger et al., 2015; Andam et al., 2017; Krüger et al., 2018
Feature modeling	Chen et al., 2005; Czarnecki et al., 2005; Batory et al., 2006
Requirements engineering	Classen et al., 2008
Release planning	Rodríguez et al., 2018

The research topics are determined based on which topics or fields in the abstract, keywords or introduction. In addition, we also considered the research fields or topics related to the journal or conference proceedings where the work was published. Some overlap between the topics is possible, since some works include a more specific topic or field than others. For example, in Table 4, feature-oriented software may also be interpreted as feature-oriented programming in some cases, but to keep it more generic, the former topic description is used instead. In addition, it is possible that a definition could fit more than one research topic, in such cases the most important or prominent one is selected. For instance, the definition by Rodríquez et al. could also fit the RE topic, but it is categorized as release planning, since this was the main topic of the work.

This categorization per research topic can be used to select a definition to utilize in the context of one of the research topics. For topics that have multiple fitting definitions, the additional factor of number of citations can suggest a preference. However, different and more extensive approaches for establishing the most suitable concept definition could be envisioned.

3.3   Abstraction Level

Some definitions might be more suitable than others given a certain context or aim. In all definitions, two aspects can be distinguished: abstraction level and viewpoint. The former aspect was inspired by the differentiation between abstract and technical feature definitions as proposed by Apel and Kästner [4]. They also recognize that features have more than one use and describe the differentiation as follows:

1. Abstract: *"features are abstract concepts of the target domain, used to specify and distinguish software systems"* (problem space)
2. Technical: *"features must be implemented in order to satisfy requirements"* (solution space)

Czarnecki and Eisenecker have separated the problem space from the solution space, in which the former focuses on domain-specific abstractions and the latter on implementation-oriented abstractions [13]. Apel and Kästner use this distinction to further define abstract and technical, relating abstract definitions to the problem space and technical definitions to the solution space [4]. In addition to this distinction between abstract and technical definitions, they have provided a list of ten definitions (all of which are also included in Table 3) and ordered them from abstract to technical. However, they have not described how they decided on which definition is more technical than another. Moreover, they identified seven abstract definitions and only three technical ones. In short, while the line between abstract and technical is clear, the gap between the two is not and the reasoning behind the order within both distinctions is vague at best.

To clarify the interpretations of abstract and technical, Sven Apel was asked to comment on the paper. He stated that the first seven definitions "take a user-centric/problem-space-centric perspective", while the eighth definition is only formal from an RE perspective. The last two definitions focus on the implementation and are thus solution-space-specific. He continues by saying that, within these categorizations, the definitions are more or less sorted by date (S. Apel, personal communication, February 12, 2019). To conclude, this approach was quite informal and therefore difficult to replicate. Moreover, it still does not solve the mystery of which definition is more abstract or technical than another. A more formal categorization is needed to tackle these challenges.

In an attempt to recreate and extend such an order based on level of abstraction (from abstract to technical), nine characteristics were extracted from the collection of 19 definitions (the other four were added later due to additional communication and refreshing of the data in 2021). The following nine characteristics were extracted:

- Abstract: characteristic, distinct (or variations thereof), aspect, abstraction
- Technical: specification, functionality (or variations thereof), requirements, behavior, unit

The identified abstract characteristics are assigned a score of 1, the technical characteristics receive a 0, then we divide this value by the number of characteristics, leading to a score between 0 and 1. In this case, 1 is the most abstract and 0 the most technical (or least abstract). A test comparing the order based on these nine characteristics and resulting score and the order of seven definitions as presented by Apel and Kästner resulted in the following findings:

- Six out of seven definitions were ordered differently.
- Two definitions were shifted three positions.
- If the line between abstract and technical is placed at 0.5, one definition shifts from abstract to technical and one is shifted the other way around.

Due to the deviation from the original order and the sensitivity of the placement of the abstract/technical line, it is concluded this approach has certain drawbacks. A second attempt, adopting a different interpretive approach, yielded better results. After analyzing the different characteristics of abstract and technical as stated by Czarnecki and Eisenecker, and Apel and Kästner, as discussed earlier in this section, the following eight characteristics were identified:

- Abstract: problem space, description of requirements, description of intended behavior and characteristic/abstract/abstraction
- Technical: solution space, satisfaction of requirements, implementation of intended behavior and functionality

Using this approach, with the same method for calculating a score, the 19 definitions were ordered once again (the results are shown in Fig. 2) with the following results:

- If the line between abstract and technical is placed at 0.5, none of the definitions shift from abstract to technical or vice versa.
- The three technical definitions are in the same order.

– Out of the seven abstract definitions, only two are out of order (and the order among those two is the same as in the order by Apel and Kästner).

To summarize, out of the ten definitions, only two were out of order (and disregarding the other definitions, those two were in the correct order). Another advantage of this approach is that it is not based on terms/characteristics extracted from the definition, but on theoretical resources by Czarnecki and Eisenecker, and Apel and Kästner. Furthermore, it should not be forgotten that it is unclear whether the original order as devised by Apel and Kästner is on an ordinal scale. It is reasonable to assume so, since the definitions are numbered. However, the reasoning behind this specific order is not thoroughly explained, apart from the descriptions of abstract and technical as stated previously. The one fully unambiguous aspect is the distinction between the abstract and technical definitions, since this was explicitly mentioned.

3.4   Viewpoint

In addition to the level of abstraction, five viewpoints were also extracted from exactly mentioned terms in the definitions:

– System
– Product
– Developer (stakeholder)
– User (stakeholder)
– Customer (stakeholder)

Firstly, system and product are considered separate viewpoints, since a system can be contained within a product, but a product can indicate more than just a system. Secondly, three stakeholders were identified and only human beings are considered a stakeholder. The developer was included, not because it was explicitly mentioned in any of the definitions, but sometimes the word stakeholder also refers to the development viewpoint. Thirdly, the user viewpoint also includes end-users and differs from the developer viewpoint, since developer do not necessarily use the product or system, but other employees of the product's or system's company might. Fourthly, customers are separated from user, since they are more specific than just any (end-) user. Finally, whenever no specific viewpoint is mentioned or can be reasonably assumed given a definition, the system is considered the viewpoint, due to features being part of the SA, which describes a system. Figure 2 shows the categorization of the definitions based on the level of abstraction score (as described previously) and the identified viewpoints.

The 19 definitions and the scoring system were also presented to a group of 26 information science students and researchers. Both expressed a difficulty in understanding what the term 'technical' was supposed to mean in this context. Given their background, they automatically assumed technical characteristics to be related to development aspects or implementations (such as code). Moreover, the level of abstraction is often seen as the level of granularity, while in this

Viewpoint

Definition	System	Product	Developer (stakeholder)	User (stakeholder)	Customer (stakeholder)
Kang et al. (1990)	✓			✓	
Bosch (2000)	✓				
Pohl et al. (2005)	✓			✓	
Chen et al. (2005)		✓		✓	✓
Guerra et al. (1996)				✓	
Kang et al. (1998)	✓				
Czarnecki & Eisenecker (2000)	✓		✓	✓	✓
Batory et al. (2004)		✓			
Czarnecki et al., 2005			✓	✓	✓
Apel et al. (2013)	✓			✓	
Rodríguez et al. (2018)			✓	✓	✓
Classen, et al. (2008)	✓				
Zave (2003)	✓				
Batory (2004)	✓				
Batory et al. (2006)		✓			
Kästner et al. (2008)			✓	✓	✓
Dit et al. (2013)			✓	✓	
Apel et al. (2007)	✓		✓	✓	✓
Apel & Kästner (2009)	✓				

(Left axis: **Level of Abstraction**, ranging from **Abstract** at top to **Technical** at bottom. Dashed line after Classen, et al. (2008) labeled "abstract - technical split".)

Fig. 2. Categorization of the 19 definitions, based on abstraction level and viewpoint.

categorization that is not the case. To make the categorization easier to read and understand, a different name and more specific minimal and maximum values would be desirable. Changing 'technical' to 'detailed' might solve the issue of misinterpreting technical characteristics, but would be an inaccurate description. The definitions do not necessarily refer to a certain level of detail and abstract definitions can still provide a detailed description of the term *feature*.

The role of RE versus SA appears crucial in this concept definition study. As Shekaran et al. explain the role of SA in RE by referring to RE as being concerned with the 'shape of the problem space', while SA focuses on the 'shape of the solution space' [29]. The distinction between problem and solution space is already present in the categorization, given the fact that the description of the problem space is considered an abstract characteristic and, on the other hand, the solution space is considered a technical characteristic [22]. To strengthen this reasoning, the Quality User Story (QUS) framework is in agreement stating that a user story (US) should be problem-oriented, meaning that *"a user story only specifies the problem, not the solution to it"* [24]. Moreover, Hofmeister et al. mention that architecture solutions help move the design from the problem space (in which architecturally significant requirements (ASRs) are formulated) to the solution space [18]. Splitting the definitions into two main categories can make selecting a definition easier, depending on the purpose for and context in which it is used. However, problem-space definitions (RE) can arguably be considered of higher quality or more useful, based on research by Berger et al. They state that *"good features need to precisely describe customer-relevant functionality"* [8]. Moreover, this would mean that definitions which include the customer viewpoints are more suitable in RE than those that do not.

4   Concept Definition Guidelines and Conclusion

Addressing our main research question *"How are features defined for different purposes in the context of information science literature?"*, we could not find a definitive answer. Many definitions of the term exist and one is not necessarily better or more accurate than the next. The selection of a definition is clearly dependent on the chosen perspective, and there is a wide difference in the adoption of a particular definition as derived from the citation count.

This paper distinguishes the definitions based on the research topics feature-oriented software, feature-oriented specifications, generative programming, software product lines, feature modeling, requirements engineering and release planning. If one definition had to be selected, it would have to be that of Kang et al. (1990) [20]. This is the oldest, it has been cited most frequently, and it is referenced more often by like-minded researchers than the other works included in this study.

However, the meaning of the term *feature* is largely dependent on its purpose, be it for requirements, architecture, development, modeling, target audience or otherwise. To complicate matters further, the viewpoint can influence the definition. Besides that, in this research a distinction was made between problem-oriented (abstract) and solution-oriented (technical). The only aid that can be provided when selecting a definition is the popularity of the definition, the research field and/or context, the intended viewpoint and audience. Even then, multiple options may be available.

With all this taken into account, the following guidelines are most fitting. First consider the research topic and select a definition that fits the topic to be written about. If that topic has multiple definitions, choose the definition with the most citations relative to its publication year.

For the topic feature-oriented software, the recommendation would be *"a prominent or distinctive user-visible aspect, quality or characteristic of a software system or systems"* from Kang et al. (1990). For feature-oriented specification, it is *"an optional or incremental unit of functionality"* from Zave (2003).

For generative programming, it is *"a distinguishable characteristic of a concept (e.g., system, component, and so on) that is relevant to some stakeholder of the concept"* from Czarnecki & Eisenecker (2000). For the research topic of software product lines, the definition *"an end-user visible characteristic of a system"* from Pohl et al. (2005) should be used. For feature modelling, *"a system property that is relevant to some stakeholder and is used to capture commonalities or discriminate among systems in a family"* from Czarnecki et al. (2005). For the topic requirements engineering, *"a triplet, f = (R, W, S), where R represents the requirements the feature satisfies, W the assumptions the feature takes about its environment and S its specification"* from Classen et al. (2008). Lastly, for the research topic of release planning, *"represent needs that are gathered via meetings with customers or other stakeholders, which, once selected, are refined during a requirements elicitation process"* from Rodríguez et al. (2018) should be used.

If the specific topic is not present in Table 4, we recommend to look at the corresponding viewpoint and to choose the most fitting definition for that viewpoint based on Fig. 2. Opt for the higher level of abstraction when talking about RE, and for the technical abstraction level when talking about SA. When there are no definitive viewpoints used, work with the definition that is most all-encompassing and relatively includes most of the most frequently used terms. So, the definition we advise would be *"a unit of functionality of a software system that satisfies a requirement, represents a design decision, and provides a potential configuration option"* by Apel and Kästner [4].

Future work could look at the use of the concept *feature* in practice rather than in the literature. Perhaps, investigating whether or not features are used differently in open source and industrial software projects or in relatively large or small software development departments could yield interesting insights.

Applications of the Concept Definition Review process to other concepts in the domain of information science, computer science, or artificial intelligence would possibly reveal the plethora of concept definitions. Agreement and standardization will assist researchers to read and understand concepts better in order to utilize them in presenting and explaining their scientific contributions.

Acknowledgements. We would like to thank Sven Apel for clarifying and discussing the categorization of definitions on the level of abstraction.

References

1. Andam, B., Burger, A., Berger, T., Chaudron, M.: FLOrIDA: Feature LOcatIon DAshboard for extracting and visualizing feature traces. In: Proceedings of the Eleventh International Workshop on Variability Modelling of Software-Intensive Systems, pp. 100–107 (2017)
2. Apel, S., Batory, D., Kästner, C., Saake, G.: Feature-Oriented Software Product Lines: Concepts and Implementation. Springer, Heidelberg (2013). https://doi.org/10.1007/978-3-642-37521-7

3. Apel, S., Lengauer, C., Batory, D., Möller, B., Kästner, C.: An algebra for feature-oriented software development. Technical Report MIP-0706, Department of Informatics and Mathematics, University of Passau (2007)
4. Apel, S., Kästner, C.: An overview of feature-oriented software development. J. Object Technol. 8(5), 49–84 (2009)
5. Batory, D., Sarvela, J., Rauschmayer, A.: Scaling step-wise refinement. IEEE Trans. Softw. Eng. 30(6), 355–371 (2004)
6. Batory, D.: Feature-oriented programming and the AHEAD tool suite. In: Proceedings of the 26th International Conference on Software Engineering, pp. 702–703 (2004)
7. Batory, D., Benavides, D., Ruiz-Cortes, A.: Automated analysis of feature models: challenges ahead. Commun. ACM 49(12), 45–47 (2006)
8. Berger, T., et al.: What is a feature? A qualitative study of features in industrial software product lines. In: Proceedings of the 19th International Conference on Software Product Lines, pp. 16–25 (2015)
9. Bosch, J.: Design and Use of Software Architectures: Adopting and Evolving a Product-Line Approach. Pearson Education, London (2000)
10. Chen, K., Zhang, W., Zhao, H., Mei, H.: An approach to constructing feature models based on requirements clustering. In: Proceedings of the 13th IEEE International Conference on Requirements Engineering, pp. 31–40 (2005)
11. Classen, A., Heymans, P., Schobbens, P.-Y.: What's in a *Feature*: a requirements engineering perspective. In: Fiadeiro, J.L., Inverardi, P. (eds.) FASE 2008. LNCS, vol. 4961, pp. 16–30. Springer, Heidelberg (2008). https://doi.org/10.1007/978-3-540-78743-3_2
12. Creswell, J.W., Creswell, J.D.: Research Design: Qualitative, Quantitative, and Mixed Methods Approaches. Sage Publications, Thousand Oaks (2017)
13. Czarnecki, K., Eisenecker, U.: Generative Programming: Methods, Tools, and Applications, vol. 16. Addison Wesley, Reading (2000)
14. Czarnecki, K., Helsen, S., Eisenecker, U.: Formalizing cardinality-based feature models and their specialization. Softw. Process Improv. Pract. 10(1), 7–29 (2005)
15. Dit, B., Revelle, M., Gethers, M., Poshyvanyk, D.: Feature location in source code: a taxonomy and survey. J. Softw. Evol. Process 25(1), 53–95 (2013)
16. Geeraerts, D.: Meaning and definition. In: van Sterkenburg, P. (ed.) A Practical Guide to Lexicography. John Benjamins Publishing Company (2003)
17. Guerra, S., Ryan, M., Sernadas, A.: Feature-oriented specifications. In: Proceedings of the ModelAge Workshop (1996)
18. Hofmeister, C., Kruchten, P., Nord, R., Obbink, H., Ran, A., America, P.: A general model of software architecture design derived from five industrial approaches. J. Syst. Softw. 80(1), 106–126 (2007)
19. Kang, K., Kim, S., Lee, J., Kim, K., Shin, E., Huh, M.: FORM: a feature-oriented reuse method with domain-specific reference architectures. Ann. Softw. Eng. 5(1), 143–168 (1998). https://doi.org/10.1023/A:1018980625587
20. Kang, K.C., Cohen, S.G., Hess, J.A., Novak, W.E., Peterson, A.S.: Feature-Oriented Domain Analysis (FODA) feasibility study. Technical report, Carnegie-Mellon University, Software Engineering Institute (1990)
21. Kästner, C., Apel, S., Kuhlemann, M.: Granularity in software product lines. In: Proceedings of the 30th International Conference on Software Engineering, pp. 311–320 (2008)
22. Kästner, C., et al.: FeatureIDE: a tool framework for feature-oriented software development. In: Proceedings of the 31st International Conference on Software Engineering, pp. 611–614. IEEE (2009)

23. Krüger, J., Gu, W., Shen, H., Mukelabai, M., Hebig, R., Berger, T.: Towards a better understanding of software features and their characteristics: a case study of Marlin. In: Proceedings of the 12th International Workshop on Variability Modelling of Software-Intensive Systems, pp. 105–112 (2018)

24. Lucassen, G., Dalpiaz, F., van der Werf, J., Brinkkemper, S.: Improving agile requirements: the quality user story framework and tool. Requir. Eng. **21**(3), 383–403 (2016)

25. Okoli, C., Schabram, K.: A guide to conducting a systematic literature review of information systems research. In: Sprouts: Working Papers on Information Systems, vol. 10, p. 26 (2010)

26. Pohl, K., Böckle, G., van der Linden, F.: Software Product Line Engineering: Foundations, Principles and Techniques. Springer, Cham (2005). https://doi.org/10.1007/3-540-28901-1

27. Rodríguez, P., Mendes, E., Turhan, B.: Key stakeholders' value propositions for feature selection in software-intensive products: an industrial case study. IEEE Trans. Softw. Eng. **46**(12), 1340–1363 (2018)

28. Shaw, M.L., Gaines, B.R.: Comparing conceptual structures: consensus, conflict, correspondence and contrast. Knowl. Acquis. **1**(4), 341–363 (1989)

29. Shekaran, C., Garlan, D., Jackson, M., Mead, N., Potts, C., Reubenstein, H.: The role of software architecture in requirements engineering. In: Proceedings of the First International Conference on Requirements Engineering, pp. 239–245 (1994)

30. Spijkman, T., Molenaar, S., Dalpiaz, F., Brinkkemper, S.: Alignment and granularity of requirements and architecture in agile development: a functional perspective. Inf. Softw. Technol. **133**, 106535 (2021)

31. UCFMapper: The various types of definitions (2021). https://www.ucfmapper.com/education/various-types-definitions/

32. Whiteley, C.: Meaning and ostensive definition. Mind **65**(259), 332–335 (1956)

33. Zave, P.: An experiment in feature engineering. In: McIver, A., Morgan, C. (eds.) Programming Methodology, pp. 353–377 (2003)

34. Zhang, J., Wang, Y., Xie, T.: Software feature refinement prioritization based on online user review mining. Inf. Softw. Technol. **108**, 30–34 (2019)

Assisted-Modeling Requirements
for Model-Driven Development Tools

David Mosquera[1]([⊠]) [iD], Marcela Ruiz[1] [iD], Oscar Pastor[2] [iD],
and Jürgen Spielberger[1] [iD]

[1] Zürich University of Applied Sciences, Gertrudstrasse 15, 8400 Winterthur, Switzerland
{mosq,ruiz,spij}@zhaw.ch
[2] PROS-VRAIN: Valencian Research Institute for Artificial Intelligence - Universitat
Politècnica de València, València, Spain
opastor@dsic.upv.es

Abstract. Model-driven development (MDD) tools allow software development
teams to increase productivity and decrease software time-to-market. Although
several MDD tools have been proposed, they are not commonly adopted by software development practitioners. Some authors have noted MDD tools are poorly
adopted due to a lack of user assistance during modeling-related tasks. This has led
model-driven engineers—i.e., engineers who create MDD tools—to equip MDD
tools with intelligent assistants, wizards for creating models, consistency checkers, and other modeling assistants to address such assist-modeling-related issues.
However, is this the way MDD users expect to be assisted during modeling in
MDD tools? Therefore, we plan and conduct two focus groups with MDD users.
We extract data around three main research questions: i) what are the challenges
perceived by MDD users during modeling for later code generation? ii) what are
the features of the current modeling assistants that users like/dislike? and iii) what
are the user's needs that are not yet satisfied by the current modeling assistants?
As a result, we gather requirements from the MDD users' perspective on how
they would like to be assisted while using MDD tools. We propose an emerging
framework for assisting MDD users during modeling based on such requirements.
In addition, we outline future challenges and research efforts for next-generation
MDD tools.

Keywords: Model-driven development · Focus group method · Framework ·
Assisted-modeling · Modeling assistants

1 Introduction

Model-driven development tools (MDD) allow software development teams to increase
productivity and decrease software time-to-market [1]. MDD tools use models for automatically generating software source code. However, MDD tools are rarely adopted by
software development practitioners. Some authors show that MDD tools have not yet
surpassed the benefits of classical approaches such as the code-centric approach [2].
This has motivated researchers to investigate and establish unaddressed challenges to

R. Guizzardi et al. (Eds.): RCIS 2022, LNBIP 446, pp. 458–474, 2022.
https://doi.org/10.1007/978-3-031-05760-1_27

improve the adoption of MDD tools [3–8]. Significantly, some authors have identified challenges about assisting MDD users during modeling, such as increasing user-centric approaches instead of technology-centric [6], understanding the modeling context [8], improving model management [5], etc.

> **MDD User:** We refer to "MDD user" in this paper to anyone who has had experience with an MDD tool to generate a software artifact.

Identified challenges as [3–8] have motivated model-driven engineers—i.e., engineers who create MDD tools—to equip MDD tools with modeling assistants. However, is this the way MDD users expect to be assisted during modeling in MDD tools? Having this question in mind, we plan and conduct two focus groups based on [9] by following the World Café [10] discussion method. We show our research overview in Fig. 1.

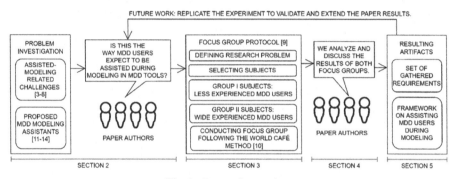

Fig. 1. Research overview.

We segmentate the focus groups as follows: the first group comprises less experienced MDD users, while the second group includes wide experienced MDD users. We aim to answer the following research questions by conducting such focus groups: i) what are the challenges perceived by MDD users during modeling for later code generation? ii) what are the features of the current modeling assistants that users like/dislike? and iii) what are the user's needs that are not yet satisfied by the current modeling assistants? We contrast the identified challenges with what related research previously observed. As a result, we gather a set of requirements of how MDD users expect to be assisted during modeling based on the challenges, features that users like/dislike, and their unsatisfied needs. Moreover, we propose an emerging framework for assisting MDD users during modeling based on such requirements. We expect researchers to propose novel modeling assistants to fulfill MDD users' requirements based on the proposed emerging framework.

The paper is structured as follows: in Sect. 2, we review identified challenges in assisting MDD users during modeling; in Sect. 3, we show the proposed focus group protocol; in Sect. 4, we present and discuss the focus group results; in Sect. 5, we gather the requirements based on the focus group results, and we propose the emerging framework for assisting MDD users during modeling; in Sect. 6, we discuss some threats to validity and limitations of our research; and, finally, in Sect. 7 we draw some conclusions and future work.

2 Background and Motivation

Some authors have been interested in the challenges surrounding MDD tools. This section explores the challenges identified by such authors, emphasizing challenges in assisting MDD users during modeling.

Abrahao et al. [6] describe User eXperience (UX) challenges in MDD tools. They represent challenges such as integrating models with user needs, identifying UX features in MDD tools, and increasing MDD tools interoperability. Regarding assisting MDD users during modeling, the work emphasizes how to transform the current MDD focus on technology to focus on the users themselves. That implies understanding the needs and contexts of the MDD users. Likewise, Aggarwal et al. [5] discuss similar challenges to those identified by Abrahao et al. [6]. However, they highlight the MDD tool customization and specific domain support as relevant challenges for assisting MDD users during modeling.

Gottardi et al. [4] perform a systematic mapping looking for general-purpose challenges in model-driven software engineering. They identify two types of challenges after reviewing 4859 studies: maintenance and methodology challenges. Additionally, they identify maintenance challenges related to assisting MDD users during modeling, such as improving debuggers, model comparators, and model version managers.

Bucchiarone et al. [3] identify several challenges in model-driven software engineering. They classify such challenges into social, foundation, domain, community, and tool challenges. Mainly, they discuss assist-modeling-related challenges into the "tool challenges" classification, such as: including human-readable requirements, integrating heterogeneous models into views, improving visualization, allowing tool scalability, and including model traceability.

Mussbacher et al. [8] explicitly identify challenges in assisted modeling in model-driven software engineering. They focus on identifying challenges regarding MDD users and their needs during modeling. Such challenges include understanding the modeling context, understanding the modeler's skills and behavior, and transferring knowledge to different domains.

Bucchiarone et al. [7] discuss the importance of modeling adoption in organizations and the progress achieved so far. They emphasize challenges, such as using artificial intelligence, including multi-paradigm modeling, and improving model management to assist users during modeling.

Up to this point, we have explored some papers comprising challenges in MDD related explicitly to assisting MDD users during modeling. Such papers have inspired novel approaches focused on developing modeling assistants for MDD users. We refer to "modeling assistants" as any software artifact intended to assist MDD users during modeling. Some examples of such novel approaches are the following: intelligent modeling assistants [11], wizards for generating models [12], and model consistency checkers [13, 14]. All these data show the perspective of researchers on how to assist software modeling. However, the researchers' perspective could differ from the way that MDD users expect to be assisted. Thus, the following main research question (MRQ) arises:

> **MRQ:** *Is this the way MDD users expect to be assisted during modeling in MDD tools?*

3   Focus Groups Protocol

We propose to conduct two focus groups [9]: a cost-efficient way of obtaining practitioner and user experience [15] to gather data around the proposed MRQ. Moreover, we specialize the MRQ into the following three research questions:

- **(RQ1)** *What are the challenges perceived by MDD users during modeling for later code generation?* We expect MDD users to describe the challenges they perceive during modeling to contrast them with the challenges devised by related works (see Sect. 2). Moreover, we expect MDD users to classify each challenge depending on the impact (high or low) if such challenge is addressed and the urgency to be addressed (urgent or not urgent).
- **(RQ2)** *What are the features of the current modeling assistants that users like/dislike?* We expect MDD users to identify the features of modeling assistants with which they have interacted and classify what they like/dislike about them. Then, we match such features with the challenges identified in RQ1. Thus, we observe which challenges have been addressed by using modeling assistants, which should remain— i.e., which MDD users like—and which should be improved—i.e., which MDD users dislike.
- **(RQ3)** *What are the user's needs that are unsatisfied by the current modeling assistants?* We expect MDD users to analyze the modeling assistants they interacted with and specify which needs are currently unsatisfied. Then, we match such unsatisfied needs with features and challenges identified in RQ2 and RQ1. Furthermore, we ask MDD users to prioritize their unsatisfied needs by using MoSCoW: a low-effort and high-consistent requirement prioritization method [16, 17]. Therefore, we observe modeling-assistance-related needs that MDD tools must/should/could/will not satisfy in the future based on the MDD users' contributions.

3.1   Selecting Subjects

We want to gather information from MDD users, so we select MDD users as the focus group subjects. However—to the best of our knowledge—a formal definition of "MDD user" has not been proposed in the literature. Indeed, some authors [6] point out that defining the MDD users is a current research challenge due to the many potential users of MDD tools. To overcome such a lack of the "MDD user" definition, we review some MDD tool-related papers to extract the authors' terms to refer to their users. As a result, we propose the following types of MDD users to select our focus group subjects[1]:

---

[1] Hereafter, we use the term "subject/subjects," referring to the focus group subject/subjects that fits/fit in one of the established MDD user types.

- *(Type A)* Non-IT related business-level users, referred to in the literature as business-level users [18], business analysts [19], end-users [19, 20], and domain users that are not necessarily computer scientists [21].
- *(Type B)* IT level users not strictly related to modeling, referred to in the literature as professional engineers with design experience [22], software developers [23–25], and IT personnel [26].
- *(Type C)* IT level users strictly related to modeling, referred in the literature as modelers [27], designers [28, 29], requirements engineers [30], model-driven developers [31], and application architects [32].

3.2 Segmentation

We conduct two focus groups composed of similar subjects—i.e., subjects with similar backgrounds and contexts—, facilitating the discussion among them [9, 33]. We show in detail the group segmentation in the following subsections.

Group I (GI)

GI comprises 11 subjects, all of whom are students of the bachelor course Rapid Software Prototyping (RASOP) at the Zürich University of Applied Sciences (ZHAW). The RASOP course is an elective 4 ECTS course offered to all engineering programs. We ask subjects to fulfill a demographic survey comprising data about their majors, their MDD tools experience, industry experience, and the type of MDD user they identify the most according to Sect. 3.1. We observe subjects are between 3rd and 4th year of their undergraduate studies (32.8 months avg. 6.0 months std. dev.), pursuing majors such as Computer Sciences (36.4%), Mechanical Engineering (27.3%), Industrial Engineering (18.2%), Electrical Engineering (9.1%), and Environmental Engineering (9.1%). Regarding their experience with MDD tools, most of them (90.9%) have three months of theoretical and practice training about MDD tools taught in the RASOP course. Only one subject (9.1%) has two months of experience using MDD tools before taking the RASOP course, having five months of experience in total. Regarding industry experience, most of the subjects (63.6%) do not have any industry experience yet. On the other hand, some subjects (36.4%) have industry experience from 0.13 to 24 months in process automation, web development, technical illustration, and software development. Finally, subjects identify themselves as Type A (45.5%) and Type B (54.5%) MDD users. The last result is consistent with what subjects answered about their majors. Most of the subjects with non-IT-related majors—e.g., Industrial Engineering—identify themselves with a non-IT-related business level MDD user—i.e., Type A MDD user. Likewise, most subjects with IT-related majors—e.g., Computer Sciences—identify themselves with an IT level MDD user not strictly related to modeling—i.e., Type B MDD user.

Group II (GII)

GII comprises three software engineering practitioners with wide experience in MDD tools working at Posity AG. Posity AG is a Swiss software development enterprise whose primary software development tool is an MDD tool named "Posity Design Studio" (PDS)[2]. Software engineering practitioners at Posity AG mainly develop data-centric

---

[2] https://posity.ch.

cloud applications by using Posity models in PDS. As we did with GI, we asked GII subjects to fulfill a demographic survey comprising data about their background, their years of industry experience, their expertise with MDD tools, and the type of MDD user with which they identify the most according to Sect. 3.1. We observe all subjects have a Computer Science background with 10 to 35 years of industry experience. Moreover, all subjects have extensive experience using MDD tools, having from 5 to 30 years of experience. Finally, subjects identify themselves as Type B (33.3%) and Type C (66.7%) MDD users.

3.3 Conducting the Focus Group Sessions

We conduct two focus group sessions from 2 to 3 h long with two moderators (the first two authors of this paper). We select the World Café method: "a simple yet powerful conversational process that helps groups of all sizes to engage in constructive dialogue" [10] to guide the discussion and interaction during the focus group session. Moreover, focus group subjects face each RQ following a brainstorming strategy [34]. First, subjects generate contributions to answer the RQ; then, subjects evaluate each contribution by discussing, improving, and refining its content. For the sake of simplicity, in this paper, we do not describe in full the setup of the World Café method, nor do we show the "raw" data obtained in the focus group. The data discussed in Sect. 4 have been prepared and refined by the authors of this paper to facilitate data presentation and analysis. However, we designed a public repository the focus group data can be consulted, including the protocol and the "raw" data[3].

4 Results and Discussion

4.1 RQ1: What are the Challenges Perceived by MDD Users During Modeling for Later Code Generation?

We identify a set of 12 challenges: six contributed by GI, three contributed by GII, and three contributed by both (see Fig. 2). According to subjects' classification, six challenges are high priority and urgent, five challenges are high priority and not urgent, no challenge is low priority and urgent, and one challenge is low priority and not urgent. We discuss the identified challenges in the following subsections.

High Priority and Urgent Challenges
GI subjects identify "decrease model and tool complexity" (C1) as a challenge since they state MDD tools are complex to use, hindering their usability. C1 agrees with Abrahao et al. [6] about the complexity of MDD tools, which negatively affects the UX during modeling. They affirm that MDD tools are much more complex than they need to be.

GII subjects identify "improve MDD tools' interoperability" (C2) as a challenge. They state MDD tools are poorly integrated with other tools, hindering import data for creating models from different sources such as existing databases, CASE (Computer-Aided Software Engineering) tools, and other MDD tools. Abrahao et al. [6] also identify

---

[3] https://github.com/DavidMosquera/RCIS2022-Focus-Group-Data.

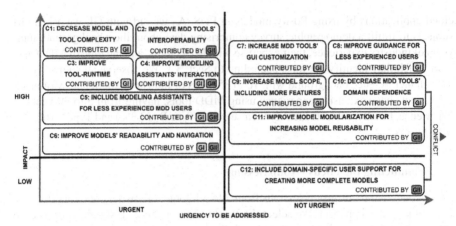

Fig. 2. Modeling challenges in MDD tools perceived by GI and GII subjects.

the interoperability of MDD tools as a challenge. They refer to this as "DSL-babel" (Domain Specific Language), since the more MDD tools are proposed with their DSLs, the more overall interoperability decrease.

GI subjects identify "improve tool-runtime" (C3) as a challenge since they state sometimes MDD tools runtime is bad, hindering a good experience during modeling. Bucchiarone et al. [3] refer to improving runtime as a challenge but mainly focus on model transformation languages and engines. C3 complements the vision of Bucchiarone et al. [3] to cover the runtime during transformations between models and the MDD tools in general.

Both groups identify two challenges regarding modeling assistants: i) "improve modeling assistants' interaction" (C4) and ii) "include modeling assistants for less experienced MDD users" (C5). In C4, subjects state modeling assistants for creating models lack usability-related features, hindering their usability, such as undo-redo commands. In C5, subjects state that facing a "blank sheet of paper" to start modeling is sometimes frustrating—especially for less experienced MDD users. So, they consider including new modeling assistants—e.g., templates—based on expert knowledge for easing model creation by less experienced MDD users as a challenge. Mussbacher et al. [8] also discuss challenges related to modeling assistants—named in their paper as intelligent modeling assistants. They coincide with C4 and C5 since modeling assistants should adapt to MDD users' context and skills to improve user interaction and ease modeling.

Both groups identify "improve models' readability and navigation" (C6) as a challenge. Subjects state that graphic models contain a lot of information, getting complex very soon and hindering their readability. They point out that such a lack of readability is due to the nature of the modeling language rather than the complexity of what they are trying to model. Moreover, they experience problems navigating between models, hindering data visualization in and between models. Bucchiarone et al. [3] also refer to readability as a challenge but mainly focus on making modeling languages "human-readable." C6 complements Bucchiarone et al. [3] since it requires improving modeling language readability and navigation through big complex models and between models.

High Priority and Not Urgent Challenges

GI subjects identify "increase MDD tools' GUI (Graphical User Interface) customization" (C7) as a challenge. GI subjects state that MDD tools' GUI does not allow them to customize shortcuts and interface element locations, negatively affecting the usability. Abrahao et al. [6] also identify MDD tools' customization as a challenge, allowing for adapting menus, pallets, and workflows for improving UX during modeling. On the other hand, Mussbacher et al. [8] highlight that modeling assistants should allow customization to increase transparency.

GI subjects identify "improve guidance for less experienced MDD users" (C8) as a challenge. They state that when they try to use a new MDD tool, there is an entry barrier, hindering the MDD tool guidance. Some authors also identify user training and guidance as a challenge since improving training and guidance to less experienced MDD users decreases the learning curve of MDD tools [4–7].

GI subjects identify "increase the model scope and include more features" (C9) as a challenge. They state that MDD tools do not yet have all the functionalities developed using programming frameworks, limiting the software development scope. Bucchiarone et al. [7] mention that several initiatives started to address C9, extending the set of features offered in MDD tools reducing the gap between programming-based tools and MDD tools. But it remains an unaddressed challenge.

GI subjects identify "decrease MDD tools domain dependence" (C10) as a challenge since MDD tool's domain dependence limits the modeling scope—mainly when the model domain differs from the MDD tool domain. C10 is opposite to challenges identified by some authors regarding MDD tools' domain dependence. Some authors specify that domain dependence is required to improve MDD users' productivity since the general-purpose tool is never really fit for purpose; one size does not provide all [3, 6, 7]. However, Mussbacher et al. [8] point out that defining appropriate, generic, domain-independent modeling interfaces and protocols is challenging for designing high integrable modeling assistants and MDD tools. Therefore, C10 and Mussbacher et al. [8] are complementary points of view.

GII subjects identify "improve model modularization for increasing model reusability" (C11) as a challenge. They state that models are poorly modularized, hindering reusing some model elements into other models. The lack of model reusing causes them to repeat information on several models, decreasing maintainability. Bucchiarone et al. [7] also highlight model reusability as a challenge that can be addressed by using AI tools. Mussbacher et al. [8] go further and include reusability in modeling assistants to reuse them in different domains, tools, and contexts.

Low Priority and Not Urgent Challenges

GII subjects identify "include domain-specific user support for creating more complete models" (C12) as a challenge. They state domain-independent modeling assistants—e.g., wizards for creating models—do not allow them to make more specific models, limiting the modeling assistants' scope. As discussed in C10, some authors agree with

C12 since domain dependence is required to improve MDD users' productivity [3, 6, 7]. However, C12 is conflictive with C10 since C10 aims to decrease MDD tools domain dependence. The conflict between C12 and C10 shows that there should be a trade-off between domain dependence and domain independence in MDD tools.

4.2 RQ2: What are the Features of the Current Modeling Assistants that Users Like/Dislike?

After discussing the challenges perceived by MDD users in Sect. 4.1, subjects identify modeling assistants' features that they like/dislike. Subjects identify a set of 11 modeling assistants' features: seven features that they like and four features that they dislike (see Fig. 3). We match them with the identified challenges and discuss the identified features in the following subsections.

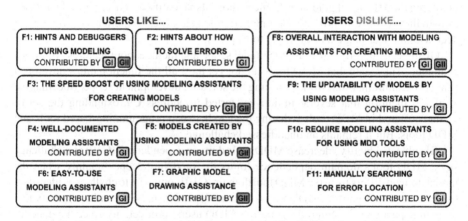

Fig. 3. Features of the current modeling assistants identified by GI and GII subjects.

Features Subjects Like

Both groups like hints and debuggers during modeling (F1). We match F1 with C1 (decrease model and tool complexity) since such debuggers and hints decrease model and tool complexity, easing error finding in models. Secondly, we observe that subjects like the following features related to C4 (improve modeling assistants' interaction): i) the speed boost during model creation by using modeling assistants (F3), ii) easy-to-use (F6) and well-documented (F4) modeling assistants, and iii) models created by using modeling assistants for creating models (F5). We match C4 with F3, F4, F5, and F6, since they aim to improve modeling assistants' interaction by bringing easy-to-use well-documented modeling assistants that boost modeling speed and create high-quality models. On the other hand, GII subjects like graphic model drawing assistance (F7) such as visual guides and entities connection. We observe such drawing assistance improves model readability, so we match F7 with C6 (improve models' readability and navigation). Finally, GI subjects like hints about how to solve errors (F2). We match F2 with C8 (improve guidance for less experienced users) since hints about how to solve errors increase the level of guidance for less experienced MDD users.

Feature Subjects Dislike

GI subjects dislike requiring modeling assistants to successfully use MDD tools (F10). They state MDD tools should be intuitive and easy to use without requiring modeling assistants. We match F10 with C1 (decrease model and tool complexity) since modeling assistants intend to decrease model tool complexity, but F10 shows subjects dislike the excessive use of modeling assistants. Secondly, both groups dislike the interaction with modeling assistants for creating models (F8). They dislike modeling assistants with many configurations, irreversible actions, required technical information, and poor dialog with the user to gather the data. We match F8 with C4 (improve modeling assistants' interaction) since F8 shows subjects dislike current modeling assistants' interaction. On the other hand, GI subjects dislike the updatability of models created by using modeling assistants (F9) since they experienced difficulties editing and completing them. We match F9 with C6 (improve models' readability and navigation) since GI subjects experienced such issues due to the lack of readability of the resulting models. Finally, GI subjects dislike manually searching for error locations (F11). We match F11 with C8 (improve guidance for less experienced users) since avoiding manually searching for error locations will improve the guidance for less experienced users.

4.3   RQ3: What are the User's Needs that are not yet Satisfied by the Current Modeling Assistants?

After identifying challenges and features of current modeling assistants, group subjects describe a set of 10 needs that are not yet satisfied by modeling assistants. We ask them to use the MoSCoW requirements prioritization method [16, 17]. As a result, group subjects classify six needs as "must have" priority, three needs as "should have" priority, one need as "could have" priority, and no need as "will not have" priority (see Fig. 4). We deeply discuss and match such needs with what groups have answered in Sects. 4.1 and 4.2 in the following subsections.

"Must have" Priority Needs

"Must have" needs are non-negotiable and mandatory to be satisfied [16, 17]. Both groups agree that MDD tools *must* improve user dialog with modeling assistants (N3) and include undo/redo commands (N1). Moreover, GII subjects state modeling assistants' interaction *must* be improved by decreasing the technical knowledge required for using them (N6).

Furthermore, GI subjects state that MDD tools *must* improve their modeling assistants' documentation (N2). We match N3, N1, N6, and N2 with C4 (improve modeling assistants' interaction), F8 (subjects dislike overall interaction with modeling assistants for creating models), and F4 (subjects like well-documented modeling assistants) since addressing such set of needs will improve the general user interaction with modeling assistants (C4 and F8) and will bring well-documented modeling assistants also improving their interaction (C4 and F4). Secondly, GI subjects state user interface aesthetics

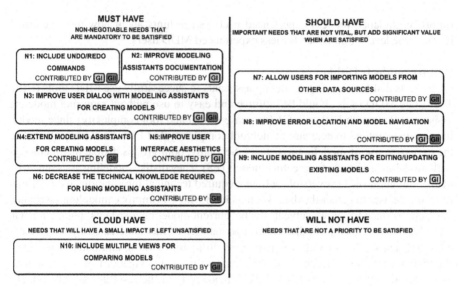

Fig. 4. Needs that are not yet satisfied by current modeling assistants; identified and prioritized by GI and GII subjects.

must be improved (N5). We match N5 with C7 (increase MDD tools' GUI customization) since increasing GUI customization implies improving the GUI aesthetics. Finally, GII subjects state that modeling assistants must be extended to create models (N4). We match N4 with C12 (include domain-specific user support for creating more complete models) since GII subjects state some extension of modeling assistants *must* be included, such as domain-specific modeling assistants for creating models.

"Should have" Priority Needs
"Should have" priority needs are important needs that are not vital but add significant value when they are satisfied [16, 17]. Both groups agree MDD tools *should* improve error location and model navigation (N8). We match N8 with F11 (subjects dislike manually searching for error location) and C8 (improve guidance for less experienced users) since avoiding manually searching for errors will improve error location, also improving guidance for less experienced users. Secondly, GI subjects state MDD tools *should* include modeling assistants for editing/updating existing models (N9). We match N9 with C5 (include modeling assistants for less experienced MDD users) since both state MDD tools should include more modeling assistants—especially for less experienced users. Finally, GII subjects identify that MDD tools *should* allow users to import models from other data sources (N7). We match N7 with C2 (improve MDD tools' interoperability) since addressing N7 will increase MDD tools' interoperability.

"Could have" Priority Needs
"Could have" priority needs will have a small impact if left unsatisfied [16, 17]. GII subjects state MDD tools *could have* multiple views for comparing models (N10). We match N10 with C6 (improve models' readability and navigation) and F7 (subjects like

graphic model drawing assistance) since including such views for comparing models will complement visual model drawing assistance, improving models' readability as a result.

5 Gathered Requirements and an Emerging Framework

We gather a set of 12 requirements based on the focus group data and matchups between modeling challenges, current modeling assistants' features, and unsatisfied needs (see Table 1). To do so, we use identified challenges and their priority as the foundation for each proposed requirement. Then, if possible, we add which modeling assistants' features should remain—i.e., those that subjects like—and those that should be improved—i.e., those that subjects dislike. Finally, if possible, we include the unsatisfied needs for identifying what MDD tools must/could/should/will not have to address their associate challenge. After analyzing the gathered requirements in Table 1, we propose an emerging framework for assisting MDD users during modeling in MDD tools (see Fig. 5).

Table 1. Proposed MDD users' requirements based on focus group data.

MDD users' requirement
R1: Improving modeling assistants' interaction is a **high** priority and **urgent** challenge (C4). From current modeling assistants for addressing C4, MDD users **like** that they are easy-to-use (F6) and well-documented (F4), boosting modeling speed (F3) and producing high-quality models (F5). But MDD users **dislike** modeling assistants' overall interaction with the user (F8). In the future, MDD tools **must** improve user dialog with modeling assistants for creating models (N3), include undo/redo commands (N1), decrease required technical knowledge to use them (N6), and improve their documentation (N2) to address C4
R2: Increasing MDD tools' GUI customization is a **high** priority and **not an urgent** challenge (C7). In the future, MDD tools **must** improve their GUI aesthetics (N5) to address C6
R3: Include modeling assistants for less experienced MDD users is a **high** priority and **urgent** challenge (C5). In the future, MDD tools **should** include modeling assistants for editing/updating existing models (N9) to address C5
R4: Improving the interoperability of MDD tools is a **high** priority and **urgent** challenge (C2). In the future, MDD tools **should** allow users to import models from other data sources (N7) to address C2
R5: Improving guidance for less experienced users is a **high** priority and **not an urgent** challenge (C8). From current modeling assistants for addressing C8, MDD users **like** hints about how to solve errors, but MDD users **dislike** manually searching for error locations. In the future, MDD tools **should** improve error location and model navigation (N8) to address C8
MDD users' requirement
R6: Improving models' readability and navigation is a **high** priority and **not an urgent** challenge (C6). From current modeling assistants for addressing C6, we **like** graphic model drawing assistance (F7), but we **dislike** the updatability of models created by modeling assistants (F9). In the future, MDD tools **could** have multiple views for comparing models (N10) to address C6

(*continued*)

Table 1. (*continued*)

R7: Including domain-specific user support for creating more complete models is a **low** priority and **not an urgent** challenge (C12). In the future, MDD tools **must** extend their modeling assistants' for creating models in specific domains (N4) to address C12

R8: Improving tools runtime is a **high** priority and **urgent** challenge (C3)

R9: Decreasing model and tool complexity is a **high** priority and **urgent** challenge (C1). From current modeling assistants for addressing C1, MDD users **like** hints and debuggers during modeling (F1), but MDD users **dislike** requiring modeling assistants for using MDD tools (F10)

R10: Increasing model scope, including more features, is a **high** priority and **not an urgent** challenge (C9)

R11: Improving model modularization for increasing model reusability is a **high** priority and **not an urgent** challenge (C11)

R12: Decreasing MDD tools' domain dependence is a **high** priority and **not an urgent** challenge (C10)

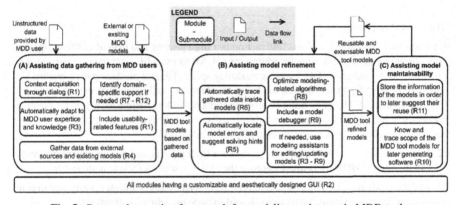

Fig. 5. Proposed emerging framework for modeling assistance in MDD tools.

The proposed emerging framework aims for allowing model-driven engineers and researchers to improve their modeling assistants based on MDD users' requirements. So, we divide such framework into three modules: A) assisting data gathering from MDD users, B) assisting model refinement, and C) assisting model maintainability. Module A aims to assist data gathering for creating models by using unstructured data or existing MDD models from external sources provided by the MDD user. Thus, Module A adapts to the user improving the user interaction and boosting modeling speed. On the other hand, Module B assists MDD users in refining the models created by Module A by tracing the data and easing error solving. Moreover, Module B includes optimized algorithms for improving tool runtime, and modeling assistants to improve model updatability. Finally, Module C allows MDD users to maintain the models through time by suggesting reusing existing models and tracing model scope into the generated software. All proposed modules must have a customizable and aesthetically designed GUI.

Some authors have already proposed frameworks for assisting modeling in MDD tools. For instance, Mussbacher et al. [8] propose a framework on intelligent modeling assistants, mainly focusing on the user interaction with the modeling assistants—such as module A of our framework. We note that our results reinforce such a research line by adding the "assisting model refinement (B)" and "assisting model maintainability (C)" modules. Thus, the frameworks can complement each other and generate modeling assistants closer to what the MDD users expect during modeling.

6 Threats to Validity and Limitations

We have identified some threats to validity and limitations during the execution of our focus groups. Regarding *conclusion validity,* we recognize that our research has a low statistical power since the population sample is small—i.e., having a sample of 14 subjects threatens our conclusion validity. Despite this, we consider our results useful and a first step to continue increasing the population sample by replicating the experiment, especially the Type C subjects, since they are a minority in our focus group segmentation (2 out of 14 subjects). Moreover, we decided to select subjects with similar backgrounds, making the focus groups homogeneous. This decision allows us to increase the *conclusion validity* since we avoid variations on the results due to individual differences among the focus group subjects—a.k.a. *random heterogeneity of subjects'* threat. However, having homogeneous groups also reduces the *external validity,* limiting our ability to generalize the focus group results. To avoid this threat arising in further replications of our focus group, we consider having more heterogeneous groups mixing them by subject types—e.g., having groups with the same number of Type A, B, and C subjects. Furthermore, the results from software engineering practitioners—i.e., GII subjects—are limited to employees from one enterprise that uses an MDD tool to develop software—i.e., Posity AG. This segmentation reduces the generality of our results since other software development enterprises do not use only MDD tools to build software. Therefore, we plan to include more software engineer practitioners from different enterprises and backgrounds in future replications, avoiding these external validity threats. Regarding *construct validity,* we decided that the focus group subjects will face each RQ "from scratch." This may have caused already identified challenges in the literature not to be discussed during the focus group sessions. However, we also avoid the subjects being biased from previously conceived challenges since we aim to gather requirements directly from MDD users and compare them with such challenges—i.e., we avoid the *interaction of testing and treatment*—increasing the *construct validity.* To overcome both limitations, we propose to use an intermediate model—e.g., the proposed framework in this paper, existing usability heuristics, among others—that allows us to discuss both subjects and literature challenges. Thus, subjects will not be biased with existing challenges, and we can identify which identified challenges are not addressed allowing us to observe unidentified challenges.

7 Conclusions and Further Work

In this paper, we have executed two focus groups based on [9] and following the World Café method [10] to answer three research questions: i) what are the challenges perceived by MDD users during modeling for later code generation? ii) what are the features of the current modeling assistants that users like/dislike? and iii) what are the user's needs that are not yet satisfied by the current modeling assistants? Such research questions aimed to collect data to gather the perspective of MDD users on how they expect to be assisted during modeling in MDD tools. After conducting the focus groups, we observed all identified challenges match or complement at least one challenge previously identified by researchers in the literature [3–8]. Moreover, we matched features that MDD users like/dislike, their unsatisfied needs, and their perceived modeling challenges to gather requirements to assist MDD users during modeling. So, we identified which features of the modeling assistants should remain and which should be improved based on what MDD users like/dislike. Furthermore, we identified which features modeling assistants must/should/could/will not have to satisfy MDD users' needs based on MoSCoW [16, 17] prioritization method. As a result, we gathered 12 requirements based on such data. Then, we proposed an emerging framework composed of three modules: A) assisting data gathering from MDD users, B) assisting model refinement, and C) assisting model maintainability. This emerging framework is a starting point for model-driven engineers and researchers to improve their modeling assistants and increase MDD tools adoption in practice. As future work, we expect to build modeling assistants following the proposed emerging framework and validate them in experiments with MDD users. Moreover, we will continue replicating our focus group, collecting more requirements to increase, improve, and validate the gathered requirements and the proposed framework. Our objective in the future is to generalize the results to a global definition of MDD users.

Acknowledgments. This work has been supported by the Zürich University of Applied Sciences (ZHAW) – School of Engineering: Institute for Applied Information Technology (InIT). Moreover, we would like to thank all RASOP course students and Posity AG practitioners for actively participating during the focus group sessions, allowing us to gather all the data we used to build our research.

References

1. Sendall, S., Kozaczynski, W.: Model transformation: the heart and soul of model-driven software development. IEEE Softw. **20**, 42–45 (2003)
2. Panach, J.I., España, S., Dieste, Ó., Pastor, Ó., Juristo, N.: In search of evidence for model-driven development claims: an experiment on quality, effort, productivity, and satisfaction. Inf. Softw. Technol. **62**, 164–186 (2015)
3. Bucchiarone, A., Cabot, J., Paige, R.F., Pierantonio, A.: Grand challenges in model-driven engineering: an analysis of the state of the research. Softw. Syst. Model. **19**(1), 5–13 (2020). https://doi.org/10.1007/s10270-019-00773-6
4. Gottardi, T., Vaccare Braga, R.T.: Understanding the successes and challenges of model-driven software engineering - a comprehensive systematic mapping. In: 2018 XLIV Latin American Computer Conference (CLEI), pp. 129–138. IEEE (2018)

5. Aggarwal, P.K., Sharma, S., Riya, Jain, P., Anupam: Gaps identification for user experience for model driven engineering. In: 11th International Conference on Cloud Computing, Data Science and Engineering, pp. 196–199. IEEE (2021)
6. Abrahao, S., et al.: User experience for model-driven engineering: challenges and future directions. In: 20th International Conference on Model Driven Engineering Languages and Systems, pp. 229–236. IEEE (2017)
7. Bucchiarone, A., et al.: What is the future of modeling? IEEE Softw. **38**, 119–127 (2021)
8. Mussbacher, G., et al.: Opportunities in intelligent modeling assistance. Softw. Syst. Model. **19**(5), 1045–1053 (2020). https://doi.org/10.1007/s10270-020-00814-5
9. Kontio, J., Bragge, J., Lehtola, L.: The focus group method as an empirical tool in software engineering. In: Guide to Advanced Empirical Software Engineering, pp. 93–116 (2008)
10. Tan, S., Brown, J.: The World Café in Singapore. J. Appl. Behav. Sci. **41**, 83–90 (2005)
11. Savary-Leblanc, M.: Improving MBSE tools UX with AI-empowered software assistants. In: 22nd International Conference on Model Driven Engineering Languages and Systems Companion, pp. 648–652. IEEE (2019)
12. ben Fraj, I., BenDaly Hlaoui, Y., BenAyed, L.: A reactive system for specifying and running flexible cloud service business processes based on machine learning. In: 45th Annual Computers, Software, and Applications Conference (COMPSAC), pp. 1483–1489. IEEE (2021)
13. Chavez, H.M., Shen, W., France, R.B., Mechling, B.A., Li, G.: An approach to checking consistency between UML class model and its Java implementation. IEEE Trans. Software Eng. **42**, 322–344 (2016)
14. Wang, C., Cavarra, A.: Checking model consistency using data-flow testing. In: 16th Asia-Pacific Software Engineering Conference, pp. 414–421. IEEE (2009)
15. Kontio, J., Lehtola, L., Bragge, J.: Using the focus group method in software engineering: obtaining practitioner and user experiences. In: International Symposium on Empirical Software Engineering, ISESE 2004, pp. 271–280. IEEE (2004)
16. Ali Khan, J., Ur Rehman, I., Hayat Khan, Y., Javed Khan, I., Rashid, S.: Comparison of requirement prioritization techniques to find best prioritization technique. Int. J. Mod. Educ. Comput. Sci. **7**, 53–59 (2015)
17. Hatton, S.: Early prioritisation of goals. In: Hainaut, J.-L., et al. (eds.) ER 2007. LNCS, vol. 4802, pp. 235–244. Springer, Heidelberg (2007). https://doi.org/10.1007/978-3-540-76292-8_29
18. Sinha, S., et al.: Auto-generation of domain-specific systems: cloud-hosted DevOps for business users. In: 13th International Conference on Cloud Computing (CLOUD), pp. 219–228. IEEE (2020)
19. Sousa, K., Mendonça, H., Lievyns, A., Vanderdonckt, J.: Getting users involved in aligning their needs with business processes models and systems. Bus. Process. Manag. J. **17**, 748–786 (2011)
20. Pérez, F., Valderas, P., Fons, J.: Towards the involvement of end-users within model-driven development. In: Costabile, M.F., Dittrich, Y., Fischer, G., Piccinno, A. (eds.) IS-EUD 2011. LNCS, vol. 6654, pp. 258–263. Springer, Heidelberg (2011). https://doi.org/10.1007/978-3-642-21530-8_23
21. Fuhrmann, H., von Hanxleden, R.: Taming graphical modeling. In: Petriu, D.C., Rouquette, N., Haugen, Ø. (eds.) MODELS 2010. LNCS, vol. 6394, pp. 196–210. Springer, Heidelberg (2010). https://doi.org/10.1007/978-3-642-16145-2_14
22. Paz, A., el Boussaidi, G., Hafedh, M.: checcsdm: a method for ensuring consistency in heterogeneous safety-critical system design. IEEE Trans. Software Eng. **47**, 2713–2739 (2020)

23. Schottle, M., Kienzle, J.: Concern-oriented interfaces for model-based reuse of APIs. In: 18th International Conference on Model Driven Engineering Languages and Systems, pp. 286–291. IEEE (2015)
24. Ohrndorf, M., Pietsch, C., Kelter, U., Grunske, L., Kehrer, T.: History-based model repair recommendations. ACM Trans. Softw. Eng. Methodol. **30**, 1–46 (2021)
25. Getir, S., Grunske, L., Bernasko, C.K., Käfer, V., Sanwald, T., Tichy, M.: CoWolf – a generic framework for multi-view co-evolution and evaluation of models. In: Kolovos, D., Wimmer, M. (eds.) ICMT 2015. LNCS, vol. 9152, pp. 34–40. Springer, Cham (2015). https://doi.org/10.1007/978-3-319-21155-8_3
26. Akiki, P.A., Bandara, A.K., Yu, Y.: Cedar studio: an IDE supporting adaptive model-driven user interfaces for enterprise applications. In: 5th ACM SIGCHI Symposium on Engineering Interactive Computing Systems, p. 139. ACM Press (2013)
27. Oberweis, A., Reussner, R.: Model validation and verification options in a contemporary UML and OCL analysis tool. In: Modellierung, pp. 203–218 (2016)
28. Danenas, P., Skersys, T., Butleris, R.: Extending drag-and-drop actions-based model-to-model transformations with natural language processing. Appl. Sci. **10**, 1–37 (2020)
29. Ameedeen, M.A., Bordbar, B., Anane, R.: Model interoperability via model driven development. J. Comput. Syst. Sci. **77**, 332–347 (2011)
30. Pires, P.F., Delicato, F.C., Cóbe, R., Batista, T., Davis, J.G., Song, J.H.: Integrating ontologies, model driven, and CNL in a multi-viewed approach for requirements engineering. Requirements Eng. **16**, 133–160 (2011)
31. Araùjo De Oliveira, R., Dingel, J., Oliveira, R.: Supporting model refinement with equivalence checking in the context of model-driven engineering with UML-RT. In: Model-Driven Engineering, Verification and Validation Workshop at the MODELS Conference (2017)
32. Ricci, L.A., Schwabe, D.: An authoring environment for model-driven web applications. In: 12th Brazilian Symposium on Multimedia and the Web, pp. 11–19. ACM Press (2006)
33. Morgan, D.: Focus Groups as Qualitative Research. SAGE Publications, Inc., Thousand Oaks (1997)
34. Paetsch, F., Eberlein, A., Maurer, F.: Requirements engineering and agile software development. In: 12th IEEE International Workshops on Enabling Technologies: Infrastructure for Collaborative Enterprises, pp. 308–313. IEEE (2003)

Model-Driven Engineering

Model-Driven Production of Data-Centric Infographics: An Application to the Impact Measurement Domain

Sergio España[✉] [ID], Vijanti Ramautar[ID], Sietse Overbeek[ID], and Tijmen Derikx

Utrecht University, Utrecht, The Netherlands
{s.espana,v.d.ramautar,s.j.overbeek}@uu.nl

Abstract. *Context and motivation*: Infographics are an engaging medium for communication. Sometimes, organisations create several infographics with the same graphic design and different data; e.g., when reporting on impact measurement. *Question/problem*: The conventional process to produce such recurrent data-centric infographics causes rework related to the disconnection between software environments. *Principal ideas/results*: This paper redesigns the process following the model-driven engineering paradigm. We present a domain-specific language to model infographics, and an interpreter that generates the infographics automatically. We have been able to model and generate infographics that report impact measurement results, which the participants of a comparative experiment have found as attractive as the original ones, and that are hard, but not impossible, to distinguish from them. *Contribution*: An innovative model-driven approach that eliminates the software environment disconnection and could facilitate the use of data-centric infographics for reporting purposes.

Keywords: Domain-specific language · Infographics · Impact measurement · Ethical social and environmental accounting · Model-driven engineering · Experiment

1 Introduction

Infographics (portmanteau for information graphic) blends data with graphic design, helping individuals and organisations concisely communicate complex information to a large audience, in a manner that can be quickly consumed and easily understood [1]. Infographics may be produced as a one-time artefact; e.g., when a consulting company summarises the results of a market research or when an instructor creates educational material for a course [2]. But infographics can also be recurringly produced and thus become a design artefact that can be reused multiple times; e.g., when an organisation produces an infographic as a companion to their annual integrated report, or when a network creates one infographic for each of its members. In this paper, we focus on the latter case, where there is typically an iterative data-centred process that starts with data requirements engineering (see Fig. 1, A1) and ends with infographics production (A7).

© The Author(s), under exclusive license to Springer Nature Switzerland AG 2022
R. Guizzardi et al. (Eds.): RCIS 2022, LNBIP 446, pp. 477–494, 2022.
https://doi.org/10.1007/978-3-031-05760-1_28

DATA ANALYST (E.G. IMPACT MANAGER OR ORGANISATIONAL ACCOUNTANT)   **ACTOR**   GRAPHIC DESIGNER
DATA ANALYSIS TOOL (E.G. IMPACT MEASUREMENT TOOL) WITH A DATABASE   **TOOL**   DESIGN APP & SPREADSHEET

Fig. 1. Reference model of a conventional process for data-centric infographic creation. It is sometimes iterative, can have many loopbacks, and not all activities are always performed. It is based on the data science process [3]. Two factors that are a source of problems are shown at the bottom; and they are instantiated for the domain of impact measurement, as an example.

Two factors are a source of complications. Firstly, due to the different skills involved during each subprocess, the actors conducting the data analysis and producing the infographic are often different; e.g., data analysts and graphic designers, respectively. Secondly, the software tools are different and have different data repository technologies, leading to an interruption in the data pipeline. This produces, at least, two problems. Any manual task involved in the migration of data from one software environment to the other might introduce errors. More importantly, changes in the data produce communication delays and require rework to update the affected infographics.

The main goal of this research is redesigning the data-centric infographics production process as a model-driven engineering process, so the infographic structure and the references to the data sources are specified in a model that a module of the data analysis software tool uses to generate the infographic, once the results are available or updated. This prevents data migration and eliminates delays. The contributions of this paper include (i) a better understanding of the current data-centric infographic production process, (ii) a domain-specific modelling language (DSL) to specify data-centric infographics, and (iii) an interpreter that renders an infographic automatically, given its model and the data. As a proof of concept, we apply our approach to the domain of impact measurement. In particular, we analyse the structure of infographics used to communicate the results of ethical, social and environmental accounting (ESEA) processes, we extend the openESEA tool [4] with the new interpreter features. We also validate the contributions by means of test cases (i.e., generating existing infographics), a comparative experiment (assessing participants perceptions of original and generated infographics), and expert assessment (by showcasing the technology to experts). Such validations allow us to produce an improvement of the DSL.

The structure of the paper is the following. Section 2 presents the research method. Section 3 provides the conceptual background for this research. Section 4 analyses existing infographics and tools used to produce them. Section 5 presents the DSL that allows modelling infographics and the interpreter that generates them. Section 6 reports on the validations of our approach and the improvements they led to. Section 7 reflects on the results. Section 8 presents conclusions and opportunities for further research.

2   Research Method

In this project, we are engineering a domain-specific language (DSL); that is, a high-level language that supports concepts and abstractions that are related to a particular application domain [5]. The research method is depicted in Fig. 2.

Fig. 2. Overview of the research method. The coarser grained activities correspond to the design cycle in [6]. We further structure our work according to the DSL engineering method proposed by Mernik et al. [5]. Inspired by [7], we have embedded SCRUM practices in activities B5 to B8.

Problem Investigation. Four types of sources inform our domain analysis: (B1) a multi-vocal literature review to better understand the infographics domain, (B2) a semi-structured interview with a stakeholder that participated in a data-centric infographics production process in the domain of ESEA, (B3) an analysis of ESEA-related infographics found online in order to identify frequent infographics component types, and (B4) an analysis of existing infographics design tools in order to identify their features. **Treatment design.** The DSL development consisted of (B5) the specification of the requirements for the DSL and its interpreter, by means of users stories, (B6) the creation of a grammar for the DSL, using the Eclipse Xtext environment [8], and (B7) the implementation of the interpreter as a module of the openESEA tool [4]. **Treatment validation.** We have validated our proposal by means of (B8) testing whether we could model and automatically generate a sample of existing infographics, (B9) conducting a comparative experiment to assess the similarity between generated infographics and the original ones, (B10) and conducting an expert assessment with 3 stakeholders. Based on the experience, we have created an improved version of the DSL and the interpreter (a quick iteration of the treatment design, not depicted in Fig. 2).

3 Conceptual Background on Infographics and on Ethical, Social and Environmental Accounting

Infographics combine graphical elements and text, to communicate facts, often serving an overarching discourse. They are used in many domains with different purposes, for instance, teaching [9], healthcare [10], political persuasion [11], research dissemination [12], and impact measurement [13]. Infographics are easy to share and tend to engage more than lengthy reports [1]. As discussed in Sect. 1, infographics are often created in a one-time effort to communicate some knowledge or message, but the model-driven approach we propose is more useful in data-centric infographics, whose design is recurrently reused on several occasions (e.g. by many organisations, by the same organisation over the years, or both). It is therefore important to decouple the data from the design specification, in order to reuse the latter. Data-centric infographics are often referred to as statistical infographics [1, 14], although we prefer the term data-centric because not all data-related components in such infographics are statistics. Effective infographics follow principles from the fields of psychology, usability, graphic design, and statistics, so as to reduce time constraints and cognitive barriers to communicating important information [15]. In this research, we focus on infographics used to communicate results of an ESEA process, which produces impact measurement data that needs to be reported to either specific stakeholder groups or the general public [4]. Many papers in computer graphics address the generation of graphics for information visualisation, ranging from model-based approaches [16] to using neural networks [17]. However, infographics design remains a broader process that has not yet been tackled, and infographics require intensive tailoring, given a domain and intended audience. In this project, we aim at infographics reporting on impact measurement results.

Ethical, Social and Environmental Accounting (ESEA) is a family of impact measurement methods intended to assess and report the performance of organisations on ethical, social or environmental topics [4]. It is also referred to by many other terms combining a qualifier (e.g., sustainability, non-financial, integrated, impact) and an action (e.g., reporting, disclosure, auditing) [18]. There are many different ESEA methods, and organisations sometimes apply more than one or want to extend an existing one. The available tools are coupled with a single method, making it difficult to tailor the tool to the needs of the organisation [4]. Examples of ESEA methods are the B Impact Assessment used by certified B Corporations [19], the Common Good Balance Sheet prescribed by the Economy for the Common Good initiative [20], the REAS Social Balance [21], and the ISO 26000 [22] and ISO 14000 [23] standards. Some proposals focus more on how to report on ESEA results than on data collection and processing; this is the case of the GRI Standards [24] and the Integrated Reporting <IR> framework [25]. While all the previously-mentioned methods are sector-independent, we are nowadays witnessing the advent of methods that are specific for a given industry sector; for instance Green IT assessment (for data centres) [26], STARS (education institutions) [27] and SAFA (agriculture) [28]. In an earlier publication, we have presented openESEA[1], an opensource, model-driven, web-based ESEA tool that intends to allow modelling any ESEA method and interpreting the models automatically to support data collection, processing, cleaning (by means of auditing and assurance), analysis, and reporting [4]. We refer to the set of indicator values collected by applying an ESEA process as (ethical, social and environmental) account.

Infographics Within the Domain of Impact Measurement. To better understand the problems related to the conventional data-centric infographics production process, we have interviewed the Social Balance coordinator of REAS-PV, a territorial network part of REAS, the Spanish Network of Networks of Alternative and Solidarity Economy[2] (*Red de Redes de Economía Alternativa y Solidaria*). REAS is a second-grade association created in 1995 grouping, in 2021, 944 organisations who commit the principles and values of the solidarity economy [29], with a joint revenue of 1007M€, employing over 23.000 people, and 25.000 volunteers. REAS is further structured in territorial networks. The member organisations apply the REAS Social Balance yearly, using an ESEA tool provided by the network [30], which supports several of the method activities (those equivalent to A2-A6 in Fig. 1). REAS often creates an infographic for each member organisation, summarising the results of their social balances[3]. They also create infographics aggregating the results per (territorial or complete) network. In 2018,

---

[1] Originally named openSEA, since 2020 the tool has been renamed openESEA, to include the business ethics dimension, also sometimes referred to as governance topics.

[2] https://reas.red.

[3] https://reasnet.com/intranet/docs/informes-e-infografias-de-la-campana-ensena-el-corazon-2020.

the Social Balance coordinator of REAS-PV commissioned an external graphic design studio to create 19 infographics reporting on the ESEA results of member organisations. The data had to be exported from the ESEA tool to an Excel spreadsheet and sent to the graphic designer. The designer received a few guidelines about the infographic appearance but was given freedom to author it to her liking. While the design activity was taking place (A7 in Fig. 1), the data of some member organisations had to be updated due to errors during the data collection or to missing data. As a result, rework was necessary to create the affected infographics again (some of them twice), and back and forth emails between the Social Balance coordinator and the graphic designer produced delays which conflicted with the communication campaign deadlines. A quick follow-up interview in 2022 confirms that they are still experiencing the same problematic issues.

4 Analyses of Infographic Component Types and Tools

Infographic Component Types. We have searched for "infographic *and* ((social or environmental accounting) *or* (environmental auditing) *or* (social auditing) *or* (non-financial report) *or* (environmental sustainability))," using Google. We have collected 58 infographics, published in the 2012–2019 year range, by organisations of diverse size and sectors, including large for-profit enterprises like Nestlé, small non-profits like the 11th Hour Racing foundation, educational institutions such as Vanderbilt University, and associations of social enterprises such as REAS. Find the infographics in a companion technical report [31]. We have analysed each infographic, identified its component types, and incrementally created the tree shown in Fig. 3.

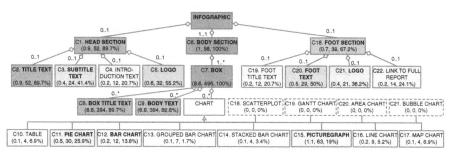

Fig. 3. Tree of infographic component types based on the analysis of 58 infographics. Aggregation relationships represent topological inclusion, and the specialisation relationships represent the existence of different types of charts. The numbers in each component type show the average number of components of such type per infographic, the total occurrences of components of such type, and the percentage of infographics having at least one component of such type, respectively. The intensity of background colour is proportional to the percentage, acting like a heatmap. We added chart component types C18–C21 after the tool analysis (Fig. 4).

Most of the infographics we have analysed have a head section located at the top, often displaying the infographic title text, subtitle text, introduction text and logo. All of them have one body section, which is the middle part displaying the text, data and chart elements. We found 8 different types of charts (for convenience, we have classified tables as a type of chart). The foot section refers to a bottom part, often having a foot title text, foot text, a logo and a link to a full report or additional information.

Analysis of Infographic Design Tool Features. A search for "Infographic *and* (tool *or* creator *or* generator)" using Google has resulted in a list of 13 tools. We have excluded 3 that require us to pay an expensive license to explore them. We have analysed the remaining 10, which are BeFunky, Canva, Easel.ly, Infogram, Lucidpress, Mind the Graph, Piktochart, Snappa, Venngage, and Visme. We have created a feature model [32] for each tool. We have identified 28 generic features (see Fig. 4). Two features in two different tools correspond to the same generic feature when they serve the same purpose, even if implemented in a slightly different way. The number of generic features per tool ranges from 10 (BeFunky) to 27 (Infogram). Some tools are stronger in terms of features (e.g., Piktochart offers all 6 features related to text and images), others in terms of chart types (e.g., Venngage offers 11 out of the 12 chart types), others in terms of target formats (e.g., Visme exports infographics to all 3 formats). The details of the analysis can be found in the technical report [31]. These analyses inform the design of the DSL and the implementation of the interpreter.

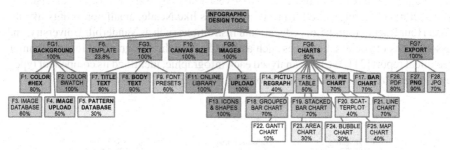

Fig. 4. Tree structuring the generic features of 10 infographic design tools. *Fi* and *FGj* are identifiers of the features and feature groups, respectively. The number at the bottom of each feature (or group) represents the percentage of tools that offer such feature (or group). The intensity of the background colour is proportional to the percentage, acting like a heatmap.

5   Domain-Specific Language for Infographics

5.1   Requirements Specification

Based on the domain analysis reported in Sects. 3 and 4, we have specified the requirements for the DSL and its interpreter, as user stories [33]. We have prioritised the requirements using the MoSCoW method [34], based on our perception of how critical each requirement is to support the infographic modelling and generation process, and taking into account that, at this stage of the research, we are aiming at a proof-of-concept tool that we can validate empirically. See a sample in Table 1.

Table 1. Sample of 4 out of 36 requirements (complete list in [31]). With respect to the abbreviations, *Pri* = priority (where *M* = must-have and *C* = could-have), *Src* = source (where *Cj* and *Fi* are component type and generic feature identifiers, *PO* = product owner), and *Imp* = implemented.

Id	Statement	Pri	Src	Imp
US2	As a user, I want to upload my own infographic specification, so that I can tweak the infographic based on my needs	M	PO	Yes
US7	As a user, I want to define the background of the infographic by selecting a colour hex code	M	F1	Yes
US16	As a user, I want to be able to create boxes, so that my infographic is divided into configurable sections	M	C7	Yes
US28	As a user, I want to create stacked bar charts on my infographic.	C	C14, F19	No

5.2 Xtext Grammar

We have implemented a textual grammar using Eclipse Xtext. The language is whitespace-aware. The grammar contains rules to define the overall infographic style and its components. As shown in Table 2, the modeller can define the infographic background colour, pattern or image (line 8) and its size (9).

Table 2. Xtext grammar excerpt containing the rule that structures the infographic model and the rule that supports piechart specification. *G#* has the line numbers corresponding to the complete grammar (presented in [31]). Lines 7, 88 and 96 were not in the first grammar version, but included after the validation (see Sect. 6.1).

G#	Grammar rule
6	`Infographic:`
7	`('type' ':' type='basic')?`
8	`& (('bgcolor' ':' bgcolor=Color)	('bgpattern' ':'` `bgpattern=Pattern)	('bgimage' ':' bgimage=ImageSrc))`
9	`& 'bgsize' ':' bgsize=SIZE_POS`
10	`& head=Head?`
11	`& texts+=Text*`
12	`& images+=Image*`
13	`& piecharts+=Piechart*`
14	`& barcharts+=Barchart*`
15	`& picturegraphs+=Picturegraph*`
16	`& foot=Foot? ;`
...	
82	`Piechart:`
83	`name=PIECHARTID ':'`
84	`BEGIN`
85	`(('bgcolor' ':' color=Color)?`
86	`& ('colors' ':' colors=COLOR_CHARTS)?`
87	`& ('data' ':' BEGIN (piedata+=ChartData)+ END)`
88	`& ('legendstyle' ':' legendstyle=LegendStyle)?`
89	`& ('padding' ':' padding=INT)?`
90	`& ('position' ':' position=SIZE_POS)`
91	`& ('showlegend' ':' showlegend=ShowOptionsOff)?`
92	`& ('showtitle' ':' showtitle=ShowOptionsOff)?`
93	`& ('showpercentage' ':' showpercentage=ShowOptionsOff)?`
94	`& ('size' ':' size=INT)?`
95	`& ('title' ':' title=STRING)?`
96	`& ('type' ':' type=PieType)?)`
97	`END ;`

The infographics component types currently supported are head section (10), text (11), image (12), pie chart (13), bar chart (14), picture graph (15), and foot section (16). As indicated by the Xtext cardinality operators, some components are optional (?), some allow zero or many occurrences (*), others allow one or many (+). The & symbol defines unordered groups of elements, making infographic specifications less rigid than with strict sequences. Each of these component types is further supported by one or several grammar rules. For the sake of brevity, we only present, also in Table 2, the rule that corresponds to the pie chart component. The other charts have a similar structure, though. For all types of chart, the modeller can specify a background colour (line 85) using a simple reserved code such as green or a hexadecimal number such as 2ca58d, the colours used in the chart itself (86), the chart data (87), padding (89), the chart position using coordinates (90), whether to show the legend (91) and the title (92) or not, the size of the chart (94), and its title (95). The chart data can be either values (e.g. 76.256) or identifiers of indicators (e.g. indscope1 is the calculation of direct CO_2 emissions produced by sources controlled or owned by an organisation). Within piecharts and barcharts, it is also possible to adjust the legend style (90). In piecharts, it is possible to define whether percentages should be shown or not (93). The definition of grids is exclusive of barcharts, and only picturegraphs can define a label. Several terminal and enum rules are required to complete the specification of the grammar, but we omit them for the sake of brevity. We disclose the complete grammar and language dictionary here [31]. The Eclipse Xtext environment can automatically generate a model editor that facilitates authoring infographic specifications by offering syntax-colouring, autocompletion, and error-checking features. Table 3 presents two fragments of the model we have created to specify one of the infographics (see Fig. 5).

Table 3. Two fragments of the specification of an infographic. *S#* has the line numbers of the specification and G# refers to the line numbers of the grammar, tracing back to Table 2.

S#	G#	Specification fragment 1	S#	G#	Specification fragment 2
1	8	bgcolor: white	423	83	piechart1:
2	9	bgsize: 1100x850	424	90	position: 488x135
3	18	head:	425	94	size: 80
4	22	position: '0x713'	426	89	padding: 10
5	21	size: 1100x137	427	85	bgcolor: ffffff
6	20	bgcolor: ffffff	428	93	showpercentage: off
7	24	title:	429	95	showtitle: off
8	34	position: 550x770	430	94	showlegend: off
9	36	value: "Vanderbilt University"	431	87	data:
10	35	maxwidth: 900	432	87	"Scope1": indscope1
11	31	align: center	433	87	"Scope2": indscope2
12	32	color: 51846e	434	87	"Scope3": indscope3
13	33	font: bold 50px Verdana	435	86	type: donut
			436	96	colors: 2ca58d,b737b2, 1982c9

5.3 Model Interpreter and Infographic Generator in OpenESEA

We have extended the openESEA tool with a module that interprets models created with the DSL, and generates the corresponding infographics. The main architectural decisions of openESEA are described in an earlier paper [4], but we summarise them here. It is a web application implemented in Type-Script using the React framework. For the backend, we use the Firestore, Authentication and Hosting services of Firebase.

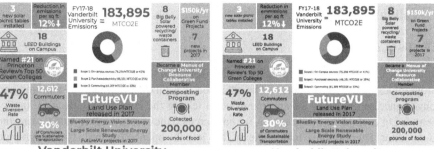

Fig. 5. Original infographic (left) and the one generated by openESEA (right), using the model excerpted in Table 3. They are infographics I_{10}^O and I_{10}^G of our test-case set (see Sect. 6.1). Some evident differences are related to font type and size; also, the pie chart is rotated $180°$. The complete set of infographics can be found in the technical report [31].

We have included a simple model repository that collects models and allows users to download them if needed, although the tool is also capable of sharing models as follows. It allows (networks of) organisations to define an official infographic by uploading a specification and making it available to their members. When a user wants to generate an infographic to communicate the results of an ethical, social and environmental account, they can decide to use the infographic specification defined by their network, by their organisation, or rather to upload a new specification. For a given account, zero or more infographics can be generated, based on different models.

To generate an infographic, the tool first parses its model with a JSON schema; since the models are YAML files, this was easy to implement. Then, it initiates an HTML5 canvas element and iterates through the infographic components, generating the corresponding graphical components in the canvas. When an infographic component presents data from an ethical, social and environmental account, the corresponding indicator values are retrieved from the account data. The user can export and download the infographic as an image. The canvas element containing the infographic is re-generated every time the user accesses the infographic, so changes to the model or the data result in an updated infographic. Finally, the tool can also generate a default infographic, that reports on the values of the account, using simple text and the organisation logo.

We have implemented 63.6% of the infographic component types (those whose name appear in bold in Fig. 3), 50.0% of the features and feature groups found on professional infographic design tools (those whose name appear in bold in Fig. 4), and 58.3% of the requirements (100% of the must-have, 50% of the should-have, 27.3% of the could-have, and 0% of the won't-have). We consider that this functional coverage is sufficient for the purpose of a proof of concept. The tool source code[4] has been released under the GNU General Public License v3.0.

6 Validation

6.1 Test Cases

To test the DSL and the interpreter, we have selected 10 infographics from the set of infographics analysed in Sect. 4. The selection is a mixture of random sampling and convenience sampling as we are interested in testing the components we have implemented. For each of the infographics, we have created the corresponding model using the Xtext editor. Then we have loaded the models into the openESEA tool. The infographic interpreter module has then rendered an infographic. See an example in Fig. 5.

Our first impression is that we have been able to generate infographics that are similar to the original ones, despite some limitations (see Sect. 7.1). For each of the models, the tool took less than 10 s to render and generate the infographic. We refer to the resulting set of test-case infographics as $I = \bigcup_{1 \leq j \leq 10} \left(I_j^O, I_j^G \right)$, where I_j^O are the original and I_j^G are the generated versions of infographic number j. We reuse this set in a comparative experiment where we further investigate the similarity.

6.2 Comparative Experiment

Objective. We want to assess the extent to which original infographics (i.e., the ones found online and analysed in Sect. 4) are indistinguishable from the corresponding generated ones (i.e., the ones that we have produced by means of creating a model and automatically generating the infographic with openESEA). Null hypotheses are:

- $H1_0$: The visual attractiveness of original infographics is similar to the visual attractiveness score of the generated infographics.
- $H2_0$: The participants are unable to distinguish whether an infographic is original or generated.

Participants. The experiment had 40 participants selected by convenience sampling. Figure 6 depicts the distribution of gender identity, age range, and level of education.

---

[4] The code of the interpreter can be found here: https://github.com/nielsrowinbik/open-sea.

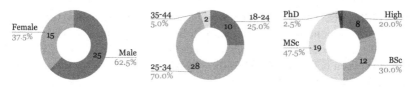

Fig. 6. Distribution of gender identity, age range, and level of education among the participants. *High, BSc, MSc* and *PhD* stand for high school-, bachelor-, master-, and doctoral-level education.

Variables. The independent variable is the *Type* of infographic. The two treatments are *Original* and *Generated* infographics. The dependent variables are the *Visual Attractiveness* (*VA*) of infographics, and the *Outcome* of guessing what type they are. To assess *VA*, we use criteria defined by Lavie and Tractinsky [35], proven to be reliable and valid to measure aesthetics of web-related content. For each infographic, participants rate the items enumerated in Table 4 on a five-point, agree-disagree, Likert scale. *VA* is then the sum of the 10 values: $VA = \sum_{1 \leq i \leq 10} VA_i$, where $10 \leq VA \leq 50$.

Table 4. Operationalisation of the Visual Attractiveness (*VA*) variable based on items from [35].

Classic aesthetics		Expressive aesthetics	
VA_1	Aesthetic design	VA_6	Creative design
VA_2	Symmetrical design	VA_7	Sophisticated design
VA_3	Pleasant design	VA_8	Original design
VA_4	Organised design	VA_9	Use of special effects
VA_5	Clean design	VA_{10}	Fascinating design

To determine whether the two types of infographics are indistinguishable, we present an infographic to the participants without disclosing its type, and we ask them to indicate whether they think it is an original or a generated infographic. Then we calculate, in a variable named *Outcome*, whether they guessed correctly (*Success*) or not (*Fail*).

Experimental Instrument. We have designed a data-collection survey with two variants: S^A and S^B. The survey starts with an introduction to the experiment. The first section contains the demographic questions. The second one contains the task instructions and 10 subsections, one for each infographic in our test-cases sample I (see Sect. 6.1). We have determined randomly what version of each infographic is presented in each survey variant, with the constraint that each should include 5 original and 5 generated infographics. S^A presents $\{I_1^O, I_2^O, I_3^O, I_4^G, I_5^G, I_6^O, I_7^G, I_8^G, I_9^G, I_{10}^O\}$ and S^B presents alternate versions $\{I_1^G, I_2^G, I_3^G, I_4^O, I_5^O, I_6^G, I_7^O, I_8^O, I_9^O, I_{10}^G\}$. All subsections contain the same 11 questions, the first 10 asking the subjects to assess the visual attractiveness of the infographic, and one asking them to guess what type of infographic it is.

Protocol. We welcome the participants with a short introduction. We obtain consent to use their data. They are randomly assigned one of the survey variants. There is no time limit for answering the survey. We thank them for participating and see them off.

Analysis of Results. We first focus on hypothesis $H1_0$. A *Shapiro-Wilk* test shows that the sample follows a normal distribution, both for the *VA* of the original infographics ($W(40) = 0.118, p = 0.221$), the generated ones ($W(40) = 0.132, p = 0.217$), and the difference between both variables ($W(40) = 0.987, p = 0.890$). There are no outliers; see the boxplots in Fig. 7.

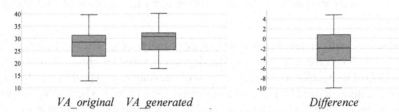

VA_original   VA_generated   Difference

Fig. 7. Boxplots of the visual attractiveness (*VA*) scores for original and generated infographics (respectively, to the left) and the difference between both variables (to the right)

To compare the *VA* between original and generated infographics, we conduct a paired-samples t-test. It is statistically significant, $t(40) = -3.41, p = 0.002$. On average, the subjects rated the *VA* of generated infographics ($M = 29.35, SD = 5.91$) higher than for original infographics ($M = 27.35, SD = 6.09$). The effect size ($d = -0.54$) is medium, according to Cohen [36]. As a result, we reject the null hypothesis $H1_0$, given that the attractiveness is not similar, but greater in the case of generated infographics.

We now focus in $H2_0$; that is, the capability of the subjects of distinguishing generated infographics from originals. We treat this as a classification problem. We assess whether there is a difference in the proportion of times that the subjects guessed the type of infographic correctly, depending on whether they were presented an original or a generated one. A chi-square test yields Table 5. The participants have been able to guess correctly whether it was generated or not, when the infographic is generated (60% of the times), but not as often when it is original (39%). This association is statistically significant, $\chi2(1) = 17.64, p < 0.001$. As an additional check, we calculate the odds ratio, which results in $OR = 2.34, p < 0.001$ (95% *CI*: 1.57, 3.50), showing again that a significant association exists. As a result, we cannot reject $H2_0$.

Table 5. Crosstabulation of *Type* of infographic and *Outcome*.

		Outcome		
		Fail	*Success*	**Total**
Type	*Generated*	80 (40.0%)	120 (60.0%)	200 (100.0%)
	Original	122 (61.0%)	78 (39.0%)	200 (100.0%)
	Total	202 (50.5%)	198 (49.5%)	400 (100.0%)

6.3 Expert Assessment

To explore whether our approach has chances of industrial adoption, we have showcased our approach to 3 independent experts. The first one is the REAS-PV Social Balance coordinator (see Sect. 3), who valued the fact that the infographics can be generated quickly and within the same tool that is used to manage the ESEA process. She expressed interest in having the features implemented in the tool used by REAS. However, she also expressed concerns about the complexity of the DSL for non-technical people. The second expert is the project manager of an ESEA tool being developed by Competa IT[5], under a proprietary licence. He was positive about the approach, highlighting that it offers a cost-effective solution for producing several infographics per each ESEA account, targeted at different stakeholder groups. He also considers such infographics a convenient way of displaying the results of certifications based on ESEA results, a functionality supported by openESEA [4] and planned in their own tool as well. One of the disadvantages he has mentioned is that, unless proper data cleansing procedures are applied, it is possible to generate an infographic with incorrect data; since the generation avoids the intervention of the graphic designer at this stage, then we miss one of the persons in the loop that could spot such errors. The third expert is the Social Balance coordinator of the Catalan Network of Solidarity Economy (XES), a territorial network of REAS that has developed the REAS ESEA tool. This tool can actually generate infographics, but their design is hardcoded rather than modelled. He expressed that other territorial networks have asked them to produce tailor-made infographics but that it is not possible, this being the reason why the involvement of graphic designers is needed. Both the second and third experts would like our approach implemented in their tools, and we are now agreeing valorisation collaborations.

7 Discussion of the Results

7.1 Reflection on the Results

Improvements After the Test Cases. We realised several limitations: (i) it was not possible to model and generate donut charts, (ii) it was not possible to adjust the legend style of bar charts, pie charts and picture graphs, and (iii) the current offering of picture graph shapes is limited to thrash bins and energy rays. We have extended the grammar with the pie charts attribute *type* to allow the definition of donut charts (line 96 in Table 2), and the attribute *legendstyle* in bar charts (line 88 in Table 2), pie charts and picture graphs. We have decided not to address the third limitation, since adding expressiveness similar to vector graphics does not pay off at this proof-of-concept stage of the project. Additionally, the product owner defined a new requirement (US17 in Table 1) that led to supporting very simple infographics by default (line 7 in Table 2). These changes required updating the interpreter accordingly.

The Generated Infographics are Sufficiently Attractive. They are perceived as more attractive than original ones. Even if the difference is statistically significant, in practical

---

[5] https://competa.com.

terms this is not relevant. Our approach does not intend to invent designs automatically. In our models, we mimicked original infographics as much as possible; so at most we intend the generated versions to be equally attractive. Anyhow, it is very positive that the generated infographics are not less attractive than the originals.

Generated Infographics are Not Indistinguishable from the Original Ones. Participants have been able to infer that some of the infographics we have presented to them are generated. However, this does not necessarily say anything bad or good about our approach. Given the positive results related to visual attractiveness and the stakeholders' intention to adopt our approach, we consider the project a success.

Our Approach can Increase the Efficiency of producing recurrent, data-centric infographics. The data pipeline is kept integrated, avoiding the disconnection mentioned in Sect. 1 and confirmed by stakeholders in Sect. 3. The ESEA accountant or network manager can trigger the generation of the infographics from the ESEA tool itself. The tool generates infographics in a reasonable lapse of time.

We Propose a New, Model-Driven Process for Producing Infographics. Graphic designers still intervene by creating a graphic design, but early in the process (A2 in Fig. 8). A remaining question is who the best candidate is to use the DSL. It is up to graphic designers to learn this skill. Otherwise, the ESEA method engineer needs to be capable of not only specifying the ESEA method (e.g., the indicators) but also to translate the infographic designs of the graphic designer into textual models, using the DSL.

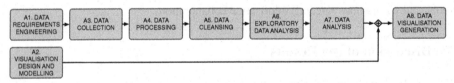

Fig. 8. Model of the redesigned process for data-centric infographic creation. It is an evolution of the one in Fig. 1, and it requires infographic modelling (A2) and generation (A8) technology.

7.2 Validity of the Results

There are a number of threats that we have had to manage or accept [37].

Conclusion Validity. We have checked the assumptions of statistical tests, obtained significant statistical results and, on top of this, we are being careful with our claims. To minimise the potential random effect of having a heterogeneous group of participants, we have designed the experiment as a balanced, paired comparison.

Internal Validity. We piloted the survey before the experiment. Also, every participant is obliged to assess whether each of the 10 infographics is original or generated, and they are not informed of the outcome. Thus, we minimise the threat of maturation and prevent that they stop halfway if they feel disappointed by failure.

Construct Validity. We rely on the opinion of participants to rate the attractiveness of infographics. While this is a threat in many situations and it would not constitute a valid measure for other types of variables, attractiveness is a subjective phenomenon and a matter of personal preference. This is also in accordance with earlier research [35]. Also, when several participants contribute to the same subjective evaluation of a stimulus, inter-subjectivity of judgment becomes intra-objectivity; see an example in aesthetics in [38]. Hypothesis guessing is a remaining threat; the participants might have assumed that we are aiming at validating an approach to generate infographics. We have set the final sample size before the experiment execution, and we have not examined the data until the experiment had terminated, to avoid experimenter biases.

External Validity. The analysis of infographic design tools showed that there are more component types than we anticipated. However, we are confident to have identified most component types that are present in infographics within the impact measurement domain (Fig. 9 shows the data saturation chart). Similarly, there are additional tools in the market that we have not analysed (e.g., Adobe Creative Cloud Express), but the main features have likely been identified (see the data saturation chart in Fig. 10).

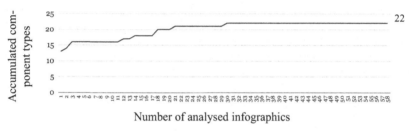

Fig. 9. Data saturation for component types was achieved after analysing the 30th infographic

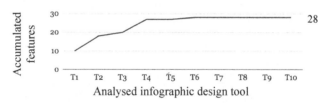

Fig. 10. Data saturation for features was achieved after analysing the 6th infographic design tool

We have focused on infographics used to communicate results of impact measurements. While this type of data-centric infographics will share many characteristics with those produced in other domains, our decision, of course, limits the generalisability of our DSL and interpreter, as well as the empirical results of the validation. However, we felt it was important to narrow down the scope of our project to produce valuable results for at least one domain. It is also relevant to discuss that, despite our desire to support a large array of chart types, infographics are often intended for a large audience that does not necessarily understand all types of charts. Limiting our efforts to the most widely used and commonly understood makes the challenge more tractable.

Finally, we have implemented a proof of concept in which engineering decisions are not always future-proof or result in the best possible performance. A couple of alternative approaches are using available JavaScript libraries for creating charts (e.g. ChartJS, rGraph), using PGF, a platform- and format-independent TeX package for generating graphics [39], or generating the infographics directly over conventional presentation software such as Microsoft PowerPoint using Visual Basic for Applications.

8   Conclusions and Future Research

In this study, we redesign the data-centric infographics production process to follow a model-driven approach (Fig. 8). The graphic design of the infographic is modelled, along with references to the data sources, using a domain-specific language (DSL). This way, the specification can be reused with different datasets; for instance, along the years or across several organisations. We refer the reader to earlier work to discover how the data (i.e., indicators of ethical, social and environmental performance) is defined and collected [4]. This approach solves the disconnection of software environments used during the conventional process (Fig. 1) by integrating the infographic generation features into the data-analysis tool. We have applied the approach to the domain of impact measurement because we had evidences of the existence of problems with the conventional process, and because it is our major area of research and we had a technology ready to be extended with the necessary features; namely, the openESEA tool [4]. Initial tests have shown the feasibility of our proposal. A later comparative experiment has proven that the generated infographics are at least as attractive as the original ones. Participants could, to some extent, guess that some of the generated infographics were not original. Overall, both the experiment results and the expert stakeholder opinions are encouraging.

In our current proposal, infographics are modelled textually, using an Xtext editor. A more elegant solution would be designing them with a what-you-see-is-what-you-get drag-and-drop editor that could later generate the resulting specification files in a transparent way. Of course, several iterations and usability studies will be needed to polish such approach. There are also known limitations in our DSL and interpreter that we plan to overcome, as well as requirements pending to be implemented (e.g., loading images directly to the tool, more options for picture graphs). We also consider supporting interactive infographics. An interesting empirical study would focus on the process rather than the outcome; e.g., asking participants to adopt our model-based approach and quantify the differences between the first usage and subsequent ones (e.g. learning curve, efficiency, quality of results). We expect that the highest gains would appear after the first modelling approach (e.g. evolving an existing model would take less time than creating it from scratch) and, especially, when reusing the same infographic over time (e.g. using the same model with different data). With such future work, we plan to contribute to the field of model-driven engineering and, at the same time, offer a suite of open-source tools for organisations that care about their impact in society and the environment.

References

1. Smiciklas, M.: The Power of Infographics: Using Pictures to Communicate and Connect with Your Audiences. Que Publishing, New York (2012)
2. Dunlap, J.C., Lowenthal, P.R.: Getting graphic about infographics: design lessons learned from popular infographics. J. Vis. Literacy **35**, 42–59 (2016)
3. O'Neil, C., Schutt, R.: Doing Data Science: Straight Talk from the Frontline. O'Reilly (2013)
4. España, S., Bik, N., Overbeek, S.: Model-driven engineering support for social and environmental accounting. In: RCIS 2019. IEEE (2019)
5. Mernik, M., Heering, J., Sloane, A.M.: When and how to develop domain-specific languages. ACM Comput. Surv. **37**(4), 316–344 (2005)
6. Wieringa, R.J.: Design Science Methodology for Information Systems and Software Engineering. Springer, Berlin (2014)
7. Karagiannis, D.: Agile modeling method engineering. In: 19th Panhellenic Conference on Informatics (PCI 2015), pp. 5–10. ACM, Athens, Greece (2015)
8. Behrens, H., et al. and Contributors: Xtext user guide v 1.0.1 (2010)
9. Martix, S., Hodson, J.: Teaching with infographics: practising new digital competencies and visual literacies (2014)
10. Arcia, A., et al.: Sometimes more is more: iterative participatory design of infographics for engagement of community members with varying levels of health literacy. J. Am. Med. Inform. Assoc. **23**(1), 174–183 (2016)
11. Amit-Danhi, E.R., Shifman, L.: Digital political infographics: a rhetorical palette of an emergent genre. New Media Soc. **20**(10), 3540–3559 (2018)
12. Thoma, B., et al.: The impact of social media promotion with infographics and podcasts on research dissemination and readership. Can. J. Emerg. Med. **20**(2), 300–306 (2018)
13. Iannaci, D.: Reporting tools for social enterprises: between impact measurement and stakeholder needs. Eur. J. Soc. Impact Circ. Econ. **1**(1b), 1–18 (2020)
14. Afrizal, D.: Statistical infographic publication: embracing the general public. In: Conference of European Statisticians. UNECE, Online (2021)
15. Otten, J.J., Cheng, K., Drewnowski, A.: Infographics and public policy: using data visualization to convey complex information. Health Aff. **34**(11), 1901–1907 (2015)
16. Mackinlay, J.: Automating the design of graphical presentations of relational information. ACM Trans. Graph. **5**, 110–141 (1986)
17. Dibia, V., Demiralp, Ç.: Data2vis: automatic generation of data visualizations using sequence-to-sequence recurrent neural networks. IEEE Comput. Graphics Appl. **39**, 33–46 (2019)
18. Gray, R., Adams, C., Owen, D.: Accountability, Social Responsibility and Sustainability: Accounting for Society and the Environment. Pearson, Boston (2014)
19. B Lab. The B Impact Assessment v6 (2018). https://bimpactassessment.net
20. Felber, C., Campos, V., Sanchis, J.R.: The common good balance sheet, an adequate tool to capture non-financials? Sustainability **11**(14), 3791 (2019)
21. Ortega, A., del Carmen Olaya, M., Bastida, R., Torrent, R., Suriñach, R.: Guia tècnica per a l'elaboraciò del balanç social. ACCID (2016)
22. ISO: ISO 26000 - Social responsibility (2010)
23. ISO: ISO 14000 - Environmental management (2015)
24. Global Reporting Initiative. GRI Standards (2018)
25. IIRC: International <IR> framework, January 2021 (2021)
26. Swiss Informatics Society. Data Center Green IT Maturity Assessment (2015)
27. Matson, L.: Sustainability Tracking, Assessment & Rating System (STARS): a tool for evaluating campus sustainability. In: UC, CSU, CCC Sustainability Conference (2008)

28. FAO: SAFA - Sustainability Assessment of Food and Agriculture systems. Guidelines version 3.0. Food and Agriculture Organization (2014)
29. REAS. A Charter for the Social Solidarity Economy (2019). https://www.guerrillatranslation.org/2019/07/01/a-charter-for-the-social-solidarity-economy
30. Xarxa d'Economia Solidària. El Balanç Social (2016). http://mercatsocial.xes.cat/ca/eines/balancsocial/
31. España, S., Ramautar, V., Overbeek, S., Derikx, T.: A domain-specific language for data-centric infographics: technical report. Utrecht University. arXiv:2203.09292 (2022)
32. Benavides, D., Trinidad, P., Ruiz-Cortés, A.: Automated reasoning on feature models. In: Pastor, O., Falcão e Cunha, J. (eds.) CAiSE 2005. LNCS, vol. 3520, pp. 491–503. Springer, Heidelberg (2005). https://doi.org/10.1007/11431855_34
33. Cohn, M.: User Stories Applied: For Agile Software Development. Addison-Wesley, Redwood City (2004)
34. Waters, K.: Prioritization using MoSCoW. Agile Planning 12, 31 (2009)
35. Lavie, T., Tractinsky, N.: Assessing dimensions of perceived visual aesthetics of web sites. Int. J. Hum Comput Stud. 60(3), 269–298 (2004)
36. Cohen, J.: Statistical Power Analysis for the Behavioral Sciences, 2nd edn. Lawrence Erlbaum Associates, Hillside (1988)
37. Wohlin, C., Runeson, P., Höst, M., Ohlsson, M.C., Regnell, B., Wesslén, A.: Experimentation in Software Engineering: An Introduction. Kluwer, Boston (2000)
38. Zen, M., Vanderdonckt, J.: Assessing user interface aesthetics based on the inter-subjectivity of judgment. In: 30th International Human Computer Interaction Conference, pp. 1–12 (2016)
39. Tantau, T.: TikZ and PGF. Manual for version 3.1.9a. Universität zu Lübeck (2021)

The B Method Meets MDE: Review, Progress and Future

Akram Idani[✉]

Univ. Grenoble Alpes, Grenoble INP, CNRS, LIG, 38000 Grenoble, France
Akram.Idani@univ-grenoble-alpes.fr

Abstract. Existing surveys about language workbenches (LWBs) ranging from 2006 to 2019 observe a poor usage of formal methods within domain-specific languages (DSLs) and call for identifying the reasons. We believe that the lack of automated formal reasoning in LWBs, and more generally in MDE, is not due to the complexity of formal methods and their mathematical background, but originates from the lack of initiatives that are dedicated to the integration of existing tools and techniques. To this aim we developed the Meeduse LWB and investigated the use of the B method to rigorously define the semantics of DSLs. The current applications of Meeduse show that the integration of DSLs together with theorem proving, animation and model-checking is viable and should be explored further. This technique is especially interesting for executable DSLs (xDSLs), which are DSLs with behavioural features. This paper is a review of our Formal MDE (FMDE) approach for xDSLs and a proposal for new avenues of investigation.

Keywords: DSLs · B method · Formal methods · MDE

1 Introduction

Developers can find a plethora of tools and approaches for building Domain-Specific Languages (DSLs) from several perspectives: design, execution, debugging and code generation. The reader can refer to [15] for a survey about DSL tools that are referenced from 2006 to 2012, and to [14] for the period between 2012 and 2019. These studies can help engineers to choose the best DSL option for a specific context and select the ones that fulfill their requirements. Unfortunately, despite that the aforementioned surveys show that DSL tools have reached a good level of maturity, they also show a major limitation of these tools; *i.e.,* the lack of formal reasoning. Kosar et al. [15] observed that only 5.7% of primary studies applied a formal analysis approach, and mentions that *"there is an urgent need in DSL research for identifying the reasons for lack of using formal methods within domain analysis and possible solutions for improvement"*. One possible reason identified by Bryant et al. [4] is that the syntactical description of DSLs is often straightforward, but specifying detailed behavior (semantics) is much more harder. This would explain why only the syntax of DSLs are formally described, but the semantics are left toward *"other less than*

R. Guizzardi et al. (Eds.): RCIS 2022, LNBIP 446, pp. 495–512, 2022.
https://doi.org/10.1007/978-3-031-05760-1_29

desirable means" [4]. The recent study of [14], appeared in 2020 (more than seven years after [15] and [4]), does not report on a better situation because it only refers to testing (the formal dimension is completely absent). This study shows that among the 59 discussed tools only 9 provide supports for testing. This is an important shortcoming because it weakens the applicability of DSLs, especially for safety-critical systems. Indeed, in these systems correctness is a strong requirement and it is often addressed by the application of formal methods.

During the last three years we investigated and developed a solution to this problem by proposing a Formal Model-Driven Engineering (FMDE) approach to create zero-fault DSLs, which are DSLs whose semantics (static and dynamic) are mathematically proved correct with regards to their structural and logical properties. We developed the Meeduse tool [10] to make the bridge between MDE and the formal B method [1]. Technically the tool is built on EMF [28] (The Eclipse Modeling Framework) and ProB [20], an animator and model-checker of the B method that is certified T2 SIL4 (Safety Integrity Level), which is the highest safety level according to the Cenelec EN 50128 standard. ProB is also used by companies such as Alstom, Thales, Siemens, General Electric, ClearSy and Systerel. In addition to the formal reasoning about the semantics of DSLs, the Meeduse tool offers execution and debugging facilities, which is a major contribution in comparison with other works built on Abstract State Machines and Maude [25]. The overall idea is that animation done in ProB is equivalently applied to the input EMF model leading to a correct model execution.

This paper surveys the current situation of xDSLs and presents the contributions of our FMDE approach for xDSLs. Note that Meeduse has been discussed in several other papers. However, every publication explored a specific facet of DSLs and the B method without a clear vision about the situation. The contributions of this paper are therefore:

1. review progress made, tool support, and case studies, which attests to the interest and power of our FMDE approach to building xDSLs.
2. put certain criticisms and received ideas concerning formal methods and translational approaches into perspectives.
3. provide useful research directions that may help gain visibility and also strengthen the application of FMDE in system design.

In Sect. 2 we present the strengths and the limitations of the two major approaches for xDSLs: translational and in-line approaches. In Sect. 3 we provide an overview about Meeduse and discuss its originality in comparison with existing works. Section 4 discusses three successful realistic applications of the tool. Finally, Sect. 5 presents the conclusion and perspectives of our works.

2   Executable DSLs: Correctness and Challenges

2.1   Observations

Several research works have been devoted by the MDE community in order to deal with executable DSLs. The major intention is to support early validation and verification in the development process. Indeed, an executable model not only represents the structural features of a system, but also deals with its behaviour. The model is therefore intended to behave like the final system should run, which provides the capability to simulate, animate and debug the system's properties before its implementation. In the literature there are two major approaches to implement the execution semantics of a DSL: translational approaches and in-place approaches. The former translate the DSL semantics into a well-established semantic domain that is assisted by numerous tools, such as MAUDE, ASM or FIACRE. The latter weave the execution semantics into meta-models, which is often done by extending the semantic domain of a DSL with action languages. Every approach has its strengths and limitations that can be summarized by [5]:

- Translational approaches:
  - *Strengths*: apply available tools, such as animators and/or model-checkers to ensure the execution capabilities of the DSL.
  - *Limitations*: first, often these approaches require complex transformations to implement the mapping from the DSL to the target semantic domain, and second, the execution results are only obtained in the target domain.
- In-place approaches:
  - *Strengths*: allow a more intuitive definition of executable DSLs since the language engineer has only to deal with concepts of the DSL and not with another target language.
  - *Limitations*: require to implement for each DSL all the execution-based tools.

Since the objective of an xDSL is to ensure early validation during the development process, the developer must have some confidence in the underlying verification tool. Nonetheless, on the one hand [14,15] show that the verification feature of existing LWBs is very limited; and on the other hand, the dependability of the resulting system is strongly related to the correctness of the DSL. In [32], the authors analysed the risks of using LWBs regarding the introduction of faults into a critical software component. The authors observed that existing LWBs have not been developed using a safety process and attested that "*particular DSLs could be, but that is only of limited use if the underlying LWB is not*". In this sense, a translational approach is much more pragmatic because it reuses well-established verification tools, that are often widely accepted by the formal methods community. In [9] we have done an empirical study with various existing xDSL platforms, and we showed that failures may originate from several concerns:

C_1: The abstract syntax of the DSL, often represented with a meta-model and the associated modeling operations such as object creation and destruction, and also the setters and getters. This kind of failures is due to the fact that the execution semantics of the DSL apply the modeling operations in order to update a given model during its execution.

C_2: The model itself, because LWBs may apply some internal choices that are not conformant to the standard semantics of meta-models. Hence, incorrect behaviours may not be exhibited while executing the model but they still possible in the final system.

C_3: The execution engine of the meta-language. For example, OCL based engines often wrongly support non-determinism as discussed in [2,31], and even if these theoretical solutions exist they are not yet integrated within tools.

2.2 Discussion

Undoubtedly, the usage of a formal method with well-established verification tools is a solution to neutralize bugs that may originate from the DSL and the modeling layer. The question is therefore "how to provide the good mixture between correctness and expressiveness?". We believe that the answer depends on the application domain of the xDSL. Indeed, formal methods are often negatively perceived by developers due to their mathematical notations, and consequently translational approaches have a limited usage for general purpose applications. Nonetheless, formal methods showed their strengths in the safety-critical domain, and they became a strong requirement to ensure zero-fault applications. Obviously, DSLs and language workbenches (LWBs) provide several benefits to this field due to their ability to share, visualize and communicate about domain concepts. Both formal methods and model-driven engineering are highly desirable in safety critical systems because domain-specific representations are omnipresent, as well as the use of provers and model-checkers. We assume that the reader may agree with this claim, even if it appears that for larger scale projects formal methods are not as widespread, because of the overhead they may create during the development activities.

In our opinion, bridging the gap between both worlds (MDE and FM) does not require innovative solutions and can be done by integrating well-established tools provided by both the Formal Methods community and the MDE community. The lack of automated formal reasoning in LWBs, and more generally in MDE, is not due to the complexity of formal methods and their mathematical background, but originates from the lack of initiatives that are dedicated to the integration of both techniques. A supporting argument could be the assertion by [15] that *"researchers within the DSL community are more interested in creating new techniques than they are in performing rigorous [empirical] evaluations"*. Moreover, the MoDeVVa Workshop (Model-Driven Engineering, Verification and Validation) collocated with the MODELS conference, and also the international conference on integrated formal methods (IFM) are good illustrations of this claim, because they have presented many approaches over the years where formal methods found their way in MDE techniques, and vice-versa. We can assume that the required material to provide a viable FMDE for DSL execution in LWBs already exists. Unfortunately, not only the integration attempts remain poor, but also the applications of existing approaches remain limited to illustrative examples without going further towards realistic safety-critical requirements. This observation may explain why existing approaches [7,23,25,30,33] are not

discussed at all in [14,15]. As far as we know, the only modeling tool with execution facilities that is proven correct is Scade [6], but this means that one
cannot extend it with domain-specific constructs without proving the correctness of these constructs and the underlying behavioural semantics. Our solution
provides a LWB (called Meeduse [22]) that allows one to formally define and
reason about the semantics (static and dynamic) of xDSLs using the formal B
method [1]. The underlying approach is a translational approach, but it goes a
step further in comparison with existing translational approaches by addressing
the three concerns discussed above and by providing realistic applications from
the safety-critical domain.

3   The Meeduse LWB

3.1   Overall View

Figure 1 summarises the main concepts of the approach. First, the MDE expert
defines the meta-model of a DSL and its underlying contextual constraints. Then,
Meeduse translates the meta-model into the B language, which produces a functional formal model describing the static semantics of the DSL (step (1) Translation).

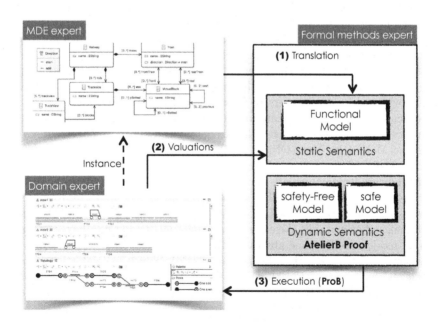

Fig. 1. The Meeduse approach.

Regarding the dynamic semantics, it is specified using additional B models
(safety-free and safe models) by means of B operations and additional invariants.

The safety-free model is an open model from which one can exhibit undesirable scenarios, and the safe model improves the latter with additional invariants and preconditions such that the undesirable behaviours are neutralized. Having these formal models, a prover (*e.g.* AtelierB) can be used in order to guarantee their correctness. This task is performed by experts in formal methods and requires skills in theorem proving.

To ensure the execution of the DSL, Meeduse applies ProB [20]. First, it injects a given model, instance of the DSL meta-model, within the B specification of the static semantics (step (2) Valuations). This step produces valuated B variables whose state transitions are managed by ProB given the B operations of the dynamic semantics machine. The tool makes the bridge between the model and the valuations of the B variables: for every animation step, the tool automatically updates the model resource such that it remains equivalent to the valuations of the B variables. Hence, traversing transitions produced by ProB result in an automatic animation of the input model (step (3) Execution).

3.2 Towards a Viable Formal Abstract Syntax

Even if Meeduse is a recent tool (its reference papers appeared in 2020), the underlying approach is not new since it provides a translation semantics to a DSL: the meta-model of the DSL is mapped in the B language through a model-to-model transformation. The limitations of translational approaches were widely discussed in the literature [4]. The first one is that it *"is very challenging to correctly map the constructs of the DSL into the constructs of the target language"*. This limitation originates from the fact that the mappings may not be at the same abstraction level and the target language may not have a simple mapping from the constructs in the source language. The use of the B method in Meeduse provides objective answers to these observations because meta-models are semantically similar to the mathematical space of the B method. Indeed, the semantics of meta-models is standardized by the Object Management Group using the MOF (the Meta-Object Facility) in which this semantics is defined by means of OCL constructs. These are built on the set theory and the first order predicate logic, which are also the foundations of the B language. Furthermore, both the MOF and the B method are state-based modeling formalisms; the semantic difference is that B applies the theory of generalized substitutions, which makes it executable. We can then assume that B and MOF have the ability to build models at the same abstraction level.

Structurally the MOF is a restriction of UML, which is comforting because this means that mapping the MOF into B is not very challenging. It can be done via a classical UML-to-B approach, which has been addressed in the past by a plethora of works and tools:

– UML2SQL [17,18]: this work provides a formal framework for the development of database applications. The B specifications are extracted from UML class diagrams, state-transition diagrams and activity diagrams. Refinement tactics are proposed in order to incrementally generate a correct implementation of a relational database.

- U2B [26,27]: this work proposes to produce a B specification, called "natural", so that the proof obligations are as simple as possible. For example, instead of generating a B machine from each UML class (of a class diagram), the authors generate a unique B machine that gathers the B constructs of the whole package.
- ArgoUML+B [19,24]: this work tried to take into account complex UML features. It proposed, on the one hand, various solutions for the translation of the UML inheritance mechanism, and on the other hand, a new formalization of behavioural UML diagrams.

Table 1 summarizes the structural features of UML class diagrams that are addressed by these approaches. The table shows that each approach has its advantages and limitations and that obviously for a better coverage of UML a combination of the various approaches is needed. The challenge is how to gather into a unified framework several UML-to-B approaches from the rich state of the art. In Meeduse the user is able to select the desired UML-to-B transformation technique and also to combine rules coming from various techniques. This is interesting because UML has been mapped into numerous state-based formal languages with similar principles (*e.g.* Z, Object-Z and ASM), and therefore the only remaining piece is to integrate these mappings within a unifying LWB such as Meeduse.

Table 1. Overview of the supported UML structural features

	UML2SQL	ArgoUML+B	U2B
Classes (undetermined instances)	+	+	+
Classes (fixed instances)	–	–	+
Class attributes	+	+	+
Distinction between multi/mono-valued attributes	+	–	–
Inheritance	+	+	+
Associations multiplicities	+	+	+
Associations navigation direction	–	+	+
Roles	+	–	+
Associations constraints	+	+	–
Distinction between fixed/non-fixed associations	+	–	–
Association	+	+	–
Associative classes	+	–	–
Parametrized classes	–	–	+

Legend: "+" (considered criterion); "–" (non considered criterion)

3.3   ProB as an Execution Engine of xDSLs

The second limitation of translational approaches, as discussed in [4], is that "*the mapping of execution results back into the DSL is not covered*". Indeed, in the existing works [7,23,25,30,33] the execution results are only obtained in the target domain because getting back the results in the source language is often claimed to be difficult and requires to extend the abstract syntax in order to model these results. The work of U. Tikhonova [29] uses the EventB language but applies classical visual animation using BMotion Studio[1] to the formal specifications. Unfortunately, this is time consuming because the DSL syntax must be redefined in BMotion Studio. Moreover, it also requires some additional verifications in order to address the compatibility between the initial DSL syntax and the graphical visualization that is managed by BMotion Studio.

None of the existing techniques offer a way to execute jointly the formal model and the domain model. They start from a DSL definition, produce a formal specification and then they "get lost" in the formal process. Meeduse gives a new vision to the integration of xDSLs and formal methods and provides solutions to this limitation (steps (2) Valuation and (3) Execution of Fig. 1). Meeduse shows that getting back the execution results in the source language is not a complicated task because, as explained in the previous section, MOF and B are both built on the set theory and the first order predicate logic. The difficulty may come from the theory on which the target language is built. By using a state-based language built on the set theory and the first order predicate logic, the semantical gap with MOF can be easily managed. We refer the reader to [10] for the technical details about how the execution results can be translated back to the original model without extending the abstract syntax of the DSL.

The proposed approach is inspired by [16], where the authors assume that a tool like an animator or model checker is able to compute all execution traces of a system and propose to implement the software system on top of programs that execute in background the formal model by choosing traversing transitions. By using xDSLs, we go a step further and reduce the effort required for writing the aforementioned programs that interact with the animator. Indeed, in our approach the developer focuses on the definition of the model's semantics and Meeduse generates a standard specification and implementation that can be executed and controlled by the animator. The solution provided by Meeduse is to guarantee the equivalence between the execution traces of the DSL and the B data provided by ProB [20] (Steps 2 and 3 in Fig. 1). As the two models, managed respectively by the LWB and ProB, evolve together during the execution and remain equivalent to each other, then debugging can be done using either the ProB tool, due to its numerous facilities for interactive animation, or an EMF-based LWB such as the Gemoc Studio [8]. Debugging with ProB (or other formal tools) may be required while developing the formal semantics of the DSL, and debugging with the Gemoc Studio (or other LWBs) may be recommended to the end users (or domain experts) who are often uncomfortable with formal tools.

Figure 2 is a screen-shot of Meeduse while running a Petri-Net model of the bounded-buffer producer/consumer synchronization [3]. The top-left side shows the graphical representation of the Petri-Net model where two processes

---

[1] http://www.stups.hhu.de/ProB/index.php5/BMotion_Studio.

share a common buffer: the producer (whose states are pending and progress) and the consumer (whose states are ready and accept). In this model, sequence (*start*; *produce*)∗ increases the number of tokens in place buffer until a fixed bound is reached; and sequence (*request*; *consume*)∗ decreases this number by five. The other views of Fig. 2 are provided by Meeduse for interactive animation and debugging: (1) the execution view from which the user can select and run possible instances of the B operation for transition firing, (2) the execution history view allowing an omniscient debugging, which is useful to go backward in time, and (3) the state view giving the various valuations of the B variables in the current state. The tool also informs the user whether the invariant is preserved or not. In addition to the animation of xDSLs, Meeduse benefits from the numerous model-checking strategies of ProB such as (without being exhaustive): heuristic-based and breadth/depth search, reachability analysis with LTL formulas and predicate decomposition.

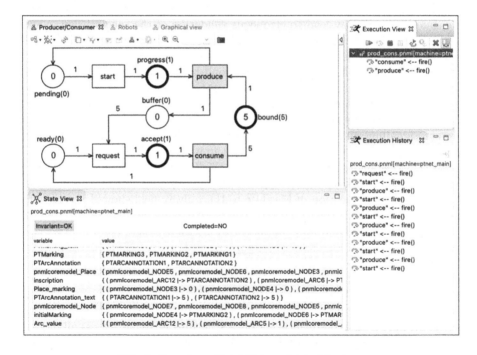

Fig. 2. Running a Petri-Net model in Meeduse

4   Applications

Meeduse has been successfully applied to realistic case studies showing the viability of our Formal MDE (FMDE) approach and the benefits of the tool to the development of correct executable DSLs:

1. To simulate safe train behaviours [11,12]. This application builds on a graphical DSL that can be used by railway experts to design railroad topologies and simulate (un)safe train movements.
2. To execute model-to-model transformations [13]. This application was carried out during the 12th edition of the transformation tool contest (TTC'19) and won the award of best verification and the third audience award.
3. To prove the execution semantics of PNML (Petri-Net Markup Language), the international standard ISO/IEC 15909 for Petri-nets, and to take benefit of the capabilities of the B method in the Petri-net field. This application led to a full-fledged tool called MeeNET (Meeduse for Petri-Nets).

4.1 FMDE for Railway Systems

It is known that the usage of formal methods in safety-critical railway systems is a strong requirement because their failures may lead to accidents and human loss. However, while formal methods provide solutions to the verification problem, human errors may lead to erroneous validation of the specification, which may produce the wrong system. Thus, involving the domain expert in the development process makes sense; first to avoid the eventual misunderstandings of the requirements by the developer and second to have a well-defined specification. Furthermore, in the railway domain, textual and graphical representations of domain concepts are widely used in the various reference documents, which allows sharing the knowledge about several railway mechanisms such as track circuits, signalling rules and interlocking systems. Formal methods and DSLs are therefore inescapable for railway systems definition and development. Unfortunately, most modeling tools (*e.g.* SafeCap, RailAid) are not built on a formal approach; and most formal solutions are not supported by domain-specific modeling tools. The application of Meeduse to the railway field showed its strength to mix both aspects in the same tool. The work was funded by the NExTRegio project of IRT Railenium and supported by SNCF Réseau. The intention was to deal with the analysis of railway signalling systems based on emergent train automation solutions, especially the European Rail Traffic Management System (ERTMS) and the underlying Train Control System (called ETCS). There are three levels of ERTMS/ETCS which differ by the used equipments and the operating mode. The first two levels are already operational, and the third one is still in design and experimentation phases: it aims at replacing signalling systems with a global European GPS-based solution for the acquisition of train positions. Our FMDE solution led to a graphical DSL equipped with proved formal semantics, which safely assists domain experts while defining the domain models and simulating the underlying operating rules. Figure 3 gives different states of the domain model (left hand side) with the list of B operations (right hand side) that can be enabled by the animator at every state.

In the first state, on top of this Figure, the domain expert can: (*i*) change the direction of trains TRAIN1 and TRAIN2; (*ii*) change the switch from straight to divergent; (*iv*) arm the auto train stop (ATS) of PORTION2; and (*v*) compose train routes using instances of operation Safe_MA_AddPortion. The

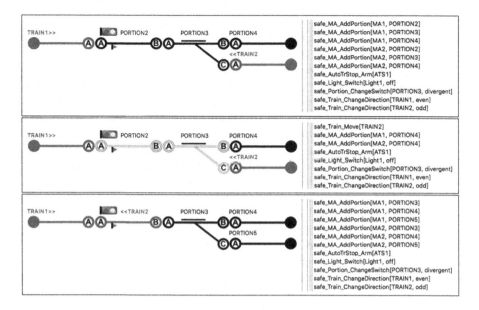

Fig. 3. Meeduse animation of a railroad model with safe execution semantics

execution of operation `Safe_MA_AddPortion(MA2,PORTION3)` followed by operation `Safe_MA_AddPortion(MA2,PORTION2)`, reaches the state presented in the middle of Fig. 3 where the color of PORTION2 and PORTION3 became orange meaning that these portions are reserved for some train. In this state, operation `safe_Train_Move` can be enabled in order to start moving TRAIN2. Operation `Safe_Train_Move(TRAIN2)` can be animated twice which leads to the state in bottom of Fig. 3 where TRAIN2 occupies PORTION2 after consuming the authorizations provided by its route.

4.2 FMDE for Model Transformations

Model transformation is a core concept in MDE. It aims to automate the extraction of platform specific representations and/or executable artifacts from high level models. Thus, the developer focuses on the modeling activities rather than on coding. After more than a decade of maturation, several tools have been developed (*e.g.* QVTo, ATL, TGG) and it became possible to leverage model transformations within software development and use them to build large-scale and complex systems. In this respect, the correctness of model transformations became crucial. Despite the advances in this field, the Verification & Validation (V&V) of model transformations still remains scattered, and perspectives on the subject are still open. Between 2012 and 2016, the international Workshop on Verification of Model Transformations (VOLT) tried to offer researchers a dedicated forum to classify, discuss, propose, and advance verification techniques dedicated to model transformations. Several solutions have been proposed:

incremental deductive verification, testing via classifying terms, applying reduction techniques, using model transformations to verify model transformations themselves. Unfortunately, these techniques were not generalized and realistic applications remain very limited or almost absent. We believe that this is due to the fact that the underlying V&V are not focused on safety-critical systems, but presented as solutions to software engineering in general – where testing remains the norm. For example, proving the transformation of UML class diagrams into RDBMS may show the complexity of a formal approach, but clearly hides how useful it is in practice, since the case does not deal with safety. A FMDE approach makes sense once critical concerns have to be taken into account. This is confirmed by our participation in the 2019 edition of the transformation tool contest (TTC'19), where the call for solutions was about a case study that is well-known in safety-critical systems: the transformation of Truth Tables (TT) into Binary Decision Diagrams (BDD). Our major observation is that among the seven participants, Meeduse was the only attempt that addressed V&V; the other solutions addressed flexibility, performance and optimality. Meeduse was not only used to verify the model transformation, but also to execute it, which is its novelty in comparison with the discussed approaches. However, the downside is its low performance when executing the transformation of huge models. Nonetheless, as the transformation is written in B, proved and attested for middle size models, it can therefore be used as a reference specification from which one can build a low-code model transformation.

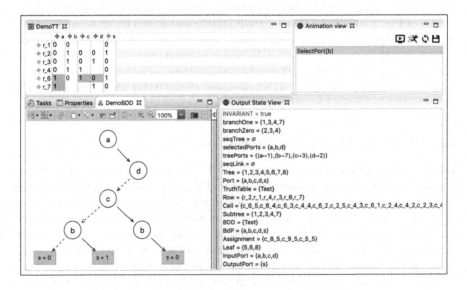

Fig. 4. Application of Meeduse to DSL transformation.

Figure 4 shows the various views of Meeduse where the modeling view deals with the representation of the input DSL. In this application, formal operational semantics, written in B, consume truth table elements and progressively produce a binary decision diagram. This application allowed us, on the one hand, to figure out how far we can push the abilities of a formal method to be integrated within model-driven engineering, and on the other hand, to present an original approach to think, reason and execute DSL transformations using a formal method.

4.3 FMDE for XML Standards

XML is used across a lot of domains because it favours readability (due to its structuring features) and interoperability between platforms (due to the availability of parsers). Several communities have developed XML-based exchange standards to structure domain concepts and improve the interoperability of the growing number of computer applications that use these domain concepts. In the railway field, we can refer to RailML (Railway Markup Language, an open XML based data exchange format for data interoperability of railway applications). An XML file is a DSL, whose static semantics (well-formedness rules) are established via an XML schema, which is itself a meta-model in the MDE jargon. Dealing with XML documents opens a broad spectrum of possibilities to FMDE, since a formal approach can be applied to any domain standard approved by experts. Having this argument, we applied Meeduse to PNML (Petri-Net Markup Language), the international standard ISO/IEC 15909 for Petri-nets. PNML provides an agreed-on interchange format that is compliant with a formal definition of Petri-nets and is managed by EMF-based platforms such as PNML Framework and The ePNK. This application was motivated by the following points:

- There exist several widely known tools that implement the Petri-net theory for verification purposes.
- Petri-nets were applied to the safety-critical domain with several success stories such as the Oslo subway.

A FMDE approach for PNML contributes towards this field by providing practitioners the possibility to benefit from the rich catalog of MDE tools without losing sight of correctness and rigorous development. This application led to a tool called MeeNET in which a PNML file can be executed, debugged, verified and translated into an implementation.

5   Discussion and Future

This paper has given an overview of our FMDE approach and presented realistic applications of our tool support, which shows that the integration of executable DSLs together with theorem proving, animation and model-checking is viable and should be explored further. However, in safety-critical systems, tools must be qualified before being used. Voelter et al., in [32] state that *"one way to qualify a given tool based on domain standards is to show that the tool has been used successfully in industrial projects built on the aforementioned standards, collect usage reports and malfunctioning reviews, and define process-based mitigation that should be used when the tool is applied"*. The formal aspects of Meeduse can be qualified in this way because it is built on the B method which is already proven in use: the B method provides a safe development process ranging from theorem proving to code generation, and both AtelierB and ProB were successfully used in several real-life safety-critical systems.

However, it is difficult to objectively apply the same principles to the non-formal aspects of Meeduse because it is an emergent approach and even if it has had several realistic applications, including a major one in the railway field, these applications were not at an industrial scale. Other ways to qualify a tool include proof and extensive validation that the tool is correct or showing that the tool itself has been developed with a formal and safe process. This is a much more realistic way because although the OMG did not establish guidelines for qualifying MOF-compliant tools, the MOF reference document provides a semi-formal specification applying XMI (for the syntax) and OCL (for the semantics). Checking that Meeduse is MOF-compliant can be done via a formal definition of the MOF semantics and the proof that data structures and algorithms used in the tool preserve these semantics all along the execution. Currently, we are working on this direction by formalizing the concepts of Meeduse using Meeduse itself, which would provide much more confidence in the tool and the underlying approach.

Furthermore, it is worthreviewing to mention that the usage of B appears as a good choice for several reasons: (1) the availability of a rich UML-to-B state of the art, that allowed us to provide a viable translation from MOF into B (considering that MOF is a restriction of UML); and (2) the usage of theorem proving, in addition to model-checking, to guarantee zero-fault DSLs. In most application domains such as Requirements Engineering, Enterprise Architectures, Business Process Management and Legal Contracts, where DSLs do exist, the commonly used verification tools are model checkers and/or SAT/SMT solvers. Meeduse may introduce theorem proving to these domains. Besides our own experience and judgement, the B Method has been compared to other state based formal methods and tools in [21] and got several good points. Regarding scalability, which is *"the ability to be well applicable to arbitrarily large and complex projects"*, the B method is ranked (Good). The work also highly ranks the verification features of B and its tool support. Nonetheless, we are aware that in order to broaden the spectrum of Meeduse several target formal approaches have to be addressed. This objective belongs to our short term perspectives (Fig. 5) and do not seem technically difficult since the power of Meeduse comes from ProB, and ProB is itself a multi-target platform covering (in addition to B) Event-B, Z, Alloy and TLA+. Translations from UML to these languages have already been investigated in the past; we just need to integrate them within Meeduse. We recall that we built this paper on the (strong) claim that FMDE does not need innovative solutions, but rather integrative initiatives of existing well-established approaches.

Figure 5 gives the evolution time-line of Meeduse with the major highlights and our current and short-term perspectives. The red time-lines refer to DSL tools that are powered by Meeduse and which have their own existence, such as MeeNET. The latter allowed us to investigate several interesting research directions such as DSL refinements and transpilation. Meeduse is a language workbench dedicated to formally instrument any EMF-based DSL with correctness concerns. We experimented it on various kinds of applications in order to evaluate its strength. For example, in the smart-home domain (project DomoSur) we focused on the execution of the DSL at run-time, being inspired by [16]. The approach also covers model transformations (M2M). Indeed, we have adapted our TTC'19 proposal to provide a LWB that is suitable for model transformations. Currently, we are working on bi-directional transformations by addressing two features: (1) proving the isomorphism of a transformation in B, and (2) updating the input model from changes done in the output model (propagation).

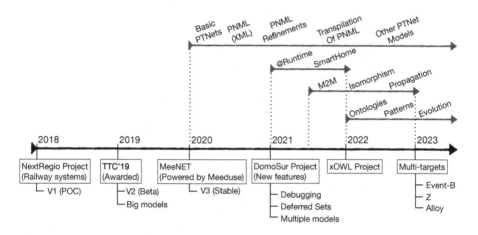

Fig. 5. Evolution time-line of Meeduse.

The applications discussed in this paper provide a narrow view of DSLs that is focused on safety-critical systems. The xOWL project (executable OWL), started this year, addresses knowledge engineering. It has a different view and studies the possibility to consider other application domains, which would allow us to analyse the implications of FMDE outside safety-critical systems. The motivation of the project is that domain ontologies evolve continuously, which leads to several problems, especially change impact analysis and resolution. Among existing works, pattern-driven techniques have been proposed to provide guidance during the ontology evolution so that it remains consistent. In the xOWL project, ontologies are described via the Ontology Web Language (OWL). Considering that OWL is a DSL, our proposal is to rethink the underlying evolution patterns by means of execution semantics that apply the expected changes to a given ontology. We have instrumented the W3C functional syntax of OWL in Meeduse, leading to a lightweight development approach. Indeed, the development effort is limited to the specification of the evolution patterns and their verification and validation. All the other features of xOWL (execution, verification and debugging) are provided by Meeduse.

Acknowledgements. The author would like to thank the LIG lab and CNRS for supporting the development of Meeduse, and he is much obliged to German Vega for his great technical contribution to the tool. This work is funded by the ANR-18-CE25-0013 PHILAE project.

References

1. Abrial, J.R.: The B-book: Assigning Programs to Meanings. Cambridge University Press, New York (1996)
2. Baar, T.: Non-deterministic constructs in OCL – what does any() mean. In: Prinz, A., Reed, R., Reed, J. (eds.) SDL 2005. LNCS, vol. 3530, pp. 32–46. Springer, Heidelberg (2005). https://doi.org/10.1007/11506843_3
3. Bobbio, A.: System modelling with petri nets. In: Colombo, A.G., de Bustamante, A.S. (eds.) Systems Reliability Assessment, vol. 7, pp. 103–143. Springer, Dordrecht (1990). https://doi.org/10.1007/978-94-009-0649-5_6
4. Bryant, B.R., Gray, J., Mernik, M., Clarke, P.J., France, R.B., Karsai, G.: Challenges and directions in formalizing the semantics of modeling languages. Comput. Sci. Inf. Syst. **8**(2), 225–253 (2011). https://doi.org/10.2298/CSIS110114012B
5. Combemale, B.: Towards Language-Oriented Modeling. Habilitation à diriger des recherches, Université de Rennes 1, December 2015
6. Dormoy, F.X.: SCADE 6: a model based solution for safety critical software development. In: Embedded Real Time Software and Systems (ERTS2008), Toulouse, France, January 2008. https://hal-insu.archives-ouvertes.fr/insu-02270108
7. Gargantini, A., Riccobene, E., Scandurra, P.: Combining formal methods and MDE techniques for model-driven system design and analysis. Int. J. Adv. Softw. 1&2 (2010)
8. Gemoc: Gemoc. http://gemoc.org/
9. Idani, A.: Formal model-driven executable DSLs: application to Petri-Nets. Int. J. Innov. Syst. Softw. Eng. (ISSE) **18**(1) (2022). https://doi.org/10.1007/s11334-021-00408-4
10. Idani, A., Ledru, Y., Vega, G.: Alliance of model driven engineering with a proof-based formal approach. Innov. Syst. Softw. Eng. **16**(3), 289–307 (2020)
11. Idani, A., Ledru, Y., Ait Wakrime, A., Ben Ayed, R., Bon, P.: Towards a tool-based domain specific approach for railway systems modeling and validation. In: Collart-Dutilleul, S., Lecomte, T., Romanovsky, A. (eds.) RSSRail 2019. LNCS, vol. 11495, pp. 23–40. Springer, Cham (2019). https://doi.org/10.1007/978-3-030-18744-6_2
12. Idani, A., Ledru, Y., Ait Wakrime, A., Ben Ayed, R., Collart-Dutilleul, S.: Incremental development of a safety critical system combining formal methods and DSMLs. In: Larsen, K.G., Willemse, T. (eds.) FMICS 2019. LNCS, vol. 11687, pp. 93–109. Springer, Cham (2019). https://doi.org/10.1007/978-3-030-27008-7_6

13. Idani, A., Vega, G., Leuschel, M.: Applying formal reasoning to model transformation: the meeduse solution. In: Proceedings of the 12th Transformation Tool Contest, co-located with STAF 2019, Software Technologies: Applications and Foundations. CEUR Workshop Proceedings, vol. 2550, pp. 33–44 (2019)
14. Iung, A., et al.: Systematic mapping study on domain-specific language development tools. Empir. Softw. Eng. **25**(5), 4205–4249 (2020). https://doi.org/10.1007/s10664-020-09872-1
15. Kosar, T., Bohra, S., Mernik, M.: Domain-specific languages: a systematic mapping study. Inf. Softw. Technol. **71**, 77–91 (2016)
16. Körner, P., Bendisposto, J., Dunkelau, J., Krings, S., Leuschel, M.: Embedding high-level formal specifications into applications. In: ter Beek, M.H., McIver, A., Oliveira, J.N. (eds.) FM 2019. LNCS, vol. 11800, pp. 519–535. Springer, Cham (2019). https://doi.org/10.1007/978-3-030-30942-8_31
17. Laleau, R., Mammar, A.: An overview of a method and its support tool for generating B specifications from UML notations. In: 15th IEEE International Conference on Automated Software Engineering, pp. 269–272 (2000)
18. Laleau, R., Polack, F.: Coming and going from UML to B: a proposal to support traceability in rigorous IS development. In: Bert, D., Bowen, J.P., Henson, M.C., Robinson, K. (eds.) ZB 2002. LNCS, vol. 2272, pp. 517–534. Springer, Heidelberg (2002). https://doi.org/10.1007/3-540-45648-1_27
19. Ledang, H.: Automatic translation from UML specifications to B. In: Automated Software Engineering, p. 436 (2001)
20. Leuschel, M., Butler, M.: ProB: an automated analysis toolset for the B method. Softw. Tools Technol. Transf. (STTT) **10**(2), 185–203 (2008)
21. Mashkoor, A., Kossak, F., Egyed, A.: Evaluating the suitability of state-based formal methods for industrial deployment. Softw. Pract. Exp. **48**(12), 2350–2379 (2018)
22. Meeduse. http://vasco.imag.fr/tools/meeduse/. Accessed 15 Dec 2020
23. Merilinna, J., Pärssinen, J.: Verification and validation in the context of domain-specific modelling. In: Proceedings of the 10th Workshop on Domain-Specific Modeling, pp. 9:1–9:6. ACM, New York (2010)
24. Meyer, E.: Développements formels par objets: utilisation conjointe de B et d'UML. Ph.D. thesis, Université de Nancy 2, March 2001
25. Rivera, J., Durán, F., Vallecillo, A.: Formal specification and analysis of domain specific models using maude. Simulation **85**, 778–792 (2009)
26. Snook, C., Butler, M.: UML-B: formal modeling and design aided by UML. ACM Trans. Softw. Eng. Methodol. **15**(1) (2006)
27. Snook, C., Butler, M.: U2B-A tool for translating UML-B models into B. In: Mermet, J. (ed.) UML-B Specification for Proven Embedded Systems Design (2004)
28. Steinberg, D., Budinsky, F., Paternostro, M., Merks, E.: EMF: Eclipse Modeling Framework 2.0, 2nd edn. Addison-Wesley Professional, Amsterdam (2009)
29. Tikhonova, U., Manders, M., Brand, van den, M., Andova, S., Verhoeff, T.: Applying model transformation and Event-B for specifying an industrial DSL. In: Workshop on Model Driven Engineering, Verification and Validation, pp. 41–50. CEUR Workshop Proceedings, CEUR-WS.org (2013)
30. Tikhonova, U.: Reusable specification templates for defining dynamic semantics of DSLs. Softw. Syst. Model. **18**, 691–720 (2017)
31. Vallecillo, A., Gogolla, M.: Adding random operations to OCL. In: Proceedings of MODELS 2017 Satellite Event, pp. 324–328. CEUR Workshop Proceedings. CEUR-WS.org (2017)

32. Voelter, M., et al.: Using language workbenches and domain-specific languages for safety-critical software development. Softw. Syst. Model. **18**(4), 2507–2530 (2018). https://doi.org/10.1007/s10270-018-0679-0
33. Zalila, F., Crégut, X., Pantel, M.: Formal verification integration approach for DSML. In: Moreira, A., Schätz, B., Gray, J., Vallecillo, A., Clarke, P. (eds.) MODELS 2013. LNCS, vol. 8107, pp. 336–351. Springer, Heidelberg (2013). https://doi.org/10.1007/978-3-642-41533-3_21

Enabling Content Management Systems as an Information Source in Model-Driven Projects

Joan Giner-Miguelez[1]($\boxtimes$) , Abel Gómez[1] , and Jordi Cabot[2]

[1] Internet Interdisciplinary Institute (IN3), Universitat Oberta de Catalunya (UOC),
Barcelona, Spain
{jginermi,agomezlla}@uoc.edu
[2] ICREA, Barcelona, Spain
jordi.cabot@icrea.cat

Abstract. Content Management Systems (CMSs) are the most popular tool when it comes to create and publish content across the web. Recently, CMSs have evolved, becoming *headless*. Content served by a *headless CMS* aims to be consumed by other applications and services through REST APIs rather than by human users through a web browser. This evolution has enabled CMSs to become a notorious source of content to be used in a variety of contexts beyond pure web navigation. As such, CMS have become an important component of many information systems. Unfortunately, we still lack the tools to properly discover and manage the information stored in a CMS, often highly customized to the needs of a specific domain. Currently, this is mostly a time-consuming and error-prone manual process.

In this paper, we propose a model-based framework to facilitate the integration of headless CMSs in software development processes. Our framework is able to discover and explicitly represent the information schema behind the CMS. This facilitates designing the interaction between the CMS model and other components consuming that information. These interactions are then generated as part of a middleware library that offers platform-agnostic access to the CMS to all the client applications. The complete framework is open-source and available online.

Keywords: Content Management System · Headless · Model-Driven Engineering · Reverse engineering · REST API

1 Introduction

Content Management Systems (CMSs) are one of the most popular options to create content across the web. These systems, such as WordPress, Drupal, or Joomla, represent nearly 61,3% in terms of published websites [3,28]. One of the main reasons for this popularity is the great user experience provided by CMSs while creating content that empowers non-technical users to be part of the content creation chain [4]. Besides, CMSs have evolved in the last years

R. Guizzardi et al. (Eds.): RCIS 2022, LNBIP 446, pp. 513–528, 2022.
https://doi.org/10.1007/978-3-031-05760-1_30

by implementing APIs to allow external apps to discover and interact with the content of these systems. This evolution, known as *headless CMSs*, has shifted the main focus of these systems, traditionally on desktop solutions, to other kinds of applications such as mobile apps and other front-end apps.

As an example, in the media industry, CMSs are widely used to build digital solutions. Sony Pictures[1], Le Figaro[2], and Syfy Channel[3] are some of the media companies powering their digital solutions with open-source CMSs. The content created inside these solutions, such as news, podcast, or video, represents the key asset of these companies. Therefore, any new software development project in these companies will likely depend on, and require, the content in the CMSs.

This integration is becoming more difficult as CMSs solutions are becoming increasingly complex as so they do the APIs they expose. Moreover, typically, large companies must deal with various deployed CMSs, some of them legacy versions, that need to be combined to satisfy the application informational needs. There is a clear need for new methods that help in managing these complex integration processes as part of new software development projects. In parallel, there is increasing adoption of the Model-Driven Engineering (MDE) practices [11,14] as MDE has been proven useful to tame the development complexity. Unfortunately, while we have several solutions to facilitate the integration of SQL [10,16] and No-SQL [1,7] backends in MDE-based processes, there is a lack of solutions to support applications that rely on headless CMSs as a content source. Therefore, the integration of headless CMSs with other apps has been done by manual solutions, being these solutions time-consuming and error-prone.

In this work, we propose a framework to enable the integration of headless CMSs in MDE-based software development processes. The framework is composed of *(i)* a core model of CMSs, *(ii)* a reverse-engineering process to extract the model from an existing deployed CMS in a UML representation, and finally, *(iii)* a code-generation process that generates a middleware library to bridge the gap between the front-end consumer app and the CMSs. The framework focuses on existing CMSs rather than on creating new ones from a UML model because we have detected that most of the CMSs are already created, and there is a need to use them as a content source.

With our framework, companies can quickly discover the information schema behind each CMS and represent it explicitly as a UML model. This model can then be integrated with other software models in the project. Finally, our framework also assists in the generation phase, simplifying the writing of all the required *glue* code between the different components. As an outcome, we have built our proposal as an Eclipse plug-in and can be found in an open repository[4].

The paper is structured as follows. Section 2 presents the general overview of the framework. Section 3 presents our *Core CMS model* and, Sect. 4 presents the reverse engineering process. Section 5 shows an example of integration and,

---

[1] https://www.sonypictures.com/tv.

[2] https://www.lefigaro.fr.

[3] https://www.syfy.com.

[4] https://hdl.handle.net/20.500.12004/1/C/RCIS/2022/001.

Sect. 6 presents the generation process. Finally, Sect. 7 presents the developed tool, Sect. 8 summarizes the related work, and Sect. 9 wraps up the conclusions and the future work.

2 Framework Overview

In this section, we provide an overview of the proposed framework. The main goals of the framework are *(i)* to extract and explicitly represent the specific model of a deployed CMS to enable the integration of CMSs as information sources within a global model based development process, and *(ii)* to generate the *glue code* that will transform any model-level interaction between the CMS model and other models (e.g. GUI models) in proper calls to the Headless CMS API.

Following Fig. 1, the first step is the *Reverse engineering* process. This process takes as input the target CMS (in particular, its URL and user credentials), and imports the predefined *Core CMS model* with basic CMS concepts. Then, creates the target CMS model as an extension to the core CMS one. The discovery process is implemented as a set of calls to the CMS API, adapted to the API offered by each CMS platform. We use UML to represent these models.

As this obtained model is a purely standard model, designers can refine it and combine it with other models. For instance, we can easily model how a User Interface model queries the CMS model to retrieve the information it needs to show to the user in a platform-independent way.

The final step is the *Code generation* process. This process takes as input the *CMS Driver* that corresponds to the CMS platform and uses it to generate a middleware library allowing consumer apps to get the information they need from the CMS. Note that the API offered by the middleware only depends on the CMS content schema, not on the underlying CMS platform. The API is consistent across any CMS technology which facilitates any future CMS migration. As part of the API we have all the usual filtering, ordering and pagination methods that may be needed to iterate on the CMS content.

In next sections we present in detail each of these steps.

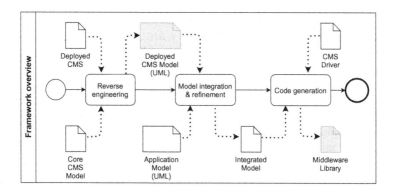

Fig. 1. Framework overview

3   The *Core CMS Model*

This section presents our core model for CMS, representing the common concepts shared by all major CMS vendors. The goal of this model is to facilitate the representation of the information schema behind specific CMSs by providing basic types that can be extended. This avoids having to start from scratch every time. Moreover the core elements also facilitate the management of CMS models when there is no need to access its detailed structure. This core model will be used by the reverse engineering process, see Sect. 4, to create the complete model-based representation of an input CMS.

We have come up with the elements in the core model after analyzing the three major CMS platforms (Drupal, WordPress and Joomla; as they together have over 70% of the market share in this domain [3,28]) and inferring the commonalities among them. Figure 2 shows a partial view of this common model. We focus on the access control and the content information parts. A more complete model, which also supports features such as revisions or translations of the content, is available in the online repository.

According to Fig. 2, we can see that many classes inherit from *ContentEntity*. This class provides the last update time of the entity and a unique identifier for it. This identifier will be used to query the CMS API and get the entity information and its linked resources.

Immediately below, we have *ContentType*, that represents any content stored in the CMS. Typically, most CMSs offer at least pages and posts as content types, but any non-trivial CMS will come with several custom types to better represent the information domain the CMS is serving. Each custom type will be a subclass of this root *ContentType* element. We can always associate a *Comment* to a content type. Comments have a hierarchical relationship, allowing different threads of comments to coexist in parallel and can be associated with a user if the comment is done by a registered one. Content types can also be classified in taxonomies. Two typical taxonomies are *Tags* and *Categories*. As we have found these taxonomies in all CMSs we have directly included them in the core model but a specific CMS could also have custom taxonomies. Taxonomies also have a hierarchical relationship.

Moreover, CMSs also have *Media* files that can be attached to the content, such as images, videos, or audio files. They are represented independently of the context where they are embedded, since media items can be reused across different content pieces. This also means that CMSs could also be used as a media repository. Finally, *Blocks* represent complementary pieces of content of a *ContentType*, for example, the ads attached to a particular content piece.

Beyond content management, CMS pay special attention to the access control and allow defining who can access the content following a standard Role-Based Access Control approach [15], where every user has a set of *Roles* attached, and every *Role* has a set of permissions attached. *Permission* represents a specific action that a user can do inside the system. There are two subtypes of permissions, the *GeneralPermission* which is not attached to a particular domain

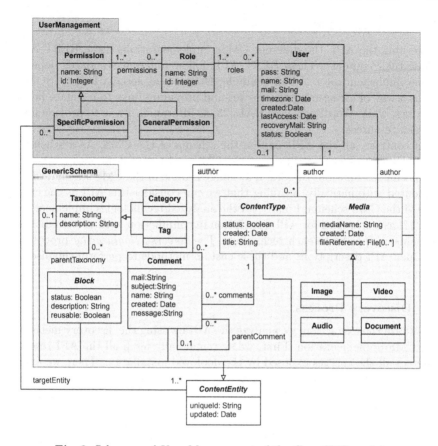

Fig. 2. Schema and User Management of the Core CMS model

entity, for instance, "access to the admin dashboard" and, the *SpecificPermission* that is related to a specific content type, for example, "edit video article".

4 The Reverse Engineering Process

In this section, we explain the *reverse engineering process*. This process is the first step in the integration framework and aims to extract the model of a deployed CMS. On one side, it starts by getting the URL and the user credentials of the target CMS to discover its API. On the other side, the process receives the *Core CMS Model* presented above to perform a set of extensions over it to extract the model of the deployed CMS. As a result, we get the model in UML facilitating the integration with other parts of the system as shown in Sect. 5.

In this section, we present a comparison of the API architecture between technologies, the discovery process followed to extract insights from the API, and the extension strategy followed over the *Core CMS Model* to fit particular data scenarios.

4.1 The CMSs API

Concerning the API, the different CMSs implementations use a REST API to expose their content[5]. These REST APIs follow different standards depending on the implementation. As an example, Drupal and Joomla follow JSON:API [12] as a way of structuring the answer and providing links to navigate between resources, while WordPress uses HAL [23] as its standard way to provide linking between resources.

In contrast, all the APIs expose the schema of the CMS. As a schema, we mean the entities and attributes of the CMS and the custom ones created over the specific installation. Besides, if a change is done in the CMS schema, the API is updated automatically, meaning that we can consider this API as the current state of the CMS schema. Finally, all the analyzed technologies implement Open API [17] as a standard for API description in conjunction with custom discovery mechanisms specific to each technology. The *Reverse Engineering* process uses the Open API descriptions and these custom methods to perform the discovery process.

4.2 Discovery Process

To perform the extraction, we analyze the API by using the discovery methods of each CMS implementation. First, we systematically fetch all the API resources to detect all the entities exposed by the deployed installation. Then, we analyze the answer to infer from which type is every detected entity and which attributes has. Once we have spotted all the entities and their attributes, we analyze, again, the answers to detect their relationships.

Listing 1. Example excerpt of API answer

```
1   "info": {
2       "title": "Journal CMS",
3   },
4   "host": "example.com",
5   "basePath": "/api",
6   "definitions": {
7       "node--videoarticle": {
8           "title": "node:videoArticle Schmea",
9           "description": "news main content",
10          "properties": {
11              "attributes": {
12                  "id": "Integer",
13                  "title": "String",
14                  "...": "..."
15              },
16              "relationships": {
17                  "0": {
18                      "type": "taxonomy--category",
19                      "link": "..."
20                  }
21              }
22          }
23  }
```

---

[5] Other API architectures as GraphQL are also implemented but not included in the standard installation. More information about the discussion can be found at https://dri.es/headless-cms-rest-vs-jsonapi-vs-graphql.

In Listing 1 we can see an example excerpt of Drupal's API answer. In this example we can see a *node--videoarticle* definition which is an entity that inherits from *ContentType*. We guess the inheritance from the term"node" which refers to *ContentType* in Drupal's terminology. Under *properties* we can see a set of attributes as *id*, *title* and others (lines from 11 to 15). Finally, we can see that our entity has a relationship with a *Taxonomy* of type *Category* and the same answer provides us a link to fetch and discover this relationship (lines from 16 to 20).

If we detect entities that are not in the proposed *core CMS* model, we can know from which types these entities inherit. For instance, in Listing 1, *VideoArticle* is a specific type of *ContentType*, and therefore, we will create a new class extending our *ContentType* class of the core model. This new class also will retain some specific annotations, such as the URL of the endpoint. It could be noted that this process is tied to every technology implementation as every technology builds the answer in different ways. Therefore, a specific extractor needs to be developed to support every CMS implementation.

4.3   Extraction Example

In Fig. 3 we have an example of an extracted model of a journal site. This example shows how the discovery process and the extensions over the *Core CMS model* can describe complex data scenarios. This example describes a journal with different types of content such as video articles, written news, and podcasts with relationships between them.

In the figure, *VideoArticle* represents the news edited as video. This class has a relationship with a *Video* and some attributes, like *likes*, which represent the number of likes made by users. In terms of Taxonomies, *VideoArticle* have a relationship with *AgeRating* which classifies the video by age recommendation. In addition, it also has a set of *Tags* to facilitate the search of this content. Finally, this class has a relationship with a set of *NewsArticle*. If a user wants further information about the news, this represents the related written news of the *VideoArticle*.

NewsArticle class, which extend from *ContentType*, represents the written news. As previous, this class has attributes, like *likes*, which represent the number of likes made by users. Also, the *NewsArticle* has a relationship with *BannerAds*, which extends from *Block*, that represents the advertisements attached to a specific news. This relationship allows configuring the advertisements policy inside the CMS, and reproduce it also in the consumer apps. In addition, *NewsArticle* can have a set of related *VideoArticle* attached to it. In the journal's example, this represents written news that have their version as video news to be shown on a streaming platform or TV. Moreover, *NewsArticle* also have media *Images* as the images of news.

Finally, another main content of the site is *Podcast*, which has a relationship with media *Audio*, which represents content edited by audio. This content audio format has been gaining popularity in the last year and is a clear example of how CMS manages different types of content.

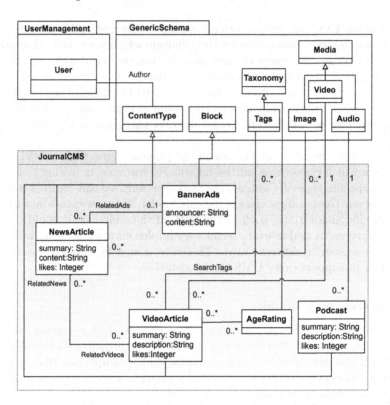

Fig. 3. Example extracted model of a journal's CMS

In conclusion, after the *Reverse Engineering* process, we obtain the model of our deployed CMS in UML. This model can work with complex data scenarios, as we can see in Fig. 3, and is ready to be integrated with other parts of the systems, as we see in the next section.

5   Model Integration and Refinement

The result of the reverse engineering process is a fully compliant UML class diagram. As such, we can use, combine, and refine that diagram as we would do with any other model of our information system. One of the most obvious integrations can be specifying how the user interface of different consumer components, such as web applications or mobile apps, query the extracted CMS model to retrieve the information they need to show to their users.

We will use this scenario to continue with our running example. Let's assume our media company is interested in building a mobile app to show a video news feed to its users. More specifically, the app will show the video feed, let users *like* some specific videos (and these *likes* will be stored the CMS for future rankings

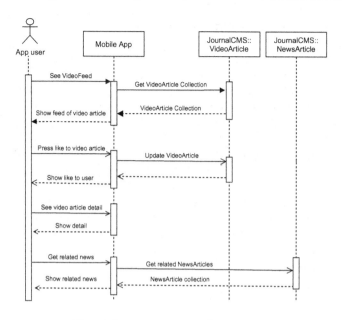

Fig. 4. Integration example sequence

and optimizations), access the details of a particular video and get related news to the ones she is watching.

All this information is available in the Journal CMS. And since we now have it explicitly represented as a model, we just need to define the interaction between the UI app model and the CMS model. A possible sequence diagram is shown in Fig. 4. In the diagram, when a user wants to see the video feed, a call to the *VideoArticle* class is performed. When a user likes a specific video, this is translated to an update call to the journal's *VideoArticle* class. Finally, when a user sees a video detail, the user can access the related content, querying, in that case, the journal's *NewsArticle* class.

The methods used in the sequence diagram are self-explanatory and, as we will see in the next section, will be made available in the next section via the code generation process. For more advanced scenario, we could also consider using a language like the Object Constraint Language [19] to more formally specify the exact queries to perform to the CMS.

Similarly, we could model other types of interactions or static diagrams with our CMS model. The key point is to realize that these interactions are modeled as any other type of interaction between two models and that when doing so the designer does not need to worry about the underlying CMS platform and version. The designer can stay at a platform-independent model. In fact, it would be possible to define a client consumer application that reads from two different CMS models without the designer not even being aware that the queried data elements belong to separate CMSs.

6   Code Generation Infrastructure

The final step of the framework is the *code generation* process. This process aims to generate a middleware library to facilitate the communication between the deployed CMS and consumer applications.

Regarding the presented integration example in Fig. 4, the middleware would be in charge of the interaction between the *Mobile app* and the content of the Journal's CMS such as *VideoArticle* and *NewsArticle*. This middleware could work with other components (e.g., generating the mobile app) by providing the glue code required to ensure the communication between the app and the content source.

In this section, following Fig. 5, we explain how this middleware is structured to be easily extensible for other technologies and how the consistent API is designed to abstract practitioners from the underlying technology. Finally, we provide a usage example of the middleware library following the integration example presented in Sect. 5.

6.1   Middleware Structure

The middleware structure is composed of a set of generated classes from the extracted model, and a *CMS Driver* composed by a set of static classes that allows us to move back and forward the data of the generated classes between the consumer app and the source CMS.

The generated classes are the dynamic part of the middleware. For each UML class of the extracted model, a java class is generated for it. This class contains the UML class attributes and relationships, and the possible sorter and filters to interact with the API. For instance, in Fig. 5, the class *VideoArticle* corresponds to the class *VideoArticle* of the example extracted model in Sect. 4. In addition to the previous ones, a *JournalSiteManager* is generated. This class is built using the singleton pattern and represents the whole CMS site, and brings access to the generated classes with specific methods for each one.

The *CMS Driver* classes are the static part of the middleware. The *Driver* class is in charge of the communication with the source CMS and is the only one tied to a particular technology. This approach means that the driver is the only class we need to reimplement to extend support for new CMS technologies. This driver has a single relationship with the *JournalSiteManager* class. In Fig. 5, we can see a *DrupalDriver* class as our journal's site is powered by Drupal.

Also, inside the *CMS Driver*, and to ensure the separation between the agnostic part and the specific part, we propose the *GenericResource* class as an interface between the generated classes and the driver. In addition, we suggest a *SearchQueryBuilder* class as a way to provide a consistent API to practitioners in terms of fetching the source CMS. At last, *SearchQuery* is the result of the query builder being an immutable class and allowing the *DrupalDriver* class to build the particular final query to the CMS API.

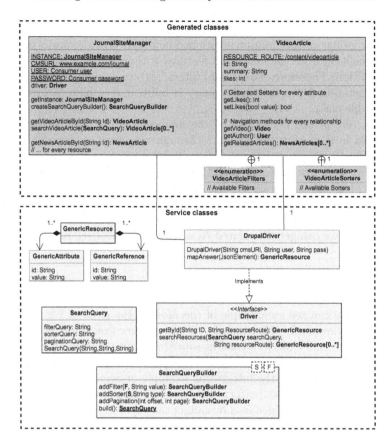

Fig. 5. Middleware's implementation diagram: *VideoArticle* example

In summary, the middleware is composed of a set of dynamic classes generated from the extracted model and a set of already developed static classes representing the *CMS Driver*.

6.2 The Middleware API

One of the key points is to facilitate the usage of this middleware without regarding the underlying technology. To do so, we have proposed a consistent API as a standard method to interact with them. We have inspired the design of the consistent API in the identified shared concepts of the different open-source CMS APIs.

The main methods to interact with the content are present in the *JournalSiteManager* of our example. This class has a *getById* and a *search* methods for every generated class from the extracted model. The first one needs a content id to return the result, and the second one needs a *SearchQuery* to do so.

The *SearchQueryBuilder* is a query builder that wraps up the concepts of filtering, ordering, and pagination mentioned before. It provides a way of building queries benefiting from the main API features without regarding its underlying implementation. Therefore, we can add filters, sorters, and pagination using the provided methods. Once the search query is built, it returns a *SearchQuery* as an immutable class.

Finally, one of the key points of the analyzed REST API is hypermedia. Hypermedia is the ability to navigate through resources to discover the API. For example, if we get a *VideoArticle* and we want to get a specific related *NewsArticle*, we know from the answer itself how to retrieve this related content. Instead of automatically getting the related content, we have decided to implement this navigation method to avoid recursive calls. In Fig. 5 we can see an example of this in the *VideoArticle* class that has navigation methods. From a *VideoArticle* we can get the related *NewsArticle* or the *User* corresponding to the author.

6.3 Middleware Usage Example

In Listing 2, we can see an example of the *Mobile App* class of Fig. 4 using the generated middleware. In this example, we can see how practitioners could easily integrate the Journal CMS as a content source without regarding the underlying technology. This example uses the java code generated by the tool presented in the next section following the integration example of Sect. 5.

In the main method, we perform the sequence represented in Fig. 4. Therefore, in line 5, we get the video feed by calling the *getVideoFeed()* method. In this method, we get an instance of *JournalSiteManager* class as it provides methods to interact with all the generated classes. Then, to get the list of videos we call the method *searchVideoArticle()* form the *JournalSiteManager* in line 18. This

Listing 2. Example MobileApp class demonstrating the use of the middleware library

```
1   class MobileApp() {
2
3     public void main(){
4       // 1 - Get the video feed
5       List<VideoArticle> feed = getVideoFeed();
6       // 2 - User press like to the first video
7       VideoArticle firstVideo = feed.get(0);
8       firstVideo.setLike(true);
9       // 3 - User wants to see the related news of the video
10      List<NewsArticle> relatedNews = firstVideo.getRelatedArticles();
11    }
12    public List<VideoArticle> getVideoFeed() {
13      // The site manager class
14      JournalSiteManager site = JournalSiteManager.getInstance();
15      // We get a search query builder
16      SearchQueryBuilder queryBuilder = site.getSearchQueryBuilder();
17      // We call the search method with an empty query
18      return site.searchVideoArticle(queryBuilder.build());
19    }
20  }
```

method requires a *SearchQuery* as an entry parameter. In our case, we get a *searchQueryBuiler*, but we do not need to set any filter or sorter configuration. So we build it directly in line 18. The method returns a list of *VideoArticles*.

Then, a user presses the *like* button of the first video of our video feed. Therefore we get the pressed video in line 7, and we call the *setlike()* method of it in the next line. Finally, in line 10, to retrieve the related content, we use the navigation methods of the *VideoArticle* class calling the *getRelatedNews()* method. This methods returns the list of *NewsArticles* related to the particular *ViodeArticle*.

7   Tool Support

We have developed a prototype implementation of our framework, available under the Eclipse Public License[6]. The source code of the project and the Eclipse update site are available in the public repository.

Our implementation is built in Java and relies on the Eclipse Modeling Framework [24] for the core modeling support. In particular, we have used Xtext and Xtend [25] to implement the code-generation process. WordPress and Drupal are the two CMSs supported by our tool, but others could be easily integrated. Further instructions about the usage of this tool are provided on the public repository.

8   Related Work

Several research works have investigated the intersection between CMS and MDE. A first group of works have focused on modeling specific CMS components or extensions/plugins. Priefer [18] proposes a model-driven approach to generate Joomla extensions. Martínez [15] presents a metamodel for the access-control rules in a CMS to be used for verification and validation purposes. Our work targets the complete CMS definition and not just specific components.

Another group of works focus on the generation of a new CMS implementation. Trías [26] propose a CMS common metamodel to capture the key concerns required to model CMS-based web applications and, Qafmolla [29] propose the generation of CMS instances from it. Similarly, Souer [22] aims to generate a CMS from an automatically generated configuration description created by business users. Modeling languages to define new CMS is also the goal of Saraiva [21]. Our work is different from these approaches in that: 1) Our model is not generic but extensible to precisely represent the information schema of the CMS (and not just global concepts as *post* or *page*) and 2) our target is the integration of the CMS with the rest of the system and therefore our framework comprises the interactions between the data consumer components and the CMS and 3) we start from an exiting CMS, which, in our opinion, is a more realistic scenario

---

[6] https://www.eclipse.org/legal/epl-2.0.

as most companies already have a CMS in place and are not looking to start from scratch but evolve it.

It is also worth mentioning works that have similar goals to ours, but that target different information sources. SQL databases (e.g. Egea [10], and Nguyen [16]), NoSQL databases (e.g. [1,5,8]), Data Warehouses (e.g. [27]) and even spreadsheets [2] have been integrated in MDE processes.

Finally, given that headless CMSs expose their data through a REST API, we comment on model-based proposals targeting REST APIs. Some relevant examples are [9], presenting an MDE-based tool to create, visualize, manage, and generate OpenAPI definitions, [6], proposing an MDE process to discover and composite JSON documents or [20], focusing on building REST APIs directly from user requirements. In our case, we do not aim to generate new APIs but to provide a specific reverse engineering process adapted to the concrete REST API endpoints and patterns offered by popular CMSs. In this sense, some concepts of these works have been integrated in our approach as building blocks on top of which we have built the specific connectors we required for the CMSs.

9   Conclusions and Future Work

In this work, we have proposed a model-based framework to integrate headless CMSs in a development process. Thanks to our framework, the conceptual schema behind the CMS can be easily exposed and reused as part of the global development process. The framework comprises a core CMS model, a reverse engineering process that extends it to precisely represent the data structure of an input CMS, and a middleware that enables modeling front-ends that need to consume that data in a platform-agnostic way.

This facilitates the reusability and evolution of the software system, also enabling potential migrations among different CMS versions and platforms, e.g., to benefit for more advanced features or scalability improvements, while minimizing the impact of such changes on all the deployed clients (e.g. mobile apps).

As further work, we will study how the reverse engineering process could take into account presentation suggestions that are now also being embedded in the CMS headless API to achieve a more homogeneous look&feel across different client front-ends. There are some important trade-offs here that deserve to be analyzed. We would also like to explore the benefits of exposing CMS data in other domains, like Machine Learning, where easy access to CMS data could be exploited for training ML models. In this context, deploying our framework as a plugin for model-based data science tools like KNIME [13] could be interesting. Finally, at the tool level, we plan to extend our support to other CMSs and to validate the tool in industrial settings.

Acknowledgements. The research described in this paper has been partially supported by the AIDOaRT Project, which has received funding from the ECSEL Joint Undertaking (JU) under grant agreement No 101007350. The JU receives support from the European Union's Horizon 2020 research and innovation programme and Sweden, Austria, Czech Republic, Finland, France, Italy, and Spain.

In addition, the research has also been partially supported by the TRANSACT project, which has received funding from the ECSEL Joint Undertaking (JU) under grant agreement No 101007260. The JU receives support from the European Union's Horizon 2020 research and innovation programme and Netherlands, Finland, Germany, Poland, Austria, Spain, Belgium. Denmark, Norway.

References

1. Abdelhedi, F., Ait Brahim, A., Atigui, F., Zurfluh, G.: MDA-based approach for NoSQL databases modelling. In: Bellatreche, L., Chakravarthy, S. (eds.) DaWaK 2017. LNCS, vol. 10440, pp. 88–102. Springer, Cham (2017). https://doi.org/10.1007/978-3-319-64283-3_7
2. Belo, O., Cunha, J., Femandes, J.P., Mendes, J., Pereira, R., Saraiva, J.: QuerySheet: a bidirectional query environment for model-driven spreadsheets. In: IEEE Symposium on Visual Languages and Human Centric Computing - VL/HCC, pp. 199–200 (2013)
3. BuildWith.com. CMS Usage Distribution in the Top 1 Million Sites (2021). https://trends.builtwith.com/cms. Accessed Jan 2022
4. Cabot, J.: Wordpress: a content management system to democratize publishing. IEEE Softw. **35**(3), 89–92 (2018)
5. Comyn-Wattiau, I., Akoka, J.: Model driven reverse engineering of NoSQL property graph databases: the case of Neo4j. In: International Conference on Big Data (Big Data), pp. 453–458. IEEE (2017)
6. Cánovas Izquierdo, J.L., Cabot, J.: JSONDiscoverer: visualizing the schema lurking behind JSON documents. Knowl. Based Syst. **103**, 52–55 (2016)
7. Daniel, G., Gómez, A., Cabot, J.: UMLto[No]SQL: mapping conceptual schemas to heterogeneous datastores. In: 2019 13th International Conference on Research Challenges in Information Science (RCIS), pp. 1–13 (2019)
8. Daniel, G., Sunyé, G., Cabot, J.: UMLtoGraphDB: mapping conceptual schemas to graph databases. In: Comyn-Wattiau, I., Tanaka, K., Song, I.-Y., Yamamoto, S., Saeki, M. (eds.) ER 2016. LNCS, vol. 9974, pp. 430–444. Springer, Cham (2016). https://doi.org/10.1007/978-3-319-46397-1_33
9. Ed-douibi, H., Cánovas Izquierdo, J.L., Bordeleau, F., Cabot, J.: WAPIml: towards a modeling infrastructure for Web APIs. In: 2019 ACM/IEEE 22nd International Conference on Model Driven Engineering Languages and Systems Companion (MODELS-C), pp. 748–752 (2019)
10. Egea, M., Dania, C., Clavel, M.: MySQL4OCL: a stored procedure-based MySQL code generator for OCL. Electron. Commun. EASST **36** (2010)
11. Hutchinson, J., Whittle, J., Rouncefield, M., Kristoffersen, S.: Empirical assessment of MDE in industry. In: Proceedings of the 33rd International Conference on Software Engineering (ICSE 2011), pp. 471–480. Association for Computing Machinery (2011)
12. JSON API Initiative. JSON:API (2020). https://jsonapi.org. Accessed Jan 2022
13. KNIME AG. KNIME's (2022). https://www.knime.com. Accessed Jan 2022
14. Liebel, G., Marko, N., Tichy, M., Leitner, A., Hansson, J.: Model-based engineering in the embedded systems domain: an industrial survey on the state-of-practice. Softw. Syst. Model. **17**(1), 91–113 (2018)

15. Martínez, S., Garcia-Alfaro, J., Cuppens, F., Cuppens-Boulahia, N., Cabot, J.: Towards an access-control metamodel for web content management systems. In: Sheng, Q.Z., Kjeldskov, J. (eds.) ICWE 2013. LNCS, vol. 8295, pp. 148–155. Springer, Cham (2013). https://doi.org/10.1007/978-3-319-04244-2_14

16. Nguyen Phuoc Bao, H., Clavel, M.: OCL2PSQL: an OCL-to-SQL code-generator for model-driven engineering. In: Dang, T.K., Küng, J., Takizawa, M., Bui, S.H. (eds.) FDSE 2019. LNCS, vol. 11814, pp. 185–203. Springer, Cham (2019). https://doi.org/10.1007/978-3-030-35653-8_13

17. OpenApi initiative. Specification (2020). https://swagger.io/specification. Accessed Jan 2022

18. Priefer, D., Rost, W., Strüber, D., Taentzer, G., Kneisel, P.: Applying MDD in the content management system domain. Softw. Syst. Model. **20**(6), 1919–1943 (2021)

19. Richters, M., Gogolla, M.: OCL: syntax, semantics, and tools. In: Clark, T., Warmer, J. (eds.) Object Modeling with the OCL. LNCS, vol. 2263, pp. 42–68. Springer, Heidelberg (2002). https://doi.org/10.1007/3-540-45669-4_4

20. Rivero, J.M., Heil, S., Grigera, J., Robles Luna, E., Gaedke, M.: An extensible, model-driven and end-user centric approach for API building. In: Casteleyn, S., Rossi, G., Winckler, M. (eds.) ICWE 2014. LNCS, vol. 8541, pp. 494–497. Springer, Cham (2014). https://doi.org/10.1007/978-3-319-08245-5_35

21. Saraiva, J.D.S., Silva, A.R.D.: Development of CMS-based web-applications using a model-driven approach. In: 2009 Fourth International Conference on Software Engineering Advances, pp. 500–505 (2009)

22. Souer, J., Kupers, T., Helms, R., Brinkkemper, S.: Model-driven web engineering for the automated configuration of web content management systems. In: Gaedke, M., Grossniklaus, M., Díaz, O. (eds.) ICWE 2009. LNCS, vol. 5648, pp. 121–135. Springer, Heidelberg (2009). https://doi.org/10.1007/978-3-642-02818-2_9

23. Stateless Group. HAL - Hypertext Application Language (2018). https://stateless.group/hal_specification. Accessed Jan 2022

24. The Eclipse Foundation. Eclipse Modeling Project - Eclipse Modeling Framework - Home (2021). www.eclipse.org/emf/. Accessed Sep 2021

25. The Eclipse Foundation. Xtend (2021). www.eclipse.org/xtend. Accessed Sep 2021

26. Trias, F., de Castro, V., Lopez-Sanz, M., Marcos, E.: Migrating traditional web applications to CMS-based web applications. Electron. Notes Theoret. Comput. Sci. **314**, 23–44 (2015)

27. Trujillo, J., Luján-Mora, S.: A UML based approach for modeling ETL processes in data warehouses. In: Song, I.-Y., Liddle, S.W., Ling, T.-W., Scheuermann, P. (eds.) ER 2003. LNCS, vol. 2813, pp. 307–320. Springer, Heidelberg (2003). https://doi.org/10.1007/978-3-540-39648-2_25

28. W3C. Usage statistics of content management systems (2021). https://w3techs.com/technologies/overview/content_management. Accessed Jan 2022

29. Qafmolla, X., Viet, C., Nguyen, K.R.: Metamodel-based generation of web content management systems. Int. J. Inf. Technol. Secur. **6**(4), 17–30 (2014)

A Global Model-Driven Denormalization Approach for Schema Migration

Jihane Mali[1,2(✉)], Shohreh Ahvar[2], Faten Atigui[3], Ahmed Azough[1],
and Nicolas Travers[1,3]

[1] Léonard de Vinci Pôle Universitaire, Research Center, Paris La Défense, France
{jihane.mali,ahmed.azough,nicolas.travers}@devinci.fr
[2] ISEP, Paris, France
{jihane.mali,shohreh.ahvar}@isep.fr
[3] CEDRIC, Conservatoire National des Arts et Métiers (CNAM), Paris, France
{faten.atigui,nicolas.travers}@cnam.fr

Abstract. With data's evolution in terms of volume, variety, and velocity, Information Systems (IS) administrators have to steadily adapt their data model and choose the best solution(s) to store and manage data in accordance with users' requirements. In this context, many existing solutions transform a source data model into a target one, but none of them leads the administrator to choose the most suitable model by offering a limited solution space automatically calculated and adapted to his needs. We propose `ModelDrivenGuide`, an automatic global approach for leading the model transformation process. It starts by transforming the conceptual model into a logical model, and it defines refinement rules that help to generate all possible data models. Our approach then relies on a heuristic to reduce the search space by avoiding cycles and redundancies. We also propose a formalisation of the denormalization process and we discuss the completeness and the complexity of our approach.

Keywords: NoSQL · MDA · Denormalization · Model refinement · Heuristic

1 Introduction

For several decades, the storage and the exploitation of data have mainly relied on relational databases. But the explosion of data, especially characterised by the 3V (Volume, Variety and Velocity), has opened up major research issues related to modeling, manipulating, and storing massive amounts of data [20].

The resulting so-called NoSQL systems correspond to *four families* of data structures: key-value oriented (KVO), column oriented (CO), document oriented (DO), and graph oriented (GO). These systems have raised new problems of transforming traditional databases into non-relational databases.

Choosing the most suitable data model family (*i.e.,* relational and NoSQL) and its implementation is a crucial task. Thus, this highly distributed context emphasizes the problem of integrating an Information System (IS) and a proper data model is essential for such systems.

Existing methods produce a single targeted data model while missing important considerations that could favor other data models with better trade-off between requirements Hence we need to produce a set of data models in order to allow the IS designer to choose the best compromise. However, generating data models that are useful for the use case among all possible ones is a hard issue. Thus, producing a reduced search space of relevant data models to help this choice becomes a real challenge.

Based on Model-Driven Architecture (MDA[1]), we propose the ModelDriven-Guide approach which starts from the conceptual model, then goes from one logical model to another thanks to refinement rules. It provides a functional decision-making process by integrating the use case composed of a set of queries, to guide the implementation choice of the SQL and NoSQL solution(s). In previous work [16], we presented an overview of our model transformation approach by focusing on the 5 *families* common meta-model to allow the specification of any data model. This work especially focuses on the following contributions:

- A complete formalization of the search space generation of data models with transformation rules and completeness, allowing to get *all* the possible models going from one logical model to another;
- A full formalism of the data model refinement problem with complexity bounds for both merge and split rules, first time in the literature;
- A recursive heuristic to reduce the search space of generated data models, based on redundancies and use cases;
- A thorough study of data model transformation by simulating schemas and use cases, and the TPC-C benchmark.

This paper is organized as follows: the related work is presented in Sect. 2, our approach is formalized in Sect. 3 and the models' generation heuristic is developed in Sect. 4. Then, we validate our approach in Sect. 5. Finally, our conclusion opens the research perspectives in Sect. 6.

2   Related Works

Model Transformation and NoSQL DB. Some of the existing approaches suggest a set of mapping rules that transform a relational model into a NoSQL one like for CO families [21] or DO families with *MongoDB* [18].

Moreover, other studies have focused on the transformation of a conceptual model into a NoSQL data model. Chebotko et al. [6] present a query-driven approach for modeling *Cassandra* starting from an ER model. They define dedicated transformation rules between these logical and physical models. De Freitas et al. [11] propose a conceptual mapping approach which transforms ER models into one of the 4 NoSQL families with an abstract formalization of the mapping

---

[1] https://www.omg.org/mda/index.htm.

Table 1. NoSQL transformation approaches

	Relational	DO	CO	KVO	GO	Approach	
[6, 14, 21]			✓			Dedicated	Manual
[18]		✓				Dedicated	Manual
[7]					✓	Dedicated	Manual
[1, 11]		✓	✓	✓	✓	Dedicated	SemiAutomated
[8]		✓	✓			Semi-Global	SemiAutomated
ModelDrivenGuide	✓	✓	✓	✓	✓	Global	Automated

rules. Also, schema evolution over time has been studied, and its relation to NoSQL data migration [19].

The following studies adopt a MDA to transform a UML class diagram into NoSQL DB. Li Y. et al. [14] propose to transform a class diagram into a CO DB with *HBase*, and Daniel G. et al. [7] into a GO DB.

Abdelhedi et al. [1] propose to transform a class diagram into a common logical model that describes the four families of NoSQL DB. Then, it's transformed into physical models related to the four families. Alfonso et al. propose the system *Mortadelo* [8] that defines a model-driven design process that generates implementations for CO and DO starting from the conceptual model.

Table 1 summarizes the main transformation strategies with dedicated conceptual/relational to NoSQL solutions or semi-global (rule-driven choice) compared to our global approach non-limited by a mapping between a source and a target concept, but covers all the possibilities using models transformation.

Reducing the Search Space. Adopting a global approach inevitably leads to the generation of a multitude of possible choices. It is therefore useful to perform a search space reduction that leads to a limited solution space from which the most optimal solution can be chosen. Statistical sampling [9] can be used to reduce the solution space, but results in approximations that may miss the optimal solution regarding the use case. Machine Learning (ML) approaches can also be useful in order to perform this task [10]. However, it requires a large amount of training data models classified according to user use case. Furthermore, ML-based approaches can be seen as a black box for IS designers.

Instead, rule-based approaches can be a good alternative to lead the reduction task in an explainable way. In our case, we have used Heuristics that have been used in the context of NoSQL DB generation. Atzeni et al. provide in [3] a set of heuristics for transforming object-oriented models into a common model before transforming it into a platform-dependent model. However those heuristics are rather driven by a query use case expressed through an affinity matrix, which makes it an approximate and probabilistic approach.

3   ModelDrivenGuide Approach

Our approach ModelDrivenGuide aims to produce a set of data models giving choices to the IS designer instead of focusing on dedicated solutions which

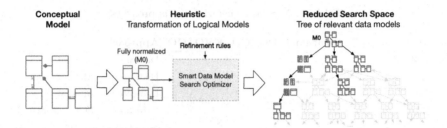

Fig. 1. ModelDrivenGuide: a heuristic to generate data models

prevents any trade-off. It is based on logical models and refinement rules as illustrated in Fig. 1.

ModelDrivenGuide begins with the fully normalized data model ($\mathcal{M}_0$) (*i.e.*, the relational model).

Then, the *Smart Data Model Search Optimizer* which is at the heart of this paper, relies on data models' denormalization with two refinement rules (Merge and Split) applied recursively on logical models (Definition 1) to obtain all possible solutions.

Especially, it is based on a heuristic that optimizes the denormalization process. This process applies refinement rules recursively without conditions to generate all models and is called *Naïve* in the following. The heuristic reduces the search space to relevant data models since some of the generated models can be redundant and even useless for the associated use case. So far, no other approach combines merge and split rules to generate *all data models* suitable for a given context of use.

Definition 1. *Let $\mathcal{M}$ be a data model conform to the 5Families meta-model [16] where $\mathcal{M} = (\mathcal{C}, \mathcal{R}, \mathcal{L}, \mathcal{K}, \mathcal{E}, \kappa)$ is composed of concepts $c(r_1, ..., r_m) \in \mathcal{C} | r_1, ..., r_m \in \mathcal{R}$, rows $r(k_1, ..., k_n) \in \mathcal{R} | k_1, ..., k_n \in \mathcal{K}$, key values $\mathcal{K}$ (Atomic Values or Complex Values), references $ref_{i \to j} \in \mathcal{L} | k_i, k_j \in \mathcal{K}$, edges $\mathcal{E} : \mathcal{C} \times \mathcal{C}$ and constraints $cons(k) \in \kappa | k \in \mathcal{K}$.*

To simplify the formalism of manipulations on data models $\mathcal{M}$, we will focus mostly on rows' transformations and references. Thus, a data model $\mathcal{M}$ can be denoted as a graph $\mathcal{M}(\mathcal{R}, \mathcal{L})$ where nodes are rows such that: $\forall i \in \{1, .., n\}, r_i \in \mathcal{R}$, and edges are references such that: $\forall j, k \in \{1, .., n\}, ref_{j \to k} \in \mathcal{L} | r_j, r_k \in \mathcal{R}, r_k = ref_{j \to k}(r_j)$. Our goal in this paper is to start from an initial $\mathcal{M}_0(\mathcal{R}, \mathcal{L})$ and to generate a set of optimal models $\mathcal{M}^{opt}$ compatible with the use case.

Driving Example: Fig. 2 gives a logical representation of a data model $\mathcal{M}$ with 4 rows **Warehouse**, **Customer**, **StoredIn** and **Order** (resp. W, C, S, and O) of 4 corresponding concepts. Four references link keys from source row (*i.e.*, primary key - bold red) to the target row (*i.e.*, foreign key).

3.1   Logical Model Refinement Rules

The Logical Model refinement is based on three transformation rules:

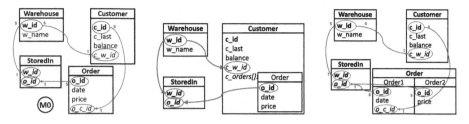

Fig. 2. Driving example   **Fig. 3.** Merge: $m(C, O, ref_{C \to O})$ **Fig. 4.** Split: $s(O, price)$

1) **merge** *rows* to produce *complex values* for nesting (CO, DO) or to merge keys for values concatenation (KVO),

2) **split** a *row* to produce two rows in a same *concept* (CO),

and 3) transform *references* into the equivalent *edges* (GO).

At logical level, models' refinement uses endogenous transformations where both source and target models are conformed to the same meta-model, here the **5Families** meta-model. We detail especially *merge* and *split* refinement rules which are the basis of the data model generation framework.

Merge is a refinement rule applied between two rows referring to each other (*i.e.,* associated classes), where the result is a single row with a complex value.

Definition 2. *Let m: $\mathcal{M} \to \mathcal{M}$ be an endogenous function that merges rows from a 5Families model $\mathcal{M}$. The merge function $m(r_i, r_j, ref_{i \to j})$ is applied on two rows $r_i, r_j \in \mathcal{R}$ (source and target rows) linked by a reference $ref_{i \to j} \in \mathcal{L}$ with corresponding keys $k_i, k_j \in \mathcal{K}$. The merge function produces a new model $\mathcal{M}'$ where r_j is a complex value of r_i, and removes $ref_{i \to j}$, denoted by:*

$$m(r_i, r_j, ref_{i \to j}) = r_i\{r_j\}$$

The merge function is a bijective function $m^{-1}(r_i\{r_j\}) = (r_i, r_j, ref_{i \to j})$ which rebuilds $ref_{i \to j}$ and non-nested rows.

This rule corresponds to the merge of two nodes in graph $\mathcal{M}$ where node j (row) is embedded in node i. Notice that $m(r_i, r_j, ref_{i \to j}) \neq m(r_j, r_i, ref_{i \to j})$ since the nested row is done in the opposite way. From reference $ref_{i \to j}$, when the target row r_j is nested into the source r_i, r_j is embedded into a list of values.

Example: applying a merge on the driving example, like $m(Customer,\text{-}\ Order, ref_{C \to O})$ results in the data model shown in Fig. 3. We notice that after applying this merge rule, the reference $ref_{C \to O}$, linking **Customer** row and **Order** row was transformed into **c_orders** containing a list of nested **Order[]**.

Cycles issue. By recursive application of merges, a problem occurs in case of data models containing cycles through references like in our driving example.

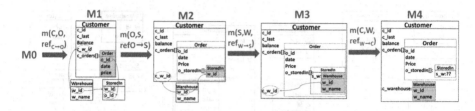

Fig. 5. Particular case: cycle

Figure 5 illustrates recursive application of four merges to obtain successively M_0 (of Fig. 2) to M_4 data models. Since the merge function removes references (Definition 2), the M_4 data model doesn't contain references anymore. However, we must notice that the M_0 contains a cycle with a reference $ref_{W \to S}$ between *Warehouse* and *StoredIn*. During the 4th merge, this reference has been transformed into a *ComplexValue* into *StoredIn[]*. But the remaining reference $ref_{C \to W}$ between *Customer* and *Warehouse* is used to merge them (data model M_4) which gets rid of the information $ref_{W \to S}$. This removal makes it impossible to get backward in the data model's generation process. This case occurs during *merges* on nested rows. We apply m^{-1} on the nested row to rebuild the reference, and apply m on the other reference.

Split is a refinement rule applied on a row containing several keys, associating several rows for the same concept (*i.e.*, CO's column family).

Definition 3. *Let $s: \mathcal{M} \to \mathcal{M}$ be an endogenous function that splits rows from a 5Families model $\mathcal{M}$. The split function $s(r_i, k)$ is applied on a row $r_i \in \mathcal{R}$ and a key value $k \in keys(r_i)$ not linked to a constraint. The split function produces a new model $\mathcal{M}'$ with two rows $\overline{r_{i_k}}$ and r_{i_k} with the same constraint key $pk \in keys(r_i)$ (i.e., primary key) where $\overline{r_{i_k}} = (pk, k_i)|\forall k_i \in keys(r_i) \wedge k_i \neq k$, and $r_{i_k} = (pk, k)$. The split function s is denoted by:*

$$s(r_i, k) = (\overline{r_{i_k}}, r_{i_k})$$

s is bijective $s^{-1}(\overline{r_{i_k}}, r_{i_k}) = (r_i)$ and merges common constraints and keys.

Example: let's take the example of the model depicted in Fig. 2, if we apply a split on Warehouse using key *price*, we obtain the model in Fig. 4. Notice that rows *Order1* and *Order2* are linked to the same concept *Order* (mixed up when a concept contains only one row).

Although splits are usually used for the CO family, they can also be useful to optimize documents size DO and thus computation time when combining only useful keys from two rows (split+merge) from useless keys (stored apart).

3.2   Inverse Functions

We should notice that *merge* and *split* refinement rules can be considered inverse functions for specific cases. In fact, if a split is applied on a key k_j as a *complex*

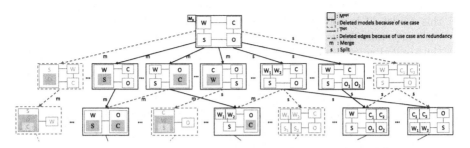

Fig. 6. Two Levels of the Graph of Possible Solutions

value r_j, it rebuilds a row with a single key containing the reference information $ref_{i \to j}$ and keys from row r_j. Then:

$$s(r_i\{r_j\}, k_j) = (r_i, r_j) = m^{-1}(r_i\{r_j\}) \tag{1}$$

At the opposite, a *merge* between two rows $\overline{r_{i_k}}$ and r_{i_k} with a same constraint pk_i can be considered to be the inverse of a *split*. Then:

$$m(\overline{r_{i_k}}, r_{i_k}, pk_i) = (r_i) = s^{-1}(\overline{r_{i_k}}, r_{i_k}) \tag{2}$$

3.3 Completeness of Our Approach

Our approach based on the previously defined refinement rules generates all possible data models by combining recursively several merges and splits. Before defining a generation strategy of data models, it is necessary to prove the completeness of our approach. It states that splits and merges are sufficient to generate all possibilities without modifying keys (*e.g.*, materialization, replication). Moreover, there is always a path between two data models.

Theorem 1. *Completeness. Merge and Split refinement rules are sufficient to generate all denormalized data models $\mathcal{M}^*$ without key modification, conformed to the 5Families meta-model, beginning with the fully normalized data model.*

Proof. We want to prove that there is at least one *path* (sequence of rules) between two denormalized models, which means there is a path between a denormalized model M_1 and the fully normalized data model M_0.

Let $step(M_1, M_2, r)$ be a rule applied between two models M_1 and M_2, where $r \in \{merge, split\}$ and $M_2 = r(M_1)$. And let $path(M_1, M_n)$ be a sequence of steps between two models M_1 and M_n, such that:

$$path(M_1, M_n) = [step(M_1, M_2, r_1), ..., step(M_{n-1}, M_n, r_{n-1})]$$

Suppose that M_0 and M_n share the same set of rows and keys and there isn't any *path* between a denormalized model M_n and the normalized one M_0.

$\implies path(M_0, M_n) = \emptyset$

$\implies \exists M, path(M, M_n) \neq \emptyset, path(M_0, M) = \emptyset$

This implies that:

1) this model M is not conformed to the *5Families* meta-model (Absurd since rules are endogenous), or

2) the applied rules can't be inverse (Absurd, from Definition 2 and 3).

Of course, it exists more than one path from a data model to another since there are several possible compositions of functions that produce a given data model. Thus, we obtain a *lattice* that represents the generated solutions where nodes are data models, and edges are applied refinement rules. The lattice reduction problem is discussed in the following section.

4   Smart Data Model Search Optimizer

Models refinement leads to the production of plenty of data models in a form of a graph. Its aim is to generate the most suitable data models and thus allow to choose the targeted solution. But the number of possibilities explodes as splits on M can be applied on each key and merges can be done in both ways.

Starting from M as $M(R, L)$, applying the ith transformation rule on the rows in R, leads to M_i. If the rule is split, it divides one row to several columns. If the rule is merge, it merges 2 rows which are a set of columns themselves.

Considering M as a set of rows, all generated M_i by the naïve process, form a graph $G(M^*, T^*)$ where M^* is the set of generated data models and T^* the transformations that link data models. This problem is a reduced problem of generating all partitioning of a set.

The total number of partitions of the set M can be calculated as the *Bell* number. *Bell* number grows exponentially by increasing the size of the set, and set partitioning has been proved to be NP-hard [13]. Moreover, splits on data models are combined with merges making the growth far more computational.

Therefore, we need a heuristic to explore only an optimal subset of state space. Considering the initial M as a state space graph of $M_0(R, L)$, and transformed graph after applying the ith transformation rule as M_i, $T(M^{opt}, T^{opt})$ is the reduced search space tree growing through transformations done by our heuristic. To obtain T we apply 1) a Depth First Search (DFS) strategy to traverse G at each step i and 2) a reduction of the graph based on the use case.

4.1   Complexity of the *Naïve* Denormalization Process

The generation of data models M^{opt} by applying refinement rules recursively on various combinations of model transformations M_i, produces a hierarchy of target data models T. At the top level of this hierarchy is the fully normalized relational model M_0 and at the lowest level are completely denormalized models.

This hierarchy's size $|T|$ depends on the size of the input data model with the number of rows $|R|$ and references $|L|$. Figure 6 shows the first two levels of the data models' generation graph (G) for our driving example, where W, C, O and S stand for *Warehouse, Customer, Order* and *StoredIn* respectively. The dashed edges show the edges that were deleted by the heuristic to avoid redundancies, cycles and transformations that do not take into consideration the use case. Consequently, dashed squares represent the lonely nodes (models) which cannot be produced since no more transformation can reach them.

The set of `merges` depends on the number of linked rows in $\mathcal{M}$ and the number of generated models is given by a *Fubini number* or *Ordered Bell number* [4]:

$$Fn_{|\mathcal{L}|} = \sum_{k=0}^{|\mathcal{L}|} k! \times \binom{|\mathcal{L}|}{k}$$

This number is an upper bound since the number of generated data models can be less depending on references' organization between rows. Our driving example can produce up to 75 models considering just merges $Fn_4 - \sum_{k=0}^{4} k! \times \binom{4}{k} = 75$.

Moreover, the number of `splits` on a row depends on the number of keys in a row $|keys(r)|$ (without primary keys) which is given by the *Bell's number* [5]:

$$B_{|keys(r)|} = \sum_{k=0}^{|keys(r)|} \binom{|keys(r)|}{k} \times B_{|keys(r)|-1}$$

Thus, in our example rows *Customer* and *Order* contain 3 non-primary keys $B_3 = 5$, for *Warehouse* $B_1 = 1$ and *StoredIn* $B_0 = 1$. And those solutions combine together producing $B_{|keys(C)|} \times B_{|keys(O)|} \times B_{|keys(W)|} \times B_{|keys(S)|} = 25$. To finish with, split and merged rows can be combined to produce all possible solutions in T. Each split row can be merged and adds new nested solutions.

Thus, we can formalize the complete denormalization problem with the combination of rows' merges and splits all together as a product. The total number of possible data models gives the following complexity measure:

$$|\mathcal{M}^*| = Fn_{|\mathcal{L}|} \times \prod_{k=1}^{|\mathcal{R}|} B_{|keys(r_k)|}$$

Our example will generate $75 \times 25 = 1,875$ data models.

As said previously, since there are several ways to produce a single $\mathcal{M}_i \in \mathcal{M}^{opt}$, it requires to prune the lattice G of solutions $\mathcal{M}^*$.

4.2   Application of the Refinement Rules

Due to the aforementioned problems, we need to provide a heuristic in order to reduce the search space and avoid cycles. The target is a subset of $\mathcal{M}^*$ called $\mathcal{M}^{opt}$ after pruning edges and nodes from T. The heuristic avoids to produce different paths between two data models. In fact, applying splits and merges in different orders will produce the same effects on the resulting data models. Moreover, every merge can be reversed by a split and produce a cycle in the generation of data models (not shown in Fig. 6). Thus, four main rules have to be adopted and are presented in the following.

A Row with Complex Values. Applying a split on a row with complex values (merged rows) gives the following result:

$$s(m(r_i, r_j, ref_{i \to j}), k_j) = s(r_i\{r_j\}, k_j) = (r_i, r_j, ref_{i \to j})$$

where r_i and r_j are two rows that are first merged and represented as $r_i\{r_j\}$, $ref_{i \to j}$ is the reference linking the two rows and used for the merge and k_j is the key on which we apply the split. We notice that this transformation takes the models' generation back in the hierarchy with the initial model (r_i, r_j) of two rows, hence generates a cycle. Thus, the first rule avoids applying a split on a row with complex values.

A Row Linked to Two or More Rows. When a row r_i is linked to two rows r_j and r_k with two references $ref_{i \to j}$ and $ref_{i \to k}$, a redundancy will occur, since:

$$m(m(r_j, r_i, ref_{j \to i}), r_k, ref_{j \to k}) = m(m(r_j, r_k, ref_{j \to k}), r_i, ref_{j \to i})$$
$$= r_j\{r_i, r_k\}$$

where $r_i, r_j, r_k \in \mathcal{R}$ and r_j is linked both to r_i $(ref_{j \to i})$ and r_k $(ref_{j \to k})$.

Since both transformations are equivalent (the result is the same data model), the heuristic discards one of the two merges.

The Use Case. By considering the use case, we can reduce the number of joins only to those used in queries that combine rows through references. Also splits should not separate keys if queries from the use case combine them. It avoids solutions which require instance reconstruction with costly joins.

Moreover, splits applied thanks to queries can generate redundancy. If queries $q(K)$ and $q(\overline{K})$ applied on r use complement keys, they will produce the same data model (i.e., $s(r, K) = s(r, \overline{K})$). To avoid this issue, we compute complementary queries to prune redundant splits.

We must notice that Theorem 1 on the completeness of our approach remains true according to the two first simplification rules. In fact, only duplicate edges in $\mathcal{M}^*$ are removed producing a DFS keeping all the nodes. Thus, it always exists a *single* path, to obtain a data model. However, this last simplification rules removes nodes since we remove edges from $\mathcal{M}^*$ if the refinement rule do not rely on a query from the use case. Commonly said in the literature [1,6,11], we assume the fact that those transformations will lead to a data model *useless* for the use case as well as the whole corresponding subtree. Consequently, the completeness remains valid for data models relying on the use case.

4.3   Heuristic of Model Generation

To formalize the generation of data models with our heuristic, Algorithm 1 gives the recursive function 5FMHeuristic which generates a list of data models $\mathcal{M}^{opt}$. It takes a relational model as an entry and produces all possible models (of all families). Starting from $i = 0$, at stage i, it processes a current data model $\mathcal{M}_i$ on which splits and merges will be applied by taking into account a set of queries $\mathcal{Q}$ from the use case. For simplicity, we consider that a query is a set of involved keys within the data model (for filters, joins, projects, aggregates, etc.).

Each time the 5FMHeuristic function is called with a new data model from the *5Families* meta model, it is added to the global output list $\mathcal{M}^{opt}$ (line 2). Then, all keys from the data model $(K \in \mathcal{K})$ are tested except those which are involved in a Primary Key constraint (line 3). To test the keys, the first check

---

Algorithm 1. 5FMHeuristic

---

global: A list of data models $\mathcal{M}^{opt}$ from *5FamiliesModel*
input: a data model $\mathcal{M}_i$ from *5FamiliesModel*, a list of input queries $\mathcal{Q}$ (a query is a set of keys from $\mathcal{M}_i$), a list of used queries $\mathcal{Q}'$
init: $\mathcal{M}_i = \mathcal{M}_0$, the relational data model, $\mathcal{M}^{opt} = \emptyset$, $\mathcal{Q}' = \emptyset$
1: **procedure** 5FMHEURISTIC($\mathcal{M}_i, \mathcal{Q}, \mathcal{Q}'$)
2:     $\mathcal{M}^{opt} := \mathcal{M}^{opt} \cup \mathcal{M}_i$
3:     **for all** key $K \in \mathcal{K} \wedge \, !Constraint(K, \text{CType=PK})$ **do**
4:         $r := Row(K)$
5:         **If** $K \notin \mathcal{Q} \cup \mathcal{Q}'$ **then**
6:             **if** $\exists k \in r | k \neq K \wedge !Constraint(k, \text{Ctype=PK})$ **then**
7:                 5FMHEURISTIC($Split(\mathcal{M}, K), \mathcal{Q}, \mathcal{Q}'$)
8:         **else**
9:             **for all** $q \in \mathcal{Q}$ **do**
10:                 **if** $\forall k \in q | Row(k) = r, \exists k \in R | k \notin q \wedge !Constraint(k, Ctype = PK) \wedge \forall q' \in \mathcal{Q} | \overline{q} \neq q'$ **then**
11:                     5FMHEURISTIC($Split(\mathcal{M}_i, q), \mathcal{Q} - q, \mathcal{Q}' \cup q$)
12:                 **else if** $\exists k \in q | Row(k) \neq r \wedge Concept(k) = Concept(K)$ **then**
13:                     continue
14:                 **else if** $\exists k_1, k_2 \in q | ref_{k_1 \rightarrow k_2} \vee ref_{k_2 \rightarrow k_1} \in \mathcal{L}$ **then**
15:                     5FMHEURISTIC($Merge(\mathcal{M}_i, q, k_1, k_2), \mathcal{Q} - q, \mathcal{Q}' \cup q$)
16:                     5FMHEURISTIC($Merge(\mathcal{M}_i, q, k_2, k_1), \mathcal{Q} - q, \mathcal{Q}' \cup q$)
17:                     5FMHEURISTIC($ReferenceToEdge(\mathcal{M}_i, q, k_1, k_2), \mathcal{Q} - q, \mathcal{Q}' \cup q$)

(lines 5–7) verifies if the key is used in a query from $\mathcal{Q} \cup \mathcal{Q}'$ (remaining and processed queries). It also checks if the key is the last remaining key from the current row R except the Primary Key (needed for instance reconstruction). In that case, the key is put in a new row by applying the rule `Split` (Definition 3) and then recursively call `5FMHeuristic` on this new data model ($\mathcal{M}_{i+1}$). If the key K belongs to at least one remaining query q from $\mathcal{Q}$ (line 9), we check the rules from the heuristic. First (lines 10–11), if all keys from q belong to the same row r and the latter can be split (except the Primary Key and non-complementary keys), we split it by taking all the keys from q in a new row, instead of the single key K. The generated data model is then recursively denormalized (line 11) without the query q (moved to $\mathcal{Q}'$).

The second rule concerns the keys from a query q (line 12) that belong to the same concept but in different rows, we then avoid applying a merge between those rows and continue to the next query (line 13).

The third rule (line 14) applies the `Merge` rule when a query q uses two keys linked by a *Reference*. In this case, we apply merges in both directions (lines 15–16). Finally, to produce GO data models, references are transformed into *Edges* (line 17) with the `ReferenceToEdge` rule (not presented in this article).

Thanks to this heuristic, the application of refinement rules will generate a tree of possibilities whose first nodes try splits and finish with merges (DFS). In addition, the use case reduces drastically the number of nodes for queries using few keys, and drives towards merges for multi-concept queries.

From our example of 4 concepts, the number of solutions, initially equal to 1,875, is reduced to only 125. This number will be reduced even more by taking into account multi-concept queries impacting the *Fubini*'s number, and far more with queries involving several keys in a same concept (*Bell*'s number).

4.4   Complexity of the Heuristic

Our problem is abstracted by a graph of solution $G(\mathcal{M}^*, \mathcal{T}^*)$, on which our heuristic has two reduction effects. First, it reduces the graph's size by pruning useless data models since it focuses only on the use case $\mathcal{Q}$, obtaining $\mathcal{M}^{opt}$. Second, it applies a DFS approach to avoid redundancies and removes all links in order to produce a tree of solutions called $\mathcal{T}^{opt}$.

Since the heuristic takes into account the distinct references used by queries $|refs(\mathcal{Q})|$, the *Fubini* number is impacted. *Bell* number with splits is also modified based on the size of distinct sets of keys from queries for a given row k: $|KeySet(\mathcal{Q}_k)|$. Thus, the upper bound of the estimated number of solutions is:

$$|\mathcal{M}^{opt}| = Fn_{|refs(\mathcal{Q})|} \times \prod_{k=1}^{|\mathcal{R}|} |KeySet(\mathcal{Q}_k)|$$

And the number of transformations to apply is: $|\mathcal{T}^{opt}| = |\mathcal{M}^{opt}| - 1$

5   Experiments

5.1   Implementation

Our `ModelDrivenGuide` approach is implemented in *Java*. It starts with a UML model automatically transformed into a relational model ($\mathcal{M}_0$).

Then, refinement rules are applied recursively on data models using the *Naïve* (generate all solutions) and the `5FMHeuristic` strategies.

In order to get the number of generated models and to see the gain obtained by our heuristic, a *signature* file has been implemented; a sorted list of concepts, rows, keys, nesting and references. It ensures their uniqueness and detects identical data models produced by two distinct paths. It will help to cut cycles in the *Naïve* strategy and to count the number of duplicated data models.

5.2   TPC-C

To illustrate our approach, we used the TPC-C[2] benchmark giving a full use case mixing at the same time transactions, joins and aggregations. For this test, we focus on the three concepts (classes in the UML model): *Customer*, *Order* and *OrderLine*.

Table 2 shows the number of generated data models, splits, merges and unique data models (thanks to signatures). The *Naïve* approach on TPC-C benchmark,

---

[2]   http://www.tpc.org/tpc_documents_current_versions/pdf/tpc-c_v5.11.0.pdf.

Table 2. Generated models

Strategy	Nb models	Nb splits	Nb merges	Nb unique models
Naïve	2,313	1,646	666	1,445
5FMHeuristic	27	2	24	27

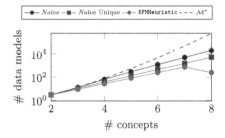

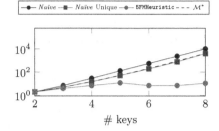

Fig. 7. Nb of data models with merges wrt. nb concepts (5 join queries)

Fig. 8. Nb of data models with splits wrt. nb keys (1 concept and 5 queries)

produces 2,313 data models by applying 1,646 splits and 666 merges (1,445 distinct data models and thus 868 redundancies).

While the 5FMHeuristic generates only 27 data models with only 2 splits (few splitting queries) and 24 merges (impacted by splits). The number of distinct data models is equal to the total number of generated models (no redundancies) which is the goal of our heuristic (DFS approach).

To compare our approach with competitors [1,8], it is important to remind that they both generate a **unique** data model that corresponds to their targeting approach optimizing one criteria. In our case, we produce this data model among 27 others offering a large choice of possible models which contains this solution. After decreasing the number of generated models noticeably, it becomes an easier task for the IS administrator to choose the most convenient one by taking into consideration either the ease of implementation, NoSQL compatibility, security policy, storage or environmental impact, etc. We must notice that among those solutions it is possible to select a data model which can be implemented in several families of databases at the same time called Polystores [2].

5.3   Data Models Generation

In order to study the behavior of the *Naïve* and the 5FMHeuristic, a generator of data models has been implemented in order to simulate the impact of denormalization strategies. It produces various data models by varying the number of concepts and keys, and also the topology of linked concepts (lines *vs* stars). We will produce the average number of generated data models per strategy. Moreover, random queries are generated to see the effect of use cases by varying the number of joins and involved keys.

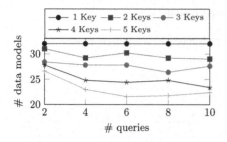

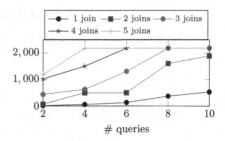

Fig. 9. 5FMHeuristic - Nb of data models (splits) wrt. Nb of filter queries (8 concepts of 5 keys each)

Fig. 10. 5FMHeuristic - Nb of data models (merges) wrt. nb joins per query (8 concepts of 5 keys each)

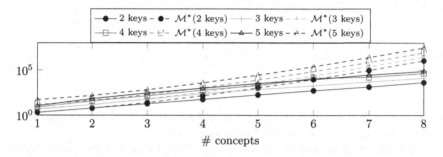

Fig. 11. 5FMHeuristic - Nb of data models wrt. nb concepts varying keys (5 queries)

Our code is available on Github[3] with a configuration file to parameter the generator.

To study the impact of the heuristic on the applied rules (Merge, Split and both of them), thanks to the aforementioned signatures, we will count the average number of generated data models from the initial data model (relational) and the generated models before and after applying the heuristic.

Figure 7 shows the evolution of the number of generated data models with the *Naïve* approach and our 5FMHeuristic. We focus only on merges by limiting keys to primary and foreign keys. The *Naïve* approach follows an exponential growth (dashed curve) but slightly lower than the *Fubini* number since simulated data models do not contain cycles (with references) nor star-shaped data models which lead to more combinations and reach this upper-bound. The *Naïve* unique approach avoids cycles and duplicates which reduces significantly the number of solutions. Our heuristic targets only required merges which can witness a decrease when it reaches bigger data models (here 7 concepts). In fact, since the number of queries/joins are a constant, their impact on the number of merges becomes noticeable. Thus the number of possibilities decreases rapidly.

Figure 8 focuses on splits varying number of keys on a single concept. The *Naïve* approach shows the distribution of the *Bell* number with duplicates

---

[3] https://github.com/leonard-de-vinci/5FM_generator/.

(dashed curve and *Naïve* unique). The number of data models produced by 5FMHeuristic is reduced sharply for splits. Here also the number of possible splits is bounded by the number of involved queries reaching a threshold based on the size of the KeySet (simplified Bell number to the KeySet).

The impact of the use case on the 5FMHeuristic is shown in Fig. 9 by varying the number of keys per query. It's applied on 8 concepts of 5 keys, each with different numbers of queries (no joins - no merges). We notice that only single-key queries produce the maximum number of splits since it offers all possibilities, while this number decreases with bigger queries. When the KeySet size is high, it implies that splits cannot be applied since all keys must be put together in a same row. Thus, we can conclude that small queries produce more data models to find more adapted solutions.

Figure 10 focuses on join queries by varying the number of joins per query. Contrary to the splits, the bigger the joins are the more data models are generated. It is due to the fact that queries allow merging more concepts with each other allowing to produce data models with a single concept.

Figure 11 plots the global impact of the 5FMHeuristic with both merges and splits by varying the number of concepts and the number of keys for each. With 5 different queries (filters and joins), it is interesting to see that the distribution remains similar for all cases. The number of keys per concept has an impact on possible splits and consequently on merges. For 8 concepts, it varies from 726 for 2 keys/concept to 16,767 data models for 4 keys/concept. We plotted the $|\mathcal{M}^*|$ complexity (dashed curves) for each key size which shows the exponential growth of the produced data models, at most 28,383,420 solutions for 8 concepts of 5 keys. We can see that the 5FMHeuristic drastically reduces the search space.

To conclude, queries with more joins and fewer keys have more impact on the number of generated data models than other ones. Moreover, our heuristic implies that splits impact possible merges due to denormalizations' combination.

6   Conclusion and Future Work

In this paper, we have proposed an IS modeling and implementation MDA-based approach that generates *all* data models by recursive denormalization to find suitable solutions. We propose the first full formalization of the problem and issues with refinement rules and data models manipulation. Then a heuristic of generations based on DFS and the use case, allows avoiding the explosion of solutions in terms of the amount of data models and paths to obtain them.

Our goal is to offer the IS designer a limited solution space of which he can choose the best model that will be adapted to his constraints (*e.g.*, efficiency, green computing, integration, polystores, security). To guide the choice of implementation, we can find performance comparisons of NoSQL DB for dedicated needs [12,17,20], studies of applicability to specific software quality choices [15].

In a future work, we would like to enrich this choice process by offering a cost model that can assist the designer in his choice by associating each model of the solution space with an implementation cost and then sort them accordingly.

We also work on the definition of the eligibility of a data model in the target DB. Indeed, the generated models must be compatible and it is necessary to define mapping rules; a key and a value for KVO, Concepts and Links for GO, etc.

References

1. Abdelhedi, F., Ait Brahim, A., Atigui, F., Zurfluh, G.: MDA-based approach for NoSQL databases modelling. In: Bellatreche, L., Chakravarthy, S. (eds.) DaWaK 2017. LNCS, vol. 10440, pp. 88–102. Springer, Cham (2017). https://doi.org/10.1007/978-3-319-64283-3_7
2. Alotaibi, R., Cautis, B., Deutsch, A., Latrache, M., Manolescu, I., Yang, Y.: Estocada: towards scalable polystore systems. VLDB Endow. **13**(12), 2949–2952 (2020)
3. Atzeni, P., Bugiotti, F., Cabibbo, L., Torlone, R.: Data modeling in the NoSQL world. Int. J. CSI **67**, 103–149 (2020)
4. Belbachir, H., Djemmada, Y., Németh, L.: The deranged Bell numbers. arXiv preprint arXiv:2102.00139 (2021)
5. Bell, E.T.: Exponential polynomials. Ann. Math. **35**(2), 258–277 (1934)
6. Chebotko, A., Kashlev, A., Lu, S.: A big data modeling methodology for apache cassandra. In: ICBD 2015, pp. 238–245. IEEE (2015)
7. Daniel, G., Sunyé, G., Cabot, J.: UMLtoGraphDB: mapping conceptual schemas to graph databases. In: Conceptual Modeling (ER 2016), pp. 430–444 (2016)
8. de la Vega, A., García-Saiz, D., Blanco, C., Zorrilla, M., Sánchez, P.: Mortadelo: automatic generation of nosql stores from platform-independent data models. Future Gen. Comput. Syst. **105**, 455–474 (2020)
9. Doshi, P., Gmytrasiewicz, P.J.: Monte Carlo sampling methods for approximating interactive POMDPS. J. Artif. Int. Res. **34**(1), 297–337 (2009)
10. Eggermont, J., Kok, J.N., Kosters, W.A.: Genetic programming for data classification: partitioning the search space. In: SAC 2004, pp. 1001–1005. New York (2004)
11. de Freitas, M.C., Souza, D.Y., Salgado, A.: Conceptual mappings to convert relational into nosql databases. In: ICEIS 2016, pp. 174–181 (2016)
12. Jatana, N., Puri, S., Ahuja, M., Kathuria, I., Gosain, D.: A survey and comparison of relational and non-relational database. IJERT **1**(6), 1–5 (2012)
13. Lewis, M., Kochenberger, G., Alidaee, B.: A new modeling and solution approach for the set-partitioning problem. COR **35**(3), 807–813 (2008)
14. Li, Y., Gu, P., Zhang, C.: Transforming UML class diagrams into HBase based on meta-model. In: ISEEE 2014, vol. 2, pp. 720–724 (2014)
15. Lourenço, J.R., Cabral, B., Carreiro, P., Vieira, M., Bernardino, J.: Choosing the right NoSQL database for the job: a quality attribute evaluation. J. Big Data **2**(1), 1–26 (2015). https://doi.org/10.1186/s40537-015-0025-0
16. Mali, J., Atigui, F., Azough, A., Travers, N.: **ModelDrivenGuide**: an approach for implementing NoSQL schemas. In: Hartmann, S., Küng, J., Kotsis, G., Tjoa, A.M., Khalil, I. (eds.) DEXA 2020. LNCS, vol. 12391, pp. 141–151. Springer, Cham (2020). https://doi.org/10.1007/978-3-030-59003-1_9
17. Raut, A.: NoSQL database and its comparison with RDBMS. Int. J. Comput. Intell. Res. **13**(7), 1645–1651 (2017)
18. Rocha, L., Vale, F., Cirilo, E., Barbosa, D., Mourão, F.: A framework for migrating relational datasets to NoSQL. Proc. Comput. Sci. **51**, 2593–2602 (2015)

19. Störl, U., Klettke, M., Scherzinger, S.: NoSQL schema evolution and data migration: state-of-the-art and opportunities. In: EDBT 2020, pp. 655–658 (2020)
20. Tauro, C.J., Aravindh, S., Shreeharsha, A.: Comparative study of the new generation, agile, scalable, high performance NoSQL databases. Int. J. Comput. Appl. **48**(20), 1–4 (2012)
21. Vajk, T., Fehér, P., Fekete, K., Charaf, H.: Denormalizing data into schema-free databases. In: CogInfoCom 2013, pp. 747–752. IEEE (2013)

State Model Inference Through the GUI Using Run-Time Test Generation

Ad Mulders[1], Olivia Rodriguez Valdes[1], Fernando Pastor Ricós[2],
Pekka Aho[1(✉)], Beatriz Marín[2], and Tanja E. J. Vos[1,2]

[1] Open Universiteit, Heerlen, The Netherlands
pekka.aho@ou.nl
[2] Universitat Politècnica de València, Valencia, Spain

Abstract. Software testing is an important part of engineering trust-worthy information systems. End-to-end testing through Graphical User Interface (GUI) can be done manually, but it is a very time consuming and costly process. There are tools to capture or manually define scripts for automating regression testing through a GUI, but the main challenge is the high maintenance cost of the scripts when the GUI changes. In addition, GUIs tend to have a large state space, so creating scripts to cover all the possible paths and defining test oracles to check all the elements of all the states would be an enormous effort. This paper presents an approach to automatically explore a GUI while inferring state models that are used for action selection in run-time GUI test generation, implemented as an extension to the open source TESTAR tool. As an initial validation, we experiment on the impact of using various state abstraction mechanisms on the model inference and the performance of the implemented action selection algorithm based on the inferred model. Later, we analyse the challenges and provide future research directions on model inference and scriptless GUI testing.

Keywords: Model inference · Automated GUI testing · TESTAR tool

1 Introduction

The world around us is strongly connected through software and information systems. Graphical User Interface (GUI) represents the main connection point between software components and the end users, and can be found in most modern applications. Testing through GUI is an important way to prevent end users from experiencing the effects of software bugs. Although GUI testing can be done manually, it is a very time consuming and costly process [14]. There are various tools to capture or manually define scripts for regression testing of GUIs, but the main challenge is the high maintenance cost of the scripts when the GUI changes [25]. In addition, GUIs tend to have a large state space, so creating scripts to cover all the possible paths and defining test oracles to check all the elements of all the states would be an enormous effort.

© The Author(s), under exclusive license to Springer Nature Switzerland AG 2022
R. Guizzardi et al. (Eds.): RCIS 2022, LNBIP 446, pp. 546–563, 2022.
https://doi.org/10.1007/978-3-031-05760-1_32

Scriptless GUI testing aims to lower the maintenance costs compared to scripted testing. In scriptless testing, there are no scripts that define the sequences of test steps prior to test execution. Instead, the testing tool decides on-the-fly during test execution which test steps are being executed, based on the action selection mechanism (ASM) being used. Although there are no test scripts to maintain, there might be some system specific instructions for the tool that require maintenance.

TESTAR (testar.org), an open-source tool for scriptless GUI testing, is based on agents that implement various ASMs. The underlying principles are very simple: generate test sequences of (state,action)-pairs by starting up the System Under Test (SUT) in its initial state, and continuously select and execute an action to bring the SUT into another state. Various case studies have shown that TESTAR effectively complements the existing manual practices and can find undiscovered failures in a SUT with reasonably low setup costs [27]. However, TESTAR suffers from not knowing exactly what has been tested, failing to give test managers the information they need for decision making.

In [13, 27], the first steps are described towards an extension of TESTAR to infer state models during GUI exploration. This feature allows creating a map of where in the SUT the tool has been and what it has done during testing. De Gier et al. [13] use the model to define offline oracles (i.e., after testing) that consisted of querying the model for accessibility information. However, inferring state models can be purposeful for TESTAR in various other ways. For example, the inferred models can be used for TESTAR's ASMs during and after the model inference, or defining various types of offline oracles that are based on comparing models between different releases or versions of a System Under Test (SUT). The models could also serve as visual reference models for testers or users, or be valuable for model-based GUI testing (MBGT) tools [3]. MBGT has not been widely adopted, because creating the models requires modelling expertise and a lot of effort [7]. If we can infer even initial models, these problems might be (partially) solved.

There are a few existing approaches for inferring models during automated exploration of the GUI that are described in Sect. 2. However, automated GUI exploration is challenging, and existing tools are mostly academic prototypes or abandoned open source projects. State space explosion is still an open challenge for the inference of state-based models through GUI. Most programs with a GUI have a huge number of possible states, and to make the size of the models manageable, some information has to be abstracted away. It is challenging to define a suitable level of abstraction and find an equilibrium between the necessary expressiveness of the extracted models and the computational complexity [19]. Abstracting away too much information might make a model unsuitable for its purpose (i.e., ASM, MBGT, oracles, etc.) and lose opportunities to discover faults and changes between versions. Abstracting away too little information might result in state space explosion, making the model less suitable for its purpose. Most of the related work does not explain in sufficient detail how they deal with the abstraction, raising questions whether their solutions are generally applicable or simply tailored for the applications used in validation.

The main contributions of this paper are:

- A description of the model inference functionality implemented into TESTAR.
- A new algorithm for ASM based on the inferred model.
- An initial validation of the test effectiveness of the new ASM in terms of code coverage and reached states.
- An initial validation of the approach by experimenting on how various abstraction mechanisms affect the inferred models.

The rest of the paper is organised as follows. Section 2 presents related work. Section 3 presents TESTAR and our approach to infer state models. Section 4 presents the use of models for action selection and experiments on the test effectiveness of the implemented ASM. Section 5 describes the experimentation to find out how various abstraction mechanisms affect the inferred models. Section 6 analyses the findings and challenges, and provides future research directions. Finally, Sect. 7 presents the main conclusions.

2 Related Work

The earliest automated GUI testing tools were so called capture and replay (C&R) tools. Using them, a test engineer manually inputs all the interactions by using mouse and keyboard while having the tool recording the test case. Afterwards, the test sequence could be executed automatically as part of a regression test set. The main advantage of C&R tools is that they are easy to use and testers do not have to write test scripts by hand. The disadvantage is that the recorded scripts are fragile for GUI changes and maintenance of the scripts is costly since the out-dated test cases have to be manually recorded again [23].

Model-based GUI testing (MBGT) [3,11,16] aims to reduce the effort for creating and maintaining the test scripts by generating the test cases from a model. MBGT approaches require that the GUI and its expected behavior is modelled on a higher level of abstraction than the GUI itself. The modelling language should be understandable by a tool that uses it to automatically generate tests. An advantage of this type of testing is that it is possible to precisely specify the exact test specifications that a GUI should conform to. Another advantage is that when the GUI changes, the test scripts do not have to be manually updated. Instead, the model is updated and the scripts/tests are generated again. The main disadvantages are that model-based GUI testing approaches require a deep knowledge of the application domain and expert knowledge of formal modelling methods and languages to manually create a model of the GUI. Modelling requires also quite a lot of time and effort.

There are a few GUI testing approaches that allow automated GUI model inference, a.k.a., GUI ripping [20] or GUI reverse engineering [15]. We can roughly distinguish 3 forms of automated model discovery: (1) static analysis, (2) dynamic analysis and (3) a hybrid combination of both static and dynamic techniques [4,17]. Model inference through static analysis uses the program's source code to infer a model of the GUI [12,26]. Static techniques concentrate

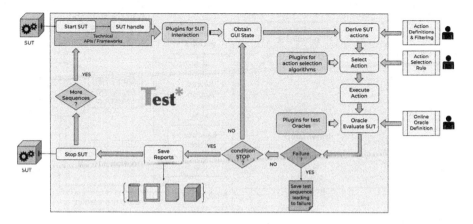

Fig. 1. TESTAR operational flow to generate sequences of (state, action)-pairs by starting up the SUT and continuously select and execute an action to bring the SUT into another state.

only on the structure of the GUI, not taking the run-time behaviour of the GUI into consideration in the model.

The dynamic model inference approaches analyse the GUI while the system is running [6]. To automatically explore the GUI, APIs or libraries are used to get access to all the GUI elements in a specific state of the application. To create a model, these tools are able to recognise whether the application is in a state that the tool has already visited before, or whether the state is being visited for the first time. Examples of tools using model inference through dynamic analysis are GUITAR [24], GUI Driver [5], Crawljax [21], Extended Ripper [8], GuiTam [22] and Murphy Tools [6]. Some approaches, e.g., [18], combine dynamic and static analysis for model extraction. As mentioned, the challenge that has to be solved for all these tools is to define a suitable level of abstraction for the model inference that ensures that the model is useful for its purpose (e.g., ASMs, offline oracles, visualisation of testing, etc.). Most of the related work does not explain in sufficient detail how they deal with abstraction. In this paper, we will make a first attempt to research how the abstraction level affects the results when using the models.

3   State Model Inference for TESTAR

The operational flow of TESTAR is shown in Fig. 1. When the SUT has *started*, TESTAR captures the current *state of the GUI* using APIs like Windows Automation API (WUIA) (for desktop), Selenium Web Driver (for web), or the Java access bridge (for Swing). This (concrete) state consists of *all* the properties (that are available through the API) of all the widgets that are part of the GUI. Subsequently, to *derive* the actions that it is able to perform in that state, it cycles through all the widgets and adds all possible actions associated with the widgets to a pool. Sometimes, if the SUT includes custom widgets and the API

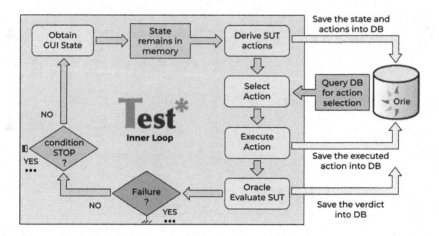

Fig. 2. TESTAR operational flow including model inference

does not detect all the widget attributes, the user has to provide TESTAR with some extra configuration to detect all the available actions correctly. From this action pool, one is *selected* by the ASM of TESTAR and *executed*. TESTAR has several ASMs available, for example based on random, prioritising new actions, or execution count [10].

After the action has been executed and the GUI has reached a new state, TESTAR will again capture the new state and derive, select and execute an action. This process repeats until the specified stop condition, for example the test sequence length or the occurrence of an error condition, is reached.

State abstraction is an important facet of scriptless GUI testing. TESTAR has an implementation to calculate state identifiers based on hashes over a selected set of widget attributes. This selected set defines the abstraction level. The abstraction level determines the number of different states TESTAR will distinguish. This can evidently influence test effectiveness and is related to the equilibrium explained in Sect. 1. To gather evidence about the suitable set of widget attributes for state abstraction, we run experiments that are described in Sect. 5.

TESTAR uses dynamic analysis techniques to infer a model. The flow for capturing the state model is depicted in Fig. 2. The state of the SUT is constantly saved into the OrientDB graph database together with available actions and the executed action. The state model can be queried by an ASM, like in Sect. 4, but also by a human, an offline oracle, or other MBGT approaches.

As indicated, the model will be built incrementally with subsequent TESTAR runs. All states (concrete and abstract) that are visited during a run are stored in the database. For analysis and reporting, the structure of the inferred model is divided into three layers as shown in Fig. 3.

The **top layer** is an abstract state model. It allows for example ASMs to use the model for action selection, or end users to analyse the behaviour of the SUT. Creating the abstract model requires thevadjust identification of unique states at a *suitable* abstraction level. As indicated, this means trying to avoid state space explosion, while simultaneously not losing the purposefulness of the model. Too abstract states can introduce non-determinism in the inferred model.

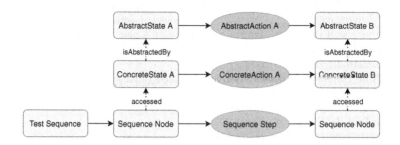

Fig. 3. Layered design of the state model

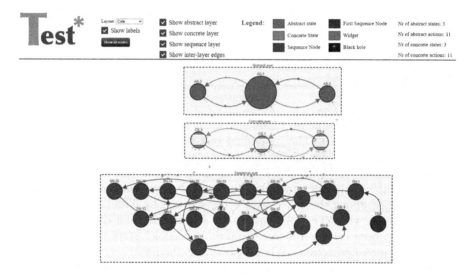

Fig. 4. Visualization of an example model inferred by TESTAR

The **mid layer** is the concrete model. This model contains all the information that can be extracted through the APIs used by TESTAR. The concrete state model will contain too many states to drive the execution of TESTAR or serve as a visual model for humans. It will serve as information storage, e.g., when a specific part of the abstract model requires deeper analysis. Each concrete state of this layer will be linked to an abstract one in the top layer, and each action will be linked to an abstract transition.

The **bottom layer** is the management layer, and its purpose will be to record meta information about the executed test sequences. Where the abstract and

---

Algorithm 1. ASM_*statemodel*: select an unvisited action

Require: s ▷ The state the SUT is currently in
Require: *State_Model*   ▷ The state model that is being inferred
Require: *path*   ▷ Path towards an unvisited action
1: **if** $path \neq empty$ **then**   ▷ ASM is following a previously determined path
2:     $a \leftarrow path.pop()$ ▷ Selected action is next one on the path
3: **else**   ▷ If the path is empty, we will create a new one to an unvisited action
4:     $reachableStates = getReachableStatesWithBFS(s, State\_Model)$
5:     $unvisitedActions \leftarrow empty$
6:     **while** ($unvisitedActions == empty \ \wedge \ reachableStates \neq empty$) **do**
7:         $s' = reachableStates.pop()$
8:         $unvisitedActions \leftarrow getActions(State\_Model, s', \textbf{unvisited})$
9:     **end while**
10:     **if** $unvisitedActions \neq empty$ **then**   ▷ Unvisited actions found with BFS
11:         $ua \leftarrow selectRandom(unvisitedActions)$ ▷ Randomly select an unvisited
12:         $path \leftarrow pathToAction(ua)$ ▷ Calculate path from s to walk to ua
13:         $a \leftarrow path.pop()$ ▷ Selected action in s is next one on the path to ua
14:     **else**   ▷ No unvisited action found with BFS
15:         $availableActions \leftarrow getActions(State\_Model, s, \textbf{all})$ ▷ Get all actions in s
16:         $a \leftarrow selectRandom(availableActions)$
17:     **end if**
18: **end if**
19: **return** a

---

concrete layers describe the SUT, the management layer describes the execution of the tests in TESTAR. The individual test sequences will be linked to the concrete states and actions of the middle layer.

Figure 4 shows an example of the layered model, where the SUT was extremely simple (only 3 abstract and 3 concrete states). The management layer has information about the exact sequence generated by TESTAR. During the model inference, when TESTAR arrives to a new state and discovers actions that have not been executed before, we use "BlackHole" state as their destination to mark unvisited actions. When a previously unvisited action is visited and TESTAR observes the SUT behaviour, the destination of the executed abstract action is updated with the observed abstract state.

4   Using the Inferred Models for Action Selection

TESTAR was extended with a new ASM (ASM_*statemodel*). The algorithm prioritises actions that have not yet been visited and can be found in Algorithm 1. The goal is to select a new action when in state s. It uses the *State_Model* and maintains a *path* of actions that leads to a specific unvisited action it wants to prioritise. If a *path* has been previously identified (i.e., *path* is not empty, line 1), the ASM just selects the next action on that path. If the *path* is empty, the ASM will try to find an unvisited action. It does so by searching (in BFS order) for *unvisitedActions* (line 8) from all the states that are reachable from s in the state model (line 4). Since s is reachable from s in 0 steps, s itself is the first state that gets checked for unvisited actions (line 7). If unvisited actions are found, it randomly selects one (ua, line 11) and updates the *path* to the state where that action can be found (line 12). Then it selects the first action that leads towards that action (line 13). If no unvisited actions are found, the ASM just randomly selects an action from those available in state s.

Table 1. Java Access Bridge properties and the possible impact of using the attribute for state abstraction in Rachota

Attribute	API	Abstract representation impact
Title	Name	Visual name of the widget. In Rachota this is a dynamic attribute because widgets update the current time
HelpText	Description	Tooltip help text of the widget. In Rachota this attribute is static
ControlType	Role	Role of the widget. Cannot always distinguish different elements, and hence can cause non-determinism
IsEnabled	States	Checks if the widget is enabled or disabled
Boundary	rect	Pixel coordinates of the widgets' position. Even one pixel change would result in a new state, so it was considered too concrete for the experiments
Path	childrenCount + parentIndex	Position in the widget-tree. Useful for differentiating states based on the structure of the widget tree

To measure the performance of the new ASM_*statemodel*, we research the following question: *How do different levels of abstraction affect the automated GUI exploration of ASM_statemodel as compared to random (ASM_random)?*

GUI exploration is measured in terms of code coverage (instruction and branch) and the number of discovered states. Code coverage is measured using JaCoCo [1]. These metrics are collected after each executed action. To be able to measure code coverage, we use open source Rachota [2] as our SUT. It is a Java Swing application for timetracking different projects. It has the following characteristics:

Java Classes	52
Methods	934
LLOC	2722
Classes incl. Inner classes	327

Each test run contains one sequence of 300 actions, which is enough [10] to show the differences between ASMs. To be able to form valid conclusions and to deal with the randomness, we repeat each test run 30 times [9].

To define different abstraction levels, we need to select attributes from the available ones. In Table 1 are the widgets attributes from the Java Access Bridge API that are implemented in TESTAR for Java Swing applications. We investigated 4 levels of abstraction:

1. Abstract: ControlType (cf. was defined in [13])
2. Intermediate: ControlType, Path
3. Dynamic: ControlType, Path, Title (including the dynamic attribute Title)
4. Customised Abstraction: ControlType, Path, HelpText, IsEnabled (this one was customised for Rachota following the impacts described in Table 1)

The results of the code coverage measurements are in Fig. 5. They clearly show that the level of abstraction affects the GUI exploration performance of the ASM_statemodel. Having too high or too low level of abstraction negatively impacts the performance. However, the ASM_statemodel outperformed ASM_random, even with a less suitable abstraction. This means that model-based ASMs are a promising way to improve effectiveness of scriptless testing.

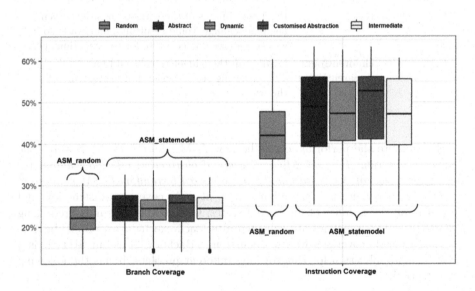

Fig. 5. The code coverage (%) that was reached when comparing ASM_random with 4 different abstraction levels of the ASM_statemodel

To investigate the impact of the changing levels of abstraction used in the experiments on the number of abstract and concrete states created, we run longer test runs of 3000 actions, again run 30 times for each configuration. The results are shown as a box plot in Fig. 6. Results show that too concrete level of abstraction creates almost as many abstract states as concrete states. As expected, the customised level creates more abstract states compared to abstract and intermediate configurations, but significantly less than the dynamic one. The customised level finds most concrete states, which indicates slightly better GUI exploration capability and matches the code coverage results.

5 Defining a Suitable Level of Abstraction

The challenge is to select a suitable subset of widget attributes for state abstraction that does not cause state-explosion. Since non-determinism of the resulting model negatively impacts its usefulness for ASMs, we run experiments to research the following question: *Can we identify widget attributes that, when used in state abstraction, generate deterministic models?*

The SUT used is the desktop application "Notepad", specifically version 1909, OS Build 18363.535. We use Notepad because it is a Windows desktop application and hence we can use the Windows Automation API that gives us over 140 attributes to choose from. That gives more choice compared to the 6 attributes of the Java Bridge from Table 1. For testing, we use the ASM_*statemodel* from Algorithm 1.

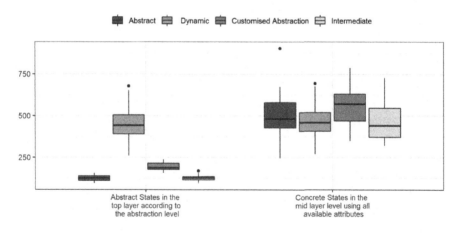

Fig. 6. The number of abstract states (top layer) and concrete states (mid layer)

We have seen in the previous section (and Fig. 6) that widget attributes used for abstraction should not be dynamic because they lead to state space explosion. Dynamic attributes are not stable because they can change their value during, or in between, runs without a detectable reason. Potentially stable attributes selected for our experiment are in the first column of Table 2.

First, we run an experiment with only one attribute in the state abstraction. A test run consisted of 4 sequences, with a maximum of 50 actions per sequence.

Running some initial tests, these values showed to be enough to detect non-determinism. We run 12 consecutive test runs for each widget attribute because we have 12 Virtual Machines (VMs) available. Table 2 shows the results. Displayed are the used widget attributes, the average number of generated test steps executed in the test for each attribute, and the total number of steps taken in each test run, before non-determinism was encountered. The results are ordered by the total number of steps executed over all 12 tests, starting with the widget attributes that "lasted the longest" before the model became non-deterministic. Although none of the models that were generated were deterministic, WidgetTitle, WidgetBoundary and WidgetHasKeyboardFocus attributes noticeably stand out from the other attributes by the average number of steps executed.

Second, we run an experiment with two attributes in the state abstraction. Combining 2 widget attributes, gives 171 possibilities. Instead of doing 4 test sequences of 50 steps each, we upgraded the number of actions per sequence to 100. The reason for the upgrade was the hypothesis that these combinations should last longer before the state model inference module encounters non-determinism. Each combination is tested 16 times, making for a total of

Table 2. Number of generated test steps before the model became non-deterministic using a single widget attribute for state abstraction.

Attribute	Avg	Total
WidgetTitle	95.5	14, 74, 74, 79, 79, 92, 92, 93, 108, 136, 145, 160
WidgetBoundary	90.6	5, 61, 64, 77, 79, 83, 84, 103, 105, 115, 155, 156
WidgetHasKeyboardFocus	82.1	38, 55, 67, 68, 70, 72, 78, 87, 100, 105, 118, 128
WidgetIsKeyboardFocusable	21.6	9, 12, , 14, 16, 17, 17, 19, 20, 26, 28, 28, 53
WidgetSetPosition	20.4	10, 11, 12, 12, 13, 16, 18, 21, 21, 22, 26, 63
WidgetIsContentElement	20.3	7, 9, 14, 15, 17, 17, 18, 21, 24, 27, 35, 39
WidgetIsOffscreen	19.7	6, 9, 11, 14, 14, 15, 15, 18, 19, 27, 29, 59
WidgetGroupLevel	19.1	7, 10, 11, 11, 12, 13, 15, 18, 22, 29, 33, 48
WidgetClassName	19.0	11, 11, 15, 16, 17, 19, 19, 19, 21, 22, 26, 32
WidgetIsControlElement	16.9	8, 11, 11, 12, 13, 13, 16, 16, 20, 24, 28, 31
WidgetIsEnabled	16.8	7, 8, 14, 15, 15, 16, 16, 19, 19, 20, 25, 28
WidgetControlType	16.3	8, 13, 13, 13, 13, 13, 17, 17, 18, 19, 25, 27
WidgetOrientationId	16.2	8, 9, 12, 13, 16, 16, 17, 19, 19, 21, 21, 23
WidgetIsPassword	15.8	6, 10, 10, 12, 14, 14, 17, 17, 19, 20, 22, 28
WidgetZIndex	15.6	9, 11, 12, 12, 13, 14, 15, 16, 16, 19, 22, 28
WidgetIsPeripheral	15.4	7, 8, 9, 13, 14, 14, 16, 19, 20, 21, 22, 22
WidgetSetSize	15.0	7, 9, 10, 12, 14, 15, 16, 17, 17, 18, 21, 24
WidgetFrameworkId	15.0	8, 11, 12, 13, 13, 14, 15, 16, 17, 18, 21, 22
WidgetRotation	14.7	8, 9, 12, 12, 13, 14, 16, 16, 16, 20, 20, 20

2736 test runs. In summary, none of the 171 combinations was able to produce a deterministic model. Moreover, after the 48 best performing combinations, the average number of steps executed per test run declines quickly. Within these 48 combinations, we find that there are 3 attributes that occur 17 times, where as the next best attribute occurs only 3 times. The 3 best performing attributes are again: WidgetTitle, WidgetBoundary and WidgetHasKeyboardFocus.

For our next experiment, we used the combination of these 3 and run this combination 16 times. Unfortunately, none of the test runs made it to the full 400 possible steps. Moreover, the average and median are lower than the highest results from the 2 attribute combinations.

In our next experiment, we tried using combinations of 5 attributes, selecting the 3 highest scoring attributes from the 2 attribute experiment again, and then adding on all the possible combinations of 2 widget attributes from the other 16 attributes. This gives us a total of 120 combinations and we run each of them 8 times. Again, no combination resulted in a deterministic model. Surprisingly, the more concrete abstraction using five attributes resulted non-determinism faster than three attributes in the previous experiment. This is probably due to dynamic nature of some of the attributes.

As the use of 5 attributes for abstraction also resulted in non-determinism, we opt to make the model even more concrete by incorporating all 32 of the control pattern properties into our tests. To make some headway, we will take the 3 high scoring general properties again (WidgetTitle, WidgetHasKeyBoardFocus and WidgetBoundary), and combine them with all the combinations of 2 control patterns. This results in 492 possible combinations, and running each one 8 times makes a total of 3936 test runs.

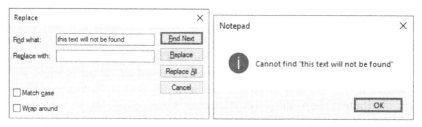

(a) Notepad 'Replace' dialog   (b) Notepad 'Cannot find' popup

Fig. 7. Notepad examples of non-determinism

Several widget combinations were able to reach the limit of 400 sequence actions without encountering non-determinism in the model, and all of these combinations included the 'Value' control pattern. Even though some combinations made it to 400 sequence steps 3 or 4 times out of the 8 test runs, they also encountered certain actions that led to non-determinism in the model. 'Value' pattern is a very 'concrete' attribute: 1) Using the 'Value' pattern can lead to models of infinite size, in the case that the application accepts text input that is not bounded. Hence, ideally, we would like to not use it in our abstraction mechanism. 2) Even while using this very concrete widget attribute, we still encountered plenty of non-determinism in the state models.

Analysing the reasons for non-determinism, we found actions that lead to different states depending on the history of actions and states traversed before. For example, in Notepad, if the 'Replace' option in the 'Edit' menu is clicked, the 'Replace' dialog is opened (see Fig. 7a). If the text written in the 'Find what' field is not found in the Notepad document, clicking 'Find Next' or 'Replace' buttons will result in the same popup dialog (see Fig. 7b), having only an 'OK' button. Clicking that button will lead back to 'Replace' dialog, but the focus remains on the button that was pressed before, and if WidgetHasKeyBoardFocus was used in the state abstraction, clicking the 'OK' button leads to 2 different states based on the action that was taken in the previous state. In this case, altering the abstraction level by adding more widget attributes would not remove the non-determinism, because the concrete states for the 2 visitations of the popup screen are also the same.

5.1   Including the Predecessor State

Another solution we can try is to incorporate the state's incoming action into our state identifier [6]. In some situations, the state could depend on the previous state, requiring taking the previous state into account in the state identification algorithm. Consequently, we decided to include the predecessor state and the incoming action in the state abstraction.

The first experiments run with all the combinations of widget attributes used in the experiments from Sect. 5 including the previous state identifier. Non-determinism related to viewing the status bar was still happening.

Subsequently, we included the incoming action in addition to the previous state in the abstraction identifier. Using the same attributes as the experiments in Sect. 5 allowed for comparison of the results. The average number of steps executed before encountering non-determinism increased significantly when using 1 or 2 widget attributes and including the incoming action. However, with more widget attributes, the results seemed to get worse, probably because in those experiments the widget attributes were selected for their good performance without incoming action. With incoming action, the best performing widget attributes were different. When executing the experiments including pattern attributes with incoming action, the combination of the ValuePattern and the 'incoming action' seems very successful. In fact, the following 3 combinations did not encounter any non-determinism during their 8 test runs of 400 actions:

1. Boundary, HasKeyboardFocus, LegacyIAccessiblePattern, Title, ValuePattern;
2. Boundary, DropTargetPattern, HasKeyboardFocus, Title, ValuePattern;
3. Boundary, ExpandCollapsePattern, HasKeyboardFocus, Title, ValuePattern.

As some of the detected cases of non-determinism were due to various length of text inputs, we run an additional experiment disabling input actions, only allowing left click actions. We found that the average number of executed steps before encountering non-determinism was increased. However, the model may be partial, and some functionality of the SUT may be excluded from the model.

After running more experiments on trying to produce a deterministic model, we had to conclude that it is not a trivial effort. Also, the quest for inferring a deterministic model by making the abstraction more concrete resulted with huge increase in the number of abstract states. We discuss the implications on possible future research directions in Sect. 6.

6   Findings, Challenges and Future Research Directions

6.1   State Abstraction

Our first finding is that tuning the abstraction level for model inference seems to be highly dependent on the specific SUT. While tuning the abstraction level, the following SUT-specific characteristics should be taken into account:

- **Dynamic increment of widgets**: In some applications, for example Rachota, we can find dynamic lists of elements on which we can constantly add new items. This constantly creates new widgets and states in the model causing a state and action explosion.
- **High number of combinatorial elements**: In some applications, for example Notepad, we can find multiple scroll lists with a large number of different elements, and from a functional point of view it is not important to cover all of these options (for example, Notepad Font selection).
- **Slide actions**: In some applications where scrolling actions are required, the exact scrolling coordinates from start to end can cause a change in the number of widgets visible in the state. Depending on the state abstraction and how the widget tree is obtained, this can create new states and cause a high number of combinatorial possibilities.
- **Popup information**: In some applications, for example Rachota, a descriptive popup message may appear for a few seconds in the GUI when the mouse is hovered over some of the widgets. This could result in a new state for the model, and it might cause non-determinism, if the hovering over a widget was not an intentional action that was executed on purpose.

A second finding is that trying to produce a deterministic state model is far from a trivial effort. There are various options to address the challenges of non-determinism in the inferred models. 1) The first could be to let the models have non-determinism and deal with it when using them. For action selection this means that we have to be able to detect when the modelled behaviour differs from the observed behaviour, and temporarily change the action selection to prevent getting stuck during GUI exploration. Another solution, and an interesting future research topic, could be using the concrete state model to navigate through states that have non-determinism in the abstract state model. 2) The second option would be to try to infer a deterministic model. This would require more SUT-specific ways to dynamically adjust the abstraction, for example based on the type of the widget, or even based on a specific widget in a specific state. TESTAR currently allows triggered behaviour that overrides normal action selection, so we plan to implement a similar way to trigger change in the calculation of state identifiers, for example ignoring a specific widget during the state abstraction. An example of a widget to be ignored from the state model could be a dynamically changing advertisement in a website. 3) A third option could be to correct non-deterministic models after run-time. Nevertheless, we have not seen this kind of technique used in a model-based testing tool yet.

Other interesting future research directions include automatically adjusting the level of abstraction, analysing the screenshots in addition to the attributes of the widget tree, and/or visualising the results of state abstraction for the user and learning from the user input to find a suitable level of abstraction.

6.2   Applying the Inferred Models in Testing

One of the core objectives for this work was to use the inferred models for a new action selection mechanism (ASM) for TESTAR. The new ASM was presented in Algorithm 1 and initial experiments show that it is better than random. Although this is a good result by itself, it is also a step towards implementing more advanced ASMs. For example ASMs based on reinforcement learning (RL) and artificial intelligence (AI) need some kind of model for learning, and the inferred model can serve that purpose.

Another advantage of the inferred state models is that human testers can use them during testing. For instance, it is interesting to have an overall view of an application's execution flow: to see the details about a certain state or executed action; to identify the path to a state where an application failed; and to obtain various metrics about the state model. Although some of this information can be obtained by querying the OrientDB database and outputting it as textual data, e.g., in tabular format, we advocate that it would be best realised by visualising the data in a way that is more intuitively understandable for humans.

Abstract state models can also allow performing conformance testing. Conformance testing is used to determine how a system under test conforms to meet the individual requirements of a particular standard. Before using inferred abstract models, the domain experts have to validate them in order to use the automatically generated test cases for conformance testing. This also requires suitable visualisation.

Another future research topic is using the inferred models for automatically finding a shortest path to reproduce a failure. This requires recognising whether a found failure is a new one or duplicate of the one we try to reproduce. Finally, the inferred models can be used for automated change detection between consequent versions of the same SUT. This kind of functionality has been evaluated with Murphy tools [6], and it is interesting as future research.

7   Conclusions

This paper describes the state model inference extension for TESTAR and reports experiments on the impact of various state abstraction mechanisms for the purpose of producing a deterministic model, and on the evaluation of the performance of an action selection algorithm using the inferred models.

The experiments on using various state abstraction mechanisms show that inferring a deterministic abstract state model is difficult, especially when trying to prevent the state space explosion. Based on our experiences, and the fact that in the literature many approaches using inferred models for GUI exploration or testing do not explain the details about state abstraction, more research and new more flexible abstraction mechanisms are needed. Also, dealing with the non-determinism in the inferred models is an important future research direction.

Based on the experiments on the impact of various levels of state abstraction for the performance of an ASM using the inferred models, we can conclude that having a suitable level of abstraction improves the performance of GUI exploration measured in code coverage. Having a too abstract or too concrete model has a negative impact on the performance. However, in our experiments, the ASM_*statemodel* performed better than the ASM_*random* with all tested abstractions.

Finally, as an immediate future work, we plan to conduct experiments on SUTs from industry in order to demonstrate the effectiveness, efficiency and usability of the TESTAR tool with the inferred models proposed on this work.

Acknowledgements. This work has been funded by ITEA3 TESTOMAT and IVVES, and H2020 DECODER and iv4XR projects. Thanks to: Arjan Hommerson, Stijn de Gouw, Stefano Schivo.

References

1. JaCoCo coverage tool. https://www.jacoco.org/jacoco/. Accessed 17 Jan 2022
2. Rachota Timetracker. http://rachota.sourceforge.net. Accessed 17 Jan 2022
3. Aho, P., Alégroth, E., Oliveira, R.A., Vos, T.E.: Evolution of automated regression testing of software systems through the graphical user interface. In: 1st International Conference on Advances in Computation, Communications and Services, pp. 16–21 (2016)
4. Aho, P., Kanstrén, T., Räty, T., Röning, J.: Automated extraction of GUI models for testing. In: Advances in Computers, vol. 95, pp. 49–112. Elsevier (2014)
5. Aho, P., Räty, T., Menz, N.: Dynamic reverse engineering of GUI models for testing. In: 2013 International Conference on Control, Decision and Information Technologies (CoDIT), pp. 441–447 (2013)
6. Aho, P., Suarez, M., Kanstren, T., Memon, A.M.: Murphy tools: utilizing extracted GUI models for industrial software testing. In: 7th ICSTW, pp. 343–348 (2014)
7. Aho, P., Suarez, M., Memon, A., Kanstrén, T.: Making GUI testing practical: bridging the gaps. In: 2015 12th International Conference on Information Technology-New Generations, pp. 439–444. IEEE (2015)

8. Amalfitano, D., Fasolino, A.R., Tramontana, P., Amatucci, N.: Considering context events in event-based testing of mobile applications. In: 6th ICSTW, pp. 126–133 (2013)

9. Arcuri, A., Briand, L.: A practical guide for using statistical tests to assess randomized algorithms in software engineering. In: 33rd ICSE, pp. 1–10. ACM (2011)

10. van der Brugge, A., Ricos, F.P., Aho, P., Marín, B., Vos, T.E.: Evaluating TESTAR's effectiveness through code coverage. In: Abrahão Gonzales, S. (ed.) XXV JISBD. SISTEDES (2021)

11. Canny, A., Palanque, P., Navarre, D.: Model-based testing of GUI applications featuring dynamic instanciation of widgets. In: 2020 IEEE International Conference on Software Testing, Verification and Validation Workshops (ICSTW), pp. 95–104 (2020). https://doi.org/10.1109/ICSTW50294.2020.00029

12. Couto, R., Ribeiro, A.N., Campos, J.C.: A patterns based reverse engineering approach for Java source code. In: 35th IEEE Software Engineering Workshop, pp. 140–147 (2012)

13. de Gier, F., Kager, D., de Gouw, S., Tanja Vos, E.: Offline oracles for accessibility evaluation with the TESTAR tool. In: 13th RCIS, pp. 1–12 (2019)

14. Grechanik, M., Xie, Q., Fu, C.: Creating GUI testing tools using accessibility technologies. In: 2009 International Conference on Software Testing, Verification, and Validation Workshops, pp. 243–250. IEEE (2009)

15. Grilo, A.M., Paiva, A.C., Faria, J.P.: Reverse engineering of GUI models for testing. In: 5th ICIST, pp. 1–6. IEEE (2010)

16. Kervinen, A., Maunumaa, M., Pääkkönen, T., Katara, M.: Model-based testing through a GUI. In: Grieskamp, W., Weise, C. (eds.) FATES 2005. LNCS, vol. 3997, pp. 16–31. Springer, Heidelberg (2006). https://doi.org/10.1007/11759744_2

17. Kull, A.: Automatic GUI model generation: state of the art. In: 2012 IEEE 23rd ISSRE Workshops, pp. 207–212. IEEE (2012)

18. Marchetto, A., Tonella, P., Ricca, F.: State-based testing of Ajax web applications. In: 2008 1st ICST, pp. 121–130 (2008)

19. Meinke, K., Walkinshaw, N.: Model-based testing and model inference. In: Margaria, T., Steffen, B. (eds.) ISoLA 2012. LNCS, vol. 7609, pp. 440–443. Springer, Heidelberg (2012). https://doi.org/10.1007/978-3-642-34026-0_32

20. Memon, A.M., Banerjee, I., Nagarajan, A.: GUI ripping: reverse engineering of graphical user interfaces for testing. In: 10th WCRE, pp. 260–269 (2003)

21. Mesbah, A., van Deursen, A., Lenselink, S.: Crawling Ajax-based web applications through dynamic analysis of user interface state changes. ACM Trans. Web 6(1), 1–30 (2012)

22. Miao, Y., Yang, X.: An FSM based GUI test automation model. In: 11th International Conference on Control Automation Robotics Vision, pp. 120–126 (2010)

23. Nedyalkova, S., Bernardino, J.: Open source capture and replay tools comparison. In: Proceedings of the International C* Conference on Computer Science and Software Engineering, pp. 117–119 (2013)

24. Nguyen, B.N., Robbins, B., Banerjee, I., Memon, A.: GUITAR: an innovative tool for automated testing of GUI-driven software. Autom. Softw. Eng. 21(1), 65–105 (2013). https://doi.org/10.1007/s10515-013-0128-9

25. Rafi, D.M., Moses, K.R.K., Petersen, K., Mäntylä, M.V.: Benefits and limitations of automated software testing: systematic literature review and practitioner survey. In: 2012 7th International Workshop on Automation of Software Test (AST), pp. 36–42. IEEE (2012)

26. Silva, J.A.C., Silva, C., Gonçalo, R.D., Saraiva, J.A., Campos, J.C.: The GUISurfer tool: towards a language independent approach to reverse engineering GUI code. In: Proceedings of the 2nd ACM SIGCHI Symposium on Engineering Interactive Computing Systems, pp. 181–186. ACM (2010)
27. Vos, T.E.J., Aho, P., Pastor Ricos, F., Rodriguez-Valdes, O., Mulders, A.: TESTAR - scriptless testing through graphical user interface. STVR **31**(3), e1771 (2021)

Machine Learning Applications

Autonomous Object Detection Using a UAV Platform in the Maritime Environment

Emmanuel Vasilopoulos$^{(\boxtimes)}$, Georgios Vosinakis⬤, Maria Krommyda, Lazaros Karagiannidis, Eleftherios Ouzounoglou, and Angelos Amditis

Institute of Communication and Computer Systems (ICCS), 15773 Athens, Greece
emmanouil.vasilopoulos@iccs.gr

Abstract. Maritime operations vary greatly in character and requirements, ranging from shipping operations to search and rescue and safety operations. Maritime operations rely heavily on surveillance and require reliable and timely data that can inform decisions and planning. Critical information in such cases includes the exact location of objects in the water, such as vessels, persons and others. Due to the unique characteristics of the maritime environment the location of even inert objects changes through time, depending on weather conditions, water currents etc. Unmanned aerial vehicles (UAV) can be used to support maritime operations by providing live video streams and images from the area of operations. Machine learning algorithms can be developed, trained and used to automatically detect and track objects of specific types and characteristics. Within the context of the EFFECTOR project we developed and present here an embedded system that employs machine learning algorithms, allowing a UAV to autonomously detect objects in the water and keep track of their changing position through time. The system is meant to supplement search and rescue, as well as maritime safety operations where a report of an object in the water needs to be verified with the object detected and tracked, providing a live video stream to support decision making.

Keywords: Maritime · UAV · UxV · Object detection · Object tracking · Machine learning · Situational awareness · Computer vision

1 Introduction

Maritime operations cover a wide variety of scenarios, including search and rescue missions and shipping management. With the increase in marine traffic and evolving maritime climate, shipping management and maritime travel safety has become a high priority issue [17]. The distinct characteristics and ever changing nature of the maritime environment present unique challenges in surveillance. These include objects drifting and changing location due to the effects of wind

© The Author(s), under exclusive license to Springer Nature Switzerland AG 2022
R. Guizzardi et al. (Eds.): RCIS 2022, LNBIP 446, pp. 567–579, 2022.
https://doi.org/10.1007/978-3-031-05760-1_33

and water currents, their position and shape altering due to rolling and pitching caused by waves and wind, while changing weather conditions can radically affect the local situation and outlook [17].

In this elusive environment timely and precise information is of utmost importance to the success of any mission. Automatic tracking systems and computer vision have been described and utilised in the detection and tracking of several types of objects in the water, from vessels to icebergs, from several platforms [20].

In the maritime field, computer vision and object detection is already being utilized in security and rescue operations [9,15], mainly running on terrestrial modules or modules carried on ships. This data can be incorporated in decision making algorithms that can increase the efficiency of utilized assets [4].

Such systems can be used in border control as well as search and rescue operations, especially since the two fields frequently overlap dynamically due to changing conditions in the maritime field. UAVs are being utilised to gather intelligence in a timely manner [5,11].

UAV systems have started to be used in terrestrial surveillance, where the video is streamed to a control station running detection and tracking through computer vision [12].

Within the context of the EFFECTOR project [3], the consortium is developing an Interoperability Framework and associated data fusion and analytics services for maritime surveillance. The aim of the project is the faster detection of new events, the enhancement of decision making and the achievement of a joint understanding and undertaking of a situation by supporting a seamless cooperation between the operating authorities and on-site intervention forces, through a secure and privacy protected network.

In our setup we present an autonomous UAV object detection and tracking solution for the maritime environment. A main advantage of our system is that all computation is taking place on the on-board the AI unit of the UAV. The UAV system we present is meant to support two distinct types of maritime operations, shipping management and search and rescue.

1.1 Contribution

Object detection algorithms based on deep learning have been trained and run on datasets gathered using UAVs [6,13]. In our system we chose to run the detection and tracking algorithm embedded in a processing unit carried on the UAV, instead of the ground control station where the video streams are sent. In this way the streamed video from the UAV already contains the bounding boxes drawn by the object detector and tracker, while a separate stream of data from the UAV transmits information on detected targets and their assigned unique IDs in JSON format.

This was done in line with an edge processing approach to reduce the required bandwidth, improve response time, and avoid delays caused by video encoding and streaming, as well as inevitable interruptions and cuts in the video stream caused by connection issues, obstacles, and weather conditions.

Since video streaming has high bandwidth requirements, any decision to reduce the streamed video resolution in case of a bad connection will not affect detection and tracking, as the detector directly receives the video feed from the UAV camera. Even if the video feed is interrupted, detection and tracking data is still transmitted via the JSON messages. If video streaming is interrupted the IDs of tracked targets are not lost. Another important aspect of this decision is the reduction of the load to the ground station, guaranteeing the solutions expandability, as adding UAVs will not greatly affect the processing load at the ground station.

2 System Design

2.1 User Requirements

As part of the EFFECTOR project we gathered user requirements from a wide variety of maritime stakeholders, including organisations and institutions involved in all categories of maritime operations.

In such operations a UAV system would be mainly utilised once an event has been detected to provide live information of the evolving incident. The main requirements of a UAV in supporting such operations were identified as:

Detection. The system will be used to identify or verify a report of an incident in the area concerned. The object detection capability allows rapid and efficient scanning of a wider area and the identification of objects of interest.

Tracking. Once a vessel of interest has been detected, live, uninterrupted information on its situation is of utmost importance in the evolution of all operations.

Timeliness. Due to the fact that conditions at seas alter quickly, detection and tracking algorithms must provide live information, or as close to live as possible with minimal delays and no interruption.

Fast Deployment. The system should be easy and fast to deploy, so that live information and tracking of a target can start as soon as possible after the incident is identified. Both in the search and rescue and maritime safety fields the initial time after the identification of an event are critical, due to their rapidly evolving character.

Interoperability. Given the abundance of legacy systems used in maritime operations and communications, any new system needs to ensure seamless interoperability with existing systems. Specifically in the EU space the Common Information Sharing Environment (CISE) [1] is being used as the basis for information exchanged between public authorities. Any messages or communications generated by the system should be compatible with CISE.

Independent and Self Sufficient Operation. The system must be able to operate independently and be self sufficient for the duration of the operation.

Cost Effectiveness. An on-board UAV surveillance system is inherently cost effective by replacing more expensive and critical airborne assets. By providing timely information of an object's location through a UAV using detection and tracking algorithms, operators will be able to more efficiently utilize all assets involved in the operation (including vessels).

2.2 Hardware Components

The entire object detection and tracking stack that will be described runs on a Jetson AGX Xavier module (see Fig. 1). This embedded processing unit is lightweight and has low energy requirements. It will compress video using dedicated hardware and perform all necessary computation. The complete technical specifications of the module are presented in Table 1. The aim is for the computing module to be carried by a UAV and directly connect to its camera feed, making the system completely autonomous.

Fig. 1. The Jetson AGX Xavier module

We run all tests with the module attached to a specially built octacopter UAV (see Fig. 2). It is equipped with a pair of daylight and thermal cameras providing a video stream for the UAV's pilot, enabling Extended Visual Line-of-Sight (EVLOS) flight capability. To provide extended capabilities of object detection and tracking, a 3-axis stabilized gimbal equipped with powerful RGB and thermal cameras is fitted under the UAV and connected to the processing unit for onboard processing. The results of the onboard processing are overlaid on the Full HD video stream and transmitted to the Intelligence Officer's workstation via a high bandwidth 2.4 GHz radio.

Fig. 2. The octacopter UAV

3   System Architecture

We designed a sequential system composed of state-of-the-art architectures, both in object detection and multiple object tracking. When an object appears in the frame of the camera, the model's purpose is to detect that object, assign an ID and keep track of its trajectory without switching IDs. Multiple object tracking (MOT) can be implemented in many ways, with the most usual being tracking-by-detection. Other methods, such as the ones provided by Open CV's Tracking API are not optimized across the tracked objects, making them unsuitable to be deployed, mainly due to their run time complexity with respect to the number of tracked objects. Such algorithms either look at neighboring pixels for every

Table 1. Jetson AGX Xavier technical specifications.

GPU	512-core Volta GPU with Tensor Cores
CPU	8-core ARM v8.2 64-bit CPU, 8 MB L2 + 4MB L3
Memory	32 GB 256-Bit LPDDR4x — 137 GB/s
Storage	32 GB eMMC 5.1
DL accelerator	(2×) NVDLA Engines
Vision accelerator	7-way VLIW Vision Processor
Encoder/Decoder	(2x) 4Kp60 — HEVC/(2x) 4Kp60 — 12-Bit Support
Size	105 mm × 105 mm × 65 mm
Deployment	Module (Jetson AGX Xavier)

tracked object making their complexity directly analogous to the number of objects tracked or have a very low accuracy. For example zooming at an object will not adjust the bounding boxes accordingly.

Considering the method used for object tracking, detection-free-tracking and tracking-by-detection were explored. Detection-free-tracking requires bounding boxes of the objects that will be tracked. Then, the tracking algorithm processes each consecutive frame and outputs their bounding boxes. In other words, tracking-by-detection algorithms processes the detections for each consecutive frame, while detection-free-tracking algorithms require one batch of the initial object detections (manual or automatic) and output the trajectories of the objects.

In detection-by-tracking, an object detection system outputs detection of all objects on a given frame. This process is repeated for every consecutive frame. New objects are detected and disappearing objects are not, automatically. It is the tracker's task to succeed where the detector fails to keep track of objects that are occluded or appear and disappear in a given frame.

The improvement, in terms of speed and accuracy, of object detection algorithms, as stated in [18], inspired an approach towards a tracking-by-detection algorithm. Tracking by detections also allows us to provide an autonomous system that does need a manual selection of targets to be tracked.

For the task of object detection, we used YOLOv5, a state-of-the-art object detector [21]. When YOLOv4 (the earlier version of the YOLO detectors) was first introduced, it was tested against EfficientDet, the state-of-the-art object detection model at that time [19]. The test was run on a V100 GPU and YOLOv4 ended up achieving the same accuracy as EfficientDet but at double the frames per second (FPS). The YOLOv5 version we opted for has a performance similar to YOLOv4, but it utilizes PyTorch instead of Darknet, greatly simplifying the process of training, testing and deployment [14].

The detection output is passed to the tracker, i.e. Deep SORT, taking into account the previous frame's output. This tracking algorithm's run time complexity is not perceivably affected by the number of objects tracked since it only processes the position and velocity of tracked objects while the output of the neural network is independent of the number of objects, greatly reducing the number of parameters that need to be processed. The overall architecture is shown in Fig. 3.

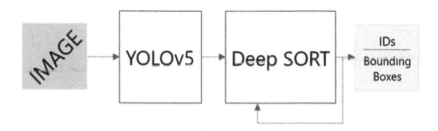

Fig. 3. The general architecture showing the distinct components of the system

3.1   Object Detection - YOLOv5

Network Architecture. YOLOv5 is a high-precision, fast-inference convolutional neural network (CNN) from the YOLO family of object detectors. Its performance is mostly related to its residual connections. The backbone and the head, as shown on Fig. 4, are its two main components.

The backbone works as a feature extractor. Starting with simple convolutional layers, features from three different stages of the component will be used on other layers of the head. This reduces the computation making the network run faster and improves accuracy. The network also uses Cross Stage Partial [23] layers (C3) to further improve efficiency in intermediate levels. Finally, the Spatial Pyramid Pooling layer (SPP) enhances the receptive field [8]. In other words, the feature produced at the end of this layer contains more information from the input.

The head is also created from convolutional layers and C3s, while using features from different stages of the backbone. Other residual connections end up at the last Detect layer, mixing up information from different scales enabling the network to more easily detect objects of various sizes.

After calculating the output, the results are only considered candidates for a detection at this stage and not actual detections. The final bounding boxes and their corresponding classes are computed by the non-maximum suppression (NMS), a post-processing algorithm responsible for merging all detections that belong to the same object [10].

The advantage of this architecture over the previous generations is its variations in sizes. The depth of the network as well as the width are configurable in pair. Four options are available: Small, Medium, Large and Extra Large.

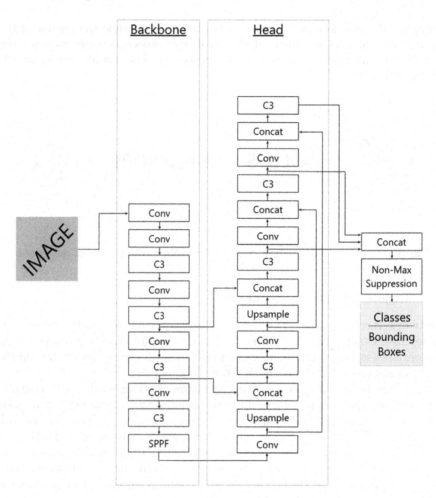

Fig. 4. YOLOv5 architecture

Each size refers to the scale of the model. The 'bigger' the size, the more the parameters of the network. More parameters means larger inference time, but better accuracy as shown in Fig. 5. The size option allows the users to deploy the model on a variety of hardware, depending on the use case. In the context of this use case, the small version of the model is trained. During inference on the Jetson AGX, the model processes a frame at 70 ms.

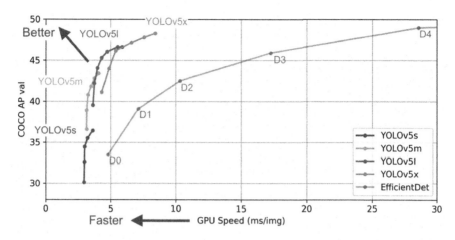

Fig. 5. YOLOv5 evaluation (source [21])

Dataset. This deep learning model calculates the positions of all the objects that appear inside the frame, while classifying the object detected. The range of sizes included in a dataset used to train this neural network is important. A dataset with only small objects or only large objects will result in corresponding inference. The model, in our use case, is trained to detect any size of objects by mixing the popular COCO [2] dataset with one that includes maritime environment images, i.e. SEAGULL [16,22]. The former is a large-scale object detection dataset containing annotated images with bounding boxes and the corresponding classes. The latter is a multi-camera video dataset for research on sea monitoring and surveillance. The videos are recorded from the point of view of a fixed wing UAV flying above the sea, over floating objects. After extracting images from the videos of the SEAGULL dataset, the two datasets were united to improve the model performance on maritime environments.

3.2 Object Tracking - Deep SORT

Multiple object tracking is the problem of keeping track of all objects of interest existing inside a frame. Each object is assigned an ID used to identify each object in the next frames. The difference with detection is that tracking stores the information of the previous time step and uses it for the current time step.

Simple online and realtime tracking (SORT) is a tracking-by-detection algorithm. It combines Kalman filtering with the Hungarian method [7]. Deep SORT refers to the addition of appearance information to the existing SORT algorithm [24]. To achieve online tracking, detection from the current and the previous frame are presented to the tracker.

Fig. 6. Detected boat with bounding box.

For every target detected the tracker keeps a state, comprising its position and velocity. The detected bounding box associated with a target updates the position of the state and a Kalman filter solves optimally its velocity components.

After calculating the detections of a frame from YOLOv5, these detections are assigned to existing targets. The detected bounding boxes (see Fig. 6) are compared to the bounding boxes of the existing targets as predicted by the tracker with the intersection-over-union (IOU) distance. The assignment is solved optimally using the Hungarian algorithm. A minimum threshold IOU rejects assignments where:

$$IOU < IOU_{min} \tag{1}$$

To get appearance information, a CNN with residual connections is trained on a re-identification dataset and its final classification layer is removed. The output of this network is a feature map, integrated into the assignment problem. The smallest cosine distance between the i-th track and j-th detection in appearance space is calculated.

Motion information is integrated with the Mahalanobis distance between predicted Kalman states and newly arrived measurements.

The two metrics are combined, and a hyperparameter adjusts the favorable metric over the final results. The first is useful for long-lasting occlusions, while the second for short-term predictions. A matching cascade algorithm makes the final associations using these metrics.

On this architecture, a model that has been pre-trained on a re-identification dataset is used directly, without re-training the model or training from scratch, with very few identity switches.

As stated in [7], the performance of the tracker is directly improved by the performance of the detector. Using a state-of-the-art object detector showed great results in both accuracy and total running time. The running time of the tracker is highly dependent on the CNN architecture. A smaller pre-trained network resulted in reduced inference time, with still very few identity switches.

3.3 CISE Compatibility

The stack includes a set of adapters, written in the Python programming language that ensure compatibility with the CISE data model. When a vessel is detected or tracked, relevant information is encoded to VESSEL entities within the CISE data model and can be transmitted directly. Directions sent to the UAV pilot by the command centre of the operation, as well as acknowledgement or modification of the directions by the pilot will adhere to the TASK entity of CISE, while the UAV will transmit its status information in the AIRCRAFT entity of the eCISE (extended CISE) data model.

3.4 Initial Testing

We ran a series of initial tests on the described architecture on the Jetson AGX described above, using a reserved part of our combined dataset (COCO and SEAGULL) that was not used in training the model.

Images from 22 videos of the SEAGULL dataset were extracted to train the object detector. The dataset consisted of around 50,000 annotated images. At first the dataset was split to train-set and test-set with a ratio of 70% and 30% correspondingly, in order to evaluate YOLOv5. After showing a COCO mean Average Precision or mAP: 0.5–0.95 greater than 0.30 after 50 epochs, the same dataset was split again with a ratio of 90% and 10%. In addition, to successfully detect a larger range of object sizes and classes, the COCO dataset was added. The deployed model was trained for 300 epochs, resulting in a COCO mAP = 0.40.

A forward pass of a small YOLOv5 lasts 70+-5 ms, when Deep SORT's neural network and the rest of the algorithm adds another 10 to 20 ms. The objects are mostly classified with a confidence greater than 80%, by the detector. The objects are correctly tracked with a minimum feed of 12 FPS.

It is important to note that this applies for a pre-trained deep SORT CNN, showing potential for fewer FPS if trained with a use-case related dataset.

4   Conclusions

The system we propose offers operational awareness by utilising a UAV with on-board detection and tracking algorithms specifically trained for the maritime environment.

During the initial stages of design we discussed with operators with experience in the field and reviewed available technologies both in the field of maritime operations as well as object detection and computer vision.

Providing an autonomous system with all computation components embedded in the UAV without limiting its operational parameters was a main concern. The system can be deployed with minimal preparation and start providing live data on reaching the area of operations without the need for any external support components.

The system is designed to be fully interoperable and compatible with the CISE information exchange environment, utilizing embedded adapters that can directly receive and provide all related information in the CISE data model.

Our solution will be extensively tested and validated in several real-time scenarios within the remit of the EFFECTOR project. Future testing will incorporate extended and new datasets, live tests in the field, as well as comparative analysis to validate the architecture's performance.

Funding Information. This work is a part of the EFFECTOR project, funded by the European Union's Horizon 2020 Research and Innovation Programme under Grant Agreement No 883374. Content reflects only the authors' view.

References

1. CISE homepage (2020). https://ec.europa.eu/oceans-and-fisheries/ocean/blue-economy/other-sectors/common-information-sharing-environment-cise_en. Accessed 18 Jan 2022
2. COCO dataset homepage (2020). https://cocodataset.org/#home. Accessed 18 Jan 2022
3. Effector project homepage (2020). https://www.effector-project.eu/. Accessed 18 Jan 2022
4. Agbissoh OTOTE, D., et al.: A decision-making algorithm for maritime search and rescue plan. Sustainability **11**(7), 2084 (2019)
5. Bauk, S., Kapidani, N., Sousa, L., Lukšić, Ž., Spuža, A.: Advantages and disadvantages of some unmanned aerial vehicles deployed in maritime surveillance. In: Maritime Transport VIII: proceedings of the 8th International Conference on Maritime Transport: Technology, Innovation and Research: Maritime Transport 2020, p. 91. Departament de Ciència i Enginyeria, Universitat Politècnica de Catalunya (2020)

6. Bazi, Y., Melgani, F.: Convolutional SVM networks for object detection in UAV imagery. IEEE Trans. Geosci. Remote Sens. **56**(6), 3107–3118 (2018)
7. Bewley, A., Ge, Z., Ott, L., Ramos, F., Upcroft, B.: Simple online and realtime tracking. In: 2016 IEEE International Conference on Image Processing (ICIP), pp. 3464–3468. IEEE (2016)
8. Bochkovskiy, A., Wang, C.Y., Liao, H.Y.M.: YOLOv4: optimal speed and accuracy of object detection. arXiv preprint arXiv:2004.10934 (2020)
9. Bürkle, A., Essendorfer, B.: Maritime surveillance with integrated systems. In: 2010 International WaterSide Security Conference, pp. 1–8. IEEE (2010)
10. Hosang, J., Benenson, R., Schiele, B.: Learning non-maximum suppression. In: Proceedings of the IEEE Conference on Computer Vision and Pattern Recognition, pp. 4507–4515 (2017)
11. Klein, N.: Maritime autonomous vehicles and international laws on boat migration: lessons from the use of drones in the Mediterranean. Mar. Policy **127**, 104447 (2021)
12. Maltezos, E., et al.: The INUS platform: a modular solution for object detection and tracking from UAVs and terrestrial surveillance assets. Computation **9**(2), 12 (2021)
13. Mittal, P., Singh, R., Sharma, A.: Deep learning-based object detection in low-altitude UAV datasets: a survey. Image Vis. Comput. **104**, 104046 (2020)
14. Nepal, U., Eslamiat, H.: Comparing YOLOv3, YOLOv4 and YOLOv5 for autonomous landing spot detection in faulty UAVs. Sensors **22**(2), 464 (2022)
15. Pelot, R., Akbari, A., Li, L.: Vessel location modeling for maritime search and rescue. In: Eiselt, H.A., Marianov, V. (eds.) Applications of Location Analysis. ISORMS, vol. 232, pp. 369–402. Springer, Cham (2015). https://doi.org/10.1007/978-3-319-20282-2_16
16. Ribeiro, R., Cruz, G., Matos, J., Bernardino, A.: A data set for airborne maritime surveillance environments. IEEE Trans. Circ. Syst. Video Technol. **29**(9), 2720–2732 (2017)
17. Rubin, A., Eiran, E.: Regional maritime security in the eastern Mediterranean: expectations and reality. Int. Aff. **95**(5), 979–997 (2019)
18. Salscheider, N.O.: Object tracking by detection with visual and motion cues. arXiv preprint arXiv:2101.07549 (2021)
19. Tan, M., Pang, R., Le, Q.V.: EfficientDet: scalable and efficient object detection. In: Proceedings of the IEEE/CVF Conference on Computer Vision and Pattern Recognition, pp. 10781–10790 (2020)
20. Tiago, A., Silva, G.: Computer-based identification and tracking of Antarctic icebergs in SAR images. Department Street of Geography, Sheffield (2004)
21. Ultralytics: YOLOv5 repository (2020). https://github.com/ultralytics/yolov5. Accessed 18 Jan 2022
22. VisLab: Seagull dataset homepage (2020). https://vislab.isr.tecnico.ulisboa.pt/seagull-dataset/. Accessed 18 Jan 2022
23. Wang, C.Y., Liao, H.Y.M., Wu, Y.H., Chen, P.Y., Hsieh, J.W., Yeh, I.H.: CSPNet: a new backbone that can enhance learning capability of CNN. In: Proceedings of the IEEE/CVF Conference on Computer Vision and Pattern Recognition Workshops, pp. 390–391 (2020)
24. Wojke, N., Bewley, A., Paulus, D.: Simple online and realtime tracking with a deep association metric. In: 2017 IEEE International Conference on Image Processing (ICIP), pp. 3645–3649. IEEE (2017)

Bosch's Industry 4.0 Advanced Data Analytics: Historical and Predictive Data Integration for Decision Support

João Galvão[1]([⊠]) (ID), Diogo Ribeiro[1] (ID), Inês Machado[1] (ID), Filipa Ferreira[1] (ID),
Júlio Gonçalves[1] (ID), Rui Faria[1] (ID), Guilherme Moreira[2] (ID), Carlos Costa[1] (ID),
Paulo Cortez[1] (ID), and Maribel Yasmina Santos[1] (ID)

[1] ALGORITMI Research Centre, University of Minho, Guimarães, Portugal
{joao.galvao,carlos.costa,pcortez,maribel}@dsi.uminho.pt,
diogo.ribeiro08@gmail.com, {a80365,a78447,a79136,
a78977}@alunos.uminho.pt
[2] Bosch Car Multimedia, Braga, Portugal
guilherme.moreira2@pt.bosch.com

Abstract. Industry 4.0, characterized by the development of automation and data exchanging technologies, has contributed to an increase in the volume of data, generated from various data sources, with great speed and variety. Organizations need to collect, store, process, and analyse this data in order to extract meaningful insights from these vast amounts of data. By overcoming these challenges imposed by what is currently known as Big Data, organizations take a step towards optimizing business processes. This paper proposes a Big Data Analytics architecture as an artefact for the integration of historical data - from the organizational business processes - and predictive data - obtained by the use of Machine Learning models -, providing an advanced data analytics environment for decision support. To support data integration in a Big Data Warehouse, a data modelling method is also proposed. These proposals were implemented and validated with a demonstration case in a multinational organization, Bosch Car Multimedia in Braga. The obtained results highlight the ability to take advantage of large amounts of historical data enhanced with predictions that support complex decision support scenarios.

Keywords: Big Data Warehousing · Advanced analytics · Machine Learning · Industry 4.0

1 Introduction

There is a huge growth in the data that is generated, being a challenge to deal with this rapid growth, as well as with the complexity of the data and its interconnection [1]. Big Data involves a large set of data in which its size and structure are not properly handled by traditional database systems, such as relational databases [2]. These large datasets usually integrate richer data for decision support, with more details about behaviours, activities, and events, providing huge diversity of data and requiring shorter response time [3]. To take full advantage of the strategic potential of the large volume of data provided by organizations, Big Data Analytics is necessary [4]. It includes procedures to extract relevant information from a large volume of data. Its main purpose is to enable organizations to make better decisions according to their mission and objectives. It also helps in solving problems quickly, providing relevant and valuable insights that can bring a competitive advantage to the organization [2].

Machine Learning is a technique for the analysis of Big Data, which consists in detecting relationships and predicting future behaviours through the modelling process based on a large set of data [2]. The use of Machine Learning algorithms with predictive capabilities in Big Data can promote the discovery of new knowledge and bring additional value to organizations [1]. The effective use of data has been increasing competitiveness and economic growth in various industries, including manufacturing [5].

This work is developed under a partnership between Bosch Car Multimediain Braga and Academia, which aims to create new scientific and technological knowledge to achieve the company's competitiveness goals by improving its main industrialization processes. This will allow a fast adaptation of the company to new market demands. Due to the complexity of both areas, Big Data Analytics and Machine Learning, and all the challenges that need to be faced to combine the analysis of huge amounts of historical data with predictions, this paper has as main goal the design, implementation, and evaluation of an advanced data analytics platform for supporting complex decision support environments. A central component in this architecture is its Big Data Warehouse (BDW), the supporting data system, ensuring the integration of all the data (historical and predictive) in a consolidated and coherent way. This architecture and the data modelling method here presented are the main contributions of this work.

This paper is structured as follows. Section 2 summarizes related work in the field, namely the development of BDWs, the adoption of Machine Learning techniques, and their integration in complex decision processes. Section 3 presents the proposed architecture for advanced data analytics, describing its various components and the supporting technologies, and the data modelling method. Section 4 describes the demonstration case, the screwing case, outlining its motivation and purpose. It also presents the supporting data model and the Machine Learning model, how the data integration and data flows were implemented, and some data visualizations for advanced data analytics. Section 5 concludes with some remarks and guidelines for future work.

2 Related Work

Since 1956, with greater emphasis on the last decade, the quantity of data has been growing exponentially and, as such, some challenges have appeared relatively to the storage and analysis of that data [6]. Important contributions in the past years had changed the databases field. Many organizations, such as Facebook, Google, and others, have had a really hard task to analyse an unprecedented amount of data that is not necessarily in a format or structure that makes it easy to analyse [7].

Big Data is mainly related with enormous amounts of unstructured data produced by high-performance applications that can range from computing applications to medical information systems. The data that is stored in this fashion, and its processing, has some specific characteristics and needs, such as: i) large-scale data, which refers to the size of the data repositories; ii) scalability issues, due to the vast amount of data and the performance concerns in its processing; iii) supporting Extraction-Transformation-Loading (ETL) processes, handling the input raw data in order to reach the required structured data; and, iv) analytical environment, designing and developing user-friendly analytical interfaces in order to extract useful knowledge from data [8].

Data-intensive systems are built to consolidate and make available relevant information for decision support. In them, data from different sources require a complex process of data integration that ensures a unified and coherent view of the organizational or application domain data [9]. As challenges such as volume, variety, or velocity emerge, Data Warehouses or other data storage systems require performant solutions able to deal with these data characteristics [10]. Big Data techniques and technologies support mixed and complex analytical workloads (e.g., streaming analysis, ad hoc querying, data visualization, data mining, simulations) in several emerging contexts [11].

Research in Big Data Warehousing [9] has proposed a structured approach for the design and implementation of BDWs, mainly focused in modelling highly autonomous objects, addressing performance issues, that integrate the relevant data to answer a specific analytical question. These objects are named Analytical Objects and can include both factual and predictive attributes. They can be integrated with Complementary Analytical Objects (to share data between several Analytical Objects), Special Objects (to normalize common Date, Time, or Spatial attributes), and Materialized Objects (to physically implement views that enhance performance).

Data Science techniques must be able to extract unknown features from data, to improve the value of the data itself, making it easier to understand behaviours, optimize processes, and improve scientific discovery. Big Data takes advantage of data analytics and Machine Learning, both being key steps for enhancing the value of data [12]. The integration of Machine Learning-based predictive applications in Big Data contexts has been proposed to address the challenges that emerge in complex decision-support environments characterized by a vast amount of data. The work of [13] aims to prevent losses caused by faults in assembly lines with a real-time monitoring system that uses data from IoT-based sensors, Big Data processing, and a hybrid prediction model.

An architecture that can automate and centralize data processing, health assessment, and prognostics is present in [14]. This architecture covers all necessary steps from acquiring data, its processing and presenting it to the users, supporting decision making. The work of [15] presents an architecture that facilitates the task of analysing and extracting value from Big Data using Hadoop-based tools for Machine Learning. This architecture supports batch and streaming processing modules, with Machine Learning tools and algorithms, so that developers take advantage of them to carry out tasks such as prediction, clustering, recommendation, or classification. Also, analytical dashboards present the results of the batch analysis and display them to the users. In [16], processing tools available in the Hadoop and Spark ecosystems, as well as optimization techniques, are combined in wind energy resource assessment and management. The work of [17] presents an architecture that includes the dimensions of data capture, processing, storage, visualization and decision-aid through Machine Learning, leaving for further research, the implementation of the proposed architecture.

In summary, several works combine Big Data with Machine Learning. However, the work presented in this paper addresses this challenge by proposing a data modelling method and an architecture that has a BDW as its central data system, integrating historical and predictive data. This integration directly supports an advanced decision-support environment that can process large datasets combining historical data with predictions obtained from models that learned from that historical data.

3 Proposed Architecture and Data Modelling Method

Research in Big Data Analytics and Machine Learning is usually done by different teams, who independently work in providing analytical means to analyse the available data. This work aims to integrate the scientific and technological contributions from both fields, supporting the integration of predictions to enrich a BDW that is used in an advanced data analytics environment that assists decision-makers for better decisions.

The proposed architecture (Fig. 1) creates a unified environment between Machine Learning processes and the BDW, as the main storage component, in Big Data contexts. Besides the storage itself, this architecture allows the monitoring of the Machine Learning models and the BDW, expanding the analytical scope beyond the decision-making at the business level - it is now possible to establish performance metrics for the BDW and the Machine Learning models and monitor them over time.

3.1 Components and Supporting Technologies

The advanced data analytics architecture (Fig. 1) is composed of three main components: *Data Sources*, *Big Data Cluster*, and *Visualization Tools*.

The Data Sources can be of different types, depending on the data they produce/handle: data can be structured, semi-structured, or unstructured. Besides this classification, the data sources may present data that is produced at different speeds, with different sizes and formats, thus justifying the context of Big Data [18].

The *Big Data Cluster* component integrates two areas, which are *Data Lake* and *Big Data Warehouse*. A distinction in data storage was made to accommodate analytical data and non-analytical data. The *Data Lake* is used to support the storage of any kind of data/processes/models such as Raw Data, Data Pipelines, Machine Learning Models, among others. The *Big Data Warehouse* is a storage ecosystem supporting the storage of data modeled as Analytical Objects, representing highly independent and autonomous entities with focus on analytical subjects in terms of decision support [9].

In the proposed architecture, the *Data Lake* has three subareas, the *Standard Raw Data*, the *Data Pipelines Repository*, and *the Machine Learning Models Repository and its Interfaces*. The purpose of the *Standard Raw Data* subarea is to standardize the data and its access, so that throughout the system data follows the same naming structure, making it easier for all users to understand and use. To be efficient and coherent, this standardization needs the definition of a set of basic rules that must be applied to all the data that is here stored. These rules should follow the DATSIS principles [19], which enable data to become Discoverable, Addressable, Trustworthy, Self-describing, Interoperable, and Secure. For example, this proposal includes rules for standardization of the attributes' name, sharing the information about the BDW and Data Lake, and ownership in the Organization Wiki, among others included in organization directives.

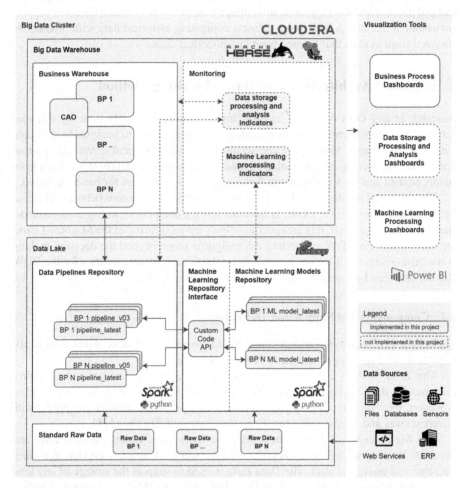

Fig. 1. Advanced data analytics architecture

The *Standard Raw Data* is used to feed two distinct, interrelated subareas, namely the *Data Pipelines Repository* and the *Machine Learning Models Repository and its Interfaces*. More specifically, data is stored in different sets linked to their Business Processes (BPs), where can be accessed in a Jupyter Notebook [20], which allows it to be read in the form of a Spark Dataframe [21]. These technologies are examples and were the ones used in this implementation. Nevertheless, other technologies with similar purposes can be used, as the technological landscape is quite diverse in this field. The same applies for the other technologies mentioned throughout this paper.

Data Pipelines in the *Data Pipelines Repository* prepare, transform and enrich data according to the defined data model, creating one or several tables in the BDW. These tables are the physical implementation of the modeled objects. BPs have their data, but they also use data that can be shared between them. This data that can be shared by several BPs is stored inside the CAO (Complementary Analytical Objects) folder of each BP. The Machine Learning models in the *Machine Learning Models Repository* provide predictions of events in the data, which brings a competitive advantage for the decision-making process. The modeled Analytical Objects integrate historical attributes with predictive attributes obtained from trained Machine Learning models. To access these predictions, an interface needs to be established between the Data Pipelines and the Machine Learning models. This interface is based on a class, developed in PySpark [22], which encodes a series of functions that allow a Machine Learning model to run on the Spark Dataframe. Thus, in the Data Pipelines development environment (in this case in Jupyter Notebooks), another notebook containing the Machine Learning Class is invoked. After invoking and correctly importing it, the functions in it are applied to the Spark Dataframe that contains the *Standard Raw Data*. The output is a Spark Dataframe, which contains the predictions for the processed events. Once the output of the Machine Learning model is obtained, it is integrated into the Data Pipeline. After integrating the predictive outputs, the Analytical Object including the historical and predictive attributes is stored in the BDW as a Hive table.

The BDW is organized according to the purpose of the data and integrates two distinct subareas: *Business Warehouse* and *Monitoring*. This division is important to efficiently store data regarding the BPs and the performance of the BDW and the Machine Learning models. Besides training Machine Learning models and using them, or creating Analytical Objects and storing them in the BDW, their evolution must be monitored over time so that these components can be improved, updated, and maintained. Otherwise, the system can become obsolete or not address performance requirements in an industrial context. For the Machine Learning models, for example, due to the volatile nature of the data in a business activity, the data that is used for training the models can change and the models need to evolve to meet the new data needs, thus obtaining more accurate predictions.

Once all the data has been integrated and properly stored into the BDW, it is possible to analyse it in the *Visualization Tools* component. This includes data visualizations that support analytical tasks with associate indicators regarding Machine Learning (such as accuracy, for instance), Data Processing (such as processing time, for instance), and Analytical Objects (such as the number of records, for instance). This component foresees analytical dashboards for the analysis of the different BPs, and for the monitoring of the Data Storage Processing and Analysis and the Machine Learning models. In the work here presented, the visualizations were implemented in PowerBI. Although the architecture presents the *Monitoring* subarea and the related visualizations, they are considered future work and for this reason are not described in this paper.

3.2 Data Modelling Method

For the design and implementation of a BDW, a data modelling approach was pursued, by following specific steps. This modelling approach extends the one presented in [9] and addresses the evolution of the BDW by integrating new BPs or domains of analysis as needed. When new domains need to be added for analysis, a set of steps must be performed. Figure 2 summarizes, with a simple example, the proposed steps (ST).

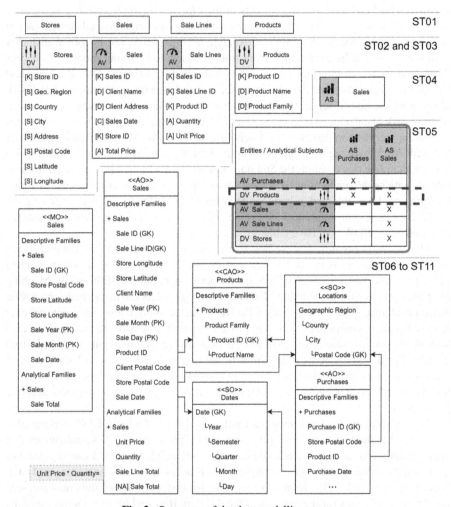

Fig. 2. Summary of the data modelling steps

ST01 – Identify the Relevant Entities of the Domain Under Analysis. These entities can be identified by domain specialists through their empirical knowledge about the organization, using existing conceptual models (e.g. ERDs), logical models (e.g. star-schemas) or others, or even directly from data sources.

ST02 – Classify Entities with Analytical Value (AV) or Descriptive Value (DV). Typically, entities with AV have two characteristics: (1) provide the main business or analytical indicators that are needed for decision support; (2) have high cardinality, as the growth rate of rows is considerably higher than the one verified in entities with DV. Entities with DV are used to provide analytical context and enable different levels of detail in the analysis. Nevertheless, it is important to mention that this method is based on a goal-driven approach, so the result of this classification depends on the domain in analysis. Thus, if the domain changes, the classification can also change.

ST03 – Identify and Classify the Attributes of the Entities. Associate the relevant attributes of each entity classified in ST02. Attributes can be classified as: Descriptive (D); Analytical (A); Date (C); Time (T); Spatial (S); Key (K). Take into consideration that some operational systems have attributes that do not add value to analytical systems. Examples of that are attributes that always have the same value, are completely null or that are not useful for analysis. Those attributes should not be considered.

ST04 – Identify the Analytical Subjects (ASs). Based on the analytical requirements of the application domain given by the decision-makers, the analytical subjects emerge from the entities with AV. They can be a subset of them, all, or a relationship between two entities with AV. For example, in the sales domain, sales and purchases (from previous iteration) were classified as entities with AV and products as an entity with DV. Figure 3 presents three different results for step ST04 according to the analytical requirements. If the focus of analysis, for now, is only about sales, then the AS will be Sales (case 1). If the focus will be the analysis of sales and purchases separately, then the result of ST04 will be the AS Sales and the AS Purchases (case 2). Another possible result is Case 3, where the focus is the sales and purchases in an integrated way. The result is an AS Sales with Purchases that, in future steps, will denormalize purchases to sales.

Fig. 3. Types of analytical subjects

ST05 – Define the Relationships Matrix. Based on the domain knowledge, the relationships matrix needs to be defined or updated, mapping each AS with the available entities (AV or DV). In this mapping, the granularity of the AS cannot be changed by the denormalization of the mapped entities.

ST06 – Identify the Special Objects (SOs). Temporal and/or spatial attributes point to the need for SOs that include the calendar, temporal or spatial descriptive attributes that are relevant in the application domain. Special care is needed in Temporal and Spatial objects. These objects are used to normalize this kind of information in the BDW. It is not recommended to use attributes with high granularity (i.e. seconds, latitude or longitude) as they will highly increase the number of records in these tables.

ST07 – Define the Analytical Objects (AOs) and Complementary Analytical Objects (CAOs). Through the relationships matrix obtained in ST05 and the mapping of attributes in ST03, AS are classified as AOs and the entities that are related to the AS are denormalized to the AOs or included in the SOs. AOs are characterized by being autonomous objects in terms of processing and by answering specific domain questions, based on a subject of interest for analytical purposes. They are highly denormalized structures that can answer queries without the constant need of joins with other data sources. The logic representation of AOs is divided into families, namely Descriptive, Analytical and Predictive Families. The attributes of the entities with AV classified as descriptive or analytical are placed inside the respective family. CAOs emerge from the relationships matrix. Without existing a strict threshold, if an entity is shared by multiple AS and that number tends to grow, then it should be considered as CAO.

ST08 – Identify the Granularity Keys (GKs). The GKs represent the level of detail of the records to be stored in an AO and integrate one or more descriptive attributes that can uniquely identify a record. Each object in the BDW needs to have a GK.

ST09 – Identify the Partition Keys (PKs). The physical partitioning scheme applied to the data is normally made through date, time or geospatial attributes, that fragment the AO into lower size files, that can be accessed individually, enabling the loading and filtering in hourly/daily batches for specific regions or countries. As an example, Hive does not properly deal with a large number of small files, so it is important to choose the appropriate PK, in other to avoid unnecessary fragmentation. Although analytical attributes can be used to form a PK, that is not recommended. These keys are relevant to increase the system performance.

ST010 – Identify the Non-Additive (NA) Analytical or Predictive Attributes. As AOs are highly denormalized structures, they can have Analytical or Predictive Attributes that do not depend of the global GK. When numerical, those attributes are classified as NA as they cannot be aggregated with a SUM in a query that uses a GROUP BY, for example.

ST011 – [Optional] Define Materialized Objects (MOs). To improve the response querying time of the BDW, sometimes is useful to create MOs. MOs are usually created to answer specific needs of the user, joining the data of one or several objects and aggregating that data by a set of attributes.

4   Industrial Demonstration Case: The Screwing Case

4.1   Motivation and Goal

The industrial facility in which this work took place is used for the development and assembly of automotive instrument clusters with the help of specialized tools and personnel. This plant focuses on optimizing assembly and testing procedures due to the critical nature of these processes for the business goals. The assembly of an instrument cluster is an extensive procedure with many checkpoints that are not the object of study in this paper. Instead, we will focus on one of the final phases where the plastic housings are combined with either printed circuit boards (PCB) or plastic parts. Bonding plastics or electronics to plastics can be achieved via a multitude of techniques that involve adhesives, welding, or the use of fasteners. Validating a fastening procedure is a difficult task as many variables are at play at the same time. The process experts develop screw tightening programs with the help of the handheld driver manufacturers to perform a fastening process and overcome most of the problems inherent to the use of this bonding technique. The settings specified on this program are used as a baseline to which we compare the actual values and assess the success of the fastening procedure. The process starts when the operator guides a handheld driver to a feeder which is always on, not controlled by the developed program. Once a screw is loaded on the screwdriver bit, the operator is guided through the tightening sequence with the aid of instructions carefully illustrated on a monitor above the station. Each inserted screw results in a Good or Fail (GoF) message on the screen which indicates whether the fastening succeeded or not. Depending on the result, two different actions are triggered: on the success, the display instructs the operator to transfer the part to the next station; on failure, the operator is instructed to stop the procedure and the process data is uploaded to a remote server where it will be thoroughly analysed by an expert tool which compares the actual results against a defect catalog. This catalog is developed and maintained by experts who constantly add new rules to accommodate new products and fault modes. One major drawback of this piece of software is its lack of scalability and its constant need for updates. With the use of Machine Learning techniques, supported by a BDW with vast amounts of historical data, we identified two models that can correctly identify good and bad screw tightening curves and provide various insights on the motives for such results.

4.2 Supporting Data and Machine Learning Models

To support the identified methods, data must be prepared and fed the models in a structured manner. The structure of data has the characteristics detailed next. For each part number p (represented by a unique identifier) we have multiple distinct serial numbers (sn). Each serial number contains a set of records regarding control procedures conducted on the shop floor, spread across multiple machines. One of these control procedures is the screw tightening procedure validation. In the data *granularity,* Fig. 4, i represents a screw fastening procedure where $i \in \{1, 2, ..., N\}$ and N represent the total number of screws for a specific p and sn pair. Each i is composed of $k \in \{1, 2, 3, ..., K\}$ observations categorized by a attributes ($a = 44$).

The focus of the analysis is on the *profile_angle* and *profile_torque* attributes as they allow us to visualize a fastening process in a 2D space (Fig. 5). Although this dataset is comprised of time-series data, we are using the angle variable ($\alpha_{i,k}$) as our sequential or temporal measure of the fastening as in most cases its values increase with the process duration.

Several machine learning models were trained for this demonstration case [23] but only the two best performing unsupervised models were selected and the predictions included in the BDW, namely the ones that used the Isolation Forest and the Autoencoder. The Isolation Forest (iForest) leverages a clear distinction of characteristics of anomalous points which are present in fewer quantities and numerically different to normal instances

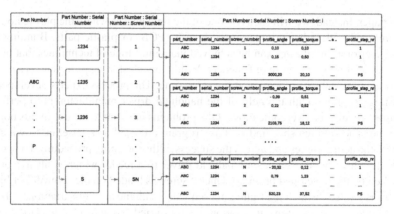

Part Number	Part Number : Serial Number	Part Number : Serial Number : Screw Number	Part Number : Serial Number : Screw Number: i						

part_number	serial_number	screw_number	profile_angle	profile_torque	.. a ..	profile_step_nr
ABC	1234	1	0,10	0,10	....	1
ABC	1234	1	0,15	0,50	....	1
....	....	....	....	....	....	....
ABC	1234	1	3000,20	20,10	....	PS

part_number	serial_number	screw_number	profile_angle	profile_torque	.. a ..	profile_step_nr
ABC	1234	2	- 0,39	0,51	....	1
ABC	1234	2	0,22	0,52	....	1
....	....	....	....	....	....	....
ABC	1234	2	2103,75	18,12	....	PS

part_number	serial_number	screw_number	profile_angle	profile_torque	.. a ..	profile_step_nr
ABC	1234	N	- 20,32	0,12	....	1
ABC	1234	N	0,79	1,23	....	1
....	....	....	....	....	....	....
ABC	1234	N	520,23	37,52	....	PS

Fig. 4. Data outlook – the screwing case

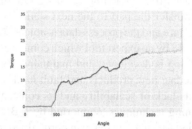

Fig. 5. Example screw tightening curve

and isolates them from normal points. Based on this principle, this anomaly detection algorithm is built upon a tree structure that attempts to isolate instances and then evaluate their normality. Anomalous instances tend to be isolated more easily as fewer features can describe them, forcing them to be closer to the root of the tree. At runtime, multiple trees are generated for a given dataset which forms an ensemble model - the iForest [24]. Normal points are isolated from anomalies that will, on average, have shorter path lengths.

Autoencoder (AE) is a type of unsupervised learning technique widely used for anomaly detection, image denoising, and feature extraction. AEs [25] are particularly strong in compressing and encoding high-dimensional data into a lower-dimensional space. This is achieved by imposing a bottleneck in its architecture which forces the neural network to create a compact representation (latent space) of the original input. Aside from this intermediate step, the AE is composed of two main stages: an encoding stage, where the input data is compressed using a specific number of features that describe the dataset, and a decoding stage where the model tries to recreate the original input with the smaller number of features present in the latent space. In this demonstration case, normal pairs of angle and torque values $(\alpha_{i,k}, \tau_{i,k})$ are provided as inputs and reconstructed $(\hat{\alpha}_{i,k}, \hat{\tau}_{i,k})$ pairs are generated by the model. Evaluating each pair (i,k) is calculated by computing the Mean Absolute Error (MAE) [26]:

$$MAE_{i,k} = (|\alpha_{i,k} - \hat{\alpha}_{i,k}| + |\tau_{i,k} - \hat{\tau}_{i,k}|)/2 \tag{1}$$

This reconstruction error ($d_{i,k} = MAE_{i,k}$) between the output and the input is then used as a decision score where greater reconstruction errors denote a higher anomaly probability.

4.3 Data Model

In this demonstration case, the Screw data model was identified following the steps presented in Subsect. 3.2 and integrates one Analytical Object (AO Screws), two Special Objects (Dates and Locations), and one materialized object (MO Screws). Due to confidentiality reasons, Fig. 6 only presents MO Screws as only this object is used to feed the dashboards here presented.

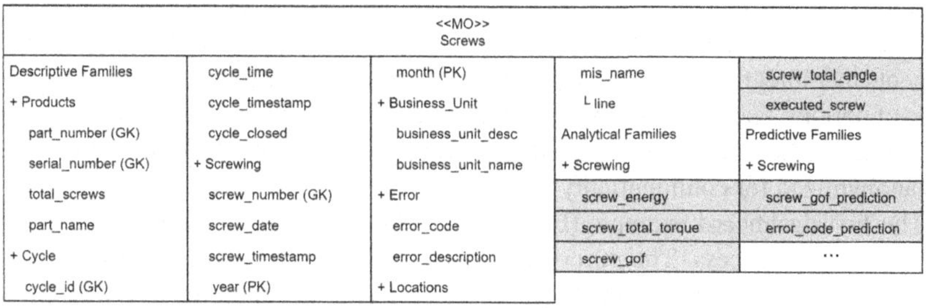

<<MO>> Screws				
Descriptive Families	cycle_time	month (PK)	mis_name	screw_total_angle
+ Products	cycle_timestamp	+ Business_Unit	∟ line	executed_screw
part_number (GK)	cycle_closed	business_unit_desc	Analytical Families	Predictive Families
serial_number (GK)	+ Screwing	business_unit_name	+ Screwing	+ Screwing
total_screws	screw_number (GK)	+ Error	screw_energy	screw_gof_prediction
part_name	screw_date	error_code	screw_total_torque	error_code_prediction
+ Cycle	screw_timestamp	error_description	screw_gof	...
cycle_id (GK)	year (PK)	+ Locations		

Fig. 6. MO screws

AO Screws will provide the necessary analytical information to MO Screws so that predictions can be made in the backend, and these predictions are stored in MO Screws along with other relevant attributes. It is important to mention that AO Screws has, for each screw, more than 400 records, and the MO Screws only has one record for each screw with the aggregated data. In the proposed model, the attributes highlighted in blue are considered NA in the AO Screws, but not in the MO Screws.

4.4 Integration and Flows

All the available data sources were integrated in the Data Pipelines that load the historical data of the screws. Once the necessary standardization has been made to the data, it is stored in the *Standard Raw Data Screw* folder. The screw data stored in the *Standard Raw Data* was used to train and optimize the prediction models. Those prediction models are stored in the *Machine Learning Models Repository* and mapped in the corresponding Application Program Interface (API).

To feed the Business Warehouse another Data Pipeline is created. This pipeline loads the screw data from the *Standard Raw Data*, and for each set of Part Number, Serial Number, Cycle ID, and Screw Number, the Machine Learning API is called to predict the result of the GoF test. With the prediction results, a Dataframe is created containing the screw data from the *Standard Raw Data* grouped by Part Number, Serial Number, Cycle ID, and Screw Number, along with the GoF prediction. After that, the Dataframe is stored in a Hive Table that matches the *Analytical Object Screw* (AO Screws), an Analytical Object modeled for this BP and that integrates all the data relevant to support the decision support needs in this industrial plant. The decision process is supported by several analytical dashboards available in the *Visualization Tools*. As proposed in the architecture, all pipelines are stored in the *Data Pipelines Repository*.

4.5 Decision Support Dashboards

Dashboards are key elements in the daily work of the decision-makers, so their design was achieved with the engagement and validation of the final users.

As a requirement for this demonstration case, decision-makers must have a set of dashboards that present macro visualizations of the screwing process, as well as more detailed ones capable to show the results of the GoF test of each screw. All the dashboards developed for this demonstration case have two main areas, an L shape bar along top and left side is dedicated to filters and a more central area with all the graphical/table elements that integrate the dashboard. In the filters area, the user can select from a wide range of options, such as temporal options, production line, or equipment, among others. Regarding all the examples presented in this paper, it is worth mentioning that all data was anonymized for confidentiality reasons.

The first dashboard example (Fig. 7) is a general dashboard, with a holistic view of the screwing process. The purpose of this dashboard is to allow the user to consult potentially important data of the business process in a fast and effective way. Starting with the first element (Fig. 7, part 1), it is possible to analyse data regarding the quantities by equipment. These quantities are related to the successful or unsuccessful production of each equipment (Screw GoF 1 and Screw GoF 0, respectively). The available data is presented in a descending order considering the produced quantity. It is possible to apply top and side filters to this same visualization, allowing, for example, a visualization of the equipment with an unsuccessful production, with the Screw GoF at 0 (Fig. 7, part 6), or filter this data by a specific equipment or production line (Fig. 7, part 4). It is important to see in the dashboards the temporal attributes, year, month, week, and day filters (Fig. 7, part 5), allowing the user to filter the data by a specific date, thus increasing the level of detail and specificity of these visualizations. It is important to see in the dashboards the temporal attributes, year, month, week, and day filters (Fig. 7, part 5), allowing the user to filter the data by a specific date, thus increasing the level of detail and specificity of these visualizations.

Figure 7, part 2, provides information about the most common errors that cause problems in a production equipment. It is possible to apply again the filters to a specific equipment, to a specific production line, to detail a specific date, search by error code and error description (Fig. 7, part 7), resulting in information regarding the equipment that most tend to suffer that specific error during the production process. In the last element (Fig. 7, part 3), a table calculates the percentage of failures relative to the production cycles, Cycle GoF, for each production line. This table shows the count of the total Cycle GoF values per production line, the count of the lines with Cycle GoF at 0 and calculates

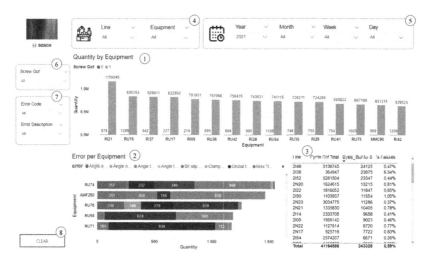

Fig. 7. Macro screw tightening dashboard.

the failure percentage. Again, the user can filter a specific production line to return the failure percentage for that same line in the chart. It is also possible to quickly clear all the filters selected by pressing the clear button (Fig. 7, part 8) which resets all the previous settings.

The dashboard with more detailed data (Fig. 8) takes as input the part number and the serial number of a product (Fig. 8, part 1). For that part and serial number, the user has an overview of the GoF test results in a bar graphic (Fig. 8, part 5). Also, the user can see the stations where the product pass considering the screws GoF test results in each station (Fig. 8, part 6). If the user selects a station, the bar graphic in Fig. 8, part 7, will highlight the screw cycles of that station and, for each Cycle ID, the number of GoF tests OK vs NOK (Not OK) is presented. Figure 8, part 8, shows in detail what are the results of the GoF test for each screw id based on a previously selected cycle id. Also, the prediction of the GoF test result is presented to the user, since in this phase decision-makers want to see the GoF test results and their prediction to evaluate if they can stop doing GoF tests or if they decide to do it by sampling. In this dashboard, a set of temporal filters can be used, Fig. 8 parts 2 and 3, and it is also possible to filter by GoF test result (part 4). Additionally, the dashboard has two cards to show the values about the percentage of failures (Fig. 8, part 9) and the total of screws (Fig. 8, part 10).

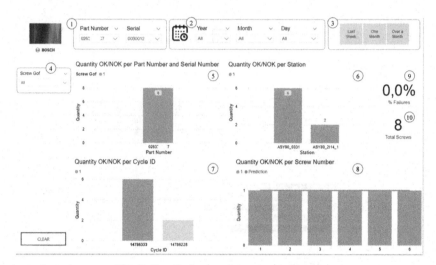

Fig. 8. Detailed screw tightening dashboard

Figure 9 presents a different serial number of the same part number presented in the dashboard of Fig. 8. For this product, in the selected cycles, 7 screws were tight (OK) and 1 is not ok (NOK), ending the process with 14,3% of failures. Also, it is possible to see that the tightening fails in the first screw and then the product changes to a different station, starting a new screw cycle that also starts the screwing process. Moreover, it is important to highlight that the prediction was capable to detect the failure in the first screw.

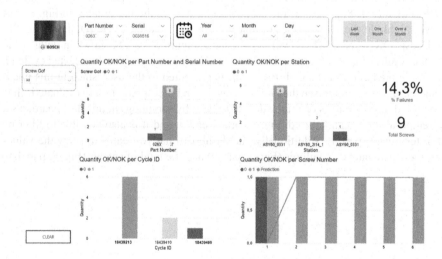

Fig. 9. Detailed screw tightening dashboard with GoF 0

5 Conclusions

This paper presented the design and implementation of an advanced data analytics environment, taking advantage of Big Data and Machine Learning techniques and technologies. The proposed architecture handles the integration of data from multiple *Data Sources* and for different business processes, providing a way to use that data to train Machine Learning models and store the predicted data. The proposed data modelling method guides practitioners through a set of steps so they can create and evolve the Big Data Warehouse physical implementation based on a logical data model.

In terms of design, the *Big Data Cluster* is divided into the Data Lake and the Big Data Warehouse areas. The Data Lake area stores the Raw Data, the Data Pipelines, and the Machine Learning models. The data in the Big Data Warehouse, the core element for analytical data storage, includes historical data and predictive data obtained using Machine Learning models. This Big Data Warehouse uses Hive tables to store objects that support all the analytical capabilities needed in the *Visualization Tools*.

The Screw Tightening demonstration case was presented, providing historical and predictive data made available throughout a set of dashboards fed by the Big Data Warehouse. In this demonstration case, the data is processed and stored on a daily basis. As future work, data needs to be processed in real-time to avoid rejection before further production steps. This presents several challenges such as applying prediction models to real-time events and the huge volume of data handled by this industrial process.

Acknowledgements. This work has been supported by FCT – *Fundação para a Ciência e Tecnologia* within the Project Scope: UIDB/00319/2020, the Doctoral scholarships PD/BDE/135100/2017 and PD/BDE/135105/2017, and European Structural and Investment Funds in the FEDER component, through the Operational Competitiveness and Internationalization Programme (COMPETE 2020) [Project nº 039479; Funding Reference: POCI-01-0247-FEDER-039479]. The authors also wish to thank the automotive electronics company staff involved with this project for providing the data and valuable domain feedback. This paper uses icons made by Freepik, from www.flaticon.com.

References

1. Wang, L., Alexander, C.A.: Machine learning in big data. Int. J. Math. Eng. Manag. Sci. **1**, 52–66 (2016)
2. Alswedani, S., Saleh, M.: Big data analytics: importance, challenges, categories, techniques, and tools. J. Adv. Trends Comput. Sci. Eng. **9**, 5384–5392 (2020)
3. Alsghaier, H.: The importance of big data analytics in business: a case study. Am. J. Softw. Eng. Appl. **6**, 111–115 (2017)
4. Rialti, R., Marzi, G., Caputo, A., Mayah, K.A.: Achieving strategic flexibility in the era of big data: the importance of knowledge management and ambidexterity. Manag. Decis. **58**, 1585–1600 (2020)
5. Gao, R.X., Wang, L., Helu, M., Teti, R.: Big data analytics for smart factories of the future. CIRP Ann. **69**, 668–692 (2020)
6. Papageorgiou, L., Eleni, P., Raftopoulou, S., Mantaiou, M., Megalooikonomou, V., Vlachakis, D.: Genomic big data hitting the storage bottleneck. EMBnet J. **24**, e910 (2018)

7. Chavalier, M., El Malki, M., Kopliku, A., Teste, O., Tournier, R.: Document-oriented data warehouses: models and extended cuboids, extended cuboids in oriented document. In: Proceedings - Conference on Research Challenges in Information Science, August 2016
8. Cuzzocrea, A., Song, I.Y., Davis, K.C.: Analytics over large-scale multidimensional data: the big data revolution! In: Conference on Information and Knowledge Management (2011)
9. Santos, M.Y., Costa, C.: Big data: concepts, warehousing and analytics. River (2020)
10. Vaisman, A., Zimányi, E.: Data warehouses: next challenges. In: Aufaure, M.-A., Zimányi, E. (eds.) eBISS 2011. LNBIP, vol. 96, pp. 1–26. Springer, Heidelberg (2012). https://doi.org/10.1007/978-3-642-27358-2_1
11. Costa, C., Santos, M.Y.: Evaluating several design patterns and trends in big data warehousing systems. In: Krogstie, J., Reijers, H.A. (eds.) CAiSE 2018. LNCS, vol. 10816, pp. 459–473. Springer, Cham (2018). https://doi.org/10.1007/978-3-319-91563-0_28
12. Elshawi, R., Sakr, S., Talia, D., Trunfio, P.: Big data systems meet machine learning challenges: towards big data science as a service. Big Data Res. **14**, 1–11 (2018)
13. Syafrudin, M., Alfian, G., Fitriyani, N.L., Rhee, J.: Performance analysis of IoT-based sensor, big data processing, and machine learning model for real-time monitoring system in automotive manufacturing. Sensors **18**, 2946 (2018)
14. Lee, J., Ardakani, H.D., Yang, S., Bagheri, B.: Industrial big data analytics and cyber-physical systems for future maintenance & service innovation. Procedia CIRP **38**, 3–7 (2015)
15. Baldominos, A., Albacete, E., Saez, Y., Isasi, P.: A scalable machine learning online service for big data real-time analysis. In: 2014 IEEE Computational Intelligence in Big Data (2014)
16. Krishnamoorthy, R., Udhayakumar, K.: Futuristic bigdata framework with optimization techniques for wind energy resource assessment and management in smart grid. In: 2021 7th International Conference on Electrical Energy Systems (ICEES), pp. 507–514 (2021)
17. Montoya-Torres, J.R., Moreno, S., Guerrero, W.J., Mejía, G.: Big data analytics and intelligent transportation systems. IFAC-PapersOnLine **54**, 216–220 (2021)
18. Cai, L., Zhu, Y.: The challenges of data quality and data quality assessment in the big data era. Data Sci. J. **14**, 1683–1470 (2015)
19. Dehghani, Z.: How to move beyond a monolithic data lake to a distributed data mesh (2019)
20. Project Jupyter: Project Jupyter | Home. https://jupyter.org/. Accessed 19 July 2021
21. Spark.apache.org: Spark SQL and DataFrames - Spark 1.5.2 Documentation. https://spark.apache.org/docs/latest/sql-programming-guide.html. Accessed 19 July 2021
22. PySpark Documentation — PySpark 3.1.2 documentation. https://spark.apache.org/docs/latest/api/python/. Accessed 19 July 2021
23. Ribeiro, D., Matos, L.M., Cortez, P., Moreira, G., Pilastri, A.: A comparison of anomaly detection methods for industrial screw tightening. In: Gervasi, O., et al. (eds.) ICCSA 2021. LNCS, vol. 12950, pp. 485–500. Springer, Cham (2021). https://doi.org/10.1007/978-3-030-86960-1_34
24. Liu, F.T., Ting, K.M., Zhou, Z.H.: Isolation forest. In: Proceedings - IEEE International Conference on Data Mining, ICDM, pp. 413–422 (2008)
25. Hinton, G.E., Salakhutdinov, R.R.: Reducing the dimensionality of data with neural networks. Science **313**, 504–507 (2006)
26. Alla, S., Adari, S.K.: Traditional Methods of Anomaly Detection. Apress, Berkeley (2019)

Benchmarking Conventional Outlier Detection Methods

Elena Tiukhova[1]([📧]) [iD], Manon Reusens[1] [iD], Bart Baesens[1,2,3] [iD],
and Monique Snoeck[1] [iD]

[1] Research Center for Information Systems Engineering (LIRIS), KU Leuven,
Naamsestraat 69, 3000 Leuven, Belgium
{elena.tiukhova,monique.snoeck}@kuleuven.be
https://feb.kuleuven.be/research/
decision-sciences-and-information-management/liris/liris
[2] Department of Decision Analytics and Risk, University of Southampton,
Southampton, UK
[3] Southampton Business School, University of Southampton,
Highfield, Southampton SO17 1BJ, UK

Abstract. Nowadays, businesses in many industries face an increasing flow of data and information. Data are at the core of the decision-making process, hence it is vital to ensure that the data are of high quality and no noise is present. Outlier detection methods are aimed to find unusual patterns in data and find their applications in many practical domains. These methods employ different techniques, ranging from pure statistical tools to deep learning models that have gained popularity in recent years. Moreover, one of the most popular outlier detection techniques are machine learning models. They have several characteristics which affect the potential of their usefulness in real-life scenarios. The goal of this paper is to add to the existing body of research on outlier detection by comparing the isolation forest, DBSCAN and LOF techniques. Thus, we investigate the research question: which ones of these outlier detection models perform best in practical business applications. To this end, three models are built on 12 datasets and compared using 5 performance metrics. The final comparison of the models is based on the McNemar's test, as well as on ranks per performance measure and on average. Three main conclusions can be made from the benchmarking study. First, the models considered in this research disagree differently, i.e. their type I and type II errors are not similar. Second, considering the time, AUPRC and sensitivity metrics, the iForest model is ranked the highest. Hence, the iForest model is the best in the cases when time performance is a key consideration as well as when the opportunity costs of not detecting an outlier are high. Third, the DBSCAN model obtains the highest ranking along the F1 score and precision dimensions. That allows us to conclude that if raising many false alarms is not an important concern, the DBSCAN model is the best to employ.

Supported by the ING Group.

Keywords: Outlier detection · iForest · Local Outlier Factor · DBSCAN

1 Introduction

It is difficult to imagine the modern world without data. Especially since the outbreak of COVID-19, the role of digital technologies has become even more crucial, and the amount of data created every minute is enormous. In 2021, the total amount of data consumed globally was 79 zettabytes, and this number is projected to grow to 180 zettabytes by 2025 [9]. Businesses try to make use of these data by building up machine learning models that facilitate decision-making processes.

For machine learning models to be unbiased and accurate, the data they utilize should be of high quality. One of the possible issues with data quality is the presence of outliers in these data. Outliers are the observations that deviate from what is considered normal and expected [15]. When data contains outliers, this may bias decisions. Outliers might signify some kind of errors, for example, errors in data capture or data processing. In such case of erroneous data, the consistency and accuracy dimensions of data quality are violated, i.e., data is incompatible with previous data and their format and does not correspond to real-world values [26]. When outliers represent correct data, they can nevertheless be of high interest in particular domains, e.g. in fraud detection, medical diagnosis, malware analysis, network intrusion detection, manufacturing quality control, etc.

Artificial intelligence and machine learning have revolutionized the way businesses operate. Similarly, the breakthrough has been made in the outlier detection domain, where thanks to machine learning, models are capable of predicting anomalous observations automatically by learning from data. However, machine learning algorithms differ in detection performance and time efficiency; hence, their business value varies significantly from one domain to another. Many benchmarking studies of outlier detection techniques exist, focusing on the technical aspects of their prediction performance (see Sect. 2.4). However, they lack the interpretation of these performance results from the business point of view. Moreover, benchmark studies rarely report on the detailed approach of the hyperparameter selection; thus, this gap is addressed in this paper as well.

The main purpose of this paper is to add to the existing body of research on outlier detection by comparing traditional (conventional) outlier detection methods, namely, the isolation forest, the density-based algorithm for discovering clusters and the local outlier factor models. Hence, we investigate the research question: which ones of these outlier detection models perform best in practical business applications. By making such benchmarking, the following contributions will be made to the outlier detection domain. First, the techniques of hyperparameters setting for the outlier detection methods are discussed. Second, the traditional machine learning methods for outlier detection are compared on benchmark outlier detection datasets. Third, the performance metrics suitable for outlier detection are utilized in order to compare the models by means of the McNemar's test and average ranking.

The remainder of this paper is structured as follows. Sections 2 gives a short overview of the traditional machine learning techniques for outlier detection and the existing benchmarking studies on these techniques. The methodology is described in Sect. 3. Sections 4 and 5 present the results of the research and discuss them. Section 6 presents the limitations and validity threats of the study. Section 7 concludes on the research.

2   Overview of Traditional Outlier Detection Methods

Originated in the 1980s in the field of intrusion detection [8], outlier detection was initially performed mainly using statistical techniques such as statistical process control tools, z-scores, box-plots, etc.

The outlier detection domain has its own assumptions that influence the way how research can be performed. Outliers are the data points that do not conform to a defined notion of a normal behavior [6]. However, the outlier detection process is not that straightforward when it comes to defining what is considered normal. The profile of normal can be defined on different levels, e.g., local, global, contextual, collective, etc. Thus, anomaly detection methods differ in a way the normal profile is defined. Also, the notion of being an outlier differs from one domain to another, making a definition of "outlierness" less straightforward. Moreover, the labelled data is not easily available in the outlier detection domain, being a major issue for the researchers [6]. That brings us to the assumption of solving the outlier detection problem using the unsupervised learning techniques. Another characteristic of the outlier detection domain is the unbalanced nature of the data: anomalous classes have fewer observations than normal classes do.

The emergence of machine learning has introduced models which are capable to learn from data and has automated the outlier detection process. The traditional machine learning models for outlier detection combine the best of two worlds: they have statistical underpinnings and at the same time learn from the data they have been trained on. In this paper, we look at the three traditional outlier detection models, namely, the isolation forest (iForest) model, the density-based algorithm for discovering clusters (DBSCAN) model and the local outlier factor (LOF) model. The former uses the notion of isolation to detect outliers, while the latter two models use the notion of density.

2.1   Isolation Forest

The isolation forest model exploits the notion of outliers being few and different from what is considered normal. Thus, the model makes two important assumptions: anomalous observations represent a minority in data and have the values that are different from normal data. In most of the cases these assumptions are realistic as they are aligned with the outlier detection domain characteristics such as the outlier definition and the nature of the data prevalent in the domain (See Sect. 2). However, these assumptions might not be realistic in the cases when anomalies constitute a substantial amount of the data.

Contrary to most other outlier detection techniques, the iForest model is optimized to detect anomalies [18]. Instead of building the profile of normal data, it uses the isolation approach that is aimed directly to find the unusual patterns in data. The iForest model builds the ensemble of binary isolation trees (iTrees) that isolate instances. Outliers are susceptible to isolation; hence, they are isolated closer to the root of the iTree. The path lengths of iTrees are averaged in the ensemble, and outliers have lower averaged path lengths than normal observations.

The isolation forest model requires specification of two training hyperparameters: the number of trees in the ensemble and the subsampling size [18]. The usage of subsamples to build an iTree allows mitigating the effects of masking and swamping, thus making isolation of outliers easier. The isolation forest model has a linear time complexity. Moreover, it has a major computational advantage over other models, as it does not calculate density or distance measures that are usually time and resource consuming. The iForest model is capable of handling large, high-dimensional datasets, so that it scales well for the big data problems.

2.2   Density-Based Algorithm for Discovering Clusters

The density-based algorithm for discovering clusters model is capable of discovering clusters of arbitrary shape, as it utilizes the density-based notion of clusters [11]. The DBSCAN model requires minimal domain knowledge because it provides a modeler with an intuitive approach of setting the model hyperparameters.

The DBSCAN model makes an assumption that clusters are dense region, so that it separates the low density regions from the high density regions. It uses the notions of core points, border points and outlier points [11]. The neighborhood of a given radius, *Eps*, must contain at least *MinPts* observations for a point to be a core point of the cluster. Border points have less than *MinPts* observation in their neighborhoods, whereas they should be reachable from the core point. The outliers are observations that do not belong to any of the clusters, i.e. they are not core points, and they cannot be reached from the core point. The assumption of clusters being dense regions allows the model to discover the clusters of any shape; however, the DBSCAN model fails to identify the clusters when clusters are of varying density.

2.3   Local Outlier Factor

The local outlier factor model utilizes the notion of density of a local neighborhood of an observation [5]. Contrary to the iForest and DBSCAN models that consider the "outlierness" as a binary property, the LOF model assigns to each observation a degree of it being an outlier. This score is called the Local Outlier Factor. The approach employing the LOF score allows to detect local anomalies that might be difficult to spot with a global approach.

The LOF score measures the observation's outlierness relative to its local neighborhood. The local neighborhood is defined by a hyperparameter *MinPts* that defines the number of nearest neighbors of an observation that form this neighborhood. The LOF score measures the density of nearest neighbors of an observation relative to the density of an observation itself. If these densities are close to each other, then the LOF score value is around 1, and the observation can be considered as an inlier. If the density of the neighbors is much higher than the density of an observation, then the LOF score is much higher than 1, and the observation is an outlier. However, there is no clear rule for deciding when the observation can be considered an outlier, so that the application of the LOF model in practice requires an intervention from the domain expert's side

2.4 Related Work

Original papers on the isolation forest algorithm and the density-based algorithm for discovering clusters provide a benchmarking of the introduced model with other outlier detection techniques. [18] provides the empirical comparison of the iForest, ORCA, SVM, LOF and Random Forests models. This benchmarking uses the AUC metric and the run time to evaluate the performance of the models. According to both AUC and run time metrics, the iForest model obtains the best performance [18]. The DBSCAN model is compared with the CLARANS model [11]: the DBSCAN model is more efficient according to the run time and is capable of both clustering and discovering noise.

A standalone benchmarking study is presented in the paper about a meta-analysis of the anomaly detection problem [10]. The comparison of the density-based (Robust Kernel Density Estimation, Ensemble Gaussian Mixture Model), model-based (One-Class SVM, Support Vector Data Description), nearest neighbors based (Local Outlier Factor, KNN Angle-based Outlier Detection) and the projection-based (Isolation Forest, Lightweight Online Detector Of Anomalies) approaches is performed using the AUC and average-precision metrics and the statistical hypothesis testing. The study reports that the isolation forest model performs best on average, while the support-vector data description and One-Class SVM models obtain a poor performance [10].

The discussed benchmarking studies lack two important things. First, they lack the contextual interpretation of the result from the business point of view, as these studies focus too much only on the technical characteristics of the performance. Second, most of them employ the AUC metric for performance evaluation that can be misleading when the data are imbalanced [13].

3 Methodology

The benchmarking experiment is designed to contrast the detection performance of outlier detection models. To that end, three traditional outlier detection models, namely the iForest, DBSCAN and LOF models, are selected. The models are trained on the twelve benchmarking datasets that are described in Sect. 3.1. Before the data can be used for training the models, they are preprocessed in a way described in Sect. 3.2 and the models are trained on these data in the way described in Sect. 3.2. For the training, these models require the hyperparameters to be specified, and the way this specification is performed is described in Sect. 3.3. Once the models are trained, we can evaluate their prediction performance and compare these models. We use five performance metrics, namely recall, precision, F1 score, AUPRC and time, along with the McNemar's test and the average ranking to contrast the models'performance. The performance metrics and comparison techniques are described in detail in Sect. 3.4. Detailed implementation of the benchmarking study is available at https://github.com/tiu-elena/benchmark-conventional-OD-methods.

3.1 Datasets

In this paper, models are trained, and the outliers are predicted using 12 benchmark outlier detection datasets. Ten of them are taken from the Outlier Detection Datasets Library [23]. The credit card dataset source is the Credit Card Fraud Detection competition from Kaggle (see the dataset's references in Table 4). The bank dataset is taken from the ADRepository. Detailed information on the dataset sources is displayed in Table 4. The datasets used in this research are diverse in the number of observations, dimensionality and in the percentage of outliers present in the data. The detailed dataset characteristics are displayed in Fig. 1.

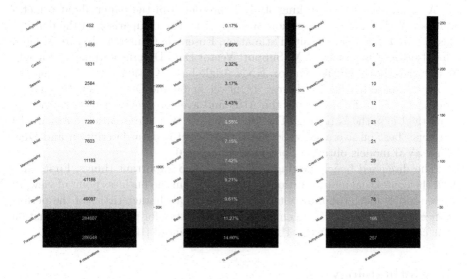

Fig. 1. Datasets characteristics

3.2 Data Preprocessing and Model Implementation

All the models are built using the *scikit-learn* library in Python [21]. The iForest model is constructed using the sklearn.ensemble.IsolationForest class. The

LOF model is built using the sklearn.neighbors.LocalOutlierFactor class. The DBSCAN model is created using the sklearn.cluster.dbscan class. The contamination parameter in the LOF model scikit-learn implementation is set to 10%. To train the models, labels are removed from the data. During the evaluation step, labels along with the models' predictions are used to calculate performance metrics.

As the DBSCAN and LOF models use the notion of density, they are sensitive to the different scales of the features. For that reason, the data is first scaled using min-max normalization. Additionally, the variables that contain only zeros are dropped. Besides, we create dummy variables to represent categorical variables.

3.3 Hyperparameters Setting

The hyperparameters of the models in this paper are set based either on the approach described in the original papers on the methods or on the research studies on the hyperparameters setting for these methods. In cases where the described approach is not feasible, the hyperparameters are set by the researchers themselves. These cases are described below in this section.

The hyperparameters of the iForest model are specified in the same way as in the original paper [18]. The subsampling size (train data size) is set to 256 as it is proved to be enough for outlier detection across a wide range of data. Another parameter is the ensemble size (the number of trees) that is set to 100 because the path lengths converge well on the ensemble of this size.

The hyperparameters of the DBSCAN model are set based on the distances to the kth nearest neighbor. The parameter k is specified in the Eq. 1 [25].

$$k = 2 \cdot \#features - 1 \tag{1}$$

Equation 2 shows the specification of the *MinPts* parameter [25].

$$MinPts = k + 1 = 2 \cdot \#features \tag{2}$$

To specify the *Eps* parameter, the distances to kth nearest neighbor for each observation are sorted and plotted. The respective distance of the point located in the first "valley" of the graph is set as an *Eps* parameter in the model (Fig. 2). On large datasets, we take a subset of the data (10% randomly) for the k-distances plot construction in order to make the distances' calculation feasible. For the Arrhythmia dataset, the approach from the Eq. 2 is not feasible: if the approach is applied, the value of k is higher than the number of observations, and the neighborhood specification is not meaningful. For the Arrhythmia dataset, the k parameter is set to *#features · 0.1* and the *MinPts* is set to *#observation · 0.1*. The *Eps* value is found using the approach illustrated in Fig. 2.

The LOF model hyperparameter, *MinPts*, specification is based on the lower-bound value of 10 from the original paper on the LOF method [5] and the upper bound value of *#observations · 0.01* that is set similarly as in the paper [27]. Equation 3 illustrates the *MinPts* parameter specification. The paper [27]

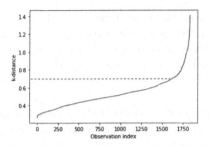

Fig. 2. *Eps* hyperparameter specification

proposes the upper bound value of *#observations · 0.03*, but we lowered the
fraction to 0.01 to make the approach work for the large datasets.

$$MinPts = max(10; \#observations \cdot 0.01) \tag{3}$$

3.4   Model Comparison

Once the models for outlier detection are built, the quality of their predictions is
evaluated using performance metrics. The class labels that are not used during
the training step are used at the model evaluation step and allow using the
metrics for classification models. The outlier detection domain is characterized
by imbalanced data, i.e., the majority of the observations are non-anomalous.
That puts some constraints on the performance metrics which can be used at the
performance evaluation step. In this paper, we use the metrics that are focused
on the quality of the positive class prediction, i.e., recall, precision, F1 score and
AUPRC. Additionally, the total time performance of the model (time to train +
time to predict) is measured and used for the models' performance evaluation.
The overview of the performance metrics is displayed in Table 1.

The AUPRC metric's thresholds are calculated using the anomaly scores
obtained from the models. For the isolation forest model, the score is the anomaly
score defined in the original paper [18]. For the LOF model, the score is the
local outlier factor defined in the original paper [5]. For the DBSCAN model,
the score is the distance to the cluster center for nonanomalous observations and
the distance to the center of the closest nonanomalous cluster for outliers.

Once the model predictions are made and the performance metrics are calculated, we need to investigate whether the models perform differently and which
model performs the best under what circumstances. In order to answer the first
question, we use McNemar's test [12]. The McNemar's test is especially useful in
the cases when the model is evaluated on a dataset only once (which is the case
in this research, as the model is evaluated on the same dataset it was trained
on but with the labels added). The McNemar's test constructs the contingency
table and tests the homogeneity of this table. In the case of classification models, this test checks whether two models disagree in the same way or not. The

Table 1. Performance metrics

Metric	Description	Formula
Recall	Shows how many of the actual outliers are captured by the model. Should be high to avoid missing the outliers	$\dfrac{TP}{TP + FN}$
Precision	Illustrates the ability of the model to precisely identify outliers. Defines how many of those predicted as outliers are actual outliers. Should be high to avoid false alarms	$\dfrac{TP}{TP + FP}$
F1 score	Illustrates the quality of the model to balance precision and recall. Works well for imbalanced datasets	$\dfrac{2 \cdot (Recall \cdot Precision)}{Recall + Precision}$
AUPRC	Shows the ability of the model to balance precision and recall on imbalanced datasets	Area under precision-recall curve across different decision thresholds
Time	Illustrates the computational costs required by the model as well as the speed of the outlier detection. Measured using the process time	Total time = Time to train + Time to predict

McNemar's test cannot comment on whether the performance of one model is better than the performance of another model.

The null hypothesis is that both models make errors in the same proportions, i.e., they have the same error rate [12]. If the test fails to reject the null hypothesis, the models have a similar proportion of errors, and we can conclude that their performances are similar. If the null hypothesis is rejected, then the proportions of errors are different and models perform differently.

4   Results

The performance metrics of the constructed models are displayed in Tables 5, 6, 7, 8, 9. The iForest model has the highest average AUPRC and recall values and the lowest average total time. It also has the lowest median total time. The DBSCAN model obtains the highest values of the F1 score and precision metrics. The average time performance of the iForest model is remarkably better than the performance of the DBSCAN model and the LOF model. The average running time of the iForest model is almost 100 times lower than the LOF performance and around 46 times lower than the DBSCAN performance.

In this paper, the McNemar's test is performed for each dataset and for each pair of the models. The p-values of the test are displayed in Table 2.

As can be seen from Table 2, for most of the datasets the null hypothesis can be rejected with more than 99% confidence. The rejection of the null hypothesis shows that models make errors differently, thus not behaving in a similar way. Also, this confirms the findings of the DBSCAN model having the highest average precision and the iForest model obtaining the highest average recall. The iForest and DBSCAN models do make mistakes, but differently: iForest tend to detect

Table 2. McNemar's test results (p-values)

Dataset	iForest vs DBSCAN	iForest vs LOF	DBSCAN vs LOF
Arrhythmia	0.54	0.003	$1.4 \cdot 10^{-5}$
Cardiocotography	0.423	$5.4 \cdot 10^{-13}$	$1.1 \cdot 10^{-17}$
ForestCover	0	0	0
Annthyroid	$8.02 \cdot 10^{-22}$	0.0002	$2.5 \cdot 10^{-34}$
Credit card	0	0	0
Mammography	$2.83 \cdot 10^{-158}$	$7.38 \cdot 10^{-13}$	$1.92 \cdot 10^{-96}$
Shuttle	0	0	$4.92 \cdot 10^{-225}$
Mnist	$2.58 \cdot 10^{-108}$	$1.25 \cdot 10^{-49}$	$9.73 \cdot 10^{-23}$
Vowels	$1.05 \cdot 10^{-23}$	0.0001	$9.41 \cdot 10^{-10}$
Seismic	0.01	$5.98 \cdot 10^{-6}$	$4.4 \cdot 10^{-9}$
Bank	0	$4.15 \cdot 10^{-103}$	0
Musk	$2.39 \cdot 10^{-36}$	$8.16 \cdot 10^{-21}$	$2.25 \cdot 10^{-86}$

as many outliers out of the true outliers as possible, but it comes at the cost of precision and results in false alarms. On the contrary, the DBSCAN model's predictions are precise and most of the predicted outliers are actual outliers. This difference in the nature of predictions results in the McNemar's test null hypothesis being rejected, as the models make errors in different proportions.

To investigate whether the performance of one model is better than the performance of another model, we calculate the average ranking of the models per performance metric. Additionally, we calculate the overall rank averaged over all the performance metrics. The results are displayed in Table 3.

Table 3. Models ranking

Model	AUPRC	F1 score	Time	Recall	Precision	Average
iForest	**1.75**	1.917	**1.83**	**1.75**	2.042	1.858
DBSCAN	2	**1.75**	1.917	2.083	**1.458**	**1.842**
LOF	2.25	2.333	2.25	2.17	2.5	2.3

As can be seen from Table 3, the overall averaged ranks of the iForest and DBSCAN models are almost the same (the DBSCAN model has slightly higher rank, but the difference is rather small). The analysis of the ranks per performance metric confirms the findings from the averaged values of the performance metrics. The iForest model is ranked the highest on the AUPRC, time and recall metrics. The DBSCAN model is ranked top 1 along the F1 score and precision dimensions. The LOF model is ranked last on all the performance metrics considered.

5   Discussion

The McNemar's test results clearly show that models make mistakes differently. However, this test does not indicate the difference in the overall error rate between the models, as it reports on the difference in proportion of error between the models. In the outlier detection domain, model errors can be of two types: type 1 errors (false positives) and type 2 errors (false negatives). The McNemar's test results show that models balance these two types differently. This fact is also illustrated by the average ranking of the models that is calculated per performance metric. This ranking shows the dominance of the iForest model for the recall metric; hence, it can better deal with false negatives and is less inclined to report that the "outlierness" is missing when it is in fact not. The DBSCAN model prevails along the precision metric. Thus, this model can better handle the false positives and raises false alarms less often.

According to the ranking, the LOF model never outperforms the iForest and DBSCAN models. It has both lower detection and time performance. However, the LOF model is the best according to the absolute value of the recall metric on large datasets with the lowest percentage of outliers (the Credit card and ForestCover datasets - see the datasets' description in Fig. 1) outperforming the iForest model. Yet, it comes at a price of having low precision values and an enormous computational time.

The iForest model is able to not miss out on the outliers at a cost of having false alarms. It obtains high recall values on large datasets (the Credit card, ForestCover and Shuttle datasets), whereas its precision, F1 score and AUPRC values on most of the large datasets (the Credit card and ForestCover datasets) are the lowest. The recall values are mostly decreasing with an increasing percentage of outliers in a dataset; thus, the model becomes less sensitive to outliers once the assumption of outliers being "few" is getting weaker. In contrast, the precision of the iForest model grows for a growing percentage of anomalies, so that the model learns to be more precise when given more anomalous data to learn from. Moreover, the precision of the iForest model mostly grows when the number of features is increasing. Thus, the model has an improved precision once it is given more features to learn from (e.g. Arrhythmia, Musk, Mnist and Bank datasets). However, it is not the case for the Cardiocotography and Shuttle datasets that obtain relatively high precision values with few features. Nevertheless, the iForest model is on average the fastest among all the models considered; hence, it requires less computational power and, therefore, it is less expensive. Also, the iForest model's high time performance allows faster detection of outliers, which is critical in industries like manufacturing where the cost of the unspotted defects is growing with time.

On the contrary, the DBSCAN model is capable of being precise. It is crucial in industries where the cost of dealing with outliers is high; thus, detecting irrelevant observations and checking them afterwards would lead to losses for the businesses. The DBSCAN model obtains the highest values for the precision metric mostly on the datasets of smaller sizes (up to 10 000 observations). Its precision is mostly decreasing for growing data size and increasing for larger numbers of features; however, some datasets do not conform to these trends, making it difficult to generalize about such dependencies.

As a result, more research is needed to investigate the dependencies of the recall and precision values on the dataset's characteristics, as in most of the cases the relationship is not straightforward.

The results show that the performance of conventional outlier detection models vary greatly depending on the datasets characteristics. The practical applicability of outlier detection models in business setting depends on the nature of the data available (its size, dimensionality and contamination level of outliers), the computing resources and their cost, the costs of a false negative and a false positive prediction and the nature of the domain itself. The DBSCAN model shows the biggest potential for the industries with high false positive costs, low computing resources costs, datasets of smaller sizes and larger number of features. It is typical for such domains as insurance fraud detection, where the investigation of fraud is time- and money-consuming. The iForest model can be advantageous in the industries where the computational resources are scarce and the costs of detecting false negatives are high and the datasets sizes are large. These characteristics are crucial in the domains like medical diagnosis, where it is vital to find out any deviations from the normal profile of a patient and the amount of the data is large.

6   Limitations and Validity Threats

This benchmark study is subject to both internal and external threats to validity. The *internal validity threats* concern the structure of the study itself. The first possible internal validity threat is the choice of the traditional outlier detection methods for this benchmarking study. The ranking of the models depends on how many models we use to obtain the performance metrics and consequently calculate this ranking. However, the three models used in this paper, i.e. the isolation forest, DBSCAN and LOF models, are the most widely used outlier detection methods in the literature nowadays [14,19,22]. Nevertheless, this threat opens up an opportunity of future research considering more outlier detection techniques for a benchmarking study. The second possible internal validity threat is how we set the hyperparameters for the models in this research. All the models require hyperparameter specification and are sensitive to these hyperparameters, thus producing different results depending on the setup. However, the hyperparameters setting in this study is based on both the original paper recommendations or the research findings that investigate the optimal ways of setting the hyperparameters for the models. Thus, the research has a solid basis of the way the hyperparameters are set up. The last possible internal validity threat is the choice of the performance metrics and tests to compare the models. The ranking averaged over the performance metrics can change depending on which performance metrics we use. However, in this research, we use a diverse set of performance metrics that measure different aspects of models' performance. These metrics are also appropriate for the outlier detection domain, thus making the research findings sound.

The *external validity threats* relate to generalizability of the study to other settings and domains. The first possible external validity threat is the choice of the datasets. The extent to which the results of this benchmarking study are generalizable is directly influenced by the diversity of the datasets that are used to build the models. This study employs twelve datasets that differ greatly in three parameters, i.e., in the number of observations and features and in the percentage of outliers. Moreover, these datasets come from different application domains. Such a variety allows for better generalization of the results, making the findings applicable to a broader context. Besides, the detailed datasets' characteristics provided in Fig. 1 allow for the transferability of the results and making it possible to apply the findings to other datasets with similar configurations. Another possible external validity threat is the choice of the business context when interpreting the performance of the models. However, the performance metrics used in this research are universal, and their interpretation can be expanded to other domains. Moreover, the datasets used in this benchmarking study are diverse and come from different domains ranging from banking (e.g., Credit Card and Bank datasets) to medicine (e.g., Annthyroid and Cardiocotography datasets).

7   Conclusion

The outlier detection is attracting more and more attention in both research and practical business worlds, as the amount of data generated on a daily basis is constantly growing. Due to the artificial intelligence and machine learning in particular, it is possible to automate the process of detecting outliers and make models learn from the data they have seen. Nevertheless, not all the models are equally efficient in doing so; hence, it is crucial to understand the strengths and weaknesses of each model in order to deploy these models in practice. Thus, it is necessary to investigate which machine learning methods perform the best when applied practically. To this end, three traditional machine learning techniques, namely, the isolation forest model, the density-based algorithm for discovering clusters model and the local outlier factor model, were compared on 12 benchmark datasets. Five performance metrics, namely, AUPRC, precision, recall, F1 score and total running time, were used to measure the performance of the models. The McNemar's test along with the average ranking were applied in order to compare the models.

The first conclusion of this paper concerns the ability of the models to detect outliers. We found that models disagree differently, i.e., their type I and type II errors are not similar. It influences the detection performance of the models: the iForest model has the highest average recall, while the DBSCAN model has the highest average precision.

The second conclusion concerns the iForest model. It obtains the best time performance and is ranked first for the recall metric. That allows us to conclude that when the cost of "not acting" is high and when time and computing resources are limited, the isolation forest model is the best choice.

The third conclusion concerns the DBSCAN model. This model is the most precise one. We conclude that the DBSCAN model is better to utilize in the cases when the cost of detecting outliers is high and false alarms are not encouraged.

Further research can be performed by exploring more outlier detection techniques, e.g., deep learning models that also allows extending the research to other data types, e.g. unstructured data. Furthermore, the comparison of the models can be extended with other techniques, e.g. the null-hypothesis testing framework.

Acknowledgements. The research was sponsored by the ING Chair on Applying Deep Learning on Metadata as a Competitive Accelerator.

A   Appendices

Table 4. Dataset sources

Dataset	Sources
Arrhythmia	[16,17,30]
Cardio	[3,24]
ForestCover	[17,28–30]
Annthyroid	[2,17,30]
Credit card	[1,7]
Mammography	[2,17,30]
Shuttle	[2,17,28–30]
Mnist	[4]
Vowels	[3,24]
Seismic	[24]
Bank	[20]
Musk	[3]

Table 5. AUPRC

Dataset	iForest	LOF	DBSCAN
Arrhythmia	0.440	0.338	0.456
Cardiocorography	0.580	0.186	0.642
ForestCover	0.062	0.248	0.025
Annthyroid	0.320	0.202	0.115
Credit card	0.104	0.453	0.390
Mammography	0.217	0.126	0.083
Shuttle	0.982	0.137	0.162
Mnist	0.269	0.313	0.334
Vowels	0.152	0.289	0.047
Seismic	0.117	0.081	0.116
Bank	0.282	0.190	0.269
Musk	0.996	0.026	1.000
Average	**0.377**	0.216	0.303

Table 6. F1 score

Dataset	iForest	LOF	DBSCAN
Arrhythmia	0.179	0.250	0.412
Cardiocotography	0.529	0.228	0.353
ForestCover	0.091	0.168	0.120
Annthyroid	0.325	0.297	0.179
Credit card	0.066	0.030	0.112
Mammography	0.196	0.199	0.218
Shuttle	0.776	0.158	0.204
Mnist	0.348	0.357	0.203
Vowels	0.179	0.418	0.554
Seismic	0.197	0.098	0.194
Bank	0.314	0.216	0.269
Musk	0.515	0.015	0.957
Average	0.310	0.203	**0.322**

Table 7. Precision

Dataset	iForest	LOF	DBSCAN
Arrhythmia	0.583	0.304	0.583
Cardio	0.535	0.224	0.677
ForestCover	0.049	0.092	0.073
Annthyroid	0.294	0.258	0.339
Credit card	0.034	0.020	0.060
Mammography	0.116	0.122	0.200
Shuttle	0.639	0.136	0.141
Mnist	0.251	0.343	0.538
Vowels	0.119	0.281	0.500
Seismic	0.138	0.081	0.132
Bank	0.363	0.229	0.189
Musk	0.346	0.010	1.000
Average	0.289	0.175	**0.369**

Table 8. Recall

Dataset	iForest	LOF	DBSCAN
Arrhythmia	0.106	0.212	0.318
Cardiocotography	0.523	0.233	0.239
ForestCover	0.689	0.956	0.332
Annthyroid	0.363	0.348	0.122
Credit card	0.833	0.910	0.841
Mammography	0.635	0.527	0.238
Shuttle	0.987	0.190	0.369
Mnist	0.566	0.373	0.201
Vowels	0.360	0.820	0.620
Seismic	0.347	0.124	0.365
Bank	0.277	0.203	0.464
Musk	1.000	0.031	0.918
Average	**0.557**	0.411	0.419

Table 9. Time

Dataset	iForest	LOF	DBSCAN
Arrhythmia	0.411	0.031	0.021
Cardiocotography	0.307	0.099	0.125
ForestCover	10.13	1153.56	119.54
Annthyroid	0.490	0.714	0.844
Credit card	17.9	2636.8	1692
Mammography	0.813	1.151	1.625
Shuttle	2.453	13.859	6.417
Mnist	1.323	2.245	1.729
Vowels	0.297	0.047	0.109
Seismic	0.427	0.255	0.198
Bank	4.927	63.714	41.938
Musk	1.047	0.448	0.438
Average	**3.377**	322.74	155.414
Median	**0.93**	0.932	1.234

References

1. Credit card fraud detection dataset (2016). https://www.kaggle.com/datasets/mlg-ulb/creditcardfraud
2. Abe, N., Zadrozny, B., Langford, J.: Outlier detection by active learning. In: Proceedings of the 12th ACM SIGKDD International Conference on Knowledge Discovery and Data Mining, pp. 504–509 (2006)
3. Aggarwal, C.C., Sathe, S.: Theoretical foundations and algorithms for outlier ensembles. ACM SIGKDD Explor. Newsl. **17**(1), 24–47 (2015)
4. Bandaragoda, T.R., Ting, K.M., Albrecht, D., Liu, F.T., Wells, J.R.: Efficient anomaly detection by isolation using nearest neighbour ensemble. In: 2014 IEEE International Conference on Data Mining Workshop, pp. 698–705. IEEE (2014)

5. Breunig, M.M., Kriegel, H.P., Ng, R.T., Sander, J.: LOF: identifying density-based local outliers. In: Proceedings of the 2000 ACM SIGMOD International Conference on Management of Data, pp. 93–104 (2000)
6. Chandola, V., Banerjee, A., Kumar, V.: Anomaly detection: a survey. ACM Comput. Surv. (CSUR) **41**(3), 1–58 (2009)
7. Dal Pozzolo, A., Caelen, O., Johnson, R.A., Bontempi, G.: Calibrating probability with undersampling for unbalanced classification. In: 2015 IEEE Symposium Series on Computational Intelligence, pp. 159–166. IEEE (2015)
8. Denning, D.E.: An intrusion-detection model. IEEE Trans. Softw. Eng. **2**, 222–232 (1987)
9. Domo: Data never sleeps 9.0. https://www.domo.com/learn/infographic/data-never-sleeps-9
10. Emmott, A., Das, S., Dietterich, T., Fern, A., Wong, W.K.: A meta-analysis of the anomaly detection problem. arXiv preprint arXiv:1503.01158 (2015)
11. Ester, M., Kriegel, H.P., Sander, J., Xu, X.: A density-based algorithm for discovering clusters in large spatial databases with noise. In: Proceedings of KDD, vol. 96, pp. 226–231 (1996)
12. Everitt, B.S.: The Analysis of Contingency Tables. Chapman and Hall/CRC, London (2019)
13. Fernández, A., García, S., Galar, M., Prati, R.C., Krawczyk, B., Herrera, F.: Learning from Imbalanced Data Sets. Springer, Cham (2018). https://doi.org/10.1007/978-3-319-98074-4
14. Hossain, F., Uddin, M.N., Halder, R.K.: Analysis of optimized machine learning and deep learning techniques for spam detection. In: 2021 IEEE International IOT, Electronics and Mechatronics Conference (IEMTRONICS), pp. 1–7. IEEE (2021)
15. Johnson, R.A., Wichern, D.W., et al.: Applied Multivariate Statistical Analysis, vol. 6. Pearson, London (2014)
16. Keller, F., Muller, E., Bohm, K.: HiCS: high contrast subspaces for density-based outlier ranking. In: 2012 IEEE 28th International Conference on Data Engineering, pp. 1037–1048. IEEE (2012)
17. Liu, F.T., Ting, K.M., Zhou, Z.H.: Isolation forest. In: 2008 Eighth IEEE International Conference on Data Mining, pp. 413–422. IEEE (2008)
18. Liu, F.T., Ting, K.M., Zhou, Z.H.: Isolation-based anomaly detection. ACM Trans. Knowl. Disc. Data (TKDD) **6**(1), 1–39 (2012). https://doi.org/10.1145/2133360.2133363
19. Nofal, S., Alfarrarjeh, A., Abu Jabal, A.: A use case of anomaly detection for identifying unusual water consumption in Jordan. Water Supply **22**(1), 1131–1140 (2022). https://doi.org/10.2166/ws.2021.210
20. Pang, G., Shen, C., Cao, L., Hengel, A.V.D.: Deep learning for anomaly detection: a review. ACM Comput. Surv. (CSUR) **54**(2), 1–38 (2021)
21. Pedregosa, F., et al.: Scikit-learn: machine learning in Python. J. Mach. Learn. Res. **12**, 2825–2830 (2011)
22. Raj, C., Khular, L., Raj, G.: Clustering based incident handling for anomaly detection in cloud infrastructures. In: 2020 10th International Conference on Cloud Computing, Data Science & Engineering (Confluence), pp. 611–616. IEEE (2020)
23. Rayana, S.: ODDS library (2016). http://odds.cs.stonybrook.edu
24. Sathe, S., Aggarwal, C.: LODES: local density meets spectral outlier detection. In: Proceedings of the 2016 SIAM International Conference on Data Mining, pp. 171–179 (2016)

25. Schubert, E., Sander, J., Ester, M., Kriegel, H.P., Xu, X.: DBSCAN revisited, revisited: why and how you should (still) use DBSCAN. ACM Trans. Database Syst. (TODS) **42**(3) (2017). https://doi.org/10.1145/3068335
26. Sidi, F., Shariat Panahy, P.H., Affendey, L.S., Jabar, M.A., Ibrahim, H., Mustapha, A.: Data quality: a survey of data quality dimensions. In: 2012 International Conference on Information Retrieval Knowledge Management, pp. 300–304 (2012). https://doi.org/10.1109/InfRKM.2012.6204995
27. Soenen, J., et al.: The effect of hyperparameter tuning on the comparative evaluation of unsupervised anomaly detection methods. In: Proceedings of the KDD'21 Workshop on Outlier Detection and Description, pp. 1–9 (2021)
28. Tan, S.C., Ting, K.M., Liu, T.F.: Fast anomaly detection for streaming data. In: Twenty-Second International Joint Conference on Artificial Intelligence (2011)
29. Ting, K.M., Zhou, G.T., Liu, F.T., Tan, J.S.C.: Mass estimation and its applications. In: Proceedings of the 16th ACM SIGKDD International Conference on Knowledge Discovery and Data Mining, pp. 989–998. Association for Computing Machinery, New York (2010). https://doi.org/10.1145/1835804.1835929
30. Ting, K., Tan, S., Liu, F.: Mass: a new ranking measure for anomaly detection. Monash University, Gippsland School of Information Technology (2009)

Forum

Handling Temporal Data Imperfections in OWL 2 - Application to Collective Memory Data Entries

Nassira Achich[1,2]([✉]) [iD], Fatma Ghorbel[1,2], Bilel Gargouri[1], Fayçal Hamdi[2] [iD], Elisabeth Métais[2], and Faiez Gargouri[1]

[1] MIRACL Laboratory, University of Sfax, Sfax, Tunisia
achichnassira@gmail.com, bilel.gargouri@fsegs.rnu.tn,
faiez.gargouri@isims.usf.tn, fatmaghorbel6@gmail.com
[2] CEDRIC Laboratory, Conservatoire National des Arts et Métiers (CNAM),
Paris, France
{faycal.hamdi,elisabeth.metais}@cnam.fr

Abstract. Dealing with imperfect temporal data entries in the context of Collective and Personal Memory applications is an imperative matter. Data are structured semantically using an ontology called "Collective Memo Onto". In this paper, we propose an approach that handles temporal data imperfections in OWL 2. We reduce to four types of imperfection defined in our typology of temporal data imperfections which are imprecision, uncertainty, simultaneously uncertainty and imprecision and conflict. The approach consists of representing imperfect quantitative and qualitative time intervals and time points by extending the 4D-fluents approach and defining new components, as well as reasoning about the handled data by extending the Allen's Interval algebra. Based on both extensions, we propose an OWL 2 ontology named "TimeOntoImperfection". The proposed qualitative temporal relations are inferred via a set of 924 SWRL rules. We validate our work by implementing a prototype based on the proposed ontology and we apply it in the context of the Collective Memory Temporal Data.

Keywords: Temporal data imperfection · Imprecision · Uncertainty · Both imprecision and uncertainty · Conflict · OWL 2 · 4D-fluents approach · Allen's interval algebra · Collective Memory Temporal Data

1 Introduction

Temporal Collective and Personal Memory Data may be affected by many types of imperfection [15]. In fact, Collective Memory is dedicated to relate historical facts like National Movement and Festivals and to describe remarkable passages from the lives of famous personalities. For instance, "Marilyn Monroe died on the night of August 4 to 5, 1962. Nearly five hours passed between the estimated time of death, around 9:30 p.m. and 10 p.m.". In this example, the imprecision is expressed in "the night of August 4 to 5, 1962" and "around 9:30 p.m. and 10 p.m.".

© The Author(s), under exclusive license to Springer Nature Switzerland AG 2022
R. Guizzardi et al. (Eds.): RCIS 2022, LNBIP 446, pp. 617–625, 2022.
https://doi.org/10.1007/978-3-031-05760-1_36

Many other kinds of imperfections that may affect temporal data are distinguished in our proposed typology [5] and the typology of Collective Memory Data imperfection proposed in [15]. Representing and reasoning about imperfect temporal data in the context of Collective Memory Model, based on an ontology named "CollectiveMemoOnto", is what we specifically addressed in this work. We reduce to four types of imperfection defined in our typology which are imprecision, uncertainty, both uncertainty and imprecision and conflict.

In the semantic web field, several approaches have been proposed to deal with perfect temporal data. However, to the best of our knowledge there is no works that deal with many temporal data imperfections at the same time.

In this paper, we propose an approach for representing and reasoning about imperfect temporal data in terms of both qualitative relations (e.g., "before") and quantitative ones (time intervals and points). It consists of: (1) Representing imperfect temporal data in OWL2. We extend the 4D-fluents approach [23] with new ontological components to represent: (1.1) imperfect quantitative temporal data, and (1.2) qualitative temporal relations between time intervals and points. Certainty degrees related to each kind of imperfection are calculated using possibility and evidence theories. (2) Reasoning about imperfect temporal data by extending the Allen's interval algebra [6]. It proposes qualitative relations only between time intervals. It is not devoted to handle imperfect time intervals. Furthermore, it is not intended to relate a time interval and a time point or two time points. We extend it by proposing qualitative temporal relations between imperfect time intervals. They preserve important properties of the original algebra. We adapt the resulting interval relations to propose temporal relations between a time interval and a time point, and two time points. (3) Proposing an OWL 2 ontology called "TimeOntoImperfection". It may be integrated in other ontologies to handle imperfect temporal data such as "CollectiveMemoOnto". It is implemented based on our extensions. Inferences are done using SWRL rules.

The structure of this paper is as follows. Preliminary concepts and related work in the fields of temporal data representation and reasoning in the Semantic Web are reviewed in Sect. 2. Section 3 introduce our proposed 4D-fluents approach extension. Section 4 introduce our proposed Allen's Interval Algebra extension. Section 5 presents our OWL 2 "TimeOntoImperfection" ontology. In Sect. 6, we present a validation in the context of Collective Memory Data.

2   Preliminaries and Related Work: Handling Temporal Data in Semantic Web

Imperfect temporal data are characterized using quantitative or qualitative terms. Imperfect quantitative temporal data means imperfect time intervals and points.

2.1   Temporal Data Representation and Reasoning About

Current technologies for the Semantic Web suffer from its lack to represent and reason about temporal data. Ontology languages such as OWL provide only binary relations and forsaken temporal data, which presents a major weakness. This explains the emergence of many researches in this context.

We classify them into two categories: *(1)* approaches that extend OWL or RDF syntax by defining new OWL or RDF operators and semantics to incorporate temporal data, which are Temporal Description Logics [7], Concrete Domains [17] and Temporal RDF [12]. *(2)* approaches that are implemented directly using OWL or RDF to represent temporal data without extending their syntax, which are Versioning [16], Reification [9], N-ary Relations [18], 4D-Fluents and Named Graphs [22]. They offer reasoning support and they can be combined with existing tools [11].

Most of these approaches handle only perfect temporal data and neglect imperfect ones and few approaches treat only some imperfections but not many imperfections at the same time. They are not intended to handle time points and qualitative temporal relations between a time interval and a time point or even two time points. Our approach should be based on existing OWL constructs. We choose to extend the 4D-fluents approach to represent imperfect quantitative temporal data and associated qualitative temporal relations since it minimizes data redundancy as the changes occur on the temporal parts and keep the static part unchanged. It maintains a full OWL expressiveness [8,13,20,24] and [14]. In 4D-fluents approach presents two classes, named "TimeSlice" and "TimeInterval". Four certain properties, named "tsTimeSliceOf", "tsTimeIntervalOf" "HasBeginnig" and "HasEnd" are introduced.

2.2   Temporal Data Reasoning: Allen's Interval Algebra

Allen proposed 13 qualitative temporal relations between perfect time intervals. He defined them in terms of the ordering of the beginning and ending bounds of the corresponding intervals. A particularity that the Allen's algebra holds, is that we can deduce new relations through the composition of other ones (e.g., "Before (A, B)" and "Equals (B, C)" give "Before (A, C)". Allen's interval algebra is not dedicated to handle uncertain time intervals and it does not relate a time point and a time interval, nor two time points. A number approaches have been extended this algebra such as [1,8,21] and [19]. These extensions are based on theories related to imprecise temporal data or uncertain temporal data. Furthermore, most of these extensions do not preserve all the properties of the original Allen's algebra. For instance, in [18], the relation "Equals" is not reflexive. However, the compositions of the resulting relations are not studied by the authors. For example, in [10], the authors do not propose the composition table of the proposed temporal relations. Most of the proposed approaches that represent and reason about imperfect temporal data, mainly deal with only imprecise temporal data or only uncertain temporal data and use fuzzy and probability theories. However, to the best of our knowledge there is no approach to deal, at the same time, with several types of imperfections in ontology.

3   Representing Temporal Data Imperfection in OWL 2

We extend the 4D-fluents approach with new ontological components and components based on OWL-Time[1] ontology to represent imperfect quantitative temporal data and associated qualitative temporal relations in OWL2. We reduce to imprecision, uncertainty, both uncertainty and imprecision at the same time and conflict as kinds of temporal data imperfections.

"TimeSlice" is the class domain for entities representing temporal parts. "time:TimeInterval" and "time:TimeInstant" are respectively the classes representing intervals and time points. "time:DateTimeDescription" is the class representing dates and time clocks. We propose an approach to deal with imprecise temporal data, specifically dates and time clocks in OWL 2 with a crisp view. We represent precise time points (dates and time clocks). For the dates, let D, Mo and Y be, respectively, precise day, month and year. We use three datatype properties from OWL-Time named "time:day", "time:month" and "time:year" to relate, respectively, "time:DateTimeDescription" and D, Mo and Y. Similarly, we represent the time clocks. We represent imprecise time points (dates and time clocks). For the dates, let D, Mo and Y be, respectively, imprecise day, month and year. We represent them by disjunctive ascending sets $\{D^{(1)}...D^{(d)}\}$, $\{Mo^{(1)}...Mo(mo)\}$ and $\{Y^{(1)}...Y^{(y)}\}$. We define for each of D, Mo and Y, respectively, two datatype properties: "HasDayFrom" and "HasDayTo", "HasMonthFrom" and "HasMonthTo", "HasYearFrom" and "HasYearTo". They are all connected to the "time:DateTimeDescription" class. Similarly, we represent the time clocks. We also represent the other types of imperfections (i.e., uncertainty, both uncertainty and imprecision, and conflict). A detailed description of this part is available in an appendix.[2]

4   Reasoning About Uncertain Temporal Data: Extending Allen's Interval Algebra

We extend the Allen's algebra to: *(1)* reason about imperfect quantitative temporal data to infer qualitative temporal relations and *(2)* to reason about the qualitative temporal relations to infer new ones.

We extend the Allen's interval algebra to reason about imperfect time intervals. When considering perfect time intervals, our approach reduces to Allen's interval algebra. We redefine the 13 Allen's relations to propose temporal relations between imperfect time intervals. For example, for uncertain time intervals, let $A = [A_{ca-}^{-}, A_{ca+}^{+}]$ and $B = [B_{cb-}^{-}, B_{cb+}^{+}]$ be two uncertain time intervals. For instance, we redefine the relation "$Before(A, B)$" as: "$Before_c(A, B)$"; where "c" is the certainty degree associated to the relation "Before" between A and B. This means that the uncertain ending bound of the interval A is less than

---

[1] https://www.w3.org/TR/owl-time/.

[2] https://cnam-my.sharepoint.com/:b:/g/personal/nassira_achich_auditeur_lecnam_net/EUCb9oFijgpJgjXEjCZmtUcBtTXfAr_t57p9YsCBnQEMtw?e=YsgVUc.

the uncertain beginning bound of B. Table 1 presents Allen's relations between uncertain intervals.

$$Before_c(A, B) \Rightarrow A^+_{ca-} < B^-_{cb+} \tag{1}$$

Table 1. Temporal relations between two uncertain time intervals A and B.

Relation(A,B)	Relations between interval bounds	Inverse(B,A)
Before$_c$(A,B)	$A^+_{ca+} < B^-_{cb-}$	After$_c$(B,A)
Meets$_c$(A,B)	$A^+_{ca+} = B^-_{cb-}$	Met-by$_c$(B,A)
Overlaps$_c$(A,B)	$(A^-_{ca-} < B^-_{cb-}) \wedge (A^+_{ca+} > B^-_{cb-}) \wedge (A^+_{ca+} < B^+_{cb+})$	Overlapped-by$_c$(B,A)
Starts$_c$(A,B)	$(A^-_{ca-} = B^-_{cb-}) \wedge (A^+_{ca+} < B^+_{cb+})$	Started-by$_c$(B,A)
During$_c$(A,B)	$(B^-_{cb-} < A^-_{ca-}) \wedge (A^+_{ca+} < B^+_{cb+})$	Contains$_c$(B,A)
Ends$_c$(A,B)	$(B^-_{cb-} < A^-_{ca-}) \wedge (A^+_{ca+} = B^+_{cb+})$	Ended-by$_c$(B,A)
Equals$_c$(A,B)	$(A^-_{ca-} = B^-_{cb-}) \wedge (A^+_{ca+} = B^+_{cb+})$	Equals$_c$(B,A)

The certainty degree "c" is inferred from the certainty degrees "c_{a+}" and "c_{b-}" using a Bayesian Network [2].

All the other tables presenting Allen's relation between imperfect time intervals of the other imperfections (imprecision, simultaneously uncertainty and imprecision, and conflict) are presented in the appendix.[3]

We adapt the qualitative temporal relations between time intervals to propose relations between a time interval and a time point as shown in Table 2 which also represents uncertainty like the last subsection. The qualitative temporal relations between time intervals are adapted to propose relations between time points, as shown in Table 3.

5   "TimeOntoImperfection": The Proposed Ontology

The temporal data imperfection ontology, named "TimeOntoImperfection".[4] it is a top-level ontology, which can be merged with other ontologies of domain that must be extended to represent and reason about imperfect temporal data. It is based on the extension of the 4D-fluent approach with predefined elements of the OWL-Time ontology, elements that we have defined to represent the targeted imperfections, and the extension of Allen's interval algebra to reason about imperfect temporal data. We create our "TimeOntoImperfection" ontology using the Protégé ontology editor. In the literature, there is not, to our knowledge, an ontology temporal data imperfection. The temporal data imperfection ontology contains 5 classes, 201 object properties and 44 data type properties that represent time interval bounds, time points, dates, and time clocks as well as all

---

[3] https://cnam-my.sharepoint.com/:b:/g/personal/nassira_achich_auditeur_lecnam_net/EeIhjd586WxGmeA_HI5OCgQBNtW1ie2ODZONZ0T0fX5oKQ?e=o7n0LQ.

[4] https://cnam-my.sharepoint.com/:u:/g/personal/nassira_achich_auditeur_lecnam_net/EWd_23zDgUVNtTUhj7uoW4QBMNh3bE3J-rt2HB60Iaa7zg?e=NKcP2z.

measures of the different types of imperfections that are the measures of certainties, the measures of possibilities, the measures of necessities and the masses of belief. We infer via a set of SWRL rules our extension of Allen's algebra.

6   Validation

In this section, we present the prototype implemented based on our ontology to explore our approach followed by a case study that we conduct in the context of the Collective Memory Data.

6.1   Prototype Implemented Based on "TimeOntoImperfection"

We propose a prototype based on our "TimeOntoImperfection" ontology to validate our work. This prototype has been implemented in Java. The main interface of the prototype allows the user to enter perfect and/or imperfect time intervals and time points and to calculate certainty degrees related to each kind of imperfection based on our proposed approach. After each new temporal data entry, the "Add Qualitative Temporal Relation" component is automatically executed to infer missing data, including associated qualitative relationships and associated metrics based on SWRL rules. Our prototype also makes it possible to perform a search on the data entered and saved by using a filter integrated in a choice bar. This is implemented with SPARQL queries.

6.2   Application to Collective Memory Data

We conduct a case study in the context of Collective Memory data whose goal is to show the interest of the current work. Collective memory data relate historical facts (e.g., "National Movement" and "Festivals") or describing remarkable passages from the lives of famous personalities (e.g., "successes", "death"). It allows the semantic representation of knowledge relating to the collective memory and individual, based on an ontology called Collective Memo Onto (CMO) that we merge with our ontology "TimeOntoIperfection" to manage the temporal dimension. Let us have the following example, "The Mona Lisa is a painting of the artist Leonardo Da Vinci. He made it between 1503 and 1506 or between 1513 and 1516, and maybe until 1519". In this example, we find two kinds of imperfections which are conflict and simultaneously uncertainty and imprecision. Let I_1 and I_2 respectively the two time intervals expressing the conflict, where $I_1 = [1503, 1506]$ and $I_2 = [1513, 1516]$. Let m_1 and m_2 the belief mass respectively associated to the I_1 and I_2. We calculate it using the evidence theory based on our approach proposed in [3]. Let P be the time point which express the uncertainty and imprecision at the same time where P = 1519. Let P_{Im} and N_{Im} the possibility and necessity degrees associated to the imprecision of P; and P_{Un} and N_{Un} the possibility and necessity degrees associated to the uncertainty of P. We calculate it using the possibility theory [4].

Table 2. Temporal relations between an uncertain time interval and an uncertain time point.

Relation(A,P)	Relations	Inverse(P,A)
Before$_c$(A,P)	$P_{cp} < A_{ca-}^-$	After$_c$(P,A)
Meets$_c$(A,P)	$P_{cp} = A_{ca+}^+$	Met-by$_c$(P,A)
Starts$_c$(A,P)	$P_{cp} = A_{ca-}^-$	Started-by$_c$(P,A)
During$_c$(A,P)	$(A_{ca-}^- < P_{cp}) \wedge (P_{cp} < A_{ca+}^+)$	Contains$_c$(P,A)
Ends$_c$(A,P)	$(P_{cp} = A_{ca+}^+)$	Ended-by$_c$(P,A)

Table 3. Temporal relations between two uncertain time points P and Q

Relation(P,Q)	Relations	Inverse(Q,P)
Before$_c$(P,Q)	$P_{cp} < Q_{cq}$	After$_c$(Q,P)
Equals$_c$(P,Q)	$(P_{cp} = Q_{cq})$	Equals$_c$(Q,P)

7   Conclusion

In this paper, we present an approach to handle several types of imperfection, which are imprecision, uncertainty, uncertainty and imprecision, and conflict, that can affect the temporal data in the context of Collective Memory Data. To represent this data, we extend the 4D-fluents; we used OWL-Time ontology and we define new ontological components and using theories of imperfection such as possibility and evidence theories. To reason about imperfect temporal data, we extend the Allen's interval algebra. Based on these extensions, we propose an OWL 2 ontology named "TimeOntoImperfection". Finally, we implement a prototype based on our ontology to validate our work. In the future, we plan to treat other kinds of imperfections defined in our typology such as redundancy.

References

1. Achich, N., Ghorbel, F., Hamdi, F., Metais, E., Gargouri, F.: Representing and reasoning about precise and imprecise time points and intervals in semantic web: dealing with dates and time clocks. In: Hartmann, S., Küng, J., Chakravarthy, S., Anderst-Kotsis, G., Tjoa, A.M., Khalil, I. (eds.) DEXA 2019. LNCS, vol. 11707, pp. 198–208. Springer, Cham (2019). https://doi.org/10.1007/978-3-030-27618-8_15
2. Achich, N., Ghorbel, F., Hamdi, F., Métais, E., Gargouri, F.: Certain and uncertain temporal data representation and reasoning in OWL 2. Int. J. Semant. Web Inf. Syst. (IJSWIS) **17**(3), 51–72 (2021)
3. Achich, N.: Traitement de l'imperfection des données temporelles saisies par l'utilisateur-Application aux logiciels destinés à des patients malades d'Alzheimer. Doctoral dissertation, HESAM Université Paris France, L'Université de Sfax Tunisie (2021)

4. Achich, N., Ghorbel, F., Hamdi, F., Metais, E., Gargouri, F.: Dealing with uncertain and imprecise time intervals in OWL2: a possibility theory-based approach. In: Cherfi, S., Perini, A., Nurcan, S. (eds.) RCIS 2021. LNBIP, vol. 415, pp. 541–557. Springer, Cham (2021). https://doi.org/10.1007/978-3-030-75018-3_35
5. Achich, N., Ghorbel, F., Hamdi, F., Metais, E., Gargouri, F.: A typology of temporal data imperfection (2019)
6. Allen, J.F.: Maintaining knowledge about temporal intervals. Commun. ACM **26**, 832–843 (1983)
7. Artale, A., Franconi, E.: A survey of temporal extensions of description logics. Ann. Math. Artif. Intell. **30**, 171–210 (2000). https://doi.org/10.1023/A:1016636131405
8. Batsakis, S., Tachmazidis, I., Antoniou, G.: Representing time and space for the semantic web. Int. J. Artif. Intell. Tools **26**(03), 1750015 (2017)
9. Buneman, P., Kostylev, E.: Annotation algebras for RDFS. In: Workshop on the Role of Semantic Web in Provenance Management (2010)
10. Gammoudi, A., Hadjali, A., Yaghlane, B.B.: Fuzz-TIME: an intelligent system for managing fuzzy temporal information. Intell. Comput. Cybern. **10**(2), 200–222 (2017)
11. Ghorbel, F., Hamdi, F., Metais, E.: Dealing with precise and imprecise temporal data in crisp ontology. Int. J. Inf. Technol. Web Eng. (IJITWE) **15**(2), 30–49 (2020)
12. Gutierrez, C.H.: Temporal RDF. In: Conference (ESWC 2005), pp. 93–107 (2005)
13. Harbelot, B.A.: Continuum: a spatiotemporal data model to represent and qualify filiation relationships. In: ACM SIGSPATIAL International Workshop, pp. 76–85 (2013)
14. Herradi, N.: A semantic representation of time intervals in OWL2. In: KEOD (2017)
15. Kharfia, H., Ghorbel, F., Gargouri, B.: Typology of data inputs imperfection in collective memory model. In: Abraham, A., Gandhi, N., Hanne, T., Hong, TP., Nogueira Rios, T., Ding, W. (eds.) ISDA 2021. LNCS, vol. 418. Springer, Cham (2022). https://doi.org/10.1007/978-3-030-96308-8_111
16. Klein, M.C.A., Fensel, D.: Ontology versioning on the semantic web. In: Semantic Web Working Symposium, pp. 75–91. Stanford University, USA (2001)
17. Lutz, C.: Description logics with concrete domains. In: Advances in M.L., pp. 265–296 (2003)
18. Noy, N., Rector, A., Hayes, P., Welty, C.: Defining N-ary relations on the semantic-web. In: W3C Working Group Note, vol. 12, no. 4 (2006)
19. Nys, G.A., Van Ruymbeke, M., Billen, R.: Spatio-temporal reasoning in CIDOC CRM: an hybrid ontology with GeoSPARQL and OWL-Time. In: CEUR Workshop Proceedings, vol. 2230. RWTH Aachen University, October 2018
20. O'Connor, M.J., Das, A.K.: A method for representing and querying temporal information in OWL. In: Fred, A., Filipe, J., Gamboa, H. (eds.) BIOSTEC 2010. CCIS, vol. 127, pp. 97–110. Springer, Heidelberg (2011). https://doi.org/10.1007/978-3-642-18472-7_8
21. Sadeghi, K.M.M., Goertzel, B.: Uncertain interval algebra via fuzzy/probabilistic modeling. In: IEEE International Conference on Fuzzy Systems, pp. 591–598 (2014)
22. Tappolet, J., Bernstein, A.: Applied temporal RDF: efficient temporal querying of RDF data with SPARQL. In: Aroyo, L., et al. (eds.) ESWC 2009. LNCS, vol. 5554, pp. 308–322. Springer, Heidelberg (2009). https://doi.org/10.1007/978-3-642-02121-3_25

23. Welty, C., Fikes, R.: A reusable ontology for fluents in OWL. In: FOIS, pp. 226–336 (2006)
24. Zekri, A., Brahmia, Z., Grandi, F., Bouaziz, R.: tOWL: a systematic approach to temporal versioning of semantic web ontologies. J. Data Semant. **5**(3), 141–163 (2016)

Integrated Approach to Assessing the Progress of Digital Transformation by Using Multiple Objective and Subjective Indicators

Daniela Borissova[1,2]([✉]) [iD], Zornitsa Dimitrova[1] [iD], Naiden Naidenov[1] [iD], and Radoslav Yoshinov[2] [iD]

[1] Institute of Information and Communication Technologies, Bulgarian Academy of Sciences, 1113 Sofia, Bulgaria
{daniela.borissova,zornitsa.dimitrova,
naiden.naidenov}@iict.bas.bg
[2] Laboratory of Telematics, Bulgarian Academy of Sciences, 1113 Sofia, Bulgaria
yoshinov@cc.bas.bg

Abstract. Digital transformation affects not only IT and innovative business models and processes, it is significantly influenced by the skills and competence of top managers and new requirements and expectations of customers. The basic driver of ongoing digital transformation in any organization is performed with the active support of the Chief Information Officer, Chief Information Security Officer, Chief Technology Officer, Chief Digital Officer, or their cooperation work. Depending on the organization/company size, the number of these chiefs could be reduced and corresponding responsibilities merged. To this end, it is necessary to identify the main responsibilities and to clarify the existing hierarchy between them. To assess the progress of the digitalization process, the article proposes a multi-criteria mathematical model, the essence of which is the consideration of both objective and subjective criteria. The realized numerical application through five objective and eight subjective criteria demonstrate the applicability of described approach.

Keywords: Digital transformation · CIO · CISO · CTO · CDO · Decision-making · Multiple criteria · Mathematical model

1 Introduction

Todays the ongoing digital transformation could be recognized not only in different areas of manufacturing companies (Lee et al. 2021) but affects also non-profit organizations (Nahrkhalaji et al. 2018), smart cities (Garvanov and Garvanova 2021), tourism (Marx et al. 2021) education (Petrova et al. 2022), etc. Regardless of the field of application, the common goal of digital transformation is to improve the efficiency, value or innovation. From the manufacturing point of view, some of the most important challenges that organizations need to be handled include: traditional processes, resistance to change, legacy business mode, limited automation, budget restrictions, absence of

© The Author(s), under exclusive license to Springer Nature Switzerland AG 2022
R. Guizzardi et al. (Eds.): RCIS 2022, LNBIP 446, pp. 626–634, 2022.
https://doi.org/10.1007/978-3-031-05760-1_37

relevant knowledge, inflexible company structure and security (Albukhitan 2020). The measures that influence the digital transformation of SMEs can be represented by the four groups concerning resources, information systems, organizational structure, and culture (Schuh et al. 2017). In this regard, it worth to mention that the cloud services are essential part of contemporary information systems (Petrov 2021). Digital transformation should be priority for top management staff and leading concept of corporate business strategy (Saarikko et al. 2020). It should be mention that digital transformation could represented via following distinguished phases: 1) digitization, 2) digitalization, and 3) digital transformation (Verhoef et al. 2021).

Digital transformation can be measured by how an organization uses IT, people and processes to realize new business models and revenues, motivated by customer expectations about products and services. This means that it is necessary to identify appropriate indicators for evaluation considering that the processes of digital transformation differ depending on the specific features of the company. The number of tasks that companies are seeking to complete is increasing extensively and this is the reason to divide it in order to be effectively accomplished. There are several key positions whose integration is key to the company's digitalization: Chief Information Officer (CIO), Chief Information Security Officer (CISO), Chief Technology Officer (CTO), and Chief Digital Officer (CDO). For different companies, some of these roles could be merged or could be considered with different importance due to the company's focus. For example, for a business-oriented company, the most important is CIO, for a technology-oriented company the most important is CTO, for a digital-oriented company the most important is CDO.

The current article is focused on: 1) highlighting the importance of CIOs, CISOs, CTOs, and CDOs as drivers of the digital transformation; 2) providing a set of different objective and subjective parameters that affect the digital transformation process; 3) providing a clear mathematical model to estimate the progress of digital transformation. The rest of the article is organized as follows: Sect. 2 contains description of roles and hierarchy of CIOs, CISOs, CTOs and CDOs; Sect. 3 describe the proposed multi-criteria decision-making model along with used criteria; Sect. 4 contains the input data for numerical testing of business activities in digital transformation along with obtained results and discussion, and conclusions are draw in Sect. 5.

2 Role and Hierarchy of CIOs, CISOs, CTOs and CDOs

The Chief Information Officer (CIO) is the executive director of the company responsible for the management, implementation, and usability of information and computer technology. As technology expands and reshapes industries worldwide, the role of information directors has grown in popularity and importance. The CIO analyzes how different technologies benefit the company or improve an existing business processes. Some authors clearly show that a CIO's background itself is not as important as to whom a CIO reports (Jones et al. 2020). An investigation about the strategic role of CIOs in the situation of information technology control weaknesses is presented (Li et al. 2021). It is found that common practices by firms to retain CIOs in information technology control weaknesses situations are ineffective. It is worth mentioning the empirical results that indicate the relationship between business strategy and CIOs of certain competencies, experiences, and personalities could lead to better organizational performance (Li and Tan 2013). Combining the COBIT framework and the AHP, the authors show how to optimize the processes to provide a sound basis for consensus in the development of the CIO function in large organizations (Shalamanov et al. 2020).

The CTO is an executive-level person who focuses on creating and implementing relevant company policies in accordance with the scientific needs to meet the business goals. The responsibilities of CTOs are focused on the development of procedures and strategies, R&D, use of technology. According to Van der Meulen, four common CTO roles could be distinguished: 1) CTO as a digital business leader; 2) CTO as a business enabler; 3) CTO as IT innovator; 4) CTO as a chief operating officer of IT (Van der Meulen 2019). In the past, the roles of CIO and CTO were performed by CIO. Empirical results reveal a positive relation between gender and CTO innovation, and companies with a stronger corporate culture supporting innovation have CTO women (Wu et al. 2021).

The main responsibilities of CISO concern to information and data security issues by providing appropriate prevention and protection against information security attacks, as well as the rapid recovery of security violation (Dhillon et al. 2021). In this regard it duties include design and implementation of security program, auditing and compliance initiatives, providing proper education and training program, working with third party security vendors. A relationship was identified between the internal audit and the information security functions, which influences the objective measures for the overall information security efficiency of the organization (Steinbart et al. 2018).

The role of CDO is associated with various activities that make it possible to turn traditional operations into digital processes. Building on a large-scale sample of firms and conducted investigations, the authors show that only about 5% of S&P firms had a CDO till the end of 2018 (Kunisch et al. 2020). As a new position, the responsibilities of CDO could be delegated to external for the company person, or to create a new CDO position. In both cases, the company's performance gains are increased substantially (Mehta et al. 2021). In addition, it is found an existing relationship between the parameters of organization design and CDO activities (Singh et al. 2020). Taking into account the main responsibilities of these chiefs, the following six relations can be identified as shown in Fig. 1.

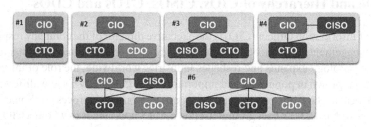

Fig. 1. Relations between CIO, CISO, CTO, and CDO

The case #1 illustrates the situation, where only positions for CIO and CTO are available in the organization. The CIO has leading role in this situation. The case #2 refers to present of three company positions for CIO, CTO and CDO. The hierarchical position in this case is taken also by CIO. Third case (#3) express also three positions for CIO, CISO and CTO where on the top in hierarchy is CIO too. Case #4 expresses the situation where CIO and CISO are at the same hierarchy level. The next two cases (#5 & #6) suppose the existence of four company positions for CIO CISO, CTO and CDO. In case #5, the CIO and CISO have equal hierarchy position while the case #6 illustrates the leading position for CIO. The common in all of these cases shown in Fig. 1 is the leading role of CIO. This means that in a micro-sized or small company situation, the CIO position must be available even if it is a part-time position.

3   Indicators and Model to Assess the Digital Transformation

To assess the success of the implementation of digital transformation, the authors propose objective and subjective indicators for measuring the results of business activities as shown in Table 1.

Table 1. Evaluation criteria to measure the performance of business activities at digitalization

Objective evaluation criteria	Subjective evaluation criteria
IT infrastructure	Strong leadership skills
Successfully implemented market innovations	Strong business communication
Return of investment	Confidence building
New customers	Problem-solving
Employee productivity	Time management
	Decision-making
	Entrepreneurial mindset
	Strategic thinking

One of the most important criteria is IT infrastructure. This is measurable objective criterion includes the development of IT infrastructure in the organization. It could be assessed as the number of introduced cloud services, digital workplaces, provision of employees with additional hardware and software. Another objective criterion is the successfully implemented market innovations that can be assessed as the number of innovations or as added business value. One of the most commonly applied market strategies is transferring to e-commerce to take the business to new geographical locations. Return of investment ratio and attracting new customers are the other two evaluation criteria of the business that affect the enterprise's progress digitalization. Last but not least is the employee productivity criterion. It can be evaluated as a profit for the company earned by one employee but this aspect can also be negatively affected by digital transformation, in case that adequate training is not provided.

The subjective criteria refer to the personal soft skills of the manager/s as CIO, CISO, CTO, CDO. These chiefs should have strong leadership and business communication skills. They need to demonstrate a confidence-building attitude in order to motivate the team that introduces the digital transformation and also to convince the other teams in the organization that digitalization will bring benefits. The C-level management including CIO, CISO, CTO, CDO, demands proving problem-solving skills and include also critical and analytical thinking. Such skills are a prerequisite that the research approach that deals with all emerging issues of the digitalization process will go smoothly. The introduction of digital tools for the application of mathematical models for group and individual decision-making, including time, and workflow management could serve as evaluation criteria. Entrepreneurial mindset and strategic thinking are also important features of success and respectively also affect the evaluation of the progress of digital transformation.

In order to measure the performance of business activities at digitalization, it is necessary to consider both criteria groups related to objective and subjective indicators. This integration requires considering these two groups separately, but within one aggregate utility function. The proposed multi-criteria model for evaluation of business activities in digital transformation is as follows:

$$DT^{performance} = \max\left\{\alpha \sum_{i=1}^{O} w_i e_i + \beta \sum_{j=1}^{S} w_j e_j\right\} \tag{1}$$

$$\alpha + \beta = 1 \tag{2}$$

$$\sum_{i=1}^{O} w_i = 1 \tag{3}$$

$$\sum_{j=1}^{S} w_j = 1 \tag{4}$$

The coefficient α expresses the importance of objective evaluation criteria while coefficient β expresses the objective ones. The coefficients w_i and w_j express the relative importance between objective and subjective criteria, e_i and e_j represents evaluation scores toward objective and subjective criteria. The range of evaluation scores expressed by e_i and e_j should have the same range as of other coefficients in the model (1)–(4). That means the acceptable range for these scores should fall between 0 and 1 to have a comparable scale. The relation (2) allows us to consider two separate parts concerning objective and subjective criteria in an integrate overall performance. The coefficients α and β make possible to realize more flexible model considering the objective and subjective criteria with different importance in the final complex evaluation.

The proposed multi-criteria model (1)–(4) can be simplified imposing the value equal to zero for the coefficient α ($\alpha = 0$) or β ($\beta = 0$). In these cases, the model (1)–(4) will rely only objective or subjective criteria in the evaluation.

4 Numerical Application

To evaluate the digitization process, the CEO was asked to set estimates for the performance of the CIO, as the test was conducted in a micro-company with the CIO. The given score toward the business activities related to the digital transformation along with weights for criteria importance are shown in Table 2.

Table 2. Evaluation score and corresponding weights for objective and subjective criteria.

Criteria		Weights for sub criteria importance		Evaluation scores	Case-1	Case-2	Case-3
Objective evaluation criteria		w_i (S-1)	w_i (S-2)	e_i	α	α	α
o-1	IT infrastructure	0.25	0.12	0.65	0.50	0.45	0.55
o-2	Successfully implemented market innovations	0.25	0.15	0.45			
o-3	Return of investment	0.25	0.26	0.54			
o-4	New customers	0.25	0.23	0.58			
o-5	Employee productivity	0.25	0.24	0.55			
Subjective evaluation criteria		w_j	w_j	e_j	β	β	β
s-1	Strong leadership skills	0.125	0.13	0.57	0.50	0.55	0.45
s-2	Strong business communication	0.125	0.15	0.78			
s-3	Confidence building	0.125	0.12	0.44			
s-4	Problem-solving	0.125	0.14	0.80			
s-5	Time management	0.125	0.11	0.56			
s-6	Decision-making	0.125	0.13	0.93			
s-7	Entrepreneurial mindset	0.125	0.09	0.78			
s-8	Strategic thinking	0.125	0.13	0.81			

Two scenarios for sub-criteria importance are presented (S-1 and S-2) and three cases that express different preferences about objective and subjective evaluation criteria. It should be mention that all estimations given in Table 2 are subjective and reflect the particular point of view of the involved CEO from a particular company. That is why this data are valid only for this company.

The comparison between described above two different scenarios under three different cases for objective and subjective criteria importance for the overall performance of company's progress is illustrated in Fig. 2.

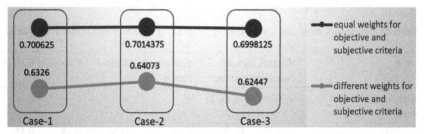

Fig. 2. Comparison of overall performance under 2 different preferences about the subjective and objective criteria weights for 3 cases

Taking into account the objective and subjective criteria with equal importance (case-1) for both scenarios (S-1 & S-2), the obtained results show a difference of 0.068025 for overall company performance. For case-2, this difference amount of 0.0607075 and for case-3 – the difference is 0.0753425 respectively.

Regardless of the variety of models that could be used, the core of these models remains in the proper identification of the evaluation criteria. It is interesting to determine the progress performance of digitalization of large-scale companies where all positions for CIO, CISO, CTO, CDO are available. This interesting direction is planned as the future development of the described in the current article approach. It is worth mentioning that the use of such an approach will not only contribute to the assessment of progress in digitalization, but will also stimulate better economic sustainability.

5   Conclusions

This article discusses the problems associated with assessing the progress of digital transformation. It has been shown that digital transformation can be measured through the IT used, the managers involved and the new business processes, as well as the revenue from improved products and services. Based on the main responsibilities of CIOs, CISOs, CTOs and CDOs, a set of different objective and subjective criteria is identified that influence the process of digital transformation. Taking into account the specifics of the described problems, a multicriteria mathematical model for estimating the progress of digital transformation is proposed. The applicability of the proposed approach has been numerically tested in the case of assessing the process of digitalization of a micro-company. The results prove the effectiveness of the proposed model, as well as the suitability of the defined two groups of objective and subjective evaluation criteria. As a future study, it is planned to determine the progress of the digitalization of large companies and, if necessary, to add new evaluation criteria.

Acknowledgment. This work is supported by the Bulgarian National Science Fund by the project *"Mathematical models, methods and algorithms for solving hard optimization problems to achieve high security in communications and better economic sustainability"*, KP-06-N52/7/19-11-2021.

References

Albukhitan, S.: Developing digital transformation strategy for manufacturing. Procedia Comput. Sci. **170**, 664–671 (2020)

Dhillon, G., Smith, K., Dissanayaka, I.: Information systems security research agenda: exploring the gap between research and practice. J. Strateg. Inf. Syst. **30**(4), 101693 (2021)

Garvanov, I., Garvanova, M.: New approach for smart cities transport development based on the internet of things concept. In: Proceedings of 17th Conference on Electrical Machines, Drives and Power Systems (ELMA), pp. 1–6 (2021)

Jones, M.C., Kappelman, L., Pavur, R., Nguyen, Q.N., Johnson, V.L.: Pathways to being CIO: the role of background revisited. Inf. Manage. **57**(5), 103234 (2020)

Kunisch, S., Menz, M., Langan, R.: Chief digital officers: an exploratory analysis of their emergence, nature, and determinants. Long Range Plan. **55**, 101999 (2020)

Lee, C.-H., Liu, C.-L., Trappey, A.J.C., Mo, J.P.T., Desouza, K.C.: Understanding digital transformation in advanced manufacturing and engineering: a bibliometric analysis, topic modeling and research trend discovery. Adv. Eng. Inform. **50**, 101428 (2021)

Li, W., Phang, S.-Y., Choi, K.W.(Stanley), Ho, S.Y.: The strategic role of CIOs in IT controls: IT control weaknesses and CIO turnover. Inf. Manage. **58**(6), 103429 (2021)

Li, Y., Tan, C.-H.: Matching business strategy and CIO characteristics: the impact on organizational performance. J. Bus. Res. **66**(2), 248–259 (2013)

Marx, S., Flynn, S., Kylanen, M.: Digital transformation in tourism: modes for continuing professional development in a virtual community of practice. Proj. Leadersh. Soc. **2**, 100034 (2021)

Mehta, N., Mehta, A., Hassan, Y., Buttner, H., RoyChowdhury, S.R.: Choices in CDO appointment and firm performance: moving towards a Stakeholder-based approach. J. Bus. Res. **134**, 233–251 (2021)

Nahrkhalaji, S.S., Shafiee, S., Shafiee, M., Hvam, L.: Challenges of digital transformation: the case of the non-profit sector. In: IEEE International Conference on Industrial Engineering and Engineering Management (IEEM), pp. 1245–1249 (2018)

Petrov, I.: Combined criteria weighting in MCDM: AHP in blocks with traditional entropy and novel hierarchy in TOPSIS evaluation of cloud services. In: Proceedings of Big Data, Knowledge and Control Systems Engineering, pp. 1–9 (2021)

Petrova, P., Kostadinova, I., Alsulami, M.H.: Embedded intelligence in a system for automatic test generation for smoothly digital transformation in higher education. In: Sgurev, V., Jotsov, V., Kacprzyk, J. (eds.) Advances in Intelligent Systems Research and Innovation. SSDC, vol. 379, pp. 441–461. Springer, Cham (2022). https://doi.org/10.1007/978-3-030-78124-8_20

Saarikko, T., Westergren, U.H., Blomquist, T.: Digital transformation: five recommendations for the digitally conscious firm. Bus. Horiz. **63**(6), 825–839 (2020)

Schuh, G., Anderl, R., Gausemeier, J., ten Hompel, M., Wahlster, W.: Industry 4.0 Maturity Index: Managing the Digital Transformation of Companies. Hacatech Study. Herbert Utz, Munich (2017)

Shalamanov, V., Sabinski, V., Georgiev, T.: Optimization of the chief information officer function in large organizations. Inf. Secur. **46**(1), 13–26 (2020)

Singh, A., Klarner, P., Hess, T.: How do chief digital officers pursue digital transformation activities? The role of organization design parameters. Long Range Plan. **53**(3), 101890 (2020)

Steinbart, P.J., Raschke, R.L., Gal, G., Dilla, W.N.: The influence of a good relationship between the internal audit and information security functions on information security outcomes. Account. Organ. Soc. **71**, 15–29 (2018)

Van der Meulen, R.: Enterprise stakeholders can realize the full value of the chief technology officer when the role is understood through four distinct personas (2019). https://www.gartner.com/smarterwithgartner/understand-the-5-common-cto-personas

Verhoef, P.C., et al.: Digital transformation: a multidisciplinary reflection and research agenda. J. Bus. Res. **122**, 889–901 (2021). https://doi.org/10.1016/j.jbusres.2019.09.022

Wu, Q., Dbouk, W., Hasan, I., Kobeissi, N., Zheng, L.: Does gender affect innovation? Evidence from female chief technology officers. Res. Policy **50**(9), 104327 (2021)

Towards Integrated Model-Driven Engineering Approach to Business Intelligence

Corentin Burnay[ID] and Benito Giunta[✉][ID]

University of Namur, Belgium, PReCISE Research Center, Namur Digital Institute,
Namur, Belgium
{corentin.burnay,benito.giunta}@unamur.be

Abstract. This paper presents a vision for an integrated and comprehensive Model-Driven Engineering (MDE) framework for Business Intelligence (BI), called BIG - Business Intelligence Generator. It starts from two observations: (i) MDE is a common approach to implement parts of a BI system and (ii) existing MDE approaches to BI are heterogeneous, not always methodologically and technically aligned, and sometimes even overlooking entire layers of the BI systems. This paper objectifies the heterogeneity of existing MDE approaches, extends on the problems it is likely to lead to, and calls for a proper end-to-end MDE-BI approach, with each layer of the MDE-BI architecture capable of proper communication and exchange with the next one. As a response, the BIG framework is introduced, under the form of a vision. The paper describes the BIG framework in general and discusses for each of its modules the benefits of the proposal. Future works required to fulfill the vision are also discussed, suggesting new avenues for research around BI and MDE.

Keywords: Business Intelligence · Model-Driven Engineering ·
Modeling · ETL · Data warehouse · OLAP · Dashboard · Automation

1 Introduction

Business Intelligence (BI hereafter) refers to the combination of technologies, softwares and models necessary to transform multiple operational data-bases – often heterogeneous in terms of structure, quality and purpose – into an integrated reporting solution. BI is an essential component of decision support in any modern organizations, and is often recognized as a critical activity [1]. It has received much research attention over the last years and can be considered as a rather mature field, well illustrated by various bibliometrics studies [2,3]. This longstanding attention to BI has resulted in a number of contributions on how to analyze, model, or otherwise specify BI systems. In particular, a number of contributions proposes the use of Model-Driven Engineering (MDE) methods as a way to deal simultaneously and more effectively with previous challenges. These MDE approaches to BI (MDE-BI) do this by defining/reusing

R. Guizzardi et al. (Eds.): RCIS 2022, LNBIP 446, pp. 635–643, 2022.
https://doi.org/10.1007/978-3-031-05760-1_38

some meta-models, each intended to model one or several layers of the BI architecture [5,6]. These layers can be summarized as (1) the integration layer (ETL), (2) the data-warehouse layer (DWH), (3) the customization layer (OLAP), and (4) the application layer (VIZ). Each meta-model comes with a set of translation rules, necessary to derive automatically the implementations of BI artifacts from model instances, thereby reducing significantly the cost of implementation, maintenance, alignment with business requirements, reuse, etc.

Despite these various advances, the proliferation of MDE-BI approaches did not completely address the BI challenge. In practice, organizations opting for MDE-BI to implement their system face an intimidating number of MDE meta-models, which are not necessarily aligned and integrated, each focusing on its own perspective on a limited number of BI layers, sometimes leaving other layers unsupported. Stated differently, the different layers of the traditional BI architecture have received unequal attention in terms of MDE-BI, while each of those parts are (i) equally important, (ii) strongly inter-dependent to deal with BI challenges and (iii) must be considered in an integrated, i.e. end-to-end, manner. This unequal and layer-centered attention to the BI technological stack is problematic in at least three regards. First, *MDE-BI models make conceptual choices and assumptions on BI layers which may not be aligned with models proposed for other layers.* For instance, a MDE model defined for the DWH layer could cover aspects such as attribute hierarchies. This very model will loose part of its value if that same concept of hierarchy is not handled downstream, through both the OLAP and the application layers. Our research to this day led to the identification of several mismatches in this regard. Second, *MDE-BI approaches are methodologically heterogeneous.* MDE is a general term, covering many different methods. Without judging the intrinsic quality of each proposal, these differences are likely to generate frictions when stacking-up the different propositions to form one single integrated MDE-BI framework. Third, *MDE-BI approaches are technically heterogeneous.* Beyond the meta-models they define, MDE-BI approaches also make choices when translating models to workable pieces of software [6]. Ultimately, this leads to a combination of several technologies required to build one single BI architecture, more likely to result in various practical difficulties when running a full BI architecture in a MDE mood.

2   Business Intelligence Generator

BI is not a sum of isolated pieces. In practice, each layer is strongly dependent on the previous ones, and influences the next ones. Assumptions and choices made in one layer, therefore, impact the entire BI system. To the best of our knowledge, this problem has never been formally taken into account and results in the inability to produce a complete BI system using MDE techniques. To fill-in this gap, our proposal is the Business Intelligence Generator (BIG) framework, depicted in Fig. 1. BIG is a combination of several MDE approaches dealing with the different layers of the BI solution in an integrated way. The result is a comprehensive MDE approach to BI, with meta-models that are connected to and aligned with each others, thereby producing models instantiations which fit together, to produce one single and coherent BI architecture. MDE's objective is to produce models – following meta-models – which can result in the automated generation of a workable solution. BIG would allow different stakeholders of the BI system to do that; with different perspectives, they will be able to specify their own part of the BI solution using visual notations, and derive most of the BI implementation automatically. BIG is still a vision, an intention. Before establishing any meta-models for each layer or formulating translation rules to automatically generate the BI artifacts, the end-to-end vision with its ultimate objectives need to be clarified, thanks this paper, and the prerequisites that would set the foundation of BIG must be identified (see Sect. 7). The vision is the result of a considerable reflection with BI researchers, practitioners and users and analysis of the state of literature. Researches about the overall architecture led us today to this proposal. The quantity of papers published on the topic of MDE for specific BI layers reassures us regarding the feasibility and relevance of such approach.

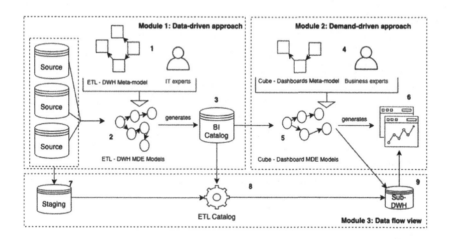

Fig. 1. Overview of the Business Intelligence Generator (BIG) framework

The overall framework is divided into three modules, each being discussed in a separate section. The first module is "data-driven module" and deals with "back-end" layers; the integration and the data warehouse (DWH) layers. The ownership of this module is clearly associated with technical experts and somehow opaque to business users. The second module is "demand-driven module" and focus on the "front-end" layers of the BI solution, including the customization and application layers. The ownership of this module goes to decision makers without entailing specific technical expertise. The last module is "data flow module" and entirely automated based on inputs from modules 1 and 2. It represents the path data would follow, from business sources to visualization applications.

It can be seen as the physical layer, the operations which will be completely managed by MDE artifacts produced by module 1 and 2.

3   Data-Driven Module in BIG

Module 1 approaches the problem of specifying the DWH of a BI system from the data sources available. It is a well recognized method [4], in which data are analyzed, transformed and pushed into the DWH regardless of the requirements of the business. The logic is that application and customization layers extract the data they need from the vast offer of the DWH. This approach has the advantage to involve only IT-experts and clearly dissociate the business from the DWH design phase. There are two essential parts in this module. As in any MDE approach, an **ETL-DWH meta-model** is necessary (Fig. 1 - item 1) to help IT-experts within the organization to specify: (i) the *tables* of the future DWH (facts and dimensions); (ii) the *fields* of each table in the DWH (attributes, measures, keys); (iii) the *data sources*; (iv) the *mappings* of each field to a data source within the ETL; (v) the *transformations* necessary on each mapping.

The instance models (Fig. 1 - item 2) produced by IT-experts based on the ETL-DWH meta-model (Fig. 1 - item 1) are then simply stored in the **BI Catalog** (Fig. 1 - item 3). At this point, there is no generation of any artifact; the BI catalog simply holds the definition of a potential future DWH as described by IT-experts. The catalog is useful in several regards: (i) stores the specification of a DWH for later code generation; (ii) exposes to downstream BI layers the list of all items available to applications in a "business-ready" fashion (data presented as a proper star-schema, leaving aside technical fields, with clean-and-ready columns, etc.); (iii) facilitates import of DWH patterns from other organisations, as a way to reduce the cost of analysis. For instance, a company willing to report on Human Resources costs could share its star schema, which could be reused by other companies.

4   Demand-Driven Module in BIG

Module 2 approaches the problem of specifying the BI applications based on business requirements, following a demand-driven approach [4]. The demand, however, would not be addressed directly to the IT-experts but would rather be expressed relative to the BI Catalog, acting as an interface between the two worlds. Business-experts would therefore produce their own requirements in terms of reporting by building on the BI Catalog, exposing in a "business-ready" fashion all business data-sources available for the application layer. To do this, BIG should propose an **OLAP-VIZ meta-model** (Fig. 1 - item 4) to help business-expert within the organization to specify (Fig. 1 - item 5): (i) the *business goals* that an expert wants to control with its applications; (ii) the *indicators* necessary to monitor these goals; (iii) the *measures* and *dimensions* required to produce the indicators; (iv) the *visualisations* expected for each indicator; (v) the *reporting features*, like drill-down, hierarchies, filters, etc.; (vi) the *links* to the BI Catalog required to feed the applications.

In this view, it is important to emphasize the complete separation of concerns between IT and business experts (respectively module 1 and 2). Technical artifacts related to the data loading and transformations are stored in the BI Catalog, and the instantiation of these loading and transformations are performed only if requested by business-experts via a *link* between the BI Catalog and their own OLAP-VIZ models. With properly defined OLAP-VIZ models, business-experts produce everything that is required to automatically generate dashboards that are aligned with their business strategy (formulated under the form of *business goals*). Most importantly, they can specify the data to be used to feed those dashboards without the intervention of IT-experts. They can easily produce new dashboards aligned with their own requirements, or update existing ones (Fig. 1 - item 6).

5   Data Flow Module in BIG

Previous claims of automated BI generation and separation of concerns significantly depend on module 3 of the BIG framework, which should be seen as the foundation – physical – layer of the framework. It is the module "where everything happens", where the data literally flows from the business operational data sources to the reports and dashboards. The data flow module calls for MDE approaches in modules 1 and 2 that are consistent with each others and that adopt compatible technologies. At this stage of the research project, we prefer to remain agnostic from a technological point of view, and only want to stress out the importance to opt for a technology that ensures sufficient flexibility to adapt to BIG agility while ensuring good performance to the OLAP and application layers. Module 3 counts three important parts.

The first part is the staging area (Fig. 1 - item 7). Staging is a common practice in most BI systems; it offers an intermediary area between the BI architecture (and its frequent heavy SQL queries) and the operational databases of the company, and helps reducing the impact of BI system on the daily activities of a business. Technically, it is nothing more than a copy of the different data sources, performed at regular time-intervals. The staging is particularly important in the BIG framework, because the automated generation of dashboards and the running of SQL queries again data-sources are likely to happen at any time, when requested by the business-experts (remember the separation of concerns between IT and business). It will therefore absorb the impact of the reporting activities, which cannot be anticipated and/or scheduled.

The second part is the **ETL Catalog** (Fig. 1 - item 8). It can be seen as the flip-side of the BI Catalog. While the BI Catalog contains a business-friendly representation of all data made available by IT-experts to business-experts, the ETL Catalog contains all the technical manipulations required to transform heterogeneous and unintegrated data sources into the data as depicted within the BI Catalog. While the BI Catalog answers the "What" question, the ETL Catalog answers the "How" question. This means any entry in the BI Catalog comes with a set of entries in the ETL Catalog specifying how the source data should be transformed and loaded when requested by business-experts. Just like its sibling, the ETL Catalog does not contain any data from the business; it only contains the specification of a potential ETL, which is not developed yet. The ETL catalog is useful in several regards: (i) stores the specification of an ETL process for later code generation; (ii) hides all technical details from the business-experts, who cannot access, modify or somehow alter the ETL catalog; (iii) dissociates data offer from its future implementation. In case of changes in the data sources, the BI Catalog remains unchanged and the IT-experts adapt the ETL Catalog to keep delivering the data; (iv) facilitates the import of ETL patterns from other organisations, together with a pattern of DWH that would be reused in the BI Catalog.

The third – and probably most critical – part is the so-called "**Sub-data warehouse**" (sub-DWH) (Fig. 1 - item 9). The need for a sub-DWH starts from a simple observation; business-experts virtually never make use of every single piece of data available in a DWH. Hence, implementing the full DWH together with the full ETL Catalog represents a huge investment for an organization, with potentially low returns. Implementing only what is needed is a way to maximize this return on investment. Sub-DWH does this in an automated way; starting from the models specified by business-experts using OLAP-VIZ meta-model, BIG will identify which pieces of data from the BI Catalog are necessary, which loading processes from the ETL Catalog are required, and will automatically produce the scripts implementing dashboards and underlying sub-DWH and ETL processes. The benefits of implementing sub-DWH are multiple: (i) optimize return-on-investment from the implementation of DWH and ETL; (ii) reduce the time to refresh the DWH, since only the necessary data are to be refreshed during the ETL; (iii) reduce the disk-space requirements of a large DWH.

The sub-DWH is a non-permanent data structure which feeds the front-end layers of the BI system on-demand. It is maintained as long as it is used by downstream OLAP cubes or dashboards, in which case BIG will put in place a mechanism for automated refresh of its data at regular intervals. It can easily be updated, based on changes on the OLAP-VIZ models by business-experts, or it can be dropped, if the sub-DWH is not used by dashboards anymore. Ultimately, the sub-DWH becomes a temporary database constantly aligned with business requirements, controlled by the business-experts at virtually no cost (beside the cost of specifying the BI and ETL Catalogs) for the IT-experts.

6   Design Process in BIG

BIG allows the participation of two different perspectives – IT and business experts – in the design process of a BI system in a continuous, independent and simultaneous manner. As already discussed, this participation happens seamlessly, without strong dependence between actors. Unlike traditional approaches,

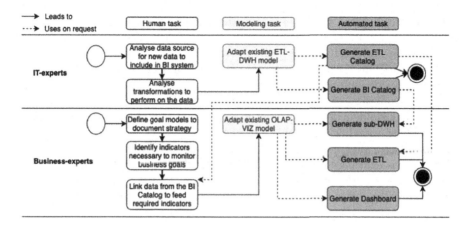

Fig. 2. Overview of the Business Intelligence Generator (BIG) process

IT-experts can specify all the data they have at hand and want to expose to business-experts in a self-service mood, without worrying about how these data will actually be used. This is possible by enriching their ETL-DWH model, which will in turn generate new offer within the BI Catalog. This process happens in a continuous way, without any strong interaction with the business itself (Fig. 2). In parallel, business-experts can specify the dashboards they need to monitor the achievement of business goals, and adapt, without delay, the indicators and underlying data necessary to operationalize such reporting. This is possible by enriching their OLAP-VIZ model, which will in turn generate new sub-DWH and related dashboards.

Ultimately, most of the interactions between the IT-expert and the business-expert modules occur in the automated tasks column, where no human input is required. The BI Catalog and ETL Catalog are automatically generated based on the ETL-DWH model. The BI Catalog is required, together with the OLAP-VIZ model of business-experts, to generate the sub-DWH. The ETL, in turn, is generated based on the ETL Catalog and the OLAP-VIZ. Dashboards can be generated based solely on the OLAP-VIZ model. Each automated task, part of the MDE-BI approach, are fed by models. The only dependence between the two modules is that business-expert are expected to refer to the BI Catalog when specifying links in their OLAP-VIZ models, and cannot proceed if necessary data are not documented yet in the BI Catalog.

7   Discussion and Future Works

BIG is an attempt to build a comprehensive and integrated MDE-BI approach to BI architectures, with the objective to automate the end-to-end implementation of BI systems. MDE approaches in general are challenging, especially with MDE-BI, considering (i) the multiple layers architecture of BI systems, (ii) the multiple stakeholders who intervene in the system definition and (iii) the decision support nature of BI, exposing the system to changes and adaptations in an uncertain world. BIG is not a disruptive approach, and does not challenge all existing contributions on the topic. On the contrary, we see it as the next logical step towards automation of BI implementation, building on various existing MDE-BI proposals. The framework clarifies the duties, use-cases and outputs for each stakeholders of a BI system. It also brings a number of benefits to these stakeholders; spending time only on what is necessary, simplified BI maintenance, reduced alignment cost, real self-service approach, reusing BI solutions. This paper is drawing up the blueprint of BIG which is still a preliminary work. We transpose the prerequisites of BIG into (i) general and (ii) module-specific future works hereafter. The aim is to clarify, for each module (Fig. 1), the next steps to handle to move forward a real and comprehensive MDE-BI framework.

(i) The General Future Works – relevant to any BIG Module. First, a *Comprehensive Literature Review* of all Model-Driven Framework for BI is necessary. Such review will allow reuse and extension of framework, will better clarify the gaps in each layer/BIG Module, and understand which adaptations are needed in the different layers in order to make them compatible. Second, there is a need for a *Business Intelligence Ontology* – such ontologies already exists, but tend to focus on particular layers of the BI architecture. Research could go on the definition of an ontology that will be unified and common to every module integrating transition from one to another. This work would contribute to one end-to-end capability of the BIG framework.

(ii) Module specific Future Works – First, *Module 1* necessitates an extension of the ontological representation of the BI catalog, with a proper mapping of the business terminology to key database-related concepts and BI concerns. Second, *Module 2* requires to derive from the ontology and meta-model of (i-2) a Dashboard Meta-Model, on top of which Model Transformation Rules should be defined to switch from a Business Model to a proper BI Dashboard implementation. Third, *Module 2* also relies on the ontological representation of BI Analytical Features and End-User Business Requirements. Such representation would offer a clear perspective on which analysis is required for which specific business need. This could be achieved by exploring well-known BI platforms's functionalities, by leading empirical studies to evaluate how each of these feature helps business users and thereby derive a graphical notation to help business specify their dashboards.

References

1. Ain, N., Vaia, G., DeLone, W.H., Waheed, M.: Two decades of research on business intelligence system adoption, utilization and success - a systematic literature review. Decis. Support Syst. **125**, 113113 (2019)
2. Chen, C.: Storey: business intelligence and analytics: from big data to big impact. MIS Q. **36**(4), 1165 (2012)

3. Liang, T.P., Liu, Y.H.: Research landscape of business intelligence and big data analytics: a bibliometrics study. Expert Syst. Appl. **111**, 2–10 (2018)
4. Ouaret, Z., Boukraa, D., Boussaid, O., Chalal, R.: AuMixDw: towards an automated hybrid approach for building XML data warehouses. Data Knowl. Eng. **120**, 60–82 (2019)
5. Schmidt, D.C.: Model-Driven Engineering. IEEE Computer Society (2006)
6. da Silva, A.R.: Model-driven engineering: a survey supported by the unified conceptual model. Comput. Lang. Syst. Struct. **43**, 139–155 (2015)

Method for Evaluating Information Security Level in Organisations

Mari Seeba[1]([✉])[ID], Sten Mäses[2][ID], and Raimundas Matulevičius[1][ID]

[1] Institute of Computer Science, University of Tartu, Tartu, Estonia
{mari.seeba,rma}@ut.ee
[2] Department of Software Science, Tallinn University of Technology, Tallinn, Estonia
sten.mases@taltech.ee

Abstract. This paper introduces a method for evaluating information security levels of organisations using a developed framework. The framework is based on Estonian Information Security Standard categories which is compatible with ISO 27001 standard. The framework covers both technical and organisational aspects of information security.

The results provide an overview of security to the organisation's management, compare different organisations across the region, and support strategic decision-making on a national level.

Keywords: Cybersecurity · Information security · Security evaluation · Maturity framework · E-ITS · Estonian case

1 Introduction

This study is motivated by the need to evaluate the security posture on a national level. For evaluating the security posture on a national level, the security levels of relevant organisations need to be measured in a standardised way. Information security standards such as ISO/IEC 27001 can provide helpful guidance for securing information assets, but lack methods for gathering quantified actionable metrics to compare different organisations and observe changes in an organisation.

This paper describes the framework for security measurement method and shows how it is used for information security assurance in Estonia. Although this paper focuses on measuring organisations information security levels in Estonia, a similar approach could be adapted in other countries. Specifically, our study investigates *how to evaluate the level of information security of an organisation*.

2 Related Work

When developing any framework for information security evaluation, several information security standards and frameworks can be used as a starting point,

R. Guizzardi et al. (Eds.): RCIS 2022, LNBIP 446, pp. 644–652, 2022.
https://doi.org/10.1007/978-3-031-05760-1_39

e.g., ISO/IEC 27k family, BSI IT-Grundschutz, CIS 18 (Center for Internet Security Critical Security Controls for Effective Cyber Defense), etc. Many standards provide maturity models, but there is a lack of a benchmark solution or model to support reliably comparable results [2]. Shukla *et al.* [9] have defined correctness, measurability, and meaningfulness as the core quality criteria of security metrics. They emphasis that systematic and complete security metrics are needed in the decision-making process. Although there are commercial tools that relate security metrics to vulnerability management and policy compliance [1], those tools tend to lack transparency and accessibility.

Le and Hoang [3] highlight that security maturity models support security posture. Such a model should (*i*) have consistent and justified maturity levels of cybersecurity across different domains; (*ii*) introduce quantitative metrics for any security assessment (in contrast to mainly qualitative metrics/processes in international standards such as ISO 27k series and NIST cybersecurity framework); (*iii*) maintain a flexible model to facilitate the inclusion of new topics connected to novel technologies. They also argue that security models could balance the qualitative assessment for management and quantitative assessment for security experts [3]. In this paper, we present a publicly accessible security measurement framework that utilises a maturity model and addresses the above challenges and shortcomings. Our framework provides measurable, systematic and transparent results, and is flexible to changes and compliant to internationally recognised standards (e.g., ISO27001).

3 Creation of Security Evaluation Framework

Our study followed Design Science Research Methodology (DSRM) [5]. Firstly, we identified the problem and elicited requirements to assess security level of organisations. Next, following the E-ITS security catalogue[1], we designed the maturity framework, which consists of ten dimensions, four maturity levels and attributes. We invited ten organisations for framework demonstration and testing experiment. Finally, we updated the framework according to the evaluation results. The resulting maturity framework is published in [8].

Problem Identification. For implementing an information security standard, an organisation needs to understand what should be changed and what is the impact of that change. Similarly, to make decisions at the state level, the government needs data to plan and estimate the security strategy. In Estonia, a new information security standard E-ITS [7] is being introduced. Hence, each implementer needs to assess its level of information security compared to set objectives for improving and monitoring its progress.

Security measuring is challenging because of dependencies, multidimensionality, dynamics, gain and loss perception biases, as well as probe effect caused by the security measurement process [6]. The problem is: *How to evaluate the*

---

[1] https://eits.ria.ee/

Table 1. Requirements for the security level evaluation

Req. 1	**Framework should cover a wide area of security-related topics**
	The tool should cover both procedural and technical measures. If a security evaluation focuses only on human aspects (e.g., HAIS-Q [4]) or only on technology, then it is likely that things remain unnoticed. Comprehensive categories should still allow minor modifications or additions to the more specific topics as the technology evolves. Specifying the aims of the security controls should be preferred, instead of defining particular technology that will likely be outdated soon. It should be possible to categorise any upcoming security control to an already existing category
Req. 2	**Framework should produce quantifiable and comparable results**
	The organisation should have the possibility to observe changes to its security level throughout regular security evaluations. Changes should be based on evidence, not gut feeling [3]
Req. 3	**Framework should be quick and easy to implement and understand**
	While the actual implementation of the security controls might take a long time, the evaluation should be intuitive to follow and take less than 1 h
Req. 4	**Framework should be aligned with a security standard**
	Following the standard structure helps to give the measurements a more coherent structure and avoids extra effort done to comply with the standard

information security level of the organisation without passing the complete gap analysis and how to evaluate the security level of the organisations comparably?

Solution Objectives Definition. To have an actionable evaluation of the information security level of the organisations, a set of requirements must be met. The requirements (Table 1) were gathered from the literature, authors' previous experience, stakeholders' expectations, and legislation. It was acknowledged that the comprehensive coverage (Req. 1) and standard-based requirements (Req. 4) are similar. We found that although comprehensive coverage might be simple to achieve with the standard, security evaluation based on a standard does not always give comprehensive coverage of topics. For example, some standards cover only technical topics.

Design and Development. Based on elicited requirements, we defined the scope of our framework design and developed the security evaluation framework. Figure 1 illustrates the process of maturity model framework design and development.

Baseline Standard. We used the Estonian information security standard (E-ITS) [7] for creating our framework. E-ITS is compliant with internationally recognised ISO27001 and has comprehensive measures, which satisfies simultaneously the Req. 1 and Req. 4. E-ITS is a baseline standard, which helps to reduce the time assigned to risk assessment for typical assets by providing ready-to-use security measures for the organisations. Only the implementation of E-ITS baseline controls is evaluated and organisation's uniqueness or special needs are not considered. Such scope produces a general E-ITS-based benchmark for evaluating the security level (Req. 2).

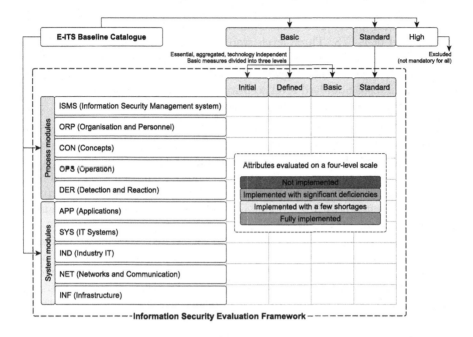

Fig. 1. Maturity framework

Dimensions of the Model. After setting the baseline standard as the base of our framework, the maturity model for security level evaluation formed naturally. The E-ITS contains measures, which are divided into ten module groups (Fig. 1) that were used as systematic dimensions, therefore ISMS, ORP, CON, OPS, DER module groups were procedural, and INF, NET, SYS, APP, IND module groups were system based technical modules. All module groups have around 90 sub-modules. For example, the OPS module converges several subtopics like outsourcing IT services, cloud service usage, and patch and change management. To meet the objectives of Req. 3, we did not transfer the sub-modules from the standard to the framework. If we go by the dimensions and predefined measures levels, this approach adds comparability to our framework and follows Req. 2.

Framework Levels. E-ITS measures are ordered Basic, Standard and High as illustrated at the top of the Fig. 1. We decided to exclude the High measures from our scope to align with Req. 2 – to limit the benchmark only on the general baseline part, which is mandatory to all organisations. Standard level measures of E-ITS we allocated directly into **Standard level** attributes of our model. E-ITS Basic level attributes we divided into three levels to enable the organisation to set reachable goals and allow more granular measurement of their improvement. We named the starting point level as **Initial Level**. At this level, the organisation solves its security issues *ad hoc* and on a need-based. The organisation is risk-sensitive and unable to deal with incidents on its own. **Defined Level** was designed to separate formal compliance documentation requirements from the processes taking place. The Basic level of our framework is complemented by **Basic Level** of E-ITS. Three levels were used (Initial, Defined and Basic) to evaluate the organisational readiness to respond to known threats and direct the organisation to the documented, trained and optimised **Standard Level** security. Achieving Standard level security allows the organisation to deal with unknown risks by significantly reducing their potential impact and loss.

Attributes of the Framework. We aimed to populate a framework with attributes so that the respondent could find evidence for each attribute implementation status if necessary. The result should be as unambiguous as possible, giving a similar outcome for the same role and information space holder. We needed both, measuring the attributes of the performance (e.g. "policy exists", "backups are done") and effectiveness (e.g. "policy is reviewed during the year", "logging is targeted and regularly monitored").

We avoided direct technology metrics to follow Req. 2. Still, we needed to keep technical measures. We customised the measures by connecting dependencies, generalising, and focusing on the essentials. For example: if there is a channel for security incident reporting, we tied the incident registration into the same attribute. We excluded all direct technology related measures or aggregated them into a general attribute (e.g. if the control was "Organisation has Office product requirements list", we aggregated it into attributes: "Requirements lists are created for applications", and included under that attribute also other applications measures like calendar, browser, etc.). Dividing attributes into levels, we followed the rules: Initial level attributes describe *ad hoc* behaviour or measures that are implemented are not managed or maintained; Defined level with formal attributes (defining policies or rules), the Basic level was populated with actual procedures and activities which were not covered by previous levels; Standard level attributes follow E-ITS Standard measures.

Evaluation Scale. We constructed the evaluation scale for attributes evaluation, which should align with Req. 3. Our first preference was the traffic light solution, which is cognitively simple to understand. The green marks attribute as fully implemented, yellow means partly implemented, and red indicates not implemented situation. The respondent's task was to colour the attributes accordingly. Based on validation feedback (see Sect.4) we decided to provide a scale with four levels.

The four-level scale supports Req. 2 to show the dynamics of organisation security even in the case of minor changes. In this case, yellow was divided into mostly implemented with some shortages (still yellow), and orange marking the significant deficiencies but still partly implemented. The results of three framework iterations are available at [8].

Demonstration. At first we received responses from 10 public or vital service providers, the size of organisations were from 30 to more than a thousand employees with subdivisions. In the second iteration we received responses from the public authority with around 350 employees. The third iteration was done by

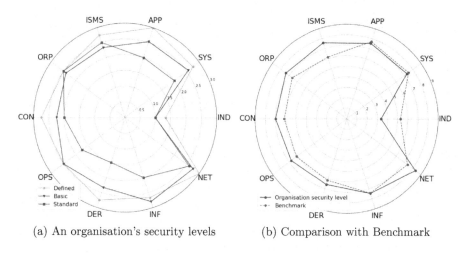

(a) An organisation's security levels   (b) Comparison with Benchmark

Fig. 2. Organisation security evaluation and comparison with benchmark (Color figure online)

expert, one author of this article, who had not participated in previous framework development phases. Each respondent was an information security officer, IT manager or data protection officer of subject organisations. Work experience in this role was from one year to more than ten years. To avoid potential leakages of vulnerabilities or security measures and due to the Public Information Act, we kept the respondents' characteristics at a general level.

Testing of the framework is conducted in three iterations. In the first and second iterations, participants received the respective file [8] in DOCX format to keep the low entrance barrier (avoiding specialised tool, login requirements or unknown formats). Each respondents task was to perform the organisation self-assessment by highlighting the maturity model framework attributes (sentences) using the evaluation scale (the traffic light colours). The data was gathered into one file, which was processed manually. At the third iteration, the expert analysed the framework structure, principles of the levels, and attributes. Each suggestion was later explored and taken into account by consensus in joint debate with the primary author of the framework.

Results Interpretation. Interpreting the results we followed only the dimensions based results, which gives us the option to change the attributes in the dimension, but the full framework still stays comparable with previous results. This approach supports the planned E-ITS yearly updates. **Use case 0.** The organisations used the coloured table to interpret organisation security using traffic light colours and the dominant visual colour to indicate the current security status before calculations. **Use case 1.** We transferred the colours into quantifiable form to perform the calculations (red equated with 1, yellow with 2 and green with 3; a four-level scale: red signifies 0, orange 1, yellow 2 and green 3). Then we calculated the average result of the organisation by each dimension and maturity level and visualised it on a radar diagram (example for illustration on Fig. 2a).

The figure indicates that the Defined level is in good shape, all dimensions values are higher than 2, except the IND dimension (it has a low value in all levels). They have not defined that it could be a security issue and have not taken industrial assets into their security management scope. Formal procedures are defined (Defined level results) but not enough training or exercises are performed (Basic level results). A high score in the Standard level gives the organisation and its partners the confidence to manage the unknown threats with minimal loss. **Use case 2.** To simplify the model, we sum each level's average value by dimension and get the information security level of organisation by dimensions (Fig. 2b blue line). I.e., a partner can not be confident about this illustrative organisation case because the DER result indicates shortages in incident management. **Use case 3.** We calculate the average value of information security level by dimension based on all organisations to get the benchmark values. The benchmark of our test-group results is shown in Fig. 2b red line. Benchmark is important for organisations to assess the status in the market or the partnership. In an illustrative case, we can say that organisation is a bit over the benchmark in the procedural dimensions, but lower values are in the technical dimension. **Use case 4.** The benchmark is also the input for state-level political and strategical decisions. Currently we estimate, based on the benchmark model, that few organisations have reached the Standard level (average dimension levels over 6) and can manage with unknown threats. Exercises and training may be needed to achieve a maturity value over 6 and 7.

Thus, our study has provide a deployment-ready model for organisations security level evaluation and an opportunity to centralise results at a state level.

4   Evaluation

We evaluated the framework compliance to stated requirements (Table 1) using semi-formal methods (feedback seminar with respondents, written feedback, interview with security expert).

Req. 1: The respondents confirmed that the framework gave an overview of the security in the scope and implementations and was still perceptible to all respondents. **Req. 2:** Few respondents noticed some duplicates in the framework, which we removed. They suggested some aesthetic issues related to the presented results. Additionally, they indicated that the results attained managers' attention and contained the motivational aspect to compare with a benchmark. One respondent provided an improvement suggestion, which was to replace the three traffic light evaluation scale with a four-level scale. This was to force the respondent to decide whether the situation is somewhat positive or rather negative. **Req. 3** was reflected with time spent (between 30 min (IT officers) to 2 h (data protection officers), average 60 min). Also, they confirmed that no entrance barriers were detected. **Req. 4** get a positive response that framework attributes gave a compact overview of E-ITS requirements: "One hour, and I got the full picture!"

The independent expert analysed the framework dimensions and levels based on literature and related work and agreed with national standard structure elements and measures. Also, attributes divided into Initial, Defined and Basic levels were double-checked to follow the internal rules (see in Sect. 3 paragraph Framework levels).

The first iterations indicated some lingual limitations. Respondents were not able to colour attributes from the initial level. To resolve the problem we revised the framework so that the higher level attributes would also contain lower level attributes. We excluded the first level results from our examples (see Fig. 2a) to avoid the possible misinterpretations.

Limitations. This work only gives insight into how it is possible to interpret the provided model data. To validate the method as an acceptable benchmark, a bigger reference group is needed. Woods and Böhme [10] have shown how indicators of security have little explanatory power alone. Furthermore, any security guideline contains the inherent weakness of generalisation. Measuring information security is a complex task, and the validity of any metrics should be considered with care and updating the attributes needs expert review.

Acknowledgement. This paper is supported in part by European Union's Horizon 2020 research and innovation programme under grant agreement No 830892, project SPARTA.

References

1. Bannam, R.: Cyber scorekeepers: a growing number of ratings firms aim to help companies and their insurers assess and manage cybersecurity risks. Risk Manage. **64**(10), 26–30 (2017)
2. Center of Internet Security: Blog — CIS Introduces v2.0 of the CIS Community Defense Model, September 2021. https://www.cisecurity.org/blog/cis-introduces-v2-0-of-the-cis-community-defense-model/
3. Le, N.T., Hoang, D.B.: Can maturity models support cyber security? In: 2016 IEEE 35th International Performance Computing and Communications Conference (IPCCC), pp. 1–7 (2016)
4. Parsons, K., McCormac, A., Butavicius, M., Pattinson, M., Jerram, C.: Determining employee awareness using the human aspects of information security questionnaire (HAIS-Q). Comput. Secur. **42**, 165–176 (2014)
5. Peffers, K., Tuunanen, T., Rothenberger, M.A., Chatterjee, S.: A design science research methodology for information systems research. J. Manage. Inf. Syst. **24**(3), 45–77 (2007)
6. Pfleeger, S.L., Cunningham, R.K.: Why measuring security is hard. IEEE Secur. Priv. Mag. **8**(4), 46–54 (2010)
7. RIA (Estonian Information System Authority): E-ITS. https://eits.ria.ee/
8. Seeba, M.: Estonian Information Security Standard (E-ITS) Based Security Level Evaluation Instrument (2021). https://doi.org/10.23673/re-298
9. Shukla, A., Katt, B., Nweke, L.O., Yeng, P.K., Weldehawaryat, G.K.: System security assurance: a systematic literature review. arXiv:2110.01904 [cs] (2021)
10. Woods, D.W., Böhme, R.: SoK: quantifying cyber risk. In: 2021 IEEE Symposium on Security and Privacy (SP), pp. 211–228 (2021)

ChouBERT: Pre-training French Language Model for Crowdsensing with Tweets in Phytosanitary Context

Shufan Jiang[1,2(✉)] , Rafael Angarita[1] , Stéphane Cormier[2] ,
Julien Orensanz[3], and Francis Rousseaux[2]

[1] Institut Supérieur d'Electronique de Paris, LISITE, Paris, France
{shufan.jiang,rafael.angarita}@isep.fr
[2] Université de Reims Champagne Ardenne, CReSTIC, Reims, France
{stephane.cormier,francis.rousseaux}@univ-reims.fr
[3] Cap2020, Gradignan, France

Abstract. To fulfil the increasing need for food of the growing population and face climate change, modern technologies have been applied to improve different farming processes. One important application scenario is to detect and measure natural hazards using sensors and data analysis techniques. Crowdsensing is a sensing paradigm that empowers ordinary people to contribute with data their sensor-enhanced mobile devices gather or generate. In this paper, we propose to use Twitter as an open crowdsensing platform for acquiring farmers knowledge. We proved this concept by applying pre-trained language models to detect individual's observation from tweets for pest monitoring.

Keywords: Transfer learning · Crowd-sensing · Plant health monitoring · Twitter

1 Introduction

Crowdsensing is a sensing paradigm that empowers ordinary people to contribute with data sensed from or generated by their sensor-enhanced mobile devices [1]. It introduces a new shift in the way we collect data by permitting to acquire local knowledge through smart devices carried by people, such as smartphones, tablets, smartwatches, among others. This allows to leverage enhanced sensors of smartphones in a fast and economical way, in contrast to more expensive traditional methods. Driven by the increasing recognition of the importance of farming to sustain humanity and the central role of farmers in the digitization of agriculture [3], we have witnessed the emergence of crowdsensing applications for smart farming [7]. Farmers are more than ever present in social media such as Facebook, WhatsApp, and Twitter [12], where they report their issues and discuss. They also search solutions to existing problems in online groups. Particularly, Twitter allows farmers to freely publish short messages called "tweets" to share their observations. Taking advantage of these observations requires of keeping track of relevant data sources among noise, extracting and organizing the information they contain and sharing it with other interested users is only possible at a high human effort, by manually inspecting, filtering and cleaning all data and connecting related entities and contexts.

Recent applications of large-scale pre-trained language models seem promising for tackling domain-specific information extraction problems from text in French [4,11]. In this work, we propose to build ChouBERT, a pre-trained language models that "learns" knowledge in plant health domain from French plant health bulletins (BSV, for *Bulletin de Santé du Végétal* in French) and recognizes similar syntax in tweets for detecting farmer's observations in French phytosanitary context. Driven by **the increasing connectivity of farmers and the emergence of online farming communities**, our goal is to explore the emerging application of on-farm observations via social networks -particularly Twitter- and propose an approach for tweet classification. We aim to answer the following research questions: *RQ1. how pre-trained language models (LM) can assist in the exploration of tweet-based crowd observations?*; and *RQ2. how to further pre-train general LMs for domain specific text classification?*

In the next section, we review related work. Then, we formalize the problem and our approach in Sect. 3. We present our solution ChouBERT in Sect. 4, we discuss the threats to validity in Sect. 5, and we give our conclusion in Sect. 6.

2   Related Work

In regard to existing works on plant health monitoring using Twitter, [16] builds different keyword-based queries to retrieve tweets about the Bogong moth and the Common Koel and compared the number of tweets with regularly planned surveys to validate the queries. This approach requires human efforts for building queries with hazard names or symptoms, and presents a problem for using Twitter to detect unfamiliar biosecurity events. [10] gathers tweet about 14 fungal diseases and proposes supervised tweet classification with Machine Learning and word embeddings. Their good accuracy proves the feasibility of categorizing tweets for monitoring known crop stresses. However, word embedding-based representations demand for disambiguation. This work also lacks in generalizability on unknown categories of tweets. In our work, we propose to apply domain-specific contextualized embedding to improve the generalizability of classifiers on unknown hazards.

3   Approach

We define individuals' observation as: a description of the presence of pest in a field in real time. However, unlike domain-specific reporting applications which frame observations in a predefined way, observations on Twitter are documented in free text or images. These observations may also exist among irrelevant tweets. These tweets may be missing essential information, such as precise location, impacted crop, the current developing status of a pest and the estimation of upcoming damage. The knowledge that helps to recognize farmers' observations can be found in: vocabularies of French crop usage [9] as formal knowledge; BSV as semi-structured domain knowledge; pre-trained LMs as knowledge of French language; tweets labelled by domain experts, containing tactical knowledge; and unlabelled tweets concerning crops or plant health issues, as a corpus of the syntax of tweets.

Figure 1 illustrates an overview of the different steps of our approach : i), data preparation: we collected tweets using keywords. Then, we invited domain experts to label a set of tweets about known issues, so that we include the experts' interest in the objective of supervised learning; ii), further pre-training: adjust the weights of pre-trained encoders using Masked Language Model (MLM) [2] with BSV and raw tweets to integrate the knowledge about plant health and the writing style of tweets in order to better project features of tweets in vectorial space; and ii), supervised tweet classification: we train classifiers with different LM representations to distinguish observations from other information.

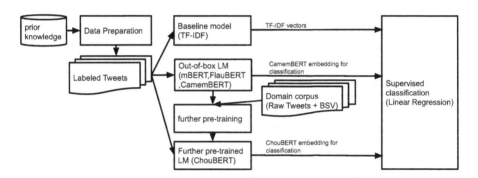

Fig. 1. Overview of the steps of the approach in the study

4   Experiments

4.1   Data Preparation

We worked in collaboration with Arvalis (www.english.arvalisinstitutduvegetal. fr) to label tweets concerning observations about 5 different natural hazards. For each of these hazards, we invited plant health researchers to label tweets according to their judgement of their pertinence. Table 1 shows the composition of our labelled set. We collect tweets for at least 2 years. We use the tweets about corn borers, corvids and barley yellow dwarf virus (JNO) to construct the training set. Of these 1358 tweets, 396 are labelled as observation (positive case). To evaluate the generalizability of our classifier on unseen hazards, we use the tweets about cording moths as supplementary training data and tweets about wireworms as supplementary test data. We chose wireworm because the word *taupin* is polysemous in French, these tweets contains many unseen noises.

Table 1. Composition of the labelled set

Hazard	French hazard name	Period	Total	Num. of observation
Corn borer	Pyrale du Maïs	2019.1–2020.12	266	56
JNO	Jaunisse Nanisante de l'Orge	2016.1–2020.9	625	229
Corvids	Corvidae	2009.8–2020.12	467	111
Coding moth	Carpocapse	2009.11–2021.9	362	49
Wireworm	Taupin	2010.3–2021.9	394	33

We downloaded 40828 BSVs [14] from data.gov.fr. 17286 BSV are in XML, and 23542 are in plain-text. A first processing is to convert XML file to plain text. A cleaning step is also necessary to remove punctuation artifacts or out-of-context agricultural information such as phone numbers. To teach the LM the characteristic of tweets, we collected tweets containing terms in a list of 669 keyword concepts in the plant health domain between January 2015 and September 2021. We used insect pest names and plant diseases names in former PestObserver website [15], and the literal value of *skos:prefLabel* and *skos:altLabel* of all nodes having type skos:Concept in FrenchCropUsage thesaurus to construct the list.

4.2 Experimental Setup

We conducted all experiments on a workstation having Intel Core i9-9900K CPU, 32 GB memory, 1 single NVIDIA GeForce RTX 3090 GPU with CUDA 10.0.130. We downloaded the LM from transformers [17]. We use fast-bert [13] wrapper for the further pre-training and the training of classifiers using linear regression. The choice of hyperparameters (see in Table 2) is based on the recommendation of BERT [2] and the configuration of our workstation. We did not do grid search for all hyperparameters on all the models for simplicity. For the further pre-training, we use implementation *CamembertForMaskedLM* in the transformer package (https://huggingface.co/transformers/v3.0.2/). We test different recipes to construct different corpus with the BSVs and tweets. We evaluate the further pre-trained LMs on the classification task.

Table 2. Hyperparameters for further pre-training and for classification

Hyperparameters	Pre-training	Classification
Batch size per GPU	[4, 8, 16]	[8, 16, 32]
Learning rate	1e−4	2e−5
Max sequence length	256	128
Epochs	[1, 2, 3, 4, 8, 16, 32]	[4, 10]
Schedule type	warmup_cosine	warmup_cosine
Optimizer type	adamw	adamw
Warm-up steps	–	300

Due to the small size of our labelled data, we perform 5-fold cross-validation keeping the same separation for our labelled set. We used these 5 labelled sets for all the classification experiments, including the experiments with baseline model and the pre-trained LMs. Figure 2 shows the number of positive labels and negative labels in each fold of training/validation set. We use the following implementations in the transformer package: *BertForSequenceClassification*, *CamemBertForSequenceClassification* and *FlauBertForSequenceClassifiaction* as classifiers, each of which is a linear layer on top of the pooled output of the LM.

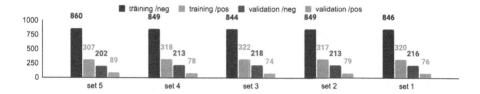

Fig. 2. Label distribution in each fold of training/validation set

To align with the linear classifiers in the transformer package, and to be differentiated from contextualized representations, we choose to fit the term frequency-inverse document frequency (TFIDF) vector of each tweets on linear regression classifier in sklearn package [8] for our baseline model. To build TFIDF feature vectors, we tokenize the tweets with or without stemming and lemmatizing, then extract all the unigrams, bigrams and trigrams, and search minimum document frequency (min-df) in $[0.005, 0.003, 0.002, 0.001]$. We find that the TFIDF vectors with stemmed tokens and min-df at 0.001 gives best average precision scores on the classification task, the average of which is 0.737186.

As presented above, there are fewer tweets about observations (positive) than non-observation (negative), and that the positives are more important, we draw the Precision-recall (PR) curve to evaluate each of the classifiers trained on the 5 folds of imbalanced data. To have a general measure of performance irrespective of any particular threshold, we use average precision score in sklearn package [8], which estimate the area under the Precision-recall curve (AUCPR) as the weighted mean of precision achieved at each threshold, with the increase in recall from the previous threshold used as the weight. In the following, we evaluate the models with the average of the 5 average precision scores.

4.3   Results and Evaluation

Choosing the Out-of-Box LM. To find the most pertinent out-of-box LM for the further pre-training on our domain corpus, we perform the classification task with the embeddings given by the following LM: CamemBERT (camembert-base and camembert-large) [6], FlauBERT (flaubert-base-uncased and flaubert-large-cased) [5], and mBERT (bert-base-multilingual-uncased) [2]. For each LM, we note the score with best average precision scores in Table 3. All the models give better representation for classification than the baseline model (0.737186), which favour contextualized embeddings. CamemBERT models (denoted by CMB in the table) outperforms FlauBERT models (denoted by FLB) . There are no significant differences between the base and large models. Thus, we choose to further pre-train Camembert-base with our corpus, and we use the classification results of CamemBERT models as our state-of-the-art models.

Table 3. Average precision scores of classification with out-of-box LMs

Model	mBERT	CMB_{large}	CMB_{base}	FLB_{base}	$FLB large$
Avg. APS	0.789853	**0.861968**	0.855935	0.845478	0.845027

The Further Pre-training. There are two mainstream strategies to pre-train domain specific LMs: either further pre-train the weights of an existing model on domain specific corpus without touching its vocabulary and tokenizer like [4], or pre-train a new LM on domain specific corpus and a tokenizer from scratch like JuriBERT [11]. In this work, we have extracted 230 MB of text from BSVs and 20 MB of tweets to construct our corpus, which is relatively small compared to those pre-trained from scratch. Thus, we decided to further pre-train existing LM on our corpus, reusing the native vocabularies and tokenizers. As we have too few tweets labelled as observation, we let BSVs teach the LM about context of observations and let unlabelled tweets to teach the LM the language style of tweets. The further pre-training is done via the Masked Language Modelling Task. Given any input sequence, 15% of the tokens are chosen randomly for prediction, of which 80% are masked, 10% are replaced with a random token and the rest 10% remain unchanged. Then the LM is trained to predict the original token, so it can learn the contextual information of the tokens, or how the tokens are organized together.

We feed the out-of-box CamemBERT base model with 3 different groups of recipes: only BSV, only tweets or both, and we note them as $ChouBERT_{BSV}$, $ChouBERT_{Tweet}$ and $ChouBERT_{BSV+Tweets}$. We fine-tune these ChouBERT models (denoted by "CHB" in the table) on the classification task and note the best performance of each model in Table 4. We can see that all three ChouBERT models have better scores than CamemBERT models, $ChouBERT_{BSV+Tweets}$ has the best results. It seems that CamemBERT do have the capacity to integrate the representation of tweets and of plant health from two different kinds of text for improving the downstream classification task when properly feed.

Table 4. Average precision scores of classification with further pre-trained LMs

Model	CHB_{Tweets}	CHB_{BSV}	$CHB_{BSV+Tweets}$	CMB_{large}	CMB_{base}
Avg. APS	0.874741	0.865134	**0.887424**	0.861968	0.855935

The Generalizability on Unseen Hazards. From the previous experiments, we selected the best hyperparameters (10 for epochs, 16 for batch size) for classification to study the effect of further-pretraining epochs on the generalizability of ChouBERT$_{BSV+Tweets}$ representation for detecting unseen hazards. Adding the tweets about coding moths to the previous training set/validation set of 3 hazards, we make a new set of 4 hazards for classification. We further pretrain ChouBERT$_{BSV+Tweets}$ for 0 (CamemBERT out-of-box model), 4, 8, 16, 32 epochs, train classifiers with 3-hazard set and 4-hazard set, test the classifiers on tweets about wireworm, so neither of the classifiers has seen the hazard during the training, and we plot the performance (the average of the 5 average precision scores) of each classifier in Fig. 3.

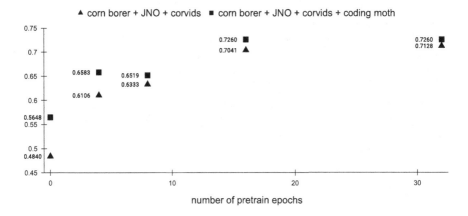

Fig. 3. Performance of different classifiers on wireworm tweets

Within the representation of each pre-trained model, the classifier trained with 4-hazard set outperforms the one trained with 3-hazard set, which implies that adding more labelled data helps improve the generalizability. Moreover, with more pretraining on text about plant health, the performance difference between the classifier trained with 4-hazard set and the one with 3-hazard set reduces. All the ChouBERT$_{BSV+Tweets}$ classifiers trained with 3-hazard set outperforms significantly the CamemBERT out-of-box classifier trained with 4-hazard set. This proves that ChouBERT$_{BSV+Tweets}$ deals better with plant health related information in tweets for classification.

5   Threats to Validity

For the labelling of the tweets about observation, we did not take into account the observations concerning the absence of hazards. Neither did we rank the pertinence or the completeness of the observations, which should be considered in future work. For the tempo-spatiality, in this study we consider all the observation tweets should be produced in real-time, if we cannot decide the temporality of a tweet, we label it as non-observation. Most of the tweets are not localized, so they can come from any francophone country.

6   Conclusion and Future Work

In this work, we presented a method to exploit crowd observations on Twitter. We built ChouBERT by applying domain adaptive pre-training to CamemBERT on BSV and tweets. We highlight the generalizability of ChouBERT representation on unseen hazards for the classification task. We can generalize this approach to improve crowdsensing based on textual content of tweets by: collecting tweets using keywords; manually labelling a small set of tweets; further pre-training language models using domain documents and tweets; and building NLP applications with the labelled set and the domain-adapted LM. For future work, we plan to evaluate our model on other NLP tasks; to study for the integration of heterogeneous text and the building of a knowledge base; and to explore other features of tweets, such as the demographic diversities in texts with contextualized embeddings. At last, our experience shows that crowd observation on Twitter is not a replacement of other monitoring paradigms, but a complementary source of information to detect weak signals, rather than quantifying the gravity of an issue.

References

1. Boubiche, D.E., et al.: Mobile crowd sensing-taxonomy, applications, challenges, and solutions. Comput. Hum. Behav. **101**, 352–370 (2019)
2. Jacob, D., et al.: BERT: pre-training of deep bidirectional transformers for language understanding. In: Proceedings of the 2019 Conference of the North American Chapter of the Association for Computational Linguistics: Human Language Technologies, vol. 1. pp. 4171–4186. ACL, June 2019
3. Klerkx, L., Jakku, E., Labarthe, P.: A review of social science on digital agriculture, smart farming and agriculture 4.0: new contributions and a future research agenda. NJAS Wageningen J. Life Sci. **90–91**, 100315 (2019)
4. Laifa, A., Gautier, L., Cruz, C.: Impact of textual data augmentation on linguistic pattern extraction to improve the idiomaticity of extractive summaries. In: Golfarelli, M., Wrembel, R., Kotsis, G., Tjoa, A.M., Khalil, I. (eds.) DaWaK 2021. LNCS, vol. 12925, pp. 143–151. Springer, Cham (2021). https://doi.org/10.1007/978-3-030-86534-4_13
5. Le, H., et al.: FlauBERT: unsupervised language model pre-training for French. In: Proceedings of The 12th Language Resources and Evaluation Conference, LREC, pp. 2479–2490. European Language Resources Association, Marseille (2020)
6. Martin, L., et al.: CamemBERT: a tasty French language model. In: Proceedings of the 58th Annual Meeting of the Association for Computational Linguistics, pp. 7203–7219. Association for Computational Linguistics, Online (2020)
7. Mendes, J., et al.: Smartphone applications targeting precision agriculture practices-a systematic review. Agronomy **10**(6), 855 (2020)

8. Pedregosa, F., et al.: Scikit-learn: machine learning in Python. J. Mach. Learn. Res. **12**, 2825–2830 (2011)

9. Roussey, C.: French Crop Usage (2021). https://doi.org/10.15454/QHFTMX

10. Shankar, P., Bitter, C., Liwicki, M.: Digital crop health monitoring by analyzing social media streams. In: 2020 IEEE/ITU International Conference on Artificial Intelligence for Good (AI4G), Geneva, Switzerland, pp. 87–94. IEEE (2020)

11. Stella, D., et al.: JuriBERT: a masked-language model adaptation for French legal text. In: Proceedings of the Natural Legal Language Processing Workshop, pp. 95–101. Association for Computational Linguistics, November 2021

12. Thollet, B.: MEDIAS SOCIAUX EN AGRICULTURE: Contribution à l'analyse des usages et de leur potentiel d'apprentissage pour la transition agroécologique. Master's thesis, AgroSup Dijon, Dijon, France, August 2020

13. Trivedi, K.: Fast-BERT (2020). https://github.com/kaushaltrivedi/fast-bert

14. Turenne, N.: Reports OCR (2016). https://www.data.gouv.fr/fr/datasets/r/c745b0bf-b135-4dc0-ba04-1e15c1b77899

15. Turenne, N., et al.: Open data platform for knowledge access in plant health domain: VESPA mining. CoRR abs/1504.06077 (2015)

16. Welvaert, M., et al.: Limits of use of social media for monitoring biosecurity events. PLOS ONE **12**(2), e0172457 (2017)

17. Wolf, T., et al.: Transformers: state-of-the-art natural language processing. In: Proceedings of the 2020 Conference on Empirical Methods in Natural Language Processing: System Demonstrations, pp. 38–45. Association for Computational Linguistics, Online, October 2020

A Conceptual Characterization of Fake News: A Positioning Paper

Nicolas Belloir[1,2(✉)], Wassila Ouerdane[3], Oscar Pastor[4], Émilien Frugier[1], and Louis-Antoine de Barmon[1]

[1] CREC St-Cyr, Académie Militaire de St-Cyr Coëtquidan, Guer, France
`nicolas.belloir@irisa.fr`
[2] IRISA, Vannes, France
[3] MICS, CentraleSupélec, Université Paris-Saclay, Gif sur Yvette, France
[4] PROS Research Group, Universitat Politècnica de València, Valencia, Spain

Abstract. Fake News have become a global phenomenon due to its explosive growth, particularly on social media. How to identify fake news is becoming an extremely attractive working domain. The lack of a sound, well-grounded conceptual characterization of what exactly a Fake news is and what are its main features, makes difficult to manage Fake News understanding, identification and creation. In this research we propose that conceptual modeling must play a crucial role to characterize Fake News content in a precise way. Only clearly delimiting what a Fake News is, it will be possible to understand and managing their different perspectives and dimensions, with the final purpose of developing any reliable framework for online Fake News detection as much automated as possible. This paper discusses the effort that should be made towards a precise conceptual model of Fake News and its relation with an XAI approach.

Keywords: Conceptual modeling · Fake news · Explainable artificial intelligence

1 Introduction

Although Fake News is not a new phenomenon [15], questions such as why it has emerged as a global topic of interest and why it is attracting increasingly more public attention are particularly relevant at this time. The leading cause is that Fake News can be created and published online faster and cheaper when compared to traditional news media such as newspapers and television [13]. In addition, recent discussions of higher education's failure to teach students how to identify Fake News have appeared in leading newspapers [9].

In a sound Information Systems Engineering context, a correct data management of Fake News strongly requires a precise conceptual characterization of "what" a Fake News is. If an information structure must represent a conceptualization, the entities that represent that conceptualization must be explicitly determined. Any information system intended to register information about

© The Author(s), under exclusive license to Springer Nature Switzerland AG 2022
R. Guizzardi et al. (Eds.): RCIS 2022, LNBIP 446, pp. 662–669, 2022.
https://doi.org/10.1007/978-3-031-05760-1_41

Fake News must identify in detail what are the relevant entities that conceptually characterize the different dimensions that must be considered to treat Fake News data correctly. Ontologically speaking, a precise ontological commitment that involves the relevant entities that constitute the conceptualization must be stated. This is the contribution that this paper addresses, by facing a fundamental question regarding the terminology and the ontology of Fake News: what constitutes and qualifies as Fake News?

To achieve that goal, in Sect. 2 we analyze related work to emphasize how ambiguous the concept of Fake News is still in practice, and how difficult is to find a unified view on what exactly a Fake News is. In Sect. 3, we propose our own classification of Fake News regarding the Information Ecosystem. On the way to propose a conceptual model of Fake News, Sect. 4 discusses the key notions that must be taken into account when the goal is to provide a sound conceptual characterization of the Fake News notion. This section ends up with a precise definition of Fake News, which in our point of view encompasses the target elements to be represented in a conceptual model. Section 5 highlights some hints on the construction of this conceptual model as an initial building block of an XAI approach.

2   Related Work

Many works focus on the study of Fake News. They are mainly interested in understanding and identifying the nature of information in Fake News to better distinguish it from real information. There has been extensive research on establishing a practical and automatic framework for online Fake News detection [20,21], intended to help online users to identify valuable information.

To develop reliable algorithms for detecting Fake News, an important ingredient is to be able to point very clearly on what is a Fake News or what are the principal features characterizing it. However, what we can notice first in the literature is that the concept of Fake News is still ambiguous, and the boundary between the definition of Fake News and other relative concepts, such as misinformation, des-information, news satire, hoax news, propaganda news, etc., is blurred. Indeed, as it is illustrated by some categorization examples [7,15,16], it is not always clear how these different concepts are related. Moreover, several definitions of Fake News have been proposed. For instance, [10] state that Fake News are "Fabricated news articles that could be potentially or intentionally misleading for the readers", or [13]: " Fake News is a news article that is intentionally and verifiably false". What we can note is that it is difficult to have a consensus and a unified vision on what exactly a Fake News is

More recently, [8] has proposed a first step towards a characterization of Fake News. It is based on seven different types of online content under the label of "Fake News" (false news, polarized content, satire, etc.), in contrast with real news by introducing a taxonomy of operational indicators in four domains. The characterization is interesting, but it is not based on a precise conceptual model. As a representation that captures someone's conceptualization of understanding of a domain, a conceptual model is the natural strategy to get a reliable domain representation that is used by human users to support communication, discussion, negotiations, etc. In the Fake News context, it allows us to define Fake News with specific and precise semantics. Moreover, the conceptual model will expose the relations between the concepts composing Fake News in a more informative and robust way, which will offer a reliable and practical means for detecting or even automatically generating Fake News. In [18] a conceptual model to examine the phenomenon of Fake News is proposed. The model focuses on the relationship between the creator and the consumer of the information, and proposes a mechanism to determine the likelihood that users will share their Fake News with others. In contrast, in our work we are particularly interested in the conceptualization of the Fake News content. A further advantage of relying upon a conceptual model is its ability to facilitate building well-justified and explainable models for Fake News detection and generation, which, to date, have rarely been available. Indeed, as it is emphasized in [19,21], despite the surge of works around the concept of Fake News, how one can automatically assess news authenticity in an *effective* and *explainable* manner is still an open issue.

We propose here to discuss and to understand what a Fake News is. The idea is to characterize Fake News content to demonstrate what is essential to consider towards building a Conceptual Model. A conceptual understanding of Fake News will help us distinguish them better and rule them from real news.

3   Fake News into the Information Ecosystem

Several concepts are close to the notion of Fake News, and it is not always clear how similar or different they are. Therefore, we propose in Fig. 1 an information categorization to situate Fake News and some closer concepts. Our categorization is based on the following observations.

On the one hand, four categories of information differ by their facticity and the author's intention to harm [17]. First, the *journalism* defined as "independent, reliable, accurate, and comprehensive information" by [6]: journalism aims to release information with a high level of facticity without intent to harm. Second, *malinformation* defined as "potentially dangerous or damaging information; inappropriate information; information people feel uncomfortable within openly accessible circulation" [12]. Thus, malinformation possess a high level of facticity released to harm. Third, *disinformation*: "includes all forms of false, inaccurate, or misleading information designed, presented and promoted to cause public harm intentionally or for profit" [4]. Thus, it has a low level of facticity released to harm someone or an institution. Finally, *misinformation* deals with "information that is false, inaccurate or misleading". We note that there is no consensus in the literature about whether the author of misinformation intends to deceive or not. In this paper, misinformation is considered as information that is false, inaccurate or misleading, released without intent to deceive. It is created when a human being misinterprets a piece of information or draw inaccurate conclusions while believing this misinterpretation or conclusion is true. Thus, it is information with a low level of facticity but no intent to harm.

On the other hand, a critical element of information is whether it is true. While journalism and malinformation present genuine information, the notions at hand don't and are consequently part of altered information. In the literature, the term "false information" is often used as opposed to "true information". However, the concept of truthfulness isn't black or white, and there is a wide range of values between true and false. Thus, using "genuine information" and "altered information" is more appropriate. As rumours can be true and not, they can only be considered information in this classification. Another element is the intent to deceive. While hoaxes and Fake News aim to deceive and thus are part of deceptive news, poor journalism, satire, and parody don't. When dealing with satire and parody, the audience is aware of the goal: to mock with evident humour the actuality [15]. This humour is how satire and parody differ from misinformation: they don't present news seriously like the former category. Consequently, they're part of "inaccurate information" along with poor journalism, but they differ by the use of humour. While satires and parodies are inaccurate because they're based on humour, poor journalism is unintentionally inaccurate information. Finally, among "misleading content" hoaxes differs from disinformation by their intent. While disinformation aims to harm, hoaxes deceive for amusement. Using this classification, we propose that Fake News are part of disinformation by their diffusion of altered information with a will to harm.

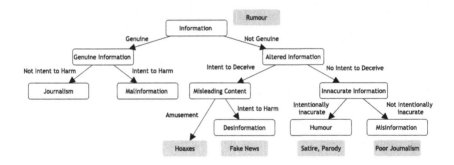

Fig. 1. Information categorization: Fake News and some related concepts

4   Characterization of the Fake News Concept

From the literature, it follows that there is no consensus on the definition of Fake News. Nevertheless, common elements can be identified. One can sum them up by saying that Fake News is fabricated to manipulate certain people. However, in general none of the proposed definitions specifies who are the people who manufacture this false information, nor how Fake News manages to convince its targets that they are true, nor who these targets are. In the following, we precise the key notions embedded into the concept of Fake News and give our definition.

4.1   Fake News are Not Created for Fun

65% of the false information about vaccination and Covid-19 came from just 12 people [2]. These are well-known figures in the American anti-vax sphere. They are conducting a coordinated campaign by fabricating Fake News about vaccines for Covid, which is part of a more extensive anti-vaccine campaign. This general pattern is typical and can be found in military operations (with a strategic level, an operational level and a tactical level). Here, the strategic level can be identified as an information war against vaccines, the operational level specifically targets vaccines against Covid and the different Fake News can be considered as a succession of tactical actions. Most Fake News fits into this three-tiered organization even if it is also possible to find isolated Fake News. Even if it is pretty easy to identify the creator of a Fake News, it is more challenging to identify people responsible for a disinformation campaign and information warfare. Nevertheless, they do exist. Thus, a conceptual model must allow to describe these protagonists.

4.2   Fake News Create a Distortion Between Real and False Facts

Most of the time, a Fake News relies upon at least one *real fact*, as pointed out by [3], and consists of at least one *false fact*. For instance, in the Fake News targeting Hillary Clinton, the real fact was that Mrs Clinton fainted during a ceremony for the 9–11 victims in New York. According to her staff, her faintness is due to a pneumonia diagnosed some days ago. The false fact was she was diagnosed with brain cancer and had only six months left to live. As illustrated, a real fact occurs in a specific context. Knowing this context is essential to understand the real part of a Fake News.

A better understanding of Fake News implies being able to identify what are the real and false facts that constitute it. Several types of true facts can be identified. For instance, a fake news can be made from a political statement, from an occurred event or from real data such as picture or video. Similarly it can be possible to identify some false fact categories. Sometimes, it can be entirely invented and has no relation to any real fact. Other times it can only be a deformation of a real fact. It can be carried out by modifying figures, falsifying papers, and retouching photos. Finally the false fact is not clear and is just a deformation of a set of real facts. The latter is pernicious and relies on several Real Facts that have no concrete link between them. They are presented in such a way that they lead to a false conclusion.

4.3   Fake News Credibility: The Authority Notion

Most readers rarely check the authenticity of information presented by a newspaper. They rather take it for granted. Fake News generally use a different vector to be propagated (e.g. social networks), and the trust in those vectors can be lower. Thus, Fake News creators need to increase Fake News credibility. One way to do so consists in invoking an authority: few news readers check the authority called upon in an article. It makes an interesting way for the creator to deceive the reader more easily and strengthen the credibility of the news. Thus, this authority notion must be incorporated in a conceptual model.

Three types of references to an authority were identified [1]: an intern authority, an external or a false one. In the first case, the source and the authority are two different entities. The authority is a renowned entity, referring to a person or an institute from whom the information released originates. By its renown and expertise, this authority provides credibility. When calling upon an external authority, the Fake News creator invokes well-known historical personalities whose words or actions are considered as a general truth. The goal is to draw a parallel between present and past by presenting the past as an unquestionable reference. This comparison reinforces the news due to the overall respect towards History. This calls upon historical figures, which generally have even more power than the first reference. Finally, the last one is the reference to a false authority, that can be either a real person with nothing to do with the fact at hand or a made-up person (a supposed expert in the area).

4.4   Fake News Target a Precise Category of People and Use a Cognitive Process Based on Emotions

Although a Fake News can influence different people in different ways, it is designed to reach a specific group of people to manipulate them in a particular way. It aims to influence the opinion of its target by generating an emotional charge. The target's reaction leads them to draw certain conclusions and change their mind about a topic. Identifying the purpose of a Fake News and who it is aimed at is therefore a challenge.

Overall, the goal of a Fake News is to make a person or a group of people change their mind about a subject. Thus, the Fake News goal is composed of three elements that would be clearly identified as part of a future conceptual model: the target, its opinion and the goal itself. The first two items are closely related since the target has an opinion on a subject. The last element embodies the influence the Fake News creator wants to have on the target. This can be either to weaken that particular opinion or to strengthen it. Thus, the goal of a Fake News can be represented by sentences according to the following scheme: "The goal of the Fake News is to" + goal + opinion + "among" + target. With the example of the Fake News regarding Hillary Clinton's health, the goal can be formulated as follows: "the goal of the Fake News is to weaken the belief that Hillary Clinton can lead the country among wavering American electors".

Behaviors and understanding of the world are both shaped by emotions, that drive or confuse people. As they emerge consciously or not, it doesn't seem easy to think and only act with reason when emotions are intense. This is why emotion loads generated in a Fake News must be considered. To be able to evaluate emotions in a future conceptual model, it is necessary to provide a clear range of values. A reasonably common characterization of emotions is given by the Plutchik wheel [11]. The latter clearly illustrates the relationship between the different emotions and their intensity.

After categorizing emotions, the conceptual model must capture the cognitive process involved in Fake News. [5] divided the mind's process into two distinct systems: (i) System 1 "is the brain's fast, automatic, intuitive approach". It is linked to the emotional and unconscious domains. Decisions are quick, impulsive, and sometimes irrational. (ii) System 2 is "the mind's slower, analytical mode, where reason dominates". It is slow and deals with the logical and conscious domain. With this two-system mind process, it is easy to understand why Fake News aim to generate emotions among readers. By doing so, they call upon the first system to overcome any attempt emerging from the second one. This emotional appeal prevents readers from awakening System 2 and forces them to think in an emotional state.

4.5   Our Fake News Definition

Based on the previous discussions, we setup the following definition of Fake News. The notions of "real" and "false" facts are at the heart of the Fake News concept and should then be considered. The emotional and cognitive process is also essential to understand the specific way the attacker influences the readers of the Fake News through its implicit conclusion. All that done having in mind to place it in a more general disinformation process. Thus, our definition is: *A Fake News is false but verifiable news composed of false facts based on real ones. Drafted in a way to trigger an emotional load, it aims to deceive its readers and influence their opinion through an implicit conclusion.*

5   Conclusions: Toward a Fake News Conceptual Model

In this article, we have shown how difficult is to clearly identify and position the concept of Fake News in the information ecosystem. Nevertheless, as the Covid pandemic and the war in Ukraine have shown, reliably classifying Fake News is an important challenge. To do so, we first classified the concept of Fake News in relation to other forms of information. Then, we identified important notions aiming at characterizing what is a Fake News, to highlight the key elements to be take into account. Our conviction is that these elements are the foundations that a conceptual model of fake news should take into account. The interest of such a model is to provide a clear characterization of what a fake news is. The applications of such a model are valuable. For example, in the challenge of identifying fake news, approaches using artificial intelligence (AI) are often proposed. But how these AI reach at these identifications is sometimes unclear. Using a conceptual model as a formal characterization model can be a solution to this problem. In this context, using a conceptual model can be seen as the first step of the XAI process proposed by [14], that it is crucial to make possible the full XAI-based process. The explainability with this approach is conceptually guided by the conceptual model which conforms the core of the contribution: to have an ontologically well-grounded definition of what a Fake News is, which is directly derived from the conceptual model. Thus, proposing a conceptual model Fake News to do so is naturally our future work.

References

1. Bondarenko, I.: Tools of explicit propaganda: cognitive underpinnings. Open J. Mod. Linguist. **10**, 23–48 (2020)
2. CCDH: The disinformation dozen - why platforms must act on twelve leading online anti-vaxxers. Technical report, Center for Countering Digital Hate (2021)
3. CCN-CERT: Disinformation in cyberspace. Technical report, CP/13 (2021)
4. European Commission, Directorate-General for Communications Networks, Content and Technology: A multi-dimensional approach to disinformation: report of the independent High level Group on fake news and online disinformation. Publications Office (2018)
5. Kahneman, D., Diener, E., Schwarz, N.: Well-Being: Foundations of Hedonic Psychology. Russell Sage Foundation (1999)
6. Kovach, B., Rosenstiel, T.: The Elements of Journalism. Three Rivers Press, New York (2014)
7. Kumar, S., Shah, N.: False information on web and social media: a survey (2018)
8. Molina, M.D., Sundar, S.S., Le, T., Lee, D.: "Fake news" is not simply false information: a concept explication and taxonomy of online content. Am. Behav. Sci. **65**(2), 180–212 (2021)
9. Ziv, W.: Op-Ed: why can't a generation that grew up online spot the misinformation in front of them? Los Angeles Times (2020)
10. Pierri, F., Ceri, S.: False news on social media: a data-driven survey. SIGMOD Rec. **48**(2), 18–27 (2019)
11. Plutchik, R.: Emotion: Theory, Research, and Experience, vol. 1. Theories of Emotion (1980)
12. Santos-d'Amorim, K., Miranda, M.: Misinformation, disinformation, and malinformation: clarifying the definitions and examples in disinfodemic times. Encontros Bibli Revista Eletrônica de Biblioteconomia e Ciência da Informação, March 2021
13. Shu, K., Sliva, A., Wang, S., Tang, J., Liu, H.: Fake news detection on social media: a data mining perspective. SIGKDD Explor. Newsl. **19**(1), 22–36 (2017)
14. Spreeuwenberg, S.: AIX: Artificial Intelligence Needs eXplanation: Why and How Transparency Increases the Success of AI Solutions. LibRT BV, Amsterdam (2019)
15. Tandoc, E., Lim, Z., Ling, R.: Defining "Fake News": a typology of scholarly definitions. Digital Journalism **6**, 1–17 (2017)
16. Wang, C.: Fake news and related concepts: definitions and recent research development. Contemp. Manag. Res. **16**, 145–174 (2020)
17. Wardle, C., Derakhshan, H.: Information disorder: toward an interdisciplinary framework for research and policy making. Council of Europe (2017)
18. Weiss, A.P., Alwan, A., Garcia, E.P., Kirakosian, A.T.: Toward a comprehensive model of fake news: a new approach to examine the creation and sharing of false information. Societies **11**(3), 1–17 (2021)
19. Zafarani, R., Zhou, X., Shu, K., Liu, H.: Fake news research: theories, detection strategies, and open problems. In: Proceedings of the 25th ACM SIGKDD, pp. 3207–3208. Association for Computing Machinery (2019)
20. Zhang, X., Ghorbani, A.: An overview of online fake news: characterization, detection, and discussion. Inf. Process. Manage. **57**(2), 102025 (2020)
21. Zhou, X., Zafarani, R.: A survey of fake news: fundamental theories, detection methods, and opportunities. ACM Comput. Surv. **53**(5), 1–40 (2020)

Combinations of Content Representation Models for Event Detection on Social Media

Elliot Maître[1,2(✉)], Max Chevalier[1], Bernard Dousset[1], Jean-Philippe Gitto[2], and Olivier Teste[1]

[1] Université de Toulouse, IRIT, Cr Rose Dieng-Kuntz, 31400 Toulouse, France
{elliot.maitre,max.chevalier,bernard.dousset,olivier.teste}@irit.fr
[2] Scalian, 22, bd Déodat de Séverac, 31770 Colomiers, France
jean-philippe.gitto@scalian.com

Abstract. Social media are becoming the preferred channel to report and discuss events happening around the world. The data from these channels can be used to detect ongoing events in real-time. A typical approach is to use event detection methods, usually consisting of a clustering phase, in which similar documents are grouped together, and then an analysis of the clusters to decide whether they deal with real-world events. To cluster together similar documents, content representation models are critical. In this paper, we individually compare the performances of different social media documents content representation models used during the clustering phase, exploiting lexical, semantic and social media specific features, like tags and URLs. To the best of our knowledge, these models are usually individually exploited in this context. We investigate their complementarity and propose to combine them.

Keywords: Information systems · Content representation models · Lexical representation · Semantic representation · Event detection · Social networks

1 Introduction

Social media are some of the main contemporary information sources, used by people but also by professionals such as journalists, business managers, politicians. They deliver information about numerous domains and can be used in multiples tasks, for instance to predict the stock market [9] and can be used in general to detect events happening around the world [4,13].

Due to the abundance of information and noise on social media, methods are needed to detect events in real-time. In previous work by McMinn [17], an event is a "significant thing that happens at some specific time and place". They identify an event by a group of entities (e.g. people, location) that are discussed in the messages. We borrow this definition for this work.

A major challenge of this task is to group documents dealing with the same event together. The text content of each document usually contains unstructured language, slang words or abbreviation but also limited context about the topic. Social media documents are also composed of other features such indexes (e.g. hashtags), metadata or links, all potentially interesting to cluster the documents.

To address this challenge, we study the performances of different document content representation models in the context of a classical event detection framework. These models exploits either lexical, semantic or social media specific features to represent the documents. To the best of our knowledge, these models, particularly text representation models, are usually individually employed in event detection methods and not combined as models encoding different aspects of the documents. Our goal is to investigate whether the information encoded by each of these models are complementary for the task of clustering event-related social media documents. To do so, we first evaluate the performances of each individual models. Then, we explore combinations of these models to obtain new similarity measures between documents and show the interest of these combinations. We focus on Twitter, the most commonly studied social media.

This article is organized as follows: Sect. 2 presents related work. Section 3 describes the event detection approach. Section 4 describes the combination method. Section 5 presents the experiments and the results.

2   Related Work

We focus on the task of open-domain event detection on social networks which consists in detecting events without prior knowledge on them [4]. Event detection methods usually fall between two categories: feature pivot or document pivot [4]. While both of these approaches are widely represented in the literature [4,13], we choose a document pivot approach to take into account more context and metadata and focus on document pivot approaches based on text.

One of the most common approach for event detection is the FSD (First Story Detection) algorithm, which was first introduced in [2]. The principle is to find the first document discussing an event and then group together new documents discussing the same event. The task is considered as a dynamic clustering task, using nearest neighbors algorithm to group the documents. Several papers improved this algorithm to speed it up [14,18], improvements being mostly focused on the nearest neighbor search. In all these papers, the tweets are represented using TF-IDF. In [15], the authors compare the performances of different text representations models. They compare TF-IDF and neural-based representation models such as Universal Sentence encoder [11]. Interestingly, they conclude that representation models based on recent architectures perform worse than TF-IDF, which is surprising considering the performances of Transformers.

In other approaches, TF-IDF is also the most common text representation model. The authors of [7] use it as well and then cluster topically similar tweets using an online incremental clustering algorithm. In [16], the authors combine TF-IDF and named entities (NE) to cluster the tweets, based on similarity criteria but also the length of the tweets. In [10], the authors propose the first method combining TF-IDF and semantic representation. They learn a representation for the words in the documents and then weight them based on their TF-IDF score, creating weighted semantic representations. They consider that two tweets are semantically related if they are generated by the same event.

In the rest of this paper, we propose to combine textual representation models to encode different dimensions of the text. Contrary to the literature, we use TF-IDF as a complementary representation to semantic representations, and not to weight semantic models.

3   Clustering for Event Detection

3.1   Description of the Framework

Figure 1 present the framework described in this section. First, tweets are retrieved from the Twitter's API. Second, they are cleaned, using methods detailed in Sect. 5.1 and the stream is discretized in time windows (e.g. 1 h). Then, the documents are clustered and the clusters are evaluated to determine whether they contain document dealing with the same event. To evaluate properly the performances of this step, we make the following hypothesis: (1) all the documents are event related, (2) each document is associated with exactly one event, (3) there is an unknown number of events. In a more realistic setup, other steps such as clusters and data filtering are needed, but they are out the scope of this study as we are interested in evaluating the performances of the clustering.

Fig. 1. The framework considered during this work.

3.2   Documents Representation

To correctly cluster related documents together, documents representation is a critical step as the performances of the clustering depend on it. We present the different content representation models we compare and combine:

Lexical similarity - We use TF-IDF, the most common text document representation model in information retrieval [5]. It represents the importance of a word in a document based on its frequency in it and its frequency in the whole corpus, under the assumption that a word that is frequent in a document but not in the corpus is representative of the document. We use an IDF calculated on the whole dataset Event2012 [17], presented in Sect. 5.1, provided by [15][1].

Semantic Similarity - Semantic representations of text documents are currently achieving state-of-the-art results in NLP, particularly those based on

---

[1] https://github.com/ina-foss/twembeddings.

Transformers [19]. In particular, we use Universal Sentence Encoder (USE) [11], which we found as the best performing on social media in previous work.

Social Network Specific Features - Hashtags are inherent to the Twitter ecosystem. They help classifying documents and assigning them to the right feed. Another interesting feature is link shared by users. Tweets are often used to react to some news published on other websites such as press websites.

Now that we know the models we use for the representation of the documents, we present the clustering algorithm.

3.3 Clustering

For each pair of documents of a window and for each document representation model, we compute its similarity to constitute a similarity matrix used to compute the clusters. For semantic and lexical representation models, we chose cosine similarity as it is the most common similarity measure in NLP [1]. For tags, we use Jaccard-Dice similarity. Finally, we determine if documents are sharing the same URL by a simple string comparison. It is a critical component of the event detection process because the performances of the clustering are directly affected by the similarity measures. To compute the clusters, we use the Louvain algorithm [8], a community detection algorithm which automatically computes the optimal number of clusters. The only parameter that this algorithm needs is a similarity threshold, which is specific to each representation model, determined during a training phase presented in Sect. 4.4.

Now that we have presented the event detection model, we detail in the following section the combination method we propose.

4   Combination Method

In this section, we first present our combination method as well as the aggregation methods and the different configurations compared in the results section.

4.1   General Description of the Method

We propose to combine different representation models using an ensemble-based similarity [12], an approach to combine different clustering methods that has also been used in related work [6]. The principle of the technique is to jointly exploit the results of different clustering processes in order to take advantage of their respective strengths. In our case, we have two majors propositions. First, we propose to combine text representation models that encode different aspects of the text and complement them with social-media specific features. Second, we propose to combine the results at the similarity level, to have more flexibility for the aggregation of the results. Figure 2 illustrates the method. First, for each representation model, a similarity matrix is computed. A model threshold is applied to this matrix to filter low similarities. Then, each similarity matrix is weighted according to the performances of its respective model for the clustering task. The matrices are aggregated to obtain a general similarity matrix, which is filtered using a general threshold. This filtered matrix is then used for the clustering. The different aggregation methods, configurations and how to obtain the thresholds and weights are described in the next sections.

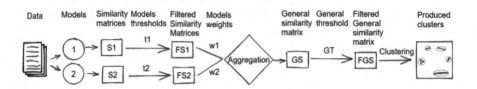

Fig. 2. Illustration of the method for the combination of two representation models.

4.2   Aggregation Methods

One of the main step of our method is to aggregate similarity matrices together. We propose three aggregations and compare their performances.

Similarity Aggregation (SA): For each representation model, for each value of the similarity matrix, if it is superior to its model threshold, then this value is used to compute the general similarity matrix. **Binary aggregation (BA):** For each representation model, for each value of the similarity matrix, if the value is superior to its model threshold the value is set to 1. If no, the value is set to 0. For SA and BA, no general threshold is applied. **General Aggregation (GA):** We directly use the similarity matrices for the aggregation without applying models thresholds. Then, we apply a general threshold to the matrix.

Now that we know how the similarity matrices are aggregated, we present the different content representation configurations we propose.

4.3   Configurations

We study two configurations: Lexical and Semantic (LS) and Lexical, Semantic and Twitter Specific (LSTS). **LS** is composed of TF-IDF, the lexical representation model, and Universal Sentence Encoder (USE), the semantic representation model. **LSTS** is composed of LS and social network specific features, namely hashtags and URLs shared in the documents.

Each of these configurations will be evaluated using the different aggregation methods presented in the Sect. 4.2.

4.4   Phases of the Experiments

We present the objectives of each phase of the experiments.

Training Phase: This phase has two objectives: first, we determine the model threshold for each representation model. We evaluate the clusters produced using each model for all threshold values, and the optimal threshold is defined as the threshold value maximizing the sum of the evaluation metrics, detailed in Sect. 5.1, as in [6]. The second objective is to determine the relative weight of each model for the combinations. For each configuration, we determine the relative importance of each model according to its performances. In practice, we compute the total of the sum of the evaluation metrics for all the models, and weight each representation model according to its contribution to the sum [6].

Validation Phase: The objective is to compute the general threshold for each configuration (i.e. LS and LSTS) for the General Aggregation (GA) method.

Testing Phase: We evaluate USE and IDF and consider these models as the baselines. We also evaluate each configuration (LS, LSTS) using each aggregation configuration (SA, BA, GA). In total, during the following experiments, we evaluate 8 representation models.

5 Experiments, Results and Analysis

5.1 Experimental Setup

Dataset: We use Event2012 [17], a corpus of 120 million tweets, collected from the 10th of October to the 7th of November 2012 using the Twitter streaming API. 159,952 tweets are labeled as event-related, distributed into 506 events. Due to the TREC policy, only tweet ids can be shared and the content of the tweets must be retrieved using the Twitter API. Some tweets are not available anymore. Thus, at the time of the experiments, we collected 69,875 labeled tweets. We sort the dataset according the date of publication of each tweet and divide it into training, validation and test sets, each one composed of 9 days of data.

Preprocessing: To clean the tweets, we remove from the text the user and retweet mentions and the URLs.

Evaluation Metrics: Each model is evaluated using B-cubed and AMI. B-cubed is a generalization of Precision, Recall, F1-score for clustering [3]. AMI measures how much information is shared between ground truth and the clustering assignment, adjusted to penalize random clusters.

5.2 Results and Analysis

The results of the training phase are presented in Table 1. For the validation phase, we calculate the thresholds for LS and LSTS in for the GA aggregation. The thresholds are **0.35** for LS and **0.30** for LSTS. The results of the testing phase are summarized in Table 2.

The combination methods can be interesting depending on the application. To favor precision, a combination using the GA aggregation can be interesting even if USE have similar performances. To favor recall, a combination using the BA aggregation is interesting. USE is the better compromise, achieving decent

Table 1. Weights and threshold for each method

Model	Threshold	Weight LS	Weight LSTS
USE	0.40	0.54	0.41
IDF	0.10	0.46	0.38
URLS	0.15	–	0.11
Hashtags	0.85	–	0.10

Table 2. Results from the Testing Phase. As a reminder, the thresholds for the GA aggregation are **0.35** for LS and **0.30** for LSTS.

Model	Aggregation method	Precision	Recall	F1-score	AMI
USE	–	0.91 ± 0.08	0.83 ± 0.19	$\mathbf{0.85 \pm 0.12}$	$\mathbf{0.76 \pm 0.22}$
IDF	–	0.85 ± 0.09	0.77 ± 0.20	0.79 ± 0.13	0.65 ± 0.24
LS	GA	0.91 ± 0.07	0.80 ± 0.20	0.83 ± 0.12	0.73 ± 0.23
	SA	0.86 ± 0.09	0.82 ± 0.20	0.82 ± 0.13	0.70 ± 0.24
	BA	0.80 ± 0.11	$\mathbf{0.86 \pm 0.19}$	0.81 ± 0.12	0.69 ± 0.24
LSTS	GA	$\mathbf{0.92 \pm 0.07}$	0.78 ± 0.21	0.83 ± 0.13	0.73 ± 0.24
	SA	0.86 ± 0.09	0.82 ± 0.21	0.82 ± 0.13	0.71 ± 0.23
	BA	0.79 ± 0.12	$\mathbf{0.86 \pm 0.19}$	0.81 ± 0.11	0.68 ± 0.23

performances in all the metrics and achieving better performances both in terms of time and computational complexity than the combinations, since only one similarity matrix has to be computed. In terms of running time, LS performs better than LSTS as computing the Jaccard similarity is a time consuming step.

6   Conclusion

In this paper, we investigate, for the clustering of event-related social media documents task, the comparison between document representation models (exploiting either lexical, semantic or specific features). To take advantage of these different models, we study the combination of multiple models, using multiple aggregation methods. Our experiments showed that some combinations can be interesting depending on the needs of the application. However, Transformer-based language models seems to have the better overall performances.

In future work, we plan to transpose these experiments to a more realistic context by including non-event related documents. A major issue in this context is to be able to evaluate the methods while most of the documents are not annotated. We plan to propose new evaluation metrics in order to facilitate the evaluation of the models and the reproducibility of the experiments.

References

1. Aggarwal, C.C., Zhai, C.X.: A survey of text clustering algorithms. In: Aggarwal, C.C., Zhai, C.X. (eds.) Mining Text Data, pp. 77–128. Springer, Boston (2012). https://doi.org/10.1007/978-1-4614-3223-4_4
2. Allan, J., Lavrenko, V., Malin, D., Swan, R.: Detections, bounds, and timelines: UMass and TDT-3. In: Proceedings of Topic Detection and Tracking Workshop, November 2000 (2000)
3. Amigó, E., Gonzalo, J., Artiles, J., Verdejo, F.: A comparison of extrinsic clustering evaluation metrics based on formal constraints. Inf. Retr. 12(4), 461–486 (2009)
4. Atefeh, F., Khreich, W.: A survey of techniques for event detection in Twitter. Comput. Intell. 31(1), 132–164 (2015)
5. Baeza-Yates, R., Ribeiro-Neto, B., et al.: Modern Information Retrieval, vol. 463. ACM Press, New York (1999)
6. Becker, H., Naaman, M., Gravano, L.: Learning similarity metrics for event identification in social media. In: Proceedings of the 3rd ACM International Conference on Web Search and Data Mining, pp. 291–300 (2010)
7. Becker, H., Naaman, M., Gravano, L.: Beyond trending topics: real-world event identification on Twitter. In: ICWSM, January 2011, vol. 11 (2011)
8. Blondel, V.D., Guillaume, J.L., Lambiotte, R., Lefebvre, E.: Fast unfolding of communities in large networks. J. Stat. Mech. Theor. Exp. P10008, 1–12 (2008)
9. Bollen, J., Mao, H., Zeng, X.: Twitter mood predicts the stock market. J. Comput. Sci. 2(1), 1–8 (2011)
10. Boom, C.D., Canneyt, S.V., Demeester, T., Dhoedt, B.: Representation learning for very short texts using weighted word embedding aggregation. CoRR abs/1607.00570 (2016)
11. Cer, D., et al.: Universal sentence encoder. CoRR abs/1803.11175 (2018)
12. Gionis, A., Mannila, H., Tsaparas, P.: Clustering aggregation. ACM Trans. Knowl. Discov. Data (TKDD) 1(1), 4-es (2007)
13. Hasan, M., Orgun, M.A., Schwitter, R.: A survey on real-time event detection from the Twitter data stream. J. Inf. Sci. 44(4), 443–463 (2018)
14. Hasan, M., Orgun, M.A., Schwitter, R.: Real-time event detection from the Twitter data stream using the Twitternews+ framework. Inf. Process. Manage. 56(3), 1146–1165 (2019)
15. Mazoyer, B., Cagé, J., Hervé, N., Hudelot, C.: A French corpus for event detection on Twitter. In: Proceedings of the 12th Language Resources and Evaluation Conference, pp. 6220–6227 (2020)
16. McMinn, A.J., Jose, J.M.: Real-time entity-based event detection for Twitter. In: Mothe, J., et al. (eds.) Experimental IR Meets Multilinguality, Multimodality, and Interaction, pp. 65–77. Springer, Cham (2015). https://doi.org/10.1007/978-3-319-24027-5_6
17. McMinn, A.J., Moshfeghi, Y., Jose, J.M.: Building a large-scale corpus for evaluating event detection on Twitter. In: Proceedings of the 22nd ACM International Conference on Information & Knowledge Management, pp. 409–418 (2013)
18. Petrović, S., Osborne, M., Lavrenko, V.: Streaming first story detection with application to Twitter. In: Human Language Technologies: The 2010 Annual Conference of the North American Chapter of the Association for Computational Linguistics, HLT 2010, pp. 181–189. Association for Computational Linguistics, USA (2010)
19. Vaswani, A., et al.: Attention is all you need. In: Guyon, I., et al. (eds.) Advances in Neural Information Processing Systems, vol. 30. Curran Associates, Inc. (2017)

Increasing Awareness and Usefulness of Open Government Data: An Empirical Analysis of Communication Methods

Abiola Paterne Chokki[✉] , Anthony Simonofski , Benoît Frénay ,
and Benoît Vanderose

University of Namur, Namur, Belgium
{abiola-paterne.chokki,anthony.simonofski}@unamur.be

Abstract. Over the past decade, governments around the world have implemented Open Government Data (OGD) policies to make their data publicly available, with collaboration and citizen engagement being one of the main goals. However, even though a lot of data is published, only a few citizens are aware of its existence and usefulness, which hinders fulfilling the purpose of OGD initiatives. The objective of this paper is to fill this gap by identifying the appropriate communication methods for raising awareness and usefulness of OGD to citizens. To achieve this goal, we first conducted a literature review to identify methods used to raise citizen awareness of OGD. Then, these identified methods were confronted with the results obtained from an online survey completed by 30 participants on their preferred methods to provide recommendations to governments. The contribution of this paper is twofold. First, it provides an inventory of communication methods identified in the literature. Second, it analyzes the gap between the use of these methods in practice and citizens' preference and uses this analysis to propose a list of methods that governments can use to promote OGD.

Keywords: Open government data · Citizens · Awareness · Communication methods

1 Introduction

Across the globe, many governments have implemented Open Government Data (OGD) policies to make their data more accessible and usable by the public. In its most common definition, OGD refers to data published by governments that can be freely used and redistributed by anyone [1]. The release of such data is most often motivated by values such as improving government transparency [2], stimulating innovation [3, 4], encouraging citizen collaboration and participation [4]. For these goals to be achieved, OGD must be used in some way, which requires that citizens know that such data exists (awareness) and know their added value (usefulness) [5]. Yet, even though a lot of data are published, only a few citizens are aware of its existence and usefulness [6, 7], which hinders achieving the goal of OGD initiatives. In this study, the term "citizens" refers to people with modest technical and data literacy.

There have been several attempts to examine whether a specific method was suitable or reports to suggest methods to increase awareness and usefulness of OGD to citizens. For instance, in [8], the author studied the contribution of the use of social media applications (Facebook, Twitter and YouTube) by Thailand's public sector to improve transparency and use of OGD. In [9], the authors proposed the use of training to promote OGD use. In [10–13], other methods in addition to the previous ones, such as OGD portals, hackathons, and newspapers, have been proposed to raise citizen awareness. However, none of these previous studies have effectively evaluated multiple methods with citizens in order to recommend to governments the appropriate methods to increase awareness and usefulness of OGD to citizens. Therefore, there is a need to conduct such a study to evaluate methods with citizens and based on this, propose appropriate methods to governments.

This paper seeks to address this gap by identifying the appropriate methods for raising awareness and usefulness of OGD to citizens. Therefore, our research question is as follows: "What are the appropriate communication methods for raising citizen awareness of the existence and usefulness of OGD?" To answer this research question, we first conducted a literature review to identify the methods used to promote OGD in previous studies. Next, these identified methods were confronted with the results obtained from an online survey completed by 30 participants on their preferred methods. The results of this survey will then be used to recommend to governments the most appropriate methods to promote OGD to citizens.

The remainder of this paper is divided into five main sections. In Sect. 2, we explain the research methodology. Section 3 explores existing methods for raising citizen awareness of the existence and usefulness of OGD and provides an overview of the survey results. In Sect. 4, we present the discussion and limitations of this study and then propose some avenues for future work. Finally, Sect. 5 provides a conclusion that summarizes the contributions of this paper.

2 Methodology

To address the research question of this study, we combined two methods: literature review and questionnaires. First, we conducted a literature review on methods used to raise citizen awareness of the existence and usefulness of OGD. The literature review was conducted using the databases "Scopus" and "Science Direct" with the keywords "open government data" or "open data", + "citizen", + "promote" or "raise awareness" + "usefulness" or "existence". We also extended our search to the grey literature and policy reports. Most of the publications found dated from 2011–2021. From these publications, an additional selection was made based on their relevance to our research, leaving altogether 15 academic articles, web pages and policy reports which were looked at more thoroughly. The retained articles were then used to collect appropriate methods for raising citizen awareness of the existence and usefulness of OGD. Second, we created an online survey[1] to collect citizens' preferred methods. The survey was pretested with two users and later shared via the following communication channels: UNAMUR

---

[1] https://forms.gle/JEnhmEKV94J3612m7.

mailbox, Facebook and Twitter to recruit citizens. In total, 30 participants completed the survey. The literature review along with the user feedback will be used to improve the current knowledge base and provide recommendations to governments.

3 Results

In this section, we first describe previous work on methods used to raise citizen awareness of the existence and usefulness of OGD. Then, we present the survey results.

3.1 Communication Methods Identified

By performing a literature review described above, we were able to identify a set of eight methods that could be used to raise citizen awareness of the existence and usefulness of OGD [10–13]. The following paragraphs explain each of these methods in more detail.

Social Media. According to [14], social media applications are new types of information and sharing tools, used in digital environments. They have been adopted by a few governments with different purposes: sharing information, interacting with citizens, promoting citizen participation in public issues or improving transparency [8, 10]. The most commonly used social media in the public sector are: **blogs, collaborative projects** (e.g., wikis, online forums), **social networking sites** (e.g., Facebook, Twitter) and **content communities** (e.g., YouTube) [8, 14]. Although social media applications offer many benefits, they can only reach specific citizens. For example, in the case of social networking sites, only the citizens who have an account and who fall within the criteria used for campaigns can be reached.

Public Outreach Campaigns. Apart from social media, a few governments have used methods such as **radio, television, newspapers, newsletters** and **poster campaigns** to inform citizens, especially of some applications built on the basis of OGD [12, 13]. The problem with these methods is that the content of the advertisement focuses on the implemented application without telling citizens that the application was implemented using open data. Therefore, citizens may use the service without knowing that it was built using open data.

Workshops and Conferences. These types of events aim to bring together various open data stakeholders to discuss the adoption and use of open data [10, 12, 15]. Two well-known examples of these types of events are the Open Government Data Camp and Open Data Day. The main advantage of these types of communication is that they help governments to have a direct discussion with citizens and also gather their feedback (e.g., needs, barriers) for improvement [16]. However, there are some drawbacks to these types of communication, such as the mandatory physical presence and limited number of participants.

Hackathons. Like the previous method, this method is an event that allows developers to design, implement and present services for a specific issue [4, 10, 12, 17]. This method allows for the promotion OGD to participants and the development of some services that can be published later to help a wider range of citizens. However, this method faces the following problems. It focuses mainly on developers and most of the results of hackathons are often not implemented or published online after the events, which does not impact the awareness of citizens, but only on the developers [11, 18].

Training and Education. This method consists of bringing together different stakeholders to inform or instruct them on a certain task with the aim of improving their performance or knowledge [10]. In [10, 12], they suggested enabling the creation of a "culture for Open Data" to students by integrating the use of open data (e.g. building apps) into academic programs. This method has been experimented in [9] and by Thessaloniki's Digital Strategy [10] but like the hackathons, this method only attracts a specific and limited part of digitally literate citizens.

Public Displays. These are mainly outdoor displays, as well as indoor displays in public spaces, which offer different benefits to users (refer to "passersby"): collecting feedback such as voting system, displaying information or accessing services [19]. The main advantage of public displays is that they help citizens to interact directly with services in real life and can be easily accessible (visible) to "passersby". However, public displays face the following problems: difficulty of interaction by a certain range of users and limited access (only available to a specific location) [20].

Applications. These include platforms developed to help users to easily access government data and also tangible examples of what can be done with published data [10, 12]. For example, **OGD portals** have offered visualizations, dashboards and success stories built from the data in addition to raw data, raising awareness of the benefits of Open Data and showing what can be done with particular datasets [10]. Apart from these features, some OGD portals have also proposed News and Events sections in their portals, which helps to increase traffic to the portal [10]. Another way to raise awareness among citizens is the development of **practical applications and services** accessible mainly on the web or mobile that use the data provided by governments and facilitate the daily life of citizens (e.g., the mobility application which helps Namur citizens to see the location of available parking in a specific area[2]). These practical applications are mainly developed either for a specific purpose or as federative applications to promote existing applications developed from open data (e.g., Datafruit[3] which summarizes in a mobile application the reuses of open datasets on the French portal). The problem with this method is that without awareness campaigns like the ones presented above, these applications cannot be acknowledged by citizens. Another problem is that each government or developer promotes their applications separately, which increases the funds used to promote the different applications and the need for citizens to go through (or install) different applications in order to use them.

Word of Mouth. This method involves citizens talking to their friends, family and other people with whom they have close relationships about a topic of open data [21, 22]. This

---

[2] https://sti.namur.be/.

[3] http://opendatatales.com/%f0%9f%a5%a5-datafruit-un-argumentaire-de-poche-de-lopen-data/.

method was less discussed in the literature of open data awareness but was proven to be one of the most powerful forms of awareness in general (e.g., e-commerce) as 92% of citizens trust their friends over traditional media [21].

3.2 Survey Results

Through questionnaires that participants completed during the rigor cycle, we were able to gather their opinions related to the research question "What are the appropriate communication methods for raising citizen awareness of the existence and usefulness of OGD?" In total, 30 participants (22 are aware of OGD and 8 are not) responded to the survey. All participants are between the ages of 18 and 50 and have at least a high school degree.

After collecting the citizens' responses on the channels through which they have been informed or wish to be informed of the existence and usefulness of OGD, we associated each citizen response with one of the methods presented in Sect. 3.1, where possible. Some citizens' responses were ambiguous (e.g., Google search or from municipality) and therefore were not considered. Figure 1 summarizes the percentage of citizens' responses on the channels through which they have been informed (A1) or wish to be informed of the existence and usefulness of OGD (A2).

Regarding the methods used in practice to raise awareness and usefulness of OGD, Fig. 1 (ref. A1) shows that the "word of mouth" channel (mainly through friends or colleagues), is the channel through which most citizens have been informed about OGD. This channel is followed by "training and education", especially "education", as many citizens indicate that they have heard about OGD in their classes. After this method comes public outreach campaigns, and after applications, workshops and conferences and hackathons. The methods "social media" and "public displays" were not mentioned by citizens. Regarding citizens' preferred methods, Fig. 1 (ref. A2) shows that citizens' preferred channel is "public outreach campaigns," followed by applications (suggested by citizens: OGD portals and OGD-based applications), training and education, and social media.

4 Discussion

This research contributes to the knowledge base in the following aspects. First, this research provides an inventory of communication methods that have been used in the literature to promote OGD to citizens. Second, unlike these previous studies that examine whether a specific method was appropriate [8] or suggest methods to increase awareness and usefulness of OGD to citizens [10–13] without providing an evaluation, this research presents the benefits and challenges of each of the methods used in the literature and evaluates each through an online survey completed by citizens. Fourth, based on the survey results, this research highlights the discrepancies between the methods used in practice and those preferred by citizens to raise citizen awareness of OGD (See Fig. 1). Fifth, based on Fig. 1, we recommend for governments to use public outreach campaigns and applications to inform citizens about the existence and the usefulness of OGD.

However, the main limitation of this research concerns the representativeness of the participants for the questionnaires. To increase this representativeness, we suggest using other communication channels or collecting data on-site in universities or public places. In this study, this was not feasible due to the COVID-19 situation.

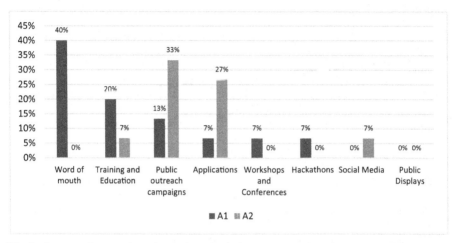

Fig. 1. Percent of respondents for each method of awareness and usefulness of OGD applied in the practice (A1) and preference by citizens (A2).

5 Conclusion and Future Work

The purpose of this paper was to identify the appropriate methods for raising awareness and usefulness of OGD to citizens. To achieve this objective, we first conducted a literature review to identify methods used to raise citizen awareness of OGD. Then, we used an online survey completed by 30 citizens to compare the results of the literature review with the citizens' perception.

The survey results, along with the literature review, allow us to make the following recommendations for governments: use public outreach campaigns and applications to inform citizens about the existence and the usefulness of OGD. The "word of mouth" method appears to be the most effective method used in practice for spreading awareness of OGD. Governments should therefore use the methods suggested above and encourage citizens to disseminate them to those around them. Since applications have proven to be one of the preferred methods and few studies have been proposed in the literature to investigate the requirements needed by a usable tool to promote OGD, the upcoming work is to fill this gap, implement them into a usable tool and evaluate the tool to see its impact on citizen awareness.

References

1. Attard, J., Orlandi, F., Scerri, S., Auer, S.: A systematic review of open government data initiatives. Gov. Inf. Q. **32**, 399–418 (2015)

2. Bertot, J.C., Jaeger, P.T., Grimes, J.M.: Using ICTs to create a culture of transparency: e-government and social media as openness and anti-corruption tools for societies. Gov. Inf. Q. **27**, 264–271 (2010)
3. Davies, T.: Open data, democracy and public sector reform: a look at open government data use from data. gov.uk (2010)
4. Johnson, P., Robinson, P.: Civic hackathons: innovation, procurement, or civic engagement? Rev. Policy Res. **31**, 349–357 (2014)
5. Zuiderwijk, A., Janssen, M., Dwivedi, Y.K.: Acceptance and use predictors of open data technologies: drawing upon the unified theory of acceptance and use of technology. Gov. Inf. Q. **32**, 429–440 (2015)
6. Toots, M., McBride, K., Kalvet, T., Krimmer, R.: Open data as enabler of public service co-creation: exploring the drivers and barriers. In: International Conference for E-Democracy and Open Government, pp. 102–112 (2017)
7. Abdelrahman, O.H.: Open government data: development, practice, and challenges. In: Open Data. IntechOpen (2021)
8. Gunawong, P.: Open government and social media: a focus on transparency. Soc. Sci. Comput. Rev. **33**, 587–598 (2015)
9. Gascó-Hernández, M., Martin, E.G., Reggi, L., Pyo, S., Luna-Reyes, L.F.: Promoting the use of open government data: cases of training and engagement. Gov. Inf. Q. **35**, 233–242 (2018)
10. Berends, J., Carrara, W., Vollers, H.: Analytical Report 6: Open Data in Cities 2 (2020)
11. Simperl, E., Walker, J.: Analytical Report 8: The Future of Open Data Portals (2020)
12. European Environment Agency: Open data and e-government good practices for fostering environmental information sharing and dissemination (2019)
13. Michael, C., Diana, F., Kate, J.: How government can promote open data and help unleash over $3 trillion in economic value (2014)
14. Mergel, I.: "A mandate for change": diffusion of social media applications among federal departments and agencies. In: Public Management Research Conference, pp. 1–27 (2011)
15. Ojo, A., Janssen, M.: Workshop about the understanding and improving the uptake and utilization open data. In: 15th Annual International Conference on Digital Government Research, pp. 350–351 (2014)
16. Cook, M., Jurkat, A.: An Open Government Research and Development Agenda Setting Workshop, Albany, NY (2011)
17. Simonofski, A., Amaral de Sousa, V., Clarinval, A., Vanderose, B.: Participation in hackathons: a multi-methods view on motivators, demotivators and citizen participation. In: Dalpiaz, F., Zdravkovic, J., Loucopoulos, P. (eds.) RCIS 2020. LNBIP, vol. 385, pp. 229–246. Springer, Cham (2020). https://doi.org/10.1007/978-3-030-50316-1_14
18. Gebka, E., Clarinval, A., Crusoe, J., Simonofski, A.: Generating value with open government data: beyond the programmer. In: 13th International Conference on Research Challenges in Information Science, pp. 1–2 (2019)
19. Clarinval, A., Simonofski, A., Vanderose, B., Dumas, B.: Public displays and citizen participation: a systematic literature review and research agenda. Transform. Gov. People Process Policy **15**(1), 1–35 (2021)
20. Coenen, J., Nofal, E., Vande Moere, A.: How the arrangement of content and location impact the use of multiple distributed public displays. In: Designing Interactive Systems Conference, pp. 1415–1426 (2019)
21. Hayes, A.: Word-of-Mouth Marketing. https://www.investopedia.com/terms/w/word-of-mouth-marketing.asp, Accessed 06 Jan 2022
22. Chen, Z., Yuan, M.: Psychology of word of mouth marketing. Curr. Opin. Psychol. **31**, 7–10 (2020)

Toward Digital ERP: A Literature Review

Benjamin De Brabander[1,2,3](✉) ⓘ, Amy Van Looy[2] ⓘ, and Stijn Viaene[1,3] ⓘ

[1] Vlerick Business School, Vlamingenstraat 83, 3000 Leuven, Belgium
Benjamin.DeBrabander@edu.vlerick.com, Stijn.Viaene@vlerick.com
[2] Faculty of Economics and Business Administration, Ghent University, Tweekerkenstraat 2, 9000 Ghent, Belgium
Amy.VanLooy@UGent.be
[3] Faculty of Business and Economics, K.U. Leuven, Naamsestraat 69, 3000 Leuven, Belgium

Abstract. Many organizations have a long history with the use of ERP. However, organizations are increasingly turning to digital capabilities to transform operational processes and business models. Extant literature has increased our understanding of ERP, but we lack comprehensive insights into the evolving nature of ERP in the context of digital transformation. Through a review of articles from the AIS Basket of Eight IT journals, we identified digital capabilities associated with contemporary ERP across five categories. The identified capabilities foreground the evolving nature of ERP, resulting in the introduction of a definition for digital ERP (D-ERP) and a call for research studying the co-evolution of D-ERP and digital transformation.

Keywords: ERP · Enterprise resource planning · Digital ERP · Digital technologies · Digital capabilities · Digital transformation

1 Introduction

In recent years, digital capabilities, such as cloud computing, low-code programming, Internet of Things (IoT), Application Programming Interface (API), and user experience journeys have become increasingly important. Some of the digital technologies and capabilities have also affected the Enterprise Resource Planning (ERP) landscape, which continues to grow between 25% and 35% yearly, with a total value of about $44 billion in 2020 (Gartner, 2021). ERP entered the digital age between 2012 and 2016, during which major ERP vendors migrated to the cloud and as-a-Service (AAS) business models [1]. However, besides the migration of ERP to the cloud [1, 2], the integration of other digital technologies and capabilities has not been explicitly researched. Digital technologies have had a significant impact on classical Information System (IS) architecture, capabilities and business models [3]. Given the continued importance of ERP, more information is needed about how the literature connects digital capabilities to ERP.

These reasons call for a closer examination of the relationship between ERP and digital capabilities. Hence, our research question is: **How does the current literature support the evolution of ERP with digital capabilities?** By combining the fields of ERP and digital capabilities, this study provides insights into how contemporary

ERP embeds digital capabilities, based on a systematic literature review (SLR) of peer-reviewed academic articles.

Section 2 presents the background and conceptual development. Section 3 elaborates on our SLR methods. Section 4 describes our findings, whereas in Sect. 5, we discuss our findings. Section 6 concludes our work.

2 Background and Conceptual Development

Prior literature on ERP research goes back to the 1990s. The term ERP was coined by the Gartner Group and included criteria for evaluating the extent to which software was actually integrated across and within the various functional silos (e.g., finance and accounting, human resources, payroll, and sales/marketing) [4]. Since then, various definitions have been used by scholars from 1998 to 2020 (Table 1).

Table 1. A non-exhaustive overview of ERP definitions over time.

Variants of definitions for ERP systems	Year	Source
Is the seamless integration of all information flowing through the company – financial and accounting information, human resource information, supply chain information, and customer information	1998	[5]
Are designed to support both the functional and operational processes of a firm's value chain, including accounting and finance, human resources, customer and sales, and supply chain management	2006	[6]
Aim to provide a unified IT architecture to enhance data consistency and integration of modular applications that support business processes	2010	[7]
Is a system of integrated software applications that standardizes, streamlines and integrates business processes across finance, human resources, procurement, distribution, and other departments	2020	[8]

Table 1 shows similarities in the various ERP definitions, such as a reference to (1) business processes, (2) integration, and (3) functions. Some components have evolved over time emphasizing a set of different applications in contrast to a single system or software. In addition, some definitions are expanding the scope from supporting functions to supporting an organization's entire value chain.

Since the dawn of the ERP systems they have been associated with transformations due to their IT capabilities to enable organizations to work according optimized processes. These transformations have been subject to intensive research and can range from straight forward IT/business process alignment [9] to complex ERP-enabled business transformations to gain a competitive advantage [10, 11]. Prior literature on the resource-based view (RBV) theory has been widely applied within the IS literature to explain how organizations gain a competitive advantage and provide superior performance. Consistent with the RBV [12] and prior literature on IT capabilities [13], we specify digital capability as "a complex bundle of digital-related resources, skills, technologies and knowledge that enable firms to coordinate activities and other resources to produce desired outcomes."

3   Methods

We conducted a systematic literature review (SLR) of peer-reviewed research published across the AIS Senior Scholars Basket of Eight Journals (AIS, 2021). We included publications from 2015 onwards, following Kranz [1], who argued that all main ERP evolved to cloud platforms by 2015. We conducted our SLR in December 2020 according to the Guidelines for performing SLR in Software Engineering [14]. In each journal, we first searched for "ERP" or "Digital" in the title, list of keywords and abstract. We reviewed each of the resulting articles to identify synonyms. We then refined our search results by including these synonyms to refine our search query. Our final inclusion and exclusion criteria combine our search query for the period 1/1/2015 to 1/1/2021. Our SLR resulted in a final selection of 30 articles (Table 2). We then manually worked through the entire set of search results to ascertain if the found publications were relevant to the discussion of the digital evolution of ERP. We identified relevant ERP or digital capabilities, mapped them to level two coded capabilities and then grouped them into five categories.

Table 2. Summary of search results (N = 30).

Description	References	Count
European Journal of Information Systems (EJIS)	[2, 15–22]	9
Information Systems Journal (ISJ)	[1, 23–26]	5
Information Systems Research (ISR)	[11, 27]	2
Journal of Information Technology (JIT)	[28]	1
Journal of Management Information Systems (JMIS)	–	0
Journal of Strategic Information Systems (JSIS)	[3, 29–32]	5
Journal of the Association for Information Systems (JAIS)	[9, 33–38]	7
MIS Quarterly (MISQ)	[39]	1

4   Findings

We analyzed the articles to find recurring topics and patterns that relate to "ERP" and "digital". We identified five categories that connected our two key topics.

4.1 Digital Capabilities

The first observed category was based on sixteen articles that name *digital capabilities*. Within this first category, we identified five recurring digital capabilities that are linked to ERP or were relevant to include in our research agenda. First, eight of them mention a factual positive influence of digital capabilities on performance that matched with the benefits that traditional ERP promise, such as increased efficiency, lower cost, increased transparency, and integration. A second identified digital capability refers to transformational power [3] and the potential of digital technologies to create new value paths and business models by combining and using them in new and innovative ways. Mendling [20] (p. 209) cites that "in digital innovation, problem and solution space emerge and co-evolve." This iterative or agile delivery approach was identified as a third digital capability (Table 3).

Table 3. An overview of digital capabilities (N = 16).

ID	Digital capabilities	References	Count
1	Improvement of performance	[1, 17–19, 21, 22, 37, 39]	8
2	Enables new business models	[3, 11, 16, 17, 20, 22, 24, 29, 32]	9
3	Iterative and agile delivery	[3, 17, 19, 20]	4
4	Modular platform	[3, 19, 22, 24, 32]	5
5	Light-touch processes	[16, 19, 38]	3

The natural synergies between multiple digital capabilities and technologies was discussed in five papers and identified as a fourth digital capability: "modular platform". This is reflected in how firms can integrate digital technologies into existing enterprise resources or as part of a broader ecosystem or platform. As a fifth topic, three papers discuss the availability of light-touch flexible processes [16].

4.2 Need for Digital Capabilities in ERP

We identified a second category, based on eleven articles that address *ERP capabilities that are improved by embedding digital capabilities*. Within this second category, we identified three recurring topics (Table 4).

Table 4. An overview of the need for digital capabilities in ERP (N = 11).

ID	Need for digital capabilities in ERP	References	Count
1	Facilitating conditions for use	[16, 20, 31, 33, 36]	5
2	Solve misfits between working and requirements	[18, 19, 25, 26, 33]	5
3	Mindfulness of novel technology aspects	[9, 20, 31, 34]	4

The first identified topic is how digital technologies enhancing facilitating conditions, such as usability and user support, have a direct positive influence on ERP feature use and overcome the effects of inertia. The second recurring topic related to deviations from standard ERP functionality and how to handle these. Malaurent and Avison [25] discussed how work-around practices can be managed and should not always be considered as deviant activities. A flexible digital layer of lightweight applications embedded in the ERP ecosystem can provide a solution for workarounds and misfits [26] without impacting the ERP core model. The third topic was observed in four papers that stated that technology adopters and designers ought to be always mindful of novel aspects of technology in relation to existing systems.

4.3 Digital Technologies in ERP

A third category was based on eight articles that name specific *digital technologies*. Four papers mentioned digital technologies that are captured in the acronym SMACIT: Social, Mobile, Analytics, Cloud, IoT [32]. Contemporary ERP-centered business application technology stacks have all five of those technologies integrated. It is interesting that IT platforms are referred to as digital and in two papers ERP and CRM systems are also referred to as digital technologies [29, 39]. Other mentioned technologies that can be related to ERP are: blockchain for SCM in ERP, digital marketplace, AI-based technology, automation, and augmented reality for production planning. These technologies are more recent and indicate the constant stream of new digital technologies each creating new opportunities for ERP to adapt and evolve. We observed from this category that modern ERP solutions have fused with multiple technologies that are referred to as digital (Table 5).

Table 5. An overview of digital technologies in ERP (N = 8).

ID	Digital technology capabilities	References	Count
1	Social, Mobile, Analytics, Cloud, IoT	[1, 20, 22, 32]	4
2	Digital business ecosystems	[19, 24]	2
3	ERP and CRM systems	[29, 39]	2
4	Blockchain; digital marketplace; and AI-based technology; augmented reality; automation	[20, 39]	2

4.4 ERP as Digital Platform

The fourth category was based on sixteen articles on *ERP within a digital platform/ecosystem context*. The first identified topic deals with introducing digital platform capabilities into or fusing with ERP (Table 6).

Table 6. An overview of platform capabilities of ERP (N = 16).

ID	Digital platform capabilities	References	Count
1	Introducing or fusing with digital platform capabilities	[2, 15, 18, 22]	4
2	Flexibility without sacrificing common core	[11, 16, 20, 28–30, 32, 35]	8
3	Re-configuration and rapid integrations	[17, 18, 22, 27, 29, 35, 37]	7
4	Resilience to cope with external shocks & trends	[20, 23, 29, 35]	4

The closely related, is the ability to maintain an operational backbone as well as a digital services platform [32]. Jain and Ramesh [19] referred to the capacity to enable flexibility without sacrificing the pursuit of common core functionalities. Research related to building digital platforms distinguishes between heavyweight and lightweight IT [28], infrastructural stability and change, and a platform core and peripheral ecosystem [11, 38]. Tallon et al. [35] confirmed the value of enriching ERP with digital capabilities by finding that IT slack was positively correlated with IT business value. A third topic regards innovative capabilities that use the efficiencies of re-configuration and rapid integrations to cope with emergent initiatives [17]. A final recurring topic focuses on creating a platform of digital options that enable the firm to sense and respond to rapidly changing market conditions [35].

4.5   ERP Processes Enabled by Digital

The fifth category was based on eight articles that name specific *processes enabled by digital technology*. We found that some typical ERP-related processes are mentioned in the literature as supported and enhanced by digital technologies. Six papers mentioned typical ERP processes, such as procurement, invoicing [20, 39], supply chain [22], planning [22] and operational [16, 19, 29] processes. In addition, typical CRM processes, such as customer engagement, sales and marketing processes, as well as e-commerce are tackled in three papers [17, 29, 37] (Table 7).

Table 7. An overview of ERP processes enabled by digital technology (N = 8).

ID	ERP processes enabled by digital	References	Count
1	Procurement, invoicing, supply chain	[20, 22, 39]	3
2	Operational processes	[16, 19, 29]	3
3	CRM processes	[17, 29, 37]	3

5 Discussion

Our findings have shown that with the rise of digital technologies and digital transformation ambitions, ERPs have integrated digital capabilities that have helped them to evolve. A most interesting observation relates to the need for ERP to "enable new business models". Notably, the literature definitions of ERP are not reflecting such evolution. Therefore, we propose a definition of digital ERP (D-ERP) as *"an integrated digital enterprise platform for configuring, operating and innovating business processes, value chains and business models end-to-end, governed in an agile way"*.

Our review highlights the need for research that addresses the transformational capabilities of D-ERP. ERP's capacity to enable business transformations has been extensively discussed in the literature [11]. Digital transformations, however, are typically associated with disruptive technologies and business models [29]. Thus, many consider digital transformation as more radical and holistic forms of business transformation [3]. However, we have seen some articles distinguish between ERP-enabled transformations and digital transformations, suggesting that ERP is a hurdle rather than an enabler of digital transformations [32]. Based on our research, that shows signs of co-evolution of ERP and digital transformation, we consider this assessment premature and call for more research on the relationship between D-ERP and digital transformation.

6 Conclusion

This literature review provides insight into the co-evolution of ERP and digital transformation. Based on our findings, we derived a definition for D(igital)-ERP. The explorative nature of our review calls for further investigation into the linkage between D-ERP and digital transformation.

References

1. Kranz, J.J., et al.: Understanding the influence of absorptive capacity and ambidexterity on the process of business model change – the case of on-premise and cloud-computing software. ISJ **26**(5), 477–517 (2016)
2. Fahmideh, M., et al.: A generic cloud migration process model. EJIS **28**(3), 233–255 (2019)
3. Chanias, S., et al.: Digital transformation strategy making in pre-digital organizations: the case of a financial services provider. JSIS **28**(1), 17–33 (2019)
4. Jacobs, F.R., et al.: Enterprise resource planning (ERP)—a brief history. J. Oper. Manag. **25**(2), 357–363 (2007)
5. Davenport, T.H.: Putting the enterprise into the enterprise system. HBR **76**(4), 121–131 (1998)
6. Ranganathan, C., et al.: ERP investments and the market value of firms: toward an understanding of influential ERP project variables. ISR **17**(2), 145–161 (2006)
7. Morris, V.: Job characteristics and job satisfaction: Understanding the role of enterprise resource planning system implementation. MIS Q. **34**(1), 143 (2010). https://doi.org/10.2307/20721418
8. Bart, P.: What is ERP? Key features of top enterprise resource planning systems. CIO (2020)
9. Sun, H., et al.: Revisiting the impact of system use on task performance: an exploitative-explorative system use framework. JAIS **20**(4), 398–433 (2019)

10. Mueller, S.K., et al.: The roles of social identity and dynamic salient group formations for ERP program management success in a postmerger context. Inf. Syst. J. **29**(3), 609–640 (2019)
11. Gregory, R.W., et al.: Paradoxes and the nature of ambidexterity in IT transformation programs. Inf. Syst. Res. **26**(1), 57–80 (2015)
12. Bharadwaj, A.S.: A resource-based perspective on information technology capability and firm performance: an empirical investigation. MIS Q. **24**(1), 169–196 (2000)
13. Dale Stoel, M., et al.: IT capabilities and firm performance: a contingency analysis of the role of industry and IT capability type. I&M **46**(3), 181–189 (2009)
14. Kitchenham, B., et al.: Systematic literature reviews in software engineering–a systematic literature review. IST **51**(1), 7–15 (2009)
15. Babaian, T., et al.: ERP prototype with built-in task and process support. Eur. J. Inf. Syst. **27**(2), 189–206 (2018)
16. Baiyere, A., et al.: Digital transformation and the new logics of business process management. EJIS **29**(3), 238–259 (2020)
17. Busquets, J.: Discovery paths: exploring emergence and IT evolutionary design in cross-border M&As. Analysing grupo Santander's acquisition of abbey (2004–2009). Eur. J. Inf. Syst. **24**(2), 178–201 (2015)
18. Ceci, F., et al.: Impact of IT offerings strategies and IT integration capability on IT vendor value creation. EJIS **28**(6), 591–611 (2019)
19. Jain, R.P., et al.: The roles of contextual elements in post-merger common platform development: an empirical investigation. EJIS **24**(2), 159–177 (2015)
20. Mendling, J., et al.: Building a complementary agenda for business process management and digital innovation. EJIS **29**(3), 208–219 (2020)
21. Nielsen, P.A., et al.: Useful business cases: value creation in IS projects. Eur. J. Inf. Syst. **26**(1), 66–83 (2017)
22. Nwankpa, J.K., et al.: Balancing exploration and exploitation of IT resources: the influence of Digital Business Intensity on perceived organizational performance. EJIS **26**(5), 469–488 (2017)
23. Heeks, R., et al.: Conceptualising the link between information systems and resilience: a developing country field study. ISJ **29**(1), 70–96 (2019)
24. Li, L., et al.: Digital transformation by SME entrepreneurs: a capability perspective. Inf. Syst. J. **28**(6), 1129–1157 (2018)
25. Malaurent, J., et al.: Reconciling global and local needs: a canonical action research project to deal with workarounds. ISJ **26**(3), 227–257 (2016)
26. van Beijsterveld, J.A.A., et al.: Solving misfits in ERP implementations by SMEs. Inf. Syst. J. **26**(4), 369–393 (2016)
27. Lee, O.-K., et al.: How does IT ambidexterity impact organizational agility? Inf. Syst. Res. **26**(2), 398–417 (2015)
28. Bygstad, B.: Generative innovation: a comparison of lightweight and heavyweight IT. J. Inf. Technol. **32**(2), 180–193 (2017)
29. Li, T., et al.: Dynamic information technology capability: concept definition and framework development. JSIS **28**(4), 101575 (2019)
30. Tallon, P.P., et al.: Information technology and the search for organizational agility: a systematic review with future research possibilities. JSIS **28**(2), 218–237 (2019)
31. Tams, S., et al.: How and why trust matters in post-adoptive usage: the mediating roles of internal and external self-efficacy. JSIS **27**(2), 170–190 (2018)
32. Vial, G.: Understanding digital transformation: a review and a research agenda. J. Strateg. Inf. Syst. **28**(2), 118–144 (2019)
33. Benlian, A.: IT feature use over time and its impact on individual task performance. J. Assoc. Inf. Syst. **16**(3), 144–173 (2015)

34. Hornyak, R., et al.: Incumbent system context and job outcomes of effective enterprise system use. JAIS **21**(2), 364–387 (2020)

35. Tallon, P.P., et al.: Business process and information technology alignment: construct conceptualization, empirical illustration, and directions for future research. J. Assoc. Inf. Syst. **17**(9), 563–589 (2016)

36. Venkatesh, V., et al.: Unified theory of acceptance and use of technology: a synthesis and the road ahead. JAIS **17**(5), 328–376 (2016)

37. Young, B.W., et al.: Inconsistent and incongruent frames during IT-enabled change: an action research study into sales process innovation. JAIS **17**(7), 495–520 (2016)

38. Widjaja, T., et al.: Monitoring the complexity of IT architectures: design principles and an IT artifact. JAIS **21**(3), 664–694 (2020)

39. Pentland, B.T., et al.: The dynamics of drift in digitized processes. MIS Q. **44**(1), 19–47 (2020)

Trust Model Recommendation Driven by Application Requirements

Chayma Sellami[1,2(✉)], Mickaël Baron[1], Stephane Jean[1], Mounir Bechchi[2], Allel Hadjali[1], and Dominique Chabot[2]

[1] LIAS - ISAE-ENSMA/University of Poitiers, Chasseneuil-du-Poitou, France
{chayma.sellami,mickael.baron,stephane.jean,allel.hadjali}@ensma.fr
[2] OCode, 322 Bis Route du Puy Charpentreau, 85000 La Roche-sur-Yon, France
{mounir,dominique}@o-code.co

Abstract. *Recommendation systems* have taken the turn of the new uses of the Internet, with the emergence of *trust-based recommendation systems*. These use trust relationships between users to predict ratings based on experiences and feedback. To obtain these ratings, many *computational models* have been developed to help users make decisions, and to improve interactions between different users within a system. Hence, choosing the appropriate model is challenging. To address this issue, we propose a two-step approach that, first, allows the user to define the requirements of his/her target system and, then, guides him/her to select the most appropriate computational model for his/her application according to the defined requirements.

Keywords: Trust · Computational model · Requirements · Modelling · Recommendation

1 Introduction

Community applications have the specificity of offering their members the tools and interfaces necessary to communicate, exchange and interact easily with each other. The functioning of these sites is also largely based on the collective sharing of information, sometimes making it possible to rapidly increase one's visibility among a whole group of contacts. These applications involve creating your own connection with the other persons by setting up actions. *Trust* is one of the key elements that allow users of these applications to interact safely. This concept is used in several fields (psychology, social sciences, politics, computer science, etc.). In computer science, several definitions have been proposed for the concept of trust. According to Gambetta in [6] "Trust (or symmetrically distrust) is a particular level of subjective probability with which an agent will perform a specific action, both before we can track each action (or independently of his ability to even trace it) and also in a context in which it affects our own action".

In the literature, there are mainly two types of trust systems: Policy-based trust systems and *Reputation-based trust systems*. The former is based on the

R. Guizzardi et al. (Eds.): RCIS 2022, LNBIP 446, pp. 694–702, 2022.
https://doi.org/10.1007/978-3-031-05760-1_45

principle of competences and authorizations required to perform an action while the latter, which we will consider in this paper, makes use of the reputation of an entity allowing to grant or not the trust towards this entity. Reputation-based trust systems use the interactions, or direct experiences, between entities and/or the experiences of others for the choice to trust another entity. Nowadays, several modern applications are based on these systems to guarantee the security within their algorithms. Examples include systems from Blablacar [13], Airbnb [7], etc. In order to guarantee and compute trust, several computational models had emerged, see [5] for a survey. The major problem in this area is how to choose the most appropriate computational model according to the user's requirements. To the best of our knowledge, there are no works in the literature that allow user to choose the computational model that best meets his/her requirements. Our contribution is divided into three steps. The first step of our approach aims at helping the user to define the expectations of their target system, and to be able to model these requirements. The second step focuses on modelling the sequence of user's actions within the system and on the way the user would like the trust evolves in the system. As for the third step of this work, and starting from the results of the first two steps, it deals with the issue related to the selection of the most adequate computational model according to the user's requirements.

The rest of the paper is made up as follows; Sect. 2 briefly describes the state of the art that contextualizes our work. Section 3 discusses the different steps of our approach for modelling trust requirements in order to help the user of trust-based systems to choose the most suitable computational model for his/her needs. To validate our proposal, we provide in Sect. 4 a complete case study with concrete examples, covering all aspects of the contribution and showing its technical feasibility. Finally, the points to remember and perspectives are presented in Sect. 5.

2   State of the Art

The internet and computer systems have become more and more important in our lives and this covers several domains, such as e-commerce, banking, health-care, etc. It is genuine that the field of use varies, but the issue that binds together all these domains is Trust. Many researchers have proposed models to represent and compute trust in computer systems. [5]. One can classify these models into: Bayesian models [1,3], Belief-based models [1,2], Discrete value models [11], Fuzzy models [4] and Flow models [14]. However, the choice of the computational model that better meets user requirements is challenging because of the variety of computational models available in the literature. Some works have been proposed to evaluate and compare computational models, see for instance [12]. [12] outlines a set of criteria that a good rating system should meet. We can cite for example : Accurate for long-term performance, Weighted toward current behavior, Robust against attacks, Amenable to statistical evaluation, etc. A good rating system should meet the above-mentioned qualities. Contrary to these studies, our aim in this paper is to propose an approach to know which computational models fulfill a set of requirements related to the trust computation. Existing studies can been seen as a static comparison of computational models whereas our proposed approach makes a dynamic comparison driven by the designer requirements.

A second issue to be addressed in this paper is the modeling of requirements based on user's activities. There are several formalisms in human-machine communication systems that allow users to define the expectations of their system, e.g.: Business Process Model and Notation (BPMN), Unified Modeling Language (UML), etc. Task Models [10] are of a great interest in system user activity and aim at producing a list of tasks, a description of the tasks and their interactions. Different task models have been developed in the literature, see (CTT, GTA, etc.). However, all these modelling methods have not been developed to deal with the trust topic. This is the first time that such methods are leveraged to deal with and model the requirements in trust-based systems.

3   An Approach to the Trust Computational Model Selection

In order to manage and compute trust, trust-based systems make use of computational models. In this work, our goal is to provide an approach to model the trust requirements in order to help the users of these systems to choose the most adequate computational model for their needs. Our approach consists of the three steps. The first step builds a model to express the application requirements of the target system. This will allow us to classify the different types of actions that can be performed by users within the system.

The second step deals with the trust requirements model. This will allow us to model the chain of the defined actions and apply constraints on the desired trust value. For example, after a single carpool, a desirable wish is the smooth evolution of trust i.e. the trust value evolves in a gradual way.

The third step is to evaluate a trust computational model against the two models defined in steps 1 and 2. The computational models will be used to compute the score, compare it with the constraints set and recommend the best suited model w.r.t. the requirements.

3.1   Application Requirements Modelling

In this section, we will discuss the first level of our approach, which defines the different types of actions of a system based on the application requirements.

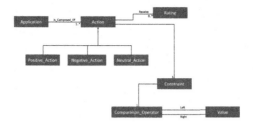

Fig. 1. Application requirements modelization

The priority for trust-based systems is to ensure safety for the user interacting with the system. Therefore, each user is characterized by reputation esteem on which cooperation choices are based. This estimate is computed from the actions made by the user. A user with a good behavior will have a great reputation value. Whereas a noxious user will have a bad reputation value. Therefore, in order to recognize a good user from a bad one, it is vital to recognize at the beginning the distinctive types of actions they can do.

Figure 1 shows the first step in our model to represent and characterize user actions. As we can see, each action in the system is linked to a rating from which we can know the type of action. In fact, each action in the system is evaluated by the other users in the interaction. Depending on the value of this evaluation, the system can distinguish whether the user has done a positive action to increase his reputation value, or a negative one to decrease it. A user can also do neutral actions that do not impact the reputation value (e.g. registering, browsing the system, etc.). Generally these actions depend on the context in which they were performed. This context represents the system application requirements.

At this step, this model makes it conceivable to know within the application which type of actions are carried out by the users. However, this step does not take into consideration the chaining of actions and does not allow to describe what is anticipated from the computational model, i.e. it does not permit to indicate requirements on the trust score. In order to be able to select the correct model for the user's requirements, it is essential to be able to depict and model the way trust evolves concurring to the user's behavior. This is the second part of our approach which will be clarified within the following section.

3.2   Model for Trust Requirements

In order to deal with the evolution of trust, it is necessary to understand the behaviour of the user and therefore the chaining of his/her actions. For this reason, we have enriched the first model, which is static, to address this issue.

In order to express the chain of actions, as mentioned in Sect. 2, we opt for the use of task models since we are interested in system user activity. The purpose of task analysis with these models is to produce a list of tasks and a description of the tasks and their interactions. Many task models have been developed, these models allow to express several characteristics about the tasks. Consequently, these models allow us to express the chain of actions, constraints, as well as a graphical and visual presentation of requirements. In contrast, the limitation of classical task models is that they do not allow to indicate constraints on the evolution of the trust score.

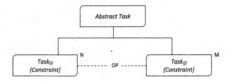

Fig. 2. A graphical example that shows our graphical proposal via Task Models principle

To enrich these models, Fig. 2 presents our proposed graphical representation of trust requirements using the principle of task models which allow us to have tasks decomposed into subtasks with a sequential time operator and information on the cardinality of tasks. As seen in Fig. 2, to describe trust requirements and the tasks that made them up, we use the concept of constraints. In this paper, the constraint is a Boolean predicate that allows to define an interval in which the value of the trust should be included. Here, we only handle simple constraints based on exact values and percentages. A percentage relies on a previous trust value to express its evolution. The integers N and M used provide information on the iteration of tasks (i.e. the number of times a task can occur). And finally, "OP" is a temporal operator to link the different tasks. Currently, only the sequence operator is managed, as it is deterministic and allows to produce a single output passing through the same tasks. Other operators are provided, but they are non-deterministic (e.g. choice, interrupt, parallel, etc.) which allow several outputs. Based on the task models, we are able to define new concepts that allowed us to express constraints on the trust requirements and to track the chaining of the user's actions. This proposal will allow us enriching our model in Fig. 1 to have a wealthier model that helps the user both to describe the sequence of human activity being on his system, to specify the type of action performed and to indicate requirements on the evolution of trust score.

From the proposed model, each trust requirement can be described by a task. The system performs tasks to compute the score, monitor the behavior, etc. The user can do different types of actions that we described earlier. Furthermore, the principle of subtasks in task models allows us to specify partial constraints for each subtask, as well as a final constraint on the task to meet the desired requirement. The partial constraint provides information on the evolution of the trust value after each subtask. While the final constraint is used to specify the desired final trust value towards the end of the task.

Using these two models, we can proceed to the last step of our approach, which is to choose the most appropriate computation model according to the user's requirements.

3.3   Trust Computational Model Recommendation

A trust-based system must, as mentioned above, have a strategy for computing the value of trust. In fact, the computation of trust must permit a choice to be made, for example, a high degree of trust in an entity allows it to be judged reliable and a choice to trust it to be made. Many authors have proposed models to represent and compute trust in systems. These models are categorized as Bayesian Models, Discrete Value Models, Belief-based Models, Fuzzy Models and Flow Models. In order to evaluate these models, this paper focuses on the criteria that permit observing the behavior of the human being within the system followed by the evolution of the trust score. The idea behind this part of the approach is to generate a scenario that the user can do in the system, put constraints on the desired way the trust score evolves (e.g. smooth evolution), and compute the value of the score using computational models developed in the literature.

As a yield of this model, we will have an estimate of the correspondence between the constraints set and the values obtained. This will facilitate the recommendation of the foremost suitable computational model for each requirement. A more complete demonstration with results will be illustrated in the next section.

4   Case Study

To validate our approach, we present in this section a complete case study that allows to cover all aspects of the contribution, to explain the complete process and to show the technical feasibility of our work. To be able to show results for our approach, a development has been made in which both models and the main computational models have been implemented. The objective is to provide a tool, called TME (Trust Model Evaluation), that can automatically assess computational models against the expressed requirements. Currently, the source code is available on the lab's forge in which you can find the complete experimentation[1]. Taking the example of carpooling systems, this concept refers to the joint and organized use of a vehicle by a driver and one or more traveling passengers for the purpose of making a common journey.

4.1   Model Requirements for Carpooling

The first part of our approach is to define the types of actions that a user can perform to subsequently monitor their sequencing, require preferences on the evolution of this value, compute a trust value and choose the appropriate computational model.

To do this, in our example, if the user receives a "Disappointing or Very Disappointing" rating, the system considers this as a negative action. If he/she receives an "Excellent or Good" rating, then he/she has taken a positive action.

---

[1] https://forge.lias-lab.fr/projects/trustmodelevaluation.

And if the user receives a "Correct" evaluation, then he/she has taken a neutral action. In order to be able to classify the type of action performed by the user, we need to associate continuous values for these discrete assessments to understand the user's behavior and score evaluation: Very Disappointing $= 0$; Disappointing $= 1$; Correct $= 2$; Good $= 3$; Excellent $= 4$.

Thus, these values allows us to put constraints on the assessments. Each user can give a $rating \in [-1; 1]$, if the $rating > 0$, the action is positive, if the $rating < 0$, the action is negative, if the $rating = 0$, the action is neutral.

Let's take the example of a passenger who made 7 trips and was evaluated after each experience as shown in Table 1.

Table 1. Example of user behavior and evaluations in the system.

Actions	Trip 1	Trip 2	Trip 3	Trip 4	Trip 5	Trip 6	Trip 7
Evaluation	Good	Good	Excellent	Disappointing	Good	Disappointing	Disappointing
Constraints	$rating > 0$	$rating > 0$	$rating > 0$	$rating > 0$	$rating < 0$	$rating > 0$	$rating < 0$
Type	Positive	Positive	Positive	Positive	Negative	Positive	Negative

4.2   Trust Requirements Model for Carpooling

At this stage, our first model is able to define the types of actions that a user can perform. The second step is to be able to model this behavior and require constraints on the trust score. The objective of this part of the approach is to be able to model what is expected of the evolution of trust in the system in relation to the actions expressed in the previous model.

Let us take the example of a forms of trust evolution among the criteria mentioned in the Sect. 2. "Weighted towards current behavior": If the user has a good reputation, i.e. he has done several positive actions (PT), and he starts doing negative actions (NT), the system should show a strong impact on the trust score (S). After the first modelling step, we can define the set of actions of the user and thus have a sequence of actions (Table 2):

Table 2. Constraints applied to the desired trust score for Weighted towards current behavior criterion.

Actions	P	P	P	P	P	P	P	N	N	N	N
Constraints	S > 22	S > 24	S > 26	S > 28	S > 30	S > 33	S < 36	S < 23	S < 14	S < 9	S < 5

4.3   Trust Computational Model Recommendation

In this example we have proposed constraints to satisfy the two criteria we have considered. For these constraints, we focused on the fact that the evolution of the trust score is progressive, to satisfy the "smooth" criterion. Thereafter, if the user behaves badly, he must lose his reputation (Weighted towards current behavior criteria).

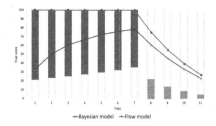

Fig. 3. Compatibility of computational models for Weighted towards current behavior criterion

Figure 3 shows results obtained from our implemented model. As we can see, our model allows to express preferences on the desired trust score value for the user (U), which can be seen in the form of bars in the figure, and to challenge existing computational models in the literature in order to know subsequently the most adequate model.

5   Conclusion and Perspectives

In this paper, we have addressed the problem of evaluating trust-based systems and choosing the most appropriate model for the application requirements. We attempted to propose an approach that helps the user to select the most fitting model for his system. We have defined and implemented a complete case study. This latter shows that our approach can be used to model different requirements, found in the literature, on the trust computation. Moreover, our approach and the associated tool compare the different computational models on the basis of a set of requirements in order to select the most appropriate one. This work can be improved by adding some features such as building a hybrid model that better meets trust requirements, i.e. if for example the designer wants to satisfy two requirements at the same time, and our approach provides him/her with a different recommendation for each requirement, merging the two recommended computational models to have a new one may be an idea.

References

1. Abdel-Hafez, A., Xu, Y., Jøsang, A.: A normal-distribution based rating aggregation method for generating product reputations. In: Web Intelligence, vol. 13, no. 1, pp. 43–51. IOS Press, January 2015
2. Acampora, G., Alghazzawi, D., Hagras, H., Vitiello, A.: An interval type-2 fuzzy logic based framework for reputation management in Peer-to-Peer e-commerce. Inf. Sci. **333**, 88–107 (2016)
3. Ashtiani, M., Azgomi, M.A.: A formulation of computational trust based on quantum decision theory. Inf. Syst. Front. **18**(4), 735–764 (2015). https://doi.org/10.1007/s10796-015-9555-4

4. Ashtiani, M., Azgomi, M.A.: A hesitant fuzzy model of computational trust considering hesitancy, vagueness and uncertainty. Appl. Soft Comput. **42**, 18–37 (2016)
5. Braga, D.D.S., Niemann, M., Hellingrath, B., Neto, F.B.D.L.: Survey on computational trust and reputation models. ACM Comput. Surv. (CSUR) **51**(5), 1–40 (2018)
6. Gambetta, D.: Can We Trust Trust. Trust Making and Breaking Cooperative Relations, vol. 13, pp. 213–237 (2000)
7. Guttentag, D. Progress on Airbnb: a literature review. J. Hosp. Tour. Technol. (2019)
8. Josang, A., Ismail, R.: The beta reputation system. In: Proceedings of the 15th Bled Electronic Commerce Conference, vol. 5, pp. 2502–2511 (2002)
9. Kamvar, S.D., Schlosser, M.T., Garcia-Molina, H.: The eigentrust algorithm for reputation management in P2P networks. In: Proceedings of the 12th International Conference on World Wide Web, pp. 640–651 (2003)
10. Limbourg, Q., Vanderdonckt, J.: Comparing task models for user interface design. In: The Handbook of Task Analysis for Human-Computer Interaction, vol. 6, pp. 135–154 (2004)
11. Lin, H., Hu, J., Huang, C., Xu, L., Wu, B.: Secure cooperative spectrum sensing and allocation in distributed cognitive radio networks. Int. J. Distrib. Sens. Netw. **11**(10), 674591 (2015)
12. Oram, A.: Peer-to-Peer: Harnessing the Power of Disruptive Technologies. O'Reilly Media, Inc., Sebastopol (2001)
13. Shaheen, S., Stocker, A., Mundler, M.: Online and app-based carpooling in france: analyzing users and practices—a study of BlaBlaCar. In: Meyer, G., Shaheen, S. (eds.) Disrupting Mobility. LNM, pp. 181–196. Springer, Cham (2017). https://doi.org/10.1007/978-3-319-51602-8_12
14. Škorić, B., de Hoogh, S.J.A., Zannone, N.: Flow-based reputation with uncertainty: evidence-based subjective logic. Int. J. Inf. Secur. **15**(4), 381–402 (2015). https://doi.org/10.1007/s10207-015-0298-5

Predicting the State of a House Using Google Street View

An Analysis of Deep Binary Classification Models for the Assessment of the Quality of Flemish Houses

Margot Geerts[1]([✉]) [ID], Kiran Shaikh[1], Jochen De Weerdt[1] [ID], and Seppe Vanden Broucke[1,2] [ID]

[1] KU Leuven, Naamsestraat 69, 3000 Leuven, Belgium
{margot.geerts,jochen.deweerdt}@kuleuven.be
[2] UGent, Tweekerkenstraat 2, 9000 Ghent, Belgium
seppe.vandenbroucke@ugent.be

Abstract. Currently, the state of a house is typically assessed by an expert, which is time and resource intensive. Therefore, an automatic assessment could have economic, social and ecological benefits. Hence, this study presents a binary classification model using transfer learning to classify Google Street View images of houses. For this purpose, a three-by-three analysis is conducted that allows to compare three different network architectures and three differently-sized data sets, using properties located in Leuven, Belgium. A DenseNet201 architecture was found to work best, as illustrated quantitatively as well as by means of state-of-the-art explainability methods.

Keywords: Google Street View · Real estate · Deep learning · Convolutional neural networks · Transfer learning

1 Introduction

Houses that are poorly isolated, or that have high carbon emissions, have a negative impact on climate change [22]. Socio-economically disadvantaged groups typically reside in houses that do not have a high degree of quality [7]. The state of a house has a big impact on the price estimation and valuation of a house [13,21]. Automated detection of properties in extremely bad state will alleviate the task of manually assessing the state of the house conducted by experts. This, in turn, would result in economic, ecological and social benefits as it facilitates gathering information about a property's condition in various scenarios. Nonetheless, no prior articles were found that focus on classifying houses based on their state using street view images. In this study, a Convolutional Neural Network (CNN) is trained on GSV images to identify houses in a bad state. The remainder of this work is organized as follows. In the next section, the related

This work was co-sponsored by a research chair ("Real Estate Analytics") granted by the Royal Federation of Belgian Notaries (Fednot).

work is further elaborated. Section 3 describes the methodology consisting of three steps: data set creation, model training and evaluation. Next, the results are presented, followed by the discussion and conclusions.

2   Related Work

Applications of machine learning (ML) techniques with Google Street View (GSV) images vary greatly. Some researchers are interested in the buildings for predicting their functionality or construction year, some focus on the vehicles parked in the street to predict socioeconomic characteristics of regions, traffic signs or street name signs [2,4,10,14,23]. To extract and use this information from the image data, either classification or object detection techniques are used. These techniques consist of training deep neural nets that need large data sets and computing resources [14]. However, as GSV images are not always usable because of low resolution and obstructions in the image, and labeling them is time intensive, it is difficult to train these models properly [2]. Therefore, with the use of transfer learning, existing, large data sets are leveraged for pretraining networks, which can then be fine-tuned for the task at hand. As several CNN architectures and large data sets exist, a common approach is to test several pretrained networks and select the model that results in the highest accuracy. Li et al. [14] find that DenseNet is a better feature extractor than AlexNet [12] and ResNet for the estimation of the construction year of a building. Kang et al. [10] prefer VGG [19] over AlexNet and ResNet to classify the functionality of buildings. For the detection of traffic signs, the pretrained SSD MobileNet is used [2].

Prior work on assessing the state of a house using images is based on the classification of interior images [16]. A pretrained VGG-16 network is used in combination with a technique for object localisation. Similar to the above, Tavakkoli et al. [21] build their own data set, in this case of street level images, which is a process that causes additional limitations. Low quality images, or images with a poor view of the house cannot be used to classify house types, but also house types that are too similar lead to misclassification. To address these issues, an object detection model was combined with a classification model to preprocess the data before it was used for image classification. Another approach is to combine several types of image data. In [13], GSV images are combined with satellite images to improve value estimation of properties.

3   Methodology

Despite the prior work, classification of houses based on their state using street view images has not been researched yet. To address this gap, a three-by-three analysis is conducted to find the most suitable CNN architecture across three differently-sized datasets. The pipeline used for the experiments consists of several steps: data set creation, model training and evaluation.

The data set was created by querying the GSV API with addresses in Leuven, Belgium. These house images were manually filtered and labeled as "good" or "bad" for state. To eliminate adjacent houses and other obstacles, bounding boxes were created in order to crop the images. This resulted in a data set of 651 images (476 "good" and 175 "bad"). Subsequently, a test set of 98 images was created. To test robustness, the remainder of the data was used to create three differently-sized training and validation sets consisting of 164, 325 and 1132 images respectively. The largest dataset was created using data set augmentation.

Three pretrained CNNs networks were selected, i.e. VGG-16 [19], DenseNet201 [8] and ResNet50V2 [6], because they emerged as either most common or best performing architectures for similar applications in related work [10,14,16]. Using the Keras API, these networks with ImageNet [3] weights were then customized. The pretrained VGG-16 network was extended with three dropout layers, a global average pooling layer, two batch normalization layers, and two dense layers. A dropout, global average pooling, and batch normalization layer were added to the ResNet50V2 model. Similarly, in the DenseNet201 model, the existing head was replaced with dropout layers, a global average pooling layer, batch normalization layers, and dense layers. The last layer in each model is the dense layer with one unit that classifies the images with the sigmoid activation function and an L2 regularizer (weight decay). Before training, the optimizer, the learning rate, the number of layers to unfreeze, and the dropout rate were tuned based on validation accuracy. Because of the rather small data sets and the significant yet doable level of class imbalance, multiple evaluation metrics were computed on the test set. The second part of model evaluation is based on the explainability technique LIME [17], which produces a mask that indicates which pixels of the image contribute most to the prediction.

4 Experimental Results

4.1 Prediction Accuracy

First, Table 1 shows the accuracy on the test set of the nine configurations. For each CNN architecture, using the largest training data set is advisable, with DenseNet201 achieving the best performance overall. This is consistent with the cross-entropy loss on the test set as reported in Table 2. When considering precision (Table 3) and recall (Table 4), the typical trade-off can be observed, with mainly the DenseNet201 being able to strike the best balance, especially when using the medium or large training sets. This is confirmed by the AUC results in Table 5.

Figure 1 shows the training and validation accuracy with increasing number of training steps for four architecture/data set combinations. Overall, training accuracy improves consistently as expected, however, fairly flat validation accuracy curves can be observed. For example, the large DenseNet201 model validation accuracy stays around 89% throughout all epochs (Fig. 1a). The small ResNet50V2 (Fig. 1c), small DenseNet201 (Fig. 1d), and medium VGG-16 model (Fig. 1b) show severe overfitting. As the CNNs are trained on very small data sets in this study,

Table 1. Accuracy on test set

	VGG-16	RESNET50V2	DENSENET201
Small	**0.8163**	0.4796	0.7959
Medium	0.2551	0.8265	**0.8776**
Large	0.8776	0.8163	**0.8980**

Table 2. Loss on test set

	VGG-16	RESNET50V2	DENSENET201
Small	0.5871	0.7988	**0.4954**
Medium	1.3817	0.4866	**0.4423**
Large	0.5406	0.5042	**0.3563**

Table 3. Precision on test set

	VGG-16	RESNET50V2	DENSENET201
Small	**0.7692**	0.3194	0.6471
Medium	0.2551	0.8333	**0.8824**
Large	0.6250	0.6667	**0.8000**

Table 4. Recall on test set

	VGG-16	RESNET50V2	DENSENET201
Small	0.4000	**0.9200**	0.4400
Medium	**1.000**	0.4000	0.6000
Large	**0.8000**	0.5600	**0.8000**

Table 5. AUC on test set

	VGG-16	RESNET50V2	DENSENET201
Small	0.8277	0.6164	**0.8296**
Medium	**0.8838**	0.8496	0.8671
Large	0.9123	0.8660	**0.9553**

while they are designed to be trained on large data sets because of their millions of parameters, this is not surprising [15]. As it has been explained above, transfer learning allows to transfer knowledge from a larger data set to a relatively smaller but similar data set. Thus, pretrained networks with ImageNet weights, which is a large data set with more than 14 million images [3], were used to avoid overfitting [18]. The other measures consisted of adding Batch Normalization layers [9], and dropout layers [15,20] to all architectures. Even though early stopping, evidenced by the small number of training steps, and several other measures were taken against overfitting, our results indicate that overfitting remains an intricate problem, especially due to the small size of the data set, and possibly also due to the class imbalance in the small data set, given that fairly high levels of accuracy can be obtained by near majority prediction models.

4.2 Model Explainability

The predictions of the best performing configuration, the DenseNet201 model with the large data set, are explained using LIME [17]. For each of the four categories of the confusion matrix, one example is analysed. Figure 2 presents the results, with for each example one image with green and red masks and one with a heatmap. Red masks indicate the pixels of the image that significantly contribute to the prediction of the label "bad", corresponding to red areas with a positive weight on the heatmap. Consequently, green areas support the prediction of the label "good" and indicate a negative weight.

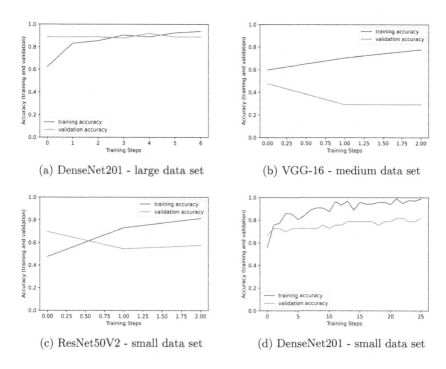

(a) DenseNet201 - large data set

(b) VGG-16 - medium data set

(c) ResNet50V2 - small data set

(d) DenseNet201 - small data set

Fig. 1. Accuracy in function of training steps for different combinations

Figure 2a shows the explanation of a house that was labeled "good" and a prediction score of 0.1187 which indicates the same label, using the threshold 0.5. The heatmap shows two areas that have a low weight, a part of the wall between the windows and the sidewalk. While the former would be important for a human labeler to assess the state of the house, the latter would not. For the true positive example (Fig. 2b), the model uses an adjacent house for the prediction, which is counterintuitive. However, the actual house carries the largest weights, producing a correct prediction. For the two wrongly predicted examples, the windows carry important weights, contributing to the correct label. In the false negative example (Fig. 2c), the pixel weights are surprising. The windows carry a negative weight while a part of the wall was attributed a large positive weight and the area with flaking paint has negligible weights. Some irrelevant parts of the image also carry relatively important weights. Lastly, the false positive example (Fig. 2d) shows that the model also recognizes the importance of the front door.

5   Discussion and Conclusions

Based on the reported performance measures on the test set, the DenseNet201 model trained on the large data set performs best compared to the other models.

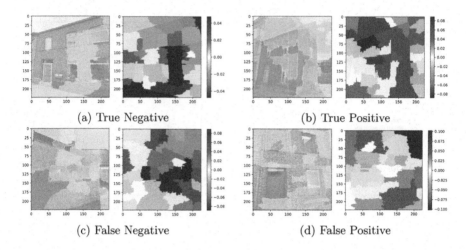

(a) True Negative   (b) True Positive

(c) False Negative   (d) False Positive

Fig. 2. LIME explanations with masks (left) and heatmap (right)

Additionally, the performance measures of the small and medium DenseNet201 models are relatively good. For this CNN architecture, an expected trend is noticeable, the accuracy, loss, precision, recall, and AUC reported on the medium data set are better than those reported on the small data set, and an even better performance is achieved on the large data set. This is expected as machine learning models tend to learn more when trained on a larger set of examples. The training and validation accuracy showed that the configurations based on the small and medium data sets overfitted severely. In addition, closer investigation revealed that the VGG-16 model only predicted the "bad" class when trained on the medium data set.

In this study, the robustness of the models is based on the performance on the three data sets. The VGG-16 and ResNet50V2 architectures are not robust as these models perform particularly bad on the medium and small data set respectively. However, DenseNet201 does perform consistently well on the three data sets, so, in this respect, DenseNet201 can be viewed as a robust architecture for this classification task. The robustness of a model can also be evaluated by comparing the performance on the test set with the training and validation performance [5]. Based on the accuracy and loss, the three CNN architectures all show consistent performance on the large data set and DenseNet201 and ResNet50V2 do so on the medium data set as well. It can be concluded that the DenseNet201 architecture is the most suitable for this task. The discrepancy in the performance between the VGG-16 models on the one hand, and the ResNet50V2 and DenseNet201 models on the other hand, implies that for this image classification task, deep networks are better suited.

Two important factors that influence these results are the size of the data sets and the class imbalance. Machine learning models typically have many parameters and are therefore trained on large data sets, but the data sets used in this study are very small. In [15] the data set CIFAR-10 is described as a small data set with 60 000 images with ten classes. However, by leveraging transfer learning to reduce the number of parameters, this problem is partially mitigated. Nevertheless, the hyperparameter tuner still optimized the models with millions of trainable parameters. Another factor that limits the performance of the models is class imbalance [1]. The minority class "bad" has a share of around 25 to 35 percent in the data sets. However, this class imbalance is not necessarily severe as other sources define ratios such as 1:100 as severe class imbalance [11].

The performance of the models is also limited by the data quality. Generalizability is the first concern as the data set's representation of the housing market is low. That is, the data set only contains houses in the specific geographical area and only terraced and detached houses. In addition, inherent to street-level images is that this type of data will never give a complete representation of a house. Specifically, regulations that limit renovations of house fronts might lead to incorrect labeling and, in consequence, incorrect predictions. Expanding the data set in several dimensions could mitigate both issues. Depending on the study area, properties in a very bad state will often be the exception. Therefore, class imbalance techniques such as the options discussed in [1], i.e. undersampling, oversampling, thresholding, cost-sensitive learning, and ensemble learning, can be used to mitigate this effect on the results.

As the described methodology involves manual work that requires quite some resources and time, further automation of the process would be extremely beneficial. A two-step method could be developed such that in the first step the raw GSV data is processed into a qualitative data set that can be used in the second step for classification. Filtering and cropping in the first step is automated by an object detection model that predicts bounding boxes enclosing the property on the image [21]. In this way, an end-to-end approach for the classification of the state of a house would be obtained that facilitates the use of this method.

References

1. Buda, M., Maki, A., Mazurowski, M.A.: A systematic study of the class imbalance problem in convolutional neural networks. Neural Netw. **106**, 249–259 (2018)
2. Campbell, A., Both, A., Sun, Q.C.: Detecting and mapping traffic signs from Google Street View images using deep learning and GIS. Comput. Environ. Urban Syst. **77** (2019). Article no. 101350
3. Deng, J., Dong, W., Socher, R., Li, L.J., Li, K., Fei-Fei, L.: ImageNet: a large-scale hierarchical image database. In: 2009 IEEE CVPR, pp. 248–255. IEEE Computer Society, Washington, DC, USA, June 2009
4. Gebru, T., et al.: Using deep learning and Google street view to estimate the demographic makeup of neighborhoods across the United States. In: Proceedings of the National Academy of Sciences, vol. 114, no. 50, pp. 13108–13113, December 2017
5. Hastie, T., Tibshirani, R., Friedman, J.H., Friedman, J.H.: The Elements of Statistical Learning: Data Mining, Inference, and Prediction, vol. 2. Springer, Cham (2009). https://doi.org/10.1007/978-0-387-21606-5
6. He, K., Zhang, X., Ren, S., Sun, J.: Identity mappings in deep residual networks. CoRR abs/1603.05027, pp. 630–645, March 2016

7. Heylen, K., Vanderstraeten, L.: Wonen in Vlaanderen anno 2018. Gompel & Svacina, Sint-Niklaas (2019)
8. Huang, G., Liu, Z., Maaten, L.V.D., Weinberger, K.Q.: Densely connected convolutional networks. CoRR abs/1608.06993, pp. 2261–2269 (2016)
9. Ioffe, S., Szegedy, C.: Batch normalization: accelerating deep network training by reducing internal covariate shift. In: 32nd International Conference on Machine Learning, ICML 2015, vol. 1, pp. 448–456, February 2015
10. Kang, J., Körner, M., Wang, Y., Taubenböck, H., Zhu, X.X.: Building instance classification using street view images. ISPRS J. Photogramm. Remote. Sens. **145**, 44–59 (2018)
11. Krawczyk, B.: Learning from imbalanced data: open challenges and future directions. Progress Artif. Intell. **5**(4), 221–232 (2016). Article no. 101350. https://doi.org/10.1007/s13748-016-0094-0
12. Krizhevsky, A., Sutskever, I., Hinton, G.E.: ImageNet classification with deep convolutional neural networks. In: Pereira, F., Burges, C.J.C., Bottou, L., Weinberger, K.Q. (eds.) Advances in Neural Information Processing Systems, vol. 25. Curran Associates Inc, Red Hook (2012)
13. Law, S., Paige, B., Russell, C.: Take a look around: using street view and satellite images to estimate house prices. ACM Trans. Intell. Syst. Technol. **10**(5), 1–19 (2019)
14. Li, Y., Chen, Y., Rajabifard, A., Khoshelham, K., Aleksandrov, M.: Estimating building age from google street view images using deep learning. In: Winter, S., Griffin, A., Sester, M. (eds.) GIScience 2018, vol. 114, pp. 1–40. Schloss Dagstuhl-Leibniz-Zentrum fuer Informatik, Dagstuhl (2018)
15. Liu, S., Deng, W.: Very deep convolutional neural network based image classification using small training sample size. In: 2015 3rd IAPR Asian Conference on Pattern Recognition (ACPR), pp. 730–734. IEEE Computer Society, Washington, DC, USA, November 2015
16. Perez, H., Tah, J.H.M., Mosavi, A.: Deep learning for detecting building defects using convolutional neural networks. Sensors (Switzerland) **19**(16), 3556 (2019)
17. Ribeiro, M.T., Singh, S., Guestrin, C.: "why should I trust you?": Explaining the predictions of any classifier. CoRR abs/1602.04938 (2016)
18. Shu, M.: Deep learning for image classification on very small datasets using transfer learning. Creative Components **345**, 14–21 (2019)
19. Simonyan, K., Zisserman, A.: Very deep convolutional networks for large-scale image recognition. In: Bengio, Y., LeCu, Y. (eds.) 3rd International Conference on Learning Representations. ICLR, La Jolla, CA, USA (2015)
20. Srivastava, N., Hinton, G., Krizhevsky, A., Sutskever, I., Salakhutdinov, R.: Dropout: a simple way to prevent neural networks from overfitting. J. Mach. Learn. Res. **15**, 1929–1958 (2014). Article no. 101350
21. Tavakkoli, V., Mohsenzadegan, K., Kyamakya, K.: A visual sensing concept for robustly classifying house types through a convolutional neural network architecture involving a multi-channel features extraction. Sensors (Switzerland) **20**(19), 1–16 (2020)
22. Ustaoglu, A., Yaras, A., Sutcu, M., Gencel, O.: Investigation of the residential building having novel environment-friendly construction materials with enhanced energy performance in diverse climate regions: Cost-efficient, low-energy and low-carbon emission. J. Build. Eng. **43**, 102617 (2021)
23. Wojna, Z., et al.: Attention-based extraction of structured information from street view imagery. In: Proceedings of the International Conference on Document Analysis and Recognition, ICDAR, vol. 1, pp. 844–850 (2017)

Towards a Comprehensive BPMN Extension for Modeling IoT-Aware Processes in Business Process Models

Yusuf Kirikkayis[✉], Florian Gallik, and Manfred Reichert

Institute of Databases and Information Systems, Ulm University, Ulm, Germany
{yusuf.kirikkayis,florian-1.gallik,manfred.reichert}@uni-ulm.de

Abstract. Internet of Thing (IoT) devices enable the collection and exchange of data over the Internet, whereas Business Process Management (BPM) is concerned with the analysis, discovery, implementation, execution, monitoring, and evolution of business processes. By enriching BPM systems with IoT capabilities, data from the real world can be captured and utilized during process execution in order to improve online process monitoring and data-driven decision making. Furthermore, this integration fosters prescriptive process monitoring, e.g., by enabling IoT-driven process adaptions when deviations between the digital process and the one actually happening in the real world occur. As a prerequisite for exploiting these benefits, IoT-related aspects of business processes need to be modeled. To enable the use of sensors, actuators, and other IoT objects in combination with process models, we introduce a BPMN 2.0 extension with IoT-related artifacts and events. We provide a first evaluation of this extension by applying it in two case studies for modeling of IoT-aware processes.

Keywords: BPMN · BPM · Internet of Things · IoT in BPM · Sensors · Actuators

1 Introduction

As electronic components have become smaller, more powerful, and less expensive, the Internet of Things (IoT) has received an upswing in recent years [1]. Many embedded components are equipped with sensors and actuators that enable collection of environmental data (sensors) as well as physical responses to specific events (actuators) [2]. IoT components can be embedded in everyday objects such as washing machines, refrigerators, vehicles, cell phones, or wearable devices. Moreover, they can be found in cyber-physical systems, smart cities, or smart logistics [3]. IoT refers to a network of physical objects or "things" being equipped with sensors, actuators and software to connect them with other devices and systems over the Internet. Such interconnected devices, in turn, constitute the basis for exchanging data [2].

While IoT allows capturing and exchanging data about the physical environment, BPM enables the analysis, discovery, implementation, execution, monitoring, and evolution of business processes [17]. BPM-enabled processes can be further enhanced by sensors, e.g., to measure the fill level of a tank and eliminate the need for manually performing this task [5]. In general, IoT technology contributes to make abstract process models real-world-aware and, thus, to align digital processes with the physical world [15]. Moreover, IoT devices can be used to automate different types of tasks, which may be physical (moving a conveyor belt) or digital (sending data or notifying a system) [14]. Furthermore, IoT enhances the monitoring, discovery, and optimization of processes, which are referred to as IoT-aware processes in the following. An important task for modeling IoT-aware processes is to properly capture IoT-related aspects. Modeling a process fosters the understanding of how the process works and allows discovering potential problems (e.g., deadlocks) before process automation. Finally, already modeled processes can be analyzed, improved, automated and optimized [17].

There are several languages for modeling business processes, such as Petri Nets, Event-driven Process Chains (EPC), Role Activity Diagrams, Resource-Event-Agent (REA), and Business Process Modeling Language (BPML) [16]. A standardized process modeling language is Business Process Model and Notation (BPMN) 2.0, which has undergone three releases since 2004 [6]. Modeling IoT-aware processes with BPMN 2.0 is a complex endeavor, and the resulting model is difficult to understand due to the potentially ambiguous use of modeling elements. As a drawback BPMN 2.0 does not allow for the explicit representation of IoT devices and IoT-related aspects, which aggravates the maintenance and servicing of IoT-aware processes significantly. In particular, the resulting process models lack structure, expressiveness, and flexibility. To overcome the lack of language elements for modeling IoT aspects, BPMN 2.0 needs to be extended. Existing BPMN 2.0 extensions for IoT-aware processes are either incomplete or do not comprehensively cover the required treatment.

In this work, we present a BPMN Extension for IoT-aware processes which enables the explicit integration of business process models with IoT devices. The approach supports the modeling of IoT-aware processes in terms of different views and levels of abstraction.

The remainder of this paper is organized as follows: In Sect. 2, we summarize the main issues that emerge when modeling IoT-aware processes with the existing BPMN 2.0 standard. Section 3 discusses existing works. In Sect. 4, we present our BPMN 2.0 extension for modeling IoT-aware processes and illustrate it along two case studies. Finally, Sect. 5 summarizes and discusses our approach.

2   Problem Statement

In order to properly model the behavior of IoT-aware business processes a multitude of input and output devices may have to be integrated with a process model. In principle, BPMN 2.0 offers various mechanisms for representing IoT devices. On one hand, script, service and business rule tasks can be used to represent IoT-related activities. On the other, resources, data objects or events may be used to model IoT-involvement [7]. However, when following such a straightforward approach, no distinction between regular BPMN tasks and IoT-related ones can be made. The familiarization with such a process may therefore take longer, as the IoT-related model elements cannot be visually distinguished from standard BPMN elements. Consequently, the modeled IoT tasks constitute a black box, i.e., it does not become transparent whether the task refers to a sensor, an actuator, or a service call [7]. On one hand, this aggravates model comprehensibility, on the other it results in a limited usability and poor maintainability of the model.

The process model, depicted in Fig. 1, deals with the treatment of the Chronic Obstructive Pulmonary Disease (COPD). COPD describes the obstruction of the lungs, which hinders the patient's breathing [7]. First, the patient's heart rhythm is checked (1). If necessary, an emergency alarm is triggered (2). Then the severity of the COPD is assessed (3) and, depending on the outcome of this assessment, either no treatment, treatment with an oxygen mask (4), or treatment with an inhaler (5) is administered. Finally, the results of the treatment are analyzed (6) and the patient record is updated accordingly (7).

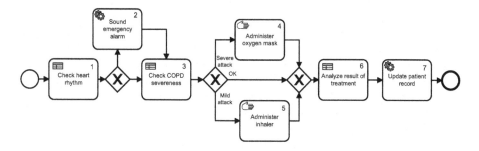

Fig. 1. Example of a process model with IoT aspects. (Adapted from [7])

When using standard BPMN 2.0 elements for modeling the physical (i.e. IoT-related) tasks of the COPD process (Fig. 1), it is unclear, which tasks are IoT-related and which are not. In addition, it is unclear which sensors and actuators, respectively, are involved in the processing of the IoT-related tasks. Instead, it becomes necessary to carefully read and understand the underlying process model in order to make assumptions whether, for example, a business rule task refers to a specific sensor or a service task represents an action of an actuator. Note that this might cause ambiguities due to labeling issues.

When representing sensors in terms of business rule tasks, the involvement of IoT devices (cf. Fig. 1, Activity (1) heart rate sensor) does not become apparent as well. Finally, there is no visual difference between an IoT-related Business Rule Task (1 & 3) and a BPMN Business Rule Task (6), or between an IoT-related Service Task (2) and a BPMN Service Task (7). This aggravates the comprehension as well as maintenance of the process model.

3   Existing Approaches

There exist several works that introduce notations, approaches, or language extensions for representing IoT devices in the context of BPMN 2.0. This section briefly describes these approaches and discusses them (Table 1).

[1] and [7] model IoT-aware processes with standard BPMN 2.0 elements. While [1] uses script tasks for integrating both sensors and actuators, [7] uses business rule tasks for representing sensors and service tasks for actuators. Cheng et al. [12] extend the BPMN 2.0 standard with a sensor task covering the following aspects: sensor device, sensor service, and sensor handler. Another approach for representing physical entities (e.g., a bottle of milk) in terms of a collapsed pool is presented in [10]. In particular, for sensing and actuation activities two new task types are introduced. Sungur et al. [13] explore the properties of wireless sensor networks (WSNs). For this purpose, they introduce a WSN Task and a WSN Pool. The WSN Task has an actionType element consisting of a question mark, an exclamation mark, and a square. It is used to specify a WSN operation as a sensing (?), actuating (!), or intermediating operation (□). A real-world temperature control scenario is suggested by [4], which enhances existing BPMN 2.0 events with a conditional event, message event, and error event. [8], extends BPMN 2.0 with a resource called ResourceExtension, included for both human and non-human resources. The ResourceExtension has some privileges (ResourcePrivileges) and types such as RFID, Sensor and Actuator (ResourceTypes). uBPMN [6] suggests additional elements for Sensor, Reader, Collector, Camera, and Microphone. Each of these elements is represented by specific task and event types. In addition, a Smart Object is introduced to represent transmitted data. In [9], an Industry 4.0 process modeling language (I4PML) extending BPMN 2.0 with the following elements is presented: Cloud app, IoT device, device data, actuation task, sensing task, human computer interface, and mobility aspect.

Though existing approaches already enable the modeling of various IoT-driven process scenarios, there remain some gaps or scenarios that cannot be fully represented. Except for [1] and [7], all other approaches extend BPMN 2.0 with specific IoT elements. While uBPMN only introduces a Start Event, none of the approaches explores the execution and/or control of an actuator in combination with an End Event. In addition, none of these approaches allows for the combined use of sensors and actuators in the context of a task. Furthermore, none of the approaches supports responses to an IoT event during task execution. Another important scenario that cannot be modeled with existing approaches is the verification of an IoT-driven condition when processing a task. There is also no concept for representing of IoT-driven processes with different levels of abstraction in the already existing approaches. Note that modeling IoT-driven processes with different abstraction levels could enable different views for various stakeholders (e.g., domain expert, BPMN expert, or IoT expert).

Table 1 provides a systematic summary of the different approaches (with ✓ indicating support of the respective feature and ✗ expressing missing support). As can be easily seen, non of the approaches comprehensively covers the treatment needed for IoT-aware processes.

Table 1. Currently supported IoT elements through the extensions

	Sensor	Actuator	Combining sensor and actuator	Start event	End event	React to IoT within a task	Intermediate event	Condition element	Physical entity	IoT data object	Abstraction level	Score
Meyer et al. [10]	✓	✓	✗	✗	✗	✗	✗	✗	✓	✗	✗	3/10
Sungur et al. [13]	✓	✓	✗	✗	✗	✗	✗	✗	✓	✗	✗	3/10
uBPMN [6]	✓	✗	✗	✓	✗	✗	✓	✓	✗	✓	✗	5/10
Cheng et al. [12]	✓	✗	✗	✗	✗	✗	✗	✗	✗	✗	✗	1/10
BPMN4WSN [11]	✓	✓	✗	✗	✗	✗	✗	✗	✓	✗	✗	3/10
Suri et al. [8]	✓	✓	✗	✗	✗	✗	✗	✗	✗	✗	✗	2/10
I4PML [9]	✓	✓	✗	✗	✗	✗	✗	✗	✓	✓	✗	4/10

4 Solution Proposal

To tackle the gaps and problems discussed in Sects. 2 and 3 respectively, we extend BPMN with artifacts and events as shown in Fig. 2. Note that these artifacts and events are not limited to specific use cases or processes, but may be used in any domain. Due to lack of space, the various elements cannot described in detail. Instead, we demonstrate the use of selected IoT artifacts and events along two IoT-aware processes from different domains. Note that all elements are decorated with a WLAN icon and labeled as "IoT". In addition, the latter in the upper left corner indicates the artifact type.

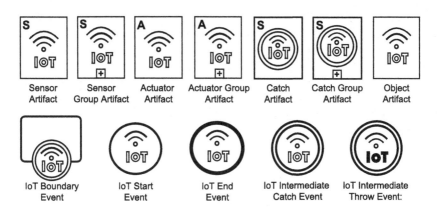

Fig. 2. Extended BPMN 2.0 elements

4.1 Case Study 1: Healthcare Process

In Case Study 1, we consider an IoT-enabled measurement process in a medical facility. The process can be modeled with our extension as shown in Fig. 3. The process begins when the physician registers the treatment with the help of an RFID scanner. The patient is then transported to the treatment room. In the treatment room, sensors are used to check whether the patient has arrived. If this does not happen within 10 min, a notification appears on a specific monitor and the process ends. If the patient has arrived within the specified time period, a measurement process is started. Afterwards, the measured data is received, the evaluation is started, and a severity score is calculated. This score serves as a basis for the next steps. If the score is above 8, an alarm is triggered with an IoT actuator artifact. If the score is 6 or 7, an indicator light is switched on. If the score is below 6 nothing happens. Finally, the process ends with a logout at the RFID sensor.

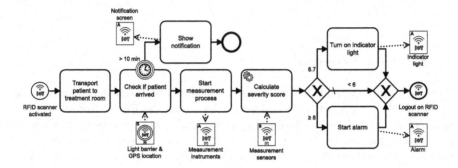

Fig. 3. Healthcare process of Case Study 1

4.2   Case Study 2: Production Process

Case Study 2 considers a process that sort workpieces based on their color in a production setting. This process could be modeled with our extension as shown in Fig. 4. Process execution starts as soon as the light barrier on the conveyor belt is triggered by an incoming workpiece. The conveyor belt is then started and remains in operation until the end light barrier is triggered or the weight on the conveyor belt exceeds 1000 kg. A light is then switched on inside the machine and a color sensor is used to determine the color of the workpiece. Finally, the workpiece is sorted according to its color.

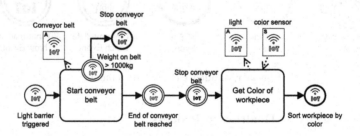

Fig. 4. Sorting process of Case Study 2

As shown, the proposed BPMN 2.0 extensions enable a proper modeling of the two IoT-aware processes analyzed in the case studies. In particular, both processes can be modeled more intuitively and specifically compared to the approaches discusses in Sects. 2 and 3.

5   Conclusions

This paper presented a BPMN 2.0 extension for modeling IoT-aware processes. Based on the problem description and a literature review, we identified gaps and problems of existing BPMN extensions for modeling IoT-aware processes. We then extended BPMN 2.0 with additional IoT artifacts and IoT-related events that address the identified gaps. In particular, the added elements enable the acquisition of physical data with the sensor artifact and the control of actuators with the actuator artifact. Furthermore, subjects and/or objects can be represented by IoT objects. IoT conditions, in turn, can be validated during task processing by the IoT intermediate catch artifacts as well as along the sequence flow by the IoT intermediate events. All artifacts can be aggregated into corresponding group artifacts to increase the abstraction level. Moreover, the process start may be triggered by an IoT condition associated with an IoT start event. In addition, a process end may execute and/or control an actuator with the IoT end event. Finally, we introduced an IoT boundary event, which allows redirecting the sequence flow based on an IoT condition. We have applied our extension in two case studies to show how the introduced artifacts and events can be used.

In future work we will perform various experiments and studies with different users such as BPMN modelers or domain experts to investigate the completeness of our extension and to study model comprehensibility. Furthermore, we integrate our extension with a process engine, i.e., the framework should support both the modeling and execution of IoT-aware processes.

Acknowledgments. This work has been funded by the Deutsche Forschungsgemeinschaft (DFG, German Research Foundation) under project number 449721677.

References

1. Domingos, D., Francisco, M.: Using BPMN to model Internet of Things behavior within business process. Int. J. Project Manag. 39–51 (2017)
2. Janiesch, C., et al.: The Internet of Things meets business process management: a manifesto. IEEE Syst. Man Cybern. Magaz. **6**(4), 34–44 (2020)
3. Chang, C., Srirama, S., Buyya, R.: Mobile cloud business process management system for the Internet of Things: a survey. ACM Comput. Surv. **49**(4), 1–42 (2016)
4. Chiu, H., Wang, M.: Extending event elements of business process model for Internet of Things. In: IEEE International Conference on Computer and Information Technology (2015)
5. Cherrier, S., Deshpande, V.: From BPM to IoT. In: Teniente, E., Weidlich, M. (eds.) BPM 2017. LNBIP, vol. 308, pp. 310–318. Springer, Cham (2018). https://doi.org/10.1007/978-3-319-74030-0_23
6. Alaaeddine, Y., Bauer, C., Saidi, R., Anind, D.: uBPMN: a BPMN extension for modeling ubiquitous business processes. In: Information and Software Technology (2016)
7. Hasić, F., Asensio, E.: Executing IoT processes in BPMN 2.0: current support and remaining challenges. In: 13th International Conference on Research Challenges in Information Science (RCIS) (2019)

8. Suri, K., Gaaloul, W., Cuccuru, A., Gerard, S.: Semantic framework for Internet of Things-aware business process development. In: 26th International Conference on Enabling Technologies: Infrastructure for Collaborative Enterprises (2017)

9. Petrasch, R., Hentschke, R.: Process modeling for Industry 4.0 applications towards an Industry 4.0 process modeling language and method. In: 13th International Joint Conference on Computer Science and Software Engineering (JCSSE) (2016)

10. Meyer, S., Ruppen, A., Hilty, L.: The things of the Internet of Things in BPMN. In: Persson, A., Stirna, J. (eds.) CAiSE 2015. LNBIP, vol. 215, pp. 285–297. Springer, Cham (2015). https://doi.org/10.1007/978-3-319-19243-7_27

11. Tranquillini, S., et al.: Process-based design and integration of wireless sensor network applications. In: Business Process Management (2012)

12. Cheng, Y., et al.: Modeling and deploying IoT-aware business process applications in sensor networks (2019)

13. Sungur, C.T., et al.: Extending BPMN for wireless sensor networks. In: IEEE 15th Conference on Business Informatics (2013)

14. Janiesch, C., et al.: The Internet-of-Things Meets Business Process Management: Mutual Benefits and Challenges (2017)

15. Gruhn, V., et al.: BRIBOT: Towards a Service-Based Methodology for Bridging Business Processes and IoT Big Data (2021)

16. Mili, G., Tremblay, G., Jaoude, G., Lefebvre, E., Elabed, L., Boussaidi, G.: Business Process Modeling Languages: Sorting Through the Alphabet Soup. Association for Computing Machinery (ACM), New York (2010)

17. Dumas, M., La Rosa, M., Mendling, J., Reijers, H.A.: Fundamentals of Business Process Management. Springer, Heidelberg (2018). https://doi.org/10.1007/978-3-662-56509-4_10

Team Selection Using Statistical and Graphical Approaches for Cricket Fantasy Leagues

S. Mohith[1]⬤, Rebhav Guha[1]⬤, Sonia Khetarpaul[1(✉)]⬤, and Samant Saurabh[2]

[1] Shiv Nadar University, Guatam Buddh Nagar, Greater Noida,
Uttar Pradesh 201310, India
sonia.khetarpaul@snu.edu.in
[2] Indian Institute of Management, Bodh Gaya, Bihar, India

Abstract. Fantasy Sports are becoming more and more popular these days, hence the race to crack it is more trending than ever. In this paper, we focus on cricket (IPL) and Dream11. Using advanced statistical and graphical models, and new performance metrics for batting and bowling we aim to build a model that can predict the top performing 11 players out of the two teams. This involves predicting the player performance and selecting the best 11 while complying with league constraints. The proposed model on an average predicts 70% of the players from the *Dream Team*.

1 Introduction

Cricket is one of the most followed team games with billions of fans all across the globe. It is a sport played by two teams with each side having eleven players. Each team is a right blend of batsmen, bowlers and all-rounders. Indian Premier League (IPL) is India's largest cricketing league and one of the most popular cricket leagues in the world. It is contested by 8 teams which represent 8 Indian cities. IPL is the most attended Cricket League in the world with major fan-following from across the globe. According to sources, the viewership count hit 380 million for IPL 2021 [13].

A fantasy sport is a type of game, where participants assemble imaginary or virtual teams composed of real players of a professional sport. Users create a team (of 11 players while complying to selection constraints) on these platforms to compete with other users. The performance of each player playing the game is converted into points based on rules set by the league that are compiled and totaled. A user's team's score is equal to the summation of points earned by all the players in the team. The more points the user's team earns, lower will be the rank of the team. In this paper our goal is to be able to create a team which is in the **winning zone** by predicting as many players as possible from the **Dream Team** (Sect. 3.2).

© The Author(s), under exclusive license to Springer Nature Switzerland AG 2022
R. Guizzardi et al. (Eds.): RCIS 2022, LNBIP 446, pp. 719–726, 2022.
https://doi.org/10.1007/978-3-031-05760-1_48

2   Related Work

Dey et al. [1] used a network approach to team formation however the approach wansn't suitable for Fantasy Sports. Although traditionally it is considered that venue plays an important role (especially for the home team), but with teams changing every year in IPL the venue is less impact-full. M.Satyam [14] states *in-strength* of a player is positively correlated with ICC rankings but these scores were not impacted by the venue. Iyer et al. [2], classified players into 3 predefined classes - Performer, Moderate, and Failure based on the set rules. Bhattacharjee et al. [3] proposed a composite index to measure the performance of cricketers irrespective of their expertise and to pick an optimal team of 11 players from a whole squad of players. Pathak et al. [4] used 3 different models to predict the winner of the match. The paper used traditional attributes of the game like average, strike rate, total runs for batting and economy rate, wickets for bowling to make the prediction. Every team was analysed against every other team. Passi et al. [5] and Lakkaraju et al. [6] introduced many new metrics like consistency, form, venue, consistency adjusted average and batting impact score and much more.

Passi et al. [7] featured player selection based on traditional Machine Learning classification algorithms. It uses various traditional attributes and derives new attributes such as form, consistency, venue and opposition score. Saurav et al. [12] discuss the integer optimisation required in such a team selection process. Hermanus [8] proposed to measure the batting performance of cricketers and use three parameters, the strike rate, the batting average and the consistency of the batsman to rank the batting performance of players. Hermanus, in [9], measured the performance of bowlers based on the wickets they took and weighted the batsmen to judge better if a bowler took an important wicket rather than an average one.

Some of the papers [5,8,9] introduced new metrics to quantify performance while other papers [2–4] have worked on predicting the player's performance using traditional metrics of the game. In this paper, we aim to predict as many players as possible from the dream team, to put the team in the winning zone. This makes the task unique from all the papers that were explored.

3   Data and Dream11

3.1   Dataset

We used the T20 format data from International cricket and IPL, collected from CrickSheet. It provides ball-by-ball data of all the cricket matches across all formats. Although the dataset had 13 years worth data from T20 matches, we chose to only use data from the last 5 years, due to reasons mentioned above.

3.2   Dream 11

Dream11 imposes a lot of constraints while selecting the 11 players. While selecting, there can be 1 to 4 Wicket-keepers, 3 to 6 Batsmen, 1 to 4 All-rounders, 3 to 6 Bowlers, maximum of 7 players can be from one team and the Total cost of selected players should be less than 100.

Wining Zone refers to the top 60% of the teams in the contest. After the match, each user's team is allotted a rank based on the total points earned by the players. Lower the rank, higher the reward.

The Dream Team is the best set of 11 players (while complying to constraints) for a particular match. The total points earned by the **dream team** is the maximum possible points that could have been earned and is announced after the match.

4 Proposed Model

The proposed model is shown in Fig. Fig. 1. Unlike in other team selection approaches, the selected 11 players here have to player with each other while in the real game they play against each other i.e., we pick players who perform strongly but not against each other since that can result in negative points for poor performance of certain players. The overall performance of a player is calculated using following equation:

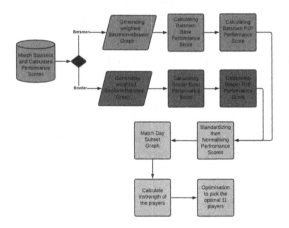

Fig. 1. Architecture of proposed model

$$S_{batsman_i} = Success\ Probability * Base\ Performance_i + In\!-\!Strength_i \quad (1)$$

4.1 Weighted Graph

The interaction a batsman has with all the bowlers and vice versa, is represented using a weighted directed graph the edges connecting them would represent the performance. An outgoing edge from a batsman to a bowler indicates the batsman's performance against the bowler and vice versa. The summation of weights of outgoing edges from a player (node) is defined as the *In-Strength*. Mathematically, the in-strength of a player i is calculated as

$$S_i^{in} = \sum_{j \neq i} W_{ji} \tag{2}$$

where W_{ji} is the weight between player i and j calculated using only the data between those two players. We call this the Player to Player or P2P metric and they are explained below for batting and bowling.

To quantify a batsman's performance against a bowler, we have defined a metric which incorporates the primary statistics of a batsman in a T20 game - runs scored (EWA), the average strike rate of the batsman and his consistency in matches (adopted from Lemmer et al. [11]). This new measure of batting performance between batsman i and bowler j is defined as

$$BP\_P2P_{i,j} = EWA(S) \times \frac{SR_{i,n}}{avg(SR_n)} \times \frac{CC_i}{avg(CC)} \tag{3}$$

where

1. EWA is exponentially weighted average with weights being $0.96, 0.96^2, 0.96^3$ ....0.96^n and S is the list of scores of batsman i against bowler j.
2. $SR_{i,n}$ is the strike rate of batsman i in the last n (10 in this case) matches and $avg(SR_n)$ is the avg strike rate of all batsmen in the last n matches.
3. CC_i is the consistency coefficient of batsman i and $avg(CC)$ is the average consistency coefficient of all batsmen.

The Consistency Coefficient measures how consistent a batsman has been throughout the time period in consideration. It is defined as:

$$CC = \frac{average\ runs}{adjusted\ standard\ deviation} \tag{4}$$

In adjusted standard deviation the scores above the average and not out scores are not taken into account since using low not out score contributes to labelling a batsman as inconsistent when he could have achieved a score closer to or above his average.

To quantify a bowler's performance against a batsman, we used a metric where for any bowler i, the P2P bowling metric against batsman j is defined as follows:

$$CBR_{i,j} = \frac{3R_{i,j}}{W_{i,j} + \frac{B_{i,j}}{6} + W_{i,j} \times \frac{R_{i,j}}{B_{i,j}}} \tag{5}$$

1. $R_{i,j}$ is the runs scored by batsman j against bowler i
2. $W_{i,j}$ is the number of times batsman j was dismissed by bowler i
3. $B_{i,j}$ is the balls bowled to batsman j by bowler i

4.2   Base Performance

Based on the overall performance (independent of opponent) we formulated a new metric called the *Base Performance* which quantifies the average performance. It is defined as

$$BP_i = ET_i \times P(Success) \tag{6}$$

where ET_i = Average batting performance (Eq. 7)

To quantify the average batting performance (ET_i) of a player we have adopted metrics from Lemmer et al. [10].

$$ET_i = \frac{(sumouta_i + (2.1 - 0.005 * avnoa_i) * sumnoa_i)}{n_i} \tag{7}$$

where n_i = number of innings played by the ith batsman

$sumouta_i$ = sum of adjusted runs in the innings where i^{th} batsman was out

$sumnoa_i$ = sum of adjusted runs in the innings where i^{th} batsman was not out

$avnoa_i$ = average of the adjusted not out scores of the ith batsman.

The adjusted runs scored by the ith player in the jth match is denoted by T_{ij} and is defined by

$$T_{ij} = Runs_{batsman_i,match_j} * (\frac{StrikeRate_{batsman_i,match_j}}{MSR_j})^{0.5} \tag{8}$$

$$StrikeRate_{batsman_i,match_j} = (\frac{Runs_{batsman_i,match_j}}{BallsFaced_{batsman_i,match_j}}) * 100 \tag{9}$$

$$MSR_j = \frac{Total \; runs \; scored \; in \; match}{Total \; balls \; bowled \; in \; the \; match} \times 100 \tag{10}$$

Success Probability for batsman is calculated as the number of times a batsman has scored more than $\frac{1}{3}^{rd}$ of his batting average where as for a bowler it is the matches where the Bowling Rate (Eq. 12) is than $\frac{1}{3}^{rd}$ of his/her average Bowling Rate

We define base performance of bowler i as:

$$BP_i = BowlingRate_i \times P(Success) \tag{11}$$

Mathematically, Bowling Rate (BR_i) is defined as:

$$BR_i = \frac{(W_{ij}) \times \frac{W_{ij}}{W_j}}{(R_{ij}) \times \frac{SR_{ij}}{SR_j}} \tag{12}$$

where W_{ij} is wickets taken by i^{th} bowler in j^{th} match, SR_j is same as Eq. 10, SR_{ij} is runs conceded by i^{th} bowler in j^{th} match

$$SR_{ij} = \frac{Runs \; Conceded \; by \; bowler}{Balls \; Bowled \; by \; the \; bowler} \times 100 \tag{13}$$

The scores calculated above are first normalised which brings the mean and standard deviation to 0 and 1 respectively. And then we bring them within the range [0,1]. The base performance and the P2P metrics are normalised and scaled separately. Using the performance metrics defined above, we have calculated the scores for a few batsmen in Table 1. On any given **Match Day** we create a subset

Table 1. Values calculated using above metrics

(a) Values calculated using above metrics

Batsmen	Base Perf	Batting Avg
AB de Villiers	35.9648	43.63
V Kohli	35.7547	40.36
MS Dhoni	28.9536	39.67
CH Gayle	27.1899	36.36
RR Pant	24.8992	33.51

(b) Batsmen against J J Bumrah (weighted graph)

Batsmen	Weight	Batting Avg
AB de Villiers	13.4625	41.66
V Kohli	6.3514	31.50
AD Russell	4.4692	20.33
MS Dhoni	4.0501	18.66

graph H using just the nodes (players) from the day's match only, ignoring all the edges to players from other teams. The in-strength of any player (node) from the graph H gives us the relative strength of the player against the players playing on that particular day. We add the base performance defined in Sect. 4.2 to the in-strength to get the final performance variable for a player.

Player	Strength	Cost	Playing	Wicketkeeper	Batsmen	Allrounder	Bowler
Samad	0.409067622	8	0	0	1	0	0
MK Pandey	1.481035807	9	1	1	0	0	0
KS Williamson	1.467014998	9	1	0	1	0	0
JM Bairstow	1.161955225	10	1	0	1	0	0
KM Jadhav	0.760470992	8	1	0	0	1	0
Sandeep Sharma	1.390342558	8.5	1	0	0	0	1
Rashid Khan	0.402781226	9.5	0	0	0	1	0
KK Ahmed	0.275347084	8	0	0	0	0	1
V Shankar	0.753843029	8.5	0	0	0	1	0
B Kumar	1.156648282	10	1	0	0	0	1
Mohammad Nabi	0.415452986	9	0	0	0	1	0
Anuj Rawat	0	8.5	0	1	0	0	0
Kartik Tyagi	0.744825911	8	0	0	0	0	1
Y Jaiswal	0	8	0	0	1	0	0
JC Buttler	1.448087544	9.5	1	1	0	0	0
SV Samson	1.049304136	9.5	0	1	0	0	0
DA Miller	0.604731822	9	1	0	1	0	0
R Parag	0.487824581	8.5	0	0	1	0	0
R Tewatia	1.016550124	9	1	0	0	0	1
CH Morris	1.121757177	9.5	1	0	0	1	0
C Sakariya	0	8.5	0	0	0	0	1
Mustafizur Rahman	0.943985952	8.5	1	0	0	0	1
Total Strength	12.55258048						
Selected		100	11	2	3	2	4
Max		100	11	4	6	4	6
Min		0	11	1	3	1	3

Fig. 2. Rajasthan Royals vs Sun Risers Hyderabad

4.3   Optimization

We used Excel's built-in tool **Solver** to solve this problem. Figure 2 shows all the variable inputs for the optimizer. The objective function is the cell *Total Strength* which is the *SUMPRODUCT* of the *Strength* (*in-strength* + the *base performance*) and the *Playing* column which the Solver tries to **maximise** by manipulating the *Playing* column which is a binary variable indicating the selection of a player in the squad. The last 3 highlighted rows show the player count and the cost constraints of the squad. The columns *Wicketkeeper, Batsmen, Allrounder* and *Bowler* are binary variables and denote the role of the player. The sum product of these columns with *Playing* column gives the number of players selected in that category. We then set the constraints as mentioned in Sect. 3.2.

5   Results

IPL takes in a lot of new players whose past data is not available in the public domain. Hence in Table 2 the column ***Possible Predictions*** contains the numbers of players from the dream team who were present in our dataset.

Considering the percentage of correct predictions out of *Possible Predictions*, it can be seen in Table 2 that on an average the model predicts **71%** of the players from the dream team. This isn't the same as saying that this model has a 71% chance of winning rewards, in fact the user's team can be almost guaranteed to stand in the winning zone every time with 6–7 players from the *Dream Team*. Having more than 7 player from the Dream Team just increases the chances of the user securing a higher rank among the participants thus getting higher rewards. Since there exists no dataset containing the match-to-match Dream11 player costs or the dream teams for previous IPL matches, it was not possible to test the model on previous year IPL matches. The model was tested on IPL 2021 matches as and when they were held. Unfortunately due to COVID-19 IPL 2021 was suspended, hence the limited results.

Table 2. Results of the model

Date	Match	Players Predicted	Possible Predictions
02 May 2021	PKBS vs DC	6	8
02 May 2021	RR vs SRH	9	11
30 April 2021	RCB vs PKBS	5	7
29 April 2021	DC vs KKR	7	10
27 April 2021	RCB vs DC	6	9
26 April 2021	PKBS vs KKR	7	9
25 April 2021	DC vs SRH	6	11
15 April 2021	RR vs DC	7	10

6   Conclusion and Future Work

The approach involved combining some earlier works to quantify player performance better and statistical/graphical approaches on a real world use case. There are a few short-comings in the proposed approach which can be improved in the future for better rewards and ranking. They are:

1. The model cannot select the **Captain** and the **Vice Captain**. Captain and Vice-captain have higher weightage to their points ($2x$ and $1.5x$).
2. As of now there is no way to include new players (players with no record in the dataset).
3. At the moment, there is no penalty for the model even it selects the least performing 7 players from the dream team.

References

1. Dey, P., Ganguly, M., Roy, S.: Network centrality based team formation: a case study on T-20 cricket. Appl. Comput. Inf. **13**(2) (2017)
2. Iyer, S.R., Sharda, R.: Prediction of athletes performance using neural networks: an application in cricket team selection. Exp. Syst. Appl. **36**(3), Part 1 (2009)
3. Bhattacharjee, D., Saikia, H.: On performance measurement of cricketers and selecting an optimum balanced team. Int. J. Perform. Anal. Sport (2014)
4. Pathak, N., Wadhwa, H.: Applications of modern classification techniques to predict the outcome of ODI cricket. Procedia Comput. Sci. **87** (2016)
5. Passi, K., Pandey, N.: Predicting players performance in one day international cricket matches using machine learning. 111–126 (2018). https://doi.org/10.5121//csit.2018.80310
6. Lakkaraju, P., Sethi, S.: Correlating the Analysis of Opinionated Texts Using SAS Text Analytics with Application of Sabermetrics to Cricket Statistics (2012)
7. Passi, K., Pandey, N.: Increased prediction accuracy in the game of cricket using machine learning. Int. J. Data Mining Knowl. Manag. Process. **8**(2) (2018)
8. Lemmer, H.H.: A measure for the batting and bowling performance of cricket players. South African J. Res. Sport Phys. Educ. Recreat. (2004)
9. Lemmer, H.H.: A method for the comparison of the bowling performances of bowlers in a match or a series of matches. South Afr. J. Res. Sport Phys. Educ. Recreat. (2005)
10. Lemmer, H.H., Bhattacharjee, D., Saikia, H.: A consistency adjusted measure for the success of prediction methods in cricket. Int. J. Sports Sci. Coach. **9**(3) (2014)
11. Lemmer, H.: A measure for the batting performance of cricket players. South Afr. J. Res. Sport Phys. Educ. Recreat. **26**, 55–64 (2004). https://doi.org/10.4314/sajrs.v26i1.25876
12. Singla, S., Shukla, S.: Integer optimisation for Dream 11 cricket team selection. Int. J. Comput. Sci. Eng. **8**, 1–6 (2020). https://doi.org/10.26438/ijcse/v8i11.16 https://doi.org/10.26438/ijcse/v8i11.16 https://doi.org/10.26438/ijcse/v8i11.16
13. https://economictimes.indiatimes.com/news/sports/big-growth-ipl-2021-viewership-count-hits-380-million-mark/the-biggest-t20-league-in-the-world/slideshow/86674520.cms
14. Mukherjee, S.: Quantifying individual performance in cricket – a network analysis of batsmen and bowlers. Phys. A Statist. Mech. Appl. **393** (2012). https://doi.org/10.1016/j.physa.2013.09.027
15. The Public Gambling Act, 1867. https://www.indiacode.nic.in/handle/123456789/2269. Accessed 12 July 2021

PM4Py-GPU: A High-Performance General-Purpose Library for Process Mining

Alessandro Berti[1,2]([✉]), Minh Phan Nghia[1,2], and Wil M.P. van der Aalst[1,2]

[1] Process and Data Science Group @ RWTH Aachen, Aachen, Germany
{a.berti,wvdaalst}@pads.rwth-aachen.de, minh.nghia.phan@rwth-aachen.de
[2] Fraunhofer Institute of Technology (FIT), Sankt Augustin, Germany

Abstract. Open-source process mining provides many algorithms for the analysis of event data which could be used to analyze mainstream processes (e.g., O2C, P2P, CRM). However, compared to commercial tools, they lack the performance and struggle to analyze large amounts of data. This paper presents PM4Py-GPU, a Python process mining library based on the NVIDIA RAPIDS framework. Thanks to the dataframe columnar storage and the high level of parallelism, a significant speed-up is achieved on classic process mining computations and processing activities.

Keywords: Process mining · Gpu analytics · Columnar storage

1 Introduction

Process mining is a branch of data science that aims to analyze the execution of business processes starting from the event data contained in the information systems supporting the processes. Several types of process mining are available, including *process discovery* (the automatic discovery of a process model from the event data), *conformance checking* (the comparison between the behavior contained in the event data against the process model, with the purpose to find deviations), *model enhancement* (the annotation of the process model with frequency/performance information) and *predictive analytics* (predicting the next path or the time until the completion of the instance). Process mining is applied worldwide to a huge amount of data using different tools (academic/commercial). Some important tool features to allow process mining in organizational settings are: the *pre-processing and transformation* possibilities, the *possibility to drill-down* (creating smaller views on the dataset, to focus on some aspect of the process), the availability of *visual analytics* (which are understandable to non-business users), the *responsiveness and performance* of the tool, and the possibilities of *machine learning* (producing useful predictive analytics and what-if analyses). Commercial tools tackle these challenges with more focus than academic/open-source tools, which, on the other hand, provide

R. Guizzardi et al. (Eds.): RCIS 2022, LNBIP 446, pp. 727–734, 2022.
https://doi.org/10.1007/978-3-031-05760-1_49

more complex analyses (e.g., process discovery with inductive miner, declarative conformance checking). The PM4Py library http://www.pm4py.org, based on the Python 3 programming language, permits to integrate with the data processing and machine learning packages which are available in the Python world (Pandas, Scikit-Learn). However, most of its algorithms work in single-thread, which is a drawback for performance. In this demo paper, we will present a GPU-based open-source library for process mining, PM4Py-GPU, based on the NVIDIA RAPIDS framework, allowing us to analyze a large amount of event data with high performance offering access to GPU-based machine learning. The speedup over other open-source libraries for general process mining purposes is more than 10x. The rest of the demonstration paper is organized as follows. Section 2 introduces the NVIDIA RAPIDS framework, which is at the base of PM4Py-GPU, and of some data formats/structures for the storage of event logs; Sect. 3 presents the implementation, the different components of the library and some code examples; Sect. 4 assess PM4Py-GPU against other products; Sect. 5 introduces the related work on process mining on big data and process mining on GPU; Finally, Sect. 6 concludes the demo paper.

2   Preliminaries

This section will first present the NVIDIA RAPIDS framework for GPU-enabled data processing and mining. Then, an overview of the most widely used file formats and data structures for the storage of event logs is provided.

2.1   NVIDIA RAPIDS

The NVIDIA RAPIDS framework https://developer.nvidia.com/rapids was launched by NVIDIA in 2018 with the purpose to enable general-purpose data science pipelines directly on the GPU. It is composed of different components: CuDF (GPU-based dataframe library for Python, analogous to Pandas), CuML (GPU-based general-purpose machine learning library for Python, similar to Scikit-learn), and CuGraph (GPU-based graph processing library for Python, similar to NetworkX). The framework is based on CUDA (developed by NVIDIA to allow low-level programming on the GPU) and uses RMM for memory management. NVIDIA RAPIDS exploit all the cores of the GPU in order to maximize the throughput. When a computation such as retrieving the maximum numeric value of a column is operated against a column of the dataframe, the different cores of the GPU act on different parts of the column, a maximum is found on every core. Then the global maximum is a reduction of these maximums. Therefore, the operation is parallelized on all the cores of the GPU. When a group-by operation is performed, the different groups are identified (also here using all the cores of the GPU) as the set of rows indices. Any operation on the group-by operation (such as taking the last value of a column per group; performing the sum of the values of a column per group; or calculating the difference between consecutive values in a group) is also performed exploiting the parallelism on the cores of the GPU.

2.2   Dataframes and File Formats for the Storage of Event Logs

In this subsection, we want to analyze the different file formats and data structures that could be used to store event logs, and the advantages/disadvantages of a columnar implementation. As a standard to interchange event logs, the XES standard is proposed https://xes-standard.org/, which is text/XML based. Therefore, the event log can be ingested in memory after parsing the XML, and this operation is quite expensive. Every attribute in a XES log is typed, and the attributes for a given case do not need to be replicated among all the events. Event logs can also be stored as CSV(s) or Parquet(s), both resembling the structure of a table. A CSV is a textual file hosting an *header row* (containing the names of the different columns separated by a separator character) and many *data rows* (containing the values of the attributes for the given row separated by a separator character). A problem with the CSV format is the typing of the attributes. A Parquet file is a binary file containing the values for each column/attribute, and applying a column-based compression. Each column/attribute is therefore strictly typed. CuDF permits the ingestion of CSV(s)/Parquet(s) into a dataframe structure. A dataframe is a table-like data structure organized as columnar storage. As many data processing operations work on a few attributes/columns of the data, adopting a columnar storage permits to retrieve specific columns with higher performance and to reduce performance problems such as cache misses. Generally, the ingestion of a Parquet file in a CuDF dataframe is faster because the data is already organized in columns. In contrast, the parsing of the text of a CSV and its transformation to a dataframe is more time expensive. However, NVIDIA CuDF is also impressive in the ingestion of CSV(s) because the different cores of the GPU are used on different parts of the CSV file.

3   Implementation and Tool

In PM4Py-GPU, we assume an event log to be ingested from a Parquet/CSV file into a CuDF dataframe using the methods available in CuDF. On top of such dataframe, different operations are possible, including:

- *Aggregations/Filtering at the Event Level*: we would like to filter in/out a row/event or perform any aggregation based solely on the properties of the row/event. Examples: filtering the events/rows for which the cost is > 1000; associate its number of occurrences to each activity.
- *Aggregations/Filtering at the Directly-Follows Level*: we would like to filter in/out rows/events or perform any sort of aggregation based on the properties of the event and of the previous (or next) event. Examples: filtering the events with activity *Insert Fine Notification* having a previous event with activity *Send Fine*; calculating the frequency/performance directly-follows graph.
- *Aggregations/Filtering at the Case Level*: this can be based on global properties of the case (e.g., the number of events in the case or the throughput time of the case) or on properties of the single event. In this setting, we need an

initial exploration of the dataframe to group the indexes of the rows based on their case and then perform the filtering/aggregation on top of it. Examples: filtering out the cases with more than 10 events; filtering the cases with at least one event with activity *Insert Fine Notification*; finding the throughput time for all the cases of the log.

- *Aggregations/Filtering at the Variant Level*: the aggregation associates each case to its variant. The filtering operation accepts a collection of variants and keeps/remove all the cases whose variant fall inside the collection. This requires a double aggregation: first, the events need to be grouped in cases. Then this grouping is used to aggregate the cases into the variants.

To facilitate these operations, in PM4Py-GPU we operate three steps starting from the original CuDF dataframe:

- The dataframe is ordered based on three criteria (in order, case identifier, the timestamp, and the absolute index of the event in the dataframe), to have the events of the same cases near each other in the dataframe, increasing the efficiency of group-by operations.
- Additional columns are added to the dataframe (including the position of the event inside a case; the timestamp and the activity of the previous event) to allow for aggregations/filtering at the directly-follows graph level.
- A *cases dataframe* is found starting from the original dataframe and having a row for each different case in the log. The columns of this dataframe include the number of events for the case, the throughput time of the case, and some numerical features that uniquely identify the case's variant. Case-based filtering is based on both the original dataframe and the cases dataframe. Variant-based filtering is applied to the cases dataframe and then reported on the original dataframe (keeping the events of the filtered cases).

The PM4Py-GPU library is available at the address https://github.com/Javert899/pm4pygpu. It does not require any further dependency than the NVIDIA RAPIDS library, which by itself depends on the availability of a GPU, the installation of the correct set of drivers, and of NVIDIA CUDA. The different modules of the library are:

- *Formatting module (format.py)*: performs the operations mentioned above on the dataframe ingested by CuDF. This enables the rest of the operations described below.
- *DFG retrieval/Paths filtering (dfg.py)*: discovers the frequency/performance directly-follows graph on the dataframe. This enables paths filtering on the dataframe.
- *EFG retrieval/Temporal Profile (efg.py)*: discovers the eventually-follows graphs or the temporal profile from the dataframe.
- *Sampling (sampling.py)*: samples the dataframe based on the specified amount of cases/events.
- *Cases dataframe (cases_df.py)*: retrieves the cases dataframe. This permits the filtering on the number of events and on the throughput time.

Table 1. Event logs used in the assessment, along with their number of events, cases, variants and activities.

Log	Events	Cases	Variants	Activities
roadtraffic_2	$1,122,940$	$300,740$	231	11
roadtraffic_5	$2,807,350$	$751,850$	231	11
roadtraffic_10	$5,614,700$	$1,503,700$	231	11
roadtraffic_20	$11,229,400$	$3,007,400$	231	11
bpic2019_2	$3,191,846$	$503,468$	$11,973$	42
bpic2019_5	$7,979,617$	$1,258,670$	$11,973$	42
bpic2019_10	$15,959,230$	$2,517,340$	$11,973$	42
bpic2018_2	$5,028,532$	$87,618$	$28,457$	41
bpic2018_5	$12,571,330$	$219,045$	$28,457$	41
bpic2018_10	$25,142,660$	$438,090$	$28,457$	51

– *Variants (variants.py)*: enables the retrieval of variants from the dataframe. This permits variant filtering.
– *Timestamp (timestamp.py)*: retrieves the timestamp values from a column of the dataframe. This permits three different types of timestamp filtering (events, cases contained, cases intersecting).
– *Endpoints (start_end_activities.py)*: retrieves the start/end activities from the dataframe. This permits filtering on the start and end activities.
– *Attributes (attributes.py)*: retrieves the values of a string/numeric attribute. This permits filtering on the values of a string/numeric attribute.
– *Feature selection (feature_selection.py)*: basilar feature extraction, keeping for every provided numerical attribute the last value per case, and for each provided string attribute its one-hot-encoding.

An example of usage of the PM4Py-GPU library, in which a Parquet log is ingested, and the directly-follows graph is computed, is reported in the following listing.

```
import cudf
from pm4pygpu import format, dfg
df = cudf.read_parquet('receipt.parquet')
df = format.apply(df)
frequency_dfg = dfg.get_frequency_dfg(df)
```

Listing 1.1: Example code of PM4Py-GPU.

4   Assessment

In this section, we want to compare PM4Py-GPU against other libraries/ solutions for process mining to evaluate mainstream operations' execution time against significant amounts of data. The compared solutions include PM4Py-GPU (described in this paper), PM4Py (CPU single-thread library for process mining in Python; https://pm4py.fit.fraunhofer.de/), the PM4Py Distributed Engine (described in the assessment). All the solutions have been run on the same machine (Threadripper 1920×, 128 GB of DDR4 RAM, NVIDIA RTX 2080). The event logs of the assessment include the Road Traffic Fine Management https://data.4tu.nl/articles/dataset/Road_Traffic_Fine_Management_Process/12683249, the BPI Challenge 2019 https://data.4tu.nl/articles/dataset/BPI_Challenge_2019/12715853 and the BPI Challenge 2018 https://data.4tu.nl/articles/dataset/BPI_Challenge_2018/12688355 event logs. The cases of every one of these logs have been replicated 2, 5, and 10 times for the assessment (the variants and activities are unchanged). Moreover, the smallest of these logs (Road Traffic Fine Management log) has also been replicated 20 times. The information about the considered event logs is reported in Table 1. In particular, the suffix (_2, _5, _10) indicates the number of replications of the cases of the log. The results of the different experiments is reported in Table 2. The first experiment is on the importing time (PM4Py vs. PM4Py-GPU; the other two software cannot be directly compared because of more aggressive pre-processing). We can see that PM4Py-GPU is slower than PM4Py in this setting (data in the GPU is stored in a way that facilitates parallelism). The second experiment is on the computation of the directly-follows graph in the four different platforms. Here, PM4Py-GPU is incredibly responsive The third experiment is on the computation of the variants in the different platforms. Here, PM4Py-GPU and the PM4Py Distributed Engine perform both well (PM4Py-GPU is faster to retrieve the variants in logs with a smaller amount of variants).

Table 2. Comparison between the execution times of different tasks. The configurations analyzed are: P4 (single-core PM4Py), P4G (PM4Py-GPU), P4D (PM4Py Distributed Engine). The tasks analyzed are: importing the event log from a Parquet file, the computation of the DFG and the computation of the variants. For the PM4Py-GPU (computing the DFG and variants), the speedup in comparison to PM4Py is also reported.

Log	Importing		DFG			Variants		
	P4	P4G	P4	P4G	P4D	P4	P4G	P4D
roadtraffic_2	**0.166 s**	1.488 s	0.335 s	**0.094 s** (3.6×)	0.252 s	1.506 s	**0.029 s** (51.9×)	0.385 s
roadtraffic_5	**0.375 s**	1.691 s	0.842 s	**0.098 s** (8.6×)	0.329 s	3.463 s	**0.040 s** (86.6×)	0.903 s
roadtraffic_10	**0.788 s**	1.962 s	1.564 s	**0.105 s** (14.9×)	0.583 s	7.908 s	**0.055 s** (144×)	1.819 s
roadtraffic_20	**1.478 s**	2.495 s	3.200 s	**0.113 s** (28.3×)	1.048 s	17.896 s	**0.092 s** (195×)	3.380 s
bpic2019_2	**0.375 s**	1.759 s	0.980 s	**0.115 s** (8.5×)	0.330 s	3.444 s	0.958 s (3.6×)	**0.794 s**
bpic2019_5	**0.976 s**	2.312 s	2.423 s	**0.156 s** (15.5×)	0.613 s	8.821 s	**0.998 s** (8.9×)	1.407 s
bpic2019_10	**1.761 s**	3.156 s	4.570 s	**0.213 s** (21.5×)	1.679 s	19.958 s	**1.071 s** (18.6×)	4.314 s
bpic2018_2	**0.353 s**	1.846 s	1.562 s	**0.162 s** (9.6×)	0.420 s	6.066 s	5.136 s (1.2×)	**0.488 s**
bpic2018_5	**0.848 s**	2.463 s	3.681 s	**0.214 s** (17.2×)	0.874 s	14.286 s	5.167 s (2.8×)	**0.973 s**
bpic2018_10	**1.737 s**	3.470 s	7.536 s	**0.306 s** (24.6×)	1.363 s	29.728 s	5.199 s (5.7×)	**1.457 s**

5   Related Work

Process Mining on Big Data Architectures: an integration between process mining techniques and Apache Hadoop has been proposed in [3]. Apache Hadoop does not work in-memory and requires the serialization of every step. Therefore, technologies such as Apache Spark could be used for in-memory process mining[1]. The drawback of Spark is the additional overhead due to the log distribution step, which limits the performance benefits of the platform. Other platform such as Apache Kafka have been used for processing of streams [5]. Application-tailored engines have also been proposed. The "PM4Py Distributed engine"[2] has been proposed as a multi-core and multi-node engine tailored for general-purpose process mining with resource awareness. However, in contrast to other distributed engines, it misses any failure-recovery option and therefore is not good for very long lasting computations. The Process Query Language (PQL) is integrated in the Celonis commercial process mining software https://www.celonis.com/ and provides high throughput for mainstream process mining computations in the cloud.

Data/Process Mining on GPU: many popular data science algorithms have been implemented on top of a GPU [1]. In particular, the training of machine learning models, which involve tensor operations, can have huge speed-ups using the GPU rather than the CPU. In [7] (LSTM neural networks) and [6] (convolutional neural networks), deep learning approaches are used for predictive purposes. Some of the process mining algorithms have been implemented on top of a GPU. In [4], the popular alpha miner algorithm is implemented on top of GPU and compared against the CPU counterpart, showing significant gains. In [2], the discovery of the paths in the log is performed on top of a GPU with a big speedup in the experimental setting.

6   Conclusion

In this paper, we presented PM4Py-GPU, a high-performance library for process mining in Python, which is based on the NVIDIA RAPIDS framework for GPU computations. The experimental results against distributed open-source software (PM4Py Distributed Engine) are very good, and the library seems suited for process mining on a significant amount of data. However, an expensive GPU is needed to make the library work, which could be a drawback for widespread usage. We should also say that the number of process mining functionalities supported by the GPU-based library is limited, hence comparisons against open-source/commercial software supporting a more comprehensive number of features might be unfair.

Acknowledgement. We thank the Alexander von Humboldt (AvH) Stiftung for supporting our research.

---

[1] https://www.pads.rwth-aachen.de/go/id/ezupn/lidx/1.

[2] https://www.pads.rwth-aachen.de/go/id/khbht.

References

1. Cano, A.: A survey on graphic processing unit computing for large-scale data mining. Wiley Interdiscip. Rev. Data Min. Knowl. Discov. **8**(1) (2018). https://doi.org/10.1002/widm.1232
2. Ferreira, D.R., Santos, R.M.: Parallelization of transition counting for process mining on multi-core CPUs and GPUs. In: Dumas, M., Fantinato, M. (eds.) BPM 2016. LNBIP, vol. 281, pp. 36–48. Springer, Cham (2017). https://doi.org/10.1007/978-3-319-58457-7_3
3. Hernández, S., van Zelst, S.J., Ezpeleta, J., van der Aalst, W.M.P.: Handling big(ger) logs: Connecting prom 6 to apache hadoop. In: Daniel, F., Zugal, S. (eds.) Proceedings of the BPM Demo Session 2015 Co-located with the 13th International Conference on Business Process Management (BPM 2015), Innsbruck, Austria, 2 September 2015. CEUR Workshop Proceedings, vol. 1418, pp. 80–84. CEUR-WS.org (2015). http://ceur-ws.org/Vol-1418/paper17.pdf
4. Kundra, D., Juneja, P., Sureka, A.: Vidushi: parallel implementation of alpha miner algorithm and performance analysis on CPU and GPU architecture. In: Reichert, M., Reijers, H.A. (eds.) BPM 2015. LNBIP, vol. 256, pp. 230–241. Springer, Cham (2016). https://doi.org/10.1007/978-3-319-42887-1_19
5. Nogueira, A.F., Rela, M.Z.: Monitoring a CI/CD workflow using process mining. SN Comput. Sci. **2**(6), 448 (2021). https://doi.org/10.1007/s42979-021-00830-2
6. Pasquadibisceglie, V., Appice, A., Castellano, G., Malerba, D.: Using convolutional neural networks for predictive process analytics. In: International Conference on Process Mining, ICPM 2019, Aachen, Germany, 24–26 June 2019, pp. 129–136. IEEE (2019). https://doi.org/10.1109/ICPM.2019.00028
7. Tax, N., Verenich, I., La Rosa, M., Dumas, M.: Predictive business process monitoring with LSTM neural networks. In: Dubois, E., Pohl, K. (eds.) CAiSE 2017. LNCS, vol. 10253, pp. 477–492. Springer, Cham (2017). https://doi.org/10.1007/978-3-319-59536-8_30

Interactive Business Process Comparison Using Conformance and Performance Insights - A Tool

Mahsa Pourbafrani[1(✉)], Majid Rafiei[1], Alessandro Berti[1,2],
and Wil M.P. van der Aalst[1,2]

[1] Process and Data Science Group @ RWTH Aachen, Aachen, Germany
{mahsa.bafrani,majid.rafiei,a.berti,wvdaalst}@pads.rwth-aachen.de
[2] Fraunhofer Institute of Technology (FIT), Sankt Augustin, Germany

Abstract. Process mining techniques make the underlying processes in organizations transparent. Historical event data are used to perform conformance checking and performance analyses. Analyzing a single process and providing visual insights has been the focus of most process mining techniques. However, comparing two processes or a single process in different situations is essential for process improvement. Different approaches have been proposed for process comparison. However, most of the techniques are either relying on the aggregated KPIs or their comparisons are based on process models, i.e., the flow of activities. Existing techniques are not able to provide understandable and insightful results for process owners. The current paper describes a tool that provides aggregated and detailed comparisons of two processes starting from their event logs using innovative visualizations. The visualizations provided by the tool are interactive. We exploit some techniques recently proposed in the literature, e.g., *stochastic conformance checking* and the *performance spectrum*, for conformance and performance comparison.

Keywords: Process mining · Event logs · Comparison visualization · Performance spectrum · Earth mover's distance

1 Introduction

Process mining [1] is a branch of data science that analyzes business processes starting from the information contained in *event logs*. Event logs store the events executed inside processes w.r.t. time, process instances, activities, and the corresponding resources. For instance, in a bank, the act of opening an account (*Activity*), for the customer number *123* (*Process Instance, or Case ID*) by John

Funded by the Deutsche Forschungsgemeinschaft (DFG, German Research Foundation) under Germany's Excellence Strategy - EXC 2023 Internet of Production- Project ID: 390621612. We also thank the Alexander von Humboldt (AvH) Stiftung for supporting our research.

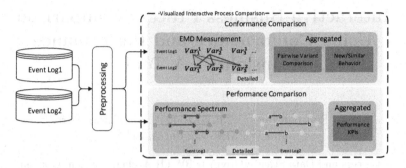

Fig. 1. The overview of the proposed tool for comparing two processes using their event logs. After preprocessing the event logs, two main modules are designed and implemented for detailed and aggregated comparisons of two processes, w.r.t. behavior and performance similarity. For instance, using *EMD* (*Earth Mover's Distance*), the cost of mapping one behavior from the first log to the second one is calculated. Using the *Performance Spectrum*, the execution times between activities a and b are compared in two event logs.

(*Resource*), at *01/10/2021 14:00:10* (*Timestamp*) is considered as an event. The sequence of events for one process instance (*Case ID*) w.r.t. their timestamps is called a *case*. A *trace* is the sequence of activities of a case. For instance, for customer 123 (a case), $\langle open\ account, deposit\ money, withdraw\ money, ...\rangle$ is the corresponding trace.

Several techniques, such as process discovery (the automatic discovery of a business process model using the event log), conformance checking (the comparison between the behavior of an event log and the corresponding process model), model enhancement (the annotation of the process model with frequency and performance information) have been provided in the process mining context. Visualizations often accompany these techniques. For instance, one of the techniques which provides an insightful visualization of event logs is the *dotted chart* visualization [17]. Such visualizations are the main resources to compare the behavior of different processes (*process comparison*) since the analyst can visually spot the differences. Process comparison is also essential to create valid simulation models, and what-if analyses [13].

To compare two processes w.r.t. their event logs, two significant aspects of the processes can be considered, the control flow (sequence of activities) and the performance patterns. In this paper, we focus on these aspects and demonstrate the features and functionalities of our proposed tool.

We elaborate on the motivation of designing and developing the proposed tool in Sect. 2. The scientific novelty and features provided by the tool are introduced in Sect. 3. In Sect. 4, we explain the tool in practice along with the technical aspects, and Sect. 5 concludes this paper by discussing the future work and limitations.

2   Motivation

In this section, the open issues are highlighted by exploring the related work. Then, the techniques that are used by the proposed tool to address these requirements are briefly explained.

Several approaches have been proposed for the comparison of processes using their event logs, e.g., in [18], a case study involves resources and activities comparisons. In [11], the authors propose a case study for process comparison among different hospitals with the focus of activity flows. In [2], the idea of process cubes is presented and applied to compare the processes in the context of education. Then, the results of queries are visualized using standard techniques such as dotted charts. In [4], the authors use process cubes to analyze and compare different aspects of business processes where they generate multidimensional processes. However, as discussed in [19], the complexity of considering all the dimensions and the effort to generate a multidimensional process is high, and it is not easy to provide an understandable visualization for the user.

Most of the current approaches for process comparison are not advanced enough in both aspects, i.e., conformance checking and performance analysis. For instance, standard comparison techniques exploit conformance checking between the event logs and the corresponding process models [5]. In addition, for performance comparison, general metrics [8] are mainly considered for the comparison. Although detailed comparison techniques such as using *Earth Mover's Distance* address this issue, e.g., in [10] and [15], there still exists a gap in transforming insights into comprehensive visualizations.

This paper proposes a tool for systematically comparing two processes or the outcomes of changed processes in different contexts. This tool supports and complements the existing comparative process mining techniques. We use a comparison technique for processes that graphically depicts the differences. Two main comparison areas are based on the distance between conformance and performance of two processes. The proposed tool visualizes the performance and compliance findings interactively. Figure 1 represents an overview of the proposed tool's architecture and modules. It includes three main modules, (A) preprocessing, (B) conformance comparison, and (C) performance comparison. The conformance comparison is inspired by the distance metrics proposed in [10] and [15]. For performance analysis, we exploit the idea of performance spectrum described in [7].

3   Approach

The comparison modules provided by our tool use different conformance and performance analyses initially proposed to analyze a single process. We adapt them for comparative purposes and create interactive visualizations for such comparisons. The modules are explained in Sects. 3.1 and 3.2.

Table 1. An example of EMD measurement for two event logs [15]. The reallocation function allocates 1 out of 50 traces $\langle a, b, c, d \rangle$ in L_1 to the same trace in L_2 and 49 traces to the trace $\langle a, e, c, d \rangle$ which is the most similar one in L_2. The sum of the table's values indicates the general EMD value, i.e., the difference between the two event logs.

A_{L_1}	A_{L_2}			
	$\langle a, b, c, d \rangle$	$\langle a, c, b, d \rangle$	$\langle a, e, c, d \rangle^{49}$	$\langle a, e, b, d \rangle^{49}$
$\langle a, b, c, d \rangle^{50}$	$\frac{1}{100} \times 0$	0×0.5	$\frac{49}{100} \times 0.25$	0×0.5
$\langle a, c, b, d \rangle^{50}$	0×0.5	$\frac{1}{100} \times 0$	0×0.5	$\frac{49}{100} \times 0.25$

3.1 Conformance Comparison

This section explains the provided method for the comparison of the control flow based on the activities and the paths recorded in the event log. We use a stochastic conformance checking technique to identify differences. A process consists of different process instances showing the possible paths that can be taken using the process model. All the possible paths that are unique traces, i.e., sequences of activities, are considered the process behaviors. Given two event logs L_1 and L_2, we denote A_{L_1} and A_{L_2} as their sets of sequences of activities. Given this information, we look for the matches and mismatches from two viewpoints; *aggregated* and *detailed*:

- *Aggregated Metrics*: we consider one of the event logs as a base and identify the non-existing behavior in another event log. For instance, if $A_{L_1} = \{\langle a, b, c \rangle, \langle a, b, e, d \rangle\}$ and $A_{L_2} = \{\langle a, b, c \rangle, \langle a, b, e, f \rangle\}$:
  - *Removed behavior* from L_1 in comparison with L_2: $A_{L_1} \backslash A_{L_2} = \{\langle a, b, e, d \rangle\}$
  - *New behavior* from L_2 in comparison with L_1: $A_{L_2} \backslash A_{L_1} = \{\langle a, b, e, f \rangle\}$

  And the measures $\frac{|A_{L_2} \backslash A_{L_1}|}{|A_{L_1} \cup A_{L_2}|}$ and $\frac{|A_{L_1} \backslash A_{L_2}|}{|A_{L_1} \cup A_{L_2}|}$ are the fraction of the new and removed behaviors, respectively. The pairwise comparison of the behaviors of processes and their frequencies, which indicate their importance in each event log, is also considered. One of the results of these metrics using an example is shown in Fig. 2.
- *Detailed Comparison*: we use the idea of Earth Mover's Distance (EMD) for the detailed comparison between traces of two event logs. EMD indicates the amount of effort required to change one pile of earth into the other. We use the conformance techniques provided in [10] to compute the EMD measurement between two event logs. The frequency of each trace is considered as the pile that needs to be moved, and the normalized edit distance (Levenshtein) is used to calculate the distance between every two traces. EMD solves an optimization problem that minimizes the cost of converting one event log to another one, i.e., it finds the best reallocation function. The outcome of applying the proposed EMD measurement to two sample event logs is shown in Fig. 3. The x-axis and y-axis represent the unique traces in the first event log (L_2) and the unique traces in the second event log (L_1), respectively.

Thus, each row is the relative effort that the first unique trace in L_1 needs to be transformed into one or more unique traces in L_2. The details of functions and formal definitions are discussed in [13].

3.2   Performance Comparison

General performance KPIs at a high level of aggregation, e.g., the average waiting time of traces, or the average service time are too abstract to be used as comparison metrics. Therefore, besides the usual metrics, we propose the usage of the performance spectrum [7]. The performance spectrum is a concept introduced to visualize the performance of a process at a detailed level. If we consider a single path between two activities a and b, the performance spectrum shows all

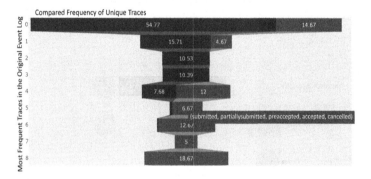

Fig. 2. The comparative frequency chart represents the similar behaviors, removed behaviors, and the new behaviors w.r.t. the second event log.

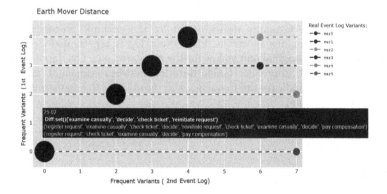

Fig. 3. Two example event logs are compared in detail. The EMD diagram depicts the differences between the two event logs in terms of the activity flow. For example, the cost of mapping each trace to the second event log is based on the activities, the order of activities, and the frequency of traces, i.e., the distance between two logs.

the temporal segments going from an event having activity a to an event having activity b in the cases of the event log. This permits to identify the time intervals with higher/lower performance, the queuing pattern (FIFO/LIFO), and other performance patterns that are useful for predictive purposes [6,9].

We use the information of the performance spectrum to calculate statistics for each segment (namely, the *average time* and the *frequency*) that are compared between two event logs. Figure 4 shows the result of the introduced performance measurement for two example processes. It represents different aspects of the results: (1) new/eliminated segments, (2) frequency of each segment, and (3) duration of each segment. For instance, given L_1 and L_2, each segment's colors refer to an event log, the size refers to the average time difference between the segments, and the transparency indicates the frequency (darker, more frequent). The gray color represents the overlapped segment in two event logs with similar performance metrics, the blue color shows the segments in the original log L_1, and the yellow points represent the new segment existing in L_2. The implementation also includes the option to display only the differences (red points).

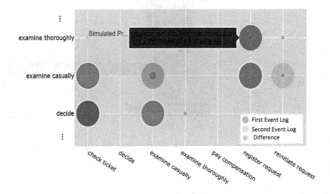

Fig. 4. Part of the performance measurement for the example process is based on the aggregated performance spectrum. Each event log is represented by a different color, i.e., blue for the original and yellow for the simulated one. Overlapping segments are represented by the gray color (same duration between segments). Each point's transparency and size indicate the frequency and duration of the segment in the event logs. (Color figure online)

4   Tool

In this section, we describe the availability, components, and maturity of the tool.

4.1   Availability

The tool is implemented as a web application. The code is publicly available.[1]
The tool is implemented in Python, and the web-based interface has been implemented using Flask. By uploading two event logs, the comparison results w.r.t.
different aspects are presented interactively. The process mining insights are on
the basis of [3,15,16]. The tool is also available as a Python library, which makes
further extension and integration for different purposes possible.

4.2   Components

The tool offers four different components. For the conformance comparison, the
EMD Comparison tab provides the Earth Mover's distance between the traces of
the first event log against the traces of the second event log. With the selection
of the variants, it is possible to focus the visualization on a given set of variants.
The *Variants Frequency Comparison* tab compares the relative frequencies of
the variants recorded in the first and second log. The *Overlap Between Logs* tab
shows how much the behavior between the two logs overlap. Finally, the *Aggregated Performance Spectrum* tab compares the performance of the segments in
the first and the second event logs. In the component, it is possible to visualize
the comparison between the performance, the aggregation of the performance in
the first event log, or the aggregation of the performance in the second one.

4.3   Maturity

The authors have used the tool in multiple projects.[2] For instance, it has been
used for assessing the quality of the simulation results in [12,14]. Moreover, a
comparison of production lines with different settings, e.g., removing one of the
stations and introducing concurrency in the process, has been done w.r.t. the
process behaviors as well as performance aspect in the *Internet of Production*
project of the RWTH Aachen University.[3]

5   Conclusion

The goal of process mining techniques is to provide insight into the processes of
organizations. Several techniques such as process model discovery, conformance
checking, and social network analysis are proposed to analyze an event log,
while some limitations exist in the comparison between two processes. Given the
complexity of the comparison task, such techniques are valuable when the result
is able to be presented comprehensively. This paper proposed innovative and
interactive visualizations to understand the differences between two processes
using their event logs. Our approach is implemented as a tool that can be used

---

[1] https://github.com/mbafrani/VisualComparison2EventLogs.
[2] https://www.researchgate.net/project/Forward-looking-in-Process-Mining.
[3] https://www.iop.rwth-aachen.de.

with other process mining and comparison techniques to capture the difference. In the current version, the tool supports comparing the traces of the two event logs and the performance comparison. As future work, we aim to support process comparison w.r.t. resources, i.e., social network analysis and roles, and embed expert knowledge.

References

1. van der Aalst, W.M.P.: Process Mining - Data Science in Action, 2nd edn. Springer, Heidelberg (2016). https://doi.org/10.1007/978-3-662-49851-4
2. van der Aalst, W.M.P., Guo, S., Gorissen, P.: Comparative process mining in education: an approach based on process cubes. In: Ceravolo, P., Accorsi, R., Cudre-Mauroux, P. (eds.) SIMPDA 2013. LNBIP, vol. 203, pp. 110–134. Springer, Heidelberg (2015). https://doi.org/10.1007/978-3-662-46436-6_6
3. Berti, A., van Zelst, S.J., van der Aalst, W.M.P.: Process mining for python (PM4Py): bridging the gap between process-and data science. In: Proceedings of the ICPM Demo Track 2019, Co-located with 1st International Conference on Process Mining (ICPM 2019), Aachen, Germany, 24–26 June 2019, pp. 13–16 (2019). http://ceur-ws.org/Vol-2374/
4. Bolt, A., van der Aalst, W.M.P.: Multidimensional process mining using process cubes. In: Gaaloul, K., Schmidt, R., Nurcan, S., Guerreiro, S., Ma, Q. (eds.) CAISE 2015. LNBIP, vol. 214, pp. 102–116. Springer, Cham (2015). https://doi.org/10.1007/978-3-319-19237-6_7
5. Carmona, J., van Dongen, B.F., Solti, A., Weidlich, M.: Conformance Checking - Relating Processes and Models. Springer, Heidelberg (2018). https://doi.org/10.1007/978-3-319-99414-7
6. Denisov, V., Fahland, D., van der Aalst, W.M.P.: Predictive performance monitoring of material handling systems using the performance spectrum. In: International Conference on Process Mining, ICPM 2019, Aachen, 24–26 June 2019, pp. 137–144. IEEE (2019)
7. Denisov, V., Fahland, D., van der Aalst, W.M. P.: Unbiased, fine-grained description of processes performance from event data. In: Weske, M., Montali, M., Weber, I., vom Brocke, J. (eds.) BPM 2018. LNCS, vol. 11080, pp. 139–157. Springer, Cham (2018). https://doi.org/10.1007/978-3-319-98648-7_9
8. Hornix, P.T.: Performance analysis of business processes through process mining. Master's Thesis, Eindhoven University of Technology (2007)
9. Klijn, E.L., Fahland, D.: Performance mining for batch processing using the performance spectrum. In: Di Francescomarino, C., Dijkman, R., Zdun, U. (eds.) BPM 2019. LNBIP, vol. 362, pp. 172–185. Springer, Cham (2019). https://doi.org/10.1007/978-3-030-37453-2_15
10. Leemans, S.J.J., Syring, A.F., van der Aalst, W.M.P.: Earth movers' stochastic conformance checking. In: BPM Forum 2019, pp. 127–143 (2019)
11. Partington, A., Wynn, M., Suriadi, S., Ouyang, C., Karnon, J.: Process mining for clinical processes: a comparative analysis of four australian hospitals. ACM Trans. Manage. Inf. Syst. 5(4) (2015). https://doi.org/10.1145/2629446
12. Pourbafrani, M., van der Aalst, W.M.P.: GenCPN: automatic CPN model generation of processes. In: 3rd International Conference ICPM 2021, Demo Track (2021)

13. Pourbafrani, M., van der Aalst, W.M.P.: Interactive process improvement using simulation of enriched process trees. In: 2nd International Workshop on AI-Enabled Process Automation (2021)
14. Pourbafrani, M., Jiao, S., van der Aalst, W.M. P.: SIMPT: Process improvement using interactive simulation of time-aware process trees. In: Cherfi, S., Perini, A., Nurcan, S. (eds.) RCIS 2021. LNBIP, vol. 415, pp. 588–594. Springer, Cham (2021). https://doi.org/10.1007/978-3-030-75018-3_40
15. Rafiei, M., van der Aalst, W.M. P.: Towards quantifying privacy in process mining. In: Leemans, S., Leopold, H. (eds.) ICPM 2020. LNBIP, vol. 406, pp. 385–397. Springer, Cham (2021). https://doi.org/10.1007/978-3-030-72693-5_29
16. Rafiei, M., van der Aalst, W.M.P.: Group-based privacy preservation techniques for process mining. Data Knowl. Eng. **134**, 101908 (2021). https://doi.org/10.1016/j.datak.2021.101908
17. Song, M., van der Aalst, W.M.P.: Supporting process mining by showing events at a glance. In: Proceedings of the 17th Annual Workshop on Information Technologies and Systems (WITS), pp. 139–145 (2007)
18. Syamsiyah, A., et al.: Business process comparison: a methodology and case study, pp. 253–267 (2017)
19. Vogelgesang, T., Kaes, G., Rinderle-Ma, S., Appelrath, H.J.: Multidimensional process mining: questions, requirements, and limitations. In: CAISE 2016 Forum, pp. 169–176 (2016). http://eprints.cs.univie.ac.at/4689/

Research Incentives in Academia Leading to Unethical Behavior

Jefferson Seide Molléri[✉][ID]

Simula Metropolitan Centre for Digital Engineering, 0167 Oslo, Norway
jefferson@simula.no

Abstract. A current practice in academia is to reward researchers for achieving outstanding performance. Although intended to boost productivity, such a practice also promotes competitiveness and could lead to unethical behavior. This position paper exposes common misconducts that arise when researchers try to game the system. It calls the research community to take preventive actions to reduce misconduct and treat such a pervasive environment with proper acknowledgment of researchers' efforts and rewards on quality rather than quantity.

Keywords: Research ethics · Researcher performance · Research quality · Incentives · Misconduct

1 Introduction

A common practice in academia nowadays is to grant incentives or rewards to researchers that achieve productivity metrics. In an ideal academic world, one expects such incentives to promote research growth in both quantity and quality. These incentives, sometimes also career development and funding opportunities, are awarded to researchers that perform better in terms of the number and impact of the publications. Therefore, the quantitative aspect plays a major role sometimes in detriment of quality. Inappropriate behavior arises from that when researchers take advantage of the system to improve their performance metrics.

2 Background

2.1 Researcher Performance Evaluation

Researcher performance is often evaluated through scientometric aspects, i.e., quantitative measures reflecting the impact of research [12]. For individual researchers, the number of publications or, more specifically, the number of publications per year, is a commonly used measure of productivity [1]. Higher publication counts reflect hard work done by a particular researcher and their students and collaborators. Moreover, by putting together individual researchers' publications in a group or department, it is possible to derive an aggregate measure, then used to compare different groups and institutions.

R. Guizzardi et al. (Eds.): RCIS 2022, LNBIP 446, pp. 744–751, 2022.
https://doi.org/10.1007/978-3-031-05760-1_51

The publications' impact is evaluated through citations, indicating the research's influence on others' works. Papers cited extensively often provide insights and experiences, describe new research directions, or summarize state-of-the-art practice in a specific field. Derivatives metrics of citation count (e.g., impact factor and h-index) are employed to assess and compare different publication venues, such as journals and conferences [1], but also researchers and organizations.

An example of such metrics' usage to assess researchers' performance, a recently published paper publicly report the 100,000 topmost scientists across all scientific fields according to their citation indexes [7]. Similarly, a bibliometric study periodically published in the Journal of Systems and Software ranks the most cited researchers and institutions in the field of Software Engineering [8]. Those papers promote a comparison based solely on quantitative performance measures.

Although those well-established measures are essential to assess the performance of researchers, they face several criticisms, see e.g. [3,10]. Quantitative measures often disregard the quality of the research produced. On the one hand, one could expect a peer review process to assess the quality and fairly reward the excellent research with a corresponding good publication, i.e., better papers published in prestigious venues. On the other hand, those qualitative measures are prone to human errors and the reviewers' subjective evaluation and require considerably more effort than collect the quantitative performance measures.

2.2   Perverse Incentives

Quantitative measures provide decision-making support for institutions hiring, promoting, or funding researchers. Those are the essential actions for researchers' career development and thus are subject to competition. Although the competition is inherited to selection processes, performance evaluation favors researchers that perform better regarding the quantity instead of quality. This skewness of the quality/quantity relation (illustrated in Fig. 1) creates a gap in the actual productivity due to unethical behavior (the difference between the solid and dotted curves).

Awards and financial incentives are a common practice to reward researchers for better performance. Such incentives intend to foster researchers to boost their productivity to match performance targets (e.g., minimum publication score). Nowadays, many universities and high-education institutes have such initiatives, and many others are designing or implementing similar programs.

The incentives, primarily intended to foster and strengthen scientific research, could produce a negative impact on researchers' behavior. Edwards and Roy [5] identify the main problems arising from the incentives practice and discusses their implications. Table 1 summarize the problems, their intended and actual effects.

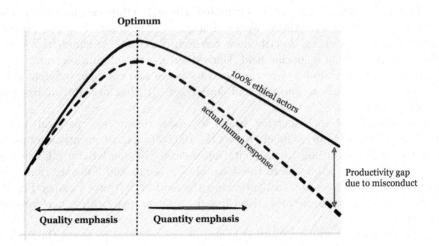

Fig. 1. True scientific productivity in relation to emphasis on research quality/quantity - adapted from Edwards and Roy [5].

Table 1. Growing Perverse Incentives in Academia - adapted from Edwards and Roy [5]

Incentive	Intended effect	Potential effect
Researchers rewarded for *increased number of publications*	• Improve research productivity • Provide a means of evaluating performance	• Avalanche of substandard, incremental papers • Poor methods • Increase in false discovery rates leading to a "natural selection of bad science" • Reduced quality of peer review
Researchers rewarded for *increased number of citations*	• Reward quality work that influences others	• Extended reference lists to inflate citations • Reviewers request citation of their work through peer review
Researchers rewarded for *increased grant funding*	• Ensure that research programs are funded • Promote growth • Generate overhead	• Increased time writing proposals and less time gathering and thinking about data • Overselling positive results and downplay of negative results
Increase *Ph.D. student productivity*	• Higher school ranking • More prestige of program	• Lower standards and create over-supply of PhDs • Postdocs often required for entry-level academic positions • PhDs hired for work MS students used to do

3   Unethical Outcomes

A scientific environment that provides rewards based only on performance metrics is prone to unethical behavior [5]. The effects of quantitative evaluations produce a competitive response among institutions and individuals. Ultimately, the competitive environment can influence those deliberately willing to benefit from the rewards to take questionable steps to maximize their productivity metrics.

3.1   Publication Misconducts

Journals and scientific publication venues are the pivots of the performance evaluation. The quantitative measures of productivity are based on scientometrics (see Sect. 2.1). To ensure a steady and progressive career development, a researcher should aim to publish their work, and even more, to foster readers to consider using it and referring back to this report in further studies. It is, therefore, vital to ensure that the publication in a high-impact venue.

Although some criticism, peer review is the most common process to assess if a candidate report is suitable for publication [1]. Most scientific publication venues have ethical guidelines to regulate the process. The guidelines cover reviewers' anonymity, conflict of interests, plagiarism, and authorship, among other topics. Despite this, inappropriate behavior and misconduct in the publication process are not unknown.

Citation Manipulation. Together with the manuscript assessment, reviewers provide comments to the authors to improve their report. Not infrequently, reviewers suggest the authors read additional material and include those references. Similarly, some scientific venues (and their editors) require self-citations before acceptance [13]. Unethical reviewers can take advantage of these practices by coercing authors to cite their own papers.

Ghost and Gift Authorship. It is difficult, if not impossible, to assess if the authors of the paper contributed to the research reported. In that sense, the scientific venues rely on the author(s) to fairly credit meaningful contributions into the report. Authorship misconducts occur both for omitting as for falsely crediting co-authors [11]. In the first case, due authors are deprived of the paper's impact, whereas in the second, guest authors benefit from the undeserved credit from the paper's impact.

Taking Undue Credit. One could particularly expect senior researchers to acknowledge all research collaborators in the submitted manuscript fairly. However, some of them deliberately use ideas and artifacts produced by other, often novice researchers (e.g., PhD candidates and research assistants) without proper acknowledgment [9,11]. Fearing reprisal from the superiors and other senior researchers, novices rarely report this exploitation to research ethics boards. The practice is particularly worrisome as the students should ideally learn ethical behavior from their seniors' exemplars.

3.2   Individual Misconduct

The exposure of individual researchers to the hypercompetitive environment has consequences on the quality of produced research. As the quantitative factors weight the career development, the "unassessed" principles of a good researcher (e.g., honesty and integrity) are often forgotten. A few borderline behaviors resulting from this trade-off relation are presented below:

Superficial Research. A large amount of small but insignificant studies increase a researcher's publication count. However, they are less likely to produce an impact on research and practice than a profound, long-term study [10]. One of the questionable practices is to produce a series of incremental papers reporting small fragments of a research topic. By choosing the easiest path to get publication numbers, researchers abdicate to contribute with the most relevant and novel results. Ultimately, this practice leads to a "bad science" environment, in which substantial research is likely to produce no impact front the vast amount of irrelevant articles.

Falsification or Fabrication of Data. Through careless or deliberated manipulation of the data, positive results are oversold on negatives [5]. Positive results are more likely to produce prestigious publications, whereas negatives struggle to be published, thus encouraging researchers to downplay or hiding them. Ideally, the scientific community should judge the negative results as important as the positive ones. Negative results can potentially refute wrong assumptions and inspire new directions for further studies.

3.3   Group Misconducts

Other undesirable effects are likely to occur on collective behavior, thus impacting researchers' groups and institutions. Productive researchers produce better results according to the quantitative metrics, therefore bringing prestige to the group. Besides that, they are expected to submit projects for funding applications. If selected, those projects often ensure grant for hiring additional researchers, often novice researchers.

In a research group, novice researchers are often perceived as a workforce. They increase the group's capacity to produce research and submit manuscripts. Senior researchers acting as advisors benefit from the novices' productivity as co-authors. Having novices responsible for the research practice, seniors move their focus to funding applications. When successful, this iterative activity increases the group's metrics (in size, performance, and financial).

Corruption in the Grant Process. The most successful groups and institutions are therefore more promising to get better grant opportunities. This, in turn, encourages the group to invest funds to offer incentives for its members based on their performance. Finally, as this cycle continues, groups with lower metrics are less motivated to compete for grants or unlikely to achieve them. Their best researchers start to search for more successful groups, where the incentives for individual researchers are plentiful.

Publishing Pact. It is an effective way to produce better productivity metrics for a research group. Members of a group are often included as co-authors in the group's publications, even when their collaboration is not substantial [10]. This inappropriate behavior is self-perpetuating, i.e., it is likely that the benefited researcher returns the favor. Mainly, novice researchers are subject to this practice starting from their senior colleagues.

Institutional Harassment. Most research institutions conduct periodic evaluations of researchers' performance based on minimum productivity targets [4]. Although these evaluations are inherited from the career development process, the results can also intimidate or penalize researchers. Cases of researchers threatened by not complying with the metrics have been reported [4]. The effects of these stressful situations impact not only the performance measures but also the morale and quality of life of the individual.

4   Reflections and Recommendations

The inappropriate behaviors and misconducts discussed herein result from a competitive environment that favors the quantitative assessment. These issues are not unheard of in information sciences research, nor are they specific to this domain. Still, it is essential to make them explicit and visible to foster the discussion and search for solutions the research community can act upon.

Changes that make the competition fairer are encouraged. As pointed out before, the changes should reduce the quantity-in-research perspective, compensating it with qualitative measures. However, a break-even between quantity and quality should be achieved, as excessive-quality assessment could ultimately hinder a researcher's productivity [5]. To achieve this optimum balance in research assessment (see Fig. 1), it is vital to:

- *Apply evaluation metrics that combine both quantitative and qualitative factors* [3]. Besides its impact, good research should be acknowledged by its credibility, integrity and accessibility. We acknowledge that measuring the quality of research is not a trivial task. The research community has yet to find a consensus about metrics to assess research quality that are both fair and feasible.
- *Reduce the importance of quantitative metrics for funding applications and career development in favor of more in-depth quality-based evaluation* [4,6, 10]. In order to lessen the adverse effects from the quantitative emphasis, it is necessary to avoid awarding prizes to researchers who achieve desired productivity figures. Only by doing so can we make room for a so expected qualitative shift.
- *Provide more stable career structures and long-term funding options* [6]. A pressing appeal for research misconduct is the fear of career stagnation or unemployment. Research institutions should concentrate efforts on assisting researchers to achieve their career goals. In addition to that, researchers could be awarded a basic fund on a regular basis rather than fostered to compete for a few multimillion-dollar research grants.

Moreover, the unethical actions are taken by individuals, i.e., researchers, reviewers, and managers of academic institutions. It is crucial to raise awareness of the problem and the implications of such misconduct to future academics and institutions' upper management. Preventive actions should be put into effect to inhibit inappropriate behavior, such as:

- *To disclaim the exact contributions of each author of the manuscript to reduce authorship misconducts* [2,11]. Many publishers have introduced disclosure of research contribution in journal articles, e.g. CRediT author statement[1]. Such initiatives also help to make explicit the role and responsibilities of each contributor.
- *In addition to the manuscript peer-review, it is important to assess the research artifacts to avoid data manipulation* [5]. Such artifacts could include digital objects that were either generated as an outcome of the research, or designed by the researchers to be used as part of the study.
- *To employ independent external reviewers in the grants selection process* [6]. It is vital to reduce biases related to prestige and networking in the grant application review process. This is particularly challenging for domains where the research community is too narrow or tightly connected. In such cases, employing reviewers with a multidisciplinary profile is encouraged.
- *Confidential report and rigorous investigation of allegations of research misconduct* [6,9]. To implement such a practice, institutions should firstly ensure an independent ethics committee (IEC) or similarly unbiased research ethics board. If such a board is locally unavailable, researchers should know which regional authorities to appeal when reporting ethics misconducts.

Finally, it would also be important that there is a genuine willingness to coach and support novice researchers in learning the skills to become strong researchers, rather than being solely used as a workforce to increase research productivity. More evenly discussions between well-established seniors and early career researchers should positively impact the research environment's ethical behavior, resulting in a better quality of research.

5 Conclusions

This position paper explores research misconducts encouraged by competitiveness in the academic environment. Researchers' performance is evaluated through a series of measures, particularly the number of publications and citations. The emphasis on the quantitative aspects provides means for gaming the system, i.e., maximizing the productivity metrics in an unappropriated manner. Several of those unethical actions and their impact on academia are discussed, followed by a few proposed solutions.

There is an urgent need to implement preventive measures in the actual scenario. Still, it is crucial to ensure long-term actions that have the potential to

---

[1] Available at https://www.elsevier.com/authors/policies-and-guidelines/credit-author-statement.

balance the ratio between quantity and quality in research. I hope that this discussion encourages readers to discuss ethical misconduct among their colleagues and organizations. By disseminating such ideas among the decision-makers, we can nurture a revolutionary change of the status quo.

Contributors

The author confirms the sole responsibility for the content in this manuscript. This position work's conceptualization and development occurred during a Ph.D. course: Research Ethics in Software Engineering.

Acknowledgement. I would like to express my gratitude to Professor Claes Wohlin for fostering discussions that led to this position paper. I am also thankful to my colleagues at BTH who participated in meaningful conversations about the topic, including Professors Emilia Mendes and Kai Petersen, and other researchers enrolled in the Research Ethics course.

References

1. Adler, R., Ewing, J., Taylor, P., et al.: Citation statistics. Statist. Sci. **24**(1), 1 (2009)
2. Bennett, D.M., Taylor, D.M.: Unethical practices in authorship of scientific papers. Emerg. Med. **15**(3), 263–270 (2003)
3. Bornmann, L., Daniel, H.D.: What do citation counts measure? a review of studies on citing behavior. J. Document. **64**(1), 45–80 (2008)
4. Colquhoun, D.: How to get good science. Physiol. News **69**, 12–14 (2008)
5. Edwards, M.A., Roy, S.: Academic research in the 21st century: maintaining scientific integrity in a climate of perverse incentives and hypercompetition. Environ. Eng. Sci. **34**(1), 51–61 (2017)
6. Gustafsson, B., Hermerén, G., Petersson, B.: Good Research Practice-What Is That? Swedish Science Research Council (2005)
7. Ioannidis, J.P., Boyack, K.W., Baas, J.: Updated science-wide author databases of standardized citation indicators. PLoS Biol. **18**(10), e3000918 (2020)
8. Karanatsiou, D., Li, Y., Arvanitou, E.M., Misirlis, N., Wong, W.E.: A bibliometric assessment of software engineering scholars and institutions (2010–2017). J. Syst. Softw. **147**, 246–261 (2019)
9. Martin, B.: Countering supervisor exploitation 1. J. Scholar. Publish. **45**(1), 74–86 (2013)
10. Parnas, D.L.: Stop the numbers game. Commun. ACM **50**(11), 19–21 (2007)
11. Solomon, J.: Programmers, professors, and parasites: credit and co-authorship in computer science. Sci. Eng. Ethics **15**(4), 467–489 (2009)
12. Van Raan, A.: Scientometrics: State-of-the-art. Scientometrics **38**(1), 205–218 (1997)
13. Wilhite, A.W., Fong, E.A.: Coercive citation in academic publishing. Science **335**(6068), 542–543 (2012)

Assessing the Ethical, Social and Environmental Performance of Conferences

Sergio España[(✉)] [ID], Vijanti Ramautar [ID], and Quang Tan Le

Utrecht University, Utrecht, The Netherlands
{s.espana,v.d.ramautar}@uu.nl

Abstract. There is a rising demand for assessing the performance of organisations on ethical, social and environmental (ESE) topics. Ethical, social and environmental accounting (ESEA) is common practice in many types of organisations and initiatives. Currently, scientific conferences are not in the spotlight nor feeling pressure to disclose their ESE accounts. However, proactively adopting these practices is an opportunity to lead the way and show commitment and responsibility. Since no existing ESEA method fits the domain of conferences well, this paper presents preliminary results on engineering such a method. We discuss material ESE topics for conferences, key performance indicators, measurement and data collection methods, and ICT infrastructure. We illustrate the method concepts by applying it to the RCIS conference series. We are confident that conference organisers and scientific communities will start assessing the performance of their conferences under many organisational sustainability dimensions and, what is more important, initiate reflection processes to improve such performance over the years.

Keywords: Ethical social and environmental accounting ·
Sustainability reporting · Ethics in information science · Scientific conferences

1 Introduction

Ethical, social and environmental accounting (ESEA) is the process of assessing the social and environmental effects of an organisation's actions and reporting them to particular interest groups and to society at large [7]. It is how responsible organisations typically assess and disclose their contributions to the community and their performance on topics related to the social dimension (e.g. social inclusion), to the environmental dimension (e.g. waste management), and to business ethics and governance (e.g. workplace democracy). The results are a valuable input for strategic managers with long-term vision.

Currently, scientific conferences are not in the spotlight nor feeling pressure to disclose their ethical, social and environmental (ESE) performance. However, proactively adopting these practices is an opportunity to lead the way and show commitment and responsibility. Most conference impact measurement initiatives

R. Guizzardi et al. (Eds.): RCIS 2022, LNBIP 446, pp. 752–760, 2022.
https://doi.org/10.1007/978-3-031-05760-1_52

have focused on bibliometric performance (e.g. [3]) and some papers have focused on CO_2 footprint and greenhouse gas emissions [13,17]. We argue that a holistic assessment of ESE topics is of paramount importance. This paper contributes the first version of an ESEA method that focuses on scientific conferences, being the foundation stone of a long-term research endeavour.

Section 2 explains the research method. Section 3 provides background and contextual knowledge on ESEA. Section 4 presents a preliminary version of the ESEA method. Section 5 proposes a list of ESE topics that we consider relevant for conferences. As a proof of concept, we have defined a few indicators that can be used to assess the performance on some topics, and calculated their values for the RCIS conference series. Section 6 concludes the paper.

2   Research Method

We aim at contributing an ESEA method that focuses on scientific conferences:

- *RQ1.* What method could assess conference performance on ESE topics?
- *RQ2.* What are material ESE topics for conferences and related indicators?
- *RQ3.* Is the method actually applicable to assess conferences?

Figure 2 presents an overview of the research method (process A). Activities with grey background are outside the scope of this paper but serve to place the project in context. For instance, we are collecting documentation of ESEA methods through a long-term multi-vocal literature review, characterising and metamodelling the methods (A1) [5]. This knowledge allows us to tailor an ESEA method that is suited for conference assessment (A2), which we metamodel with the Process Deliverable Diagram (PDD) technique [20]. To better understand the domain of conferences, we conduct an analysis of their stakeholders (A3), following the analytical framework in [9]. We propose a set of topics that are material to conferences, and some indicators that can be used to assess conference performance on those topics (A4). To do this, we have carried out a brainstorming session, we have reviewed ESEA frameworks and methods (e.g. Global Reporting Initiative[1], and we have elicited topics from other academics through informal conversations. We then define the ICT requirements to support the method (A5). As a proof of concept, we apply some of the indicators to earlier editions of the RCIS conference series[2] (A6). As future work, the method will be completed. This implies designing the stakeholder surveys (A7). Then, during the ESEA4RCIS participatory workshop[3], we plan to validate the topics and indicators with stakeholders (A8), and apply the method in order to perform the ESEA of RCIS 2022 (executing process B, discussed in Sect. 4). We also hope that this workshop will initiate deeper stakeholder engagement actions, such as reflecting on the results and improvement planning.

---

[1] https://globalreporting.org/standards.
[2] https://www.rcis-conf.com.
[3] https://esea4rcis.sites.uu.nl.

3   Background Information

3.1   The Overarching Improvement Cycle

ESEA processes typically provide key performance indicators to guide a continuous ESE improvement cycle [1], such as the one we now describe.

Materiality Assessment (P1) allows organisations identifying and prioritising the ESE topics that are relevant for them [21]. These depend on organisational characteristics, strategy, sector, and stakeholders. Sometimes, materiality assessment is performed by ESEA method engineers, who prescribe, along with the method activities and deliverables, the topics they find to be relevant for the intended method users. This is the approach used by networks such as B Corporations, and Economy for the Common Good[4]. Following the same approach, we propose a set of material topics for conferences in Sect. 5.

ESEA methods (P2) typically define a set of indicators to assess organisational performance on a set of ESE topics. Terminology in this domain is very diverse, but we have created the so called openESEA metamodel, which offers a conceptual framework [5]. The values of direct indicators can be provided by a handful of organisational managers with access to the organisational information systems and policies, through stakeholder surveys or, if necessary, by other data collection means (e.g. scraping). Collecting evidences is necessary if later auditing and assurance are required. Indirect indicators require formulas or algorithms to produce their value.

Improvement planning (P3) typically starts with a gap analysis between the current performance revealed by the ESE account and either (i) the sustainability strategy goals of the organisation, (ii) the performance of peers according to industry benchmarks, or (iii) the requirements of a certification the organisation aims at. Once the set of topics whose performance they wish to increase are identified, a set of improvement actions are defined, either by means of brainstorming, top-down or bottom-up feedback, or advice from external consultants.

Organisational re-engineering (P4) is the execution of the improvement plan. Changes should also be reflected in the enterprise models, and their effects are assessed in the next iteration of the cycle.

Organisations often disclose the results of the ESEA or the whole cycle in a non-financial report (a.k.a. sustainability report, social balance, integrated report) [16], alongside narrative and examples of good business practices.

3.2   The Need for Stakeholder Management and Engagement

Timely stakeholder participation is essential for the effective enactment of the improvement cycle in general, and of ESEA in particular [6]. Such participation results in more influential assessment processes and increases the positive impact of subsequent actions, reducing their social conflicts and costs [19]. We have identified relevant groups in the context of scientific conferences (see Fig. 1).

---

[4] https://www.bcorporation.net, https://www.ecogood.org.

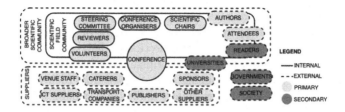

Fig. 1. Main conference stakeholder groups

We plan to climb the 5-level stakeholder participation scale from Pretty [14]. Our paper presents a proposal engineered with limited participation from peers; i.e. level 2. It is meant to trigger discussions in our scientific community and kick-start greater engagement. During the upcoming ESEA4RCIS workshop, we will continue the materiality assessment process with the participants, and we will collaboratively perform the ESEA of RCIS 2022; this jumps to level 4. During the later reflection moments, we hope to engage additional stakeholder groups, such as the RCIS Steering Committee and part of the RCIS community. By reaching level 5, where self-organisation is achieved, we expect to minimise resistance to change the status quo of conference management.

4   ESEA Method Overview

Figure 2 shows our first attempt at producing an ESEA method for assessing the ESE performance of conferences. Process A, at the top, depicts the research method; find the explanation in Sect. 2. Process B, at the bottom, shows the actual ESEA method process and data structure. It starts with the ESEA accountants collecting data for the direct indicators (activity B1), refined as a partial order of finer-grained activities. Collection can be done by manually entering known data, scraping available data, or sending out stakeholder surveys. If the ESEA accountants are working on behalf of the conference organisers, they will likely have access to information they can manually enter themselves (B1.1). Scraping (B1.2) allows gathering vasts amounts of data (e.g. author names, affiliations), but this data should then be cleansed (B1.3). Surveys (B1.4) can collect more personal data (e.g. ages, perceptions, opinions). Indirect indicators can be calculated automatically (B2). Once all the indicators are calculated, the ESE account can be submitted (B3). Auditing the collected data (B4) is optional and it depends on the needs for accuracy or assurance; auditing is a complex process that we leave out of the scope in this paper. To produce a comprehensive report (B5), the following activities are often performed in an intertwined way. The results need to be interpreted (B5.1). Visualising results with the proper tabular or graphical formats (B5.2) helps with the interpretation. Writing the report (B5.3) often requires additional context-setting, story-telling, explanations, and examples. Disclosing the results (B5.4) is up to the conference steering committee and organisers, but it is convenient for accountability and transparency purposes, as well as an enabler for collective reflection. To support the ESEA of RCIS 2022 we will use openESEA, an open-source model-driven tool we are developing [5]; it supports the manual collection of data (B1.2). For automated collection (B1.1) we resort to Python scripts, publishing house APIs[5] and a spreadsheet. The gender of authors has been guessed from their full names using an online service[6], and we have manually audited the results until achieving a probability of accuracy of 97.2%.

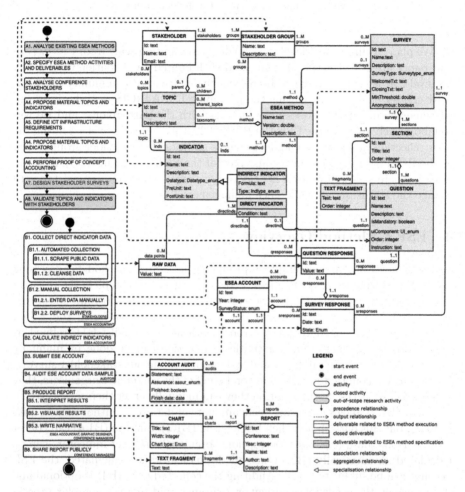

Fig. 2. An overview of an ESEA method for conferences (process B), along with the research method followed to produce it (process A), expressed as a PDD [20]

---

[5] https://developer.ieee.org and https://dev.springernature.com.

[6] https://namsor.app python SDK (we acknowledge that gender is a social and personal construct, that we have artificially restricted it to a binary classification, and that the only way to determine it accurately is by asking the persons themselves).

5   Material Topics for Conferences

Let this initial proposal trigger discussions within our scientific community, conference organisers and steering committees. We calculate some indicators for RCIS conference editions from 2007 to 2021, purposely avoiding interpretation.

Climate Change and Carbon Emissions. Academics show interest in estimating the CO_2 footprint of conferences. Sources of emissions are travelling [2,4,17], proceedings [15], catering and additional conference material (e.g. complementary gift bag) [13]. Conference organisers could also resort to renewable energy sources.

Data Privacy and Cybersecurity. Conference organisers manage financial information of attendees and other data considered sensitive by the GDPR. For instance, dietary constraints can reveal religion or ethnic origin. They are expected to protect this data properly and make sure that any outsourced service providers (e.g. event management organisation) do the same.

Research Topics. Conference calls for papers often list themes and topics of interest. Whether these encourage research on ESE topics is an indication of the awareness of the chairs towards such topics. For instance, 3 out of 15 RCIS editions (2016, 2017, 2021) have a theme related to an ESE topic (e.g. *Information science for a socially and environmentally responsible world*, in 2016). The extent to which submitted and accepted papers actually address such ESE topics is an indication of the real commitments of the underlying scientific community. This could be assessed using a sustainability maturity framework [11].

Diversity, Inclusion and Equity. Diversity refers to all aspects of human difference (e.g. socio-economic status, sexual preference, race, ethnicity, gender identity). Inclusion refers to creating a space where all individuals have a sense of belonging and are able to achieve to their potential. Equity acknowledges that there are unique experiences and barriers that not everyone faces [12]. Equity can be expected at the level of all stakeholder groups. An occasional practice is offering grants to authors with low incomes (e.g. from developing countries). Across the 2008–2019 RCIS editions, 31.8% of General Chairs and 45.5% of Programme Chairs are persons identified as having a feminine gender, and the ratio is for paper authors is 35.0%. The gender ratio of first authors within each edition ranges from 24.1% in 2011 to 57.8% in 2018. From 2007 until 2018, an average of 8.6 developing countries (those with an economy either developing or in transition, according to United Nations) were represented by accepted paper authors; from 2019 to 2021, only 5 developing countries were represented, on average.

Waste Management. Conferences produce waste [10] and need to manage materials and goods that are no longer of direct use to them. They need to comply with all applicable legal criteria for disposal of waste, including e-waste. To act responsibly, they need to proactively reduce the amount of waste, while also optimising and researching new ways to recycle.

Responsible Conference Management. Conventional concerns refer to proper financial management and offering the attendees the expected services, but we should be more ambitious. Volunteers should be treated fairly (e.g. duties aligned with interests, ensure their learning experience, free registration, acknowledgements). Reviewers should receive a reasonable number of reviews aligned with their preferences, be given sufficient time avoiding periods in which the reviewing can hinder work-life balance (end-of-year vacations). Registration and other attendance costs should be affordable (wide range of accommodation offerings). The early-bird (or lowest) full-conference registration fees for paper authors of earlier in-person RCIS conference editions from 2009 to 2019 range from €390, in 2013, to €595, in 2019. Since conferences receive public money (directly as subsidies, indirectly as travel expense reimbursements), society could demand that their financial accounts are publicly disclosed. To the best of our knowledge, no edition has done this so far. And how the surplus (if any) is used, seems a matter of concern; i.e., whether benefits are reinvested in the community (e.g. as research grants) or instead shared among a few stakeholders (e.g. a private organisation). As a scientific community, we could also consider whether we would like that certain decisions about our conferences are made democratically.

Open Science. This calls for transparent and accessible knowledge, shared and developed through collaborative networks [18]. E.g. whether conference proceedings are open publications, whether reviewing systems align with open reviews, whether the conference promotes or prescribes the publication of software artefacts under open-source licences, and the publication of FAIR and open data. Of course, conference management systems could also be free software. From 2009 to 2019, RCIS proceedings have been published by IEEE and, from 2020 onwards, by Springer; none of these books are offered as open publications.

Attendee Experience. Attendees need to be satisfied with the scientific and organisational aspects. But their work-life balance should also be considered (e.g. schedules compatible with family matters, availability of milking and relaxation rooms, childcare possibilities). Lastly, outreach by means of science communication in layman terms returns back to society.

Responsible Supply Chain Management. Responsible procurement trades off economic social and environmental criteria and it is important for many organisations [8], but is rarely heard of in scientific conference management. Conferences could support social entrepreneurship and NGOs by hiring or procuring services and products from them. Catering could offer organic and local food.

6   This Is Just the Start

In this paper, we make an initial proposal for an ethical, social and environmental accounting method for conferences. We also propose a list of topics that goes beyond the usual suspects (e.g. CO_2 footprint) and intends to be thought-provoking. We expect that the discussions will span further than the

RCIS Forum. When applying the method later on, it is not the actual accounting results what is most important, but the process of obtaining those results, the reflections that they can produce within the scientific community, and the effort of jointly tackling the subsequent improvement actions. Together, we can contribute to more ethical, socially inclusive, and environmentally friendly conferences.

References

1. Arnold, V., Lampe, J.C., Sutton, S.G.: Understanding the factors underlying ethical organizations: Enabling continuous ethical improvement. J. Appl. Bus. Res. **15**(3), 1–20 (1999)
2. Biørn-Hansen, A., Pargman, D., Eriksson, E., Romero, M., Laaksolahti, J., Robért, M.: Exploring the problem space of CO2 emission reductions from academic flying. Sustainability **13**(21), 12206 (2021)
3. Cash, P., Škec, S., Štorga, M.: A bibliometric analysis of the DESIGN 2012 conference. In: ICED 2013. The Design Society (2013)
4. Coroama, V.C., Hilty, L.M., Birtel, M.: Effects of internet-based multiple-site conferences on greenhouse gas emissions. Telemat. Inform. **29**(4) (2012)
5. España, S., Bik, N., Overbeek, S.: Model-driven engineering support for social and environmental accounting. In: RCIS 2019, pp. 1–12. IEEE (2019)
6. Gao, S.S., Zhang, J.J.: Stakeholder engagement, social auditing and corporate sustainability. Bus. Process Manag. J. (2006)
7. Gray, R., Owen, D., Maunders, K.: Corporate Social Reporting: Accounting and Accountability. Prentice-Hall (1987)
8. Hoejmose, S.U., Adrien-Kirby, A.J.: Socially and environmentally responsible procurement: a literature review and future research agenda of a managerial issue in the 21st century. J. Purch. Supply Manag. **18**(4), 232–242 (2012)
9. Madsen, H., Ulhøi, J.P.: Integrating environmental and stakeholder management. Special Issue: Netw. Environ. Manag. Sustain. Develop. **10**(2), 77–88 (2001)
10. Mankaa, R.N., Bolz, M., Palumbo, E., Neugebauer, S., Traverso, M.: Walk-the-talk: sustainable events management as common practice for sustainability conferences. In: 12th Italian LCA Network Conference, pp. 11–12 (2018)
11. Mann, S., Bates, O., Maher, R.: Shifting the maturity needle of ICT for sustainability. In: ICT4S 2018, pp. 209–226. EasyChair (2018)
12. McCleary-Gaddy, A.: Be explicit: defining the difference between the office of diversity & inclusion and the office of diversity & equity. Med. Teach. **41**(12), 1443–1444 (2019)
13. Neugebauer, S., Bolz, M., Mankaa, R., Traverso, M.: How sustainable are sustainability conferences?-Comprehensive life cycle assessment of an international conference series in Europe. J. Clean. Prod. **242**, 118516 (2020)
14. Pretty, J.N.: Participatory learning for sustainable agriculture. World Dev. **23**(8), 1247–1263 (1995)
15. Spinellis, D., Louridas, P.: The carbon footprint of conference papers. PloS One **8**(6) (2013)
16. Stolowy, H., Paugam, L.: The expansion of non-financial reporting: an exploratory study. Acc. Bus. Res. **48**(5), 525–548 (2018)
17. Stroud, J.T., Feeley, K.J.: Responsible academia: optimizing conference locations to minimize greenhouse gas emissions. Ecography **38**(4) (2015)

18. Vicente-Saez, R., Martinez-Fuentes, C.: Open science now: a systematic literature review for an integrated definition. J. Bus. Res. **88**, 428–436 (2018)
19. Voinov, A., Bousquet, F.: Modelling with stakeholders. Environ. Model. Softw. **25**(11), 1268–1281 (2010)
20. van de Weerd, I., Brinkkemper, S.: Meta-modeling for situational analysis and design methods. In: Handbook of Research on Modern Systems Analysis and Design Technologies and Applications, pp. 35–54. IGI Global (2009)
21. Whitehead, J.: Prioritizing sustainability indicators: using materiality analysis to guide sustainability assessment and strategy. Bus. Strat. Env. **26**(3), 399–412 (2017)

ANGLEr: A Next-Generation Natural Language Exploratory Framework

Timotej Knez[(✉)], Marko Bajec, and Slavko Žitnik

University of Ljubljana, Ljubljana, Slovenia
`timotej.knez@fri.uni-lj.si`

Abstract. Natural language processing is used for solving a wide variety of problems. Some scholars and interest groups working with language resources are not well versed in programming, so there is a need for a good graphical framework that allows users to quickly design and test natural language processing pipelines without the need for programming. The existing frameworks do not satisfy all the requirements for such a tool. We, therefore, propose a new framework that provides a simple way for its users to build language processing pipelines. It also allows a simple programming language agnostic way for adding new modules, which will help the adoption by natural language processing developers and researchers. The main parts of the proposed framework consist of (a) a pluggable Docker-based architecture, (b) a general data model, and (c) APIs description along with the graphical user interface. The proposed design is being used for implementation of a new natural language processing framework, called ANGLEr.

Keywords: Natural language processing · Graphical framework · Language understanding

1 Introduction

Recently we have seen a lot of interest in the natural language processing tools with Uima [5] receiving around 6 thousand downloads and GATE [3] receiving over 400 thousand downloads since 2005[1]. NLP tools are used by people from various backgrounds. For the users outside the computer science area to use the available tools effectively, we have to provide them with an intuitive graphical user interface that allows a simplified construction of text processing pipelines. In this paper, we design an architecture for a new framework that enables quick and simple construction of natural language processing pipelines - ANGLEr. In addition to that, the framework features a way to add new tools that is simple for developers to implement. The researchers could showcase their projects by including them in our framework with very little additional effort. In the past, several frameworks for simplifying NLP workflows were designed. However, they

---

[1] The number of downloads was recorded by https://sourceforge.net.

R. Guizzardi et al. (Eds.): RCIS 2022, LNBIP 446, pp. 761–768, 2022.
https://doi.org/10.1007/978-3-031-05760-1_53

all fall short either in the expandability or the ease of use. In our work we identify the key components of such framework and improve upon the existing frameworks by fixing the identified flaws.

The main contributions of our proposed framework are the following:

a) An architecture that allows simple inclusion of new and already existing tools to the system.
b) A data model that is general enough to store information from many different tools, while enabling compatibility between different tools.
c) A graphical user interface that speeds up the process of building a pipeline and makes the framework more approachable for users without technical knowledge.

The rest of this paper is organised as follows: in Sect. 2 we present the existing frameworks and compare them to our proposed framework. In Sect. 3 we present our framework and its most important parts. We conclude the paper in Sect. 4.

2   Existing Frameworks Review

We examined the existing frameworks and selected the ones that we believe to be most useful for users with limited programming knowledge. Analyzing the strengths and weaknesses of the existing tools helps us to define a list of features that could be improved by introducing a new framework. The comparison is summarized in Table 1. We compare the frameworks on a set of features that were presented by some of the frameworks as their main selling points while others were selected as we find them important for usability. We compare the tools on (1) their graphical user interface (GUI), which affects usability for new users, (2) data model, which affects expandability, and (3) the language required for creation of plugins, which is important for plugin developers.

GATE: One of the best-known frameworks for text processing is GATE [3]. It was designed to feature a unified data model that supports a wide variety of language processing tools. While the GATE framework provides a graphical way for pipeline construction, its interface has not been significantly updated in over 10 years. The program is thus difficult to use for new users. Another limitation of the GATE framework is that it requires all of its plugins to be written in the Java language. Because of that, it is difficult to adapt an already existing text processing algorithm to work with the framework.

UIMA: Uima [5] is a framework for extracting information from unstructured documents like text, images, emails and so on. It features a data model that combines extracted information from all previous components. While Uima provides some graphical tools, they are not combined into a single application and do not provide an easy way for creating processing pipelines. The primary way of using UIMA still requires the user to edit XML descriptor documents, which limits usability for new users and slows down the development. New components for the framework can be written in the Java or the C++ programming languages.

Orange: Another popular program for graphical creation of machine learning pipelines is Orange [4]. The main feature of the orange framework is a great user interface that is friendly even for non technical users. Orange provides a variety of widgets designed for machine learning tasks and data visualisation. The framework was designed primarily to work with relational data for classic machine learning. It supports two extensions for processing natural language which allow us to use the existing machine learning tools on text documents. The largest drawback when using Orange for text processing is that the tabular representation, used for representing orange data, is not well suited for representing information needed for text processing. The two extensions tackle this problem in different ways.

Orange Text Mining: The Orange text mining extension implements a data model based on the tabular representation used by the existing tools. This representation allows integration with the existing machine learning and visualization tools, however, it also limits the implementation of some text processing algorithms. Its biggest limitation is that it only works with features on a document level.

Textable: On the other hand, the Textable extension uses a significantly different approach to data representation. In Textable, different tools produce different data models. This allows for more flexibility when developing new tools, however, it also limits the compatibility between different types of tools. Ideally, we would want a single data model that is capable of supporting every tool. This way we could reach high compatibility between tools while keeping the flexibility when creating new ones.

Libraries: In the past multiple libraries that support text processing have been developed. These libraries are commonly used by experts in the field of natural language processing. They are not well suited for use by other people that might also be interested in language processing. One of such libraries is called OpenNLP [1]. It supports the development of natural language applications in the Java programming language. There are also multiple Python libraries, such as NLTK [2], Gobbli [6], and Stanza [7]. To use any of these libraries, the user is required to have some programming knowledge. Writing a program is also a lot slower than constructing a pipeline through a graphical interface.

Table 1. Comparison of different natural language processing frameworks.

Framework	Graphical UI	Unified data model	Plugin language
GATE	Yes (native)	Yes (too general)	Java
UIMA	Yes (limited)	Yes	Java or C++
Orange	Yes (native)	Diverse	Python
Orange textable	Yes (native)	Diverse	Python
Orange text mining	Yes (native)	Tabular data only	Python
Libraries	No	Diverse	Java
ANGLEr	**Yes (Web)**	**Yes (versioned)**	**N/A (Docker packaged)**

3   ANGLEr: A Next-Generation Natural Language Exploratory Framework

Based on the problems we identified during the review of the existing frameworks we identified the following important elements: (a) common and versioned data model, (b) extensible Docker-based architecture with API-based communication, and (c) unified and pluggable user interface (see Table 1). The data model and APIs will ease the creation of new functionalities for the community. On the other hand, the user interface and simple installation or public hosting are important for the general audience who need to use the language processing tools but do not have the programming skills.

3.1   Data Model

The data model defines a structure that is used to transfer the analysis results between the tools. We aim to support all text processing tools that are available in the existing frameworks as well as new types of algorithms that have not been identified yet. A general view of the data model is presented in Fig. 1.

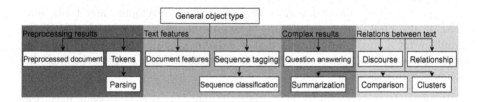

Fig. 1. The upper-level hierarchy of ANGLEr object types.

The data model is a central part of the proposed system. It has to provide enough versatility to support every processing tool while keeping a well defined structure to enable compatibility between different tools. We propose a model comprised of two parts. The first part is a corpus of all documents that we want to process. We store the basic metadata about the corpus and each of the documents captured in it. Each document can also be separated into sentences and tokens in case this information was provided in the loaded corpus. The second part of the model stores all of the processor outputs. Outputs and inputs to available processors are encoded in the same schema - i.e., a processor's output can be directly another algorithm's input. We define a list of object types that can be used for representing results as follows:

Preprocessed document type stores a new version of the documents in a corpus after a preprocessing algorithm has been applied.
Document features type stores the values of the features that represent each document in a corpus.

Sequence classification type is used to store information about the result of the classification of an entire document or a part of a document.

Sequence tagging type is used to store a set of tags for different parts of the documents. This type could be used for instance to store named entity recognition results.

Relationship type is used to represent relations between two entities from the source documents.

Discourse type is used to store entities that appear in the documents. Each entity also contains a list of all entity mentions that correspond to this entity. This object would be used for instance to store coreference information.

Parsing type stores parts of the document that are linked together. For example, this type would be used to store phrases detected using a chunking algorithm.

Tokens type stores the information about tokens that appear in the documents. This type would be used to represent for instance the result of a tokenizer or n-gram generator.

Summarization type is used to store summaries of a document.

Question answering type is used to store questions and their answers based on the provided text.

Comparison type is used to store similarity between different parts of the documents.

Clusters type is used to store clusters of similar documents.

Type hierarchy is proposed to improve backward compatibility when adding new object types. This way each new specific object type contains the attributes of its parent as well as some of its own attributes (each level also allows for key-value metadata storage). An algorithm that was designed to work with a general object type can also work with all of its descendants. For example, a tool for performing named entity recognition would accept the *token* type to get the tokens on which to perform named entity recognition. The user could also provide the *parsing* type since it is a descendant of the *token* type.

3.2 High-Level Architecture

The goal of our framework architecture (presented in Fig. 2) is that adding a new module would be as simple as possible. In addition to that, we want to make sure that the framework and its modules can be also used by third-party applications. Based on this, we have decided that each module should run as a Docker container, which simplifies the inclusion of already existing tools into our framework. The modules are connected to the ANGLEr backend, which is responsible for managing and running the pipeline. The user interacts with the graphical user interface that runs as a web page in a web browser.

3.3 REST Interface

All of the communication between the parts of the framework is done over REST application programming interfaces (API). This also allows third-party applications to access the modules and the ANGLEr backend.

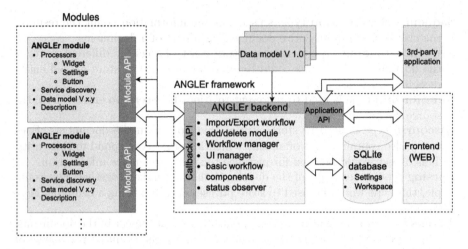

Fig. 2. ANGLEr high-level system architecture. The communication between all of the modules an the backend is done using the versioned data model.

```
[{"input_name": "Training data", "types": [
    {"type_name": "Tokens", "type": "tokens"},{"type_name": "Part of speech", "type": "sequenceTagging"},
    {"type_name": "Named entities", "type": "sequenceTagging"}
    ]},
    {"input_name": "Validation data", "types": [
    {"type_name": "Tokens", "type": "tokens"},{"type_name": "Part of speech", "type": "sequenceTagging"}]}]
```

Fig. 3. An example of the input attribute for a named entity recognition module.

The framework features multiple REST APIs shown in Fig. 2. The *Module API* (Table 2) is responsible for commands that are sent from the ANGLEr backend to a module. It allows the ANGLEr backend to gather information about the processors that are running in a module. It also allows the backend to send the data and start its processing. After the processing is done, the module sends the results to the backend using the *Callback API*. The ANGLEr backend also exposes the *Application API*, which is responsible for communication with the graphical user interface, as well as with any third-party applications that might want to use the ANGLEr functionality. Since the *Module API* has to be implemented by each module, we present its endpoints in Table 2.

Once a module is added to the framework, the ANGLEr backend creates a request to the *about* endpoint, which provides the basic information about the entire module. After that, it requests the *processors* endpoint, which lists all processors that are contained in the module. This information allows the framework to visualise the processor in the graphical interface. It also provides the endpoints that the ANGLEr backend uses to send data for processing and to show configuration and visualisation pages.

Table 2. Endpoints in the *Module API*. The underlined attributes are required. The processors endpoint returns a list of objects where each object has the presented attributes.

Endpoint	Parameters	Description
/about	<u>UUID</u>	Identifier of the module
	<u>name</u>	Name of the module
	<u>version</u>	Module version
	<u>data_model</u>	Version of the data model used
	Desc	Module description
	Authors	List of authors
	Organisation	Organisation of the authors
	url	URL address of the page about the module
/processors	<u>name</u>	Name of a processor
	short_name	Short version of the name
	data_endpoint	Endpoint where the data for processing can be sent
	settings_endpoint	An address of a page containing processor settings
	ui_endpoint	An address of a page for visualization
	<u>icon</u>	An address of the icon to be used to represent the processor
	category	A category of the processors menu that should contain this processor
	<u>inputs</u>	A list of objects representing different inputs. Each object should contain a name and a list of types required. An example is shown in Fig. 3
	<u>outputs</u>	A list of objects representing different outputs. Each object should contain a name and a type of the output
/docs	html_page	A web page containing the documentation

3.4 Graphical User Interface

The user interface is a key component for making the framework simple to use. It allows the users without any programming knowledge to use the framework. This is especially important since language processing tools are very useful for linguists, who typically do not have advanced computer knowledge. We propose a simple design for the user interface, which is shown in Fig. 4. The tabs at the top of the page organise the tools into groups based on their purpose. For example, the groups contain the tools for text preprocessing, for semantic and syntactic analysis etc. A user can get a widget for each tool by dragging and dropping. The widget has a page for setting its parameters and a dialog for selecting the input. A widget can optionally also provide a page for data visualisation (see the Fig. 4 right). The user connects the widgets into a pipeline that can be stored to a file and loaded by any user. The ANGLEr framework then executes the pipeline to process the data.

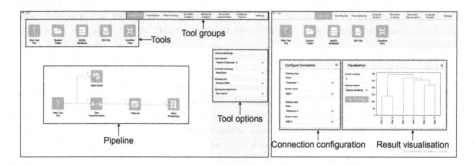

Fig. 4. The main parts of the ANGLEr user interface.

4 Conclusion

We describe the main components for implementing a new natural language processing framework - ANGLEr. We believe that a new framework based on our proposal would provide a large improvement over the existing frameworks and would greatly benefit users that are working with natural language processing. The framework would provide a fast way for prototyping when developing text processing pipelines. It would also allow users with no programming knowledge to build advanced NLP pipelines. In addition to that, the new framework would provide the researchers with a great way for showcasing their work in the NLP area. Since the tools can be implemented in any programming language, their inclusion is much less complicated than with existing frameworks.

References

1. Apache: Opennlp (2010). http://opennlp.apache.org
2. Bird, S., Loper, E.: NLTK: the natural language toolkit. Association for Computational Linguistics (2004)
3. Cunningham, H.: Gate, a general architecture for text engineering. Comput. Humanit. **36**(2), 223–254 (2002)
4. Demšar, J., et al.: Orange: data mining toolbox in Python. J. Mach. Learn. Res. **14**(1), 2349–2353 (2013)
5. Ferrucci, D., Lally, A.: UIMA: an architectural approach to unstructured information processing in the corporate research environment. Nat. Lang. Eng. **10**(3–4), 327–348 (2004)
6. Nance, J., Baumgartner, P.: gobbli: a uniform interface to deep learning for text in Python. J. Open Source Softw. **6**(62), 2395 (2021)
7. Qi, P., Zhang, Y., Zhang, Y., Bolton, J., Manning, C.D.: Stanza: a Python natural language processing toolkit for many human languages. In: Proceedings of the 58th Annual Meeting of the Association for Computational Linguistics: System Demonstrations, pp. 101–108 (2020)

Doctoral Consortium

Towards a Roadmap for Developing Digital Work Practices: A Maturity Model Approach

Pooria Jafari[✉] [iD]

Department of Business Informatics and Operations Management, Faculty of Economics and Business Administration, Ghent University, Tweekerkenstraat 2, 9000 Ghent, Belgium
Pooria.Jafari@UGent.be

Abstract. The digital economy brings us a wide range of services and products assisted by emerging technologies in Industry 4.0. Nevertheless, already since the 1990s and early 2000s, IT has had a huge impact on the digitalization and digitization of businesses. This phenomenon of advancing in digital-oriented work practices is not only affecting the customer side, but is also changing the way of working within organizations. Although employees are one of the crucial elements in each organization and their level of work satisfaction is critical to the efficiency of a business, their work impression is still undergoing scrutiny when it gets to digital process innovations. Moreover, organizations still benefit from assistance in their adoption of digital-oriented work practices, for which a related maturity model (MM) can be one of the solutions. The current Ph.D. plan consists of three research projects that follow a mixed-method approach with a combination of quantitative and qualitative designs, within an overall design of Design-Science Research (DSR). Project 1 has two subprojects (i.e., a Systematic Literature Review (SLR) and a data-driven analysis). It starts with analyzing the relevant literature from a people–process–technology (PPT) perspective to extract relevant factors when digitalizing business processes. Afterwards, it investigates a representative set of European employee data using statistical data analysis (e.g., factor analysis and ANOVA) and data mining (e.g., clustering) techniques to delve into the impact of digital-oriented work practices on work satisfaction. After this artefact identification, we continue with Project 2 (i.e., expert panel, case study) to add evidence for a maturity-based gradation along with the identified clusters of digital-oriented work practices. Finally, Project 3 helps concretize the intended MM by focusing on the relationships with employee satisfaction and the relevant factors. Our findings will assist organizations to upgrade their work practices (i.e., including assessment and improvement advice), while simultaneously empowering their employees.

Keywords: People-process-technology · Maturity model · Work satisfaction

1 Research Problem and State of the Art

In the new era of information technology, almost all services and business activities are benefiting from technology advances. In the field of business, the concept of using digital innovation has become a primary trend, taking opportunities from emerging technologies

R. Guizzardi et al. (Eds.): RCIS 2022, LNBIP 446, pp. 771–778, 2022.
https://doi.org/10.1007/978-3-031-05760-1_54

and triggering organizations to work differently in a digital economy. As a result, the work practices have to adapt to the new technologies, and also employees are affected [1]. Similarly, business processes (i.e., as being part of every business and innovation) are no exception to this shift, and are fully part of this evolution from traditional to digitalized work. Moreover, digital transformations in work practices are assisted by new technologies (i.e., Internet of Things, robotics, blockchain) to drastically change or even potentially disrupt the way of working for employees [2, 3]. Such digital transformations have potential for both employees and their organizations to perform more efficiently than in a traditional or non-digital way of organizing work [4]. Additionally, the development of emerging technologies and the way of managing these new disruptive innovations require new aspects to be added as main primary factors (e.g., differences in terms of management functions, innovation, customer focus) [5]. All above-mentioned factors help increase operational efficiency and improve resource utilization under the umbrella of "value stream management", namely with business processes being the value chains within organizations [6]. For instance, their importance have been highlighted during the recent disaster management plans due to Covid-19, including restrictions in task management within organizations.

Our underlying theoretical perspective is called people-process-technology (PPT). Its roots go back to the early 1960s, when Leavitt proposed a model for creating change in organizations. The final version of this model was formed by the three main factors (i.e., people-related, process-related, and technology-related factors) as well as the common zone between them (i.e., relationships between people-process, people-technology, and process-technology). This theoretical view claims that an organization's top management decides and acts in a common zone between the people and process dimensions in order to empower employees to do their tasks more efficiently [5].

Meanwhile, most of the existing studies have been focusing on customer satisfaction, the market, and innovation management [7, 8]. As an extension to earlier views, this doctoral journey aims to primarily focus on the employee-related aspects by considering their work satisfaction while supporting the interaction between the PPT factors. Our objective is to gradually build and test a maturity model (MM) with different adoption degrees of digital-oriented work practices in order to enable organizations to assess their current level of digital work and by evolving accordingly. This self-assessment instrument will give an opportunity to organizations to know about their current level of maturity and either stay and develop within the same level, or to clime up the maturity ladder to get to the higher levels of digitalization.

The intended MM will be created based on our findings along three research projects, each with their own research question:

- Main RQ1. Which PPT (i.e., people, process, technology) subfactors are most decisive for characterizing the different degrees of digital-oriented work practices? And, what is the expected link with work satisfaction?
- Main RQ2. How can the PPT subfactors of RQ1 be combined in maturity levels to build and test a MM for assessing an organization's digital-oriented work practices?
- Main RQ3. How can the PPT subfactors be measured along the different levels of a MM for digital-oriented work practices? And, to what extent does higher maturity in digital-oriented work practices correspond to higher work satisfaction?

To tackle these research questions, our overall approach is situated in the Design-Science Research (DSR) methodology [9–11], and combining quantitative with qualitative methodologies.

2 Research Methodology

Design-Science Research (DSR) is an established paradigm in the field of information systems research that aims at constructing a solution via artefact development instead of explaining an existing phenomenon [9–11]. DSR is based on iterative cycles that refine artefacts in each build-test iteration. In this study, we will design our artefacts in three iterations with the aim of MM development. First, the maturity levels will be extracted by a data-driven analysis and literature review in a first iteration. In a second iteration, we will validate the proposed maturity levels with subject matter experts and in an illustrative case study. Finally, in a third iteration, we will build a measurement tool to assess an organization's current level of maturity by means of our proposed MM. Thus, we defined three projects to tackle each of these three iterations. An overview of these three projects is shown in Fig. 1. The Ph.D. is currently situated in the middle of Project 2. We subsequently discuss the methodology of each project.

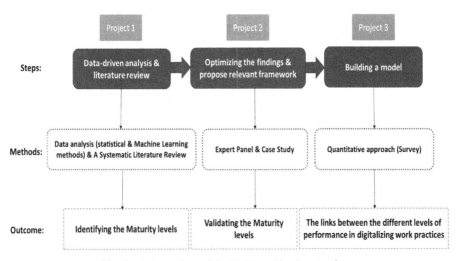

Fig. 1. An overview of the Ph.D. and its three projects.

2.1 Project 1. Systematic Literature Review and Data-Driven Analysis for Level Identification

To get a better comprehension of the PPT sub factors to be involved in our analysis, we started with a systematic literature review (SLR). An SLR is *"a means of evaluating and interpreting all available research relevant to a particular research question, topic area, or phenomenon of interest*" (p. vi) [12]. This SLR method[1] determines the areas for collecting compatible sources based on a protocol with inclusion and exclusion criteria [13, 14]. As a result of the related filtering approach, 45 relevant papers were included in our literature-based analysis. To assign the potentially relevant PPT factors concerning digital-oriented work practices, we counted the sampled papers in one or more compatible groups of variables (Table 1). As a result, we extracted a total of nineteen independent variables, of which four consisted of individual and organizational characteristics and 15 PPT subfactors were representations of digital work.

Table 1. Selection of variables in the dataset, derived from the literature study.

Digital-oriented work characteristics along PPT		
PPT pillar	Main variables	Sub variables
Technology	IT use	
Process	Quality control	
	Pace of work	
	Work interruptions	Work interruptions (frequency)
		Work interruptions (effect)
	Cognitive work	Problem-solving tasks
		Task complexity
		Task repetitiveness
		Task novelty
		Multi-skilled tasks
People	Employee involvement	
	Decision autonomy	
	Skills match	
	On-the-job training	
	Payment for team performance	

For instance, payment for team performance (as an organizational characteristic) is important for process work when employees work together in multi-disciplinary teams. Ideally, employees (as process workers) are also paid as team members with the aim of team work development. Although such variables do not directly affect the Process factor, they enact organizational advancements from the perspective of HR, and are thus organization capabilities from within the People factor [19]. The seven dependent variables (i.e., perceived work performance, work usefulness, perceived working conditions, career advancement, work motivation, work satisfaction, work security) were related to

---

[1] https://drive.google.com/file/d/1x6MvIp6M_szsk24xnDQqgdJ15f4VSETS/view?usp=sharing.

work satisfaction and included in our final list to examine work satisfaction from an employee's perspective.

In a second phase, we applied our findings to the most recent working conditions dataset of the European Foundation for the Improvement of Living and Working Conditions [15]. This dataset is publicly available and was chosen because of its representativeness in terms of the number of respondents, and because of its matching variables with the PPT subfactors supported by our SLR outcomes. More specifically, the 2015 dataset consisted of 43,850 European employees. We first prepared our data by doing data cleansing, and removed missing values. All statistical analyses were conducted using SPSS (version 26). The 15 digital work variables were classified using a cluster analysis (i.e., K-means) [16]. For our independent variable of work satisfaction, we conducted a factor analysis and extracted two main factors among seven dependent variables: "Current Work Perception" (CWP) and "Future Work Perception" (FWP). In order to find any difference in work satisfaction among the obtained clusters, ANOVA-based testing was conducted.

The clustering output will be input for the remaining projects, during which we identify any logical progress over time in order to turn them into gradual maturity levels. Thus, the clustering exercise in Project 1 was our first step to derive the intended MM.

2.2 Project 2. Validation Study with Expert Panel and Illustrative Case Study

An iterative validation of the MM and the related maturity levels was the aim of Project 2, with the PPT-related subfactors acting as "construct" artefacts. Finding a maturity ladder approach among the earlier identified clusters served as our "model" artefact. We therefore decided to carry out an expert panel and then a case study to evaluate the proposed "model" artefact, and to examine the validation feedback in a second DSR iteration [17]. Thus, the proposed framework in Project 2 was constructed by the extracted artefacts from the first iteration. Similarly, the output of this iteration in Project 2 will be a validated model for the third iteration (Project 3), which then aims at defining the measurable relationships in our proposed model.

The ultimate purpose of the expert panel was to redesign the model that presents the relationships among the PPT constructs, and to gradually demonstrate its positive impact on business value [18]. The expert panel was conducted with 12 participants (i.e., 10 CIOs, 2 CTOs) from 12 countries across three continents (i.e., 1 from North America, 2 from the Middle East, 7 from Western Europe, and 2 from Eastern Europe). They gained their work experience in 10 sectors and in different organizational sizes (i.e., 1 micro, 2 small, 2 medium, 7 large-sized). On average, the interviews took 36 min, during which we asked the experts to evaluate our proposed framework based on four traditional DSR criteria (i.e., usefulness, quality, efficiency, effectiveness) using a 5-point Likert scale (1 = fully disagree; 5 = fully agree), plus an "I don't know" option. We therefore visualized our proposed maturity levels along a ladder, and by explaining their main differences in terms of the five most differentiating PPT subfactors as derived from Project 1 (i.e., IT use, decision autonomy, employee involvement, task complexity, and task repetitiveness). Additionally, we asked the experts about their points of view to further extend and update each of the five subfactors with an eye on: (1) emerging technologies and (2) the impact of the global COVID-19 pandemic on the digitalization efforts of organizations.

Currently, this is the actual stage of our research work in the entire Ph.D. path. Within Project 2, we also plan to conduct one illustrative case study to actually demonstrate an organization's longitudinal journey along the different maturity levels. We will therefore translate and apply our findings of the previous iterations together with employees of a large-sized organization that is situated in one of the two highest levels of our suggested MM. The illustrative case study will thus show the digitalization journey of an organization that has well advanced in terms of digitalizing its work practices. The longitudinal focus will especially be on the different stages that the organization has experienced over time, and its triggers to evolve from lower to higher digital-oriented work practices. The three departments related to PPT (i.e., people, process, and technology) will be involved (i.e., HR, operations and/or a process-related center of excellence, and IT). We plan to have interviews with two roles per department (i.e., one employee, one manager) on order to understand their opinions about their organization's evolution along the three PPT pillars. Furthermore, we will collect relevant documents to obtain data triangulation (e.g., mission statement, digitalization plans).

2.3   Project 3. Generalization Study with a Quantitative Approach (Survey)

While Project 1 and Project 2 aimed at identifying and validating our proposed MM, Project 3 continues with the "method" artefact for assessing and improving maturity. More specifically, each MM needs its own measurement tool to enable its users to assess their organization based on the respective model stages (i.e., maturity levels). In this iteration, we will conduct an international survey to attach a measurement to each maturity level. In other words, after a verification and validation in the first and second iterations, we will now focus on building a measurement instrument to make our suggested MM more applicable for digital-oriented work practices within organizations.

We plan to target at least 300 managers from three continents (i.e., North America, Europe, Asia). The survey questions will be derived from the survey questions in Project 1 and the extensions gathered from the expert feedback in Project 2. The PPT subfactors will act as latent variables, each of which will have three to five statements on a 5-point Likert scale (1 = most unimportant; 5 = most important). Managers will be asked about their degree of importance (weight) for their organizations.

Before the large-scale survey will be launched, we plan to organize a pilot survey with 50 Belgian managers. These results can be evaluated using a factor analysis per latent construct. Afterwards, the large-scale survey results will be analyzed using Partial Least Squares - Structural Equation Modelling (PLS-SEM). This statistical analysis will also allow investigating the relationships between the maturity levels and the related performance outcomes (and particularly work satisfaction) [19].

3   Intermediate Results

The Ph.D. journey is currently in the middle of Project 2 (i.e., after 2.5 years of research), and ends by September 2023. Meanwhile, we have been able to select 15 PPT-related sub factors (Table 1) along five differentiators that best describe the maturity level transitions:

- Main people-related differentiators: decision autonomy, employee involvement
- Main process-related differentiators: task complexity, and task repetitiveness
- Main technology-related differentiators: different level of IT use ranging from simple producting tools to more advanced tools

The clustering method in Project 1 resulted in five types of digital-oriented work practices. A deeper understanding of these types has brought us to a MM for gradually adopting digital-oriented work practices, and for which the cluster types serve as maturity levels with an underlying gradation of the main PPT differentiators:

- Initial: slightly digitized and little empowerment work with unambiguous tasks
- Developed: highly digitized and little empowered work with difficult tasks
- Optimized: slightly digitized and greatly empowered work with difficult tasks
- Advanced: somewhat digitized and empowered work with challenging tasks
- Excellent: greatly digitized and greatly empowered work with challenging tasks

In Project 1, we uncovered significant performance differences between the clustered digital-oriented work practices in terms of one's "Current Work Perception" (CWP) and "Future Work Perception" (FWP). The results showed that employees in organizations with an "optimized" maturity level (i.e., in the middle of our MM) were most satisfied with their current work. For future work, we observed a stronger positive relationship (i.e., the higher the maturity level, the more satisfied about the future).

Based on Project 2, most experts considered the DSR evaluation criteria higher than average (i.e., circa 35% with "very agree" and 40% with "agree"). The experts' priorities were, from higher to lower: (1) IT use, (2) employee involvement, (3) task complexity, (4) decision autonomy, and (5) task repetitiveness. The experts considered that the model would enhance once more details and measurements are developed in order to create a more useful and efficient MM. This motivates us to continue with Project 3.

4   Conclusion and Future Steps

This doctoral consortium report has described three DSR-related iterations to gradually build and test a MM for digital-oriented work practices, underpinned by the theory of PPT. The findings were first verified by a Systematic Literature Review (SLR) and proven by a data analysis round in Project 1. In Project 2, we gathered extra validation by means of an expert panel and case study round, which further enhanced the practical implications of our intended MM. In Project 3, we will develop a measurement instrument for the MM and enrich this PPT association between the journey of digital-oriented work practices and work satisfaction. In the end, the findings can be applied as a roadmap for different organizations to determine a digitalization path for digital work.

Acknowledgments. This Ph.D. is organized by Ghent University (Belgium) under the supervision of Prof. Dr. Amy Van Looy. I acknowledge the financial support of BOF, grant 01N14219.

References

1. Van Looy, A.: A quantitative study of the link between business process management and digital innovation. In: Carmona, J., Engels, G., Kumar, A. (eds.) BPM 2017. LNBIP, vol. 297, pp. 177–192. Springer, Cham (2017). https://doi.org/10.1007/978-3-319-65015-9_11
2. Braa K., Rolland K.H.: Horizontal information systems. In: Organizational and Social Perspectives on Information Technology, pp. 83–101 (2000)
3. Schwab, D.K.: The Fourth Industrial Revolution. Crown, New York (2017)
4. vom Brocke, J., Zelt, S., Schmiedel, T.: On the role of context in business process management. Int. J. Inf. Manage. **36**(3), 486–495 (2016)
5. Prodan, M., Prodan, A., Purcarea, A.A.: Three new dimensions to people, process, technology improvement model. In: Rocha, A., Correia, A.M., Costanzo, S., Reis, L.P. (eds.) New Contributions in Information Systems and Technologies. AISC, vol. 353, pp. 481–490. Springer, Cham (2015). https://doi.org/10.1007/978-3-319-16486-1_47
6. Muñoz, M., Rodríguez, M.N.: A guidance to implement or reinforce a DevOps approach in organizations: A casestudy. J. Softw. Evol. Proc. (2021). https://doi.org/10.1002/smr.2342. e2342
7. Darwin, J.R.: Drivers of employee engagement and innovation in information technology industry. IOSR J. Bus. Manag. **20**, 38–46 (2018)
8. Trkman, P., Mertens, W., Viaene, S., Gemmel, P.: From business process management to customer process management. Bus. Process. Manag. J. **21**(2), 250–266 (2015)
9. Hevner, A., March, S.T., Park, J., Ram, S.: Design science in information systems research. Manag. Inf. Syst. Q. **28**(1), 75–106 (2004)
10. Peffers, K., Tuunanen, T., Rothenberger, M.A., Chatterjee, S.: A design science research methodology for information systems research. J. Manag. Inf. Syst. **24**(3), 45–77 (2007)
11. Reubens, R.: To craft, by design, for sustainability: Towards holistic sustainability design for developing-country enterprises, Thompson, New Delhi, pp. 31–44 (2016)
12. Kitchenham, B.: Procedures for Performing Systematic Reviews, vol. 33. Keele University, Keele (2004)
13. Boellt, S.K., Cecez-Kecmanovic, D.: On being 'systematic' in literature reviews in IS. J. Inf. Technol. **30**, 161–173 (2015)
14. King, W.R., He, J.: Understanding the role and methods of meta-analysis in IS research. Commun. Assoc. Inform. Sys. **16**, 665–686 (2005)
15. EuroFound: European Foundation for the Improvement of Living and Working Conditions. Retrieved from EuroFound: https://www.eurofound.europa.eu/surveys/european-wor king-conditions-surveys-ewcs. Accessed 1 Nov 2019
16. Embrechts, M.J., Gatti, C.J., Linton, J., Roysam, B.: Hierarchical clustering for large data sets. In: Georgieva, P., Mihaylova, L., Jain, L. (eds.) Advances in Intelligent Signal Processing and Data Mining. Studies in Computational Intelligence, vol. 410. Springer, Heidelberg (2013). https://doi.org/10.1007/978-3-642-28696-4_8
17. Bandara, W., Van Looy, A., Merideth, J., Meyers, L.: Holistic guidelines for selecting and adapting BPM Maturity Models (BPM MMs). In: Fahland, D., Ghidini, C., Becker, J., Dumas, M. (eds.) BPM 2020. LNBIP, vol. 392, pp. 263–278. Springer, Cham (2020). https://doi.org/10.1007/978-3-030-58638-6_16
18. Becker, J., Knackstedt, R., Pöppelbuß, J.: Developing maturity models for IT management. Bus. Inf. Syst. Eng. **1**(3), 213–222 (2009)
19. Van Looy, A.: Capabilities for managing business processes: a measurement instrument. Bus. Process. Manag. J. **25**(1), 287–311 (2020)

Evolutionary Scriptless Testing

Lianne Valerie Hufkens[(✉)]

Open Universiteit, Heerlen, The Netherlands
lianne.hufkens@ou.nl

Abstract. Automated scriptless testing approaches use Action Selection Rules (ASR) to generate on-the-fly test sequences when testing a software system. Currently, these rules are manually designed and implemented. In this paper we present our research on how to automatically create ASRs by evolving them using an evolutionary algorithm. Expected results are an automated system for Evolutionary Scriptless Testing containing a representation of ASRs, different fitness functions and manipulation operators.

Keywords: Automated GUI testing · Scriptless testing · Evolutionary testing

1 Introduction

TESTAR [1,2] implements a scriptless approach for automated testing at the Graphical User Interface (GUI) level. It is based on agents that implement Action Selection Rules (ASR). The underlying principle is very simple: generate test sequences of (state, action)-pairs by starting up the System Under Test (SUT) in its initial state, and continuously select an action to bring the SUT into another state.

Selecting actions characterizes the fundamental challenge of intelligent systems, i.e. what to do next [15]. Currently, TESTAR offers a few default rules for action selection. One of these rules, i.e. Random, has already shown to be a valuable addition to the testing process in industrial contexts [3,4,12]. If a tester wants to add new more sophisticated rules, these need to be *manually* designed and implemented.

The hypothesis of this research is that we can *automate* the creation of effective new rules by evolving them using an Evolutionary Algorithm (EA). Using EA on software engineering problems and reformulating them as optimisation problems is known as Search-Based Software Engineering (SBSE). According to Harman et al. [8], three things are needed for such a reformulation: a representation of individuals, a fitness function, and a set of manipulation operators.

In this paper, we briefly discuss SBSE for scriptless testing in Sect. 2. Section 3 explains the working process of the TESTAR tool. Section 4 describes the research needed to obtain the three parts to use EA to automate the creation of ASRs, and Sect. 6 presents the expected contributions. Finally, Sect. 6 presents the plan for validation and evaluation.

R. Guizzardi et al. (Eds.): RCIS 2022, LNBIP 446, pp. 779–785, 2022.
https://doi.org/10.1007/978-3-031-05760-1_55

2   Related Work

Using EAs in software testing is not new, an extensive amount of literature exists [9,10]. However, most existing work concentrates on test input data or sequence generation. Evolving complete ASRs for scriptless testing, as we propose in this work, is a new research direction that has some initial results that we will describe in continuation.

The first work that defined a preliminary and simple representation and fitness function is [5]. Subsequently, [6] presents results on using both in TESTAR, showing an increase in test effectiveness. Then a more complex representation is researched in [7] and [14]. After that [13] makes a first attempt to define more complex fitness functions. However, the efficiency and effectiveness of the complex representation and fitness functions have not yet been validated. The research of this paper aims to do that by implementing the automatic creation of effective ASRs through an Evolutionary Algorithm and validate them with a corpus of fresh SUTs.

3   TESTAR

A test run with TESTAR works as follows (see Fig. 1): start the SUT, detect the current state (i.e. all available control elements (widgets)) of the GUI, derive all possible actions for these widgets, select one action using the current ASR, execute that action, and observe the effects. This flow is repeated until a stop criterion is met, e.g. when a previously defined sequence length has been reached or a crash has been detected.

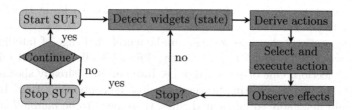

Fig. 1. Test sequence execution flow in TESTAR.

The default ASRs currently in TESTAR are Random (RND), Least Executed Actions (LEA), Prioritize New Actions (PNA) and Unvisited Action First (UAF).

State abstraction (StAb) is an important facet of scriptless GUI testing. TESTAR has an implementation to calculate state identifiers based on hashes over a selected set of widget attributes. This selected set defines the abstraction level.

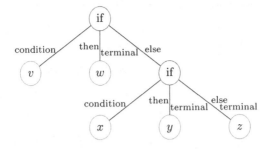

Fig. 2. The structure of an Action Selection Rule tree.

4   Towards Evolutionary Scriptless Testing

The goal of this research is to evolve new ASRs for TESTAR using EA. We will call this Evolutionary Scriptless Testing (EST). As indicated, for this we need to research three components such that we can formulate an optimisation problem and use EA to solve the problem:

1. a representation of the individuals (i.e. the ASRs)
2. a fitness function
3. a set of manipulation operators

Subsequently, to solve the optimisation problem, the EA mimics evolution in nature. It manages a population of individuals (candidate solutions) whereby the most suitable individuals (as judged by the fitness function) survive to the next generation. Every round the manipulation operators diversify the population and replace the individuals that did not make the cut with new individuals. This process is repeated until termination.

In the next subsections, we will describe the research that is needed for each of these three components.

4.1   Representation: ASRs as Individuals

The EA variant used in this research is Genetic Programming (GP), which represents individuals as trees. This gives the ASRs an *if-then-else*-like structure, as is shown in Fig. 2. Every set of children consists of 1 condition (a state predicate over the state of the GUI) and 2 ASRs. The ASRs can be subtrees or terminals. In [14] a first version of a context free grammar is presented for ASRs. Examples of conditions are 'state has not changed' and 'number of actions available' of a specified type. Examples of terminals are 'random unexecuted action' and 'random drag action'. Towards our EST implementation this grammar will be extended with more conditions and terminals.

Listing 1.1 shows a simple ASR example. It encodes the following strategy: if there are any click actions available, pick one of the not executed actions among those click actions. If not, randomly select an action that has been executed the least number of times.

```
IF  available−actions−of−type  click−action
THEN  random−unexecuted−action−of−type  click−action
ELSE  random−least−executed−action
```

Listing 1.1. An example of an ASR.

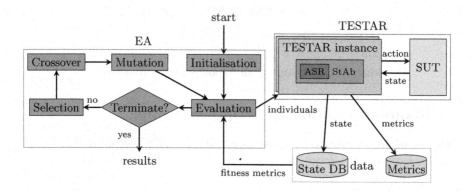

Fig. 3. The proposed evolutionary system

4.2   Fitness Function

A fitness function judges the performance of individuals (i.e. the ASRs) [8]. Only the most suitable individuals are kept to form the basis for the next generation. The fitness should be based on a suitable metric that leads to test effectiveness. In [6,7,14] the used fitness was "the number of different (abstract) states visited during testing". In [13] other fitness functions have been defined that focus on introducing "new actions" more often, achieving high "state model coverage" or "code coverage".

Whether these fitness functions lead to ASRs that do better testing is still something that needs to be investigated on realistic SUTs. Defining the right fitness function is not an easy task; the goal of testing is to find faults, but not finding any is not necessarily a proof that the testing process was adequate. Other challenges will be researched in this work towards ETS. For instance, finding a suitable level of abstraction for states, which is an important issue that can influence the fitness [15]. We need to find an equilibrium between the necessary expressiveness of the states and the computational complexity [11]. Also, we need to research how we can include more metrics to judge fitness, and whether these can be used in a multi-objective search.

Returning to the previous ASR example (Listing 1.1), let us say that this ASR has been loaded into a TESTAR instance and used for a full run. The resulting code coverage is 76% and it reached 54 states out of approximately 80 states total. The used fitness function weights the results of this ASR and awards it a score of 37. All other ASRs have also received a score, with which the

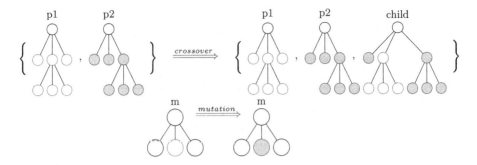

Fig. 4. The manipulation operators *crossover* and *mutation* for tree-structures.

Evolutionary Algorithm can judge which are more 'fit', and thus which ASRs to keep.

4.3   Manipulation Operators: Crossover and Mutation

Manipulation operators (see Fig. 4) control how much the individuals (i.e. ASRs) of the population change with every generation. The two operators *crossover* and *mutation* determine the next generation of individuals.

Crossover. Two ASRs act as 'parents' and reproduce. Their child is randomly assembled from copies of the components of their parents.

Mutation. Randomly apply a change to one node of a chosen individual, in keeping with the constraints. This can change a condition or a terminal to another component of the same type. Alternatively, the tree can gain or lose a set of children at that position.

Typically, one or more applications of *mutation* follow the *crossover* phase.

5   Expected Contributions

In order to be able to execute EST with the resulting three components (individuals, fitness and operators) and validate our hypothesis we will develop an evolutionary system to execute the tests.

The system will have 3 main components: the EA system, TESTAR and a database. Figure 3 provides an high-level overview of the system.

After initialisation, the *EA system* will evaluate every ASR of the newly generated population. This evaluation starts with *TESTAR* that uses the ASR during a complete test run. During and after the run, TESTAR will collect state data and metrics (e.g. percentage of code covered) and save this in a *database*. Subsequently, the ASRs are evaluated using the fitness function and the collected metrics. The results are fed back into the EA system and a new population is built up from the old in the following steps (selection, crossover, mutation),

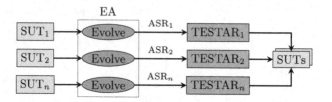

Fig. 5. The setup for validating the evolved ASRs.

whereby the best performing individuals form the basis. Starting from the evaluation step this process repeats. After a set number of generations the EA process terminates.

As mentioned in Sect. 4, our contribution will be a suitable representation definition, an updated fitness function, and our choice of manipulation operators. For this we must define and collect suitable data for building fitness functions and measuring test effectiveness.

6   Plan for Validation and Evaluation

To validate our approach we need to select SUTs and need to define a set of metrics to measure test effectiveness (preferably these should not be exactly the same as those used for the fitness function.) On one hand, we will select (or create) a set of toy programs to use for initial exploratory experiments to establish the groundwork for further research and validation with more complex SUTs. On the other hand, we will select realistic open-source SUTs and closed-source SUTs obtained from our IVVES[1] project partners. These SUTs can be web applications, desktop applications or mobile apps, since TESTAR can handle all of them due to its plug-in architecture [15].

Figure 5 outlines the approach for validating the evolved ASRs. For every SUT in the training corpus, the EA will have evolved an ASR to match. These 'optimised' ASRs are then each loaded into a TESTAR instance and tested on all SUTs from the training corpus. Naturally, every ASR should perform well on the SUT it was evolved on, but the results with other SUTs will indicate if it also works well on similar SUTs.

As part of the evaluation, all ASRs (earlier results, new results, as well as the current ones in TESTAR) will be tested on a validation corpus (a fresh set of SUTs). This should prove whether the ASRs work well on a new SUT.

Acknowledgements. This work has been funded by the ITEA3 IVVES project, with contract number 18022.

---

[1] www.ivves.eu.

References

1. Testar on github. https://github.com/TESTARtool/TESTAR_dev
2. Testar project download page. https://testar.org/download/
3. Bauersfeld, S., Vos, T.E., Condori-Fernández, N., Bagnato, A., Brosse, E.: Evaluating the testar tool in an industrial case study. In: Proceedings of the 8th ACM/IEEE International Symposium on Empirical Software Engineering and Measurement, pp. 1–9 (2014)
4. Chahim, H., Duran, M., Vos, T.E.J., Aho, P., Condori Fernandez, N.: Scriptless testing at the GUI level in an industrial setting. In: Dalpiaz, F., Zdravkovic, J., Loucopoulos, P. (eds.) RCIS 2020. LNBIP, vol. 385, pp. 267–284. Springer, Cham (2020). https://doi.org/10.1007/978-3-030-50316-1_16
5. Esparcia-Alcázar, A.I., Almenar, F., Rueda, U., Vos, T.E.J.: Evolving rules for action selection in automated testing via genetic programming - a first approach. In: Squillero, G., Sim, K. (eds.) EvoApplications 2017. LNCS, vol. 10200, pp. 82–95. Springer, Cham (2017). https://doi.org/10.1007/978-3-319-55792-2_6
6. Esparcia-Alcázar, A.I., Almenar, F., Vos, T.E.J., Rueda, U.: Using genetic programming to evolve action selection rules in traversal-based automated software testing: results obtained with the TESTAR tool. Memetic Comput. **10**(3), 257–265 (2018). https://doi.org/10.1007/s12293-018-0263-8
7. de Groot, M.: Using evolutionary computing to improve black box monkey testing on a Graphical User Interface. Master's thesis, Open University of the Netherlands, Heerlen, Netherlands (2018)
8. Harman, M., Jones, B.F.: Search-based software engineering. Inf. Softw. Technol. **43**, 833–839 (2001). https://doi.org/10.1016/S0950-5849(01)00189-6
9. Khari, M., Kumar, P.: An extensive evaluation of search-based software testing: a review. Soft Comput. **23**(6), 1933–1946 (2019). https://doi.org/10.1007/s00500-017-2906-y
10. McMinn, P.: Search-based software testing: past, present and future. In: 2011 IEEE Fourth International Conference on Software Testing, Verification and Validation Workshops, pp. 153–163 (2011). https://doi.org/10.1109/ICSTW.2011.100
11. Meinke, K., Walkinshaw, N.: Model-based testing and model inference. In: Margaria, T., Steffen, B. (eds.) ISoLA 2012. LNCS, vol. 7609, pp. 440–443. Springer, Heidelberg (2012). https://doi.org/10.1007/978-3-642-34026-0_32
12. Ricós, F.P., Aho, P., Vos, T., Boigues, I.T., Blasco, E.C., Martínez, H.M.: Deploying TESTAR to enable remote testing in an industrial CI pipeline: a case-based evaluation. In: Margaria, T., Steffen, B. (eds.) ISoLA 2020. LNCS, vol. 12476, pp. 543–557. Springer, Cham (2020). https://doi.org/10.1007/978-3-030-61362-4_31
13. Stoyanov, N.: Strategy based genetic algorithms approach in automated GUI testing. Master's thesis, Eindhoven University of Technology, Eindhoven, Netherlands (2020)
14. Theuws, G.: Random action selection vs genetic programming: a case study in TESTAR. Master's thesis, Open University of the Netherlands, Heerlen, Netherlands (2020)
15. Vos, T.E.J., Aho, P., Ricos, F.P., Rodriguez-Valdes, O., Mulders, A.: Testar - scriptless testing through graphical user interface. Softw. Test. Verification Reliabil. **31**(3), e1771 (2021). https://doi.org/10.1002/stvr.1771, https://onlinelibrary.wiley.com/doi/abs/10.1002/stvr.1771

Scriptless Testing for Extended Reality Systems

Fernando Pastor Ricós[✉][iD]

Universitat Politècnica de València, València, Spain
fpastor@pros.upv.es

Abstract. Extended Reality (XR) systems are complex applications that have emerged in a wide variety of domains, such as computer games and medical practice. Testing XR software is mainly done manually by human testers, which implies a high cost in terms of time and money. Current automated testing approaches for XR systems consist of rudimentary capture and replay of scripts. However, this approach only works for simple test scenarios. Moreover, it is well-known that the scripts break easily each time the XR system is changed. There are research projects aimed at using autonomous agents that will follow scripted instructions to test XR functionalities. Nonetheless, using only scripted testing techniques, it is difficult and expensive to tackle the challenges of testing XR systems. This thesis is focus on the use of automated scriptless testing for XR systems. This way we help to reduce part of the manual testing effort and complement the scripted techniques.

Keywords: Scriptless testing · State model inference · Extended reality

1 Introduction

XR systems have been on the rise in recent years to allow users to interact with simulated environments. XR software systems have emerged in different domains, ranging from medicine, for marketing purposes, to computer games [19]. The latter sector accounts for about 50% of the Virtual Reality (VR) software market. The complexity of the navigation and interaction in 3D spaces, the increasing importance of user experience, the existence of randomness and non-determinism in VR games, together with the agile development requirements of the market [9], makes XR testing a challenging and critical task [10].

Nowadays, XR testing practice is mainly done manually. This implies high costs in terms of time and money. The sector lacks automation processes, frameworks, and tools [13]. Even though automated software testing has been significantly researched to reduce the problems of manual testing (time, cost, etc.), for instance, as shown for testing through the Graphical User Interface (GUI) [16], the existing testing tools do not consider the complexity of testing XR systems.

R. Guizzardi et al. (Eds.): RCIS 2022, LNBIP 446, pp. 786–794, 2022.
https://doi.org/10.1007/978-3-031-05760-1_56

Automated scripted testing is based on scripts that are manually crafted, generated from models, or recorded with a tool to replay them later [15]. These scripts contain the interactions that the automated tool must execute to validate some functional requirements. However, the current automated testing approaches for XR systems are often based on the rudimentary capture and replay of scripts, which only works for simple test scenarios. Furthermore, the scripts break easily when the XR system is changed.

In order to tackle the problems of applying automated scripted testing to XR systems, the Intelligent Verification/Validation for Extended Reality Based Systems (IV4XR, 2019–2022) project is developing a framework to allow the observation of XR entities and the use of autonomous Functional Test Agents (FTAs) to achieve testing goals [14]. These FTAs follow scripted tactics and imply the need to manually create and maintain scripts.

Manually defining and maintaining scripts to cover all possible functional System Under Test (SUT) dependencies would mean investing a lot of money and time. Therefore, using only scripted techniques, it is difficult and expensive to tackle the challenges of testing XR systems. To reduce the effort related to test scripts, additional and complementary testing approaches are required.

Scriptless testing is an outstanding approach intended to automatically interact and validate the software without the use of scripts. These tools automatically explore and generate test sequences by selecting and executing the available actions in the discovered states. The use of scriptless testing tools has proven to be complementary to scripted tools for covering different parts of the SUT and for detecting unexpected software failures [11]. Moreover, scriptless testing can be beneficial to check offline oracles [7], to visualize model transitions and changes between different versions of the same software [2], to automatically generate test cases from the model [1], or to automatically apply regression testing to detect changes [8]. Despite all these benefits, there is a lack of research to apply the scriptless approach for XR environments.

The goal of this research is to evaluate the benefits of using scriptless testing for two VR software systems. To do that, we select the scriptless open source tool TESTAR [22]. TESTAR is an actively maintained tool that tests desktop, web and mobile applications through the GUI. In this research, TESTAR must be evolved to be an intelligent exploratory agent that tests XR environments. Thus, we advocate that scriptless testing can help developers and testers at XR companies by reducing manual effort and complementing scripted techniques.

The rest of this paper is structured as follows. Section 2 presents the related work. Section 3 introduces the main characteristics of TESTAR. Section 4 exposes research challenges for XR systems. Section 5 presents the completion plan for this thesis research, including the validation plan. Section 6 concludes.

2 Related Work

Testing XR systems is difficult. The development of XR systems differs from traditional systems due to less clear requirements, high level of interactivity and increased realism, entities behavior dependent on the context, among others; making testing a challenging work [17]. Testing XR systems is mainly performed manually, because it is needed to deal with interfaces in 3D spaces, human behavior is difficult to reproduce, and entities are highly dependent on the context.

Little research has been performed on automated testing of XR systems. In [12], the difference in the evolution of the requirements between traditional desktop, web or mobile software, and XR video game open source projects is presented. This study shows that, as a XR video game project evolves, the development effort is reduced from code functionality and focuses more on the multimedia features. It also shows that testing objectives of XR systems evolve differently than those of traditional software, and that malfunctions of XR systems are more related to the interface than the programming errors.

In [4], a script based approach for automated functional testing of XR systems is proposed. It offers a semi-formal language to describe the requirements specification and then automates the generation of test cases using scene graph concepts to represent the virtual environment. Although this proposal can be a useful scripted approach, the effort to maintain the scripts is problematic.

The iv4XR project uses autonomous agents that follow declarative tactics to test XR systems [14]. Artificial Intelligent (AI) techniques are applied to have robustness tactic-based tests and reduce the maintenance effort [18]. MBT is used to generate tactic-based test cases automatically [5]. However, manual effort is still required if the XR testing needs to cover a high amount of system paths, either to create tactic tests or to craft a model of an existing level.

The scriptless approach and the use of AI agents has not been thoroughly investigated in terms of test effectiveness, efficiency and usefulness. Consequently, the motivation of this work is to research the benefits of using the scriptless tool TESTAR as an exploratory agent within the iv4XR project.

3   TESTAR Tool

TESTAR is a scriptless tool that, while testing, infers a state model that is stored in an OrientDB graph database. TESTAR connects and interacts with the SUT using an API to obtain the current state and execute actions. Windows Automation and Java access bridge can be used for desktop applications, Selenium WebDriver for web pages, and Appium for mobile applications. To be able to connect and interact with XR systems, we use the open source iv4XR Java software plugin that allows TESTAR to interact with two iv4XR use cases: LabRecruits[1] and Space Engineers (SE)[2]. LabRecruits is a VR open source demo game and SE is an industrial VR game with millions of players.

The logical flow of TESTAR together with the integration of the iv4XR plugin is shown in Fig. 1. The World Object Model (WOM) interface allows TESTAR to observe the information of the virtual entities through a set of properties such as the entity type, the position and the size. Thus, the XR state

---

[1] https://github.com/iv4xr-project/labrecruits.
[2] https://github.com/iv4xr-project/iv4xr-se-plugin.

consists on the set of entities that exist in the observation range. Because the type of entities and their properties are different depending on the virtual SUT, TESTAR needs to define the XR state differently for LabRecruits and SE. For example, SE entities contain properties that indicate the actual integrity and 3D position of blocks, which is something that does not exist for LabRecruits game.

XR actions are the available interactions that TESTAR can execute in each XR state. We derive two types of XR actions: basic commands and compound tactics. A basic command action is the most basic event that TESTAR can execute, e.g. move or rotate one step, equip a tool and start or stop using a tool. However, due to the essence of XR systems, most of the time it is necessary to execute a compound tactical action that contains several basic commands. For example, to interact with a SE block entity TESTAR needs to rotate to aim the block, move to reach the block, equip a tool and start using this tool.

Based on the available actions in a state s, TESTAR selects and executes an action a and obtains a new state s'. The transition (s, a, s') is then stored into the state model and TESTAR continues with state s'. It generates test sequences until the STOP condition is met. Thus, the state model is a knowledge graph that contains information about the elements in the states, as well as the executed and non-executed actions.

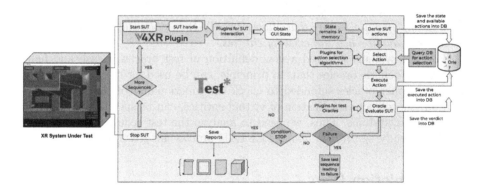

Fig. 1. TESTAR operational flow

During the integration of the iv4XR plugin, TESTAR was able to discover two failures. In LabRecruits, we detected a hang exception when interacting with a non-interactive entity. In SE, we discovered a plugin exception when TESTAR took off his helmet and died while exploring SE system. For this reason, we expect to use possible failures found by TESTAR as a measure of test effectiveness.

4   XR Research Challenges

To evaluate scriptless testing for XR systems, we work to solve four challenges that are described in this section.

When TESTAR interacts with desktop, web or mobile SUTs, most of the GUI actions are based on mouse movements in the 2D screen followed by click, drag or type events. However, for XR systems there is greater complexity when executing actions. An XR action may require move and orientate TESTAR in 3D environments and deal with obstructive objects to reach the entities. Therefore we have to research: *1. How to develop an action derivation mechanism to perceive the environment and realize smart interaction movements.*

TESTAR maintains two types of states in its model: concrete and abstract states. Concrete states are created using all the properties of all the entities. Abstract states creation can be customized by selecting which properties are used for state abstraction. A suitable level of abstraction allows TESTAR to select non-dynamic properties to create a traceable model avoiding a state explosion. However, for XR systems, some properties, such as the position of the agent, are important but too concrete to determine the XR state. Thus, we need to analyze: *2. How to define a suitable approach to abstract the area of the TESTAR agent.*

For GUI applications, when TESTAR obtains the GUI state, the tool detects the available widgets to interact with. However, for XR systems, the observation about the reachable entities is restricted to the observation range and obstructive objects. A basic movement can modify the observed entities and discover that new entities can be reached by navigating to certain positions. Since TESTAR does not contain such a complex navigation feature, we need to investigate: *3. How to implement a feature that allows TESTAR to navigate the XR space and remember how to reach the existing entities.*

For XR systems we need a new definition of test oracles. Crash and hang failures are generic oracles that, in principle, can be used for most XR systems. However, other oracles intended to check the functionality of the XR interactions or the correct visualization of virtual entities, can be specific and different between XR systems. Thus, we need to analyze the test requirements for each SUT and study: *4. How to adapt TESTAR oracles for XR systems.*

5   Completion Plan

This thesis aims to improve four functionalities of TESTAR to tackle the challenges presented above. It is the first time presented in a doctoral consortium.

5.1   Smart Interaction Movements

TESTAR needs to be able to observe the environment, select an entity to interact with, and perceive which virtual objects obstruct the movement to the entity. For LabRecruits, TESTAR can use tactical actions, together with the automatic NavMesh [21] map created by Unity to follow the NavMesh positions to reach the desired entity. However, in other XR systems such as SE VR game, this NavMesh map does not exist by default. We are working on creating a 2D as well as 3D pathfinding (with jetpack) algorithm for SE. This constructs a sparse grid on-the-fly as the agent moves around and avoids obstacles.

5.2   XR State Abstraction

Current abstraction mechanisms are not feasible for some XR systems. In SE, TESTAR movements change the observation range and, therefore, the SE blocks of the XR state. This implies the constant creation of new abstract states in the model. A possible solution for SE is to use the location area of TESTAR together with the SE grid entities. One grid is a group of blocks that generally represents a structure such as a spatial base. The exploration movements along these grids will modify the observed blocks but not the grid structure.

5.3   Navigable State Model Layer

In XR systems, especially in virtual games, the user has the possibility to move around the virtual world or a navigable area, in order to reach the interactive entities. To determine which are the reachable entities, TESTAR needs to explore the navigable areas and store the position of the discovered entities.

A navigable state has been implemented in the TESTAR state model. Figure 2 shows an example of this feature with LabRecruits. TESTAR prioritizes the exploratory movements to discover the reachable entities of the navigable state. After fully exploring the navigable state, TESTAR selects an interactive action not executed previously. Then, TESTAR starts a new exploration to be able to map which interactive actions of the SUT connects the existing navigable states.

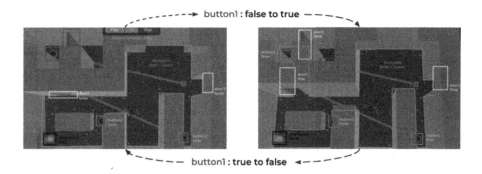

Fig. 2. TESTAR navigable state transition

5.4   XR Test Oracles

Although there are generic oracles that can be applied to most systems to detect crashes, hangs or exception messages, the definition of what a failure is and the oracles to detect it, depends on each system. In the development of XR systems, a lot of effort is dedicated to the visualization of virtual objects. Thus, the oracles aiming to verify the correct visualization of these objects are of great interest in the development processes of virtual systems.

IV4XR project has a trained model that helps to verify the correct visualization of SE blocks according to their integrity values. TESTAR is able to detect the type and the integrity of the blocks using the IV4XR plugin and take screenshots of the SE system using the Windows Accessibility API. Thus, we aim to research the integration of the model in TESTAR to be used as a visual oracle while TESTAR explores the virtual environment.

5.5 Validation Plan

LabRecruits and SE partners are interested in conducting this research and publishing the validation results. We will evaluate TESTAR effectiveness, efficiency, and the usefulness of the test results. Let us look into each of these separately.

– RQ1: How does TESTAR improve test effectiveness of XR systems?

On non-XR systems, TESTAR test effectiveness can be measured by the usual code coverage, GUI coverage and failures found. Although these are still desired (two interesting system failures were already found during the TESTAR integration), the usual metrics are not enough. This is because, for XR systems, test effectiveness also depends on game-specific properties like the entities, the interactions, the reality, the high level game objectives, the levels and the players.

Therefore, to answer RQ1, first we need to discuss with the product owners to determine what they consider to be effective for their particular system. Then, we need to execute a series of experiments to obtain and evaluate these metrics.

For LabRecruits, it is considered effective to test the transition coverage of the buttons and doors at different levels. Also, because this system is developed with Unity, it is possible to obtain code coverage metrics [20]. For SE, it is considered effective to test the integrity and the visualization of the blocks when using a tool to interact with it, or the correct use of materials to build blocks.

In this first PhD year, we expect to execute a series of experiments with LabRecruits to measure how effective is TESTAR in terms of transition and code coverage. Since LabRecruits can have different types of levels, and because TESTAR always contains a certain level of randomness, we need to make sure we obtain significant evidence. For this we will realize an empirical evaluation and use for example the Bayesian statistical analysis [6]. Furthermore, for SE, we need to research and determine which metrics to use to measure effectiveness.

For the second PhD year, we plan to conduct an industrial experiment with SE. We expect to use TESTAR to explore SE and interact with different blocks to validate properties such as integrity, and research the use of an image recognition model as an oracle to validate blocks visualization. Then, realize an empirical evaluation by using the effectiveness metrics defined during this PhD year.

– RQ2: How useful are the test results of TESTAR?

TESTAR creates logs, HTML reports with screenshots, and the state model, as test result artifacts. Logs and HTML reports are useful to visualize how a test sequence was executed, especially if a failure has been found. The state model is an artifact that can be useful for more purposes, as was stated in Sect. 1.

One of the research topics of the IV4XR project is the use of a MBT tool to generate button-interaction test cases. To use the MBT tool with an existing LabRecruits level, it is necessary to manually craft the finite interaction model before being able to generate test cases [5]. TESTAR navigable state model contains the necessary information that the MBT tool needs to automatically generate these test cases. For this reason, we expect to analyze the useful complementarity of TESTAR and the MBT tool before the end of this PhD year.

– RQ3: How efficient is the TESTAR exploration process on XR systems?

Besides investigating effectiveness and usefulness of the TESTAR results, we still need to research the third important property of testing: efficiency [3].

Because TESTAR does not test a specific set of actions but explores the existing amount of available actions, it needs more time than scripted approaches to reach significant effectiveness. For the third PhD year, we expect to research the integration of a distributed framework that allows the execution of multiple TESTAR instances that use the state model as a central knowledge database.

6 Conclusion

In this doctoral consortium paper we have presented the challenges for testing XR systems, the research performed so far, the goal of our research, and the completion activities and validation plan that will be carried on during the PhD.

Acknowledgement. This work has been funded by iv4XR (H2020, 856716) project.

References

1. Aho, P., Alégroth, E., Oliveira, R., Vos, T.: Evolution of automated regression testing of software systems through the graphical user interface. In: 1st ACCS, pp. 16–21 (2016)
2. Aho, P., Suarez, M., Kanstrén, T., Memon, A.M.: Murphy tools: utilizing extracted GUI models for industrial software testing. In: ICSTW, pp. 343–348 (2014)
3. Böhme, M., Paul, S.: A probabilistic analysis of the efficiency of automated software testing. IEEE Trans. Softw. Eng. **42**(4), 345–360 (2015)
4. Correa Souza, A.C., Nunes, F.L., Delamaro, M.E.: An automated functional testing approach for virtual reality applications. STVR **28**(8), e1690 (2018)
5. Ferdous, R., Kifetew, F., Prandi, D., Prasetya, I.S.W.B., Shirzadehhajimahmood, S., Susi, A.: Search-based automated play testing of computer games: a model-based approach. In: O'Reilly, U., Devroey, X. (eds.) SSBSE 2021. LNCS, vol. 12914, pp. 56–71. Springer, Cham (2021). https://doi.org/10.1007/978-3-030-88106-1_5
6. Furia, C., Feldt, R., Torkar, R.: Bayesian data analysis in empirical software engineering research. IEEE TSE **47**(9), 1786–1810 (2021)
7. de Gier, F., Kager, D., de Gouw, S., Vos, T.E.J.: Offline oracles for accessibility evaluation with the Testar tool. In: 13th RCIS, pp. 1–12. IEEE (2019)
8. Grilo, A.M., Paiva, A.C., Faria, J.P.: Reverse engineering of GUI models for testing. In: 5th Iberian Conference on Information Systems and Technology, pp. 1–6. IEEE (2010)

9. Kropp, M., Meier, A., Anslow, C., Biddle, R.: Satisfaction, practices, and influences in agile software development. In: 22nd EASE, pp. 112–121 (2018)

10. Lin, D., Bezemer, C.P., Hassan, A.E.: Studying the urgent updates of popular games on the steam platform. Emp. Softw. Eng. **22**(4), 2095–2126 (2017)

11. Machiry, A., Tahiliani, R., Naik, M.: Dynodroid: an input generation system for android apps. In: 9th FSE, pp. 224–234 (2013)

12. Pascarella, L., Palomba, F., Di Penta, M., Bacchelli, A.: How is video game development different from software development in open source? In: 2018 IEEE/ACM 15th MSR, pp. 392–402 (2018)

13. Politowski, C., Petrillo, F., Guéhéneuc, Y.G.: A survey of video game testing. arXiv preprint arXiv:2103.06431 (2021)

14. Prasetya, I., Dastani, M., Prada, R., Vos, T.E.J., Dignum, F., Kifetew, F.: Aplib: tactical agents for testing computer games. In: 8th EMAS, pp. 21–41 (2020)

15. Rafi, D.M., Moses, K.R.K., Petersen, K., Mäntylä, M.V.: Benefits and limitations of automated software testing: systematic literature review and practitioner survey. In: 7th International Workshop on Automation of Software Test (AST), pp. 36–42. IEEE (2012)

16. Rodríguez-Valdés, O., Vos, T.E.J., Aho, P., Marín, B.: 30 Years of automated GUI testing: a bibliometric analysis. In: Paiva, A.C.R., Cavalli, A.R., Ventura Martins, P., Pérez-Castillo, R. (eds.) QUATIC 2021. CCIS, vol. 1439, pp. 473–488. Springer, Cham (2021). https://doi.org/10.1007/978-3-030-85347-1_34

17. Santos, R., Magalhães, C., Capretz, L., Correia-Neto, J., da Silva, F., Saher, A.: Computer games are serious business and so is their quality: particularities of software testing in game development from the perspective of practitioners. In: 12th ESEM, pp. 1–10 (2018)

18. Shirzadehhajimahmood, S., Prasetya, I., Dignum, F., Dastani, M., Keller, G.: Using an agent-based approach for robust automated testing of computer games. In: 12th A-TEST, pp. 1–8 (2021)

19. Stanney, K., Nye, H., Haddad, S., Hale, K., Padron, C., Cohn, J.: Extended reality (xr) environments. Handbook of Human Factors and Ergonomics, pp. 782–815 (2021)

20. Unity. Code coverage (2019). https://docs.unity3d.com/Packages/com.unity.testtools.codecoverage@1.0. Accessed 18 Feb 2022

21. Unity. Navigation system in unity (2021). https://docs.unity3d.com/Manual/nav-NavigationSystem.html. Accessed 18 Feb 2022

22. Vos, T., Aho, P., Pastor Ricos, F., Rodriguez-Valdes, O., Mulders, A.: Testar-Scriptless testing through graphical user interface. STVR **31**(3), e1771 (2021)

Artificial Intelligence: Impacts of Explainability on Value Creation and Decision Making

Taoufik El Oualidi[⊠] [iD]

University of Paris-Saclay, University of Evry, IMT-BS, LITEM, 91025 Evry, France
`taoufik.el-oualidi@universite-paris-saclay.fr`

Abstract. Over the last few years, companies' investment in new AI systems has seen a strong and constant progression. However, except for the Big Tech, the use of AI is still marginal at this stage, and seems to spark cautiousness and apprehension. A potential reason for this hesitation may be linked to a lack of trust related in particular to the so-called black box AI technologies such as deep learning. This is why our research objective is to explore the effects of explainability on trust in these new AI-based digital systems with which the users can either interact or directly accept its results in case of fully autonomous system. More precisely, in the perspective of an industrialized use of AI, we would like to study the role of explainability for stakeholders in the decision-making process as well as in value creation.

Keywords: Artificial Intelligence · Explainability · Trust · Value creation · Decision making · Machine learning

1 Introduction

Idealized or handled with apprehension [1, 2], companies see AI as a key investment for its value creation potential [3]. This explains why between 2015 and 2020, all sectors combined, investment in AI has seen increased by more than 500% reaching thereby above 67 billion dollars [4]. And nothing to date suggests a downward trend in the coming years. Estimated to be worth more than $300 billion by 2021, the global AI market is growing rather exponentially [5]. Despite this craze, the large-scale use of AI-based systems is still a minority in most companies [6]. This reveals that the use of AI is still in its infancy and is becoming a strategic priority.

Nonetheless, beyond the technological challenge and the quest for cost reduction or quality improvement, AI brings forth complex and unprecedented modalities of human-machine interaction [7, 8]. Undoubtedly more than for any other digital system, the trust that is required at the heart of this interaction is a crucial factor for the successful large-scale integration of AI [9]. As a true pillar of trust, explainability manifests itself as a promising tool in the construction of an AI-based organizational and digital device.

Our research work, started this year, is taking place in an IT department of the French company La Poste Group where we will be involved in the management of AI during the PhD research endeavor. By placing the concept of explainable AI at the heart of

R. Guizzardi et al. (Eds.): RCIS 2022, LNBIP 446, pp. 795–802, 2022.
https://doi.org/10.1007/978-3-031-05760-1_57

this study, we are interested in its impact on the decision-making and value creation processes. This study is set to contribute to a better understanding of the role that AI explainability can play as well as ways in which it can be taken into consideration.

2   The Problem

By promoting the development of decision support or prediction algorithms, AI systems give access to a complexity that the user did not integrate until now in his work. However, what appears to be an important technical advance is being challenged by the opacity of recent AI techniques known as Machine Learning [10–12]. Indeed, the functioning of these AI-based systems such as neural networks or Support Vector Machines (SVM) are generally opaque and perplexing to both the data scientist and the end user. The said opacity constitutes a major obstacle to the efficient use of AI systems [6, 9]. Technical quality does not always guarantee usage or adoption, and human factors play a vital role in this matter [13, 14].

Among the factors that drive adoption and value creation, trust is of utmost importance [9, 15]. In fact, Trust allows the user to act with confidence based on recommendations from an automated and artificially intelligent decision support tool, even in situations of uncertainty [16, 17]. That being said, one must also acknowledge the fact that over confidence, i.e., trust without limit, can have harmful effects.

To foster a trusting relationship with new AI systems, explainability appears to be a promising approach. Graphical representation is not always the most suitable medium to facilitate explainability, and some forms may contribute more than others to accelerate decision-making [18, 19].

While research in the area of Explainable AI (XAI) is growing [12, 20, 21], research on its adoption and impact remains limited [9]. To date, no informed consensus has emerged on this topic. Moreover, some researchers not only defend the idea that the explainability requirement cannot be systematic, but they also question its usefulness [22].

Thus, making the most opaque AI techniques more explainable is a challenge for companies that want to gain the trust of their users to stay successful. In particular, we focus on the issues related to selecting the most useful manner to convey the explanation, and its potential impact and adoption by end-users at the individual and organizational levels.

3   State of the Art

This questioning about the explainability of AI-based systems is not new and dates back to the early 1970s [23]. If progress has been made to make AI models and their results more understandable, there remain some shortcomings. These shortcomings are accentuated today with the advent of Big Data [24] and ever more powerful machine learning techniques. Thus, deep learning algorithms resemble black boxes, and once again trigger the need for explainability. Faced with performance, legal, and ethical requirements, companies see the explainability of AI as a way to acquire the confidence necessary to grow their business. We are then witnessing a significant revival of interest in explainable AI that materializes in the academic literature starting in 2015 [21] and at the end of 2017 with DARPA's Explainable AI (XAI) project [25]. Explainable AI thus sets an ambitious goal to make all algorithms as powerful as they are explainable.

In parallel to this, AI-based solution providers are starting to offer integrated functionalities that offer explanations for each result computed by an AI [26]. We are also witnessing a data scientist community that is very active on the subject and that is developing algorithms to try to meet this need for explainability [27]. Although companies are at a stage where they have not yet acquired sufficient maturity on the use of AI, they are well aware that explainable AI is a valuable asset.

To conclude, at this stage of our research, we have not yet found any research works that addresses explicitly the impact and the adoption of AI explainability in a business context similar to ours.

4 Research Objectives and Methodology

This research effort aims to explore the organizational and economic effects of new AI-based digital systems in companies. The aim is to study more specifically the role of explainability for stakeholders in the decision-making process and in the creation of value. On the basis of empirical data, the goal is to gain a better understanding of ways in which integration of explainability in the construction of a system (organizational and technical) can ensure better performance. More precisely, the topic of this thesis is articulated around four main questions:

1. How does the need for explainability materialize and position itself in the course of an AI-based system implementation project?
2. Does the incorporation of mechanisms for interpreting and explaining the results provided by AI-based organizational and digital devices contribute to an increase in user confidence, and ultimately, to the adoption of these devices and to value creation?
3. What forms and types of explanation are more likely to foster user confidence in their decision-making process?
4. Are there moderating factors that interfere during the explainability process (e.g., age, gender, level of maturity in digital use, hierarchical position, etc.)?

To carry out this study, we chose a hybrid exploration that uses an abductive approach in order to take into account empirical observations and theoretical knowledge [28–30]. The intervention in the field, within the framework of AI-based systems implementation projects, allows us to study the process of taking into account the explainability in its actual context [31]. With regard to this intervention context, which is favorable to the interaction between research and practice, our ambition is to conduct controlled experiments combined with a Design Science approach [32–34].

5   Theoretical Background

We rely on three theoretical fields. First, management information system (MIS) theories on the impact of IS, such as Delone and McLean's theory [10, 11], and how information quality in AI-based systems can be impacted by explainability.

Secondly, theories in organization sciences [35, 36] allow us to address the complexity of decisional [37–39] and structural [40] dimensions. In particular, studies show that AI has the capacity to influence decisions in companies [41]. This leads us to assume that it contributes to the co-construction of meaning by interfering in the course of vertical and horizontal interactions in the company.

Thirdly, the theoretical field of Knowledge Management on organizational learning [42–45], and explainability as a contributor to the justification of the resulting information. With the help of new learning technologies, such as deep learning, we assume that AI plays a role in the production, dissemination and updating of knowledge.

6   Expected Tangible Results

For a supply chain topic, we are currently developing a proof of concept (POC) with AI-based technology whose objective is twofold: to be a valuable tool both for decision support and for material transportation expenses reducing. Currently, in order to ensure the postal mail processing and distribution, material transportation management is daily performed by La Poste teams according to a regional division of the territory. Based on a daily transportation plan consisting of approximately 20,000 routes, each team is responsible for identifying and choosing the best transportation methods (type of truck, day, schedule, etc.) to meet the supply needs of materials. This supply process is characterized by a strong expertise of the teams who have developed routines allowing them to provide answers within three days. Identifying the best way to meet requests is handled by each team after matching material requirements with routes that have sufficient carrying capacity. The AI algorithm is designed to predict the carrying capacities on all the routes planned for D, D + 1 and D + 2. With this prediction, AI contributes to the supply process by enabling this automated matching task every morning that the teams will be in charge for validating. The confidence of the teams is essential for this validation step. The explainable AI is a real challenge to obtain this trust and generate gains.

The outcomes expected by this POC will offer a better understanding of how the company is impacted by the use of AI. In a decision-making point of view, AI capabilities interact with teams and their managers. The economic evaluation should indicate if these decisions have a positive effect or not. So much information to analyze to feed our research work. The two following decision-making process indicates the existing one without AI (see Fig. 1) and this with AI as it is planned through the POC (see Fig. 2). Here we have the opportunity to further analyze the impact of AI on decision-making and value creation.

At this stage, the chosen way consists first in producing an analytical framework in order to identify if a typology of explainability exists according to specific AI algorithms. Furthermore, we assume that depending on the nature and the level of project requirements (performance, ethics, legal and regulatory…), the chosen AI algorithm will affect the explainability in a specific way. Such a framework could help us, for example on specifying the expected formalism (visual representation, text…) and to evaluate how it contributes to an increase in user confidence.

In terms of risk, we realize that La Poste Group is a large historical French company evolving in a complex environment. Then, we are mindful of the risks inherent in such an environment that could interfere with our work (access to sensitive data, organizational or strategic changes…). Therefore, we will make a particular effort to better control the threats to validity throughout our research.

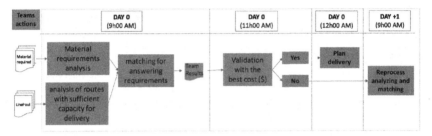

Fig. 1. Existing decision-making process without AI used

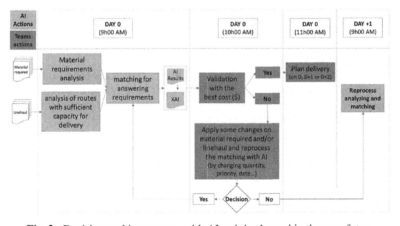

Fig. 2. Decision-making process with AI as it is planned in the near future

7   Conclusion

This study aims to contribute to a better understanding of the role and impact of the explainability of AI-based systems in decision-making and value creation processes in companies. On the basis of initial evidence from field experience as well as from scientific and professional literature, this study shows that the need for explainability is likely to increase with regard to expected investments in the implementation of AI-based digital systems.

If explainability can play an important role in the confidence in these systems, we have underlined the need to make this concept operative. It is in this sense that avenues of reflection are proposed in order to identify different types of explainability with a view to their articulation with another concept, that of interpretability. It would then seem that the construction of new organizational and digital devices is to be imagined. The creativity of organizations will be an essential asset to meet the challenge.

References

1. Ferguson, Y.: 1. Ce que l'intelligence artificielle fait de l'homme au travail. Visite sociologique d'une entreprise. In: Les mutations du travail, pp. 23–42. La Découverte (2019)
2. Mateu, J.-B., Pluchart, J.-J.: L'économie de l'intelligence artificielle. Rev. D'econ. Financ. **135**, 257–272 (2019)
3. Jöhnk, J., Weißert, M., Wyrtki, K.: Ready or not, AI comes—an interview study of organizational AI readiness factors. Bus. Inf. Syst. Eng. **63**, 5–20 (2021)
4. Zhang, D., et al.: The AI index 2021 annual report. arXiv preprint arXiv:2103.06312 (2021)
5. Le marché de l'IA et son périmètre. SAY. **5**, 144–145 (2021)
6. Wang, R.: How AI Changes the Rules: New Imperatives for the Intelligent Organization. https://sloanreview.mit.edu/offer-sas-how-ai-changes-the-rules-2020/
7. Haenlein, M., Kaplan, A.: A brief history of artificial intelligence: on the past, present, and future of artificial intelligence. Calif. Manage. Rev. **61**, 5–14 (2019)
8. Schroder, A., Constantiou, I., Tuunainen, V., Austin, R.D.: Human-AI Collaboration – Coordinating Automation and Augmentation Tasks in a Digital Service Company (2022)
9. Glikson, E., Woolley, A.W.: Human trust in artificial intelligence: review of empirical research. Acad. Manag. Ann. **14**, 627–660 (2020)
10. Chakraborti, T., Kulkarni, A., Sreedharan, S., Smith, D.E., Kambhampati, S.: Explicability? legibility? predictability? transparency? privacy? security? the emerging landscape of interpretable agent behavior. In: Proceedings of the International Conference on Automated Planning and Scheduling, pp. 86–96 (2019)
11. Champin, P.-A., Fuchs, B., Guin, N., Mille, A.: Explicabilité: vers des dispositifs numériques interagissant en intelligence avec l'utilisateur. In: Atelier Humains et IA, Travailler en Intelligence à EGC (2020)
12. Jouis, G., Mouchère, H., Picarougne, F., Ardouin, A.: Tour d'horizon autour de l'explicabilité des modèles profonds. In: Rencontres des Jeunes Chercheur· ses en Intelligence Artificielle (RJCIA 2020) (2020)
13. Petter, S., DeLone, W., McLean, E.R.: Information systems success: the quest for the independent variables. J. Manag. Inf. Syst. **29**, 7–62 (2013)
14. Dwivedi, Y.K., et al.: Research on information systems failures and successes: status update and future directions. Inf. Syst. Front. **17**, 143–157 (2014)
15. Höddinghaus, M., Sondern, D., Hertel, G.: The automation of leadership functions: would people trust decision algorithms? Comput. Hum. Behav. **116**, 106635 (2021)
16. Madsen, M., Gregor, S.: Measuring human-computer trust. In: 11th Australasian Conference on Information Systems, pp. 6–8. Citeseer (2000)
17. Hoff, K.A., Bashir, M.: Trust in automation: integrating empirical evidence on factors that influence trust. Hum. Factors **57**, 407–434 (2015)

18. Gedikli, F., Jannach, D., Ge, M.: How should I explain? A comparison of different explanation types for recommender systems. Int. J. Hum. Comput. Stud. **72**, 367–382 (2014)
19. Longo, L., Goebel, R., Lecue, F., Kieseberg, P., Holzinger, A.: Explainable artificial intelligence: concepts, applications, research challenges and visions. In: Holzinger, A., Peter Kieseberg, A., Tjoa, M., Weippl, E. (eds.) CD-MAKE 2020. LNCS, vol. 12279, pp. 1–16. Springer, Cham (2020). https://doi.org/10.1007/978-3-030-57321-8_1
20. Gunning, D., Stefik, M., Choi, J., Miller, T., Stumpf, S., Yang, G.-Z.: XAI—explainable artificial intelligence. Sci. Robot. **4** (2019)
21. Adadi, A., Berrada, M.: Peeking inside the black-box: a survey on explainable artificial intelligence (XAI). IEEE Access **6**, 52138–52160 (2018)
22. Robbins, S.: A misdirected principle with a catch: explicability for AI. Mind. Mach. **29**, 495–514 (2019)
23. Preece, A.: Asking 'Why' in AI: explainability of intelligent systems–perspectives and challenges. Intell. Syst. Account. Financ. Manage. **25**, 63–72 (2018)
24. Duan, Y., Edwards, J.S., Dwivedi, Y.K.: Artificial intelligence for decision making in the era of Big Data – evolution, challenges and research agenda. Int. J. Inf. Manage. **48**, 63–71 (2019). https://doi.org/10.1016/j.ijinfomgt.2019.01.021
25. Gunning, D.: Explainable artificial intelligence (xai). Defense Advanced Research Projects Agency (DARPA), nd Web. 2 (2017)
26. Explainability with Dataiku. https://www.dataiku.com/product/key-capabilities/explainability/. Accessed 16 Feb 2022
27. Bhatnagar, P.: Explainable AI(XAI) - A guide to 7 packages in Python to explain your models. https://towardsdatascience.com/explainable-ai-xai-a-guide-to-7-packages-in-python-to-explain-your-models-932967f0634b
28. Koenig, G.: Production de la connaissance et constitution des pratiques organisationnelles. Rev. Gest. Ressour. Hum. 4–17 (1993)
29. Thiétart, R.-A.: Méthodes de Recherche en Management, 4th edn. Dunod (2014)
30. Dumez, H.: Méthodologie de la Recherche Qualitative: Les Questions Clés de la Démarche Compréhensive. Vuibert (2016)
31. Argyris, C.: Action science and organizational learning. J. Manager. Psychol. **10**(6), 20–26 (1995)
32. Zelkowitz, M.V., Wallace, D.R.: Experimental models for validating technology. Computer **31**, 23–31 (1998)
33. Hevner, A., Chatterjee, S.: Design science research in information systems. In: Hevner, A., Chatterjee, S. (eds.) Design Research in Information Systems, pp. 9–22. Springer, Boston (2010). https://doi.org/10.1007/978-1-4419-5653-8_2
34. Pascal, A.: Le design science dans le domaine des systèmes d'information: mise en débat et perspectives. Syst. D'inf. Manag. **17**, 7–31 (2012)
35. Rojot, J.: Théorie des Organisations. Editions ESKA, Paris (2016)
36. Perrow, C.: Complex organizations; a critical essay (1972)
37. March, J.G., Simon, H.A.: Organizations (1958)
38. Weick, K.E.: Enacted sensemaking in crisis situations [1]. J. Manage. Stud. **25**, 305–317 (1988)
39. Weick, K.E.: Sensemaking in Organizations. Sage (1995)
40. Mintzberg, H.: Structure et Dynamique des Organisations. Ed. d'Organisation, Paris (1982)
41. Benbya, H., Pachidi, S., Jarvenpaa, S.: Special issue editorial: artificial intelligence in organizations: implications for information systems research. J. Assoc. Inf. Syst. **22**, 10 (2021)
42. Nonaka, I.: A dynamic theory of organizational knowledge creation. Organ. Sci. **5**, 14–37 (1994)
43. Nonaka, I., Von Krogh, G., Voelpel, S.: Organizational knowledge creation theory: evolutionary paths and future advances. Organ. Stud. **27**, 1179–1208 (2006)

44. Argyris, C., Schön, D.A.: Organizational learning: a theory of action perspective. Reis **77/78**, 345–348 (1997)
45. Deken, F., Carlile, P.R., Berends, H., Lauche, K.: Generating novelty through interdependent routines: a process model of routine work. Organ. Sci. **27**, 659–677 (2016)

Towards Empirically Validated Process Modelling Education Using a BPMN Formalism

Ilia Maslov[(⊠)] [iD]

LIRIS Department, KU Leuven, 1000 Brussels, Belgium
`ilia.maslov@kuleuven.be`

Abstract. "A picture is worth more than a thousand words" may be said of process models, using the words of business and IT leaders. Business Process Model and Notation (BPMN) is a de facto standard used for business process modelling that helps flexible and responsive understanding, analysis and communication of business processes and inter-organisational collaborations. Despite the importance, there are significant gaps in providing empirically justified and systematic pedagogy in teaching process modelling. The research seeks to cover the gap through the systematic literature review of the broader area of conceptual modelling, analysis of novice modellers' common errors and patterns, design of learning goals and course design of process modelling and validating the result using experiments.

Keywords: Business process modelling · Process modelling education · BPMN · Course design · Formative assessment · e-learning

1 Context and Motivation

Conceptual modelling (thereon, CM for short) is the process and discipline of describing aspects of the physical and social world via simplifying it through abstraction and conceptualisation for understanding, communication and managing complexity (Mylopoulos 1992; Buchmann et al. 2019). CM is broad and is used in areas such as software engineering and business process management, and there are particular idiosyncratic differences (Guarino et al. 2020). For example, there are many modelling languages used for different purposes in enterprise modelling – an activity used to capture, represent and capitalise basic facts and knowledge about an enterprise(s) how it's structured, organised and operated. It's contrary to, say, IS modelling, where the purpose is to model an IS, its functionalities, implementation and so on. BPMN (Business Process Model and Notation) is a de facto industry standard for process modelling (thereon, PM for short; Vernadat 2002; Linden and Proper 2014). PM is used to support strategic and operational tasks in organisations, bridge the communication gaps between business and IT and facilitate the engineering of information systems. Hence, it is essential to deliver effective teaching of PM to business students (Dumas et al. 2018). Proficient PM requires the translation of verbal descriptions and non-explicit requirements into formal and visual diagrams of conceptual models and coping with ambiguity, versatility, open-ended problems, making it a challenging activity for novices and experts alike (Claes

R. Guizzardi et al. (Eds.): RCIS 2022, LNBIP 446, pp. 803–810, 2022.
https://doi.org/10.1007/978-3-031-05760-1_58

et al. 2017). To this task, usually, the pedagogy is not systematised, primarily based on the teacher's personal experience, while also imbalanced in terms of different levels of remembering, understanding and applying the knowledge, with the eminent gaps in scaffolding and overall evaluation of student's knowledge, leading to a wide diversity of pedagogy methods (Bogdanova and Snoeck 2017 and 2018a). Moreover, even though the quality of process models may impact the outcome of business processes and high demand is placed on the high-quality process models, the general quality of such models is low in many cases. Assisting in training modellers may help improve process modelling skills and subsequent process models (Claes et al. 2017). Furthermore, despite the extant quality frameworks and studies about process model quality, which also focus more on the syntactical rather than semantic quality metrics, there are currently no concrete frameworks applicable to effective teaching of process modelling (de Oca et al. 2015; Claes et al. 2017; De Meyer and Claes 2018). Out of 87 studies analysed by Avila et al. (2020), only a third provide empirical evidence for process modelling guidelines, warranting consistent definitions, empirical evidence and feedback about guidelines to modellers. Given all that, researchers in CM and PM education call for better, more systematic and effective approaches to deliver such education. It is also suggested that CM education may not always corroborate between fields of application, e.g., process modelling and IS modelling (Buchmann et al. 2019; Rosenthal et al. 2019). This research project will seek to systematically analyse findings from the CM education and apply them in grounding research on PM education, as the latter is conceived partially as a type of the former in the literature. The project will also seek to establish process quality frameworks, which are specific to teaching PM. Optimal learning objectives and course design qualities shall be designed using the framework, which are then to be validated through the use of experiments and assignments, hence aiming towards promoting scientifically rigorous theory within PM education discipline. Consequently, the sections will introduce state of the art, research objectives and planned research methodology to attain the objectives.

2   State of the Art

In this section, I shall briefly review the concepts relevant to the research project and state of the art literature discussions. Figure 1 below summarises the research framework of concepts and relationships relevant to the project. It is based on the preliminary results from a systematic literature review and helps frame the research, even though it is still work-in-progress and may not be considered as very precise, with further potential to rethink the proposed conceptualization and abstraction of phenomena. Due to the space limitations, the framework description will be briefly sketched using the evidence examples from the literature, primarily via the explanation of *Relationships* (***R***) between *Concepts* (***C***), which are numerically referenced.

There are multiple individual factors (***C1***) that have an impact on the learning outcomes and constitute an important theme of learner differences (***R10***) (Lim and Morris 2009). For example, the learner's working memory's two functions (holding and processing of information and ability to build new relations between elements) were found to influence the process and hence, the model quality (***R1, R6***) (Martini et al. 2016).

Intelligence is strongly correlated to the ability of individuals to learn and apply new information and the structuredness and self-directedness of learning (*R2*) (Gottfredson 1997). The cognitive styles of people correlate with the way process models are constructed (*R1*) (Figl and Recker 2016). Personality and intrinsic students' motivation to learn are very reliable and highly important in determining the learning outcome and how a student learns (*R2, R10*) (Zhou 2015). Academic discipline may affect how students view the process of CM and understand the course content through their discipline lens (*R1, R2*) (Buchmann et al. 2019). Educators are suggested to target individual learners' motivation by suggesting learning application or how useful it would be to the learners in the future. It is also recommended to match students' knowledge level, learning needs, cognitive style to the learning difficulty and variation in delivering instructions, hence promoting (at least partially) individually personalized course design (*R2, R5*) (Lim and Morris 2009; Cope and Kalantzis 2016).

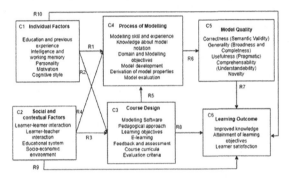

Fig. 1. Proposed (work-in-progress) research framework.

In addition, there are relevant social and other contextual factors (*C2*). Even though individual factors better predict motivation to study and learning outcomes, social factors were found more frequent. Social factors can be conceptualised as social punishment avoidance, social expectations by parents and teachers, material rewards and social privileges. Hence, both the classroom social environment and broader socio-economic environment significantly impact learning outcomes and are suggested to keep in mind when viewing the educational system at large (*R3, R9*) (Marić and Sakač 2014). Per se, collaborative CM in higher education may facilitate learning outcomes, mediate the interaction between individual and social factors, and promote transformation and reflection between various forms of knowledge (*R4, R6*) (Kosonen et al. 2010). In collaborative PM within corporate settings, modellers may adapt their modelling style, with stronger modellers being able to create both concrete and abstract models and focus more on the comprehensibility of the models. In contrast, weaker modellers can create only concrete models (*R4, R6*) (Wilmont 2020).

Recent research indicates the potential to adapt course design (*C3*) to deliver more effective training of PM skills and achieve better learning outcomes (*C6*). Buchmann et al. (2019) viewed the course design of teaching CM as a design problem and using workshops while bearing in mind students' individual factors (*R3, R4*). Claes et al.

(2017) have developed a method that adapts to the student's cognitive profile and trains with the most fitting approach to model (*R2, R5, R8*). Bogdanova and Snoeck (2018b) trained students' CM through a Massive Open Online Course's formative assessment using a learning ontology that connects learning items, learning objectives and errors, and such feedback is well-received by the students (*R3, R5, R8*). Sanchez-Ferreres et al. (2020) developed an automated software, which allowed to make BPMN models and automatically provide immediate feedback about quality issues in the model (*R5, R8*). Software systems overall play an important part both in training and in future work related to process modelling (*R5, R8*) (Dumas et al. 2018), as part of an e-learning trend, which is using ICT to deliver information in training and education, often in the blended learning context (*R5, R6*) (e.g., Lim and Morris 2009).

There are several frameworks to evaluate conceptual models' qualities and guidelines to follow during process modelling to improve the quality (*C4* and *C5*). Process model quality frameworks (*C4*) may include 3QM (Lindland et al. 1994), Guidelines of Modelling (Uthmann and Becker 1999), a taxonomy of quality components (Thalheim 2010). The qualities of conceptual models may be used to evaluate the errors made during the teaching and learning process by the novice modellers (*R6, R7*) (Bogdanova and Snoeck 2018). There are also PM guidelines used to improve the process of PM and novice modellers' skills (*R5, R6*), e.g., 7PMG is a set of seven process modelling heuristic guidelines (Mendling et al. 2010). There are currently no directly applicable process model quality frameworks and guidelines in the context of teaching novice modellers' because the existing ones are too abstract to apply in the educational context (*R7, R8*) (de Oca et al. 2015; De Meyer and Claes 2018; Avila et al. 2020).

3   Research Objectives and Methodology

The PhD research will be conducted for four years. In this work, the ultimate goal is to create empirically validated learning goals, course design and modelling proficiency assessment methods to provide effective and systematic PM education, thus extending the theory of PM education. Since BPMN is an industry-standard and used by companies (in and outside the IT sector), teaching BPMN and quantifying modelling quality and modellers' proficiency is thus also relevant for training programs within organisations. Hence, the research project leans towards a pragmatic research philosophy and research design methodology (e.g., Buchmann et al. 2019). The research stages within a research design methodology may have to be iterative, following the four research objectives as roughly the four stages: (1) to explore the problem, (2) to analyse the problem, (3) to design the solution, (4) to validate the solution. The research will promote reproducible research via research data management plans making collected data available online on free repositories and following other ethical research guidelines (Arnold et al. 2019). The research project will seek to achieve the research objectives (ROs) via the following approaches and methodologies:

RO1. *To explore the state of the art in the literature of teaching methods and student behaviour in the field of conceptual and process modelling.*

RO1 is to give research context to the PhD project, and will mainly be achieved by answering the **RQ**: "*How can the research findings from CM education help to extend the* *research of process modelling education?"* Given the early literature review findings, there are some significant gaps in connecting research theories of CM education to PM education, which may differ concerning educational context, students' background, intended use, etc. Covering the research gap can help promote the understanding of PM education based on the findings in CM education. On this basis, the first paper is currently in progress, aiming to answer RQ1. The review is based on the bibliometric methods (e.g., see Zupic and Čater 2015) of analysing bibliographical data of 2000–2021 years with the help of statistical methods, which is an uptrending technique facilitated by the availability of powerful software and hardware. Bibliometrics promote research replicability and facilitate easy analysis of the broader field of research through literature analysis. If combined with qualitative content analysis, it further promotes in-depth analysis of emerging themes and research topics. At later stages, the research will continue to explore literature at need.

RO2. To analyse students' novice modeller's behaviour of process modelling, using a dataset of BPMN models, focusing on typical types of errors (patterns).

RO2 does raise an important **RQ**: *"Can we find error patterns and types that typically reflect novice modeller's behavior and expose novices's miscomprehension of PM in particular?"* Analysis of modeller's behaviour will primarily deal with produced process models and qualities (*C5*). The research project will deal with a previously collected extensive dataset of students' BPMN models. The dataset contains over 1000 BPMN models already. Data is collected from three campuses, three different teachers, and three different programs across two universities in Belgium. Each academic year new data can be added. Attaining RO2 will involve assessing the models via qualitative manual analysis and potentially quantitative software-facilitated analysis. In such a way, an empirically evidenced taxonomy of common novice modellers' errors could be constructed, like in Bogdanova and Snoeck (2018a). However, feedback and review from the course teachers will be required. Such a qualitative approach to systematisation of empirically-based taxonomies and categories is an inductive approach, which allows building a theory of common errors in the learning of PM ground-up, which can later be used for course design and be tested in experiments. The approach also follows the ideas of the past research (e.g., 3QM) but aims to extend existing taxonomies with more semantic metrics and guidelines. The analysis could be combined with the research of process patterns (Fellmann et al. 2018), but in the context of education. The analysis may be done also at the modelling process level (*C4*). For example, using eye-tracking data from the way students interact with the process models could give insights into cognitive styles of process modelling (Tallon et al. 2019). This can help move towards a broader theory of PM education, with early attempts at hypothesizing different concepts and relationships shown in Fig. 1.

RO3. To design and validate pedagogical approaches and theories (process pattern learning, a taxonomy of errors, formative testing and (semi-)automated feedback) for PM, through experiments, surveys, interviews and (formative) test results.

The final RO raises an important, pragmatically valuable **RQ**: *"What scientifically rigorous pedagogical approaches are most effective to training process modelling, and among novice modellers in particular?"* Designing systematic learning goals and course content (*C3* and *C6*) of PM education may be accomplished using design methodology (Buchmann et al. 2019). Within the research scope of this project, there already exist some pedagogical solutions, suggesting some course design (e.g., see Bogdanova and Snoeck 2018a). Past literature has already tried to analyse these solutions based on the classical learning philosophies: behaviourism, constructivism, cognitivism (e.g., see Rosenthal 2019). This research project will seek to extend the existing literature, developing PM-specific learning goals and course design. Notably, Cope and Kalantzis (2016) have suggested moving the pedagogy from the traditional teacher-led, didactic pedagogy towards more reflexive pedagogy, which is an adapted pedagogy with effective use of digital media, not just the traditional pedagogy done over the Internet. They suggested using seven affordances provided by the new e-learning technologies, including recursive feedback throughout the learning, multimodal meaning of knowledge, active knowledge-making by the students, collaborative learning, and personalised learning. This theoretical framework may provide invaluable guidelines to frame the design methodology. Furthermore, designing an adapted Bloom's learning framework, like Bogdanova and Snoeck (2019a), can help arrive at appropriate learning goals and outcomes in education. For example, previously established error taxonomies and patterns in the educational context can further improve PM pedagogy (Bogdanova and Snoeck 2018a). Only recently, researchers started to systematise patterns of processes (Fellmann et al. 2018), which is a potential to research the yet underexplored field of pattern-based learning of process models. Designing and utilising formative assessment IT systems can help with personalized teaching of CM in different areas, and in particular in PM, even though it is still an underexplored topic of research (Claes et al. 2017; Bogdanova and Snoeck 2018b; Bogdanova and Snoeck 2019b; Sanchez-Ferreres et al. 2020). The potential is to replicate and extend the experiment in Bogdanova and Snoeck (2018b), as they have constructed a MOOC and suggested improving the learning outcomes via formative assessment in the course design methodology. IT and software are thus an essential theme in the research project as a critical component. ICT is the main mode through which conceptual models of all types are generally created in academia and the workplace and ICT facilitate learning under an emerging e-learning paradigm.

Conducting experiments in the learning context is one of the most popular approaches to validate learning goals and course design hypotheses. It is empirically evidenced, and there is a lack of empirical studies in the research topic area (Avila et al. 2020). The primary goal of the RO3 is to validate the hypotheses concerning the implementation of the improved pedagogical approach (*C3, C4* and *C6*). There are multiple ways to validate the hypotheses through the experiments. Experiments could be based around specific cognitive tasks involving certain experiment conditions with learning outcomes as outcome variables while controlling for independent variables (Martini et al. 2016). For example, Claes et al. (2017) suggested analysing the two student groups with and without new method implementation to teach process modelling before and after administration, showing potential for their method. RO4 will seek to validate the designed course instructions. Another approach to validation is to scrutinise learning outcomes and monitor their evolution. Equally, student's learning progress can be analysed through formative assessment and continuous feedback.

4   Conclusions and Future Work

The topic of process modelling education is a complex phenomenon, drawing on knowledge from different disciplines. The research project has a pragmatic orientation, with the ultimate goal of moving towards an empirically validated effective PM education pedagogical framework. To this objective, a research design methodology is a practical approach. Within this methodology, there will be an emphasis on getting empirical data and validating the hypotheses using experiments. The project has been focused on RO1 with the ongoing SLR insofar. The required data is available for RO2, and for RO3, there are valuable frameworks to be used to design appropriated learning goals and contents, such as ideas suggested by Cope and Kalantzis (2016) or Bogdanova and Snoeck (2019a). RO2 and RO3 will be achieved in parallel and iteratively. Several potential approaches have been discussed, such as an experiment to analyse the impact of real-time process model simulation on learning outcomes.

References

Arnold, B., et al.: The turning way: a handbook for reproducible data science. Zenodo (2019)

Avila, D.T., dos Santos, R.I., Mendling, J., Thom, L.H.: A systematic literature review of process modeling guidelines and their empirical support. BPM J. (2020)

Bogdanova, D., Snoeck, M.: Learning from errors: error-based exercises in domain modelling pedagogy. In: IFIP Working Conference on the Practice of Enterprise Modeling, pp. 321–334. Springer, Cham (2018). https://doi.org/10.1007/978-3-030-02302-7_20

Bogdanova, D., Snoeck, M.: Using MOOC technology and formative assessment in a conceptual modelling course: an experience report. In: Proceedings of the 21st ACM/IEEE International Conference on Model Driven Engineering Languages and Systems: Companion Proceedings, pp. 67–73 (2018). https://doi.org/10.1007/978-3-030-02302-7_20

Bogdanova, D., Snoeck, M.: CaMeLOT: an educational framework for conceptual data modelling. Inf. Softw. Technol. **110**, 92–107 (2019)

Bogdanova, D., Snoeck, M.: Use of personalised feedback reports in a blended conceptual modelling course. In: 2019 ACM/IEEE 22nd International Conference on Model Driven Engineering Languages and Systems Companion, pp. 672–679. IEEE (2019)

Buchmann, R.A., Ghiran, A.M., Döller, V., Karagiannis, D.: Conceptual modeling education as a "design problem." Complex Syst. Inf. Model. Q. **21**, 21–33 (2019)

Claes, J., Vanderfeesten, I., Gailly, F., Grefen, P., Poels, G.: The structured process modeling method (SPMM) what is the best way for me to construct a process model? Decis. Support Syst. **100**, 57–76 (2017)

Cope, B., Kalantzis, M.: E-learning Ecologies. Routledge, Nova Iorque (2016)

De Meyer, P., Claes, J.: An overview of process model quality literature-the comprehensive process model quality framework. arXiv preprint arXiv:1808.07930 (2018)

de Oca, I.M.M., Snoeck, M., Reijers, H.A., Rodríguez-Morffi, A.: A systematic literature review of studies on business process modeling quality. Inf. Softw. Technol. **58**, 187–205 (2015)

Dumas, M., La Rosa, M., Mendling, J., Reijers, H.A.: Introduction to business process management. In: Fundamentals of BPM, pp. 1–33. Springer, Heidelberg (2018). https://doi.org/10.1007/978-3-642-33143-5_1

Fellmann, M., Koschmider, A., Laue, R., Schoknecht, A., Vetter, A.: Business process model patterns: state-of-the-art, research classification and taxonomy. BPM J. (2018)

Figl, K., Recker, J.: Exploring cognitive style and task-specific preferences for process representations. Requirements Eng. **21**(1), 63–85 (2016). https://doi.org/10.1007/s00766-014-0210-2

Gottfredson, L.S.: Why g matters: the complexity of everyday life. Intelligence **24**(1), 79–132 (1997)

Guarino, N., Guizzardi, G., Mylopoulos, J.: On the philosophical foundations of conceptual models. Inf. Modell. Knowl. Bases **31**(321), 1 (2020)

Kosonen, K., Ilomäki, L., Lakkala, M.: Collaborative conceptual mapping in teaching qualitative methods. Blended learning in Finland, pp. 138–153 (2010)

Lim, D.H., Morris, M.L.: Learner and instructional factors influencing learning outcomes within a blended learning environment. J. Educ. Technol. Soc. **12**(4), 282–293 (2009)

Linden, D.V.D., Proper, H.A.: On the accommodation of conceptual distinctions in conceptual modeling languages. Modellierung 2014 (2014)

Lindland, O.I., Sindre, G., Solvberg, A.: Understanding quality in conceptual modeling. IEEE Softw. **11**(2), 42–49 (1994)

Marić, M., Sakač, M.: Individual and social factors related to students' academic achievement and motivation for learning. Suvremena psihologija **17**(1), 63–79 (2014)

Martini, M., Pinggera, J., Neurauter, M., Sachse, P., Furtner, M.R., Weber, B.: The impact of working memory and the "process of process modelling" on model quality: Investigating experienced versus inexperienced modellers. Sci. Rep. **6**(1), 1–12 (2016)

Mylopoulos, J.: Conceptual modelling and Telos. Conceptual modelling, databases, and CASE: an integrated view of information system development, pp. 49–68 (1992)

Rosenthal, K., Ternes, B., Strecker, S.: Learning Conceptual Modeling: structuring overview, research themes and paths for future research. In: ECIS (2019)

Sanchez-Ferreres, J., et al.: Supporting the process of learning and teaching process models. IEEE Trans. Learn. Technol. **13**(3), 552–566 (2020)

Tallon, M., et al.: Comprehension of business process models: Insight into cognitive strategies via eye tracking. Expert Syst. Appl. **136**, 145–158 (2019)

Thalheim, B.: Towards a theory of conceptual modelling. J. Univers. Comput. Sci. **16**(20), 3102–3137 (2010)

Uthmann, C.V., Becker, J.: Guidelines of modelling (GoM) for business process simulation. In: Scholz-Reiter, B., Stahlmann, HD., Nethe, A. (eds.) Process Modelling, pp. 100–116. Springer, Heidelberg (1999). https://doi.org/10.1007/978-3-642-60120-0_7

Vernadat, F.B.: Enterprise modelling and integration. In: Kosanke, K., Jochem, R., Nell, J.G., Bas, A.O. (eds.) Enterprise Inter- and Intra-Organizational Integration. ITIFIP, vol. 108, pp. 25–33. Springer, Boston, MA (2003). https://doi.org/10.1007/978-0-387-35621-1_4

Wilmont, I.: Cognitive aspects of conceptual modelling: exploring abstract reasoning and executive control in modelling practice (Doctoral dissertation, Sl: sn) (2020)

Zhou, M.: Moderating effect of self-determination in the relationship between big five personality and academic performance. Personal. Individual Diff. **86** (2015)

Zupic, I., Čater, T.: Bibliometric methods in management and organisation. Organisat. Res. Methods **18**(3), 429–472 (2015)

Multi-task Learning for Automatic Event-Centric Temporal Knowledge Graph Construction

Timotej Knez[✉]

University of Ljubljana, Ljubljana, Slovenia
timotej.knez@fri.uni-lj.si

Abstract. An important aspect of understanding written language is recognising and understanding events described in a document. Each event is usually associated with a specific time or time period when it occurred. Humans naturally understand the time of each event based on our common sense and the relations between the events, expressed in the documents. In our work we will explore and implement a system for automated extraction of temporal relations between the events in a document as well as of additional attributes like date, time, duration etc. for placing the events in time. Our system will use the extracted information to build a graph representing the events seen in a document. We will also combine the temporal knowledge over multiple documents to build a global knowledge base that will serve as a collection of common sense about the temporal aspect of common events, allowing the system to use the gathered knowledge about the events to derive information not explicitly expressed in the document.

Keywords: NLP · Knowledge graphs · Temporal relations · Information retrieval

1 Introduction

Computer understanding of written language is a crucial part of various applications. It enables the use of vast resources available in the unstructured text form, as well as a more natural interaction between humans and computers. In order for computers to understand natural language, they have to extract parts of the document that are of interest and often also recognise the relations that hold between those parts. One of the common ways for achieving this is entity and relation extraction. The goal of this process is to recognize entities described in a document and also recognise the relations between them. The relations extracted in this way are then usually represented in the form of a knowledge graph. This is a graph where entities are represented by vertices and the relations between them are represented by edges. When trying to understand the content of a document, an important aspect is the time and order of the events that are described. The process of recognising temporal relations that hold between events described in a document

R. Guizzardi et al. (Eds.): RCIS 2022, LNBIP 446, pp. 811–818, 2022.
https://doi.org/10.1007/978-3-031-05760-1_59

is often referred to as temporal relation extraction. The goal is to find the events in a text document and recognise the temporal relations that hold between them. For example, we want to recognise the sequence of events. In addition to the temporal relations between the events, we would also like to extract additional attributes about the time of the event. For example, the time when the event occurred and its duration. These attributes can help us to gain a better understanding of how the events transpired. Our goal is to (a) create and evaluate a neural network architecture for recognising relations between the events in a document and their temporal features. The model will take advantage of a pretrained transformer-based language model. Apart from the text document, the model will also use an external source of common sense knowledge. This should help with the events where the temporal relations or attributes are not expressed explicitly. (b) The general temporal knowledge about the observed events will get stored in a global time-based knowledge base. This knowledge will be used as an additional source of common sense for extracting information about the following events. The entities stored in the global knowledge base will be also linked to an existing knowledge base, for example to Wikidata. This will make the knowledge base more useful to our future projects and to other researchers. (c) The final component of our system will combine the events and information about them, extracted from a document, and present that in the form of a single time-centric knowledge graph. By representing the relations between events in a graph we will enable inference of document level relations and the relations that are not expressed directly. This component will also visualise the final graph showing the events and entities in a document.

2   Proposed End-to-End Architecture

The main goal of our research is to create a system for the automatic construction of knowledge graphs focused on events and their temporal attributes. The temporal attributes like time, date and duration will help place the events at specific points in time. The system will consist of multiple parts shown in Fig. 1. In the first part, the document will get preprocessed and the events and entities will get extracted. The preprocessing will be done using existing tools, while the entity and event extraction will likely need some modifications from the existing models. One option for the implementation of the event and entity extraction model is to use the work done by Ro et al. [14], as they have built a model for open relation extraction, which solves a similar problem. Another promising method for entity extraction is genre [3]. This method also performs entity linking, which would be useful for later stages of the pipeline.

The second part of the system (number 1 in Fig. 1) will be responsible for identifying relations between the events and the event temporal attributes. To solve this task we will design a new neural model architecture. The model will use the tokens, recognised events, and entities from the previous part. In addition to that, the model will take advantage of additional common-sense knowledge like (a) statistics about the common relations between the events [12], and (b) the existing information about the events captured in a knowledge graph like the

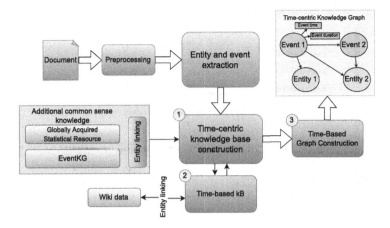

Fig. 1. Proposed system architecture. The parts shown in blue will be based on already existing architectures, while the parts shown in red will be designed from scratch and the parts in gray are external resources.

one presented in [6]. In order to take advantage of these external resources, the extracted entities will need to be linked to the ones present in existing knowledge graphs. We plan on implementing the extraction of multiple temporal features using a multi-task learning approach where a part of the network will be shared between the models for recognising each attribute. The relations and attributes extracted in the second part will also be used to build and update a global knowledge base (number 2 in Fig. 1) containing information about the common events. The information captured in this global knowledge base will be also used as a source of common sense for the following relation and attribute predictions. By using entity linking it will also be possible to reference information captured in an external knowledge base, like for example in Wikidata. In the third part (number 3 in Fig. 1) the recognised relations and attributes are combined into a single knowledge graph and visualised.

We present an example of the data after each step of the pipeline in Fig. 2. The system starts with a sentence containing two events and three entities. These get grouped into two groups containing an event and its corresponding entities and one group containing the two events. After that, the system extracts event attributes and relation between the events. Finally, the extracted information gets combined into a time-based knowledge graph.

3   Research Questions

Research Question 1: Which temporal attributes are most relevant for language understanding?

The goal of our research is to design a system that can automatically extract temporal attributes about events described in a document. By extracting more attributes we are increasing the amount of information that is captured in the

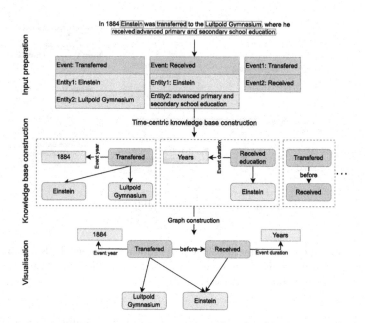

Fig. 2. An example of the data after each processing step in our system.

final graph; however, we are also increasing the complexity of the task, which will likely result in lower accuracy. Because of that, it is important to focus on the attributes that will be most helpful for understanding the document. The attributes that we will be recognising will be based on the ones described in the time ontology in owl [8].

Research Question 2: Which additional knowledge is useful for recognising temporal attributes of the events?

One of the advantages of our system for building a time-centric knowledge base is that it takes advantage of external sources of common sense in order to improve the quality of the extraction process. In our work, we will try using multiple common sense resources and determine which provide the most benefit.

Research Question 3: To what extent can the temporal attributes be extracted?

The main component of the system will be performing recognition of multiple event attributes. In order for the final knowledge base to be useful, extracted attributes have to be reliable enough. We will explore the quality of the attributes that can be extracted when using unstructured documents as a source for our pipeline.

4   Related Work

The idea of building a knowledge graph containing events and their temporal features has already been somewhat explored. One example of an event-based temporal knowledge graph was proposed by Gottschalk and Demidova [6]. They built a knowledge graph containing temporal information about historic events. The knowledge graph was built by aggregating information from other already existing knowledge graphs and semi-structured documents. This is a major limitation since most of the information available is presented in the form of unstructured documents. In order to take advantage of the information contained in unstructured text documents, we can use temporal relation extraction. In the area of temporal relation extraction, a number of approaches have already been described and tested. Traditionally the researchers used a number of hand-engineered features to predict the temporal relations between different events. This approaches were the ones performing best in the i2b2 challenge [16] and the clinical TempEval tasks [1,2]. In recent years the use of neural networks has gained popularity as neural models provided better results than hand-engineered features. One of the early approaches using neural networks for temporal relation extraction has been proposed by Dligach et al. [5]. They compared the use of LSTM and convolutional neural network architectures for extracting temporal relations between medical events. In their tests, the convolutional network outperformed the LSTM network. They also propose the use of specialized tokens for marking the events in a sentence. Their work focused on data from the medical domain. The use of neural networks for extracting temporal relations between events has been further explored by Ning et al. [11]. They use an LSTM network to extract temporal relations. They also use statistical features about the common relations between two verbs. The use of static resource to aid temporal relation extraction has been presented in their earlier work [12]. Since some verb pairs might not be present in the precomputed list of frequencies, they introduce the idea of encoding the triplets containing the two verbs and a relation between them. The encoding is done in a way that allows them to predict probabilities even for unseen triplets. That way they are able to improve over the previous state of the art performance on the MATRES dataset. On the other hand, Vashishtha et al. [17] aim to predict more precise temporal relations between the events. They use the neural networks for fine-grained predictions of temporal relations, where the aim is not only to detect coarse relations like which event happened before the other, but also how long the events were, how much time has passed between them and so on. Using their methods they are able to generate timelines of events described in a pair of sentences. The timelines capture the event sequence as well as the relative duration of each event. They also released their dataset to the public.

In recent years the use of large transformer-based language models [18] has become very popular in the area of relation extraction. One of the most well-known transformer-based models called Bidirectional Encoder Representations from Transformers (BERT) was proposed by Devlin et al. [4]. The use of the BERT model improved the results of many natural language tasks including the task of relation extraction. The use of BERT for relation extraction was explored in the work by Ro et al. [14]. They use a multi-head attention network on top of a BERT model to label the parts of the sentence corresponding to the entities and their relations. This provides a way to achieve open information extraction. In contrast with using a transformer model to label a sentence, recent approaches for relation extraction seem to also focus on generative architectures that use an approach similar to machine translation in order to translate a sentence into a set of relation triplets. This approach was described by Han et al. [7] and by Zhang et al. [19] Josifoski et al. [9] present a way to use a generative entity and relation extraction in a way that limits the extracted relations to a set of relations that we are interested in. This approach allows building highly structured knowledge graphs that are more useful for most computer applications.

Since transformer based model architectures have shown large improvements for the tasks of open and closed relation extraction, we can expect that they would also achieve performance gain for the task of temporal relation extraction. This idea has been researched by Lin et al. [10] as they used a neural network architecture using a BERT pretrained language model for extracting the contains relation in the medical domain. They use their model on the THYME corpus [15]. Zhang et al. [20] also develop a temporal relation extraction model based on BERT, however, their model is not limited to a single domain as it is able to predict the before, after, during and is_included relations from the MATRES corpus [13]. In our work, we would like to take this idea even further by not only extracting the temporal relations but also multiple temporal attributes like time, date, duration and similar, as well as recognising entities that participate in the event. The extracted information will get combined into a single temporal knowledge graph that will contain information about the entire document.

5   Data Construction

In our work, we plan on developing a neural network architecture capable of extracting temporal relations between events. For training the model, we will need enough labelled data. We will likely use a combination of existing datasets. The first promising dataset is the fine-grained relation extraction dataset published by Vashishtha et al. [17]. The dataset contains 91.000 event pairs that have been manually placed on relative timelines. Another dataset commonly used for temporal relation extraction is the MATRES dataset [13]. This dataset contains documents with annotated events and time expressions as well as temporal relations between the events. The dataset only features the simple temporal relations like *before, after* and *simultaneous*. Since our model is designed to solve multiple tasks, we can use data from multiple similar datasets by simply training each task on a separate dataset. We believe the existing datasets contain enough data for training our model.

6   Scientific Contributions

By researching the proposed topic we will make multiple scientific contributions. The main contributions of our work are the following:

- An algorithm for temporal attribute and relation extraction using external semantic resources (common sense knowledge and temporal knowledge base).
- A new architecture for end-to-end temporal data extraction.
- Temporal centric knowledge base along with visualisation techniques and entity links to existing knowledge bases.

Additionally, there are also some smaller contributions, like the use of existing architectures, designed for relation extraction on the task of event extraction. Another smaller contribution is the exploration of external resources that can be used to improve temporal information extraction.

7   Conclusion

In our work, we will design a system capable of automatically building a temporal-centric knowledge graph from a document. The main part of the research will be focused on the extraction of temporal attributes and relations about the events. The process will take advantage of the external resources providing common sense as well as a global knowledge base that will get automatically updated over time. The global temporal knowledge base that we will create will be also available to other researchers to use in their projects or to enrich the existing knowledge bases with temporal information. The code developed as a part of our research will be published in a public repository, allowing anyone to use our work in their future research. We also plan on publishing all models that we will train to make our work more easily reproducible.

The automatic construction of event-centric temporal knowledge graphs enables a variety of applications. For example, it could be used directly as a technique for summarising a document by providing a series of events and their corresponding entities. For instance, we could summarize a cooking recipe to form a list of steps. Constructing such knowledge graphs could also be used as an approach for classifying documents based on their contents.

References

1. Bethard, S., Derczynski, L., Savova, G., Pustejovsky, J., Verhagen, M.: Semeval-2015 task 6: clinical tempeval. In: Proceedings of the 9th International Workshop on Semantic Evaluation (SemEval 2015), pp. 806–814 (2015)
2. Bethard, S., et al.: Semeval-2016 task 12: clinical tempeval. In: Proceedings of the 10th International Workshop on Semantic Evaluation (SemEval-2016), pp. 1052–1062 (2016)
3. Cao, N.D., Izacard, G., Riedel, S., Petroni, F.: Autoregressive entity retrieval. In: International Conference on Learning Representations (2021). https://openreview.net/forum?id=5k8F6UU39V

4. Devlin, J., Chang, M.W., Lee, K., Toutanova, K.: BERT: pre-training of deep bidirectional transformers for language understanding. In: Proceedings of NAACL-HLT, pp. 4171–4186 (2019)
5. Dligach, D., Miller, T., Lin, C., Bethard, S., Savova, G.: Neural temporal relation extraction. In: Proceedings of the 15th Conference of the European Chapter of the Association for Computational Linguistics: Volume 2, Short Papers, pp. 746–751 (2017)
6. Gottschalk, S., Demidova, E.: EventKG: a multilingual event-centric temporal knowledge graph. In: Gangemi, A., et al. (eds.) ESWC 2018. LNCS, vol. 10843, pp. 272–287. Springer, Cham (2018). https://doi.org/10.1007/978-3-319-93417-4_18
7. Han, J., Wang, H.: Generative adversarial networks for open information extraction. Adv. Comput. Intell. **1**(4), 1–11 (2021). https://doi.org/10.1007/s43674-021-00006-8
8. Hobbs, J.R., Pan, F.: Time ontology in owl. W3C working draft, vol. 27, no. 133, pp. 3–36 (2006)
9. Josifoski, M., De Cao, N., Peyrard, M., West, R.: GenIE: generative information extraction. arXiv preprint arXiv:2112.08340 (2021)
10. Lin, C., Miller, T., Dligach, D., Bethard, S., Savova, G.: A BERT-based universal model for both within-and cross-sentence clinical temporal relation extraction. In: Proceedings of the 2nd Clinical Natural Language Processing Workshop, pp. 65–71 (2019)
11. Ning, Q., Subramanian, S., Roth, D.: An improved neural baseline for temporal relation extraction. In: Proceedings of the 2019 Conference on Empirical Methods in Natural Language Processing and the 9th International Joint Conference on Natural Language Processing (EMNLP-IJCNLP), pp. 6203–6209 (2019)
12. Ning, Q., Wu, H., Peng, H., Roth, D.: Improving temporal relation extraction with a globally acquired statistical resource. In: Proceedings of the 2018 Conference of the North American Chapter of the Association for Computational Linguistics: Human Language Technologies, Volume 1 (Long Papers), pp. 841–851 (2018)
13. Ning, Q., Wu, H., Roth, D.: A multi-axis annotation scheme for event temporal relations. In: Proceedings of the 56th Annual Meeting of the Association for Computational Linguistics (Volume 1: Long Papers), pp. 1318–1328 (2018)
14. Ro, Y., Lee, Y., Kang, P.: Multi2oie: multilingual open information extraction based on multi-head attention with BERT. In: Findings of ACL: EMNLP 2020 (2020)
15. Styler, W.F., et al.: Temporal annotation in the clinical domain. Trans. Assoc. Comput. Linguist. **2**, 143–154 (2014)
16. Sun, W., Rumshisky, A., Uzuner, O.: Evaluating temporal relations in clinical text: 2012 i2b2 challenge. J. Am. Med. Inf. Assoc. **20**(5), 806–813 (2013)
17. Vashishtha, S., Van Durme, B., White, A.S.: Fine-grained temporal relation extraction. In: Proceedings of the 57th Annual Meeting of the Association for Computational Linguistics, pp. 2906–2919 (2019)
18. Vaswani, A., et al.: Attention is all you need. In: Advances in Neural Information Processing Systems 30 (2017)
19. Zhang, N., et al.: Contrastive information extraction with generative transformer. IEEE/ACM Trans. Audio Speech Lang. Process. **29**, 3077–3088 (2021)
20. Zhang, S., Huang, L., Ning, Q.: Extracting temporal event relation with syntactic-guided temporal graph transformer. arXiv preprint arXiv:2104.09570 (2021)

Tutorials

Information Security and Risk Management: Trustworthiness and Human Interaction

Stephen C. Phillips[1] (iD), Nicholas Fair[1(✉)] (iD), Gencer Erdogan[2] (iD),
and Simeon Tverdal[2] (iD)

[1] IT Innovation Centre, University of Southampton, Southampton, UK
{S.C.Phillips, N.S.Fair}@soton.ac.uk
[2] Sustainable Communication Technologies, SINTEF Digital, Oslo, Norway
{gencer.erdogan, simeon.tverdal}@sintef.no

1 Tutorial Abstract

As digital information has come to underpin the majority of modern systems in almost all domains (e.g. business, finance, government, education, health, third sector), increasingly sophisticated cybersecurity attacks have become an unavoidable reality of modern life. In the face of this, regulation and best practice are increasing moving from simplistic security control tick-lists towards risk management frameworks (such as recommended in the EU's GDPR and NIS directive and described in standards such as ISO 27005). Consequently, it is highly relevant for students, practitioners, and researchers alike to understand risk management, systems modelling, attack paths, and human interactions and risks in order to understand the central value and importance of cybersecurity risk management in supporting trustworthiness in information systems.

As part of the H2020 CyberKit4SME project, this interactive, hands-on tutorial will explore state-of-the-art approaches to trustworthy cybersecurity risk management that is able to effectively and sufficiently account for the risks that humans introduce into any information system [1]. After establishing the basic concepts around cybersecurity, trustworthiness, system modelling, risk management and socio-technical theory, an exploration of the importance and role of visualised attack paths in providing easily understood risks, thereby ensuring intelligent risk management tools do not become 'black boxes' to their users, will be undertaken. Alongside this, how attack paths help support human decision-making by pinpointing the most effective risk mitigation strategies will be investigated. In addition, the tutorial will explore human interaction flows and how they can combine with attack paths to empower comprehensive cybersecurity risk assessments and help guide holistic mitigations. In the final part of the tutorial, there will be an opportunity to get practical experience of modelling an information system and identifying and mitigating the cybersecurity risks to it using two tools: the System Security Modeller [2, 3] (University of Southampton) and the Human and Organisational Risk Modelling framework (SINTEF) which is derived from the Customer Journey Modelling Language [4, 5] (CJML).

Acknowledgements. This work has received funding from the European Union's Horizon 2020 research and innovation programme under grant agreement No. 883188.

R. Guizzardi et al. (Eds.): RCIS 2022, LNBIP 446, pp. 821–822, 2022.
https://doi.org/10.1007/978-3-031-05760-1

Learning Goals

By the end of the tutorial attendees will have a general understanding of risk management; of what is meant by trustworthiness in cybersecurity system modelling and of the risk impact of humans in information systems. A specific understanding of the role of visualised attack paths in promoting trustworthiness and human choice in cybersecurity system modelling; of how to conceptualise and model human interaction flows and of how attack paths and human interaction flows interact when assessing cybersecurity risk and risk mitigation strategies will be imparted along with experience using a system modeller and interaction flow charts to identify cybersecurity risk and risk mitigation strategies.

Presenters

Dr. Stephen C. Phillips, Principal Research Engineer, technical coordinator of the H2020 CyberKit4SME project, System Security Modeller product manager.
Dr. Gencer Erdogan, Research Scientist, technical lead H2020 CyberKit4SME, developer of the Human and Organisational Risk Modelling framework.
Dr. Nic Fair, Research Engineer, digital education and learning expert, contributor to the System Security Modeller.
Simeon Andersen Tverdal, Researcher, developer of Human and Organisational Risk Modelling framework.

References

1. Boletsis, C., Halvorsrud, R., Pickering, J.B., Phillips, S.C., Surridge, M.: Cybersecurity for SMEs: introducing the human element into socio-technical cybersecurity risk assessment. In: Proceedings of the 16th International Joint Conference on Computer Vision, Imaging and Computer Graphics Theory and Applications (VISIGRAPP 2021), vol. 3, pp. 266–274
2. Surridge, M., et al.: Modelling compliance threats and security analysis of cross border health data exchange. In: Attiogbé, C., Ferrarotti, F., Maabout, S. (eds.) MEDI 2019. CCIS, vol. 1085, pp 180–189 (2019). Springer, Cham. https://doi.org/10.1007/978-3-030-32213-7_14
3. Mohammadi, N., Goeke, L., Heisel, M., Surridge, M.: Systematic risk assessment of cloud computing systems using a combined model-based approach. In: Proceedings of the 22nd International Conference on Enterprise Information Systems, vol. 2, pp. 53-66
4. Halvorsrud, R., Kvale, K., Følstad, A.: Improving service quality through customer journey analysis. J. Serv. Theory Pract. **26**, 840-867 (2016)
5. Halvorsrud, R., Boletsis, C., Garcia-Ceja, E.: Designing a modeling language for customer journeys: lessons learned from user involvement. In: Proceedings of the 24th ACM/IEEE International Conference on Model Driven Engineering Languages and Systems (MODELS), pp. 239–249 (2021)

The Challenge of Collecting and Analyzing Information from Citizens and Social Media in Emergencies: The Crowd4SDG Experience and Tools

Barbara Pernici[1]([✉])(ID), Carlo Bono[1](ID), Jose Luis Fernandez-Marquez[2](ID), and Mehmet Oğuz Mülâyim[3](ID)

[1] Politecnico di Milano, DEIB, Via Giuseppe Ponzio, 34, 20133 Milan, Italy
{barbara.pernici,carlo.bono}@polimi.it
[2] University of Geneva, Geneva, Switzerland
joseluis.fernandez@unige.ch
[3] Artificial Intelligence Research Institute (IIIA-CSIC), Cerdanyola del Vallès, Spain
oguz@iiia.csic.es

Tutorial Abstract

Every year more than 150 million people worldwide are affected by natural disasters. As declared by the United Nations Office for the Coordination of Humanitarian Affairs, "The first 72 h after a disaster are crucial; response must begin during that time to save lives". Social media has been demonstrated to be a potential data source to provide actionable data just as a disaster happens and develops, thus allowing emergency responders to better coordinate their activities. However, social media data also presents many challenges regarding data quality and geolocation (i.e., the geographical location of a post). Over the years, several technologies enabled the retrieval and processing of high volumes of data, with artificial intelligence often employed as a replacement for human intelligence for data classification tasks. Nevertheless, the need to deliver high-quality results within a critical response time is still a major challenge.

In this tutorial, we will see how crowdsourcing assisted by artificial intelligence can make a significant contribution, especially where critical thinking and decision making are needed, in extracting valuable information from unconventional data sources. The tutorial will introduce the basics for extracting and analyzing information from social media, with a specific focus on retrieving images in an emergency after a natural disaster. We will provide the basics about social media crawling and analysis. A specific focus will be given to fine-grained geolocalization of tweets and the combination of AI and crowdsourcing to filter relevant images and confirm or improve geolocations, which are needed to deliver high-quality information. Our experiences with social media analysis (e.g., [5]), geolocalization (e.g., [1, 6]), and crowdsourcing (e.g., [2, 3]) obtained

This work was funded by the EU H2020 project Crowd4SDG "Citizen Science for Monitoring Climate Impacts and Achieving Climate Resilience", #872944.

R. Guizzardi et al. (Eds.): RCIS 2022, LNBIP 446, pp. 823–824, 2022.
https://doi.org/10.1007/978-3-031-05760-1

in a recently concluded H2020 project E2mC (Evolution of Emergency Copernicus services) [4] and in the on-going H2020 project Crowd4SDG (Citizen Science for Monitoring Climate Impacts and Achieving Climate Resilience, https://crowd4sdg.eu) will be illustrated.

The objective of the tutorial is to provide an introduction and hands-on experience in some of the tools available in the field of emergency information systems. In particular, we focus on the tools that enable the search and analysis of social media posts, mainly on Twitter but also on other social media. The analysis of posts includes approaches for selecting relevant images based on image contents and text analysis techniques for information extraction. We also show how we could leverage citizen scientists by setting up a crowdsourcing environment, based on the PyBossa open-source platform (https://pybossa.com), and we demonstrate how we evaluate the quality of crowdsourcing results. We will also discuss the methods and processes for using such tools in a sudden emergency to gather different types of information to support first responders and decision makers.

The tutorial is intended for participants who represent organizations looking for emergency data, who can benefit from collective intelligence, especially where there is a data gap in their research using traditional data sources, but are skeptical in reliability of this kind information; communities and agencies looking for tools to analyze the data; individuals interested in learning about available tools that can enrich and ensure reliability and usability of data obtained from social media.

References

1. Bono, C., Pernici, B., Fernandez-Marquez, J.L., Shankar, A.R., Mülâyim, M.O., Nemni, E.: TriggerCit: Early Flood Alerting using Twitter and Geolocation - a comparison with alternative sources (2022). https://doi.org/10.48550/arxiv.2202.12014

2. Cerquides, J., Mülâyim, M.O.: Crowdnalysis: A software library to help analyze crowdsourcing results (2022). https://doi.org/10.5281/zenodo.5898579

3. Cerquides, J., Mülâyim, M.O., Hernández-González, J., Ravi Shankar, A., Fernandez-Marquez, J.L.: A Conceptual Probabilistic Framework for Annotation Aggregation of Citizen Science Data. Mathematics 9(8) (2021). https://doi.org/10.3390/math9080875

4. Havas, C., et al.: E2mC: improving emergency management service practice through social media and crowdsourcing analysis in near real time. Sensors 17(12), 2766 (2017). https://doi.org/10.3390/s17122766

5. Negri, V., et al.: Image-based Social Sensing: Combining AI and the Crowd to Mine Policy-Adherence Indicators from Twitter (2020). https://doi.org/10.48550/ARXIV.2010.03021

6. Scalia, G., Francalanci, C., Pernici, B.: CIME: context-aware geolocation of emergency-related posts. GeoInformatica 26(1), 125–157 (2021). https://doi.org/10.1007/s10707-021-00446-x

Information Science Research with Machine Learning: Best Practices and Pitfalls

Andreas Vogelsang$^{(\boxtimes)}$

University of Cologne, Cologne, Germany
`vogelsang@cs.uni-koeln.de`

1 Tutorial Abstract

More and more research on Information Science is based on the use of techniques from machine learning (ML). Supervised machine learning, where ML models are trained on labeled datasets, has been used for categorizing requirements [1, 7, 11], for information retrieval from large documents [3], or for data-driven risk management [10]. Unsupervised machine learning, where ML models are trained on unlabeled data to identify recurring patterns has been used to cluster user feedback and requirements [6, 8]. Advances in ML have further enhanced the possibilities to process large amounts of natural language text [2]. This enables new areas for IS research such as user feedback mining [5], app store analytics [9], or crowd-based requirements engineering [4]. ML becomes so prevalent in IS research because of the ever growing availability of data and the ease of using ML algorithms out of the box based on frameworks and libraries. Although ML algorithms are so approachable, researchers can still make a lot of methodological mistakes that may invalidate a study or, if these flaws are not detected by unaware reviewers, lead to invalid conclusions in published IS research papers. The aim of this 90-min tutorial is to:

- provide an overview of ML techniques and their capabilities for IS research
- Describe the typical steps of an ML pipeline
- Make participants aware of best practices and the most common pitfalls when applying ML for IS research

We will walk the audience through a typical ML pipeline and discuss pitfalls and best practices in each step. We will focus on pitfalls prevalent to IS research scenarios such as reliability of data labeled by humans, imbalanced data sets, lack of baseline comparisons, lack of clear problem description, or lack of hyperparameter optimization. The tutorial is designed for researchers interested in using ML techniques in their research with the goal to increase the quality and reliability of IS research that uses ML techniques.

© The Author(s), under exclusive license to Springer Nature Switzerland AG 2022
R. Guizzardi et al. (Eds.): RCIS 2022, LNBIP 446, pp. 825–826, 2022.
https://doi.org/10.1007/978-3-031-05760-1

References

1. Dalpiaz, F., Dell'Anna, D., Aydemir, F.B., Cevikol, S.: Requirements classification with interpretable machine learning and dependency parsing. In: IEEE International Requirements Engineering Conference (RE). pp. 142–152 (2019). https://doi.org/10.1109/RE.2019.00025
2. Dalpiaz, F., Ferrari, A., Franch, X., Palomares, C.: Natural language processing for requirements engineering: the best is yet to come. IEEE Softw. 35(5), 115–119 (2018). https://doi.org/10.1109/MS.2018.3571242
3. Frattini, J., Fischbach, J., Mendez, D., Unterkalmsteiner, M., Vogelsang, A., Wnuk, K.: Causality in requirements artifacts: prevalence, detection, and impact. Requirements Eng. (2022). https://doi.org/10.1007/s00766-022-00371-x
4. Groen, E.C., et al.: The crowd in requirements engineering: the landscape and challenges. IEEE Softw. 34(2), 44–52 (2017). https://doi.org/10.1109/MS.2017.33
5. Guzman, E., Maalej, W.: How do users like this feature? A fine grained sentiment analysis of app reviews. In: IEEE International Requirements Engineering Conference (RE), pp. 153–162 (2014). https://doi.org/10.1109/RE.2014.6912257
6. Gülle, K.J., Ford, N., Ebel, P., Brokhausen, F., Vogelsang, A.: Topic modeling on user stories using word mover's distance. In: IEEE Seventh International Workshop on Artificial Intelligence for Requirements Engineering (AIRE), pp. 52–60 (2020). https://doi.org/10.1109/AIRE51212.2020.00015
7. Hey, T., Keim, J., Koziolek, A., Tichy, W.F.: NoRBERT : transfer learning for requirements classification. In: IEEE International Requirements Engineering Conference (RE), pp. 169–179 (2020). https://doi.org/10.1109/RE48521.2020.00028
8. Maalej, W., Nayebi, M., Johann, T., Ruhe, G.: Toward data-driven requirements engineering. IEEE Softw. 33(1), 48–54 (2016). https://doi.org/10.1109/MS.2015.153
9. Pagano, D., Maalej, W.: User feedback in the appstore: an empirical study. In: IEEE International Requirements Engineering Conference (RE), pp. 125–134 (2013). https://doi.org/10.1109/RE.2013.6636712
10. Wiesweg, F., Vogelsang, A., Mendez, D.: Data-driven risk management for requirements engineering: an automated approach based on Bayesian networks. In: IEEE International Requirements Engineering Conference (RE), pp. 125–135 (2020). https://doi.org/10.1109/RE48521.2020.00024
11. Winkler, J., Vogelsang, A.: Automatic classification of requirements based on convolutional neural networks. In: IEEE International Requirements Engineering Conference Workshops (REW), pp. 39–45 (2016). https://doi.org/10.1109/REW.2016.021

Author Index

Printed in the United States
by Baker & Taylor Publisher Services